COSMETIC APPLICATIONS OF LASER AND LIGHT-BASED SYSTEMS

PERSONAL CARE AND COSMETIC TECHNOLOGY

Series Editor: Meyer Rosen

President, Interactive Consulting, Inc., NY, USA

The aim of this book series is to disseminate the latest personal care and cosmetic technology developments with a particular emphasis on accessible and practical content. These books will appeal to scientists, engineers, technicians, business managers, and marketing personnel.

For more information about the book series and new book proposals please contact William Andrew at info@williamandrew.com.

http://www.williamandrew.com/PersonalCareCosmetic.php

COSMETIC APPLICATIONS OF LASER AND LIGHT-BASED SYSTEMS

Edited by
Gurpreet S. Ahluwalia

William Andrew
Norwich, NY, USA

ISBN: 978-0-8155-1572-2

Library of Congress Cataloging-in-Publication Data

Cosmetic applications of laser & light-based systems / edited by Gurpreet S. Ahluwalia.
p. ; cm. -- (Personal care and cosmetic technology)
Includes bibliographical references and index.
ISBN 978-0-8155-1572-2 (alk. paper)
1. Skin--Laser surgery. 2. Skin--Diseases--Phototherapy. 3. Hair--Diseases--Treatment. 4. Cosmetic delivery systems. 5. Surgery, Plastic. I. Ahluwalia, Gurpreet S. II. Title: Cosmetic applications of laser and light-based systems. III. Series.
[DNLM: 1. Cosmetic Techniques. 2. Laser Therapy--methods. 3. Hair--physiology. 4. Photochemotherapy--methods. 5. Skin Physiology. WR 650 C8327 2009]
RL120.L37C72 2009
617.4'770598--dc22
2008038597

Printed in the United States of America

This book is printed on acid-free paper.

10 9 8 7 6 5 4 3 2 1

Published by:
William Andrew Inc.
13 Eaton Avenue
Norwich, NY 13815
1-800-932-7045
www.williamandrew.com

ENVIRONMENTALLY FRIENDLY
This book has been printed digitally because this process does not use any plates, ink, chemicals, or press solutions that are harmful to the environment. The paper used in this book has a 30% recycled content.

To my mother and father, Surinder and Surjeet Ahluwalia for teaching me the virtues of life and providing unconditional love and support

To my wife Gail for her encouragement, patience and understanding

To my son Sean Preet and daughter Anjuli for their love and support

To my mentor David A. Cooney from National Cancer Institute, NIH who taught me the fundamentals of scientific investigation

Contents

Contributors

Gurpreet S. Ahluwalia PhD
The Gillette Company
A wholly owned subsidiary of The P&G Company
Needham, MA
USA

(Senior Director, Dermatology Clinical R&D, at Allergan, Inc., Irvine, CA, USA, as of November 2008)

Murad Alam MD
Northwestern University Feinberg School of Medicine
Associate Professor Dermatology and Otolaryngology, and Surgery
Chief, Section of Cutaneous and Aesthetic Surgery
Department of Dermatology
Northwestern University
Chicago, IL
USA

Macrene Alexiades-Armenakas MD, PhD
Assistant Clinical Professor
Department of Dermatology
Yale University School of Medicine

Director
Dermatology and Laser Surgery Private Practice
New York, NY
USA

Gregory B. Altshuler PhD
Vice President, Research and Development
Palomar Medical Technologies, Inc.
Burlington, MA
USA

Todd J. Banks PharmD, RPh
Regulatory Affairs Manager
The Procter and Gamble Company
Cincinnati, OH
USA

Natalia V. Botchkareva MD, PhD
The Gillette Company
A wholly owned subsidiary of The P&G Company
Needham, MA
USA

(At School of Life Sciences, The University of Bradford, Bradford, UK, as of November 2008)

Roy G. Geronemus MD
Director
Laser & Skin Surgery Center of New York
New York, NY
USA

David J. Goldberg MD, JD
Clinical Professor of Dermatology
Mount Sinai School of Medicine
New York, NY
USA

Director
Skin Laser & Surgery Specialists of New York and New Jersey
New York, NY
USA

Jillian Havey
Research fellow
Northwestern University Feinberg School of Medicine
Department of Dermatology
Northwestern University
Chicago, IL
USA

James Henry PhD
The Procter and Gamble Company
Cincinnati, OH
USA

Mary P. Lupo MD
Department of Dermatology
Tulane Medical School

Lupo Center for Aesthetic and General Dermatology
New Orleans, LA
USA

David H. McDaniel MD
Director
Laser Skin & Vein Center of Virginia
Virginia Beach, VA
USA

Cheri Millikin
The Procter and Gamble Company
Cincinnati, OH
USA

Serge Mordon PhD
Research Director
INSERM & Lille University Hospital
Lille
France

Michael Moretti
Editor/Publisher
Medical Insight, Inc.
Aliso Viejo, CA
USA

Vic A. Narurkar MD
Director and Founder
Bay Area Laser Institute
San Francisco, CA
USA

J. Frank Nash PhD
The Procter and Gamble Company
Cincinnati, OH
USA

Keyvan Nouri MD, FAAD
Professor of Dermatology and Otolaryngology
Director of Mohs Micrographic Surgery, Dermatologic and Laser Surgery Unit
Director of Surgical Training
Department of Dermatology & Cutaneous Surgery
University of Miami Miller School of Medicine
Miami, FL
USA

John E. Oblong PhD
The Procter and Gamble Company
Cincinnati, OH
USA

Rita V. Patel
Department of Dermatology
University of Miami
Miller School of Medicine
Miami, FL
USA

Valerie Anne Randall PhD
Professor
School of Life Sciences
The University of Bradford
Bradford
UK

Neil S. Sadick MD, FAAD
Clinical Professor of Dermatology
Weill Medical College of Cornell University
New York, NY
USA

Cosmetic, Laser and Dermatologic Surgery
New York, NY
USA

Douglas Shander PhD
The Gillette Company
A wholly owned subsidiary of The P&G Company
Needham, MA
USA

(At Trichoresearch, Gaithersburg, MD, USA, as of November 2008)

Anita Singh
Research Fellow
Department of Dermatology & Cutaneous Surgery
University of Miami
School of Medicine
Miami, FL
USA

David Sliney PhD
Consulting Medical Physicist
Fallston, MD
USA

Retired from US Army Center for Health Promotion
and Preventive Medicine
Aberdeen Proving Ground, MD
USA

Pete Styczynski PhD
The Procter and Gamble Company
Cincinnati, OH
USA

Valery V. Tuchin PhD
Institute of Optics and Biophotonics
Saratov State University
Saratov
Russia

Institute of Precise Mechanics and Control
Saratov
Russia

Voraphol Vejjabhinanta MD
Senior Clinical Research Fellow
Department of Dermatology & Cutaneous Surgery
University of Miami Miller School of Medicine
Miami, FL
USA

Clinical Instructor
Suphannahong Dermatology Institute
Bangkok
Thailand

Susan Walgrave MD
Zel Skin & Laser Specialists
Edina, MN
USA

Melea Ward
The Procter and Gamble Company
Cincinnati, OH
USA

Robert A. Weiss MD
Director
MD Laser Skin & Vein Institute, LLC
Hunt Valley, MD
USA

Dale Wiener
Palomar Medical Technologies, Inc.
Burlington, MA
USA

Paul Wiener
Palomar Medical Technologies, Inc.
Burlington, MA
USA

Brian Zelickson MD
Associate Professor of Dermatology
Department of Dermatology
University of Minnesota
Minneapolis, MN
USA

Zel Skin & Laser Specialists
Edina, MN
USA

Preface

Though cosmetic science dates back nearly 4000 years, it is in the last two to four decades that the industry has made the most progress by coming up with high potency bioactive ingredients now part of cosmeceuticals, innovative topical drugs for beauty treatments, minimally invasive injectables such as Botox® Cosmetic and dermal fillers, and non-invasive, non-ablative laser and light-based systems for cosmetic dermatology. The laser and light-based systems are preferred by the consumer who demands more than what creams and topical drugs can deliver and thinks that injectables and surgery are a step too far. The cosmetic targets for these systems are diverse and include the removal of unwanted hair; the treatment of photodamaged and unevenly pigmented skin to improve tone, texture, and imperfections similar to what is achieved with aggressive peels and exfoliants; and the treatment of fine lines, wrinkles, and laxity to improve skin appearance and give it a rejuvenated look. The treatment of acne, vascular disorders, cellulite, pseudofolliculitis barbae (PFB), and the removal of tattoos and benign pigmented lesions are some additional conditions targeted by laser and light systems.

All laser and light-based systems for cosmetic dermatology are regulated by the FDA as medical devices. The FDA clears (not approves) these devices for marketing based on a determination of their substantial equivalence to a predicate marketed device under the Agency's 510(k) provisions. This has allowed for technology advancements to rapidly enter the marketplace without having to go through a lengthy regulatory approval process. This has resulted in the introduction of a large number of laser and light systems in the past two decades for a broad array of skin conditions. As good as these systems work in terms of their effectiveness and safety, there are certain limitations imposed by the physiological and biochemical makeup of their biological targets. Moreover, there are marked inter-individual differences between subjects in their response to the benefits and side effects of laser and light system treatments. Understanding the causes of this variability can go a long way toward individualizing treatment regimens and identifying synergistic combinations for providing desired benefits to the consumer.

The purpose of this book is to provide the research community a comprehensive review of the technology, from the basic biology of the involved target to the efficacy and safety that are specific to the device and the cosmetic dermatology indication. The text is organized into six parts and 25 total chapters. Each chapter is dedicated to a specific topic authored by experts in their field. Part 1 covers the technology fundamentals related to the physiology and biochemistry of skin and hair along with the biophysical principles of laser technology that are relevant to understanding specific light-tissue interactions. Part 2 covers available hair management options including various laser and light-based technologies and the laser effects on hair follicle biology at the molecular level. Available options for enhancing skin appearance, including microdermabrasion, cosmeceuticals, topical drugs, and combination treatments with a focus on various light-based systems are discussed in Part 3. Laser treatment of diverse skin conditions, including cellulite, acne, and PFB, and for wound healing, creating synergies with topical drugs, and the use of photodynamic therapy for enhanced cosmetic benefits are discussed in Parts 4 and 5. Part 6 is dedicated to the safety, including dermal and eye, and the regulatory aspects of gaining marketing clearance, of laser and light-based systems.

The next frontier in the quest for beautiful skin and youthful appearance is likely to be the combination of topical chemistry and medical devices. It is likely that the light-based devices being developed for the aesthetic home-use market will be complemented by cosmeceuticals and topical drug products to provide consumers with a complete beauty solution in the privacy of their homes.

I would like to thank all the contributors to this work, each of whom devoted their time and effort to reviewing the available literature, to sharing their personal experiences in cosmetic dermatology procedures, and to sharing their clinical and basic research findings.

Gurpreet S. Ahluwalia
Irvine, California
October 2008

PART 1
BASIC TECHNOLOGY AND TARGETS FOR LIGHT-BASED SYSTEMS

1

The Biology of Hair Growth

Valerie Anne Randall and Natalia V. Botchkareva

Centre of Skin Sciences, School of Life Sciences, University of Bradford, Bradford, UK

Gurpreet S. Ahluwalia (ed.), *Cosmetic Applications of Laser and Light-Based Systems*, 3–35,

1.1 Introduction

The hair follicle is a highly dynamic organ found only in mammals. Although frequently overlooked, the follicle is fascinating from many viewpoints. For cell and developmental biologists it has an almost unique ability in mammals to regenerate itself, recapitulating many embryonic steps en route [1,2]. For zoologists, it is a mammalian characteristic, significant for their evolutionary success and crucial for the survival of many mammals-loss of fur or faulty colouration leads to death from cold or predation. Human follicles also pose a unique paradox for endocrinologists as the same hormones, androgens, cause stimulation of hair growth in many areas, while simultaneously inhibiting scalp follicles causing balding [3,4]. In contrast, hair is often seen as rather irrelevant medically, as human hair loss is not life threatening. Nevertheless, hair is very important for most people [5]. Many men spend significant time shaving daily and vast amounts are spent on hair products; a 'bad hair day' is a common expression for days when everything goes wrong! This reflects the important role hair plays in human communication in both social and sexual contexts and explains why hair disorders such as hirsutism (excessive hair growth) or alopecia (hair loss/balding) cause serious psychological distress [6].

Hair growth is co-ordinated by hormones, usually in parallel to changes in the individual's age and stage of development or environmental alterations like day-length [7]. Hormones instruct the follicle to undergo appropriate changes so that during the next hair cycle, the new hair produced differs in colour and/or size. This chapter will review the functions of hair, its structure and the processes occurring during the hair growth cycle, the changes which can occur with the seasons, and the importance of the main regulator of human hair growth, the androgens. Throughout the chapter, the main emphasis will be on human hair growth.

1.2 The Functions of Hair

Mammalian skin produces hair everywhere except for the glabrous skin of the lips, palms, and soles. Although obvious in most mammals, human hair growth is so reduced with tiny, virtually colourless *vellus* hairs in many areas, that we are termed the "naked ape". Externally hairs are thin, flexible tubes of dead, fully keratinised epithelial cells; they vary in colour, length, diameter, and cross-sectional shape. Inside the skin hairs are part of individual living hair follicles, cylindrical epithelial downgrowths into the dermis and subcutaneous fat, which enlarge at the base into the hair bulb surrounding the mesenchyme-derived dermal papilla (Fig. 1.1) [8].

In many mammals, hair's important roles include insulation for thermoregulation, appropriate colour for camouflage [9], and a protective physical barrier, for example, from ultraviolet light. Follicles also specialise as neuroreceptors (e.g. whiskers) or for sexual communication like the lion's mane [10]. Human hair's main functions are protection and communication; it has virtually lost insulation and camouflage roles, although seasonal variation [11–13] and hair erection when cold indicate the evolutionary history. Children's

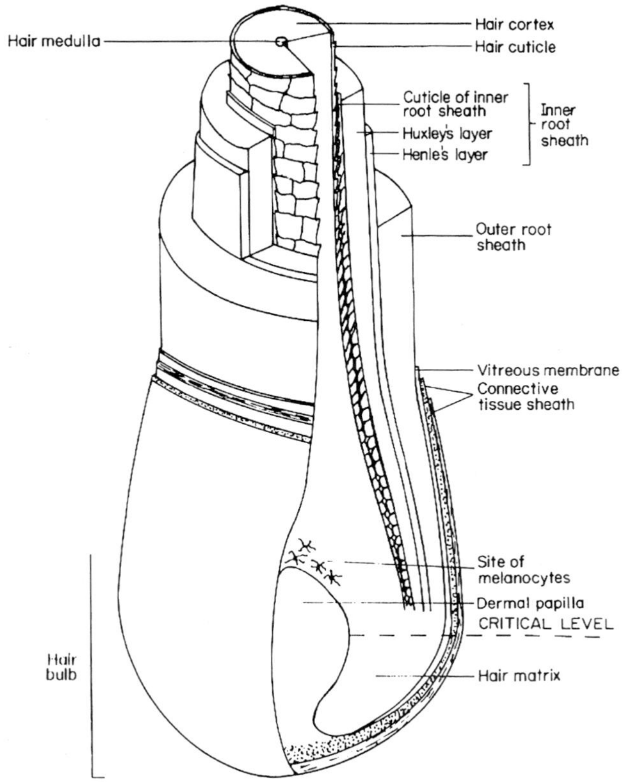

Figure 1.1 The hair follicle. The right-hand side of this diagram shows a section through the lower hair follicle while the left represents a three-dimensional view cut away to reveal the various layers. Drawing by Richard J. Dew. Reproduced from Randall [3].

hairs are mainly protective; eyebrows and eyelashes stop things entering the eyes, while scalp hair probably prevents sunlight, cold, and physical damage to the head and neck [14]. Scalp hair is also important in social communication. Abundant, good-quality hair signals good health, in contrast to sparse, brittle hair indicating starvation or disease [15]. Customs involving head hair spread across many cultures throughout history. Hair removal generally has strong depersonalising roles (e.g. head shaving of prisoners and Christian/Buddhist monks), while long uncut hair has positive connotations like Samson's strength in the Bible.

Other human hair is involved in sexual communication. Pubic and axillary hair development signals puberty in both sexes [16–18], and sexually mature men exhibit masculinity with visible beard, chest, and upper pubic diamond hair (Fig. 1.2). The beard's strong signal and its potential involvement in a display of threatening behaviour, like the lion's mane, [5,10,14] may explain its common removal in "Westernised" countries. This important communication role explains the serious psychological consequences and impact on quality of life seen in hair disorders like hirsutism, excessive male pattern hair growth in women, and hair loss, such as alopecia areata, an autoimmune disease affecting both sexes [19]. Common balding, androgenetic alopecia or male pattern hair loss [20], also causes negative effects, even among men who have never sought medical help [6]. Its high incidence in Caucasians and occurrence in other primates suggest a natural phenomenon, a secondary

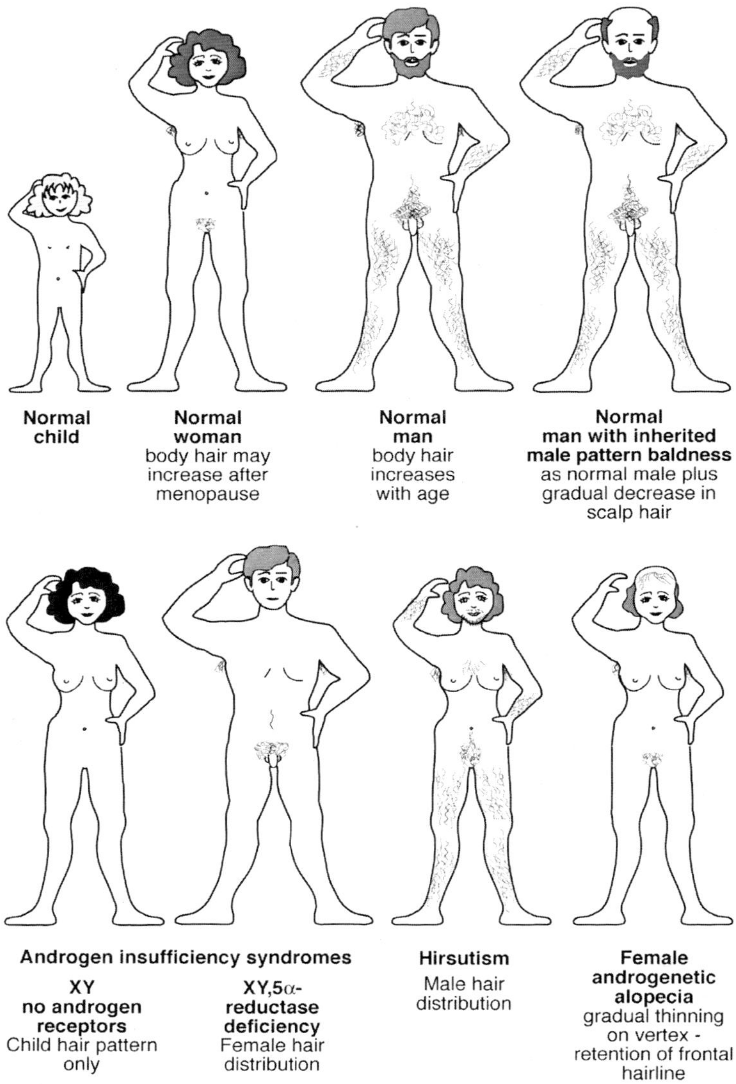

Figure 1.2 Human hair distribution under differing endocrine conditions. Normal patterns of human hair growth are shown in the upper panel. Visible (i.e. terminal) hair with protective functions normally develops in children on the scalp, eyelashes, and eyebrows. Once puberty occurs, further terminal hair develops on the axilla and pubis in both sexes and on the face, chest, limbs, and often back in men. In people with the appropriate genetic tendency, androgens may also stimulate hair loss from the scalp in a patterned manner causing androgenetic alopecia. The various androgen insufficiency syndromes (lower panel) demonstrate that none of this occurs without functional androgen receptors and that only axillary and female pattern of lower pubic triangle hairs are formed in the absence of 5α-reductase type-2. Male pattern hair growth (hirsutism) occurs in women with abnormalities of plasma androgens or from idiopathic causes and women may also develop a different form of hair loss, female androgenetic alopecia. Reproduced from Randall [221].

sexual characteristic, rather than a disorder. Marked balding would identify the older male leader, like the silver-backed gorilla or the senior stag's largest antlers. Other suggestions include advantages in fighting, as flushed bald skin would look aggressive or offer less hair for opponents to pull [14]. If any of these were evolutionary pressures to develop balding, the lower incidence among Africans [21] suggests that any possible advantages were outweighed by hair's important protection from the tropical sun. Whatever the origin, looking older is not beneficial in the industrialised world's current youth-orientated culture.

1.3 Hair Follicle Anatomy

The hair follicle can be divided into three anatomical compartments: the infundibulum, isthmus, and the inferior segment. The upper follicle is permanent, whereas the lower follicle, the inferior segment, regenerates with each hair follicle cycle. The infundibulum extends from the skin surface to the sebaceous duct. The isthmus, the permanent middle portion, extends from the duct of sebaceous gland to the exertion of arrector pilli muscle. The inferior segment consists of the suprabulbar area and the hair bulb. The hair bulb consists of extensively proliferating keratinocytes and pigment-producing melanocytes of the hair matrix that surround the pear-shaped dermal papilla, which contains specialised fibroblast-type cells embedded in an extracellular matrix and separated from the keratinocytes by a basement membrane [22]. The hair matrix keratinocytes move upwards and differentiate into the hair shaft, as well as into the inner root sheath; the melanocytes transfer pigment into the developing hair keratinocytes to give the hair its colour. The epithelial portion of the hair follicle is separated from the surrounding dermis by the *perifollicular connective tissue* or *dermal sheath*. This consists of an inner basement membrane called the hyaline or glassy membrane and an outer connective tissue sheath. The major compartments of the hair follicle from the innermost to the outermost include the hair shaft, the inner root sheath, the outer root sheath, and the connective tissue sheath (Fig. 1.1).

1.3.1 The Hair Shaft

The hair shaft consists of the medulla, cortex. Immediately above the matrix cells, hair shaft cells begin to express specific hair shaft keratins in the prekeratogenous zone [23]. The medulla is a central part of larger hairs, such as beard hairs, and a specific keratin expressed in this layer of cells can be controlled by androgens [24]. The cortex is composed of longitudinally arranged fibres. The hair shaft cuticle covers the hair, and its integrity and properties have a great impact on the appearance of the hair. It is formed by a layer of scales that interlock with opposing scales of the inner root sheath, which allows the hair shaft and the inner root sheath to move upwards together.

1.3.2 The Inner Root Sheath

The inner root sheath consists of four layers: the cuticle, Huxley's layer, Henle's layer, and the companion layer. The cells of the inner root sheath cuticle partially overlap with the cuticle cells of the hair shaft, anchoring the hair shaft tightly to the follicle. Inner root sheath cells produce keratins 1/10 and trichohyalin that serve as an intracellular "cement" giving strength to the inner root sheath to support and mould the growing hair shaft, as well as guide its upward movement. The transcription factor GATA-3 is critical for inner root sheath differentiation and lineage. Mice lacking this gene fail to form an inner root sheath [25]. The inner root sheath separates the hair shaft from the outer root sheath, which forms the external concentric layer of epithelial cells in the hair follicle.

1.3.3 The Outer Root Sheath

The outer root sheath contains a heterogeneous cell population including keratinocytes expressing keratins 5 and 14, keratinocyte and melanocyte stem cell progeny migrating downward to the hair matrix, and differentiating melanocytes [26–29]. Between the insertion of the arrector pili muscle and duct of the sebaceous gland the outer root sheath forms a distinct bulge, which has been identified as a reservoir of multipotent stem cells [30]. These cells are biochemically distinct and can be identified by long-term retention of BrdU or by immunodetection of cytokeratins 15 and 19, CD 34 (in mice), and CD 200 (in humans) [31–34]. In addition, these cells are characterised by their low proliferative rate and their capacity for giving rise to several different cell types including epidermal keratinocytes, sebaceous gland cells, and the various different types of epithelial cells of the lower follicle [35]. This area also contains melanocyte stem cells [36]. Moreover, recently nestin, the neural stem cell marker protein, was also shown to be expressed in the bulge area of the hair follicle. Nestin-positive stem cells isolated from this area could differentiate into neurons, glia, smooth muscle cells, and melanocytes in vitro. Experiments in mice confirmed that nestin-expressing hair follicle stem cells can differentiate into blood vessels and neural tissue after transplantation to the subcutis of nude mice [37]. These experiments suggest that hair-follicle bulge-area stem cells may provide an accessible source of undifferentiated multipotent stem cells for therapeutic applications [37].

1.3.4 The Dermal Papilla

The hair bulb encloses the follicular dermal papilla, which comprises a group of mesenchyme-derived cells, the dermal papilla cells, mucopolysaccharide-rich stroma, nerve fibres, and a single capillary loop. The follicular papilla is believed to be one of the most important drivers to instruct the hair follicle to grow and form a particularly sized and pigmented hair shaft. Several experiments have shown that the dermal papilla has powerful inductive properties. Dermal papilla cells transplanted into non-hair-bearing epidermis are able to induce the formation of new hair follicles [38,39]. The dermal papilla is an essential source of paracrine factors critical for hair growth and melanogenesis; it is believed to be the interpreter of circulating signals such as hormones to the follicle (discussed in Section 1.7). Specific examples of factors produced by the dermal papilla that influence hair growth include noggin, which exerts a hair growth-inducing effect by antagonising bone morphogenetic protein (BMP) signalling and activation of the BMP receptor IA expressed in the follicular epithelium [40]. Keratinocyte growth factor (KGF) is also produced by the anagen dermal papilla, and its receptor, FGFR2, is found predominantly in the matrix keratinocytes. The activation of this pathway by injections of KGF into nude mice induces hair growth at the site of injection [41]. Dermal papilla cells also express hepatocyte growth factor (HGF) [42]. Transgenic mice overexpressing HGF display accelerated hair follicle development [42]. Insulin-like growth factor-I (IGF-I) found in the dermal papilla also serves as an important morphogen in the hair follicle [43]. In addition, stem cell factor (SCF) produced by the dermal papilla [44] is essential for proliferation, differentiation, and melanin production by follicular melanocytes expressing its receptor c-kit [26]. The dermal papilla also displays unusually strong alkaline phosphatase activity during the entire hair cycle [45]. Although a role for alkaline phosphatase remains obscure, hair growth is reduced when inhibitors of alkaline phosphatase are applied [46].

Interestingly, recent studies suggested that follicle dermal papilla and connective (or dermal) sheath cells may act as stem cells for both follicular and interfollicular dermis. Moreover, the stem cell potential of follicle dermal cells extends beyond the skin. Jahoda and colleagues have demonstrated that rodent hair follicle dermal cells have haematopoietic stem cell activity [47] and can also be directed towards adipocyte and osteocyte phenotypes (reviewed in [48]).

1.4 Changing the Hair Produced by a Follicle via the Hair Growth Cycle

To fulfil all the roles described in Section 1.2, the hair produced by a follicle often needs to change and follicles possess a unique mechanism for this, the hair growth cycle [1,2] (Fig. 1.3). This involves destruction of the original lower follicle, and its regeneration to form another, which can produce hair with different characteristics. Thus, post-natal follicles retain the ability to recapitulate the later stages of follicular embryogenesis throughout life. Exactly how differently sized a hair can be to its immediate predecessor is currently unclear because many changes take several years (e.g. growing a full beard) [49]. Hairs are produced in anagen, the growth phase. Once a hair reaches full length, a short apoptosis-driven involution phase, catagen, occurs, where cell division and pigmentation stops, the hair becomes fully keratinised with a swollen "club" end and moves up in the skin with the regressed dermal papilla. After a period of rest, telogen, the dermal papilla cells and associated keratinocyte stem cells reactivate and a new lower follicle develops downwards inside the dermal sheath which surrounded the previous follicle. The new hair then grows up into the original upper follicle (Fig. 1.3). The existing hair is generally lost; although previously thought to be due to the new hair's upward movement, a further active shedding stage, exogen, is now proposed [50–53].

Hair follicle regeneration is characterised by dramatic changes in its microanatomy and cellular activity. Hair follicle transition between distinct hair cycle stages is governed by epithelial–mesenchymal interactions between the keratinocytes of the follicular epithelium and the dermal papilla fibroblasts. Cell fate during hair follicle growth and involution is controlled by numerous growth regulators that induce survival and/or differentiation or apoptosis. During hair follicle active growth and hair production, the activity of factors promoting proliferation, differentiation, and survival predominates, while hair follicle regression is characterised by activation of various signalling pathways that induce apoptosis in hair follicle cells [53–55].

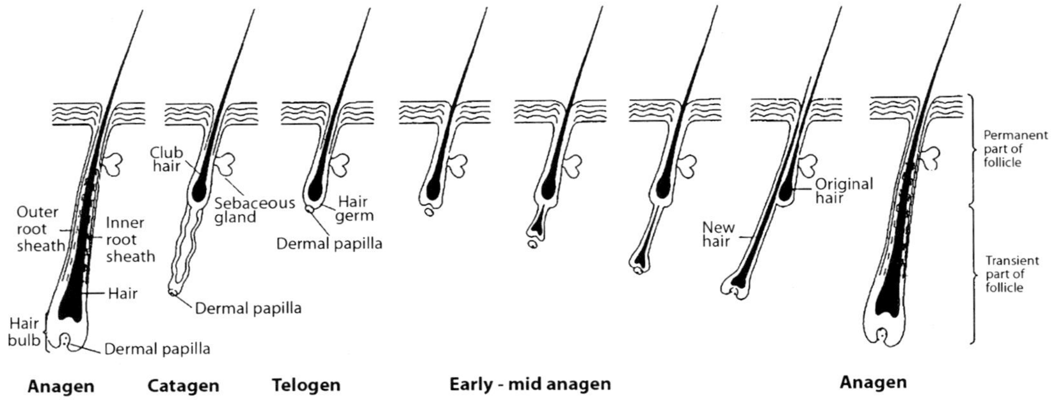

Figure 1.3 The hair follicle growth cycle. Hair follicles go through well established repeated cycles of development and growth (anagen), regression (catagen), and rest (telogen) [1,2] to enable the replacement of hairs, often by another of differing colour or size. An additional phase, exogen, has been reported where the resting club hair is released [87,88]. Modified from Randall [3].

1.4.1 Telogen—The Resting Phase

Telogen hair follicles are very short in length. They are characterised by a lack of pigment-producing melanocytes and the inner root sheath. Their compact ball-shaped dermal papilla is closely attached to a small cap of secondary hair germ keratinocytes containing hair follicle stem cells. A balance of local growth stimulators and inhibitors in the proximal part of the telogen hair follicle appears to be critical for the initiation of the telogen–anagen transition. In particular, activation of the Shh pathway induces hair follicle transition from telogen to anagen [56]. The high sensitivity of telogen hair follicles to Shh pathways was confirmed by the initiation of anagen by a single topical application of synthetic, non-peptidyl small molecule agonists of the Hh pathway [57]. On the other hand, telogen skin has been suggested to contain inhibitors of hair growth [58]. Bone morphogenetic protein 4 (BMP4) has been identified as one of these inhibitors, as antagonising the BMP4 pathway by its endogenous inhibitor, noggin, induces active hair growth in post-natal telogen skin in vivo [26]. Interestingly, noggin increased Shh mRNA in the hair follicle, while BMP4 downregulated Shh [26].

Cell proliferation in the germinative compartment of the telogen hair follicle can also be activated by applying mechanical or chemical stimuli. For instance, removing the hair shaft from telogen follicles by epilation results in a new hair growth wave [59]. The molecular mechanisms underlying this induction remain largely unknown. However, plucking-induced anagen is widely used as a model for studying the hair cycle in mice to evaluate the expression pattern of genes of interest at distinct hair cycle stages, although there is always the possibility of abnormal effects due to the wounding caused by plucking. In addition, telogen–anagen transformation of mouse hair follicles can also be induced by the administration of immunosuppressants such as cyclosporin A and FK506 [60,61]. Indeed, the stimulation of unwanted hair growth is one of the most common dermatological side effects of immunosuppressive cyclosporine A therapy, seen in transplantation medicine and in the treatment of autoimmune diseases [62].

1.4.2 Anagen—The Growth Phase

Anagen can be divided into six stages. During early phases, hair progenitor cells proliferate, envelope the growing dermal papilla, grow downwards into the skin and begin to differentiate into the hair shaft and inner root sheath. In mid anagen, melanocytes located in the hair matrix show pigment-producing activity, and the newly formed hair shaft begins to develop. In late anagen, full restoration of the hair fibre-producing unit is achieved, which is characterised by the formation of the epithelial hair bulb surrounding the dermal papilla, located deep in the subcutaneous tissue, and the new hair shaft emerges from the skin surface [59,63,64]. During anagen, active signal exchanges occur between the epithelial cells of the hair bulb and the fibroblasts of the dermal papilla. Actively proliferating and postmitotic keratinocytes of the hair matrix express receptors and/or intracellular signalling components of a variety of signalling pathways (β-catenin/Lef-1, c-kit, c-met, FGFR2, IGF-IR), while the corresponding ligands are expressed in the dermal papilla (Wnt5a, SCF, HGF, FGF7, IGF-1) (reviewed in [54,63]).

In addition to hair follicle tissue remodelling, skin innervation and vascular networks also undergo substantial changes with the progression of the anagen stage [65,66]. Perifollicular vascularisation is significantly increased during anagen. It correlates with the upregulation of the expression of vascular endothelial growth factor (VEGF) mRNA, a potent angiogenic growth factor, produced by keratinocytes of the outer root sheath. In transgenic mice overexpressing VEGF, perifollicular vascularisation was strongly induced, which resulted in accelerated hair growth and increased size of hair follicles and hair shafts [67]. In contrast, application of suppressors of angiogenesis leads to hair growth reduction [68]. Therefore, cutaneous vasculature may have a great impact on the hair shaft producing activity of hair follicle cells.

1.4.3 Catagen—The Regressive Phase

Anagen is followed by a phase of hair follicle involution, catagen. Catagen was first characterised in detail by Kligman [69] and Straile [70]. At the beginning of catagen, proliferation and differentiation of hair matrix keratinocytes reduces dramatically, the pigment-producing activity of melanocytes ceases, and hair shaft production is completed. During catagen, the follicle compartments involved in hair production are reduced to sizes that allow them to regenerate in the next hair cycle after receiving the appropriate stimulation. The hair follicle shortens in length by up to 70%. Although catagen is often considered a regressive event, it is an exquisitely orchestrated, energy-requiring remodelling process, whose progression assures renewal of a further generation of the hair follicle. Morphologically and functionally, catagen is divided into eight sub-stages [59]. During catagen, a specialised structure, the club hair is formed. The keratinised brush-like structure at the base of the club hair is surrounded by epithelial cells of the outer root sheath and anchors the hair in the telogen follicle. During catagen, the dermal papilla is transformed into a cluster of quiescent cells that are closely adjacent to the regressing hair follicle epithelium and travel from the subcutis to the dermis/subcutis border to maintain contact with the distal portion of the hair follicle epithelium including the secondary hair germ and bulge. Catagen is characterised by several simultaneously occurring and tightly coordinated cellular programs. The most important characteristic feature is a well-coordinated apoptosis occurring in the proximal part of the hair follicle. Apoptosis is regulated differently in each follicle compartment and distinct cell populations show different abilities to undergo apoptosis [55]. The majority of the follicular epithelial cells and melanocytes are very susceptible to apoptosis, while dermal papilla fibroblasts and the populations of keratinocytes and melanocytes selected for survival display a high resistance [71,72]. The physiological involution of the hair follicle may be triggered by the withdrawal of dermal papilla-derived growth factors that maintain cell proliferation and differentiation in the anagen hair follicle, and by a variety of stimuli, including signalling via death receptors (Fig. 1.4).

One of the candidate molecules mediating apoptosis in hair matrix keratinocytes after growth factor withdrawal is p53. Mice lacking p53 showed significantly retarded catagen progression, compared with control mice confirming a pro-apoptotic role for p53 in the hair follicle [26]. The delicate proliferation-apoptosis balance, essential for follicle cyclic behaviour, can also be controlled by survivin [73]. Survivin, a member of the apoptosis inhibitor protein family, is implicated in the control of cell proliferation as well as the inhibition of apoptosis [74]. Survivin, expressed in the proliferating keratinocytes of the anagen hair matrix and outer root sheath, disappears with the progression of catagen [73]. Before or during catagen, outer root sheath keratinocytes produce several important catagen-promoting secreted molecules: fibroblast growth factor-5 short isoform, neurotrophins, transforming growth factor-β1/2 (TGF-β1/2), IGF binding protein 3, and thrombospondin-1 [75–78]. Several important growth factors were discovered as modulators of catagen development

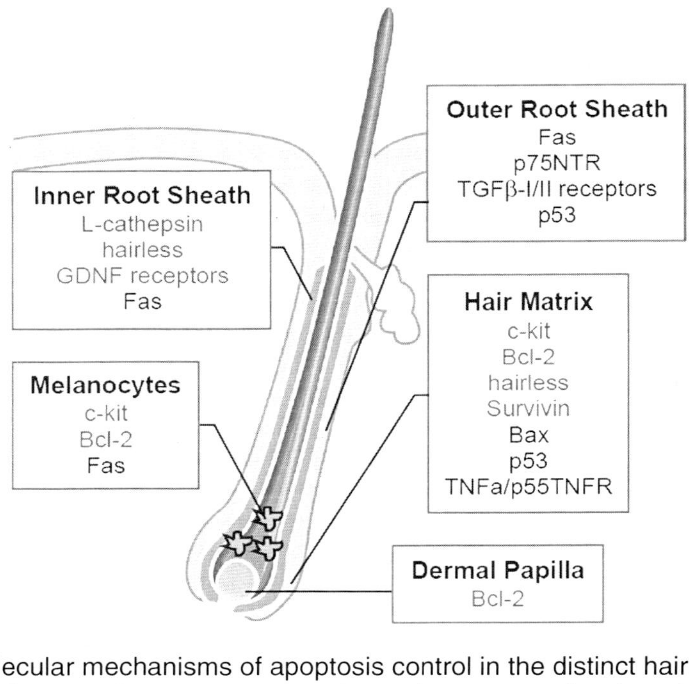

Figure 1.4 Molecular mechanisms of apoptosis control in the distinct hair follicle compartments. Scheme demonstrates the expression pattern of anti- and pro-apoptotic molecules (shown in brown and black respectively) in the hair follicle.

by gene knockout studies. The most remarkable phenotype was seen in mice lacking the fibroblast growth factor-5 (*Fgf5*) gene whose hair was 50% longer than their wild type littermates, giving an "angora-like" phenotype [79]. Neurotrophins and TGF-β1 also induce premature catagen onset. Mice overexpressing distinct members of the neurotrophin family (BDNF, NT-3) show premature catagen development in part by stimulation of pro-apoptotic signalling through the p75 kD neurotrophin receptor in the outer root sheath [75]. TGF-β1 knockout mice display delayed catagen onset [76]. Neurotrophins and TGF-β2 also exert catagen-promoting effects on human hair follicles in organ culture [80,81]. Catagen can also be initiated by several other molecules, such as endothelin-1, insulin-like growth factor binding proteins-3/4/5, interleukin-1, vitamin D receptor (reviewed in [82]), prolactin [83,84], endocannabinoids [85], or thrombospondin-1 [78].

1.4.4 Exogen—Hair Shedding

An additional phase of the hair cycle called exogen was recently recognised; this involves hair shaft shedding from the telogen follicle [86], an active process, accompanied by the activation of proteolytic processes in the follicular root [87]. Exogen was also recently characterized in human follicles. It was shown that while anagen and telogen hairs are firmly anchored to the follicle, exogen hairs are passively retained within the follicles. In addition, exogen clubs do not retain remnants of the outer root sheath, in contrast to plucked telogen hairs [88].

The new hair formed during the next anagen may resemble its predecessor, like most human scalp hair, or may differ markedly like the brown summer and white winter hairs of Scottish hares [9]. The type of hair produced depends on the regulatory dermal papilla [89,90] although the cell biology and biochemistry of their mechanisms are not fully understood. The duration of hair cycle stages varies in different body areas. Human scalp hair follicles have the longest anagen phase, which can last up to several years; they also display a relatively short catagen phase (1–2 weeks) followed by a telogen phase lasting several months. The majority of scalp hair follicles are in anagen (80–85%), with the rest either in catagen (2%) or telogen (10–15%). The anagen phase of follicles in other body regions is substantially shorter, for example on the arms, legs, and thighs it ranges from 3 to 4 months [26]. It is clear that anagen length generally determines hair length; long scalp hairs are produced by follicles with anagens over 2 years, while short finger hairs only grow for around 2 months [91].

1.5 Hair Pigmentation

The colour of hair is variable. It is important in many mammals for camouflage and in human beings for making hair visible, such as the increased colour of sexually related hair after puberty [17,18]. Loss of hair pigment resulting in greying and whitening of hair is one of the first characteristics of ageing. Within the hair follicle, neural crest-derived melanocytes in the hair bulb produce and transport melanin to the keratinocytes of the precortical zone that differentiate to form the pigmented hair shaft. The hair follicle pigmentary unit in the bulb cyclically regenerates synchronously with the hair follicle during the hair cycle. The melanogenic activity of the follicular melanocytes is strictly coupled to the anagen stage, decreases during late anagen and early catagen, and ceases during late catagen and telogen [26,92,93].

In the anagen hair follicle, melanocytes are divided into three distinct subpopulations. The first population is located in the hair follicle bulge and represents melanocyte stem cells that repopulate the melanocytes in the new hair bulb formed at the onset of anagen [26,36,94]. The second population is located in the hair follicle outer root sheath and represents differentiating melanocytes. The third is located in the hair matrix above the dermal papilla and actively produces melanin [26,93] (Fig. 1.5). Melanogenesis is controlled by several key enzymes that are uniquely expressed in the melanocytes (reviewed in [95]). Tyrosinase catalyses the rate-limiting initial events of melanogenesis, and mutations in tyrosinase gene lead to loss of pigment [96]. Tyrosinase-related proteins (TRP) 1 and TRP2 share 40–45% amino acid identity with tyrosinase and are also critically important for melanogenesis, functioning as downstream enzymes in the melanin biosynthetic pathway [97].

Hair pigmentation is tightly regulated by several hormones and growth factors. Androgens play a major role in causing alterations of human hair colour, including increase of pigment during vellus to terminal hair switches in many regions such as the beard after puberty, or the converse on the scalp during male pattern balding [98]. Changes in anagen-associated melanogenesis are accompanied by changes in the gene expression of melanocortin 1 receptor (MC1-R) activated by POMC-derived ACTH and MSH peptides [99], and ACTH and α-MSH are able to promote human follicular melanocyte differentiation by

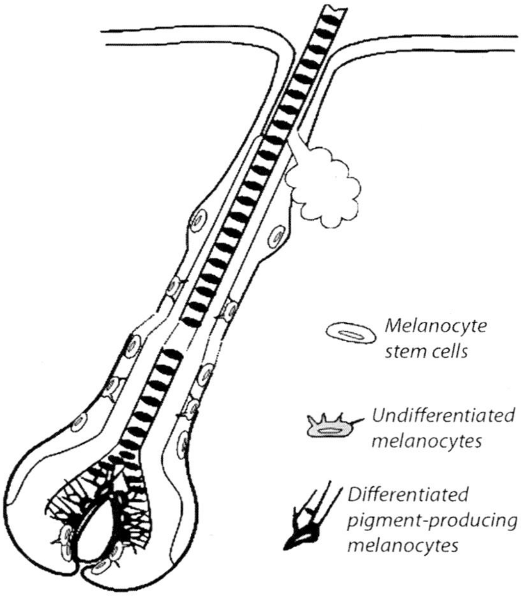

Figure 1.5 Hair follicle melanocyte distribution. Schematic drawing represents localisation of different subpopulations of melanocytes in the anagen hair follicle. Melanocyte stem cells are located in the bulge, the differentiating melanocyte are mostly located in the outer root sheath, while differentiated melanogenically active melanocytes are present in the hair bulb.

up-regulating melanogenesis, dendricity, and proliferation in less differentiated melanocyte subpopulations [100]. SCF/c-kit signalling is required for cyclic regeneration of the hair pigmentation unit. Pharmacological inhibition of SCF/c-kit signalling in vivo leads to the production of depigmented hairs in rodents [26]. In addition, other proteins known to be involved in melanocyte biology, including agouti signal protein, the endothelin family, fibroblast growth factor 2, and hepatocyte growth factor may be important for modulating the activity of hair follicle melanocytes during the hair cycle (reviewed in [101,102]).

1.6 Seasonal Changes in Hair Growth

Hair follicles are under hormonal regulation due to the importance of coordinating alterations in insulative and colour properties of a mammal's coat to the environment or visibility to changes in sexual development. Seasonal changes usually occur twice a year in temperate regions with coordinated waves of growth and moulting to produce a thicker, warmer winter coat and shorter summer pelage. These are linked to day-length, and to a lesser extent to temperature, like seasonal breeding activity [7,103]; nutrient availability can also affect hair type because of the high metabolic requirements of hair production [104].

1.6.1 Hormonal Coordination of Seasonal Changes in Animals

Studies in many species, including sheep, hamsters, mink, and ground squirrels [105,106], show that long daylight hours initiate short periods of daily melatonin secretion by the pineal gland and summer coat development, while short (winter) day-length increases melatonin secretion and stimulates a longer, warmer pelage [7,103]. The pineal gland acts as a neuroendocrine transducer converting nerve impulses stimulated by daylight to reduced secretion of melatonin, normally secreted in the dark. Melatonin signals are generally translated to the follicle by the hypothalamus-pituitary route; for example, melatonin administration into the sheep hypothalamus stimulates short day responses [107]. However, although disconnecting the hypothalamus and pituitary removes seasonal changes in body weight and the wool's normal cycling pattern, long days stimulate a minor moult [103]. Prolactin levels continue to cycle, suggesting melatonin also acts directly on the pituitary prolactin secretion. Since both growth hormone and IGF-1 levels are also reduced, this may prevent prolactin's full effect as IGF-1 receptors are present in goat follicles [108] and IGF-1 can stimulate human hair growth in vitro [109].

There is strong evidence for prolactin's involvement in seasonal coat changes in Djungarian hamsters [106], goats [108], mink [110], sheep [110,111], and deer [112]. Increased prolactin levels in long daylight correspond to low summer growth and low prolactin concentrations during short days with increased winter growth; moulting occurred in sheep after maximal prolactin levels [103]. Prolactin infusion inhibits goat hair growth locally [113] and prolactin receptors are located in rodent [114,115] and mink [116] skin and the dermal papilla and epithelial compartments of sheep follicles [117]. Interestingly, sheep [111], mink [116], and non-seasonal laboratory rodent [115] follicles also express prolactin mRNA.

Other hormones implicated in regulating mammalian hair growth cycles include the sex steroids, oestradiol and testosterone, and the adrenal steroids; these delay anagen in rats [7,118], while gonadectomy in rats and adrenalectomy in rats and mink [7,118,119] advance it. Topical application of 17β-oestradiol to mice skin inhibits hair growth and accelerates catagen, while antioestrogens promote early anagen [120–124]. Rat dermal papillae take up oestradiol [125] and both oestrogen receptors α (ERα) and β (ERβ) are detected in human follicles [126] and cultured dermal papilla cells [127]. Testosterone also delays seasonal hair growth in badgers [128], while urinary cortisol levels are negatively correlated with hair loss in rhesus macaque monkeys [129]. In contrast, thyroid hormones advance anagen while thyroidectomy or propythiouracil delay it [7,118]. How these circulating hormones interact is still unclear, but the main drivers in seasonal coat changes are light, melatonin, and prolactin.

1.6.2 Seasonal Variation in Human Hair Growth

Seasonal changes are much less obvious in human beings, where follicle cycles are generally unsynchronised after age one, except in groups of three follicles called Demeijère trios [26]. Regular annual cycles in human scalp [11–13], beard, and other body hair [11] were only recognised relatively recently.

Seasonal changes in hair growth were evident in 14 healthy Caucasian men aged 18–39 years studied for 18 months in Sheffield, UK (latitude 53.4°N); these men also showed pronounced seasonal behaviour, spending much more time outside in summer, despite their indoor employment [11]. Scalp hair showed a single annual cycle with over 90% of follicles in anagen in the spring falling to around 80% in the autumn; the number of hairs shed in the autumn also more than doubled [11] (Fig. 1.6). Similar increased head-hair shedding in New York women [12] indicates an autumnal moult. Since scalp hair usually grows for at least 2–3 years [91], detection of an annual cycle indicates a strong response of any follicles able to react, presumably those in later stages of anagen.

Changes also occurred in male characteristic, androgen-dependent body hair [11]. Winter beard and thigh hair growth rate were low, but increased significantly in the summer (Fig. 1. 6). French men showed similar summer peaks in semen volume, sperm count, and mobility [130] suggesting androgen-related effects; their luteinising hormone (LH), testosterone, and 17β-oestradiol levels showed autumnal peaks. Low winter testosterone and higher summer levels were also reported in European men [131,132] and pubertal boys [133]. Testosterone changes probably alter beard and thigh hair growth rate, but they are less likely to regulate scalp follicles as seasonal changes also occur in women. However, androgens do inhibit some scalp follicles in genetically susceptible individuals causing balding [134] and dermal papilla cells derived from non-balding scalp follicles contain low levels of androgen receptors making such a response possible [135]. Annual fluctuations of thyroid hormones, with peaks of T3 in September and free T4 in October [136], could also influence scalp growth, but hypothyroidism is normally associated with hair loss [137].

In contrast to these single cycles, thigh follicles showed biannual changes in anagen, with 80% of follicles growing in May and November, falling to around 60% in March and August [11] (Fig. 1. 6).This pattern is similar to the spring and autumn moults of many temperate mammals [7] and may reflect such seasonal moulting from our evolutionary past. Presumably these cycles are controlled like those in Section 1.6.1. Human beings can respond to altered day-length by changing melatonin, prolactin, and cortisol secretion, but

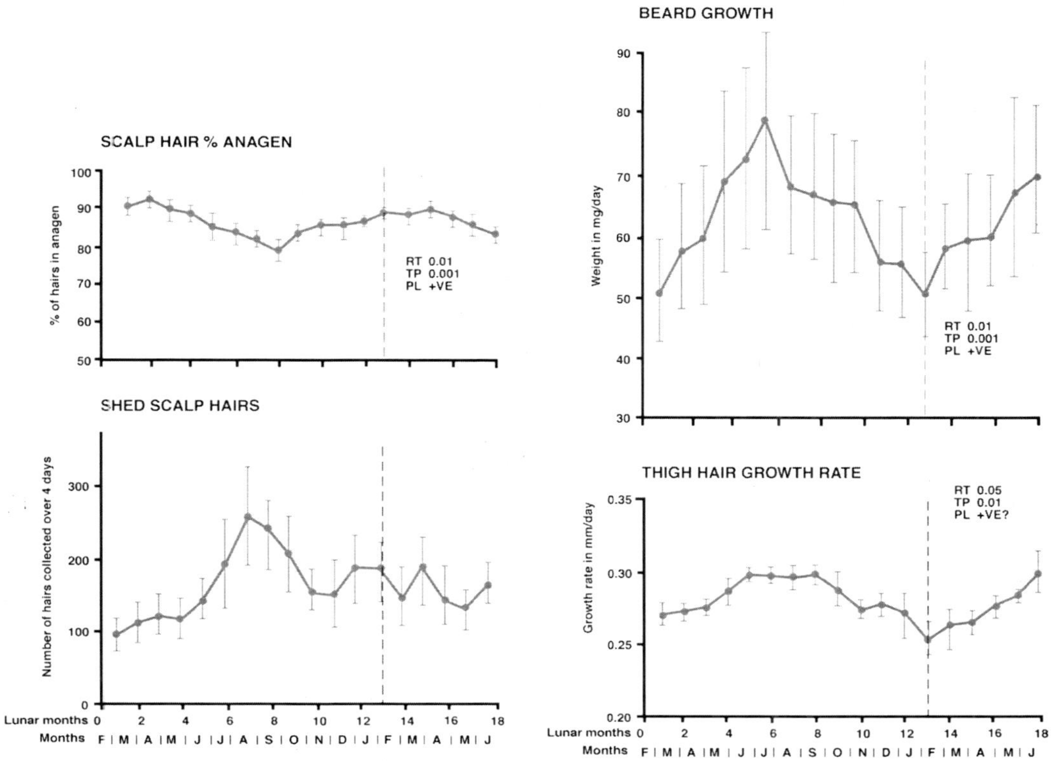

Figure 1.6 Seasonal changes in human hair growth. Hair follicles on the scalp (left) and body (right) of British men with indoor occupations living in the north of England show significant seasonal variation. Scalp hair (upper panel) has a single annual cycle with most follicles in anagen in spring, with anagen numbers falling in autumn; the number of hairs shed (lower panel) paralleled this. Facial (upper panel) and thigh hair (lower panel) grows significantly faster in the summer months and more slowly in the winter. Measurements are mean ± SEM for Caucasian men (13 scalp and beard, 14 thigh); there is wide variation in beard heaviness in individual men [49]. Statistical analysis was carried out using runs (RT), turning points (TP), and phase length (PL) tests. Data from Randall and Ebling [11], redrawn from Randall VA [221].

the artificially manipulated light of urban environments suppress these responses [138]. Nevertheless, people in Antarctica [139] and those with seasonal affective disorder [140] maintain melatonin rhythms and Randall and Ebling's study population definitely exhibited seasonal behaviour despite indoor occupations [11].

These annual changes are important for any investigations of scalp or androgen-dependent hair growth, particularly in individuals living in temperate zones. For hair loss patients, any condition may be exacerbated during the increased autumnal shedding. They also have important implications for any assessments of new therapies or treatments to stimulate, inhibit or remove hair; to be accurate measurements need to be carried out over a year to avoid natural seasonal variations.

1.7 Hormonal Regulation of Human Hair Growth

Apart from seasonal changes (Section 1.6), the most obvious regulators of human hair growth are androgens, as long as individuals have good nutrition [15,141] and normal thyroid function [137,142]. Pregnancy hormones also effect hair growth causing diffuse hair loss post-partum.

1.7.1 Pregnancy

Lynfield [143] found more scalp follicles were in anagen during the second and third trimesters (95%) and for about a week after birth; by six weeks this fell to about 76%, remaining low for 3 months. Pregnancy hormones maintain follicles in anagen, but after birth many enter catagen and telogen, causing a synchronised partial shedding or moult. This may be particularly noticeable in autumn due to seasonal shedding (Section 1.6.2). Which hormones are involved is uncertain, although oestrogen and prolactin are possibilities. Human follicles have prolactin [144] and 17β-oestradiol [126,127] receptors, but 17β-oestradiol inhibits cultured human follicles [145], and rodent hair growth, accelerating catagen onset [121–123], the opposite of the pregnancy effect. Prolactin reduces human follicular growth in vitro [144] supporting a role in post-partum shedding.

1.7.2 Androgens

1.7.2.1 Human Hair Follicles Show Paradoxically Different Intrinsic Responses to Androgens

Androgens' dramatic stimulation of hair growth is seen first in puberty with pubic and axillary hair development in both sexes [16–18]. These changes parallel the rise in plasma androgens, occurring later in boys than girls [146,147]. Testosterone stimulates beard growth in eunuchs and elderly men [148] and castration inhibits beard growth [49] and male pattern baldness [149], but individuals with complete androgen insufficiency (i.e. without functional androgen receptors) highlight the essential involvement of androgens [150]. As they cannot respond to androgen, these XY individuals develop a female-type phenotype, but without any pubic or axillary hair or any androgenetic alopecia (Fig. 1. 2). Growth hormone is also required for the full androgen response as sexual hair development is inhibited in growth hormone deficiency [151].

Androgens stimulate tiny *vellus* follicles producing fine, virtually colourless, almost invisible hairs to transform into larger, deeper follicles forming longer, thicker, more pigmented hairs (Fig. 1.7). Follicles must pass through the hair cycle, regenerating the lower follicle to carry out such changes (Section 1.4). Although androgens stimulate hair growth in many areas, causing greater hair growth on the face, upper pubic diamond, chest, etc. in men [49], they can also have the opposite effect on specific scalp areas, often in the same individual, causing balding [57]. This involves the reverse transformation of large, deep follicles producing long, often heavily pigmented *terminal* scalp hairs to miniaturised vellus follicles forming tiny, almost invisible hairs (Fig. 1.7).

During puberty, the hairline is usually straight across the top of the forehead. In many men this frontal hairline progressively regresses in two wings and thinning occurs

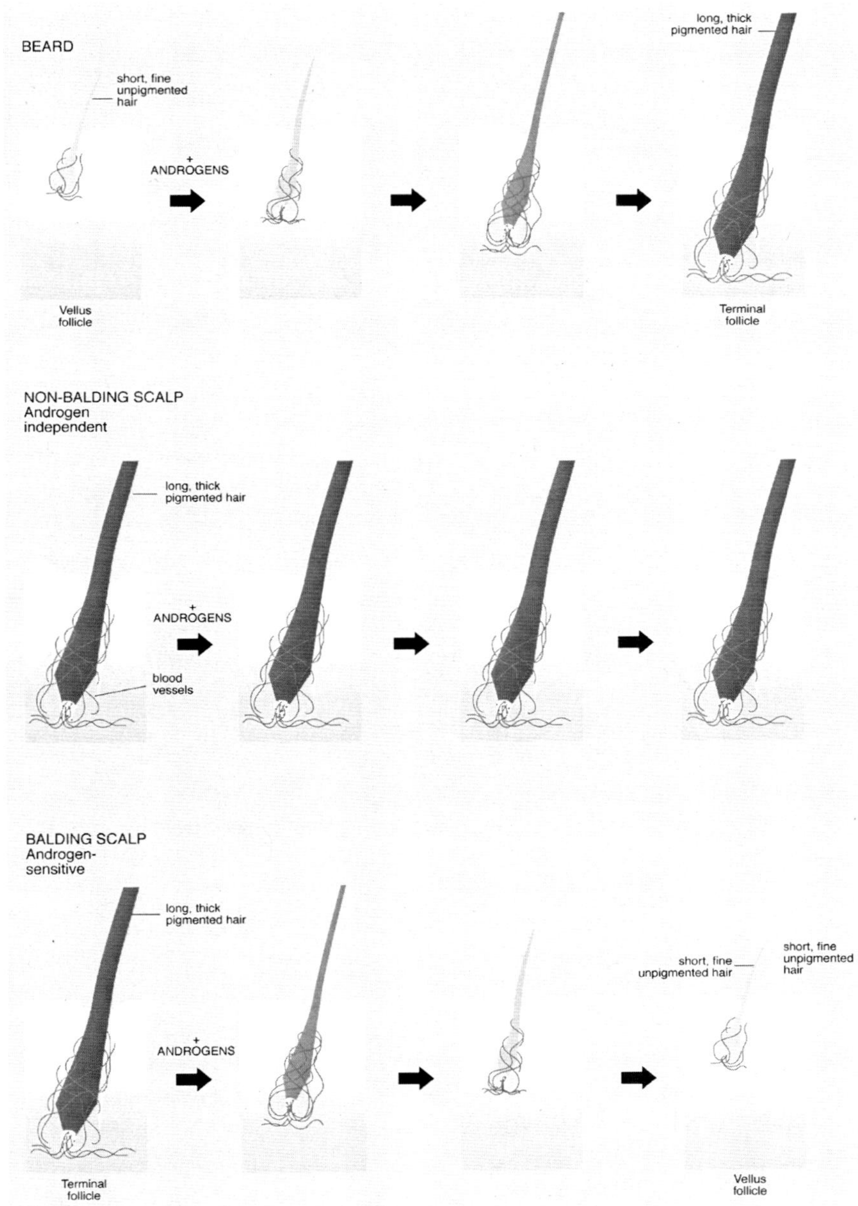

Figure 1.7 Androgens have paradoxically different effects on human hair follicles depending on their body site. In many areas, androgens stimulate the gradual transformation of small follicles producing tiny, virtually colourless, *vellus* hairs to terminal follicles producing longer, thicker, and more pigmented hairs during and after puberty (upper panel) [49]. These changes involve passing through the hair cycle (see Fig. 1.3). At the same time many follicles in the scalp and eyelashes continue to produce the same type of hairs, apparently unaffected by androgens (middle panel). In complete contrast, androgens may cause inhibition of follicles on specific areas of the scalp in genetically susceptible individuals causing the reverse transformation of terminal follicles to *vellus* ones and androgenetic alopecia [134]. Diagram reproduced from Randall [221].

mid-vertex [134]. These areas gradually expand in a precise pattern exposing 'bare' scalp [134,152]; the lower sides and back normally retain terminal hair (Fig. 1.2). Androgenetic alopecia is reviewed thoroughly elsewhere [20,153]. Similar hair loss, considered androgen-dependent, can occur in women, but the pattern differs; the frontal hairline is normally retained while generalised thinning progresses on the vertex until it appears bald [154].

In contrast, androgens appear to have no effect on other hairs like the eyelashes (Fig. 1.7). This is an intriguing and unique biological paradox. How does one hormone stimulate an organ, the hair follicle, in many areas, but have no effect in another, while at the same time, cause inhibition in the same organ in another part of the body, often in the same individual?

There are also significant differences between androgen-stimulated follicles. Axillary and lower pubic follicles enlarge in response to female levels of androgens, while other follicles require male levels [146,147]. Follicles also differ in their sensitivity, or speed of response. Facial follicles enlarge first above the mouth (moustache) and on the chin in boys and hirsute women; this spreads gradually over the face and neck [18]. This progression resembles the patterned inhibition during balding [134,152]. Many androgen responses are gradual, with some follicles taking years to show the full response. Beard weight increases dramatically during puberty but continues rising until the mid-thirties, while terminal hairs may only be visible on the chest and ear canal years later [49] and the miniaturisation processes of androgenetic alopecia continue well into old age [134,152]. This delay parallels the late onset of androgen-dependent benign prostatic hypertrophy and prostatic carcinoma [135].

Another demonstration of the intrinsic behaviour of human follicles is the contrast between beard and axillary hair growth. Although both increase rapidly during puberty, beard growth remains heavy, while axillary hair is maximal in the mid-twenties before falling rapidly in both sexes [49].This is another paradox; why do follicles in some areas no longer show their androgenic responses, while in many others they maintain or extend them?

These contrasts are presumably due to differential gene expression within individual follicles, since all follicles are exposed to the same circulating hormones and, from the complete androgen insensitivity syndrome, require the same receptor. [150]. Follicles' retention of their original androgen response when transplanted, the basis of corrective cosmetic surgery confirms this [156]. Presumably, this genetic programming occurs, in the patterning processes during development. Interestingly, the dermis of the chick's frontal parietal scalp, which parallels human balding regions, develops from the neural crest, while the occipital-temporal region, our non-balding area, arises from the mesoderm [157]. The molecular mechanisms involved in forming different types of follicles during embryogenesis are unclear, but secreted signalling factors, such as Eda, sonic hedgehog, Wnt, and various growth factor families (e.g. BMPs, nuclear factors), including various homeobox genes, and others such as Hairless and Tabby, plus transmembrane and extracellular matrix molecules are all implicated [158,159].

Human follicles require androgens not only for their initial transformation, but also need them to maintain many of the effects. If men are castrated after puberty neither beard growth nor male pattern balding return to prepubertal levels [22,134] suggesting that some altered gene expression does not require androgens for maintenance or lower levels can maintain some effect. Nevertheless, beard growth increases in the summer [11] (Fig. 1.6), probably in response to increased circulating androgens (Section 1.6), antiandrogen treatment reduces hair growth in hirsutism [160] and more selective blockers of androgen action, 5α-reductase inhibitors such as finasteride, can cause regrowth in androgenetic alopecia [161,162]. This suggests that androgens are required to maintain most of the responses, as well as initiating progression.

These intrinsic differences in hair follicle androgen responses have important consequences for anyone wishing to investigate androgen action. It is essential to study follicles which respond appropriately in vivo for the question being addressed. Unfortunately, this means that the most available human material, non-balding scalp, is often inappropriate.

Genetics also appears important in androgen-dependent hair growth. Male pattern baldness [149,163,164] and heavy beard growth [49] run in families, Caucasian men and women generally have greater hair growth than Japanese [49], despite similar testosterone levels [165], and African men exhibit much less baldness [21]. Several genes have been investigated for association with androgenetic alopecia. Interestingly, women with polycystic ovaries and their brothers with early balding exhibit links to one allele of the steroid metabolism gene, CYP17 [166]. No association was found with neutral polymorphic markers of genes for testosterone metabolising enzymes 5α-reductase type-1 or -2 in balding [167,168]; however, *Stu* I restriction fragment length polymorphism (RFLP) in exon 1 of the androgen receptor was present in young (98%) and older (92%) balding men, although also in 77% of older controls [169]. Although single triplet repeats of CAG or GAC were unaltered, short/short polymorphic CAG/GGC haplotypes were significantly higher in balding subjects. Interestingly, Spanish girls with precocious puberty (i.e. before 8 years) showed smaller numbers of CAG repeats [170] and shorter triplet repeat lengths are associated with another androgen-dependent condition, prostate cancer [171]. Whether this has functional significance like increased androgen sensitivity or simply reflects linkage disequilibrium with a causative mutation is unclear. However, increased sensitivity is not supported by the similarity of steroid binding capability between androgen receptors from balding and non-balding follicle dermal papilla cells [172].

1.7.2.2 The Mechanism of Androgen Action in Hair Follicles

Specific effects of androgens on hair follicle cells. Androgens must alter many aspects of follicular cell activity to cause these changes in follicle and hair type. They must alter the ability of epithelial matrix cells to divide, determine whether they should differentiate into medulla (found in some large hairs), and regulate the pigment produced and/or transferred by follicular melanocytes. They must also alter dermal papilla size which has a constant relationship with the hair and follicle size [173,174], and ensure the dermal sheath surrounding follicles expands to accommodate larger follicles. These responses are also quite complex; for example, altering hair length could involve changing cell division rate, that is, hair growth rate, and/or the actual growing period, anagen. Anagen length seems the most important. Thigh hair is three times longer in young men than women, but grows only slightly faster for a much longer period [175]. Androgens do cause such alterations as antiandrogen treatment reduces hair diameter, growth rate, length, pigmentation, and medullation in hirsute women [176], while blocking 5α-reductase activity increases many of these aspects in alopecia [161]. This raises the question: are androgens acting on each target cell individually or operating through one coordinating system with indirect effects on other cell types?

General mechanism of action of androgens. Androgens, like other steroid hormones, diffuse through cell membranes to act on target cells by binding to specific intracellular receptors. These hormone-receptor complexes undergo conformational changes exposing DNA binding sites and bind to specific hormone response elements (HRE) in the DNA, often in combination with accessory (coactivating) proteins, promoting expression of specific, hormone-regulated genes [177]. Androgen action is more complex than other steroids. Testosterone, the main male circulating androgen, binds receptors in some tissues (e.g. skeletal muscle). However, in others, including secondary sexual tissues like the prostate, testosterone is metabolised intracellularly by 5α-reductase enzymes to 5α-dihydrotestosterone, a more potent androgen, which binds more strongly to the androgen receptor to activate gene expression [178].

Androgen-dependent follicles require androgen receptors to respond as highlighted by the absence of adult body hair in complete androgen insensitivity (Fig. 1.2) [150], but the need for 5α-reductase varies with body region. Men with 5α-reductase type-2 deficiency only produce female patterns of pubic and axillary hair growth, although their body shapes become masculinised [179] (Fig. 1.2). Therefore, 5α-dihydrotestosterone appears necessary for follicles characteristic of men, including beard, chest, and upper pubic diamond, while testosterone itself can stimulate the axilla and lower pubic triangle follicles also found in women. Since androgenetic alopecia is not seen in 5α-reductase type-2 deficient men and the 5α-reductase type-2 inhibitor, finasteride, can restore hair growth [85,86], 5α-reductase type-2 also seems important for androgen-dependent balding.

Why some follicles need 5α-dihydrotestosterone and others testosterone to stimulate the same types of cell biological changes that lead to larger hairs is unclear; presumably, the cells use different intracellular coactivating proteins to act with the receptor.

Current model for androgen action in hair follicles. Hair follicle growth is complex but rarely abnormal, indicating a highly controlled system. This suggests that androgen action is coordinated through one part of the follicle. The current hypothesis, proposed in 1990 by Randall et al. [180], focuses on the dermal papilla with androgens acting directly on dermal papilla cells where they bind to androgen receptors and then initiate the altered gene expression of regulatory factors which influence other target cells (Fig. 1.8). These factors could be soluble paracrine factors and/or extracellular matrix factors; extracellular matrix forms much of the papilla volume, and dermal papilla size corresponds to hair and follicle size [173,174]. In this model the dermal papilla is the primary direct target, while other cells such as keratinocytes and melanocytes are indirect targets.

This hypothesis evolved from several concepts reviewed elsewhere [3,180] including dermal papilla determination of the type of hair produced [89]; adult follicle cycles partially recapitulating their embryogenic development; strong parallels in androgen dependency and age-related changes between hair follicles and the prostate; and androgens acting on embryonic prostate epithelium through the mesenchyme [155]. There is now strong experimental support for this model. Androgen receptors are found in the dermal papilla [126,181] and in cultured dermal papilla cells derived from androgen-sensitive follicles including beard [135], balding scalp [172], and deer manes [182]. Cells from androgen-sensitive sites contain higher levels of specific, saturable androgen receptors than androgen-insensitive non-balding scalp in vitro [135,172,183]. Importantly, beard, but not pubic or non-balding scalp cultured dermal papilla cells metabolise testosterone to 5α-dihydrotestosterone in vitro [184–186] reflecting hair growth in 5α-reductase deficiency;

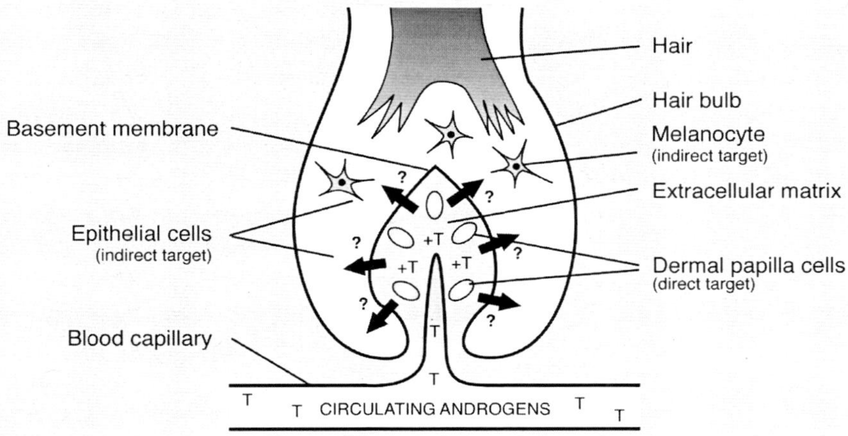

Figure 1.8 The current model for androgen action in the hair follicle. In this model androgens from the blood enter the hair follicle via the dermal papilla's blood supply. They are bound by androgen receptors in the dermal papilla cells causing changes in their production of regulatory paracrine factors; these then alter the activity of dermal papilla cells, follicular keratinocytes, melanocytes, etc. T = testosterone; ? = unknown paracrine factors. Reproduced from Randall [221].

5α-reductase type-2 gene expression also supports this [183]. These results led to wide acceptance of this hypothesis.

However, some recent observations suggest minor modifications. The dermal sheath, which isolates the follicle from the dermis, now seems to have other important roles as well, as it can form a new dermal papilla and stimulate follicle development [187]. Cultured dermal sheath cells from beard follicles contain similar levels of androgen receptors to dermal papilla cells (personal observations) and balding dermal sheath and dermal papilla express mRNA for 5α-reductase type-2 [188]. This indicates that the dermal sheath can respond directly to androgens without the dermal papilla acting as an intermediary. The sheath may be a reserve to replace a lost dermal papilla's key roles because of hair's essential role for mammalian survival and/or dermal sheath cells may respond directly to androgens to facilitate alterations in sheath, or even dermal papilla, size in forming a differently sized follicle.

Recently, a very specialised keratin, hHa7, was found in the medulla of hairs from beard, pubis, and axilla [189]. The medulla is formed by central hair cells which develop large air-filled spaces. Beard medulla cells showed coexpression of keratin hHa7 and the androgen receptor. Since the hHa7 gene promoter also contained sequences with high homology to the androgen response element (ARE), keratin hHa7 expression may be androgen-regulated. However, no stimulation occurred when the promoter was transfected into prostate cells and keratin hHa7 with the same promoter is also expressed in androgen-insensitive body hairs of chimpanzees [190] making the significance unclear. Nevertheless, the current model needs modification to include possible specific, direct action of androgens on lower dermal sheath and medulla cells.

The alteration of signalling molecules in the hair follicle by androgens. The final part of the mechanism of androgen action involves the alteration of paracrine signalling factors produced by dermal papilla cells. There is great interest in paracrine signalling in developing and cycling follicles, aiming to understand hair follicles as dynamic organs (see Sections 1.2 and 1.3) [90,190]. Unfortunately, there are few practical animal models for studying androgen effects [191] because of the special effects of androgens on human follicles. Fortunately, cultured dermal papilla cells from follicles with different sensitivities to androgens offer a useful model in which to study androgen effects due to the dermal papilla's central role, their abilities to be grown from small skin samples, to stimulate hair growth in vivo at low passage numbers [89,90], and to retain characteristics in vitro which reflect their androgen responses in vivo [191] (discussed earlier). They secrete both extracellular matrix [192] and soluble, proteinaceous factors which stimulate growth in other dermal papilla cells [180,193], outer root sheath cells [194,195], and transformed epidermal keratinocytes [196]. Soluble factors from human cells can cross species affecting rodent cell growth in vitro and in vivo [197], paralleling the ability of human dermal papillae to induce hair growth in vivo in athymic mice [198].

Importantly, physiological levels of testosterone in vitro increase the ability of beard cells to promote increased growth of other beard dermal papilla cells [193], outer root sheath cells [195], and keratinocytes [196] in line with the hypothesis. Interestingly, testosterone had no effect on non-balding scalp cells and only beard cells responded to the soluble factors produced [193], suggesting they have different receptors to non-balding scalp cells. This implies that an autocrine mechanism is involved in androgen-stimulated beard cell growth; androgen-mediated changes do involve alterations in dermal papilla cell numbers as well as the amount of extracellular matrix [174]. A need to modify the autocrine production of growth factors could contribute to the slow androgenic response, which often takes many years to reach full effect [22,134]. In contrast to the beard cell stimulation, testosterone decreased the mitogenic capacity of androgenetic alopecia dermal papilla cells from both men [196] and stump-tailed macaques [199]. All these results support the dermal papilla based model and demonstrate that the paradoxical androgen effects observed in vivo are reflected in vitro, strengthening the use of cultured dermal papilla cells as a model system for studying androgen action in vitro.

The main priority now is to identify the factors that androgens alter. So far, only IGF-1 is identified as secreted by beard cells under androgens in vitro [181]. IGF-1 is a potent mitogen which maintains anagen in cultured human follicles [109,200] and abnormal hair growth occurs in the IGF-I receptor deficient mouse [201] supporting its importance. Beard cells also secrete more SCF than non-balding scalp cells, although this is unaltered by androgens in vitro [202]. Since SCF plays important roles in epidermal [203] and hair pigmentation development [204], the dermal papilla probably provides local SCF for follicular melanocytes [202]. Androgens in vivo presumably increase *scf* expression by facial dermal papilla cells to cause hair darkening when boys' vellus hairs transform to adult beard. Recently DNA microarray methods also revealed that three genes, *sfrp-2, mn1*, and *atp1β1*, were expressed at significantly higher levels in beard than normal scalp cells, but no changes were detected due to androgen in vitro [205].

Although androgenetic alopecia dermal papilla cells are even more difficult to culture than normal follicles [206], androgens inhibit their expression of protease nexin-1, a potent inhibitor of serine proteases, which regulate cellular growth and differentiation in many tissues [207]. Androgens also stimulate their production of TGF-β and TGF-β2 [208,209]. TGF-β is a strong candidate for an inhibitor of keratinocyte activity in alopecia because it inhibits human follicle growth in vitro promoting catagen-like changes in human beings [111,210] and mice [211]; a probable TGF-β1 suppressor delays catagen in mice [212] and follicular keratinocytes have receptors for TGF-β [213]. However, in a limited DNA macroarray analysis TGF-β2 and TNF-α were actually slightly reduced in balding cells [214]. Balding scalp-cell conditioned media also inhibits human and rodent dermal papilla cell growth in vitro and delays mouse hair growth in vivo suggesting active secretion of inhibitory factors [197]. This is unlikely to involve TGF-β which is associated with the transition from anagen to catagen [210,211] and whose receptors are only detected on keratinocytes [213]. Thus, studying dermal papilla cells implicates several factors already: IGF-1 in enlargement, SCF in increased pigmentation, and nexin-1 and TGF-β in miniaturisation. Alterations in several factors are probably necessary to precisely control the major cell biological rearrangements required when follicles change size. Further research into such factors should help clarify the complex follicular cell interactions and the pathogenesis of androgen-dependent disorders.

1.8 Treatment of Hair Growth Disorders

Because human hair plays important roles in social and sexual communication (discussed in Section 1.2), hair where it is unwanted or hair loss is a source of embarrassment and psychological distress. A variety of methods are available to help control both excess hair growth and hair loss. The earliest methods used to remove hair were physical means such as shaving, followed by depilatory creams, waxes, or sugars; new developments include the use of lasers (see Chapter 10), and chemical inhibitors of hair growth such as Vaniqua [216,215]. Many substances have been suggested to stimulate hair growth over the years [20,217] with one of the most recent also being laser treatment. However, the most established promoters are topical applications of minoxidil (Regaine) or oral finasteride (Propecia) a 5α-reductase inhibitor used to block androgen effects in androgenetic alopecia [161]. The mechanism of action of minoxidil, an antihypertensive agent that promoted hair growth as an unacceptable side effect, has been a mystery despite its use for over 20 years; recent research supports action via potassium channels in the dermal papilla [218,219]. The most effective method remains transplanting androgen-independent hair follicles from the base of the scalp to the affected areas where they retain their intrinsic independence to androgens and maintain terminal hair [156]. Current research includes attempts to culture cells from hair follicles to amplify the individual's donor follicles.

Despite this range of treatments, neither excess hair growth nor hair loss are fully controlled; since much unwanted hair growth or hair loss is potentiated by androgens, any treatment has to be applied frequently and continually to counteract the constant supply of hormonal stimulation. Recently, successful clinical response to finasteride was related to increased dermal papilla expression of IGF-1 [220], confirming the importance of dermal papilla-produced paracrine factors and emphasising the dermal papilla's key role in androgen action. Greater understanding should lead to exciting new ways to treat hair disorders, as molecular pharmacology can devise very specific drugs and transport through the skin can target particular areas.

References

1. Dry FW. The coat of the mouse (*Mus musculus*). J Genet 1926; 16: 32–35.
2. Kligman AG. The human hair cycle. J Invest Dermatol 1959; 33: 307–316.
3. Randall VA. Androgens and human hair growth. Clin Endocrinol 1994; 40: 439–457.
4. Randall VA. Androgens and hair: a biological paradox. In: Nieschlag E, Behre HM, editors, Testosterone: Action, Deficiency, Substitution, 3rd Ed. Berlin: Cambridge University Press, 2004, pp. 207–231.
5. Jansen VAA, van Baalen M. Altruism through beard chromodynamics. Nature 2006; 440: 663–666.
6. Girman CJ, Rhodes T, Lilly FRW, Guo SS, Siervogal RM, Patrick DL, Chumlea WC. Effects of self-perceived hair loss in a community sample of men. Dermatol 1998; 197: 223–229.
7. Ebling FG, Hale PA, Randall VA. Hormones and hair growth. In: Goldsmith LA, editor, Biochemistry and Physiology of the Skin, 2nd .Ed. Oxford: Clarendon Press, 1991, pp. 660–690.
8. Montagna W, Van Scott EJ. The anatomy of the hair follicle. In: Montagna W, Ellis RA, editors, The Biology of Hair Growth. New York: Academic Press, 1958, pp. 39–64.
9. Flux JEC. Colour change of mountain hares (*Lepus timidus scotius*) in north-east Scotland. Zoology 1970; 162: 345–358.
10. West PM, Packer C. Sexual selection, temperature and the lion's mane. Science 2002; 297: 1339–1343.
11. Randall VA, Ebling EJG. Seasonal changes in human hair growth. Br J Dermatol 1991; 124: 146–151.
12. Orentreich N. Scalp hair replacement in men. In: Montagna W, Dobson RL, editors, Advances in Biology of Skin. Vol. 9: Hair Growth. Oxford: Pergamon Press, 1969, pp. 99–108.
13. Courtois M, Loussouarn G, Howseau S, et al. Periodicity in the growth and shedding of hair. Br J Dermatol 1996; 134: 47–54.
14. Goodhart CB. The evolutionary significance of human hair patterns and skin coloring. Adv Sci 1960; 17: 53–58.
15. Bradfield RB. Protein deprivation: comparative response of hair roots, serum protein and urinary nitrogen. Am J Clin Nutr 1971; 24: 405–410.
16. Reynolds EL. The appearance of adult patterns of body hair in man. Ann NY Acad Sci 1951; 53: 576–584.
17. Marshall WA, Tanner JM. Variations in pattern of pubertal change in girls. Arch Dis Child 1969; 44: 291–303.
18. Marshall WA, Tanner JM. Variations in the pattern of pubertal changes in boys. Arch Dis Child 1970; 45: 13–23.
19. Randall VA. Is alopecia areata an autoimmune disease? Lancet 2001; 358: 1922–1924.
20. Randall VA. Physiology and pathophysiology of androgenetic alopecia. In: Degroot LJ, Jameson JL, editors, Endocrinology, 5th Ed. Section XIV, Burger HG, editor, Male Reproduction. Philadelphia: WB Saunders Co., 2005, pp. 3295–3309.
21. Setty LR. Hair patterns of the scalp of white and negro males. Am J Phys Anthrop 1970; 33: 49–55.
22. Nutbrown M, Randall VA. Differences between connective tissue-epithelial junctions in human skin and the anagen hair follicle. J Invest Dermatol 1995; 104: 90–94.
23. Langbein L, Schweizer J. Keratins of the human hair follicle. Int Rev Cytol 2005; 243: 1–78.
24. Jave-Suarez LF, Langbein L, Winter H, Praetzel S, Rogers MA, Schweizer J. Androgen regulation of the human hair follicle: the type I hair keratin hHa7 is a direct target gene in trichocytes. J Invest Dermatol 2004; 122: 555–564.
25. Kaufman CK, Zhou P, Pasolli HA, Rendl M, Bolotin D, Lim KC, Dai X, Alegre ML, Fuchs E. GATA-3: an unexpected regulator of cell lineage determination in skin. Genes Dev 2003; 17: 2108–2122.
26. Botchkareva NV, Khlgatian M, Longley BJ, Botchkarev VA, Gilchrest BA. SCF/c-kit signaling is required for cyclic regeneration of the hair pigmentation unit. FASEB J 2001; 15: 645–658.

27. Byrne C, Fuchs E. Probing keratinocyte and differentiation specificity of the human K5 promoter in vitro and in transgenic mice. Mol Cell Biol 1993; 13: 3176–3190.
28. Coulombe PA, Kopan R, Fuchs E. Expression of keratin K14 in the epidermis and hair follicle: insights into complex programs of differentiation. J Cell Biol 1989; 109: 2295–2312.
29. Oshima H, Rochat A, Kedzia C, Kobayashi K, Barrandon Y. Morphogenesis and renewal of hair follicles from adult multipotent stem cells. Cell 2001; 104: 233–245.
30. Cotsarelis G, Sun TT, Lavker RM. Label-retaining cells reside in the bulge area of pilosebaceous unit: implications for follicular stem cells, hair cycle, and skin carcinogenesis. Cell 1990; 61: 1329–1337.
31. Commo S, Bernard BA. Immunohistochemical analysis of tissue remodelling during the anagen-catagen transition of the human hair follicle. Br J Dermatol 1997; 137: 31–38.
32. Lyle S, Christofidou-Solomidou M, Liu Y, Elder DE, Albelda S, Cotsarelis G. The C8/144B monoclonal antibody recognizes cytokeratin 15 and defines the location of human hair follicle stem cells. J Cell Sci 1998; 111: 3179–3188.
33. Michel M, Torok N, Godbout MJ, Lussier M, Pierrete G, Royal A, Germain L. Keratin 19 as a biochemical marker of skin stem cells in vivo and in vitro: keratin 19 expressing cells are differentially localized in function of anatomic sites, and their number varies with donor age and culture stage. J Cell Sci 1996; 109: 1017–1028.
34. Ohyama M, Terunuma A, Tock C, Radonovich M, Pise-Masison C, Hopping S, Brady J, Udey M, Vogel JC. Characterization and isolation of stem cell-enriched human hair follicle bulge cells. J Clin Invest 2006; 116: 249–260.
35. Cotsarelis G. Epithelial stem cells: a folliculocentric view. J Invest Dermatol 2006; 126: 1459–1468.
36. Nishimura EK, Jordan SA, Oshima H, Yoshida H, Osawa M, Moriyama M, Jackson IJ, Barrandon Y, Miyachi Y, Nishikawa S. Dominant role of the niche in melanocyte stem-cell fate determination. Nature 2002; 416: 854–860.
37. Hoffman RM. The potential of nestin-expressing hair follicle stem cells in regenerative medicine. Expert Opin Biol Ther 2007; 7: 289–291.
38. Jahoda CA, Horne KA, Oliver RF. Induction of hair growth by implantation of cultured dermal papilla cells. Nature 1984; 311: 560–562.
39. Reynolds AJ, Lawrence C, Cserhalmi-Friedman PB, Christiano AM, Jahoda CA. Trans-gender induction of hair follicles. Nature 1999; 402: 33–34.
40. Botchkarev VA, Botchkareva NV, Nakamura M, Huber O, Funa K, Lauster R, Paus R, Gilchrest BA. Noggin is required for induction of hair follicle growth phase in postnatal skin. FASEB J 2001; 15: 2205–2214.
41. Danilenko DM, Ring BD, Yanagihara D, Benson W, Wiemann B, Starnes CO, Pierce GF. Keratinocyte growth factor is an important endogenous mediator of hair follicle growth, development, and differentiation. Normalization of the nu/nu follicular differentiation defect and amelioration of chemotherapy-induced alopecia. Am J Pathol 1995; 147: 145–154.
42. Lindner G, Menrad A, Gherardi E, Merlino G, Welker P, Handjiski B, Roloff B, Paus R. Involvement of hepatocyte growth factor/scatter factor and met receptor signaling in hair follicle morphogenesis and cycling. FASEB J 2000; 14: 319–332.
43. Rudman SM, Philpott MP, Thomas GA, Kealey T. The role of IGF-I in human skin and its appendages: morphogen as well as mitogen? J Invest Dermatol 1997; 109: 770–777.
44. Hibberts NA, Messenger AG, Randall VA. Dermal papilla cells derived from beard hair follicles secrete more stem cell factor (SCF) in culture than scalp cells or dermal fibroblasts. Biochem Biophys Res Commun 1996; 222: 401–405.
45. Handjiski BK, Eichmuller S, Hofmann U, Czarnetzki BM, Paus R. Alkaline phosphatase activity and localization during the murine hair cycle. Br J Dermatol 1994; 131: 303–310.
46. Styczynski P, Ahluwalia G. Reduction of hair growth, USA, 2001, US6299865.
47. Lako M, Armstrong L, Cairns PM, Harris S, Hole N, Jahoda CA. Hair follicle dermal cells repopulate the mouse haematopoietic system. J Cell Sci 2002; 115: 3967–3974.
48. Richardson GD, Arnott EC, Whitehouse CJ, Lawrence CM, Reynolds AJ, Hole N, Jahoda CA. Plasticity of rodent and human hair follicle dermal cells: implications for cell therapy and tissue engineering. J Invest Dermatol Symp Proc 2005; 10: 180–183.

49. Hamilton JB. Age, sex and genetic factors in the regulation of hair growth in man: a comparison of Caucasian and Japanese populations. In: Montagna W, Ellis RA, editors, The Biology of Hair Growth. New York: Academic Press, 1958, pp. 399–433.
50. Chase HB. Growth of the hair. Physiol Rev 1954; 34: 113–126.
51. Chase HB, Eaton GJ. The growth of hair follicles in waves. Ann NY Acad Sci 1959; 83: 365–368.
52. Stenn KS, Combates NJ, Eilertsen KJ, Gordon JS, Pardinas JR, Parimoo S, Prouty SM. Hair follicle growth controls. Dermatol Clin 1996; 14: 543–558.
53. Kligman AM. The human hair cycle. J Invest Dermatol 1959; 33: 307–316.
54. Botchkarev VA, Kishimoto J. Molecular control of epithelial-mesenchymal interactions during hair follicle cycling. J Invest Dermatol Symp Proc 2003; 8: 46–55.
55. Botchkareva NV, Ahluwalia G, Shander D. Apoptosis in the hair follicle. J Invest Dermatol 2006; 126: 258–264.
56. Sato N, Leopold PL, Crystal RG. Induction of the hair growth phase in postnatal mice by localized transient expression of Sonic hedgehog. J Clin Invest 1999; 104: 855–864.
57. Paladini RD, Saleh J, Qian C, Xu GX, Rubin LL. Modulation of hair growth with small molecule agonists of the hedgehog signaling pathway. J Invest Dermatol 2005; 125: 638–646.
58. Paus R, Stenn KS, Link RE. Telogen skin contains an inhibitor of hair growth. Br J Dermatol 1990; 122: 777–784.
59. Muller-Rover S, Handjiski B, van der Veen C, Eichmuller S, Foitzik K, McKay IA, Stenn KS, Paus R. A comprehensive guide for the accurate classification of murine hair follicles in distinct hair cycle stages. J Invest Dermatol 2001; 117: 3–15.
60. Iwabuchi T, Maruyama T, Sei Y, Adachi K. Effects of immunosuppressive peptidyl-prolyl cis-trans isomerase (PPIase) inhibitors, cyclosporin A, FK506, ascomycin and rapamycin, on hair growth initiation in mouse: immunosuppression is not required for new hair growth. J Dermatol Sci 1995; 9: 64–69.
61. Paus R, Stenn KS, Link RE. The induction of anagen hair growth in telogen mouse skin by cyclosporine A administration. Lab Invest 1989; 60: 365–369.
62. Lutz G. Effects of cyclosporin A on hair. Review. Skin Pharmacol 1994; 7: 101–104.
63. Paus R, Cotsarelis G. The biology of hair follicles. N Engl J Med 1999; 341: 491–497.
64. Stenn KS, Paus R. Controls of hair follicle cycling. Physiol Rev 2001; 81: 449–494.
65. Botchkarev VA, Eichmuller S, Johansson O, Paus R. Hair cycle-dependent plasticity of skin and hair follicle innervation in normal murine skin. J Comp Neurol 1997; 386: 379–395.
66. Mecklenburg L, Tobin DJ, Muller-Rover S, Handjiski B, Wendt G, Peters EM, Pohl S, Moll I, Paus R. Active hair growth (anagen) is associated with angiogenesis. J Invest Dermatol 2000; 114: 909–916.
67. Yano K, Brown LF, Detmar M. Control of hair growth and follicle size by VEGF-mediated angiogenesis. J Clin Invest 2001; 107: 409–417.
68. Ahluwalia G, Styczynski P, Shander D. Inhibition of hair growth. USA, 2000, US6093748.
69. Kligman AG. The human hair cycle. J Invest Dermatol 1959; 33: 307–316.
70. Straile WE, Chase HB, Arsenault C. Growth and differentiation of hair follicles between periods of activity and quiescence. J Exp Zool 1961; 148: 205–221.
71. Lindner G, Botchkarev VA, Botchkareva NV, Ling G, van der Veen C, Paus R. Analysis of apoptosis during hair follicle regression (catagen). Am J Pathol 1997; 151: 1601–1617.
72. Seiberg M, Marthinuss J, Stenn KS. Changes in expression of apoptosis-associated genes in skin mark early catagen. J Invest Dermatol 1995; 104: 78–82.
73. Botchkareva NV, Kahn M, Ahluwalia G, Shander D. Survivin in the human hair follicle. J Invest Dermatol 2007; 127: 479–482.
74. Altieri DC. Survivin, versatile modulation of cell division and apoptosis in cancer. Oncogene 2003; 22: 8581–8589.
75. Botchkarev VA, Botchkareva NV, Peters EM, Paus R. Epithelial growth control by neurotrophins: leads and lessons from the hair follicle. Prog Brain Res 2004; 146: 493–513.
76. Foitzik K, Lindner G, Mueller-Roever S, Maurer M, Botchkareva N, Botchkarev V, Handjiski B, Metz M, Hibino T, Soma T, Dotto GP, Paus R. Control of murine hair follicle regression (catagen) by TGF-β1 in vivo. FASEB J 2000; 14: 752–760.

77. Suzuki S, Ota Y, Ozawa K, Imamura T. Dual-mode regulation of hair growth cycle by two Fgf-5 gene products. J Invest Dermatol 2000; 114: 456–463.
78. Yano K, Brown LF, Lawler J, Miyakawa T, Detmar M. Thrombospondin-1 plays a critical role in the induction of hair follicle involution and vascular regression during the catagen phase. J Invest Dermatol 2003; 120: 14–19.
79. Hebert JM, Rosenquist T, Gotz J, Martin GR. FGF5 as a regulator of the hair growth cycle: evidence from targeted and spontaneous mutations. Cell 1994; 78: 1017–1025.
80. Tsuji Y, Denda S, Soma T, Raftery L, Momoi T, Hibino T. A potential suppressor of TGF-β delays catagen progression in hair follicles. J Invest Dermatol Symp Proc 2003; 8: 65–68.
81. Peters EM, Stieglitz MG, Liezman C, Overall RW, Nakamura M, Hagen E, Klapp BF, Arck P, Paus R. p75 Neurotrophin receptor-mediated signaling promotes human hair follicle regression (catagen). Am J Pathol 2006; 168: 221–234.
82. Stenn KS, Paus R. Controls of hair follicle cycling. Physiol Rev 2001; 81: 449–494.
83. Foitzik K, Krause K, Conrad F, Nakamura M, Funk W, Paus R. Human scalp hair follicles are both a target and a source of prolactin, which serves as an autocrine and/or paracrine promoter of apoptosis-driven hair follicle regression. Am J Pathol 2006; 168: 748–756.
84. Pearson AJ, Ashby MG, Wildermoth JE, Craven AJ, Nixon AJ. Effect of exogenous prolactin on the hair growth cycle. Exp Dermatol 1999; 8: 358–360.
85. Telek A, Bíró T, Bodó E, Tóth BI, Borbíró I, Kunos G, Paus R. Inhibition of human hair follicle growth by endo- and exocannabinoids. FASEB J 2007; 21: 3534–3541.
86. Stenn KS, Parimoo S, Prouty SM. Growth of the hair follicle, a cycling and regenerating biological system. In: Chuong C-M, editor, Molecular Basis of Epithelial Appendage Morphogenesis. Austin, TX: RG Landes Co., 1998, pp. 111–130.
87. Milner Y, Sudnik J, Filippi M, Kizoulis M, Kashgarian M, Stenn K. Exogen, shedding phase of the hair growth cycle: characterization of a mouse model. J Invest Dermatol 2002; 119: 639–644.
88. Van Neste D, Leroy T, Conil S. Exogen hair characterization in human scalp. Skin Res Technol 2007; 13: 436–443.
89. Reynolds AJ, Jahoda CAB. Cultured human and rat tooth papilla cells induce hair follicle regeneration and fibre growth. Differentiation 2004; 72: 566–575.
90. Waters JM, Richardson GD, Jahoda CA. Hair follicle stem cells. Semin Cell Dev Biol 2007; 18: 245–254.
91. Saitoh M, Sakamoto M. Human hair cycle. J Invest Dermatol 1970; 54: 65–81.
92. Slominski A, Paus R, Costantino R. Differential expression and activity of melanogenesis-related proteins during induced hair growth in mice. J Invest Dermatol 1991; 96: 172–179.
93. Tobin DJ, Slominski A, Botchkarev V, Paus R. The fate of hair follicle melanocytes during the hair growth cycle. J Invest Dermatol 1999; 4: 323–332.
94. Osawa M, Egawa G, Mak SS, Moriyama M, Freter R, Yonetani S, Beermann F, Nishikawa S. Molecular characterization of melanocyte stem cells in their niche. Development (Cambridge, England) 2006; 132: 5589–5599.
95. Hearing VJ. Biochemical control of melanogenesis and melanosomal organization. J Invest Dermatol Symp Proc 1999; 4: 24–28.
96. Beermann F. The tyrosinase related protein-1 (Tyrp1) promoter in transgenic experiments: targeted expression to the retinal pigment epithelium. Cell Mol Biol 1999; 45: 961–968.
97. Winder A, Kobayashi T, Tsukamoto K, Urabe K, Aroca P, Kameyama K, Hearing VJ. The tyrosinase gene family—interactions of melanogenic proteins to regulate melanogenesis. Cell Mol Biol Res 1994; 40: 613–626.
98. Randall VA. Hormonal regulation of hair follicles exhibits a biological paradox. Semin Cell Dev Biol. 2007; 18(2): 274–285.
99. Ermak G, Slominski A. Production of POMC, CRH-R1, MC1, and MC2 receptor mRNA and expression of tyrosinase gene in relation to hair cycle and dexamethasone treatment in the C57BL/6 mouse skin. J Invest Dermatol 1997; 108: 160–165.
100. Kauser S, Thody AJ, Schallreuter KU, Gummer CL, Tobin DJ. A fully functional proopiomelanocortin/melanocortin-1 receptor system regulates the differentiation of human scalp hair follicle melanocytes. Endocrinology 2005; 146: 532–543.

101. Slominski A, Wortsman J, Plonka PM, Schallreuter KU, Paus R, Tobin DJ. Hair follicle pigmentation. J Invest Dermatol 2005; 124: 13–21.
102. Tobin DJ, Hordinsky M, Bernard BA. Hair pigmentation: a research update. J Investig Dermatol Symp Proc. 2005; 10:275-9. Review
103. Lincoln GA, Richardson M. Photo-neuroendocrine control of seasonal cycles in body weight, pelage growth and reproduction: lessons from the HPD sheep model. Comp Biochem Physiol Part C 1998; 119: 283–294.
104. Johnson E. Seasonal changes in the skin of mammals. Symp Zool Soc Land 1977; 39: 373–404.
105. Santiago-Moreno J, Lopez-Sebastian A, del Campo A, Gonzalez-Bulnes A, Picazo R, Gomez-Brunet A. Effect of constant-release melatonin implants and prolonged exposure to a long day photoperiod on prolactin secretion and hair growth in mouflon (*Ovis gmelini musimon*). Domest Anim Endocrinol 2004; 26: 303–314.
106. Duncan MJ, Goldman BD. Hormonal regulation of the annual pelage color cycle in the Djungarian hamster, *Phodopus surgorus*. II. Role of prolactin. J Exp Zool 1984; 230: 97–103.
107. Lincoln GA. Effects of placing micro-implant of melatonin in the pars tuberalis, pars distalis and the lateral septum of the forebrain on the secretion of follicle stimulating hormone and prolactin and testicular size in rams. J Endocrinol 1994; 142: 267–276.
108. Dicks P, Morgan CJ, Morgan PJ, Kelly D, Williams LM. The localisation and characterisation of insulin-like growth factor-1 receptors and the investigation of melatonin receptors on the hair follicles of seasonal and non-seasonal fibre-producing goats. J Neuroendocrinol 1996; 151: 55–63.
109. Philpott M. The roles of growth factors in hair follicles: investigations using cultured hair follicles. In: Camacho FM, Randall VA, Price VH, editors, Hair and Its Disorders: Biology, Pathology and Management. London: Martin Dunitz, 2000, pp. 103–113.
110. Rougeot J, Allain D, Martinet L. Photoperiodic and hormonal control of seasonal coat changes in mammals with special reference to sheep and mink. Acta Zoologica Fennica 1984; 171: 13–18.
111. Nixon AJ, Ford CA, Wildermouth JE, Craven AJ, Ashby MG. Regulation of prolactin receptor expression in ovine skin in relation to circulating prolactin and wool follicle growth status. J Endocrinol 2002; 172: 605–614.
112. Curlewis JD, Loudon AS, Milne JA, McNeilly AS. Effects of chronic long-acting bromocriptine treatment on liveweight, voluntary food intake, coat growth and breeding season in non-pregnant red deer hinds. J Endocrinol 1988; 119: 413–420.
113. Puchala R, Pierzynowski SG, Wuliji T, Goetsch AL, Soto-Navarro SA, Sahlu T. Effects of prolactin administered to a perfused area of the skin of Angora goats. J Anim Sci 2003; 81: 279–284.
114. Outit A, Morel G, Kelly PA. Visualisation of gene expression of short and long forms of prolactin receptor in the rat. Endocrinol 1993; 133: 135–144.
115. Foitzik K, Krause K, Nixon AJ, Ford CA, Ohnemus U, Pearson AJ, Paus R. Prolactin and its receptor are expressed in murine hair follicle epithelium, show hair cycle-dependent expression, and induce catagen. Am J Pathol 2003; 162: 1611–1621.
116. Rose J, Garwood T, Jaber B. Prolactin receptor concentrations in the skin of mink during the winter fur growth cycle. J Exp Zool 1995; 271: 205–210.
117. Choy VJ, Nixon AJ, Pearson AJ. Distribution of prolactin receptor immuno-reactivity in ovine skin and changes during the wool follicle growth cycle. J Endocrinol 1977; 155: 265–275.
118. Johnson E. Qualitative studies of hair growth in the albino rat. II. The effects of sex hormones. J Endocrinol 1958; 16: 351–359.
119. Rose J. Bilateral adrenalectomy induces early onset of summer fur growth in mink (*Mustela vison*). Comp Biochem Physiol C Parmacol Toxicol Endocrinol 1995; 111: 243–247.
120. Oh HS, Smart RC. An estrogen receptor pathway regulates the telogen-anagen hair follicle transition and influences epidermal cell proliferation. Proc Natl Acad Sci USA 1996; 93: 12525–12530.
121. Smart RC, Oh HS, Chanda S, Robinette CL. Effects of 17-β-estradiol and ICI 182 780 on hair growth in various strains of mice. J Invest Dermatol Symp Proc 1999; 4: 285–289.
122. Chanda S, Robinette CL, Couse JF, Smart RC. 17β-estradiol and ICI-182780 regulate the hair follicle cycle in mice through an estrogen receptor-α pathway. Am J Physiol Endocrinol Metab 2000; 278: E202–E210.

123. Movérare S, Lindberg MK, Faergemann J, Gustafsson JA, Ohlsson C. Estrogen receptor alpha, but not estrogen receptor beta, is involved in the regulation of the hair follicle cycling as well as the thickness of epidermis in male mice. J Invest Dermatol. 2002;119:1053-8.
124. Ohnemus U, Uenalan M, Conrad F, Handjiski B, Mecklenburg L, Nakamura M, et al. Hair cycle control by estrogens: catagen induction via ERα is checked by ERβ signalling. Endocrinol 2005; 145: 1214–1225.
125. Bidmon HJ, Pitts JD, Solomon HF, Bondi JV, Stumpf WE. Estradiol distribution and penetration in rat skin after topical application, studied by high resolution autoradiography. Histochem 1990; 95: 43–54.
126. Thornton MJ, Taylor AH, Mulligan K, Al-Azzawi F, Lyon CC, O'Driscoll J, et al. The distribution of estrogen receptor β is distinct to that of estrogen receptor α and the androgen receptor in human skin and the pilosebaceous unit. J Invest Dermatol Symp Proc 2003; 8: 100–103.
127. Thornton MJ, Nelson LD, Taylor AH, Birch MP, Laing I, Messenger AG. The modulation of aromatase and estrogen receptor α in cultured human dermal papilla cells by dexamethasone: a novel mechanism for selective action of estrogen via estrogen receptor β? J Invest Dermatol 2006; 126: 2010–2018.
128. Maurel D, Coutant C, Boissin J. Thyroid and gonadal regulation of hair growth during the seasonal moult in the male European badger, *Meles meles L.* Gen Comp Endocrinol 1987; 65: 317–327.
129. Steinmetz HW, Kaumanns W, Dix I, Heistermann M, Fox M, Kaup F-J. Coat condition, housing condition and measurement of faecal cortisol metabolites—a non-invasive study about alopecia in captive rhesus macaques (*Macaca mulatta*). J Med Primatol 2006; 35: 3–11.
130. Reinberg A, Smolensky MH, Hallek M, Smith KD, Steinberger E. Annual variation in semen characteristics and plasma hormone levels in men undergoing vasectomy. Fertil Steril 1988; 49: 309–315.
131. Reinberg A, Lagoguey M, Chauffourinier JM, Cesselin F. Circannual and circadian rhythms in plasma testosterone in five healthy young Parisian males. Acta Endocrinol 1975; 80: 732–743.
132. Smals AGH, Kloppenberg PWC, Benrad THJ. Circannual cycle in plasma testosterone levels in man. J Clin Endocrinol Metabol 1976; 42: 979–982.
133. Bellastella A, Criscuoco T, Mango A, Perrone L, Sawisi AJ, Faggiano M. Circannual rhythms of LH, FSH, testosterone, prolactin and cortisol during puberty. Clin Endocrinol 1983; 19: 453–459.
134. Hamilton JB. Patterned loss of hair in man; types and incidence. Ann NY Acad Sci 1951; 53: 708–728.
135. Randall VA, Thornton MJ, Messenger AG. Cultured dermal papilla cells from androgen-dependent human hair follicles (e.g. beard) contain more androgen receptors than those from non-balding areas of scalp. J Endocrinol 1992; 3: 141–147.
136. Pasquali R, Baraldi G, Casimirri F, Mattioli L, Capelli M, Melchionda N, Capani F, Labo G. Seasonal variations of total and free thyroid hormones in healthy men: a chronobiological study. Acta Endocimol (Copenh) 1984; 107: 42–48.
137. Eckert J, Church RE, Ebling FJ, Munro DS. Hair loss in women. Br J Dermatol 1967; 79: 543–548.
138. Wehr TA. Effects of seasonal changes in daylength on human neuroendocrine function. Horm Res 1998; 49: 118–124.
139. Yoneyama S, Hashimoto S, Honma K. Seasonal changes of human circadian rhythms in Antarctica. Am J Physiol 1999; 227: R1091–R1097.
140. Wehr TA, Duncan WC Jr, Sher L, Aeschbach D, Schwartz PJ, Turner EH, Postolache TT, Rosenthal NE. A circadian signal of change of season in patients with seasonal affective disorder. Arch Gen Psychiatry 2001; 58: 1115–1116.
141. Rushton HD. Commentary: Decreased serum ferritin and alopecia in women. J Invest Dermatol 2003; 121: xvii–xviii.
142. Jackson D, Church RE, Ebling FJG. Hair diameter in female baldness. Br J Dermatol 1972; 87: 361–367.
143. Lynfield YL. Effect of pregnancy on the human hair cycle. J Invest Dermatol 1960; 35: 323–327.

144. Fiotzik K, Krause K, Conrad F, Nakamura M, Funk W, Paus R. Human scalp hair follicles are both a target and a source of prolactin, which serves as an autocrine and/or paracrine promoter of apoptosis-driven hair follicle regression. Am J Pathol 2006; 168: 748–756.
145. Conrad F, Ohnemus U, Bodo E, Bettermann A, Paus R. Estrogens and human scalp hair growth-still more questions than answers. J Invest Dermatol 2004; 122: 840–842.
146. Winter JSD, Faiman C. Pituitary-gonadal relations in male children and adolescents. Paed Res 1972; 6: 125–135.
147. Winter JSD, Faiman C. Pituitary-gonadal relations in female children and adolescents. Paed Res 1973; 7: 948–953.
148. Chieffi M. Effect of testosterone administration on the beard growth of elderly males. J Gerontol 1949; 4: 200–204.
149. Hamilton JB. Effect of castration in adolescent and young adult males upon further changes in the proportions of bare and hairy scalp. J Clin Endocrinol Metabol 1960; 20: 1309–1318.
150. McPhaul MJ. Mutations that alter androgen function; androgen insensitivity and related disorders. In: Degroot LJ, Jameson JL, editors, Endocrinology, 5th Ed. Section XIV, Burger HG, editor, Male Reproduction. Philadelphia: WB Saunders Co., 2005, pp. 3139–3157.
151. Blok GJ, de Boer H, Gooren LJ, van der Veen EA. Growth hormone substitution in adult growth hormone-deficient men augments androgen effects on the skin. Clin Endocrinol 1997; 47: 29–36.
152. Norwood OTT. Male pattern baldness. Classification and incidence. South Med J 1975; 68: 1359–1370.
153. Randall VA. The biology of androgenetic alopecia. In: Camacho FM, Randall VA, Price VH, editors, Hair and Its Disorders: Biology, Pathology and Management. London: Martin Dunitz, 2000, pp. 123–136.
154. Ludwig E. Classification of the types of androgenic alopecia (common baldness) arising in the female sex. Br J Dermatol 1977; 97: 249–256.
155. Hayward S, Donjacour AA, Bhowmick NA, Thomson AA, Cunha GR. Endocrinology of the prostate and benign prostatic hyperplasia. In: Degroot LJ, Jameson JLB, editors, Endocrinology, 5th Ed. Section XIV, Burger HG, editor, Male Reproduction. Philadelphia: WB Saunders Co., 2005, pp. 3311–3324.
156. Orentreich N, Durr NP. Biology of scalp hair growth. Clin Plast Surg 1982; 9: 197–205.
157. Ziller C. Pattern formation in neural crest derivatives. In: Van Neste D, Randall VA, editors, Hair Research for the Next Millennium. Amsterdam: Elsevier Science, 1996, pp. 1–5.
158. Wu-Kuo T, Chuong C-M. Developmental biology of hair follicles and other skin appendages. In: Camacho FM, Randall VA, Price VH, editors, Hair and Its Disorders: Biology, Pathology and Management. London: Martin Dunitz, 2000, pp. 17–37.
159. Mou C, Jackson B, Schneider P, Overbeek PA, Headon DJ. Generation of the primary hair follicle pattern. Proc Natl Acad Sci USA 2006; 103: 9075–9080.
160. Fruzetti F. Treatment of hirsutism: antiandrogen and 5α-reductase inhibitor therapy. In: Azziz R, Nestler JE, Dewailly D, editors, Androgen Excess Disorders in Women. Philadelphia: Lippincott-Raven, 1997, pp. 787–797.
161. Kaufman KD, Olsen EA, Whiting D, Savi R, De Villez R, Bergfeld W, the Finasteride Male Pattern Hair Loss Study Group. Finasteride in the treatment of men with androgenetic alopecia. J Am Acad Dermatol 1998; 39: 578–589.
162. Whiting DA, Olsen EA, Savin R, Halper L, Rodgers A, Wang L, Hustad C, Palmisano J, Male Pattern Hair Loss Study Group. Efficacy and tolerability of finasteride 1 mg in men aged 41 to 60 years with male pattern hair loss. Eur J Dermatol 2003; 13: 150–160.
163. Birch MP, Messenger AG. Genetic factors predispose to balding and non-balding in men. Eur J Dermatol 2001; 11: 309–314.
164. Ellis JA, Harrap SB. The genetics of androgenetic alopecia. Clin Dermatol 2001; 19: 149–154.
165. Ewing JA, Rouse BA. Hirsutism, race and testosterone levels: comparison of East Asians and Euro-Americans. Hum Biol 1978; 50: 209–215.
166. Carey AH, Chan KL, Short F, White D, Williamson R, Franks S. Evidence for a single gene effect causing polycystic ovaries and male pattern baldness. Clin Endocrinol 1993; 38: 653–658.

167. Ellis JA, Stebbing M, Harrap SB. Genetic analysis of male pattern baldness and the 5α-reductase genes. J Invest Dermatol 1998; 110: 849–853.
168. Ha SJ, Kim JS, Myung JW, Lee HJ, Kim JW. Analysis of genetic polymorphisms of steroid 5α-reductase type 1 and 2 genes in Korean men with androgenetic alopecia. J Dermatol Sci 2003; 31: 135–141.
169. Ellis JA, Stebbing M, Harrap SB. Polymorphism of androgen receptor gene is associated with male pattern baldness. J Invest Derm 2000; 116: 452–455.
170. Ibanez L, Ong KK, Mongan N, Jaaskelainen J, Marcos MV, Hughes IA, De Zegher F, Dunger DB. Androgen receptor gene CAG repeat polymorphism in the development of ovarian hyperandrogenism. J Clin Endocr Metab 2003; 88: 3333–3338.
171. Stanford JL, Just JJ, Gibbs M, Wicklund KG, Neal CL, Blumenstein BA, Ostrander EA. Polymorphic repeats in the androgen receptor gene: molecular markers of prostate cancer risk. Cancer Res 1997; 57: 1194–1198.
172. Hibberts NA, Howell AE, Randall VA. Dermal papilla cells from human balding scalp hair follicles contain higher levels of androgen receptors than those from non-balding scalp. J Endocrinol 1998; 156: 59–65.
173. Van Scott EJ, Ekel TM. Geometric relationships between the matrix of the hair bulb and its dermal papilla in normal and alopecic scalp. J Invest Dermatol 1958; 31: 281–287.
174. Elliot K, Stephenson TJ, Messenger AG. Differences in hair follicle dermal papilla volume are due to extracellular matrix volume and cell number: implications for the control of hair follicle size and androgen responses. J Invest Dermatol 1999; 113: 873–877.
175. Seago SV, Ebling FJG. The hair cycle on the thigh and upper arm. Br J Dermatol 1985; 113: 9–16.
176. Sawers RA, Randall VA, Iqbal MJ. Studies on the clinical and endocrine aspects of antiandrogens. In: Jeffcoate JL, editor, Androgens and Antiandrogen Therapy. Current Topics in Endocrinology, Vol. 1. Chichester: John Wiley, 1982, pp. 145–168.
177. Handelsman DJ. Androgen action and pharmacologic uses. In: DeGroot LJ, Jameson JL, editors, Endocrinology, 5th Ed. Section XIV, Burger HG, editor, Male Reproduction. Philadelphia: WB Saunders Co., 2005, pp. 3121–3138.
178. Randall VA. The role of 5α-reductase in health and disease. Baillières Clin Endocrinol Metabol 1994; 8: 405–431.
179. Wilson JD, Griffin JE, Russell DW. Steroid 5α-reductase 2 deficiency. Endocr Rev 1993; 14: 577–593.
180. Randall VA, Thornton MJ, Hamada K, Redfern CPF, Nutbrown M, Ebling FJG, Messenger AG. Androgens and the hair follicle: cultured human dermal papilla cells as a model system. Ann NY Acad Sci 1991; 642: 355–375.
181. Itami S, Kurata S, Takayasu S. Androgen induction of follicular epithelial cell growth is mediated via insulin-like growth factor I from dermal papilla cells. Biochem Biophys Res Commun 1995; 212: 988–994.
182. Thornton MJ, Hibberts NA, Street T, Brinklow BR, Loudon AS, Randall VA. Androgen receptors are only present in mesenchyme-derived dermal papilla cells of red deer (*Cervus elaphus*) neck follicles when raised androgens induce a mane in the breeding season. J Endocrinol 2001; 168: 401–418.
183. Ando Y, Yamaguchi Y, Hamada K, Yoshikawa K, Itami S. Expression of mRNA for androgen receptor, 5α-reductase and 17β-hydroxysteroid dehydrogenase in human dermal papilla cells. Br J Dermatol 1999; 141: 840–845.
184. Itami S, Kurata S, Takayasu S. 5α-Reductase activity in cultured human dermal papilla cells from beard compared with reticular dermal fibroblasts. J Invest Dermatol 1990; 94: 150–152.
185. Thornton MJ, Liang I, Hamada K, Messenger AG, Randall VA. Differences in testosterone metabolism by beard and scalp hair follicle dermal papilla cells. Clin Endocrinol 1993; 39: 633–639.
186. Hamada K, Thornton MJ, Liang I, Messenger AG, Randall VA. Pubic and axillary dermal papilla cells do not produce 5α-dihydrotestosterone in culture. J Invest Dermatol 1996; 106: 1017–1022.

187. Reynolds AJ, Lawrence C, Cserhalmi-Friedman PB, Christiano AM, Jahoda CAB. Trans-gender induction of hair follicles. Nature 1999; 402: 33–34.
188. Asada Y, Sonoda T, Ojiro M, Kurata S, Sato T, Ezaki T, Takayasu S. 5α-Reductase type 2 is constitutively expressed in the dermal papilla and connective tissue sheath of the hair follicle in vivo but not during culture in vitro. J Clin Endocrinol Metab 2001; 86: 2875–2880.
189. Jave-Suarez LF, Langbein L, Winter H, Praetzel S, Rogers MA, Schweizer J. Androgen regulation of the human hair follicle: the type 1 hair keratin hHa7 is a direct target gene in trichocytes. J Invest Dermatol 2004; 122: 555–564.
190. Rendl M, Lewis L, Fuchs E. Molecular dissection of mesenchymal–epithelial interactions in the hair follicle. PLoS Biol 2005; 3(11): e331.
191. Randall VA, Sundberg JP, Philpott MP. Animal and in vitro models for the study of hair follicles. J Invest Dermatol Symp Proc 2003; 8: 39–45.
192. Messenger AG, Elliott K, Temple A, Randall VA. Expression of basement membrane proteins and interstitial collagens in dermal papillae of human hair follicles. J Invest Dermatol 1991; 96: 93–97.
193. Thornton MJ, Hamada K, Messenger AG, Randall VA. Beard, but not scalp, dermal papilla cells secrete autocrine growth factors in response to testosterone in vitro. J Invest Dermatol 1998; 111: 727–732.
194. Limat A, Hunziker T, Waelti ER, Inaebrit SP, Wiesmann U, Brathen LR. Soluble factors from human hair papilla cells and dermal fibroblasts dramatically increase the clonal growth of outer root sheath cells. Arch Dermatol Res 1993; 285: 205–210.
195. Itami S, Kurata S, Sonada T, Takayasu S: Interactions between dermal papilla cells and follicular epithelial cells in vitro: effect of androgen. Br J Dermatol 1995; 132: 527–532.
196. Hibberts NA, Randall VA. Testosterone inhibits the capacity of cultured cells from human balding scalp dermal papilla cells to produce keratinocyte mitogenic factors. In: Van Neste DV, Randall VA, editors, Hair Research for the Next Millennium. Amsterdam: Elsevier Science, 1996, pp. 303–306.
197. Hamada K, Randall VA. Inhibitory autocrine factors produced by the mesenchyme-derived hair follicle dermal papilla may be a key to male pattern baldness. Br J Dermatol 2006; 154: 609–618.
198. Jahoda CA, Oliver RF, Reynolds AJ, Forrester JC, Gillespie JW, Cserhalmi-Friedman PB, Christiano AM, Horne KA. Trans-species hair growth induction by human hair follicle dermal papillae. Exp Dermatol 2001; 10: 229–237.
199. Obana N, Chang C, Uno H. Inhibition of hair growth by testosterone in the presence of dermal papilla cells from the frontal bald scalp of the post-pubertal stump-tailed macaque. Endocrinol 1997; 138: 356–361.
200. Philpott MP, Sanders DA, Kealey T. Effects of insulin and insulin-like growth factors on cultured human hair follicles; IGF-1 at physiologic concentrations is an important regulator of hair follicle growth in vitro. J Invest Dermatol 1994; 102: 857–861.
201. Liu JP, Baker J, Perkins AS, Robertson EH, Efstratiadis A. Mice carrying null mutations of the genes encoding insulin-like growth factor 1 (IGF-1) and type 1 IGF receptor (IGF 1r). Cell 1993; 75: 59–72.
202. Hibberts NA, Messenger AG, Randall VA. Dermal papilla cells derived from beard hair follicles secrete more stem cell factor (SCF) in culture than scalp cells or dermal fibroblasts. Biochem Biophys Res Commun 1996; 222: 401–415.
203. Williams DE, de Vries P, Namen AE, Widmer MB, Lyman SD. The steel factor. Dev Biol 1992; 151: 368–376.
204. Fleischman RA, Saltman DL, Stastry V, Zneimer S. Deletion of the c-kit proto-oncogene in the human developmental defect piebald trait. Proc Natl Acad Sci USA 1991; 88: 10885–10889.
205. Rutberg SE, Kolpak ML, Gourley JA, Tan G, Henry JP, Shander S. Differences in expression of specific biomarkers distinguish human beard from scalp dermal papilla cells. J Invest Dermatol 2006; 126: 2583–2595.
206. Randall VA, Hibberts NA, Hamada K. A comparison of the culture and growth of dermal papilla cells derived from normal and balding (androgenetic alopecia) scalp. Br J Dermatol 1996; 134: 437–444.

207. Sonada T, Asada Y, Kurata S, Takayasu S. The mRNA for protease nexin-1 is expressed in human dermal papilla cells and its level is affected by androgen. J Invest Dermatol 1999; 113: 308–313.
208. Inui S, Fukuzato Y, Nakajima F, Yoshikawa K, Itami S. Androgen-inducible TGF-β1 from balding dermal papilla cells inhibits epithelial cell growth: a clue to understand paradoxical effects of androgen on human hair growth. FASEB J 2002; 16: 1967–1969.
209. Hibino T, Nishiyama T. Role of TGF-β2 in the human hair cycle. J Dermatol Sci 2004; 35: 9–18.
210. Soma T, Tsuji Y, Hibino T. Involvement of transforming growth factor-β2 in catagen induction during the human hair cycle. J Invest Dermatol 2002; 118: 993–997.
211. Soma T, Dohrmann CE, Hibino T, Raftery LA. Profile of transforming growth factor-β responses during the murine hair cycle. J Invest Dermatol 2003; 121: 969–975.
212. Tsuji Y, Denda S, Soma T, Raferty L, Momoi T, Hibino T. A potential suppressor of TGF-β delays catagen progression in hair follicles. J Invest Derm Symp Proc 2003; 8: 65–68.
213. Wollina U, Lange D, Funa K, Paus R. Expression of transforming growth factor beta isoforms and their receptors during hair growth phases in mice. Histol Histopathol 1996; 11: 431–436.
214. Midorikawa T, Chikazawa T, Yoshino T, Takada K, Arase S. Different gene expression profile observed in dermal papilla cells related to androgenic alopecia by DNA macroarray analysis. J Dermatol Sci 2004; 36: 25–32.
215. Azziz R. The evaluation and management of hirsutism. Obstet Gynecol 2003; 101: 995–1007.
216. Ross EK, Shapiro J. Management of hair loss. Dermatol Clin 2005; 23: 227–243.
217. Randall VA, Lanigan S, Hamzavi I, Chamberlain James L. New dimensions in Hirsutism. Lasers Med Sci 2006; 21: 126–133.
218. Davies GC, Thornton MJ, Jenner TJ, Chen YJ, Hansen JB, Carr RD, Randall VA. Novel and established potassium channel openers stimulate hair growth in vitro: implications for their modes of action in hair follicles. J Invest Dermatol 2005; 124: 686–694.
219. Shorter K, Farjo NP, Picksley SM, Randall VA. The human hair follicle contains two forms of ATP-sensitive potassium channels, only one of which is sensitive to minoxidil. FASEB J 2008; 22: 1725–1736.
220. Tang L, Bernardo O, Bolduc C, Lui H, Madani S, Shapiro J. The expression of insulin-like growth factor 1 in follicular dermal papillae correlates with therapeutic efficacy of finasteride in androgenetic alopecia. J Am Acad Dermatol 2003; 49: 229–233.
221. Randall VA. Androgens: the main regulator of human hair growth. In: Camacho FM, Randall VA, Price VH, editors, Hair and Its Disorders: Biology, Pathology and Management. London: Martin Dunitz, 2000, pp. 69–82.

2

Skin Biology: Understanding Biological Targets for Improving Appearance

John E. Oblong and Cheri Millikin

The Procter and Gamble Company, Cincinnati, OH, USA

2.1 Introduction

The human skin is the body's largest organ and serves critical functions such as acting as the first line of defense from daily exposure to environmental insults (ranging from microorganisms to irradiation to pollutants), helping to regulate the body's internal core temperature and water content, as well as providing rudimentary support and sensory interface with the outside world. This totally integrated structure serves the body's unique needs for maintaining its integrity, functionality, and defenses from the environment. However, in this capacity, the skin undergoes some of the most challenging conditions in the body, and its ability to respond to these challenges highlights the unique properties that it possesses. This chapter will provide a general overview of skin physiology and biochemical processes that regulate skin health and appearance. In addition, the changes that take place in skin as a function of aging and environmental insults from chronic UV damage are described.

Gurpreet S. Ahluwalia (ed.), *Cosmetic Applications of Laser and Light-Based Systems*, 37–48,

2.2 Basics of Skin Physiology

The morphology of skin is comprised of two primary layers of viable tissue that covers nearly the entire surface of the body [1,2]. These layers include the epidermis and dermis interspersed by a basement membrane, all of which reside on a subcutaneous fat layer or hypodermis. Residing in the skin are numerous structural appendages, including the hair follicle, eccrine, and sebaceous glands as well as capillary networks (Fig. 2.1).

The epidermis is the upper (or outermost) layer of the skin and comprises a cellular continuum from the underlying viable cell layers up through to the stratum corneum at the surface. This layer can be further subdivided into four layers: the basal, spinous, granular, and cornified layers (Fig. 2.2). Each cellular layer in the epidermis represents various stages along a process in which basal epidermal keratinocytes undergo a continuous cycle of proliferation, differentiation, and apoptosis moving upward from the basal layer to finally yield corneocytes that make up the stratum corneum. Basal keratinocytes reside at the lower portion of the epidermis supported on the basement membrane that separates the epidermis from the dermis. These mitotically active cells undergo a proliferative cycle to generate daughter cells that are physically dislocated upward into the spinous and granular layers and undergo the process of differentiation into corneocytes. During this process, the cells are attached to each other through desmosome connections via cadherins and

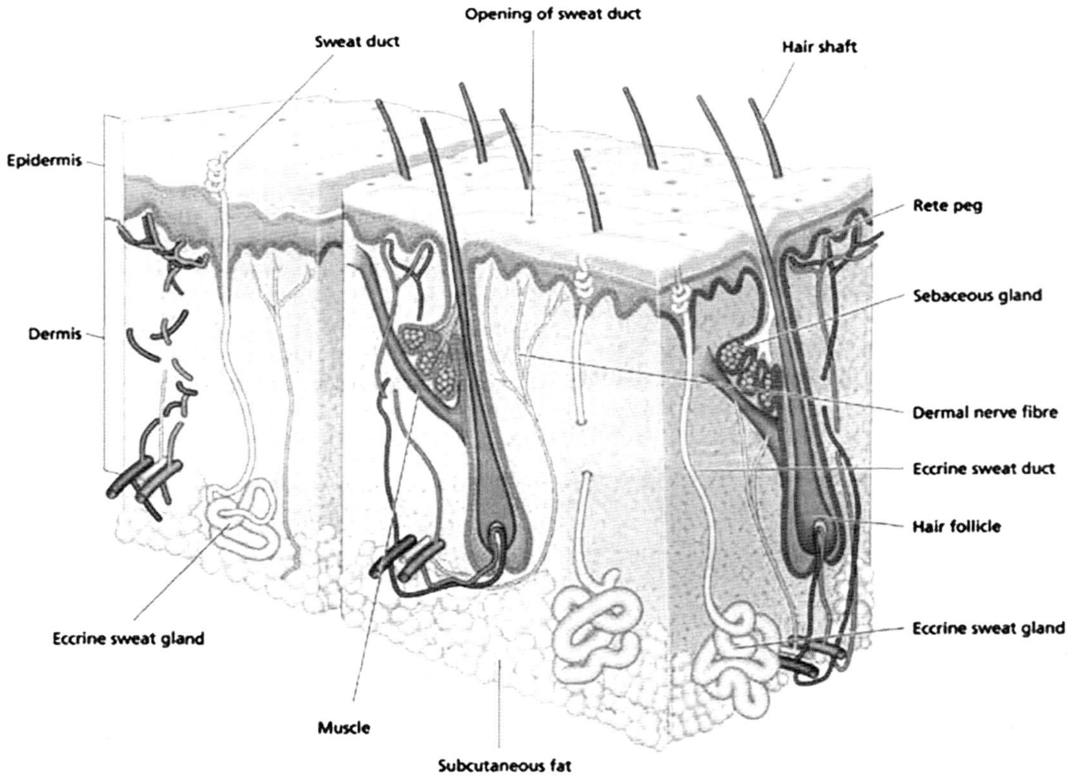

Figure 2.1 General schematic of skin's architecture.

intracellular keratin proteins become complexed with filaggrin, a large molecular weight protein that is posttranslationally processed (for a general review, see [3]). On passing through the spinous and granular layers, the cells undergo morphological changes that render them flatter in structure as they lose their cellular viability, undergo alternate keratin expression profiles, and transform into cellular remnants. A younger-aged epidermis turns over on average in 28–30 days, and this can rise to 40+ days in older-aged skin. Structurally, the resulting corneocytes remain connected to each other via integrins concentrated in surface desmosomes [4] and are interspersed with lipid bilayer lamellar structures, the latter of which also help provide part of the water barrier properties of skin. The layers of corneocytes in the stratum corneum averages 18–25 and the resulting barrier that is formed provides up to 98% of the water retention ability of the skin. The filaggrin present in corneocytes can be proteolytically degraded into small peptide fragments and ultimately into individual amino acids, depending on the relative water state of the upper layers of the epidermis. These amino acids along with urea and lactic acid in the stratum corneum are referred to as the skin's natural moisturizing factor (NMF). As external humidity changes impact the skin's water content, the keratin–filaggrin complexes in the corneocytes serve as a repository to generate more NMF via proteolysis to help counterbalance any trans-epidermal water loss. In contrast, exposure of skin to excessive humidity can be deleterious as well, leading to swelling of the stratum corneum and disruption of the lamellar structure bodies.

Other types of cells present in the epidermis include antigen presenting Langerhans, Merkel, and melanocytes. Langerhans cells are macrophages that serve as a primary defense to help prevent infection as well as block aberrant cellular growth as in the case of transformed tumor cells. These cells, along with macrophages in the dermis, are the main reason for the skin to be considered an immunologically related organ, because they function as the initial response when the skin comes into contact with a foreign substance. Merkel cells are essentially modified keratinocytes that are connected via desmosomes to surrounding keratinocytes and also serve the main role of mechanosensory detection via connections to nerve endings.

The melanocytes are specialized dendritic-like cells interspersed amongst basal keratinocytes and serve the primary function of producing melanin that is distributed to surrounding keratinocytes. Each melanocyte is in contact with upward of 30 keratinocytes via dendritic processes [5]. Melanin itself is comprised of eumelanin and pheomelanin, two pigmentary components that generate the diversity of coloration observed amongst the global population [6]. The melanins are complex polymers derived from tyrosine, which is converted to DOPA and dopaquinone by tyrosinase, a critical enzyme that is one of the regulatory points for melanogenesis [7]. At the molecular level, the chemistry of melanogenesis is a multistep process that involves a series of oxidative and complexation reactions, including complexation of dopaquinone with cysteine (derived from glutathione), which leads to the production of various forms of pheomelanin and is responsible for yellow or red pigment colors [8]. Pheomelanin is the primary pigment observed in red hair and light-skinned individuals. Alternatively, dopaquinone is converted into dopachrome, which can take two pathways to eumelanin production. Eumelanin is the primary pigment observed in darker-skinned individuals.

The general pigmentation of skin occurs via the active transport of melanin granules called melanosomes to neighboring keratinocytes via the melanocyte's dendritic processes [9] (Fig. 2.2). Melanosomes are lysosome-like structures whose characteristics

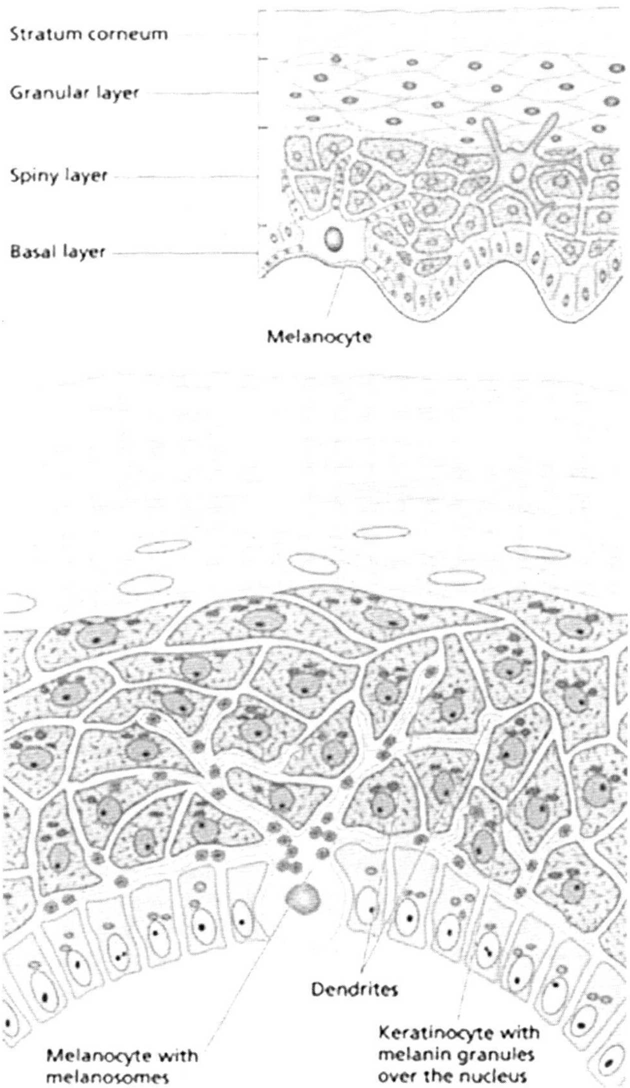

Figure 2.2 The primary layers present in the epidermis and melanin distribution from melanocytes.

differ depending on the type of melanin produced. Pheomelanosomes are primitive spherical structures whereas eumelanosomes are oval structures and express three times more tyrosinase than pheomelanosomes. The regulation of melanin production is very complex and involves upward of 80 genes [10,11]. The synthetic process has been found to be regulated by various extracellular signaling components that trigger a signal transduction cascade via melanocortin receptors [12,13]. While the baseline state of melanin in each individual's skin is dictated by genetic composition, external triggers such as UV irradiation and other forms of stress can lead to significant alterations in net synthesis of the melanins [14].

Underlying the epidermal layer is the basement membrane, a complex cutaneous network comprised of varying collagens, including types IV, VII, and XVII and anchoring fibrils. Published work examining the basement membrane zone (BMZ) has shown that complex hemidesmosome attachments present in the BMZ are critical for maintaining a stable integration between the epidermis and dermis, which is evident in some genetic disorders that leads to blistering of the skin [15]. These hemidesmosomes extend from the basal keratinocytes into the basement membrane. On the dermal side, anchoring fibrils allow for anchorage into the papillary region of the dermis.

The BMZ interface resides on the surface of the dermis, the thickest portion of the skin that makes up 90% of the total skin thickness, and is comprised primarily of various extracellular matrix (ECM) components, which render the skin's resilience and elasticity. The dermis can be divided into two layers, the papillary and reticular layers. The papillary layer resides immediately under the epidermis and helps provide some of the support to the basement membrane interface as well as extensions into the epidermis that are called rete ridges. These in turn help to maintain the skin's integrity via a better physical interface between the epidermis and dermis at a macro level. In contrast to the epidermis, the papillary layer is relatively sparse in total cellular content, yet contains mesenchymal-derived dermal fibroblasts. These cells serve the primary function of synthesizing ECM components, which include types I and III collagen, elastin, glycosaminoglycans (GAGs), and fibronectin, of which type I and III collagen make up greater than 85% of the total dermal ECM protein content. Collagen synthesis involves various posttranslational modification steps, including proteolytic removal of N- and C-terminal peptide fragments, arrangement into 3- and 4-triple helix complexes that finally assemble into regularly arranged fibrillar structures.

This blend of the collagen fibrillar network with the other ECM components provides the skin's strength, elastic, and turgor properties. The ECM content of the papillary layer is relatively densely packed in irregular distributions. Dermal fibroblasts help regulate the overall content, and thereby structural integrity, of the dermis by regulating the resupply of new collagen and the turnover and removal of older or damaged collagen. This occurs via a balance of new collagen synthesis and altered expression patterns of matrix metalloproteases (MMPs) and tissue inhibitor of matrix metalloproteases (TIMPs). The ability to turnover damaged collagen and repair with newly synthesized collagen is particularly relevant during photodamage, wound healing, and responses to other environmental insults. The reticular layer is sparser in cell content and is comprised of loosely held coarse fibers of collagen and other ECM components such as elastin. It serves as a main structural component to the skin because of its overall thickness and as an anchoring connective tissue for such appendages as sweat glands and hair follicles. Finally, the capillary vasculature that helps supply nutrients and waste removal from the skin resides in the dermis. Thus, when there is an injury that leads to blood loss, there is clear damage that extends through the epidermis into the dermis.

2.3 Changes in Skin Structure and Integrity as a Function of Environment and Aging

Over the course of an individual's life, human skin undergoes a steady process of morphological, structural, and biochemical alterations that are characterized as fine lines/wrinkles, texture, uneven skin tone, hyperpigmented spots, and loss of elasticity and resilience [16]. There are several working theories on the key causative scenarios to help explain changes observed in the aging process in general [17], ranging from oxidative stress and mitochondrial efficiency [18], telomere shortening [19], to hormonal changes [20]. One of the more germane theories for facial skin is the free radical theory of aging, which proposes that a lifetime of exposure to oxidative damage from intra- and extracellular radical oxygen species (ROS) will lead to an accumulation of damage that ultimately limits a cell's ability to function at its proper capacitance in maintaining intracellular homeostasis and proper communication with the ECM. Compounding this is a reduction in the redox status and antioxidant defenses of cells, which limits the ability to neutralize ROS as they are continually generated from metabolic processes. ROS can be viewed as causing aberrant chemical, and thereby physical, changes to proteins and lipids, and also trigger specific molecular signaling pathways in response to the damage [21]. Relative to aging, diminished redox status in cells as well as mitochondrial efficiency and oxidative radical generation are generally key sources of ROS in human tissue. In skin, environmental challenges can dramatically increase transient or acute levels.

While several aging theories are applicable to changes in human skin, the two primary drivers of these changes that have been studied extensively include photodamage from chronic UV exposure and intrinsic (or chronological) aging [22]. While research over the past few decades has found that photoaging and intrinsic aging can be, to a certain extent, superimposed upon each other [23,24], it is quite clear that UV damage elicits the greatest changes that are observed in skin and can accelerate the processes that are already being impacted by chronological aging [25,26]. These changes include decreased ECM content by both stimulation of the degradation process and reduction in new synthesis as well as more direct chemical damage to the ECM. A combination of various cutaneous changes in the skin leads to the general observation of fine lines and wrinkles, which may be further exaggerated by muscular contractions and connections to the underlying hypodermis [27]. Other environmental insults do have an effect on the aging process of skin [28], but they are not as significantly impactful as chronic UV exposure.

The observation of gross morphological changes in skin as a function of photodamage have been extensively studied and noted. The actual pathogenic agents that drive these changes are UV-generated radical oxygen species and hydrogen peroxides, which can trigger a cascade of biophysical and biochemical processes, some of which are understood and others which are speculated (Fig. 2.3). More recently, the usage of molecular techniques has begun to shed light on the actual mechanistic changes that occur down to the gene expression level [29]. However, even at the molecular level, it remains difficult to separate the effects of chronological aging in addition to chronic UV exposure because the common denominators of ROS and oxidative stress have implications in both phenomena (albeit the radical formation is initiated from alternate sources).

Acute changes in the environment, particularly humidity and temperature, can quickly lead to changes in the overall appearance of the skin. In skin that undergoes acute photodamage, the epidermis is often in a hyperplastic condition, which can be observed as a thickened epidermis. In the stratum corneum, the normal exfoliation of the corneocytes on the surface occurs via proteolytic cleavage of the desmosome connections, leading to sloughing of individual or small numbers of connected corneocytes. Disruption of this process when humidity conditions are altered can lead to an aberrant removal of the

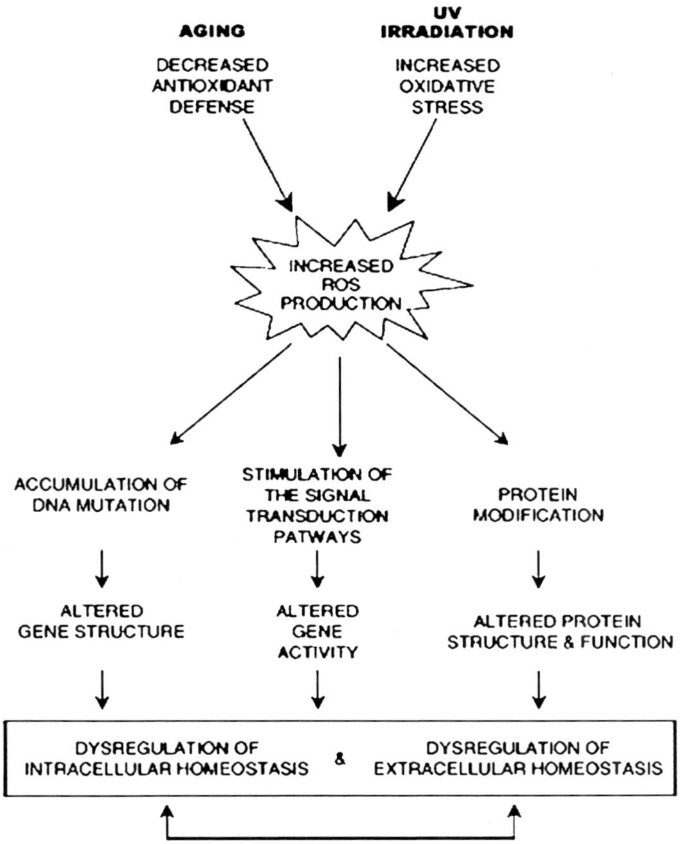

Figure 2.3 Schematic of oxidative stress via ROS as generated by intrinsic aging and UV exposure.

corneocytes and in some instances to a thickening of the epidermis. In addition, as a function of age, alterations in the turnover rate as well as expression patterns of filaggrin can lead to further disruption of the stratum corneum and its barrier properties. Thus, elderly skin tends to appear more dry, rough, and less translucent than younger-aged skin [30]. In the event of a wound or UV damage to the skin, the overall repair process and heightened sensitivity to sunburn is delayed in older-aged skin as well.

2.4 Photodamage to Skin

It is well-accepted that chronic UV exposure is one of the primary drivers for changes in the structure and function of skin that can be visualized as increased fine lines and wrinkles, altered pigmentation, and physical property changes. These changes, particularly in facial skin, can be detected at the structural, cellular, and molecular levels. Biophysical changes caused by ROS include protein degeneration that impacts structure and function as evidenced by cross-linking and glycation of ECM proteins including collagen and elastin [31] and lipid peroxidation. In addition, there is an unbalancing of the cellular redox status which in turn can impact the expression of stress response genes.

In general, there is a marked decrease in the levels of new collagen synthesized in the papillary layers, which reside closer to the surface and presumably sustain more UV damage than the underlying reticular layers [32]. The effect on net collagen is exaggerated by an increase in the expression levels of MMPs, including MMP-1, MMP-3, and MMP-9. The elevation of MMP protein levels in turn leads to a steady degradation of existing collagen fibril networks. This overall thinning process of the dermis is thought to lead to the sagging and furrowing of skin, causing wrinkles [33]. The degeneration and gross alterations in the elastin fiber network, referred to as elastosis, produce a thickened mass that is presumed to impact the skin's elastic properties, rendering it less resilient in comparison to nonphotodamaged skin. However, it is unclear whether any newly synthesized elastin by fibroblasts leads to functionally relevant elastin, as can be measured by restoration of skin's elasticity. To date, this does not appear to be the case. Levels of GAGs, another major EMC component, are elevated in the upper dermis [34]. The increase in GAGs are thought to occur in response to UV-induced damage to collagen, the likely function of GAG being to protect collagen from potential degradation by endogenous proteases. However, excessive levels are probably deleterious to the visible appearance of skin. In vitro cell culture experiments [35] indicate that the addition of GAGs inhibits collagen bundle assembly and thus would be expected to interfere with the dermal repair processes. The combination of these various cutaneous changes in skin contributes to the appearance of fine lines and wrinkles. The overall decrease in collagen content in skin, elevated levels of aberrantly cross-linked collagen, and increased GAG levels lead to the strong appearance of fine lines and wrinkles in human facial skin.

At the molecular level, it has been proposed that the generation of ROS inside the skin leads to the activation of specific signaling pathways that are induced [29]. Upon UV exposure, several cytokine and growth factor signaling pathways are activated, including EGF, TNFα, PDGF, and IL-1 and activate a MAP kinase-mediated cascade of signal transduction. In addition, UV irradiation of cells can lead to an increase in hydrogen peroxide production, which in turn may further stimulate or enhance the signaling pathways, particularly the G-protein-coupled protein-kinase-mediated ones. Upon transduction to the nucleus, there is further activation of the AP-1 transcription factor, which in turn regulates several stress response genes as well as collagen synthesis. Of particular relevance is type I collagen and the MMP family. However, there is a divergence of regulation in that UV-induced AP-1 upregulates MMP expression but suppresses type I collagen (both COL1A1 and COL1A2) gene expression. The overall mechanism of UV-generated ROS and hydrogen peroxide is not currently understood. However, the net result on alterations in gene expression patterns strongly supports the lowered responsiveness of damaged keratinocytes and fibroblasts to the environmental insult and the skin's ability to repair itself. This is further compounded in aged skin in contrast to younger-aged skin.

UV exposure can also cause hyperpigmentation of the skin, which is observed as a darkening of an area of skin caused by increased concentrations of melanin [36], which in turn acts as a natural sun screen to help protect the skin from further damage [37]. This phenomenon can also be elicited by aging and skin injury and results in the appearance of age spots, also known as lentigines. Age spots are harmless, flat, brown discolorations of the skin that usually appear on sun-exposed areas of the hands, neck, and face of people older than 40 years of age. However, pigmented lesions can also be the hallmark for early stages of melanoma. Long-term chronic exposure to UV damage has clearly been connected not only to premature aging and skin discoloration but is also one of the primary inducers of this deadly form of cancer.

2.5 Intrinsic Aging of Skin

Intrinsic or chronological aging occurs in all organs and cells in the body, including skin, and some of the major processes that are involved span cellular senescence, decreased metabolic capacitance, and diminished repair processes, including DNA repair as well as stress response [38].

The summation of these alterations in skin leads to clear physiological changes that can be observed as fine wrinkles, dryness, sallowness, and loss of elasticity. At the cellular level, there is a decreased proliferation pattern of epidermal keratinocytes and reduction in thickness of the dermis, which correlates with reduced collagen synthesis by dermal fibroblasts and aberrant melanogenesis from melanocytes. These changes not only lead to a general thinning of the skin but also impact the response to external insults and changes in the skin's physical properties, including loss of elasticity. For example, elderly skin is more susceptible to sunburn from acute UV exposure and wound healing is significantly impaired, rendering the skin more prone to injuries such as tears, ulcerations, and infections. Underlying these changes is a general onset of senescence in fibroblasts—one of the primary theories of aging—that may be correlated to telomere length, which serves as a "mitotic clock" [39]. Compounding these changes in female skin is also the effect of hormonal changes that can impact the general physiology of skin. Changes in estrogen levels have been associated with decreased collagen synthesis and these changes, including wound healing response, may be overcome by hormone replacement therapy in the elderly [40].

In the epidermis, the turnover rate is typically found to be extended, increasing up to 40 days in the elderly and a general thinning of the skin is observed. In younger-aged skin, the thickness of the epidermis is 35–50 μm, whereas in elderly skin it decreases to 25–40 μm. However, the relative number of cellular layers remains intact. Water retention potential is also diminished in elderly skin, which can be observed as having more incidence of dry skin conditions than younger-aged skin [41]. The number of active melanocytes decreases by about 10–20% per decade, probably explaining in part the increased vulnerability to UV radiation in old age [42]. The remaining melanocytes are sometimes observed to increase in size and are irregularly dispersed, potentially explaining the appearance of age spots in elderly skin, particularly in photodamaged skin [43]. The number of Langerhans cells decreases as a function of age but activity is also affected by free radical damage from UV exposure, further compromising wound healing and prevention of infections.

In older-aged skin, the undulations of the rete ridges in the BMZ extending from the dermis into the epidermis are less noticeable, leading to a general flattening of the basement membrane interface. This is thought to be part of the reason that elderly skin is more fragile to tears and blistering as well as overall wound- healing responses.

In the dermis, one of the most significant changes that can be observed as a function of aging is a general thinning of the ECM content. As the ECM is such a large percentage of the thickness of the dermis, this can quite easily be observed at a gross morphological level.

The overall thickness is estimated to decrease by 20% on average in nonphotodamaged areas. In areas with photodamage, this decrease is often higher due to the combination of the aging process and UV-induced chronic changes. The loss of ECM content in the dermis is primarily due to a decrease in the synthesis of new type I and III collagen from fibroblasts as well as an increase in MMP expression levels and aberrant TIMP regulation [44]. This net imbalance would cause the loss of collagen content, particularly in the papillary region where dermal fibroblasts undergo a steady progression into senescence [45]. There is also a loss of blood vessel networks, which can impact the coloration of skin and limit metabolic processes due to nutrient distribution and waste removal. The decreased levels of new collagen synthesis and the increase in degradation via proteolytic processes and nonenzymatic cross-linking modifications such as glycation contribute to the appearance of fine lines and wrinkles as well as sallowness. Of more relevance in elderly skin, the ability to repair following a wound event is compromised as well.

2.6 Treatment Effects

The fundamental understanding of the effects of photodamage and aging on skin that has occurred over the past few decades has led to the identification and commercialization of various technologies that have a mechanistic rationale for potential treatment effects, including reversal of the appearance of fine lines and wrinkles and pigmentary disorders. These range from actual physical agents that either remove varying layers of the stratum corneum and epidermis to stimulate a wound repair process or directly restore the ECM content in thinned dermis of photodamaged skin. Hydroxy acids, such as salicylic, lactic, and phenol, have been used in ranges of concentrations to remove varying depths of the epidermis. Under extreme situations, in which a significant portion of the epidermis is removed, there are dramatic effects in reducing the amount of fine lines and wrinkles and the evening of texture and pigmentation. However, there can be significant side effects of skin burns, thereby limiting the actual physical levels of these materials in cosmetic products and thus rendering them somewhat effective but limited. Other types of materials that act in a physical and acute manner (benefits observed immediately) are dermal fillers. These types of materials are injected directly into the dermis of facial skin and include collagen, hyaluronic acid, and microspheres, the last of which provides more permanent effects. Another type of injectionable agent is Botox, the bacterial botulinum neurotoxin. Upon injection, the neurotoxin causes a temporary relaxation of muscles in the skin by blocking neural stimulation, thereby allowing the skin to relax and essentially makes fine lines and wrinkles unnoticeable.

More selective technologies that are connected to aspects of collagen synthesis include retinoids, ascorbic acid, and peptides, all of which have been shown to stimulate or enhance collagen synthesis from dermal fibroblasts in vitro and/or in vivo. In addition, there are numerous antioxidant-based technologies that in theory would impact the oxidative stress generated in the skin from UV and intrinsic aging. A more specific example includes niacinamide, which was identified based on its ability to restore the imbalance in key redox regulators that occur in aged skin. Relative to pigmentation, the usage of bleaching agents such as hydroxyquinone can be very effective in gross alteration of pigmentation but more selective agents such as retinoids, vitamin C analogues, and glucosamine derivatives can be used cosmetically with reduced negative side effects.

The last decade has seen a literal explosion of knowledge and development in the safe usage of devices in the professional marketplace to treat various skin ailments, including photodamage. This area is reviewed in more detail in Chapter 15 but essentially this methodology has significantly impacted the professional aesthetics marketplace financially as well as by providing to patients treatments that are both less invasive and deliver significant efficacy.

In the long run, continued understanding of the mechanisms and signaling pathways that are impacted in skin as a function of photodamage and intrinsic aging will provide additional mechanistic insight that should lead to more effective technologies. Equally important is that these technologies would be selective in their action, thereby in theory allowing for safe usage by consumers of both cosmetic and drug products. A separate but related, aspect is the continued development of home-use energy emitting devices that, either alone or in combination with topical agents, will provide to the consumer an even better arsenal to combat the ravages of time and environmental damage.

References

1. Montagna W. (1962) *The Structure and Function of Skin*, 2nd edition. New York: Academic Press.
2. Greaves MW. (1976) Physiology of skin. *J Invest Dermatol.* **67**:66–69.
3. Dale BA, Holbrook KA. (1987) Developmental expression of human epidermal keratins and filaggrin. *Curr Top Dev Biol.* **22**:127–151.
4. Furukawa F, Takigawa M, Matsuyoshi N, Shirahama S, Wakita H, Fujita M, Horiguchi Y, Imamura S. (1994) Cadherins in cutaneous biology. *J Dermatol.* **21**:802–813.
5. Hoath SB, Leahy DG. (2003) The organization of human epidermis: functional epidermal units and phi proportionality. *J Invest Dermatol.* **121**:1440–1446.
6. Hunt G, Kyne S, Ito S, Wakamatsu K, Todd C, Thody A. (1995) Eumelanin and phaeomelanin contents of human epidermis and cultured melanocytes. *Pigment Cell Res.* **8**:202–208.
7. Riley PA. (1993) Mechanistic aspects of the control of tyrosinase activity. *Pigment Cell Res.* **6**:182–185.
8. Land EJ, Ramsden CA, Riley PA. (2001) Pulse radiolysis studies of ortho-quinone chemistry relevant to melanogenesis. *J Photochem Photobiol B.* **64**:123–135.
9. Boissy RE. (2003) Melanosome transfer to and translocation in the keratinocyte. *Exp Dermatol.* **12**:5–12.
10. Hearing VJ. (1999) Biochemical control of melanogenesis and melanosomal organization. *J Investig Dermatol Symp Proc.* **4**:24–28.
11. Schallreuter KU. (2007) Advances in melanocyte basic science research. *Dermatol Clin.* **25**:283–291.
12. Abdel-Malek Z, Suzuki I, Tada A, Im S, Akcali C. (1999) The melanocortin-1 receptor and human pigmentation. *Ann NY Acad Sci.* **885**:117–133.
13. Kauser S, Schallreuter KU, Thody AJ, Gummer C, Tobin DJ. (2003) Regulation of human epidermal melanocyte biology by beta-endorphin. *J Invest Dermatol.* **120**:1073–1080.
14. Costin GE, Hearing VJ. (2007) Human skin pigmentation: melanocytes modulate skin color in response to stress. *FASEB J.* **21**:976–994.
15. Uitto J, Pulkkinen L. (1996) Molecular complexity of the cutaneous basement membrane zone. *Mol Biol Rep.* **23**:35–46.
16. Lavker RM, Zheng PS, Dong G. (1986) Morphology of aged skin. *Dermatol Clin.* **4**:379–389.
17. Viña J, Borrás C, Miquel J. (2007) Theories of ageing. *IUBMB Life.* **59**:249–254.
18. Wei YH, Lu CY, Wei CY, Ma YS, Lee HC. (2001) Oxidative stress in human aging and mitochondrial disease-consequences of defective mitochondrial respiration and impaired antioxidant enzyme system. *Chin J Physiol.* **44**:1–11.
19. Harley CB, Vaziri H, Counter CM, Allsopp RC. (1992) The telomere hypothesis of cellular aging. *Exp Gerontol.* **27**:375–382

20. Morley JE, Unterman TG. (2000) Hormonal fountains of youth. *J Lab Clin Med.* **135**:364–366.
21. Bulteau AL, Moreau M, Nizard C, Friguet B. (2007) Proteasome and photoaging: the effects of UV irradiation. *Ann NY Acad Sci.* **1100**:280–290.
22. Fenske NA, Lober CW. (1986) Structural and functional changes of normal aging skin. *J Am Acad Dermatol.* **15**:571–585.
23. Gilchrest BA, Yaar M. (1992) Ageing and photoageing of the skin: observations at the cellular and molecular level. *Br J Dermatol.* **41**:25–30.
24. Rabe JH, Mamelak AJ, McElgunn PJ, Morison WL, Sauder DN. (2006) Photoaging: mechanisms and repair. *J Am Acad Dermatol.* **55**:1–19.
25. Kligman LH, Kligman AM. (1986) The nature of photoaging: its prevention and repair. *Photodermatol.* **3**:215–227.
26. Kang S, Fisher GJ, Voorhees JJ. (2001) Photoaging: pathogenesis, prevention, and treatment. *Clin Geriatr Med.* **17**:643–659.
27. Pierard GE, Lapiere CM. (1989) The microanatomical basis of facial frown lines. *Arch Dermatol.*, **125**:1090–1092.
28. Morita A. (2007) Tobacco smoke causes premature skin aging. *J Dermatol Sci.* **48**:169–175.
29. Fisher GJ, Kang S, Varani J, Bata-Csorgo Z, Wan Y, Datta S, Voorhees JJ. (2002) Mechanisms of photoaging and chronological skin aging. *Arch Dermatol.* **138**:1462–1470.
30. Hashizume H. (2004) Skin aging and dry skin. *J Dermatol.* **31**:603–609.
31. Alpermann H, Vogel HG. (1978) Effect of repeated ultraviolet irradiation on skin of hairless mice. *Arch Dermatol Res.* **262**:15–25.
32. Bernstein EF, Chen YQ, Kopp JB, Fisher L, Brown DB, Hahn PJ, Robey FA, Lakkakorpi J, Uitto J. (1996) Long-term sun exposure alters the collagen of the papillary dermis. Comparison of sun-protected and photoaged skin by northern analysis, immunohistochemical staining, and confocal laser scanning microscopy. *J Am Acad Dermatol.* **34**:209–218.
33. Contet-Audonneau JL, Jeanmaire C, Pauly, G. (1999) A histological study of human wrinkle structures: comparison between sun-exposed areas of the face, with or without wrinkles, and sun-protected areas. *Brit J Dermatol.* **140**:1038–1047.
34. Gonzalez S, Moran M, Kochevar IE. (1999) Chronic photodamage in skin of mast cell-deficient mice. *Photochem Photobiol.* **70**:248–253.
35. Guidry C, Grinnell F. (1987) Heparin modulates the organization of hydrated collagen gels and inhibits gel contraction by fibroblasts. *J Cell Biol.* **104**:1097–1103.
36. Ortonne JP. (1990) The effects of ultraviolet exposure on skin melanin pigmentation. *J Int Med Res.* **18**:8C–17C.
37. Kollias N, Sayre RM, Zeise L, Chedekel MR. (1991) Photoprotection by melanin. *J Photochem Photobiol B.* **9**:135–160.
38. Makrantonaki, E, Zouboulis CC. (2007) Molecular mechanisms of skin aging: state of the art. *Ann NY Acad Sci.* **1119**:40–50.
39. Allsopp RC, Harley CB. (1995) Evidence for a critical telomere length in senescent human fibroblasts. *Exp Cell Res.* **219**:130–136.
40. Castelo-Branco C, Duran M, González-Merlo J. (1992) Skin collagen changes related to age and hormone replacement therapy. *Maturitas.* **15**:113–119.
41. Cua AB, Wilhelm KP, Maibach HI. (1990) Frictional properties of human skin: relation to age, sex and anatomical region, stratum corneum hydration and transepidermal water loss. *Br J Dermatol.* **123**:473–479.
42. Ortonne JP. (1990) Pigmentary changes of the ageing skin. *Br J Dermatol.* **122**:21–28.
43. Haddad MM, Xu W, Medrano EE. (1998) Aging in epidermal melanocytes: cell cycle genes and melanins. *J Investig Dermatol Symp Proc.* **3**:36–40.
44. Hornebeck W. (2003) Down-regulation of tissue inhibitor of matrix metalloprotease-1 (TIMP-1) in aged human skin contributes to matrix degradation and impaired cell growth and survival. *Pathol Biol.* **51**:569–573.
45. Campisi J. (1998) The role of cellular senescence in skin aging. *J Investig Dermatol Symp Proc.* **3**:1–5.

3

Physics Behind Light-Based Systems: Skin and Hair Follicle Interactions with Light

Gregory B. Altshuler[1] *and Valery V. Tuchin*[2,3]

[1]*Palomar Medical Technologies, Inc., Burlington, MA, USA*
[2]*Institute of Optics and Biophotonics, Saratov State University, Saratov, Russia*
[3]*Institute of Precise Mechanics and Control of RAS, Saratov, Russia*

Gurpreet S. Ahluwalia (ed.), *Cosmetic Applications of Laser and Light-Based Systems*, 49–123,

3.1 Introduction

Skin as a biological tissue is an optically inhomogeneous and absorbing medium whose average refractive index is higher than that of air. This is responsible for the partial reflection of radiation at the skin/air interface, while the remaining part penetrates the skin. Multiple scattering and absorption are responsible for laser beam broadening and eventual decay as the radiation travels through the skin, whereas bulk scattering within skin dermis and underlying tissues is a major cause of the dispersion of a large fraction of radiation in the backward direction. Therefore, light propagation within the skin depends on the scattering and absorption properties of its compartments: cells, cell organelles, and various fiber structures [1–11]. The size, shape, and density of these structures, their refractive index, relative to the interstitial ground substance, and the polarization state of the incident light all play important roles in the propagation of light in tissues.

Light interaction with a multilayer and multicomponent skin is a very complicated process [1–11]. The horny skin layer (stratum corneum) reflects about 5–7% of the incident light. A collimated light beam is transformed to a diffuse one by microscopic inhomogeneities at the air/horny layer interface. A major part of reflected light results from backscattering in different skin layers (stratum corneum, epidermis, dermis, blood, and fat). The absorption of diffuse light by the major skin pigments, such as melanin, hemoglobin, and its oxygenated form, is an informative feature for diagnosis and monitoring of skin pathology and aging.

Light-induced thermal effects in skin are important for diagnostics, therapy, and surgery. The optothermal diagnostic methods are based on detection of the time-dependent heat generation, induced in skin by a comparably low-intensive pulsed or modulated optical radiation. They allow one to estimate optical, thermal, and acoustic properties of skin and underlying tissues that depend on peculiarities of tissue structure.

For thermal phototherapy and surgery, much higher light intensities are used. In these cases, controllable temperatures rise, and thermal and/or thermo-mechanical damage (coagulation, vaporization, vacuolization, pyrolysis, ablation) of skin are important.

In this chapter we discuss the basic physics of light and light interaction with skin that defines light propagation in skin and light photothermal action. Light refraction, scattering, absorption, as well as spectral and polarization properties are analyzed. Different light sources and fibers for light delivery are briefly described. Skin's optical properties, its penetration depth, transmittance, reflectance, and fluorescence spectra formation are also discussed. The prospective use of skin optical clearing technology for more effective applications of various optical methods is also presented. Mechanisms of light tissue interaction of inducing photochemical, photothermal, and photomechanical reactions are discussed in the framework of skin selective photothermolysis and extensions of this technology.

3.2 What is Light?

3.2.1 Electromagnetic Waves and Photons

Light is the common name of electromagnetic radiation (EMR) that we can see. Lamps, lasers, and light emitting diodes (LEDs) that generate light can also emit EMR, which is not visible. However, they have specific features characteristic to the visible light, such as photochemical action for the shorter wavelengths (violet color) and thermal action for the longer ones (red color). Thus, two neighboring regions of EMR (i.e., ultraviolet (UV) and infrared (IR)) also belong to light. UV and IR light can be seen (visualized) with the help of matrix photodetectors such as CCD or IR thermal cameras. Electromagnetic radiation propagates in vacuum or different media in the form of electromagnetic waves, which are periodical oscillations of electrical and magnetic fields in time and space.

Light can be also described as a stream of photons. Photon is a quantum of EMR, usually considered as an elementary particle that has energy $E_{ph} = h\nu$ or $h(c/\lambda)$, where h is the Planck's constant (a physical constant that is used to describe the sizes of quanta; it plays a central role in the theory of quantum mechanics, and is named after Max Planck, one of the founders of quantum theory), ν is the frequency of light, c is the speed of light, and λ is the wavelength of light. Wavelength is the distance between two adjacent peaks in electric or magnetic fields of EMR of a light wave, measured typically in nanometers (nm) or micrometers (μm): 1μm is 10^{-9} meter (m) and 1 μm is 10^{-6} m; $h = 6.626 \times 10^{-34}$ joules × seconds; $c = 3 \times 10^{8}$ m/s in a free space, ν is expressed in hertz (Hz), 1 Hz is s^{-1}, because of high frequency of light its frequency typically expressed in THz: 1 THz is 10^{12} Hz. For example, IR radiation with the wavelength of 10 μm is oscillating with the frequency of 30 THz.

To evaluate the total energy of the light beam, we need to account for each photon that was detected or interacted with the target; if the total number of photons is N, then $E_{total} = N \times E_{ph} = N \times h\nu$.

3.2.2 Wavelength Range

Physicians who apply light in phototherapy or vision science classify the whole light spectrum (i.e., from 100 nm to 1000 μm) based on its major mechanism of interaction with biological cells and tissues. In particular, light spectral ranges are described as: *ultraviolet* (UV) light—UVC, 100–280 nm; UVB, 280–315 nm; and UVA, 315–400 nm; *visible*—400–780 nm (violet, 400–450 nm; blue, 450–480 nm; green, 510–560 nm; yellow, 560–590 nm; orange, 590–620 nm; and red, 620–780 nm); *infrared* (IR) light—IRA, 0.78–1.4 μm; IRB, 1.4–3.0 μm; and IRC, 3–1000 μm. However, physicists who consider light's interaction with and propagation in a biological media (atmosphere, ocean, etc.) classify light spectrum as UV (100–400 nm), visible (400–800 nm), near IR (NIR) (0.8–2.5 μm), middle IR (MIR) (2.5–50 μm), and far IR (FIR) (50–2000 μm). Presently, as light is more and more widely and effectively used in medicine, both classifications and terminologies are in use in the biomedical optics world. For example, because of a great success in tissue spectroscopy and imaging in the near infrared range the term NIR is often used now by physicians. A current interest and future perspective of the terahertz range of electromagnetic radiation in biomedical applications is the spreading of the light wavelength range used in medicine to the 2000 μm that physicists use.

3.2.3 Energy and Power

To characterize the efficiency of light interaction with biological tissue (inducing a photochemical reaction, temperature increase, evaporation, thermal mechanical breaking, etc.) besides choosing the wavelength of light, its energetic parameters are also important. Two major parameters are typically used: energy and power. Energy is the ability of light (as well as other forms of energy, such as mechanical, thermal, electrical, chemical, and nuclear) to produce some work; energy E is measured in joules (J). Power is the rate of delivery of energy; it is normally measured in watts (W) (i.e., joules per second (J/s)). The smaller and bigger energy units are in use in biomedical optics and photomedicine to characterize light sources and delivery optics; typically they are: energy—microjoule (μJ), 10^{-6} *J*; millijoule (mJ), 10^{-3} J; kilojoule (kJ), 10^3 J; power (P)—microwatt (μW), 10^{-6} W; milliwatt (mW), 10^{-3} W; kilowatt (kW), 10^3 W; megawatt (MW), 10^6 W; and gigawatt (GW), 10^9 W.

At light interaction with tissues, produced photophysical, photochemical, or photobiological effects depend on energy density and/or power density that was provided within the target area. Energy density or fluence is the energy of the light wave that propagates through a unit area which is perpendicular to the direction of propagation of the light wave. Fluence is measured in J/cm^2 or J/m^2. Power density is the power of the light wave that propagates through a unit area which is perpendicular to the direction of propagation of the light wave. Power density or intensity is measured in W/cm^2 or W/m^2. The relationship between fluence (f) and intensity (I) is given by: $F = I \cdot \tau_p$, where τ_p is the length of pulse (pulsewidth) or exposure time. In inhomogeneous light scattering media to which tissues belong, the following parameter is often used: fluence rate (or total radiant energy fluence rate), that is, the sum of the radiance over all angles at a point $\bar{r}$; the quantity that is typically measured in irradiated tissues in units watts per square meter or centimeter (W/m^2 or W/cm^2).

Several measures of light are commonly known as intensity: radiant intensity is a radiometric quantity, measured in watts per steradian (W/sr); luminous intensity is a photometric quantity, measured in lumens per steradian (lm/sr), or candela (cd); radiance (irradiance) is commonly called "intensity" or "quantum flux" measured in W/m^2 or W/cm^2.

3.2.4 Light Beam and Divergence

In general, a light beam is a slender stream of light. Light is an electromagnetic wave. Electromagnetic waves can be characterized by wave fronts. A wave front is a surface where the electromagnetic field of light is oscillating in the same phase. In the geometric optics approximation, light can be presented as a family of rays. These rays are always perpendicular to the wave front. Often the user needs a collimated beam—a beam of light in which all rays are parallel to each other and the wave front is a plane. Such a beam, in some cases, can be provided automatically by using an appropriate laser or can be formed by special optics (with possible significant loss of light energy) using conventional light sources such as lamps.

A laser beam is a group of nearly parallel rays generated by a laser; a light beam with a Gaussian shape for the transverse intensity profile: if the intensity at the center of the beam is I_0, then the formula for a Gaussian beam is $I = I_0 \exp(-2r^2/w_0^2)$, where r is the radial distance from the axis and w_0 is the beam "waist" (the narrowest part of a Gaussian beam). The intensity profile of such a beam is said to be bell-shaped. Besides laser beam, a single-mode fiber with a core diameter of several microns also creates a Gaussian beam at its output.

The "spreading" of a light beam in general, and in particular of a laser beam as it moves away from the laser or light source is called beam divergence. The initial beam divergence of the light source is important for a light beam focusing on the target and for controlling the light spot diameter on the target surface. Typically, in order to control light-treatment effects, the light beam is focused onto the target surface by a lens, and the distance between output lens and target surface is varied to provide the needed light spot size and power density within the area of treatment. The radius of the beam in the focal plane of the lens with a focal length f is given by: $w=f{\cdot}\theta$, where θ is the beam divergence. Single-mode lasers or single-mode fibers have a minimal beam divergence and can provide minimal light spot size. Minimal light spot size in the focal plane of the aberration free optical system can be close to the wavelength λ.

3.2.5 Continuous Wave and Pulsed Light

Evidently, light-tissue interaction depends on temporal parameters of the light, whether it is continuous wave (CW) or pulsed. A CW mode means that emitted waves are not intermittent or broken up into damped wave trains, but unless intentionally interrupted, follow one another without any interval of time between them. Pulsed light can be produced as a single pulse of duration τ_p (pulsewidth), measured in seconds (s), or as successive trains of pulses with some repetition frequency (rate) f_p, measured in hertz (1/s). Lamps can generate light pulses of duration τ_p in millisecond (ms) (10^{-3} s), microsecond (μs) (10^{-6} s), or nanosecond (ns) (10^{-9} s) ranges, and only lasers can generate more shorter pulses, that is, in picosecond (ps) (10^{-12} s) and femtosecond (fs) (10^{-15} s) ranges with a high repetition rate f_p up to 100 MHz. A laser with Q-switching produces the so-called giant pulses, as the mode-locked laser produces ultrashort pulses with a high repetition rate. In dependence of technology used, the form of pulses can be different: rectangular, triangle, or Gaussian.

To describe energetic properties of pulsed light, a few more characteristics should be introduced, such as pulse energy E_p, peak power P_p (power within the individual pulse) and average power for a train of pulses. Peak power is calculated as $P_p = E_p/\tau_p$. Thus, for ultrashort pulses, peak power can be extremely high even for low or moderate light energies, and tissue breakdown can be expected; however the average power of light, calculated as $P_{ave} = E_p \times f_p$, cannot be very high. For example, if a light source generates pulses with an energy of $E_p = 0.1$ J, at a rate of $f_p = 1$ Hz (1/s) and duration of $\tau_p = 10$ ns, then $P_p = E_p/\tau_p = 10^7$ W, or 10 MW, as the average power is only $P_{ave} = E_p \times f_p = 0.1$ W or 100 mW.

3.2.6 Coherence and Monochromaticity

Coherent light that is typically produced by lasers is light in which the electromagnetic waves maintain a fixed phase relationship over a period of time, and in which the phase relationship remains constant for various points in the plane that is perpendicular to the direction of propagation.

Coherence length of a light source characterizes the degree of temporal coherence of the emitted light, $l_C = c\tau_C$, where c is the light speed and τ_C is the coherence time, which is approximately equal to the pulse duration of the pulsed light source or inversely proportional to the wavelength bandwidth $\Delta\lambda$ of a CW light source, $\tau_C \sim \lambda^2/(c\Delta\lambda)$. A single frequency CW gas-discharge He–Ne laser with a narrow bandwidth $\Delta\lambda = 10^{-6}$ nm and wavelength $\lambda = 632.8$ nm has a coherence length $l_C \approx 400$ m; a multimode diode laser with $\Delta\lambda = 30$ nm and $\lambda = 830$ nm has $l_C \approx 23$ μm. For a titanium sapphire laser with $\lambda = 820$ nm, the bandwidth may be as big as 140 nm; therefore, coherence length is very short $l_C \approx 2$ μm. The shortest $l_C \approx 0.9$ μm is for a white light source ($\Delta\lambda = 400$ nm). Coherence length is a fundamental parameter for optical coherence tomography (OCT); lower the l_C value, the better is the image resolution that can be achieved. OCT systems based on a titanium sapphire laser or a white light source allows one to image skin with a subcellular resolution of 1–2 μm.

Monochromatic light is light of one color (wavelength) only, ideally produced by a CW single-frequency well-stabilized laser.

Quasi-monochromatic light is light that has a very narrow but nonzero wavelength (frequency) bandwidth; it can be presented as a group of monochromatic waves with a slightly different wavelength.

Nonmonochromatic light has a broad wavelength bandwidth and can be presented as many groups of monochromatic waves with different wavelengths.

3.2.7 Light Refraction

Light refraction is the change in direction of a ray of light when passing obliquely from one medium into another in which the light speed is different. Light refraction is characterized by the index of refraction-a number (n) indicating the speed of light in a given medium, as either the ratio of the speed of light in a vacuum to that in the given medium (absolute index of refraction), or the ratio of the speed of light in a specified medium to that in the given medium (relative index of refraction), $m = n_1/n_2$. For different human skin components, refractive index (RI) in the visible/NIR wavelength range varies from a value a little bit higher than for water due to influence of some organic components ~1.35 for interstitial fluid to 1.55 for the stratum corneum.

3.2.8 Polarization

Polarization of light is a state in which rays of light exhibit different properties in different directions. When the electric field vector of the EMR oscillates in a single, fixed plane all along the beam, the light is said to be linearly (plane) polarized; when the plane of the electric field rotates, the light is said to be elliptically polarized because the electric field vector traces out an ellipse at a fixed point in space as a function of time; and when the ellipse happens to be a circle, the light is said to be circularly polarized.

The degree of polarization is the quantity that characterizes the ratio of the intensity of polarized light to the total intensity of light, $P_L = (I_{\parallel} - I_{\perp})/(I_{\parallel} + I_{\perp})$, where $I_{\parallel}$ is the intensity of light polarized in parallel and $I_{\perp}$ perpendicular to polarization plane. When polarized light traces a tissue, its depolarization (destruction of light polarization) happens because of the complex character of light's interaction with the inhomogeneous (scattering) medium (i.e., tissue). As a characteristic of such an interaction, the depolarization length is introduced. The depolarization length is the length of light beam transport in a scattering-depolarizing medium in which the polarization degree decays to a definite level compared to the totally polarized incident light.

Many tissues, including skin, feature polarization anisotropy which is an inequality of polarization properties along different axes. Polarization properties of light propagating within a tissue are sensitive to changes in tissue morphology, for instance, due to skin collagen aging. On this basis, a number of polarization-gating techniques were designed. These techniques provide a selection of diffuse photon groups with different path lengths, in particular ballistic or least-scattering photons that carry information about tissue structure. Skin polarization optical imaging and spectroscopy techniques were recently suggested.

A polarizer is a device, often a crystal or prism, which produces polarized light from unpolarized light of a conventional light source. Laser light is principally polarized, and depending on laser construction it can provide a high degree of linear or circular polarization.

3.3 Light Sources

3.3.1 Spontaneous and Stimulated Emission

Light is emitted by atoms (molecules) of the light source material that can be a gas, a liquid, or a solid. Atoms can be in different excited states when electrons possess energy according to their position in relation to the nucleus of an atom. The closer the electron is to the nucleus, the lower is the energy. When the energy of an electron changes, it must do so in certain definite steps, and not in a continuous manner. The positions in which electrons may be found according to their energy are called energy levels and sublevels. These levels are counted by their steps outward, and the numbers allotted to them are their quantum numbers. Excitation of atoms (molecules) can be provided by different ways: by heating, by electrical discharge, or by optical pumping.

An excited atom (molecule) is able to relax with time. In other words, the excited state has a lifetime that refers to the time the atom (molecule) stays in its excited state before emitting a photon spontaneously (spontaneous emission) or Lose energy nonradiatively (by collisions with the other atoms). Thus, the lifetime is related to the rates of excited state decay, to the facility of the relaxation pathway, radiative and nonradiative. If the rate of spontaneous emission or any of the other rates are fast, the lifetime is short (for commonly used fluorescent compounds typical excited state decay times are within the range 1 ns–1 ms). An atomic optical transition is typically an electronic transition, where energy is given out as electromagnetic radiation in the optical range. Direction of spontaneously emitted photons is random, and the frequency (wavelength) is also random in the limits of the bandwidth of luminescence of the excited transition. As a result, most spontaneous emitting light sources have an isotropic direction of emission and a wide range of frequencies (polychromatic).

Intensive stimulated emission of light by a group of atoms (molecules) is possible when the higher energy levels of these atoms are populated more intensively than the lower ones (inversed population). Such inversion of population can be done by different methods including two component gas systems with coinciding energetic levels with different relaxation times or optical pumping. Stimulated emission is characterized by the generation of a new photon which is identical to the excitation photon that initially interacted with the atom. As a result, we receive two photons with the same wavelength, phase, and direction of propagation, instead of one. Stimulation emission is the basic concept for lasing. Stimulated emission in an active medium with inversed population is developed as a photon avalanche with identical directions of propagation, frequencies, and polarizations. A laser is an active medium with inversed population that is placed between two paralleled mirrors. During lasing, the photon avalanche is propagating between two paralleled mirrors and is amplified during each time of intersection with the active medium. As a result, the laser beam is formed with a very low divergence and single wavelength (monochromatic beam).

3.3.2 Heat Sources

All bodies emit electromagnetic radiation in the wavelength spectral range and with intensities that correspond to their temperature. This is called heat radiation or blackbody radiation. Spontaneous and stimulated emission by atoms are the basis for the Planck curve (function), which gives the intensity radiated by a blackbody as a function of wavelength for a definite body temperature. The normalized spectral irradiancy of heat sources versus wavelength shows that the emitting spectrum is very broad, and is shifted to IR range by decreasing temperature (see Fig. 3.1). A blackbody is an object that absorbs all the electromagnetic energy that falls on the object, no matter what the wavelength of the radiation. Many objects made from condensed materials (for instance, metals, tissues) can be considered as blackbodies. The area under the Planck curve increases as the temperature is increased (the Stefan–Boltzmann law); the peak in the emitted energy moves to the shorter wavelengths as the temperature is increased (Wien's law). Follow Wien's low maximum of heat radiation spectrum, is the function of body temperature as power four. Thus, strongly heated metals are used in light sources with a broadband visible and infrared radiation. For example, filament lamp with a temperature of 2000K emits light with peak wavelength at approximately 1400 nm.

3.3.3 Halogen Lamps

A halogen lamp has an iodine-cycle tungsten incandescent lamp as the visible/near infrared (360 nm to >1 μm) light source for spectrophotometry and phototherapy, and recently

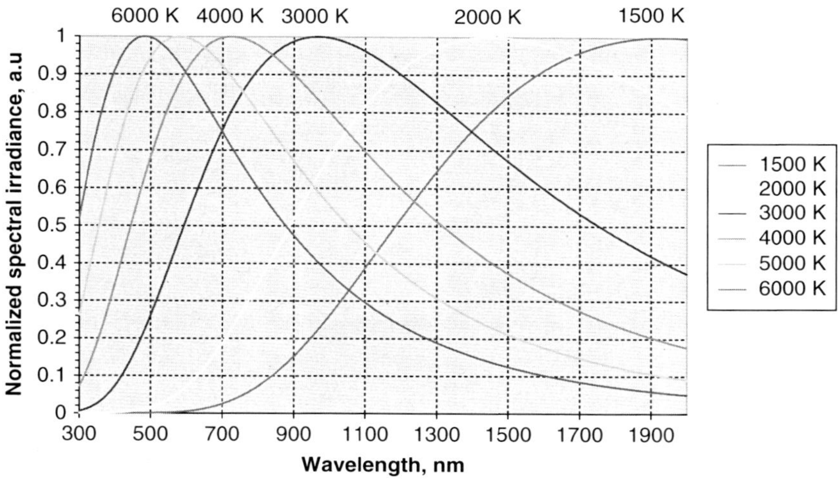

Figure 3.1 The normalized spectral irradiancy of heat sources vs. wavelength.

it has also been used in IR phototherapy. As it is gas-filled, the tungsten electrode can be heated up to 4000 K. At this temperature, the maximum spectra of emission is about λ_{max} =750 nm (see Fig. 3.1). About 30% of the total power is emitted at wavelengths shorter than λ_{max} and the remaining 70% is emitted at wavelengths longer than λ_{max}.

3.3.4 Arc Lamps

An arc or flash lamp is a lamp that uses a luminous and heat emission of plasma bridge formed in a gap between two conductors or terminals when they are separated. A mercury arc lamp is a discharge arc lamp filled with mercury vapor at high pressure; it gives out a very bright UV and visible light at some wavelengths including 303, 312, 365, 405, 436, 546, and 578 nm. Another arc lamp producing the so-called IPL (intensive pulse light) and often used in tissue spectroscopy and dermatology is a xenon or krypton lamp that is filled with xenon or krypton. It gives out a very bright UV and visible light in the range from 200 nm to >3.0 μm. The output spectrum of an arc lamp is a mixed emission spectrum of plasma as a heat source and spontaneous fluorescence of plasma ions. For a high energy short pulse, the temperature of arc lamp plasma can be very high (6000–10000 K) and the dominant emission is provided by heat sources with spectrum close to blackbody (see Fig. 3.1). For CW or long pulse mode, the temperature of arc lamp plasma is relatively low (3000–6000 K) and the emission spectrum has a significant portion of fluorescence light in the red and NIR spectral range.

3.3.5 Light Emitting and Superluminescent Diodes

An LED is a semiconductor device that emits light when the forward-directed current passes the *p–n* junction. LEDs are the light sources with a wide range of selected wavelengths from UV to IR. A typical wavelength bandwidth is 20–30 nm for LEDs working in the visible range. An LED's light power ranges from a few milliwatts to a few watts. Their light beam divergence is of a few tenths of a degree. They can be used as prospective light sources for many applications because of their high efficiency (conversion of electrical to light energy), long life time (more than 10^5 hours), ability to emit many different wavelengths (colors), and high brightness.

A superluminescent diode (SLD) is a very bright diode light source with a broad bandwidth. It is usually manufactured using a laser diode technology (heterostructure, waveguide, etc.), but without reflecting mirrors (there is an antireflection coating at the diode faces, or their out-of-parallelism is provided). Its main difference from a LED is that it has a uniform wavefront of the output radiation which allows one to couple its radiation into a single mode fiber. The SLDs are used in different medical OCT systems.

New semiconductor technologies that are based on heterostructure concept are important for fabrication of short-wavelength LEDs, SLDs, and diode lasers. Heterostructure is a semiconductor junction which is composed of layers of dissimilar semiconductor materials with nonequal band gaps. The quantum heterostructure has a size that restricts the movements of the charge carriers and forces them into a quantum confinement that leads to the formation of a set of discrete energy levels with sharper density, than that for structures of more conventional sizes.

3.3.6 Lasers: Gas, Solid-State, and Diode

Laser is acronym for *light amplification by the stimulated emission of radiation*. Laser is a device that generates a beam of light that is collimated, monochromatic, and coherent. Laser radiation is characterized by its wavelength, power, and pulse- or continuous wave-mode of generation. Any CW lasers can work in the pulse mode by a switch-on and switch-off pumping power but many pulse lasers cannot work in CW. Normally, lasers are characterized by the output wavelength (nm or μm), spectral bandwidth (nm), energy characteristics such as power (mW, W, kW) for CW laser, and energy per pulse (J), pulsewidth (ns, μs, ms, s), repetition rate (Hz), and average power (mW, W, kW) for pulse lasers. An important practical characteristic of a laser is its efficiency, which is the ratio of the output laser power to the input electrical power of laser pumping and expressed in percentage. Lasers of higher efficiency normally have the smallest size and lower cost.

To provide a high precision of laser beam focusing and to transport its radiation through the single-mode fibers, single-mode lasers are used. Such lasers produce a light beam with a Gaussian shape of the transverse intensity profile without any spatial oscillations, the so-called single transverse mode lasers. In general, such lasers generate many optical frequencies (so-called longitudinal modes), which have the same transverse Gaussian shape.

A pulse laser is a laser that generates a single pulse or a set of pulses. Laser pulses can be produced by simple switching the pumping power on and off . However, two technologies are typically used to produce a special laser pulsing, they are: Q-switching and mode-locking. The Q-switching, sometimes known as giant pulse formation, is a technique by which a laser can be made to produce a pulsed output beam with a very high power. The technique allows for the production of light pulses with extremely high (gigawatt) peak power, much higher than would be produced by the same laser if it is operating in a CW mode.

A mode-locked laser is a multimode laser with synchronously irradiating modes, and the regime is obtained by applying an intracavity high-frequency modulator, with typical pulse duration of up to a subpicosecond range, and a repetition frequency of dozens of megahertz.

A cavity-dumped mode-locked laser is a laser that uses a specific technology for producing high-energy ultrashort laser pulses by decreasing the pulse repetition rate. The laser output mirror is replaced by an optical selector consisting of a couple of spherical mirrors and an acousto- or electrooptical deflector, which extracts a pulse from the cavity after it has passed over a few dozen cavity lengths. The pulse energy is accumulated between two sequential extractions: the pulse repetition rate can be tuned in to the range from dozens of hertz to a few megahertz.

Compared to mode-locking, Q-switching leads to much lower pulse repetition rates, much higher pulse energies, and much longer pulse durations; both techniques are sometimes applied simultaneously.

Most lasers emit at a particular wavelength, in tunable lasers, one can vary the wavelength over some limited spectral range.

There are a huge variety of lasers and laser systems available in the market. Lasers can be classified in accordance with the active media used, such as gas, solid state, liquid, and semiconductor (diode) lasers. For example, a gas laser is a laser whose active medium is a gas or mixture of gases. We briefly present the most popular lasers used in medicine.

CO_2 (carbon dioxide) laser—a laser in which the lasing medium is CO_2 gas with an IR emission from 9.2 to 11.1 μm with the maximal efficiency at 10.6 μm and a power from a few watts to a few kilowatts. Both CW and pulsed regimes are available. Lasers are tunable in the limits of CO_2 molecules spectral range (9.2–11.1 μm). CO_2 laser has a very high efficiency, up to 40%. Because of a high absorption of tissues in this wavelength range, CO_2 laser is mostly used for tissue ablation.

Excimer laser—a laser whose lasing medium is an excited molecular complex, an excimer (molecule-dimer). The emission is in the UV range. Examples are: ArF laser, 193 nm; KrF laser, 248 nm; XeCl laser, 308 nm; and XeF laser, 351 nm. These lasers are tunable in some limits (10–20 nm). Because of a high absorption by tissues in the UV range, excimer lasers are widely used for tissue ablation with a high precision in both directions: in tissue depth and transversely. Eye refractive surgery technologies are based on these lasers.

Dye laser—a laser in which the laser medium is a liquid dye. Dye lasers emit in a broad spectral range (e.g., in the visible), and are tunable. Wavelengths range is from 340 to 960 nm, at optical frequency doubling—from 217 to 380 nm, and at parametric conversion—from 1060 to 3100 nm. Its emitted energy is from 1 mJ to 50 J in periodic pulse mode. The mean power is from 0.06 to 20 W. Pulse duration is from several nanoseconds to several microseconds and pulse frequency from a single pulse to 1 kHz. Train of microsecond pulses can be used to generate millisecond pulses. It is used in spectroscopy and photochemistry of biological molecules and is one of the best lasers for blood vessels coagulation.

A solid-state laser has an active medium as a matrix of crystal, glass or ceramic doped by active ions. Different crystal matrices, such as sapphire, yttrium aluminum garnet (YAG), alexandrite, yttrium scandium gallium garnet (YSGG) and others are used in lasers. Active ions can be Nd (neodymium), Cr (chromium), Er (erbium), Ho (holmium), Tm (thulium) and others. Active ions in different matrices have different laser wavelengths. For example Cr^{3+} doping sapphire (ruby laser) provides laser wavelength of 694 nm, but the same ions doping alexandrite crystal (alexandrite laser) give laser wavelength of 755 nm. Solid-state lasers are pumped by optical radiation from a flash (arc) lamp or from other laser, for example, a diode laser. Efficiency of flash lamp pumped laser is about 0.1–5%. Diode laser pumped solid-state lasers have an efficiency in the range of 10–50%.

Nd:YAG (neodymium:yttrium aluminum garnet) laser is one of the most efficient solid-state lasers whose lasing medium is the crystal Nd:YAG with emission in the NIR at 1064 nm; other less intensive lines at 946, 1319, 1335, 1338, 1356, and 1833 nm are also available. This laser is often used at optical frequency doubling—532 nm, the third harmonic of the radiation (355 nm) is also widely applicable in photomedicine. Both CW and pulsed regimes are available. Typical power of the main harmonic (1064 nm) is from a few watts to a few hundred watts in CW mode. Pulsed lasers are characterized by a high repetition rate, up to 300 Hz; their pulse duration varies from a few nanoseconds to hundred milliseconds, and the pulse energy is 0.05–100 J; the pulse power amounts up to several megawatts, the average power up to 1000 W. Several others lasers are based on active Nd ions, for example, neodymium:yttrium aluminum perovskite laser (Nd:YAP). This laser emits at wavelengths λ = 1054, 1341 nm, and other wavelengths. Nd laser is one of most popular lasers in photomedicine.

Er:YAG (erbium:yttrium aluminum garnet) laser—a solid-state laser whose lasing medium is the crystal Er:YAG with emission in mid-IR at 2.79–2.94 μm—is one of the most effective lasers for ablation of different tissues, including skin and hard tissues, because of its unique wavelength that coincides with the strongest absorption band of water (normal oscillatory modes of water molecules, λ = 2.94 μm). Typical power range is from a few watts to a few tenths of watts. For miniature systems (a crystal 4 mm in diameter and 75 mm long), the pulse duration in the free-run regime is in the microsecond range with the pulse-repetition rate of 25 Hz, pulse energy of a few joules and average power of a few watts. In the Q-switching regime the pulse duration is in the nanosecond range with a pulse energy of ~100 mJ.

A diode laser is a semiconductor injection laser. This laser is pumped by electrical current through a multilayered semiconductor structure (heterostructure), including a so-called quantum well heterostructure that maximizes a laser's optical mode overlap and injected carriers. The optical mode overlap is optimized with the gain to produce lasers with lower threshold currents. One of the widely used diode lasers is GaAs (gallium arsenide) laser—a laser based on the semiconductor material GaAs; the emission is in the NIR, at about 830 nm. More complex compositions allowing one to have the desired wavelength and output power are also designed: GaP_xAs_{1-x} lasers [emit light from 640 nm (x = 0.4) to 830 nm (x = 0)]. $Ga_xIn_{1-x}As_yP_{1-y}$ lasers, at $y = 2.2x$ and for different values of x, emit in the range 920–1500 nm. These lasers emit light in range up to 2000 nm. Pb_xS_{1-x}, $Sn_xPb_{1-x}Te$ and $Sn_xPb_{1-x}Se$ lasers, for different values of x, emit in the range 2.5–49 μm. GaN (gallium nitride) laser emits in the short wavelength range from 360 to 450 nm.

A single diode laser emitter has a typical size of laser aperture of 1.5–100 μm and cavity length of 0.5–3 mm. The maximum output power of a single diode laser emitter is in the range from 0.5 to 10 W. High-power diode laser is usually a plate planar array of laser bar with laser aperture up to 10 mm. One laser bar comprises 10–90 single laser emitters. Maximum output power of a diode laser bar is in the range from twenty to several hundred watts. Diode lasers are the most efficient lasers with efficiency up to 70%. Diode lasers have a very high beam divergence: 50°–90° in a fast axis and 5°–20° in a slow axis. Special micro optics is necessary to form a low divergence beam or for coupling diode laser power into the fiber. Diode laser can work in the CW mode or in pulse mode by pulsing pumping electrical current.

Diode lasers are used for pumping of other lasers, that allows one to produce very robust and compact totally solid-state systems, such as a diode-pumped Nd:YAG, which is an integrated solid-state laser with a Nd:YAG crystal as a lasing medium and optical pumping provided by a single-diode lasers or by a diode bars. Another example of such a system is a so-called fiber laser. In fiber laser active medium is glass or crystal fiber with core doped by active ions such as Nd, Yt, Er, Tm. Diode laser power is injected in the cladding of such fiber for pumping of active ions. Fiber lasers have a very high efficiency and the best quality of laser beam.

3.3.7 Light Delivery Fibers

The success of light treatment is largely dependent on the light-delivery system used. Generally, a light guide, which delivers light energy to a target to illuminate it, consists of an assembly of optical fibers that are bundled but not ordered. A fiber is an optical waveguide that uses the phenomenon of total internal reflection for light transportation with low losses and is made from transparent glass, quartz, polymer, or crystal, usually with a circular cross section. It consists of at least two parts, an inner part or core, with a higher refractive index and through which light propagates, and an outer part or cladding, with a lower refractive index and which provides a totally reflecting interface between the core and the cladding.

A multimode fiber is a single fiber that allows the excitation (direction) of many modes (rays); for example, for a fiber with a core diameter of 50 μm, numerical aperture, NA = 0.2, and an excitation wavelength of 633 nm, the number of excited modes is equal to 1250.

A single-mode fiber is a fiber in which only a single mode can be excited; for a fiber with a numerical aperture NA = 0.1 and wavelength 633 nm the single mode can be excited if the core diameter is less than 4.8 μm.

The numerical aperture is characteristic of the light-gathering power of an objective or optical fiber; it is proportional to the sine of the acceptance angle, a higher NA more light is collected by a fiber.

A fiber bundle is a flexible bundle of individual optical fibers arranged in an ordered or disordered manner and correspondingly named regular and irregular bundles. The irregular fiber bundles are used for illumination and collecting light from a tissue, as well as a regular bundle provides transportation of tissue image.

Fiber-optics is a well-developed industrial field, where various fiber instrumentation is available, such as connectors, couplers, GRIN- or selfoc-lenses, focons, fiber multiplexes (dividers), as well as fiber-optic catheters—a flexible single fiber or a fiber bundle used to move light into body cavities and back. For example, GRIN (gradient index) lens focuses light through a precisely controlled radial variation of the lens material's index of refraction from the optical axis to the edge of the lens; this allows a GRIN lens with flat or angle-polished surfaces to collimate light emitted from an optical fiber or to focus an incident beam into an optical fiber; end faces can be provided with an antireflection coating to avoid unwanted back reflection.

3.3.8 Laser versus Noncoherent Light Sources

Individual photons from laser or noncoherent light sources are identical, and a tissue does not know how photon was born. The choice between laser and noncoherent light sources is strictly dependent on the particular application. In therapeutic applications, a laser is almost just one choice for short pulse (<0.1 ms), small spot size or microbeam (<1 mm), fiber delivery, narrow absorption spectrum bandwidth of chromophore (<50 nm), and focusing beams in tissues. In other cases, noncoherent sources can be very competitive with laser, especially if cost and size of equipment are important parameters. For example, for light-based hair removal systems the target is the melanin in hair follicles. For deep penetration into the skin, very broad spectrum of wavelengths can be used, from 600 to 1200 nm. Melanin absorbs light in this entire spectrum. For greater safety, the ratio of irradiance of melanin in the hair bulb (0.5–1.5 mm depth) and bulge (2–5 mm depth) areas to irradiance of melanin in epidermis (0.05–0.1 mm depth) must be maximal. This condition can be achieved just for a large beam size on the surface of the skin, when diffusion of light in the skin due scattering in the follicle depth is smaller than the initial beam size. Based on these criteria, the beam size on the skin surface must be about 2 cm. At the same time, for permanent hair removal the fluence on the skin surface must be in the range 30–100 J/cm^2. The total energy of the light pulse must be of 100–300 J. A laser with such high energy is very expensive. At the same time, a Xe arc lamp with a filtered 600–1200 nm spectrum has an efficiency of about 15 %; it is small and low-cost.

3.4 Light Propagation in Skin

3.4.1 Light Absorption and Scattering

Absorbing medium is the medium that absorbs light at certain wavelengths or wavelength bands. The absorption band is a range of wavelengths for which a medium absorbs more strongly than at adjacent wavelengths. The process of light absorption is the transformation of light (radiant) energy to some other form of energy, usually heat, as the light transverses tissue. Commonly, an absorbing medium consists of absorption centers that are particles or molecules that absorb light. To characterize the absorption of a medium, an absorption coefficient μ_a is introduced; in a nonscattering sample it is defined as the reciprocal of the distance x_a over which light of intensity $I(x = 0) = I_0$ is attenuated (due to absorption) to $I(x_a) = I_0/e \approx 0.37I_0$; the units are typically cm^{-1} or mm^{-1}. Behind this definition is a fundamental process of light wave or photon absorption that is characterized by a cross section, that is, the ability of a molecule to absorb a photon of a particular wavelength and polarization. Although the units are given as an area, it does not refer to an actual area size, at least partially because the density or state of the target molecule will affect the probability of absorption. Quantitatively, the number dN of photons absorbed between the points x and $x + \mathrm{d}x$ along the path of a light beam is the product of the number N of photons penetrating to depth x times the number ρ of absorbing molecules per unit volume times the absorption cross section σ_{abs}: $\mathrm{d}N/\mathrm{d}x = -\rho\sigma_{abs}N$, $\mu_a = \rho\sigma_{abs}$.

In spectroscopy, a few terms are commonly used, such as absorbance that is the ratio of the absorbed light intensity to the incident intensity, thus, it is a dimensionless quantity and absorption spectrum, that is, the spectrum formed by light that has passed through a medium in which light of certain wavelengths was absorbed.

Absorption spectra can be also expressed in terms of absorption coefficient. Such spectra for skin and its components: water (75%), epidermis, melanosome, vessel wall, and whole blood are presented in Fig. 3.2. The diagnostic/therapeutic lasers and their wavelengths are presented on the crossings of water absorption curve, and the corresponding laser

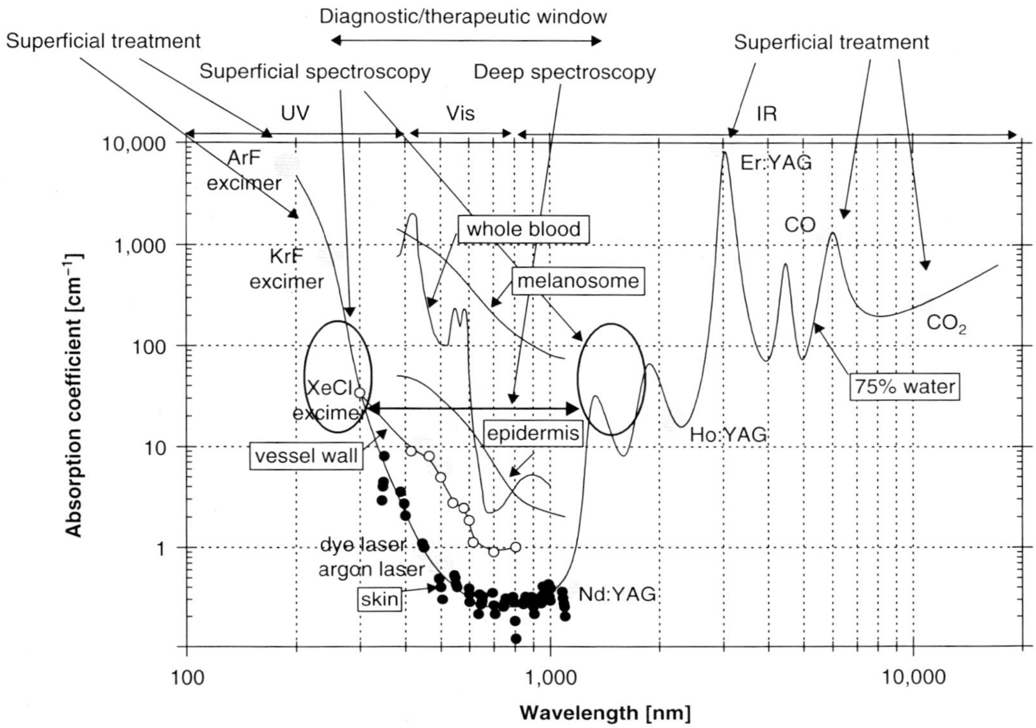

Figure 3.2 Absorption spectra of skin and skin components: water (75%), epidermis, melanosome, vessel wall, and whole blood; diagnostic/therapeutic lasers and their wavelengths as well as diagnostic/therapeutic window and wavelength ranges suitable for superficial and deep spectroscopy or treatment are also shown (adapted from ref. [1]).

wavelengths show what absorption coefficient is expected at absorption in water. Because water is the major component of any tissue, including skin, presented absorption coefficients will be only slightly corrected for real tissues. As seen in Fig. 3.2, for bloodless skin in the visible range, absorption is equal to water (75%), because dermis as the main skin component that is well supplied by water defines the absorption coefficient in this range.

In photomedicine, many diagnostic and treatment technologies use endogenous or exogenous agents—photosensitizers that are substances that increase the absorption of a tissue at a particular wavelength band, and may significantly accelerate photothermal or photochemical treatments.

Tissues are not only absorbing, but also an inhomogeneous media with different levels of organization that include cells, fibers, and macroinhomogeneities, such as, for instance, skin appendages or tumors. The sizes of cells and tissue structure elements vary in size from a few tenths nanometers to hundreds of micrometers [1]. Mammalian cells have diameters in the range of 5–75 μm. In the epidermal layer, the cells are large (with an average cross-sectional area of about 80 μm^2 for living epidermis and 1300 μm^2 for the outmost horny layer) and quite uniform in size. Fat cells, each containing a single lipid droplet that nearly fills the entire cell and therefore results in eccentric placement of the cytoplasm and nucleus, have a wide range of diameters from a few microns to 50–75 μm. Fat cells may reach a diameter of 100–200 μm in pathological cases. There are a wide variety of structures within cells that determine tissue light scattering. Cell nuclei are on the order of 5–10 μm in diameter; mitochondria, lysosomes and peroxisoms have dimensions of 1–2 μm; ribosomes are on the order of 20 nm in diameter; and structures within various organelles can have dimensions up to a few hundred nanometers.

In the course of transport (travel) in an inhomogeneous medium with absorption a photon changes its direction due to reflection, refraction, diffraction, or scattering and can be absorbed by an appropriate molecule on its way. Light scattering means a change in the direction of propagation of light in a turbid medium caused by reflection and refraction by microscopic internal structures. Such small structures that are smaller or comparable with the wavelength of propagating light are commonly called scatterers. A scatterer is an inhomogeneity or a particle of a medium that refracts or diffracts light; light is diffused or deflected as a result of collisions between the wave and particles of the medium.

There are a number of parameters that describe the scattering process. Scattering angle is related to photon scattered by a particle, so that its trajectory is deflected by a deflection (scattering) angle θ in the scattering plane and/or by azimuthal angle of scattering φ in the plane perpendicular to the scattering plane. Scattering plane is a plane defined by positions of a light source, a scattering particle, and a detector.

To characterize the scattering efficiency of a medium, a scattering coefficient μ_s is introduced; in a nonabsorbing sample it is defined as the reciprocal of the distance x_s over which the light of intensity $I(x = 0) = I_0$ is attenuated (due to scattering) to $I(x_s) = I_0/e \approx 0.37I_0$; the units are typically cm^{-1} or mm^{-1}. Behind this definition is also a fundamental process of photon scattering that is characterized by a photon scattering cross section, that is, the ability of a particle to scatter a photon of a particular wavelength and polarization; although the units are given as an area, it does not refer to an actual size area; quantitatively, the number dN of photons scattered between the points x and $x + dx$ along the path of a light beam is the product of the number N of photons penetrating to depth x times the number ρ of scattering particles per unit volume times the scattering cross section σ_{sca}: $dN/dx = -\rho\sigma_{sca}N$. The scattering coefficient μ_s (cm^{-1}) describes a medium containing many scattering particles at a concentration described as a volume density ρ (cm^3); the scattering coefficient is essentially the cross-sectional area σ_{sca} (cm^{-1}) per unit volume of medium: $\mu_s = \rho\sigma_{sca}$.

A collimated (laser) beam is attenuated in a thin tissue layer of thickness x in accordance with the exponential law, Bouguer—Beer—Lambert law,

$$I(x) = I_0 \exp(-\mu_t x), \tag{3.1}$$

where $I(x)$ is the intensity of transmitted light measured using a distant photodetector with a small aperture (on-line or collimated transmittance), W/cm^2; I_0 is the incident light intensity, W/cm^2;

$$\mu_t = \mu_a + \mu_s \tag{3.2}$$

is the extinction coefficient (interaction or total attenuation coefficient). Therefore, attenuation is a decrease in energy per unit area of a wave or beam of light: it occurs as the distance from the source increases and is caused by absorption or scattering, or both. The attenuation (extinction) coefficient is the reciprocal of the distance over which light of intensity I_0 is attenuated to $I_0/e \approx 0.37I_0$; the units are typically cm^{-1}. The greatest extinction of the forward scattered light is due to particles with dimensions between λ and 10λ. The particles with diameters between $\lambda/4$ and $\lambda/2$ are the dominant backscatterers.

Besides scattering coefficient, scattering process is also characterized by a so-called scattering phase function—the function that describes the scattering properties of the medium, and is in fact the probability density function for a photon traveling in some direction to be scattered in some new direction [1,3]. It characterizes an elementary scattering act: if scattering is relatively symmetric to the direction of the incident wave, then the phase function depends only on the scattering angle θ (angle between two directions, the new and former one), $p(\theta)$. Scattering anisotropy factor, g, is a major parameter of $p(\theta)$ and is a measure of the amount of forward direction retained after a single scattering event. If a photon is scattered by a particle so that its trajectory is deflected by a scattering angle θ, then the component of the new trajectory which is aligned in the forward direction is presented as $\cos\theta$. There is an average scattering angle and the mean value of $\langle\cos\theta\rangle$ is defined as the anisotropy factor $g \equiv \langle\cos\theta\rangle$. The value of g varies in the range from -1 to 1: $g = 0$ corresponds to isotropic (Rayleigh) scattering, $g = 1$ to total forward scattering (Mie scattering at large particles), and -1 to total backward scattering.

Rayleigh scattering relates to scattering by small particles (with respect to the wavelength of the incident light) when the scattered irradiance is inversely proportional to λ^4 and increases as a^6 (a is the radius of a particle), and the angular distribution of the scattered light is isotropic. Mie scattering relates to scattering by comparatively large spherical particles, which are of the order of the wavelength, and based on an exact solution of Maxwell's electromagnetic field equations for a homogeneous sphere. Typical tissues contain both types of scatterers small and large (for instance, cell components and collagen fibers of connective tissues).

Often, such characteristic of turbid materials and tissues as albedo can be useful for the prediction of light propagation in a tissue. Albedo is the ratio of the scattering to extinction cross section (or coefficient); ranges from zero for a completely absorbing medium to unity for a completely scattering medium.

All the above-mentioned issues are related to a single scattering—the scattering process that occurs when a wave undertakes no more than one collision with the particles of the medium in which it propagates. In contrast, in many tissues we have multiple scattering—a scattering process in which, on average, each photon undertakes many scattering events. However, for some tissues or tissue thin slices, single scattering approximation which assumes that the tissue is sufficiently thin for single scattering, accurately estimates the reflection and transmission for the slab and may be valid.

To evaluate what kind of scattering regime is realized, a so-called mean free path length (MFP) may be estimated. MFP is the mean distance between two successive interactions with scattering or absorption which a photon traveling in a scattering-absorption medium is experiencing, MFP = $1/\mu_t$. Thus, one can differentiate optically thin (transparent) or optically thick sample that is under investigation. Optical thickness is the depth of a material or medium in which the intensity of light of a given wavelength is reduced by a factor of $1/e$ ($e = 2.7$) because of absorption and/or scattering [see Eq. (3.1)]. A sample with high thickness and/or high turbidity that correspond to a few optical thickness depths is optically thick, as well as a sample with a low thickness and/or low turbidity that correspond to one or less than one optical thickness depth is optically thin.

The radiation transfer theory (RTT) is the basic theory allowing one to calculate light distributions in the multiple scattering media with absorption, such as tissues. The heart of this theory is the radiation transfer equation (RTE)-the Boltzmann or linear transport equation, which is a balance equation describing the flow of particles (e.g., photons) in a given volume element that takes into account their velocity c, location r, and changes due to collisions (i.e., scattering and absorption). The basic parameter for this theory is the reduced scattering coefficient that is a lumped property incorporating the scattering coefficient μ_s and the scattering anisotropy factor g: $\mu_s' = \mu_s(1 - g)$ (cm^{-1}); μ_s' describes the diffusion of photons in a random walk of step size of $1/\mu_s'$ (cm), where each step involves isotropic scattering; this is equivalent to description of photon movement using many small steps $1/\mu_s$ that each involve only a partial (anisotropic) deflection angle if there are many scattering events before an absorption event, that is, $\mu_a << \mu_s'$ (diffusion regime); μ_s' is useful in the diffusion regime which is commonly encountered when treating how visible and near-infrared light propagates through tissues[1,3].

The transport mean free path (TMFP) of a photon (cm) is defined as $l_t = (1/\mu_t') = (\mu_a + \mu_s')^{-1}$, where $\mu_t' = \mu_a + \mu_s'$ is the transport coefficient. The TMFP in a medium with anisotropic single scattering significantly exceeds the MFP, $l_t >> l_{ph}$. The l_t is the distance over which the photon loses its initial direction.

A power law for dependence of the scattering coefficient on the wavelength is typical for many tissues: $\mu_s \propto \lambda^{-h}$ [1]. For instance, for different tissue structures, associated with skin aging or rejuvenation, parameter h ranges from 1 to 2. The reduced scattering coefficient also obeys a power law, $\mu_s' \propto \lambda^{-h}$, that was experimentally demonstrated for normal and dehydrated rat skin in an in vitro study for the wavelength range of 500–1200 nm [4]. Parameter h was equal to 1.12 for normal skin and was decreased at topical application of glycerol (mostly dehydration effect): $h = 1.09$ for 5 min in glycerol, 0.85 for 10 min, 0.52 for 20 min, and 0.9 for the rehydrated sample.

In vivo backscattering measurements for human skin have also demonstrated the power law for the reduced scattering coefficient [5]: $\mu_s' = q\lambda^{-h}$ (cm^{-1}, λ in µm). In particular, for reflectance spectra from the human forearm in the wavelength range 700–900 nm, constants q and h were determined as 5.50 ± 0.11 and 1.11 ± 0.08, respectively. From Mie theory, it follows that the power constant h is related to an averaged size of the scatterers with the Mie-equivalent radius a_M. Once h is determined, this radius can be derived from equations [5]:

$$h = -1109.5a_M^3 + 341.67a_M^2 - 9.36961a_M - 3.9359 \quad (a_M < 0.23\,\mu\text{m})$$
$$h = 23.909a_M^3 - 37.218a_M^2 + 19.534a_M - 3.965 \quad (0.23 < a_M < 0.60\,\mu\text{m}).$$

These relations are valid for a relative refractive index between equivalent spheres and the surrounding medium equal to $m = 1.037$.

To describe pulsed light propagation in a tissue, the time-dependent radiation transfer theory is used. This theory is based on the time-dependent linear transport equation, which is a balance equation describing the time-dependent flow of particles (e.g., photons) in a given volume element that takes into account their velocity c, location $\bar{r}$, and changes due to collisions (i.e., scattering and absorption).

The time-resolved methods that use pulsed or modulated laser beams for irradiating tissues under study can separate different components of scattering photons from a sample in forward (transillumination) or backscattering operating modes. These groups of photons are so- called ballistic (coherent) photons—a group of unscattered and strictly straightforward scattered photons; quasi-ballistic (snake or zigzag) photons—photons that migrate within a scattering medium along trajectories that are close but not the same as for ballistic photons; and diffusion photons, which are typically the largest group of photons that migrate for a longer time in a tissue along multistep random trajectories. Each of these groups carries information about optical (morphological) properties of a tissue. Ballistic photons are good for getting precise tissue images similar to X-ray computer tomography; however, in many tissues because of strong scattering this group of photons is typically negligibly small. Snake photons, which undergo a few scattering events that are all in the forward or near-forward direction, retain the image bearing characteristics to some extent and are detectable. Due to the high intensity of the diffusion component, it is much more practical to use diffusion photons to estimate optical properties of tissues; however, the spatial resolution may not be very high. To improve the spatial resolution of diffusion methods, various approaches for selective detection of informative photons are suggested, such as spatially-resolved, angle-resolved, and polarization-sensitive gating [1].

One of the parameters which are typically used in tissue study is a scattering spectrum—the spectrum of scattered light; it can be differential, measured or calculated for a certain scattering angle, or integrated within an angle (field of view) of the measuring spectrometer.

The modeling of light propagation in a tissue by taking into account the experimental geometry, source and detector characteristics, and the known optical properties of a sample, and prediction of the measurements and associated accuracies that result are often needed for planning the strategy of light treatment or diagnosis. Such modeling and predictions are classified as a solution of the forward scattering problem. In contrast, the inverse scattering problem solution is the attempt to take a set of measurements and error estimates, and only a limited set of parameters describing the sample and experiment, and to derive the remaining parameters. Usually the geometry is known, intensities or their parameters are measured, and the optical properties or sizes of tissue scatterers need to be derived. If these properties are considered to be spatially varying, then the resultant solutions can be presented as a 2D or 3D function of space, that is, as an image.

All the above-mentioned points are characteristic of elastic (static) light scattering, where light is scattered elastically by static (motionless) objects without changes in photon energy or light frequency. In contrast, the dynamic (quasi-elastic) light scattering involves scattering by a moving object which causes a Doppler shift in the frequency of the scattered wave relative to the frequency of the incident light. A Doppler shift is the apparent change in the frequency of a wave, such as a light wave or sound wave, resulting from a change in the distance between the source of the wave and the receiver.

Inelastic light scattering is a basis for very powerful spectroscopy of biological molecules and tissues, Raman spectroscopy. From quantum theory it follows that an individual light scattering event is considered as an absorption by a particle of the scattering medium of the incident photon with energy h and then emission of the photon with energy $h\nu'$, at $\nu \neq \nu''$ light scattering is accompanied by redistribution of energy between the radiation and the medium and is called inelastic; at $\nu = \nu'$ scattering is elastic.

3.4.2 Skin Chromophores and Fluorophores

Absorption is only one way by which light can interact with the skin. Absorption of the UV and visible light in skin is due to electronic excitation of aromatic or conjugated unsaturated chromophores. A chromophore is a chemical that absorbs light with a characteristic spectral pattern. There are many kinds of chromophores in the skin, but a few major chromophores predominantly determine the optical absorption within each skin layer [6,7]. Spectral ranges of absorption of the main skin chromophores are presented in Fig. 3.3.

Proteins found in the epidermis contain the aromatic amino acids tryptophan and tyrosine which have a characteristic absorption band near 270–280 nm; urocanic acid and the nucleic acids also contribute to this absorption band with a maximum near 260–270 nm (Fig. 3.4). Epidermal melanin plays an important role in limiting the penetration depth of light in the skin [8]: it effectively absorbs at all wavelengths from 300 to 1200 nm, but the strongest absorption occurs at shorter wavelengths, in the near-UV spectral range (Figs. 3.4 and 3.5).

In the IR spectral range, the skin absorption spectrum is essentially determined by the absorption of water in the skin (Fig. 3.2).

Some of the major dermal chromophores are oxyhemoglobin, deoxyhemoglobin, bilirubin, carotenoids, and porphyrins (Fig. 3.5). Both the oxygenated and deoxygenated forms of hemoglobin absorb light. Oxyhemoglobin has its strongest absorption band at 415 nm (Soret band), and it has two secondary absorption bands at 542 and 577 nm (Q bands). Deoxyhemoglobin has primary absorption band at 430 nm and it has a single secondary absorption band at 555 nm. Both hemoglobins exhibit the lowest absorption at wavelengths longer than 620 nm. Bilirubin has two relatively broad absorption bands near 330 and 460 nm.

The absorption of diffuse light by skin pigments is a measure of bilirubin content, hemoglobin concentration and its saturation with oxygen, and the concentration of pharmaceutical products in blood and tissues; these characteristics are widely used in the diagnosis of various diseases (Fig. 3.5). Certain phototherapeutic and diagnostic modalities take advantage in dependence of transdermal penetration of visible and near infrared (NIR)

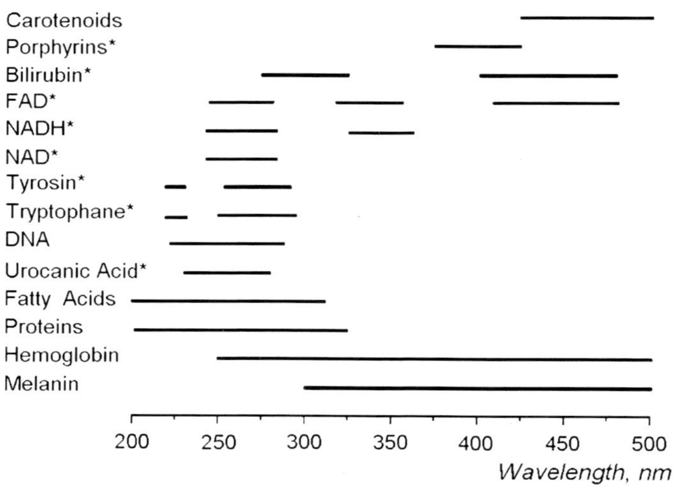

Figure 3.3 Spectral ranges of absorption of the main skin chromophores. For chromophores marked with (*), range indicated is a half-width of the band [7].

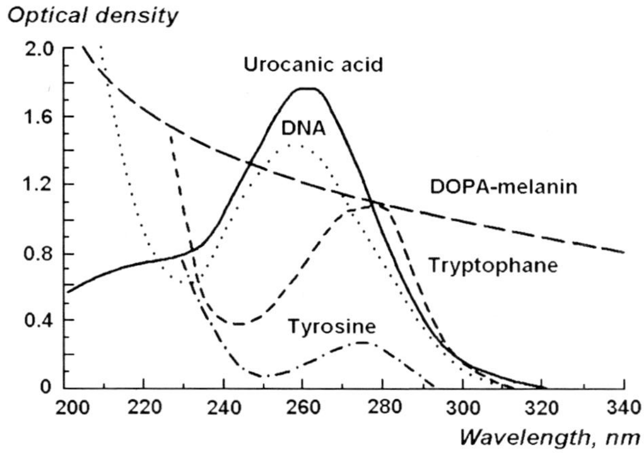

Figure 3.4 UV absorption spectra of major chromophores of human skin [DOPA-melanin, 1.5 mg % in H_2O; urocanic acid, 10^4 M in H_2O; DNA, calf thymus, 10 mg % in H_2O (pH 4.5); tryptophan, 2 × 10^4 M (pH 7); tyrosine, 2 × 10^4 M (pH 7)] [6].

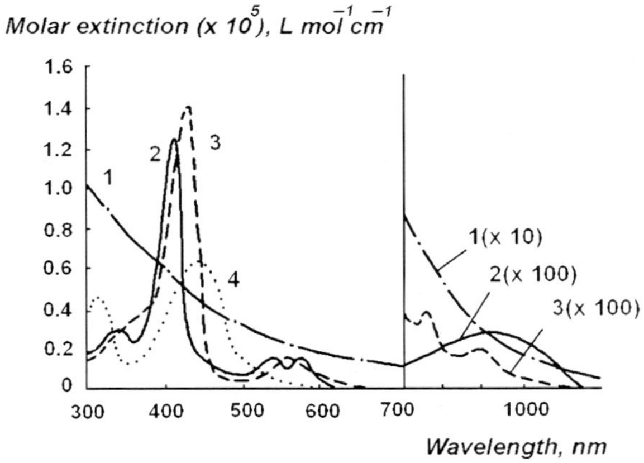

Figure 3.5 Molar attenuation spectra for solutions of major visible light-absorbing human skin pigments: 1, DOPA-melanin (H_2O); 2, oxyhemoglobin (H_2O); 3, hemoglobin (H_2O); 4, bilirubin ($CHCl_3$) [6].

light inside the body in the wavelength region corresponding to the therapeutic or diagnostic window (600–900 nm) (Fig. 3.2).

Human skin contains various types of native fluorophores (chromophores that emit light with a characteristic spectral pattern) at its excitation by a proper wavelength with unique absorption and emission spectra, different fluorescence quantum efficiency, different fluorescence decay time, and different distribution within the skin. Some fluorophores have similar absorption and fluorescence spectra, and typically, fluorescence spectra measured on the skin surface are the result of the overlapping bands of various such fluorophores. The skin also contains nonfluorescent chromophores, such as hemoglobin and melanin. These chromophores may absorb fluorescence light emitted by the other fluorophores, and thus may modify initial fluorescence spectra due to filtering properties of nonfluorescing chromophores.

When the excitation wavelength is increased, new fluorophores are involved in the formation of the shape of fluorescence spectrum. The closer the excitation wavelength to the center of the so-called therapeutic/diagnostic window (600–900 nm), the larger the penetration depth of the excitation light in tissue, and the larger is the tissue volume probed by the excitation light. As a result, new kinds of fluorophores located in deeper skin layers contribute to the total tissue fluorescence measured.

The spectral ranges of fluorescence of the main skin fluorophores are presented in Fig. 3.6 [7]. It can be seen that the skin autofluorescence (AF) (natural fluorescence of a tissue) in the UVA range is dominated by the fluorescence bands of aromatic amino acids, namely tyrosine and tryptophan. Tyrosine and tryptophan content in epidermis is more than twice that of the whole skin, and this is why epidermis has high AF in the UVA range. This also explains why the fluorescence of psoriatic stratum corneum is significantly higher than that for normal stratum corneum.

Among the endogenous skin fluorophores are the different forms of nicotinamide adenine dinucleotide (NAD) and keratin located in the epidermis and dermal collagen. The reduced (NADH) and oxidized (NAD^+) forms of NAD take part in cellular metabolism, and the intensity of their specific fluorescence (fluorescence maxima near 460 nm and 435 nm, respectively) is used for the quantitative NADH detection and differential diagnostics of the metabolism dysfunction. A similarity between the AF spectrum of the human skin in vivo and the emission spectrum of the keratin (maximum near 450 nm) was found.

Collagen is one of the most important skin fluorophores. Approximately, 75% of the dry weight of the dermal tissue is composed of the collagen fibers. Collagen is the main structural component of the connective tissue and accounts for about 90% of protein in human dermis. Dermis holds a thin fibrillar network mainly composed of Types I (about 80%) and III (about 20%) collagen. Collagen of Type IV is found in the cellular basement membrane,

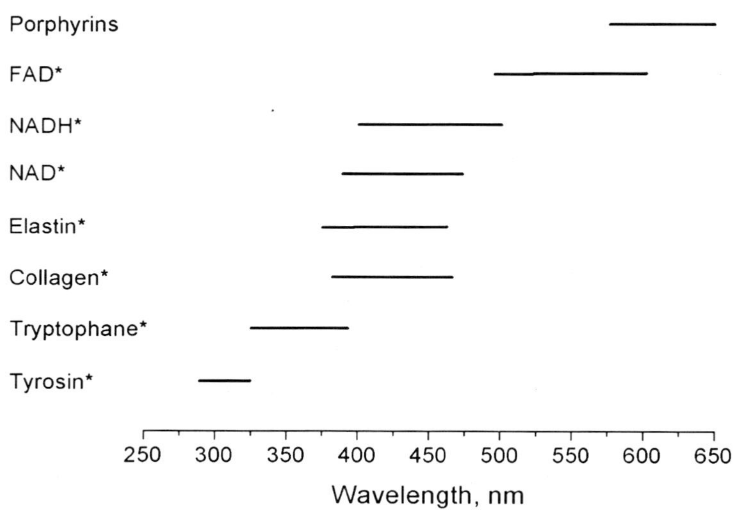

Figure 3.6 Spectral ranges of fluorescence of the main skin fluorophores. For fluorophores, marked with (*), range indicated is a half-width of the band [7].

which is connected to the collagen network by anchoring fibers containing collagen VII [12]. Collagen fibers exhibit a constant density throughout all the dermal layers.

All absorption spectra of different chromophores can be the function of temperature and light intensity. For example, absorption spectrum of whole blood can be dramatically changed in the range from normal body temperature to temperature of water vaporization. It can be explained by methemoglobin formation and coagulation of blood. Another example is shifting of water absorption spectrum to the shorter wavelength range at temperature increase [13] (Fig. 3.7).

3.4.3 Refractive Index Variations in Skin

Soft tissue is composed of closely packed groups of cells entrapped in a network of fibers through which water percolates. At a microscopic scale, the tissue components have no pronounced boundaries. They appear to merge into a continuous structure with spatial variations in the refractive index. It has been shown that the tissue components that contribute most to the local refractive-index variations are the connective tissue fibers (bundles of elastin and collagen), cytoplasmic organelles (mitochondria, lysosoms, and peroxisomes), cell nuclei, and melanin granules [1]. For different parts of a biological cell, values of refractive index in the NIR range can be estimated as follows: extracellular fluid, $\bar{n}$ = 1.35–1.36; cytoplasm, 1.360–1.375, cell membrane, 1.46, nucleus, 1.38–1.41; mitochondria and organelles, 1.38–1.41; melanin, 1.6–1.7.

Scattering arises from mismatches in refractive index of the components that make up the cell. Organelles and subcomponents of organelles having indices different from their surroundings are expected to be the primary sources of cellular scattering. For instance, the nucleus is a significant scatterer because it is often the largest organelle in the cell, and its size increases relative to the rest of the cell throughout neoplastic progression. Mitochondria (0.5–1.5 μm in diameter), lysosomes (0.5 μm), and peroxisomes (0.5 μm) are very important scatterers whose size relative to the wavelength of light suggests that they must make a significant contribution to backscattering. Granular melanin, traditionally thought

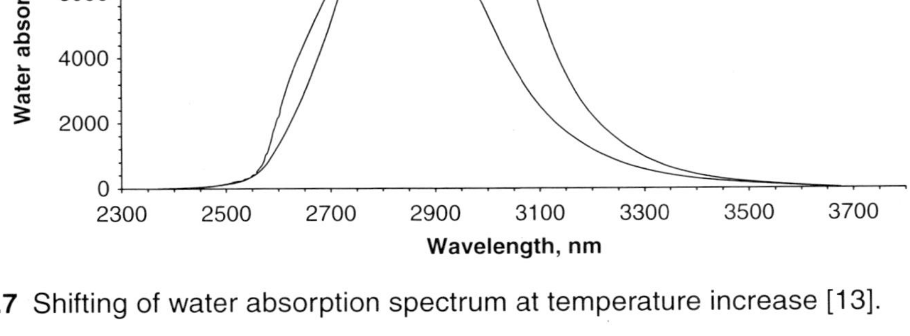

Figure 3.7 Shifting of water absorption spectrum at temperature increase [13].

of as an absorber, must be considered an important scatterer because of its size and high refractive index. Besides cell components, fibrous tissue structures such as collagen and elastin must be considered important scatterers.

The refractive indices of tissue structure elements, such as the fibrils, the interstitial medium, nuclei, cytoplasm, organelles, and the tissue itself can be derived using the law of Gladstone and Dale, which states that the resulting value represents an average of the refractive indices of the components related to their volume fractions as [1]:

$$\bar{n} = \sum_{i=1}^{N} n_i f_i, \quad \sum_i f_i = 1, \tag{3.3}$$

where n_i and f_i are the refractive index and volume fraction of the individual components, respectively, and N is the number of components.

The average background index is defined as the weighted average of refractive indices of the cytoplasm and the interstitial fluid, n_{cp} and n_{is}, as

$$\bar{n}_0 = f_{cp} n_{cp} + (1 - f_{cp}) n_{is}, \tag{3.4}$$

where f_{cp} is the volume fraction of the fluid in the tissue contained inside the cells. Literature data presented allows one to estimate $n_{cp} = 1.367$ and $n_{is} = 1.355$. Since approximately 60% of the total fluid in soft tissue is contained in the intracellular compartment, it follows from Eq. (3.4) that $\bar{n}_0 = 0.6{\cdot}(1.367) + 0.4{\cdot}(1.355) = 1.362$. The refractive index of scatterers can be defined as the sum of the background index and the mean index variation, $\bar{n}_0 = \bar{n}_0 + \langle\Delta n\rangle$, which can be approximated by another volume-weight average,

$$\langle\Delta n\rangle = f_f(n_f - n_{is}) + f_{nc}(n_{nc} - n_{cp}) + f_{or}(n_{or} - n_{cp}). \tag{3.5}$$

Here the subscripts f, is, nc, cp, and or refer to the fibers, interstitial fluid, nuclei, cytoplasm, and organelles, respectively, which were identified above as the major contributors to index variations. The terms in parentheses in this expression are the differences between the refractive indices of the three types of tissue component and their respective backgrounds; the multiplying factors are the volume fractions of the elements in the solid portion of the tissue. The refractive index of the connective-tissue fibers is about 1.47, which corresponds to about 55% hydration of collagen, its main component. The nucleus and the cytoplasmic organelles in mammalian cells that contain similar concentrations of proteins and nucleic acids, such as mitochondria and the ribosomes, have refractive indices that lie within a relative narrow range (1.38–1.41). Taking this into account and assuming that $n_{nc} = n_{or} = 1.40$, the mean index variation can be expressed in terms of the fibrous-tissue fraction c_f only: $\langle\Delta n\rangle = f_f(n_f - n_{is}) + (1 - f_f)(n_{nc} - n_{cp})$.

Collagen and elastin fibers compose approximately 70% of the fat-free dry weight of the dermis. Therefore, for dermis $f_f = 0.7$, $n_f - n_{is} = 1.470–1.355 = 0.115$ and $n_{nc} - n_{cp} = n_{or} - n_{cp} = 1.400–1.367 = 0.033$, the mean index variations $\langle\Delta n\rangle = 0.7{\cdot}(0.115) + (1–0.7){\cdot}(0.033) = 0.09$ and $\bar{n}_s = \bar{n}_0 + \langle\Delta n\rangle \approx 1.36 + 0.09 = 1.45$. The mean refractive index of skin or any other tissue can be evaluated using the law of Gladstone and Dale with respect to a two-phase system: scatterers with volume fraction of f_s and background material with volume fraction $(1 - f_s)$

$$\bar{n} = f_s \bar{n}_s + (1 - f_s)\bar{n}_0. \tag{3.6}$$

In practice, the mean refractive index of a tissue is a measurable value, thus volume fraction of scatterers could be determined from Eq. (3.6). For example, for skin dermis RI measured in vivo on the wavelength of He–Ne laser (633 nm) $\bar{n} = 1.396$ [1]; thus, using the estimates presented above for $\bar{n}_s = 1.45$ and $\bar{n}_0 = 1.36$, we can determine the volume fraction of scatterers as $f_s \approx 0.4$. This parameter is very important for evaluation and monitoring of skin scattering, that is, skin appearance, especially at its aging or influence of bad ecology or in a course of its rejuvenation. Scatterer volume fraction determination is also important for evaluation of efficiency of skin sun (UV) protective screens based on scattering particles.

3.4.4 Optical Properties and Penetration Depth of Skin

For skin, in vivo measurements of optical properties are possible only in the geometry of backscattering. The spatially resolved reflectance $R(r_{sd})$ is defined as the power of the backscattered light per unit of area detected by a receiver at the surface of the skin at a distance r_{sd} from the source, r_{sd} is the source–detector separation. $R(r_{sd})$ depends on the optical properties of the sample, that is, the absorption coefficient μ_a, the scattering coefficient μ_s, and the anisotropy factor g, the refractive index n, and the numerical aperture NA of the receiving system [1,5,9–11,14–26].

When optical parameters of skin are under investigation, the small source-detector separations should be used, where the diffusion approximation is not valid due to its proximity to the tissue boundary. In that case, more sophisticated approximations of the RTE solution should be employed; in particular, a numerical solution of the inverse problem by the Monte Carlo (MC) method is prospective.

In the six-detecting-fiber system made of 0.4 mm-core diameter optical fiber that is described in ref. [23], source-detector distances were: $r_{sd} = 0.44$, 0.78, 0.92, 1.22, 1.40, and 1.84 mm. In the course of their in vivo studies, temperature dependences of the absorption and the reduced scattering coefficients of human forearm skin have been determined.

Two precise optical systems, the fiber optic spectrometer yielding spatially resolved backreflectance spectra and the single wavelength fiber-optic-CCD tissue imager, were used for in vivo measurements of anisotropy of scattering and absorption coefficients of human skin at different body locations [22]. The source-detector distances between 0.33 and 10.0 mm for 18 detecting 0.2 mm core diameter fibers linearly aligned with a central illuminating fiber provided a 2D mapping of reflected intensity by rotation of the detecting fiber system around the illuminating fiber. The MC code accounting for a two-layered tissue model (skin itself and a semi-infinite subcutaneous fat layer) with two groups of scatterers, one of randomly distributed scatterers and another of infinite dielectric cylinders (dermal collagen fibers) aligned along one of the principle Cartesian axes parallel to the skin surface, was designed to evaluate scattering and absorption coefficients distributions. In the skin layer, the scattering coefficient was recalculated before each interaction event according to the current direction of photon propagation as $\mu_s = \mu_{s0}\,[1 + f_s(0.5 - |\cos\psi|)]$, where μ_{s0} is a base scattering coefficient, f_s is the fraction of scatterers oriented in the preferential direction, and ψ is the angle between the current photon direction and the cylinder axis.

Optical coherence tomography (OCT) is a newly developed modality that allows one to evaluate the scattering and absorption properties of tissue in vivo within the limits of an

OCT penetration depth of 1–3 mm [1,24]. In its simplest form, this method assumes that backscattered light from a tissue decreases in the intensity according to $I_b \cong I_0\exp[-2(\mu_a + \mu_s)z]$, where $2z$ is the round-trip distance of light backscattered at depth z. For most tissues in the NIR, $\mu_a << \mu_s$; thus, μ_s can be estimated roughly as $\mu_s \cong 1/2z\{\ln[I_b(z)/I_0]\}$.

Direct measurement of the scattering phase function $p(\theta)$ is important for the choice of an adequate model for the tissue being examined. The scattering phase function or g-factor is usually determined from in vitro goniophotometric measurements [10,11]. For human dermis and epidermis in the wavelength range from 300 to 1300 nm, g-factor is well-described by the empirical formula [10]: $g_e \sim g_d \sim 0.62 + \lambda \times 0.29 \times 10^{-3}$, where the wavelength λ is given in nanometers.

The values of absorption and scattering coefficients and scattering anisotropy factor for the human skin measured in vitro, ex vivo, or in vivo are presented in Table 3.1.

The direct measurement of the penetration depth of a tissue at some specific wavelength is important for providing of laser phototherapy strategy. For example, at attenuation of a wide laser beam of intensity I_0 in a thick tissue at depths $z > l_d = 1/\mu_{eff}$ may be described instead of Eq. (3.1) by the following equation [1]:

$$I(z) \approx I_0 b_s \exp(-\mu_{eff} z), \tag{3.7}$$

where b_s accounts for additional irradiation of upper layers of a tissue due to multiple backscattering (photon recycling effect) and $\mu_{eff} = [3\mu_a(\mu_s' + \mu_a)]^{1/2}$ photon diffusion due to multiple scattering. Respectively, the depth of light penetration into a tissue is

$$l_e = l_d[\ln b_s + 1]. \tag{3.8}$$

Typically, for tissues b_s =1–5 for beam diameter of 1–20 mm. Thus, when wide laser beams are used for irradiation of highly scattering tissues with low absorption, CW light energy is accumulated in the tissue due to high multiplicity of chaotic long-path photon migrations. A highly scattering medium works as a random cavity providing the capacity of light energy. The light power density within the superficial tissue layers may substantially (up to threefold for the human skin [10]) exceed the incident power density and cause the overheating at laser photothermolysis or the overdosage during photodynamic therapy. On the other hand, photon recycling effect can be used for more effective irradiation of undersurface lesions at relatively small incident power densities.

Figure 3.8 illustrates results of the reconstruction of the spectral dependences of the absorption (a) and reduced scattering coefficients (b) of the human skin [16]. The reconstruction was done on the basis of in vitro measurements of skin samples transmittance and reflectance. Using received data for $\mu_a(\lambda)$ and $\mu_s'(\lambda)$, calculations of the skin penetration depth ($l_e \approx l_d$) was performed (c). The maximal penetration depth of 3.5 mm was found at the wavelength 1090 nm. At some other wavelengths, such as 600, 633, 660, 700, 750, 800, 850, and 900 nm, the penetration depth was correspondingly equal to 1.5, 1.7, 1.8, 2.0, 2.2, 2.3, 2.4, and 2.5 mm.

It is important to note that the penetration depth shown in Fig. 3.8 is defined as depth at which an initially collimated large diameter beam is attenuated e = 2.7 times. The attenuation profile of light in the skin is approximately an exponential function and photons with low density can be found at depths much deeper than defined by Fig. 3.8. Smaller diameter

Table 3.1 Optical Properties of the Human Skin, Its Components, and Underlying Tissues Measured In Vitro, Ex Vivo, and In Vivo (rms values are given in parentheses)

Tissue	λ (nm)	μ_a (cm^{-1})	μ_s (cm^{-1})	μ_s' (cm^{-1})	g	Remarks
In vitro measurements						
Stratum corneum	193	6000	–	–	–	Frozen sections [9]
	250	1150	2600	260	0.9	Data from graphs of ref. [10]; μ_s' is calculated
	308	600	2400	240	0.9	
	337	330	2300	230	0.9	
	351	300	2200	220	0.9	
	400	230	2000	200	0.9	
Epidermis	250	1000	2000	616	0.69	Data from graphs of ref. [10]; μ_s' and g are calculated
	308	300	1400	407	0.71	
	337	120	1200	338	0.72	
	351	100	1100	306	0.72	
	415	66	800	206	0.74	
	488	50	600	143	0.76	
	514	44	600	139	0.77	
	585	36	470	99	0.79	
	633	35	450	88	0.80	
	800	40	420	62	0.85	
Dermis	250	35	833	257	0.69	Data from graphs of ref. [10]; values are transformed in accordance with data for = 633 nm [11] (bloodless tissue, hydration—85%), μ_s' and g are calculated
	308	12	583	170	0.71	
	337	8.2	500	141	0.72	
	351	7	458	127	0.72	
	415	4.7	320	82	0.74	
	488	3.5	250	60	0.76	
	514	3	250	58	0.77	
	585	3	196	41	0.79	
	633	2.7	187.5	37	0.80	
	800	2.3	175	30	0.85	

(Continued)

Table 3.1 Optical Properties of the Human Skin, Its Components, and Underlying Tissues Measured In Vitro, Ex Vivo, and In Vivo (rms values are given in parentheses) (Continued)

Tissue	λ (nm)	μ_a (cm^{-1})	μ_s (cm^{-1})	μ_s' (cm^{-1})	g	Remarks
Dermis	749	0.24 (0.19)	–	23.1 (0.75)	–	Frozen sections, double integrating sphere (DIS) [14]
	789	0.75 (0.06)	–	22.8 (1.29)	–	
	836	0.98 (0.15)	–	15.9 (2.16)	–	
Caucasian male skin ($n = 3$)	500	5.1	–	50	–	Integrating sphere (IS), inverse adding-doubling (IAD) method; sample thickness: 0.40, 0.23, 0.25 mm [15]
	810	0.26	–	15.8	–	
Caucasian male skin ($n = 3$), external pressure 0.1 kg/cm^2	500	15.3	–	167.4	–	IS, IAD, sample thickness: 0.15, 0.05, 0.13 mm [15]
	810	0.63	–	52.7	–	
Caucasian male skin ($n = 3$), external pressure 1 kg/cm^2	500	13.6	–	156.7	–	IS, IAD; sample thickness: 0.12, 0.05, 0.13 mm [15]
	810	0.57	–	53.4	–	
Caucasian female skin ($n = 3$)	500	5.2	–	23.9	–	IS, IAD; sample thickness: 0.42,0.50, 0.50 mm [15]
	810	0.97	–	8.2	–	
Caucasian female skin ($n = 3$), external pressure 0.1 kg/cm^2	500	7.4	–	31.5	–	IS, IAD; sample thickness: 0.30, 0.30, 0.34 mm [15]
	810	1.4	–	11.3	–	
Caucasian female skin ($n = 3$), external pressure 1 kg/cm^2	500	10.0	–	40.2	–	IS, IAD; sample thickness: 0.27, 0.20, 0.23 mm [15]
	810	1.7	–	13.1	–	
Hispanic male skin ($n = 3$)	500	3.8	–	24.2	–	IS, IAD; sample thickness: 0.70, 0.78, 0.63 mm [15]
	810	0.87	–	7.5	–	

Hispanic male skin (*n* = 3), external pressure 0.1 kg/cm^2	500	5.1	–	37.6	–	IS, IAD; sample thickness: 0.35, 0.62, 0.48 mm [15]
	810	0.93	–	11.4	–	
Hispanic male skin (*n* = 3), external pressure 1 kg/cm^2	500	6.2	–	40.4	–	IS, IAD; sample thickness: 0.28, 0.48, 0.33 mm [15]
	810	0.87	–	10.2	–	
Caucasian skin (*n* = 21)	400	3.76 (0.35)	–	71.79 (9.42)	–	IS, IAD; tissue slabs, 1–6 mm; post-mortem; <24 hr after death; stored at 20°C in saline; measurements at room temperature; in the spectral range 400–2000 nm: $\mu_s' = 1.1 \times 10^{12}\lambda^{-4} + 73.7\lambda^{-0.22}$, [λ] = nm [16]
	500	1.19 (0.16)	–	32.46 (4.21)	–	
	600	0.69 (0.13)	–	21.78 (2.98)	–	
	700	0.48 (0.11)	–	16.69 (2.27)	–	
	800	0.43 (0.11)	–	14.02 (1.89)	–	
	900	0.33 (0.02)	–	15.66 (2.06)	–	
	1000	0.27 (0.03)	–	16.83 (2.77)	–	
	1100	0.16 (0.04)	–	17.11 (2.69)	–	
	1200	0.54 (0.04)	–	16.71 (2.89)	–	
	1300	0.41 (0.07)	–	14.69 (2.59)	–	
	1400	1.64 (0.31)	–	14.28 (3.69)	–	
	1500	1.69 (0.35)	–	14.41 (3.75)	–	
	1600	1.19 (0.22)	–	14.16 (3.41)	–	
	1700	1.55 (0.28)	–	14.71 (3.51)	–	
	1800	1.44 (0.22)	–	13.36 (2.91)	–	
	1900	2.14 (0.28)	–	12.15 (3.05)	–	
	2000	1.74 (0.29)	–	12.01 (2.91)	–	

(Continued)

Table 3.1 Optical Properties of the Human Skin, Its Components, and Underlying Tissues Measured In Vitro, Ex Vivo, and In Vivo (rms values are given in parentheses) (Continued)

Tissue	λ (nm)	μ_a (cm^{-1})	μ_s (cm^{-1})	μ_s' (cm^{-1})	g	Remarks
Epidermis	370	1.35 (0.16)	–	11.56 (1.25)	0.8	IS, inverse Monte Carlo (IMC) technique, slabs [17]
	470	0.84 (0.06)	–	7.96 (0.82)	0.8	
	570	0.39 (0.08)	–	5.52 (0.55)	0.8	
	670	0.26 (0.08)	–	4.48 (0.43)	0.8	
	770	0.19 (0.06)	–	3.79 (0.37)	0.8	
	870	0.10 (0.05)	–	3.41 (0.34)	0.8	
	970	0.06 (0.03)	–	3.15 (0.34)	0.8	
	1070	0.02 (0.02)	–	2.97 (0.32)	0.8	
	1170	0.06 (0.04)	–	2.71 (0.31)	0.8	
	1270	0.06 (0.04)	–	2.62 (0.31)	0.8	
	1370	0.56 (0.14)	–	2.50 (0.31)	0.8	
	1470	2.96 (0.42)	–	3.08 (0.45)	0.8	
	1570	1.01 (0.20)	–	2.39 (0.34)	0.8	
Dermis	370	0.98 (0.14)	–	8.76 (1.36)	0.8	IS, IMC, slabs [17]
	470	0.43 (0.06)	–	5.36 (0.60)	0.8	
	570	0.22 (0.03)	–	3.50 (0.31)	0.8	
	670	0.15 (0.02)	–	2.78 (0.26)	0.8	
	770	0.13 (0.02)	–	2.33 (0.24)	0.8	
	870	0.09 (0.02)	–	2.05 (0.22)	0.8	
	970	0.08 (0.02)	–	1.90 (0.22)	0.8	
	1070	0.05 (0.02)	–	1.79 (0.21)	0.8	
	1170	0.12 (0.02)	–	1.69 (0.20)	0.8	
	1270	0.10 (0.02)	–	1.63 (0.20)	0.8	
	1370	0.48 (0.04)	–	1.66 (0.19)	0.8	
	1470	2.19 (0.20)	–	2.13 (0.21)	0.8	
	1570	0.85 (0.07)	–	1.65 (0.19)	0.8	

Subcutaneous fat	370	1.18 (0.21)	–	5.27 (0.69)	0.8	IS, IMC, slabs [17]
	470	0.75 (0.09)	–	3.92 (0.50)	0.8	
	570	0.31 (0.09)	–	2.89 (0.36)	0.8	
	670	0.13 (0.03)	–	2.40 (0.27)	0.8	
	770	0.11 (0.02)	–	2.07 (0.21)	0.8	
	870	0.09 (0.02)	–	1.89 (0.19)	0.8	
	970	0.09 (0.03)	–	1.76 (0.18)	0.8	
	1070	0.07 (0.02)	–	1.68 (0.15)	0.8	
	1170	0.14 (0.03)	–	1.63 (0.15)	0.8	
	1270	0.10 (0.03)	–	1.59 (0.14)	0.8	
	1370	0.27 (0.04)	–	1.60 (0.15)	0.8	
	1470	1.08 (0.18)	–	1.81 (0.19)	0.8	
	1570	0.43 (0.07)	–	1.60 (0.16)	0.8	
Ex vivo measurements						
Caucasian dermis (n = 12)	633	0.33 (0.09)	–	27.3 (5.4)	0.9	A single integrating sphere "comparison" method, IMC; samples from abdominal and breast tissue obtained from plastic surgery or post-mortem examinations, g = 0.9 is supposed value in calculations [18]
	700	0.19 (0.06)	–	23.2 (4.1)	0.9	
	900	0.13 (0.07)	–	16.3 (2.5)	0.9	
Negroid dermis (n = 5)	633	2.41 (1.53)	–	32.1 (20.4)	0.9	
	700	1.49 (0.88)	–	26.8 (14.1)	0.9	
	900	0.45 (0.18)	–	18.1 (0.4)	0.9	
Subdermis (primarily globular fat cells) (n = 12)	633	0.13 (0.05)	–	12.6 (3.4)	0.9	
	700	0.09 (0.03)	–	12.1 (3.2)	0.9	
	900	0.12 (0.04)	–	10.8 (2.7)	0.9	

(Continued)

Table 3.1 Optical Properties of the Human Skin, Its Components, and Underlying Tissues Measured In Vitro, Ex Vivo, and In Vivo (rms values are given in parentheses) (Continued)

Tissue	λ (nm)	μ_a (cm^{-1})	μ_s (cm^{-1})	μ_s' (cm^{-1})	g	Remarks
Sample/subject—01/01; female (F), age = 51 yr; back of knee, left leg; moderate inflammation in dermis; SC = 40–70 μm; E = 40–150 μm; D = 300 μm	1460 1600 2200	17.88 (1.12) 5.35 (0.24) 7.46 (0.56)	– – –	10.74 (0.49) 8.06 (0.29) 7.17 (0.26)	– – –	DIS, IAD; slabs containing stratum corneum (SC), epidermis (E), and dermis (D), taken from 14 subjects; measured within 24 hr of excision; heated to 37°C; three measurements on each side of the sample; 2.5-cm diameter sample ports on the setup, for small sample size reduced to 1.3 cm*; totally data for 52 wavelengths in the range 1000–2200 nm are available [19]
02/01; F, age = 51 yr; back of knee, left leg; moderate inflammation in dermis; SC = 40–70 μm; E = 40–140 μm; D = 300 μm	1460 1600 2200	18.70 (1.13) 5.46 (0.27) 8.86 (0.46)	– – –	11.39 (0.65) 8.62 (0.34) 8.15 (0.26)	– – –	
03/02; F, age = 66 yr; lower back, right side; mild solar damage; SC = 20–50 μm; E = 30 μm; D = 200 μm	1460 1600 2200	16.01 (0.56) 4.91 (0.10) 10.94 (0.23)	– – –	9.83 (0.59) 6.78 (0.45) 9.00 (0.54)	– – –	

04/02; F, age = 66 yr; lower back, right side; mild solar damage; SC = 20–50 μm; E = 30 μm; D = 200 μm	1460 1600 2200	12.65 (0.96) 3.86 (0.28) 8.58 (0.55)	– – –	8.61 (0.63) 6.04 (0.29) 7.74 (0.23)	– – –
05/03; F, age = 67 yr; shin, right leg; mild solar damage, chronic inflammation; SC = 20–50 μm; E = 30–50 μm; D = 200 μm	1460 1600 2200	16.58 (3.26) 5.15 (0.60) 9.65 (1.17)	– – –	11.68 (1.41) 8.89 (1.11) 10.31 (0.81)	– – –
06/03; F, age = 67 yr; shin, right leg; mild solar damage, chronic inflammation; SC = 20–50 μm; E = 30–50 μm; D = 200 μm	1460 1600 2200	18.07 (0.42) 5.60 (0.17) 11.26 (0.16)	– – –	13.13 (0.63) 10.34 (0.52) 12.20 (0.88)	– – –
07/04; M, age = 64 yr; thigh, right leg; mild chronic dermatitis; SC = 20–30 μm; E = 50–90 μm; D = 300 μm	1000 1460 1600 2200	0.69 (0.01) 16.64 (0.95) 4.96 (0.27) 13.04 (2.36)	– – – –	10.45 (0.61) 10.75 (0.81) 7.72 (0.40) 9.42 (1.57)	– – – –

(Continued)

Table 3.1 Optical Properties of the Human Skin, Its Components, and Underlying Tissues Measured In Vitro, Ex Vivo, and In Vivo (rms values are given in parentheses) (Continued)

Tissue	λ (nm)	μ_a (cm^{-1})	μ_s (cm^{-1})	μ_s' (cm^{-1})	g	Remarks
08/05; M, age = 75 yr; lower thigh, left leg; normal skin; SC = 8–12 μm; E = 20–60 μm; D = 200 μm	1000	0.83 (0.03)	–	12.25 (1.20)	–	
	1460	19.06 (1.22)	–	11.46 (1.09)	–	
	1600	5.75 (0.27)	–	8.31 (0.76)	–	
	2200	11.92 (0.41)	–	10.34 (0.76)	–	
09/05; M, age = 75 yr; lower thigh, left leg; normal skin; SC = 8–12 μm; E = 20–60 μm; D = 200 μm	1000	0.85 (0.02)	–	11.66 (0.96)	–	
	1460	18.03 (2.01)	–	11.19 (1.51)	–	
	1600	5.61 (0.56)	–	7.87 (0.81)	–	
	2200	11.85 (0.83)	–	10.03 (0.90)	–	
10/06; F, age = 42 yr; groin, left side; mild chronic inflammation; SC = 5 μm; E = 25–30 μm; D = 200 μm	1000	0.80 (0.01)	–	14.17 (0.71)	–	
	1460	20.49 (0.89)	–	13.64 (1.44)	–	
	1600	5.85 (0.14)	–	10.05 (0.55)	–	
	2200	12.46 (0.42)	–	11.79 (0.69)	–	
11/06; F, age = 42 yr; groin, left side; mild chronic inflammation; SC = 5 μm; E = 25–30 μm; D = 200 μm	1000	0.77 (0.03)	–	13.95 (1.12)	–	
	1460	20.24 (1.04)	–	13.18 (1.72)	–	
	1600	5.76 (0.28)	–	9.48 (0.91)	–	
	2200	12.71 (0.58)	–	10.89 (1.20)	–	

12/07; M, age = 33 yr; posterior thigh, right side; mild chronic dermatitis; SC = 2–5 μm; E = 5–10 μm; D = 300 μm	1000	0.82 (0.02)	–	14.35 (0.81)	–
	1460	19.01 (1.28)	–	13.30 (0.91)	–
	1600	5.81 (0.33)	–	10.14 (0.49)	–
	2200	11.13 (1.21)	–	9.00 (0.33)	–
13/08; F, age = 52 yr; axillary, right side; mild perivascular chronic inflammation; SC = 5–7 μm; E = 25 μm; D = 100 μm	1000	0.97 (0.08)	–	13.70 (0.35)	–
	1460	21.39 (1.25)	–	12.54 (0.72)	–
	1600	6.17 (0.30)	–	9.94 (0.78)	–
	2200	12.53 (0.84)	–	9.45 (0.84)	–
14/09; M, age = 37 yr; back of thigh, upper left; mild chronic dermatitis; SC = 3 μm; E = 13 μm; D = 300 μm	1000	0.82 (0.02)	–	15.00 (0.49)	–
	1460	23.31 (0.71)	–	12.32 (0.51)	–
	1600	6.68 (0.11)	–	10.01 (0.37)	–
	2200	15.19 (1.37)	–	8.54 (0.52)	–
15/10; M, age = 70 yr; scalp; mild chronic dermatitis w/solar elastosis; SC = 4–15 μm; E = 8–10 μm; D = 200 μm	1000	1.04 (0.02)	–	12.26 (0.44)	–
	1460	15.95 (0.99)	–	10.75 (1.20)	–
	1600	5.09 (0.23)	–	8.83 (0.92)	–
	2200	12.65 (0.52)	–	8.83 (1.94)	–

(Continued)

Table 3.1 Optical Properties of the Human Skin, Its Components, and Underlying Tissues Measured In Vitro, Ex Vivo, and In Vivo (rms values are given in parentheses) (Continued)

Tissue	λ (nm)	μ_a (cm^{-1})	μ_s (cm^{-1})	μ_s' (cm^{-1})	g	Remarks
16/11*; M, age = 61 yr; scalp; mild chronic dermatitis w/solar elastosis; SC = 2–4 μm; E = 6 μm; D = 300 μm	1000	0.79 (0.02)	–	13.11 (0.61)	–	
	1460	16.47 (1.05)	–	12.45 (0.56)	–	
	1600	5.11 (0.24)	–	10.43 (0.57)	–	
	2200	13.30 (1.48)	–	9.89 (0.79)	–	
17/12*; F, age = 68 yr; scalp/facial tissue; mild solar damage, chronic inflammation; SC = 2 μm; E = 8–10 μm; D = 200 μm	1000	1.06 (0.03)	–	8.79 (1.18)	–	
	1460	12.81 (1.84)	–	9.60 (0.57)	–	
	1600	4.26 (0.50)	–	6.93 (0.75)	–	
	2200	11.32 (1.52)	–	8.14 (0.81)	–	
18/12*; F, age = 68 yr; scalp/ facial tissue; sever solar damage, mild chronic inflammation; SC = 2μm; E = 8–10 μm; D = 150 μm	1000	1.32 (0.05)	–	8.63 (1.91)	–	
	1460	12.68 (5.07)	–	8.74 (1.26)	–	
	1600	4.31 (1.34)	–	6.60 (0.92)	–	
	2200	11.33 (3.05)	–	7.30 (0.24)	–	

19/13*; F, age = 53 yr; scalp/facial tissue; mild chronic inflammation; SC = 4 μm; E = 10 μm; D = 200 μm	1000	1.55 (0.02)	–	11.96 (0.65)	–
	1460	16.13 (1.38)	–	11.52 (0.64)	–
	1600	5.38 (0.31)	–	8.65 (0.54)	–
	2200	13.84 (1.02)	–	9.67 (0.65)	–
20/13; F, age = 53 yr; scalp/facial tissue; mild solar damage; SC = 4 μm; E = 10 μm; D = 200 μm	1000	1.53 (0.02)	–	12.89 (0.77)	–
	1460	16.82 (1.13)	–	12.01 (0.81)	–
	1600	5.57 (0.19)	–	9.47 (0.60)	–
	2200	13.46 (0.58)	–	10.41 (0.71)	–
21/14; F, age = 52 yr; abdomen; mild chronic inflammation; SC = 4–5 μm; E = 10 μm; D = 200 μm	1000	0.88 (0.03)	–	14.96 (1.28)	–
	1460	18.21 (2.51)	–	14.20 (0.71)	–
	1600	5.74 (0.68)	–	10.58 (0.44)	–
	2200	11.33 (0.76)	–	10.40 (0.47)	–
22/14; F, age = 52 yr; abdomen; mild chronic inflammation; SC = 4–5 μm; E = 10 μm; D = 200 μm	1000	0.94 (0.02)	–	15.26 (0.63)	–
	1460	18.46 (1.64)	–	15.10 (1.01)	–
	1600	5.76 (0.31)	–	11.05 (0.39)	–
	2200	13.72 (0.52)	–	13.72 (0.42)	–

(Continued)

Table 3.1 Optical Properties of the Human Skin, Its Components, and Underlying Tissues Measured In Vitro, Ex Vivo, and In Vivo (rms values are given in parentheses) (Continued)

Tissue	λ (nm)	μ_a (cm^{-1})	μ_s (cm^{-1})	μ_s' (cm^{-1})	g	Remarks
Subcutaneous fat	400	2.26 (0.24)	–	13.39 (2.78)	–	IS, IAD; tissue slabs, 1–3 mm; <6 hr after surgery; stored at 20°C in saline; measurements at room temperature; in the spectral range 600–1500 nm: $\mu_s' = 1.05\times10^3\lambda^{-0.68}$, [λ] = nm [16]
(n = 6)	500	1.49 (0.06)	–	13.82 (4.00)	–	
	600	1.18 (0.02)	–	13.39 (4.65)	–	
	700	1.11 (0.05)	–	12.17 (4.41)	–	
	800	1.07 (0.11)	–	11.62 (4.63)	–	
	900	1.07 (0.07)	–	9.97 (3.42)	–	
	1000	1.06 (0.06)	–	9.39 (3.32)	–	
	1100	1.01 (0.05)	–	8.74 (3.28)	–	
	1200	1.06 (0.07)	–	7.91 (3.17)	–	
	1300	0.89 (0.07)	–	7.81 (3.19)	–	
	1400	1.08 (0.03)	–	7.51 (3.31)	–	
	1500	1.05 (0.02)	–	7.36 (3.42)	–	
	1600	0.89 (0.04)	–	7.16 (3.21)	–	
	1700	1.26 (0.07)	–	7.53 (3.33)	–	
	1800	1.21 (0.01)	–	7.50 (3.48)	–	
	1900	1.62 (0.06)	–	8.72 (4.15)	–	
	2000	1.43 (0.09)	–	8.24 (4.03)	–	
In vivo measurements						
Dermis	660	0.07–0.2	–	9–14.5	–	Spatially-resolved reflectance (SRR) technique, IMC [20]
Skin	633	0.62	–	32	–	Ref. [5]
	700	0.38	–	28.7	–	

Skin (0–1 mm)	633	0.67	–	16.2	–	Ref. [21]
Skin (1–2 mm)	633	0.026	–	12.0	–	
Skin (>2 mm)	633	0.96	–	5.3	–	
Forearm:						
Epidermis	633	8^{Δ}	–	17.5	0.9^{Δ}	SRR; ($^{\Delta}$) from literature [21]
Dermis	633	0.15	–	17.5	0.9^{Δ}	
Epidermis and	750	0.375	–	15	–	SRR; diffusion approximation [22]
dermis	750	0.03	–	10	–	
Subcutaneous fat						
Arm	633	0.17 (0.01)	–	9.08 (0.05)	–	SRR, 9 detecting 600-μm fibers; mean separation 1.7 mm; Mie phase function [5]
	660	0.128 (0.005)	–	8.68 (0.05)	–	
	700	0.090 (0.002)	–	8.14 (0.05)	–	
Foot sole	633	0.072 (0.002)	–	11.17 (0.09)	–	
	660	0.053 (0.003)	–	10.45 (0.09)	–	
	700	0.037 (0.001)	–	9.52 (0.08)	–	
Forehead	633	0.090 (0.009)	–	16.72 (0.09)	–	
	660	0.052 (0.003)	–	16.16 (0.08)	–	
	700	0.0240 (0.002)	–	15.38 (0.06)	–	
Abdominal skin:						
Chosen direction	810	0.014	–	20	–	SRR; CCD detector, r_{sd}≤10 mm; diffusion approximation [22]
perpendicular direction (along collagen fibers)	810	0.07	–	10	–	
Forearm (light skin, $n = 7$):						
Skin	590	2.372 (0.282)	–	9.191 (0.931)	–	SRR; MC-generated grid [23]
temperature—22°C	750	0.966 (0.110)	–	7.340 (0.901)	–	
	950	0.981 (0.073)	–	6.067 (0.847)	–	
Skin	590	2.869 (0.289)	–	9.613 (0.894)	–	
temperature—38°C	750	1.157 (0.106)	–	7.649 (0.971)	–	
	950	1.135 (0.123)	–	6.234 (0.928)	–	

(Continued)

Table 3.1 Optical Properties of the Human Skin, Its Components, and Underlying Tissues Measured In Vitro, Ex Vivo, and In Vivo (rms values are given in parentheses)

Tissue	λ (nm)	μ_a (cm^{-1})	μ_s (cm^{-1})	μ_s' (cm^{-1})	g	Remarks
Dermis of a lower	1300	–	47	–	–	OCT [24]
arm	1300	–	12	–	–	
Stratum corneum						
of finger						
Volar side of lower						
arm:	1300	–	15–20	–	–	OCT [25]
Epidermis	1300	–	80–100	–	–	
Upper dermis						
Palm of hand:	1300	–	10–15	–	–	
Stratum corneum						
Epidermis:	1300	–	60–70	–	–	
Grandular layer	1300	–	40–50	–	–	
Basal layer	1300	–	50–80	–	–	
Upper dermis						
Volar side of lower arm (epidermis and dermis)	1300	–	140	–	–	OCT [26]; depth up to 350 μm; skin treated with a detergent solution (2% of anionic tensides in water)
Normal	1300	–	80	–	–	
Treated						
Forearm (5 subjects, 14 measurements)	800	0.23 (0.04)	–	6.8 (0.8)	–	Time-domain technique, μ_s' (cm^{-1}) ≈ $11–5.1\times10^{-3}\lambda$, λ = 760–900 nm [14]

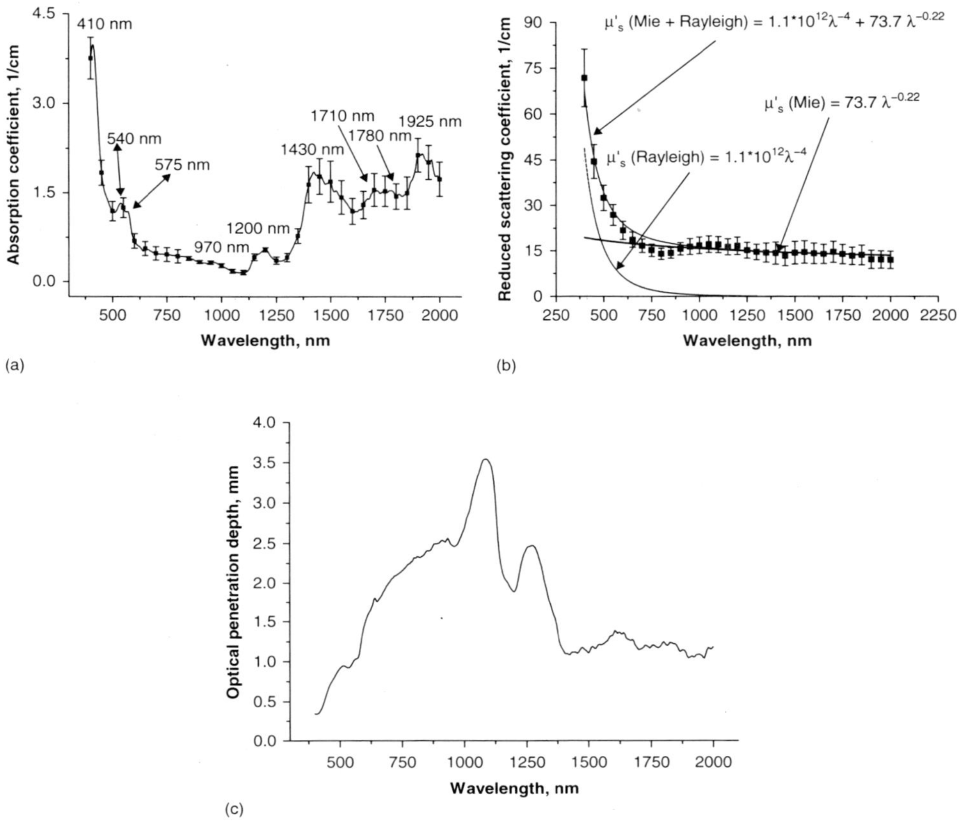

Figure 3.8 Results of in vitro measurements of optical properties of the human skin samples ($N = 21$) using integrating sphere (IS) technique for measurements and inverse adding-doubling method (IAD) for the reconstruction of the optical properties [16]. Tissue slabs of 1–6 mm in thickness taken post-mortem (<24 hours after death) and stored at 20°C in saline were investigated. Measurements were taken at room temperature. (a) The spectral dependence of absorption coefficient μ_a, the vertical lines show the standard deviation values. (b) The spectral dependence of reduced scattering coefficient μ_s', and its approximation by the power law; the symbols correspond to the averaged experimental data and the vertical lines show the standard deviation values; the bold and dashed lines show the contribution of the Mie and Rayleigh scattering, respectively; the solid line shows the combination of the Mie and Rayleigh scattering. (c) The optical penetration depth of light $l_e \approx l_d$ into skin over the wavelength ranges from 400 to 2000 nm.

beams always have low penetration depth than large diameter beams. The penetration depth can be increased by focusing beam into the skin. Mechanical skin deformation by compression or vacuum allows for increase of penetration depth due to scattering coefficient reduction by skin deformation and absorption coefficient decrease by blood displacement to the neighboring tissue regions. Another method of penetration depth increase is optical clearing that is based on tissue impregnation by a biocompatible solution and causes matching of refractive indices of scatterers and tissue ground material which may dramatically decrease tissue scattering coefficient [1].

The quantity usually measured in dosimetry is the irradiance $F(\bar{r})$, which is defined as the power per receiving area of a flat detector. For such definition, light entering differently from perpendicular incidence contributes with reduced impact, and light from below does not contribute at all. Light-induced tissue heating or any photobiological effect in tissue or cells depends on light absorption. For isotropic media, absorption is not sensitive to the angle of irradiation; thus, an adequate light dosimetric quantity should be the total radiant energy fluence rate $U(\bar{r})$ or the space irradiance, defined as the light power hitting a sphere divided by sphere's cross-section [27]. For an isotropic space distribution of light intensity, the space irradiance is four times the irradiance measured for the same point within tissue ($\bar{r}$).

The mean refractive index $\bar{n}$ of a tissue is defined by the refractive indices of its scattering centers material n_s and ground (surrounding) matter n_0 [Eq. (3.6)]. The refractive index variation in tissues, quantified by the ratio $m \equiv n_s/n_0$, determines light scattering efficiency. Measuring refractive indices in tissues and their constituent components is an important focus of interest in tissue optics because index of refraction determines light reflection and refraction at the interfaces between air and tissue, detecting fiber and tissue, and tissue layers; it also strongly influences light propagation and distribution within tissues, defines speed of light in tissue, and governs how the photons migrate.

Indeed, the optical properties of tissues, including refractive indices, are known to depend on water content. The refractive indices of water over a broad wavelength ranging from 200 nm to 200 μm have been reported in ref. [28]. Specifically, $n_w = 1.396$ for $\lambda = 200$ nm, 1.335 for $\lambda = 500$ nm, 1.142 for $\lambda = 2{,}800$ nm, 1.400 for $\lambda = 3{,}500$ nm, 1.218 for $\lambda = 10{,}000$ nm, and 2.130 for $\lambda = 200$ μm.

To model tissue by a mixture of water and a bioorganic compound of a tissue is more adequate. For instance, the refractive index of human skin can be approximated by a 70/30 mixture of water and protein [19]. Assuming that protein has a constant refractive index value of 1.5 over the entire wavelength range, the authors of ref. [19] have suggested the following expression for estimation of skin index of refraction:

$$n_{skin}(\lambda)=0.7(1.58 - 8.45 \times 10^{-4}\lambda + 1.10 \times 10^{-6}\lambda^2 - 7.19 \times 10^{-10}\lambda^3 + 2.32 \times 10^{-13}\lambda^4 - 2.98 \times 10^{-17}\lambda^5) + 0.3 \times 1.5, \quad (3.9)$$

where the wavelength λ is in nanometers.

A more precise semiempirical dispersion formula for a whole human skin, derived from in vitro experimental data for normal and immersed (optically cleared) skin, has a view [29]:

$$n_{skin}(\lambda)=1.3090 - 4.3460 \times 10^2\lambda^{-2} + 1.6065 \times 10^9\lambda^{-4} - 1.2811 \times 10^{14}\lambda^{-6}. \quad (3.10)$$

For skin optics, this is of great importance to know the dispersion properties of melanin, which is contained in skin and hair. Melanin granules are the major back-reflecting particles in OCT and small-scale spatially resolved spectroscopy of skin. The wavelength dependence of the refractive index of melanin particles in the range from 350 to 800 nm was reconstructed on the basis of spectroscopic and electronic microscopy studies of water suspensions of natural melanin [29,30]:

$$n_M(\lambda)=1.6840 - 1.8723 \times 10^4\lambda^{-2} + 1.0964 \times 10^{10}\lambda^{-4} - 8.6484 \times 10^{14}\lambda^{-6}. \quad (3.11)$$

3.4.5 Transmittance and Reflectance Spectra of Skin

A few important definitions first:

1. Optical transmittance is the ratio of the light intensity $I(d)$ transmitted through a sample of thickness d to the incident intensity I_0, $T = I(d)/I_0$; it is a dimensionless quantity; from Eq. (3.1) it follows that $I(d)/I_0 = \exp(-\mu_t d)$;
2. Transmittance spectra show the dependence of transmittance on the wavelength: $T(\lambda)$;
3. Reflectance (reflection coefficient) is the ratio of the light intensity reflected from a sample I_r to the incident intensity I_0, $R = I_r/I_0$; it is a dimensional quantity;
4. Reflectance spectra show the dependence of reflectance on the wavelength: $R(\lambda)$;
5. Reflectance depends on the angle of view of a detector, when as a detector an integrating sphere is used diffusion reflectance R_d is measured.

For skin it is not easy to provide transmittance measurements in vivo, thus such measurements is used more often in in vitro studies for the prediction of skin optical parameters. Two types of transmittance are typically introduced—the collimated transmittance that is based on collection of light transmitted only in the direction of the incident beam, T_c, which can be measured by a distant detector with a small aperture, and the total transmittance that is based on a collection of all light transmitted (scattered) within the hemisphere in the forward direction, T_t, which can be measured by a detector with an integrating sphere. Both quantities can be measured for the definite wavelength or for a whole spectrum of interest, typically in the visible/NIR range.

The reflectance spectroscopy is much more suitable for skin measurements. A few different methods are available; each of them solving a particular problem in skin appearance or pathology monitoring and, thus, has its own algorithm of operation and corresponding hardware. One of them is a spatially-resolved reflectance technique (SRR)—a technique that uses two or more fibers to illuminate skin and collect the back-reflected light; the positions of the illuminating and light-collecting fibers can be fixed or scanned along the skin surface perpendicularly or having some angle to the surface. Usually, a grating spectrograph at the output of receiving fiber (fibers) in combination with an optical multichannel analyzer (cooled CCD or photodiode array) as a detector is used for such measurements.

Due to the lower thickness of the epidermis compared with the dermis, scattering in the epidermis is of less importance than dermal scattering. Dermal tissue is practically entirely responsible for the majority of light scattering that takes place in the skin, and also determines the diffuse pattern of light distribution within the skin and the formation of the back-scattered diffuse reflectance. Scattering is generally stronger in the UV spectral range, but the strong absorption of epidermal melanin and dermal blood is an important factor responsible for the reduction of the back-scattered light and the generation of the skin reflectance spectrum [6]. Thus, absorption and scattering determine the amount of light emerging from the skin surface, which is closely related to the diffuse reflectance R_d.

Representative diffuse reflectance spectra of human skin are shown in Fig. 3.9. In the UV spectral range (<300 nm) the reflectance R_d is generally very small due to strong epidermal absorption, which reduces the amount of backscattered light to the level comparable

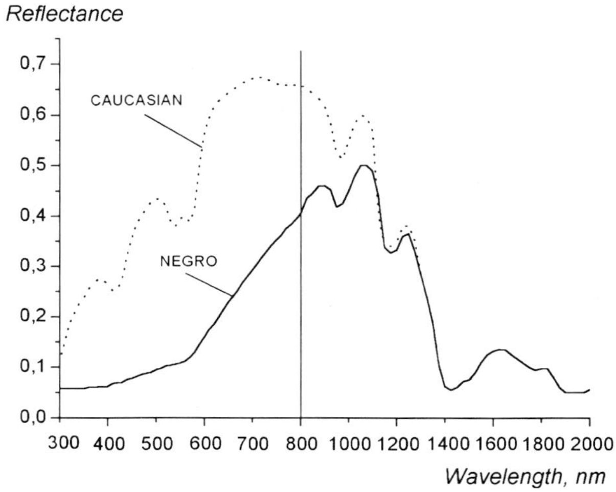

Figure 3.9 Typical diffuse reflectance spectra of white (Caucasian) and black (African) human skin [6].

with Fresnel's reflectance. The penetration depth of optical radiation within the epidermis does not exceed a few cell layers, and epidermal chromophores have a small effect on the diffuse reflectance spectrum.

In the UVA spectral range (315–400 nm) the skin reflectance exceeds Fresnel's reflectance, which indicates an increase in the back-scattered radiation. The penetration depth of optical radiation increases up to hundreds of micrometers, and epidermal chromophores affect the shape of the reflectance spectrum.

In the visible spectral range (400–800 nm) the penetration depth is between 0.5–2.5 mm (see Fig. 3.8). In this case, both absorption and scattering play a dominant role in the formation of the diffuse reflectance spectrum. The fraction of back scattered light increases due to multiple scattering within the skin. The value of R_d is between 15% and 70%, and the reflectance spectrum has a sharp minimum in the spectral range 415–430 nm, due to hemoglobin absorption in the dermis. The reflectance spectrum has the characteristic dips in the spectral range 540–580 nm which are due to the Q-absorption bands of hemoglobin. Additional weak minima in the reflectance may be noted due to absorption of carotene (480 nm) and bilirubin (460 nm).

In the spectral range 600–1500 nm, absorption is even lower, scattering dominates absorption, and the penetration depth is of 3.5 mm. Light within the skin is entirely diffuse, thus the diffuse reflectance is increased up to 35–70%. In the near IR spectral range, the skin reflectance increases up until 800–900 nm, and then decreases due to water absorption bands.

In the study of the perception of skin color, the chromaticity coordinates are used. One of the first mathematically defined color spaces was the CIE XYZ color space (also known as CIE 1931 color space), created by the International Commission on Illumination (Commission Internationale de l'Éclairage—CIE) in 1931. The human eye has receptors for short (S), middle (M), and long (L) wavelengths, also known as blue, green, and red receptors; this means that one, in principle, needs three parameters to describe a color sensation. A specific method for associating three numbers (or tristimulus values) with each color is called a color space: the CIE XYZ color space is one of many such spaces; however, the CIE XYZ color space is special, because it is based on direct measurements of the human eye, and serves as the basis from which many other color spaces are defined; the CIE1964 standard observer is based on the mean 10-degree color matching functions.

3.4.6 Polarization Anisotropy

Skin, as many other tissues is a polarization anisotropic (i.e., birefringent) medium [1]. Birefringence is the phenomenon exhibited by certain materials in which an incident ray of light is split into two rays, called an ordinary ray and an extraordinary ray, which are plane-(linear) polarized in mutually orthogonal planes, or circular-polarized in opposite directions (left and right). Skin linear birefringence results primarily from the linear anisotropy of a dermis fibrous structure. The refractive index of dermis is higher along the length of fibers than across them. A specific dermis structure can be modeled as a system composed of parallel cylinders that create a uniaxial birefringent medium with the optic axis parallel to the cylinder axes. This is called birefringence of form that arises when the relative optical phase between the orthogonal polarization components is nonzero for forward-scattered light. After multiple forward scattering events, a relative phase difference accumulates and a phase delay (δ_{oe}) similar to that observed in birefringent crystalline materials is introduced between orthogonal polarization components. For organized linear structures, an increase in phase delay may be characterized by a difference ((Δn_{oe}) in the effective refractive index for light polarized along, and perpendicular to, the long axis of the linear structures. The effect of tissue birefringence on the propagation of linearly polarized light is dependent on the angle between the incident polarization orientation and the tissue axis. Phase retardation δ_{oe} between orthogonal polarization components, is proportional to the distance d traveled through the birefringent medium

$$\delta_{oe} = \frac{2\delta d \Delta n_{oe}}{\lambda_0} \tag{3.12}$$

A medium of parallel cylinders is a positive uniaxial birefringent medium [$\Delta n_{oe} = (n_e - n_o) > 0$]. Therefore, a case defined by an incident optical field directed parallel to the cylinder axes will be called "extraordinary," and a case with the incident optical field perpendicular to the cylinder axes will be called "ordinary". The difference ($n_e - n_o$) between the extraordinary index and the ordinary index is a measure of the birefringence of a medium comprised of cylinders. For the Rayleigh limit ($\lambda >>$ cylinder diameter), the form birefringence becomes [1].

$$\Delta n_{oe} = (n_e - n_o) = \frac{f_1 f_2 (n_1 - n_2)^2}{f_1 n_1 + f_2 n_2}, \tag{3.13}$$

where f_1 is the volume fraction of the cylinders; f_2 is the volume fraction of the ground substance; and n_1, n_2 are the corresponding indices. For a given difference of indices n_1 and n_2, maximal birefringence is expected for approximately equal volume fractions of thin cylinders and ground material.

Form birefringence is used as an instrument for studying tissue composition. If n_1 and n_2 are known, the measured phase shift δ_{oe} and evaluation of the corresponding birefringence Δn_{oe} allows one to assess the volume fraction occupied by the particles [see Eqs. (3.12) and (3.13)]. The reported value of the human skin birefringence Δn_{oe} is of the order of 10^{-3} [1].

A new technique—polarization-sensitive optical coherence tomography (PS OCT)—allows for a high precision measurement of linear birefringence in a turbid tissue [1]. For example, the porcine skin birefringence measured by PS OCT is of $\Delta n_{oe} = 1.5 \times 10^{-3} - 3.5 \times 10^{-3}$. Such birefringence provides up to 90% phase retardation at a depth on the order of several hundred micrometers.

The magnitude of birefringence is related to the density and refractive properties of the collagen fibers, whereas the orientation of the fast axis indicates the orientation of the collagen fibers. The amplitude and orientation of birefringence of the skin and cartilage are not as uniformly distributed as in a tendon. In other words, the densities of collagen fibers in skin and cartilage are not as uniform as in a tendon, and the orientation of the collagen fibers is not distributed in as orderly a fashion.

It was experimentally demonstrated that in a turbid tissue, laser radiation retains linear polarization on the level of $P_L \leq 0.1$ within a few TMFP l_t, that is, $2.5l_t$. Specifically, for skin irradiated in the red and NIR ranges, $\mu_a \cong 0.4$ cm^{-1}, $\mu_s' \cong 20$ cm^{-1}, and correspondingly $l_t \cong 0.48$ mm. Consequently, light propagating in skin can retain linear polarization within a length of about 1.2 mm. Such an optical path in a tissue corresponds to a time delay on the order of 5.3 ps, which provides an opportunity to produce polarization images of macro-inhomogeneities in a tissue with a spatial resolution equivalent to the spatial resolution that can be achieved by the selection of photons using more sophisticated time-resolved techniques. In addition to the selection of diffuse-scattered photons, polarization imaging makes it possible to eliminate the specular reflection from the surface of a tissue, which allows one to use this technique to image microvessels in facile skin and detect birefringence and optical activity in superficial tissue layers [1,31–34].

Polarization imaging is a new modality in tissue optics [1,31–34]. The most prospective approaches for polarization tissue imaging are: linear polarization degree mapping, two-dimensional backscattering, PS OCT, and full-field polarization-speckle technique. The most robust and cheap is a linear polarization degree (P_L) mapping technique which is based on registration of two-dimensional polarization patterns for the backscattering of a polarized incident narrow laser beam [33]. As an illustration in Fig. 3.10, a scheme of experimental setup for polarization imaging and three different types of images of skin burn lesion are shown. Two images within the imaging area (x, y) are acquired: one "parallel" $[I_{\parallel}(x, y)]$ and one "perpendicular" $[I_{\perp}(x, y)]$. These images are algebraically combined to yield: $P_L(x, y) = (I_{\parallel} - I_{\perp})/(I_{\parallel} + I_{\perp})$. The numerator rejects randomly polarized diffuse reflectance. Normalization by the denominator cancels common attenuation due to melanin pigmentation. The copolarized surface image is characteristic by a clearly seen superficial skin papillary pattern, as well as cross-polarized image gives more information about the status of subsurface skin vessels.

A similar camera system, but one that uses an incoherent white light source such as xenon lamp, is described in ref. 33, where results of a pilot clinical study of various skin pathologies using polarized light are presented. The polarization images of pigmented skin sites (freckles, tattoos, pigmented nevi) and unpigmented skin sites (nonpigmented intradermal nevi, neurofibromas, actinic keratosis, malignant basal cell carcinomas, squamous

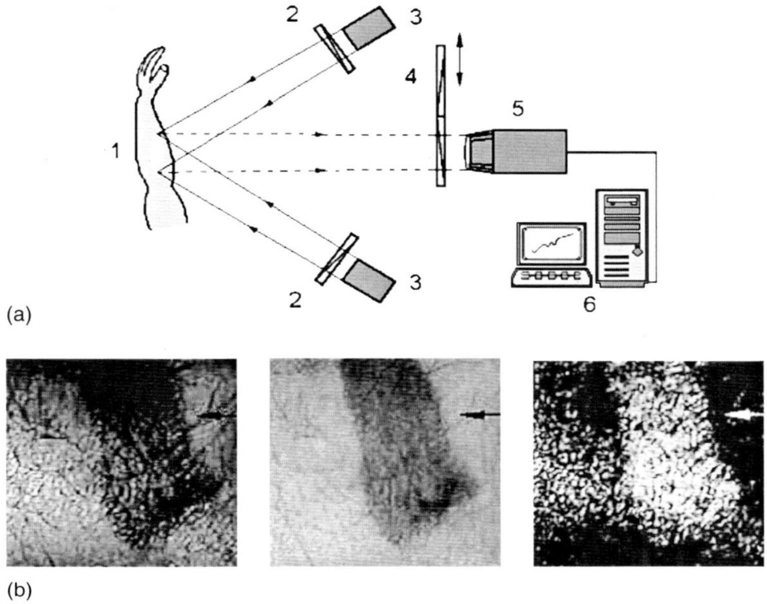

Figure 3.10 Polarization imaging of skin in vivo.35 (a) 1, Skin site; 2, polarization filters; 3, light sources; 4, polarization and interferential filters; 5, monochrom CCD camera; 6, PC. (b) Polarization images (λ = 550 nm) of skin burn lesion of the volunteer: from the left to the right: a co-polarized [$I_{\parallel}(x, y)$], a crossed-polarized [$I_{\perp}(x, y)$], and degree of polarization image [$P_L(x, y)$].

cell carcinomas, vascular abnormalities (venous lakes), and burn scars) are analyzed to find the differences caused by various skin pathologies (see some examples in Fig. 3.11) [33]. A comparative analysis of polarization images of normal and diseased human skin has shown the ability of the aforementioned approach to emphasize image contrast based on light scattering in the superficial layers of the skin. The polarization images can visualize disruption of the normal texture of the papillary and upper reticular layers caused by skin pathology. Polarization imaging can be considered as an adequately effective tool for identifying skin cancer margins, and for guiding surgical excision of skin cancer. Various modalities of polarization imaging are also considered in ref. [35].

3.4.7 Fluorescence

Fluorescence, more generally luminescence, is light not generated at high temperatures alone. It is different from incandescence, in that it usually occurs at low temperatures and is thus a form of cold body radiation. It can be caused by, for example, chemical or biochemical reactions, optical energy absorption; many kinds of luminescence are known: fluorescence, phosphorescence, bioluminescence, chemoluminescence, electroluminescence, radioluminescence, photoluminescence, and etc. Fluorescence is a property of emitting light of a longer wavelength on absorption of light energy, essentially occurs simultaneously with the excitation of a sample.

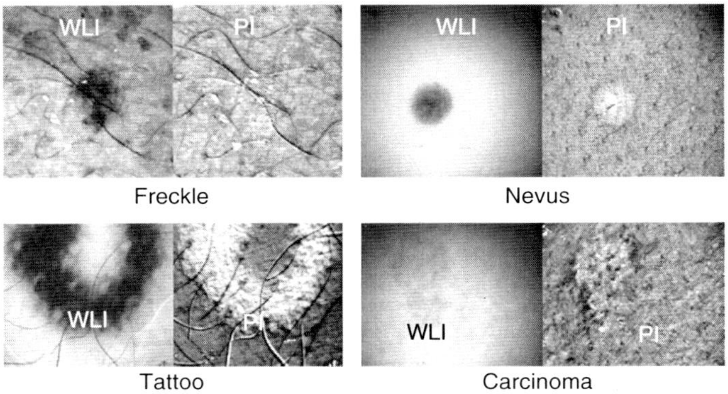

Figure 3.11 Comparison of white light (WLI) versus polarization (PI) images [33]. A freckle: polarization image removes the melanin from a freckle. A benign pigmented nevus: polarization image removes the melanin and shows apparent scatter, the drop of polarized light reflectance from epidermis lining the hair follicles is seen. Tattoo: polarization image lightens the "blackness" of the tattoo, specular reflectance of polarized light off the carbon particles yields a strong image. Malignant basal cell carcinoma: white light image underestimates the extent of the skin cancer.

Fluorescence is characterized by the emission spectrum that is the emission obtained from a luminescent material at different wavelengths when it is excited by a narrow range of shorter wavelengths, as well as by excitation spectrum, that is, the emission spectrum monitored at one wavelength and the intensity at this wavelength is measured as a function of the exciting wavelength. Autofluorescence (AF) is a natural fluorescence of a material (a tissue) due to excitation of the endogenous fluorophores in contrast to fluorescence of a stained material (a tissues or a cell) when exogenous fluorophores are excited.

Human skin contains various types of native fluorophores with unique absorption and emission spectra (Fig. 3.6). The observations regarding the central role of the epidermal chromophores, such as keratin and NADH, in the formation of the AF spectrum of the human skin is based on a fact that the in vitro fluorescence spectra of keratin and NADH are very similar to the in vivo AF spectra of the human skin [36].

In the case of collagen and elastin, which are located predominantly within the papillary and reticular layer of dermis, the situation is a bit different. Both excitation and emission light is attenuated because of absorption by melanin. In addition, fluorescence intensity in 400–480 nm range is subject to attenuation by the other skin chromophores, such as hemoglobin, porphyrins, carotenoids etc. (Fig. 3.3). Both the total intensity and the spectral features may be affected [7,37,38].

The fluorescence intensity on excitation and emission wavelengths can best be depicted with a 3D plot (Fig. 3.12). A simple inspection of the presented spectra leads to two basic observations: the human skin exhibits a rather characteristic AF pattern, and the skin AF intensity is subject to marked individual variations.

A 2D contour plot of a 3D skin AF pattern, usually referred to as the fluorescence excitation-emission matrix (EEMs,) is shown in Fig. 3.13. One of the goals of fluorescence spectroscopy is the identification of excitation wavelengths suitable for the differentiation of various pathological conditions. Most of the biological components, which are either related

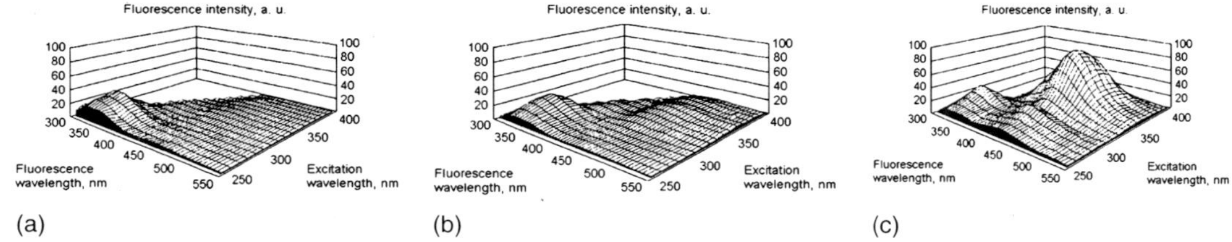

Figure 3.12 3D plots of the AF spectra of the human skin measured ex vivo at different excitation wavelengths: (a) 40-year-old man; (b) and (c) 60- and 87-year-old women, respectively [7]. Measurements were performed for skin samples of 20 × 20 mm size with subcutaneous fat obtained from the patients in the course of skin plastic surgery (abdominal and lower extremities regions).

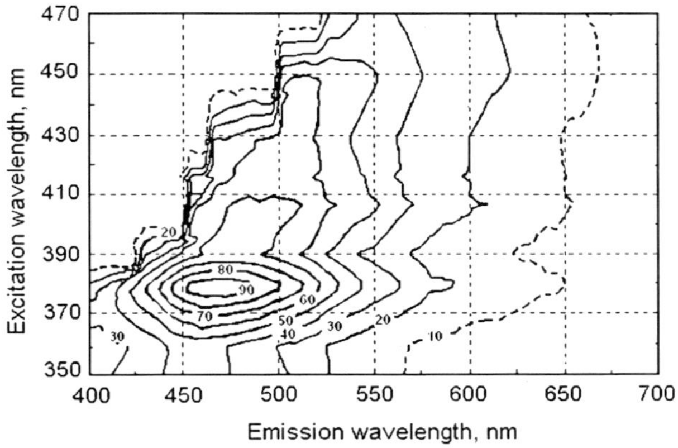

Figure 3.13 The excitation-emission maps (EEMs) of the in vivo human skin AF emission [39].

to the skin tissue structure, or are involved in metabolic and functional processes, generate fluorescence emission in the UV-visible spectral region. As a result, different morpho-functional conditions of the skin, related to biochemical and physiochemical alterations, can be characterized on the basis of information available in the fluorescence EMMs [7,39].

3.4.8 Skin Optical Clearing

The refractive index variation in tissues, quantified by the ratio $m \equiv n_s/n_0$, determines light scattering efficiency. For example, in a simple tissue model, such as dielectric spheres of equal diameter $2a$, the reduced scattering coefficient is described by [40]:

$$\mu_s' \equiv \mu_s(1-g) = 3.28\ \pi a^2\ \rho(2\pi a/\lambda)^{0.37}\ (m-1)^{2.09}, \tag{3.14}$$

where ρ is the volume density of the scatterers, g is the scattering anisotropy factor, and λ is the light wavelength in the scattering medium. At equalizing (matching) of refractive indices (RI) of the interstitial fluid, n_0, and scatterers, n_s, m $\rightarrow$ 1, $\mu_s' \sim \mu_s \rightarrow 0$. Skin transmittance, $T = I(d)/I_0 \sim \exp\{-[3\mu_a(\mu_s' + \mu_a)]^{1/2}d\}$ [Eq. (3.7)], can be increased substantially by reduction of the scattering coefficient, because for a native skin in the visible/NIR range $\mu_s' >> \mu_a$. One of the ways to decrease the scattering coefficient of skin is by impregnating it with a solution [optical clearing agent (OCA)] with RI, n_{OCA}, higher than n_0 [1,41]. If a hyperosmotic OCA is applied topically to the skin, besides its diffusion into skin, tissue water will flow outside from a tissue and corresponding tissue dehydration will take place, this action will lead to additional matching of RI of scatterers relative to the background and more effective packing of scatters (tissue shrinkage). Both processes provide more effective light transport through skin.

The excellent diffusional resistance of the skin stratum corneum (SC) makes the transdermal delivery of immersion agents and water lost by skin difficult [42]. The diffusion of water across the SC is a passive process that can be modified during the application of hyperosmotic OCAs. The water content of the innermost layer of the SC is in equilibrium with the adjacent moist granular layer. The outside cell layer, however, is in equilibrium with the environment, and it is certainly drier than the innermost cornified layer. Dermis is the thicker layer of the skin, which is mostly fibrous tissue well-supplied by blood, and thus can be easily impregnated by exogenous or endogenous liquids (immersion agents). Subcutaneous tissue contains a big portion of fat cellular layer, which is much less penetrative for diffusing molecules than dermis. Such specific structure of skin defines the methodology of its effective optical clearing, which is related to the immersion of refractive indices of scatterers (keratinocytes components in epidermis, collagen and elastin fibers in dermis) and ground matter [41].

Experimental studies of optical clearing of skin using glycerol, glycerol–water solutions, glucose, propylene glycol, polyethylene glycol, DMSO, sunscreen creams, cosmetic lotions, gels, and pharmaceutical products were recently overviewed [1,41,43]. In vivo topical application of these agents made human skin more optically translucent within a time period, from a few minutes to a few hours.

To enhance OCA permeation through SC, a number of specific physical procedures, such as heating, electrophoresis, sonophoresis, and laser-induced stress, as well as chemical enhancers, such as oleic acid and DMSO, are usually applied. A method of accelerating penetration of the index-matching compounds by enhancing skin permeability through creating a lattice of micro-zones (islets) of limited thermal damage in the SC was recently proposed [44]. A combination of a flashlamp system (EsteLux, Palomar Medical Technologies, Inc.) and a specially designed appliqué with a pattern of absorbing centers (center size ~75 μm, lattice pitch ~450 μm) has been used to create the lattice of islets of damage (LID). Several index-matching agents, including glucose and glycerol, have been tested. A high degree of optical clearance of a full-thickness pig, rat, chicken, and human skin in vitro and in vivo has been demonstrated with 40% glucose and 88% glycerol solution after creating a LID with a few optical pulses (fluence 14–36 J/cm^2, 20 ms pulse duration).

To provide faster and more effective skin optical clearing, an intradermal injection can be applied. Figure 3.14 shows the reflectance spectra and the corresponding time-dependent reflectance for a few spectral components measured for a human healthy volunteer at intradermal injection of 40% glucose solution [45]. The reflectance spectra are determined by the diffusion reflection of the skin layers with the well-pronounced bands caused by blood absorption. Within one hour after glucose injection, the skin reflection coefficient decreases in average by a factor of 3.8 and then exhibits a slow increase, which indicates that glucose is eliminated from the observation area, and the skin reflectance tends to restore

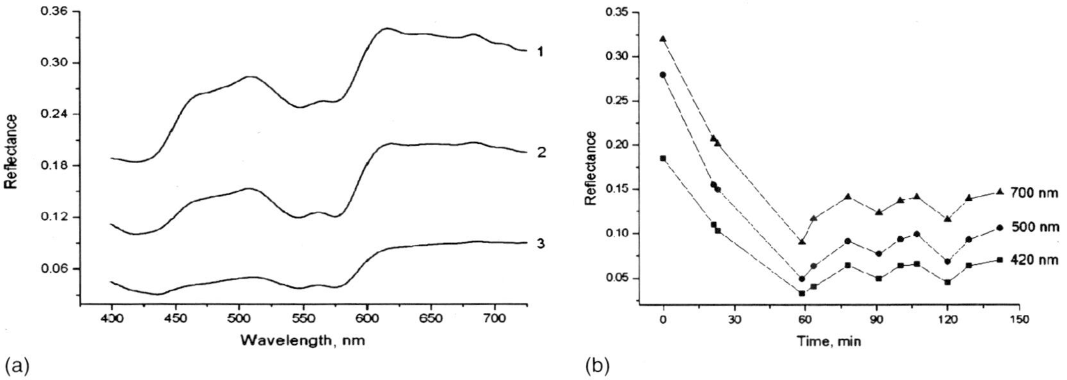

Figure 3.14 (a) The reflectance spectra and (b) the time-dependent reflectance at three wavelengths (420, 500, and 700 nm) of the human skin measured at hyperdermal injection of 0.1 ml of 40% glucose into the internal side of the forearm of the male volunteer for different time intervals; (1) intact skin, (2) at 23 min and (3) at 60 min after injection [45].

itself to the initial level. Basing on this results and skin model, it was suggested that the main contribution to clearing in the first stage (first hour) is due to the RI matching between collagen fibrils of the dermis (n = 1.46) and the interstitial space (initially n = 1.36) to which glucose (n = 1.39) diffuses.

For applications, it is important that skin preserves transparency (low reflectance) for a few hours after injection, which is defined by predominant diffusion of glucose along the skin surface, because the upper and lower layers of the skin—epidermis and fat—have much lower (a few orders) permeability for glucose than dermis. It is seen from Fig. 3.14 that at dermal clearing the contrast of hemoglobin absorption bands is significantly higher than for normal skin, but for prolonged immersion (curve 3) the contrast is again not very high. This is important for the optimization of clearing time at imaging of tissue abnormalities associated with hemoglobin or other absorbers.

Because of the limitation of probing the depth of OCT imaging (1–2 mm for skin), its combination with OCA immersion can be a useful technology for skin diagnosis and monitoring. This is illustrated by the OCT images of human skin with psoriatic erythrodermia acquired before, and in some time after application of glycerol (Fig. 3.15) [46]. In one hour

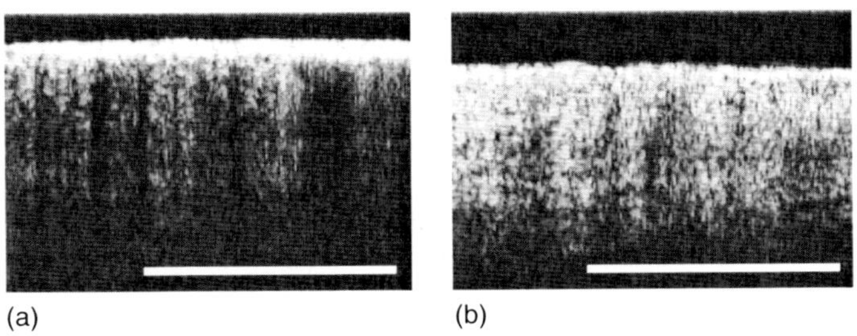

Figure 3.15 OCT images of skin with psoriatic erythrodermia: (a) before topical application of glycerol; (b) 60 min after application of glycerol [46].

of glycerol application, OCT image differs from the initial image in greater penetration depth and better contrast. These image improvements facilitate identifying of important morphological phenomenon of acanthosis.

Squeezing (compressing) or stretching of skin produces a significant increase in its optical transmission [1,41,43]. The major reasons for that are the following: (1) increased optical tissue homogeneity due to removal of blood and interstitial fluid from the compressed site; (2) more close packing of tissue components causes less scattering due to cooperative (interference) effects; and (3) less tissue thickness.

Spectral properties of skin can be effectively controlled by applying an external localized pressure when UV induced erythema (skin redness) is developed [7]. The intensity of skin reflectance and autofluorescence is well-controlled at pressure applied to the skin site. Due to more effective fluorescence attenuation by blood hemoglobin at more intensive erythema, skin compression more effectively increases fluorescence output.

The light propagation in human skin at mechanical tension was studied in vivo using diffuse reflectometry [22]. It was found that intact skin has its own anisotropy which is believed to be caused by the preferential orientation of collagen fibers in the dermis, as described by Langer's skin tension lines, and at skin external stretching, scattering coefficient and corresponding light back-reflectance and transmittance can be effectively controlled. At external forced tension, more significant damping of scattering along the direction of mechanical stress was determined. The reduced scattering coefficient varied by up a factor of two between different directions of light propagation at the same position.

The measurements of the deformations and applied loads and estimating the biomechanical properties of tissue are critical to many areas of the health sciences, including monitoring of the tension in wound closures, skin flaps, and tissue expanders [47]. Such measurements which can be provided by detection of the polarized light reflectivity will allow surgeons to treat wounds more successfully by minimizing scar tissue and maximizing the speed of treatment, by letting them know how much the skin can be stretched at each treatment step. In vivo human experiments showed that the specular reflection from skin changes with stretch [47]. For small values of stretch, the specular reflectivity measured for He–Ne laser ($\lambda = 633$ nm) beam with the 45° angle of incidence increases linearly with strain. The linear relationship between applied stretch and polarized reflectivity can be understood if the skin surface is approximated by a sinusoidal profile in the resting stage. Stretching reduces amplitude and increases spatial scale of skin profile, thereby making it smoother and flatter, resulting in a corresponding increase of reflectivity. For larger stretches [for strains above 8.8% (5-mm stretch)] for the human subject tested, the dependence is saturated and even goes down. The stretches in two perpendicular directions (parallel and perpendicular to the long axis of the forearm) yield good correlation between stretch and reflected light intensity and shows that skin has anisotropic properties, which can be detected by light reflection [47].

A reproducible effect of temperature between 25 and 40°C on the reduced scattering coefficient of human dermis and subdermis was found in ex vivo study in the NIR [18]. For dermis, the relative change in the reduced scattering coefficient showed an increase [$(4.7 \pm 0.5) \times 10^{-3}$°C^{-1}] and for subdermis a decrease [$(-1.4 \pm 0.28) \times 10^{-3}$°C^{-1}]. It was hypothesized that the observed positive and negative temperature coefficients of scattering for dermis and subdermis are connected with differences in their structural components.

3.5 Mechanisms of Light Tissue Interaction

3.5.1 Photochemicals

Photochemical interaction of light and tissue is of great interest for the study of tissue damage induced by solar radiation, in particular in skin-aging process, as well as for the designing of controllable technologies for tissue repairing and rejuvenation. Such interaction depends on the type of endogenous or exogenous chromophore (photosensitizer) involved in photochemical reaction, oxygen tension, and light wavelength, intensity, and exposure. To characterize a photochemical reaction a quantum yield is introduced. For a radiation-induced process, quantum yield is the number of times that a defined event (usually a chemical reaction step) occurs per photon absorbed by the system. It is a measure of the efficiency with which absorbed light produces some effect. Since not all photons are absorbed productively, the typical quantum yield is less than 1. Quantum yields greater than 1 are possible for photo-induced or radiation-induced chain reactions, in which a single photon may trigger a long chain of transformations.

3.5.2 Photothermal and Photomechanical Mechanisms

When photons traveling in a tissue are absorbed, heat is generated. Generated heat induces several effects in tissue which can be presented in the order of amount of heat deposition: temperature increase and reversible and irreversible alterations in tissue. The following types of irreversible tissue damage are expected as tissue temperature rises past T_{crit}: *coagulation* (denaturation of cellular and tissue proteins) is the basis for *tissue welding*; *vaporization* [tissue dehydration and vapor bubbles formation (*vacuolization*), $T \geq 100°C$)] is the basis for tissue mechanical destruction; *pyrolysis* ($T \approx 350$–$450°C$). For short light pulse, all these processes develop as explosion or thermal ablation. All these phenomena are named as photothermal mechanism. During ablation, high pressure is developing in tissue, which can be a reason for shock wave formation and mechanical damage of tissue. This phenomenon is named photomechanical mechanism.

The generated heat, described by the heat source term S at a point $\mathbf{r}$ is proportional to the fluence rate of light $\phi(\mathbf{r})$ (mW/cm^2) and absorption coefficient $\mu_a(\mathbf{r})$ at this point [48–52]:

$$S(\mathbf{r}) = \mu_a(\mathbf{r})\phi(\mathbf{r}). \quad (3.15)$$

The traditional *bioheat equation* originated from the energy balance describes the change in tissue temperature over time at point $\mathbf{r}$ in the tissue

$$\rho c \frac{\partial T(r,t)}{\partial t} = \nabla[k_m \nabla T(r,t)] + S(r) + \rho C w (T_a - T_v) \quad (3.16)$$

where ρ is the tissue density (g/cm^3), C is the tissue specific heat (mJ/g°C), $T(\mathbf{r},t)$ is the tissue temperature (°C) at time t, k_m is the thermal conductivity (mW/cm°C), $S(\mathbf{r})$ is the heat source term (mW/cm^3), w is the tissue perfusion rate (g/cm^{3}s), T_a is the inlet arterial temperature (°C), and T_v is the outlet veinual temperature (°C), all at point $\mathbf{r}$ in the tissue. In this equation convection, radiation, vaporization, and metabolic heat effects are not accounted for, because of their negligible effect in many practical cases. The source term is assumed to be stationary over the time interval of heating. The first term to the right of the equal sign describes any heat conduction (typically away from point **r**), and the source term accounts for heat generation due to photon absorption. In most cases of light (laser) tissue interaction, the heat transfer caused by perfusion (last term) is negligible.

To solve this equation, initial and boundary conditions must be accounted for. The initial condition is the tissue temperature at $t = 0$ and the boundary conditions depend on tissue structure and geometry of light heating. Methods of solving of the *bioheat equation* can be found in refs. [48–50].

Damage to a tissue results when it is exposed to a high temperature for a long time [48–52]. The damage function is expressed in terms of an Arrhenius integral:

$$\Omega(\tau) = \ln\left(\frac{C(0)}{C(\tau)}\right) = A\int_0^{\tau} e^{-\frac{E_a}{RT(t)}}\mathrm{d}t, \tag{3.17}$$

where τ is the total heating time (s); $C(0)$ is the original concentration of undamaged tissue; $C(\tau)$ is the remaining concentration of undamaged tissue after time τ ; A is an empirical determined constant (s^{-1}); E_a is an empirically determined activation energy barrier (J/mole); R is the universal gas constant (8.32 J/mole·K); and T is the absolute temperature (K).

At noninvasive optical diagnostic and some photochemical applications of light, one has to keep tissue below the damaging temperature so-called the critical temperature T_{crit}. This temperature is defined as the temperature where the damage accumulation rate, $\mathrm{d}\Omega/\mathrm{d}t$, is equal to 1.0 [51]:

$$T_{crit} = \frac{E_a}{R\ln(A)} \tag{3.18}$$

The constants A and E_a can be calculated on the basis of experimental data when tissue is exposed to a constant temperature [49]. For example, for pig skin, $A = 3.1 \times 10^{98}$ and $E_a = 6.28 \times 10^5$, that gives $T_{crit} = 59.7°C$.

With CW light sources due to the increase of the temperature difference between the irradiation and the surrounding tissue, conduction of heat away from the light absorption point and into surrounding tissue increases. In dependence of light energy, large tissue volumes may be damaged, or losing of heat at the target tissue component may be expected. For pulsed light, a little heat is usually lost during the pulse duration since light absorption is a fast process while heat conduction is relatively slow; therefore, more precise tissue damage is possible.

The disadvantage of thermal ablation with CW light sources is undesirable damage to surrounding tissue via its coagulation. Pulsed light can deliver sufficient energy to ablate tissue in each pulse, but in a short enough time, that tissue is removed before any heat is transferred to the surrounding tissue. To achieve a precise tissue ablation, lasers with a very short penetration depth, like excimer ArF laser or Er:YAG, are used (Fig. 3.2).

For skin as a turbid medium irradiated with wide laser beams (>0.1mm), the effect of backscattering causes a higher subsurface fluence rate compared with the incident laser fluence [Eq. (3.7)]. Therefore, the z-axial light distribution in tissue and the corresponding stress distribution have a complex profile, with a maximum at a subsurface layer. The stress amplitude adjacent to the irradiated surface $\delta p(0)$ and the stress exponential tail into the depth (z) of the tissue sample are proportional to tissue absorption coefficient μ_a and the incident laser pulse energy E_0 [53,54]:

$$\delta p(0) = \Gamma \mu_a E(0), \text{ at surface } (z = 0), \quad (3.19)$$

$$\delta p(z) = \Gamma \mu_a b_s E_0 \exp(-\mu_{eff} z), \text{ for } z > 1/\mu_{eff}, \quad (3.20)$$

where $\Gamma = \beta v_a^2 / c_T$, c_T is the specific heat of the tissue, b_s is the factor that accounts for the effect of backscattered irradiance that increases the effective energy absorbed in the subsurface layer, μ_{eff} is defined early, $E(0)$ is the subsurface irradiance, and E_0 is the incident laser pulse energy at the sample surface (J/cm^2). For optically thick samples [53, 54]:

$$E(0) \approx (1 + 7.1 R_d) E_0, \quad (3.21)$$

where R_d is the total diffuse reflection.

The Grüneisen parameter Γ is a dimensionless, temperature-dependent factor proportional to the fraction of thermal energy converted into mechanical stress. For water it can be expressed with an empirical formula as [53]:

$$\Gamma = 0.0043 + 0.0053T, \quad (3.22)$$

where temperature T is measured in degrees Celsius; for $T = 37°C$, $\Gamma \approx 0.2$.

Equations (3.19) and (3.20) are strictly valid only when the heating process is much faster than expansion of the medium. The stress is temporarily confined during laser-heat deposition when the duration of the laser pulse is much shorter than the time of stress propagation across the depth of light penetration in the tissue sample. Such conditions of temporal pressure confinement in a volume of irradiated tissue allow for the most efficient pressure generation [53,54].

3.6 Theory of Photothermal Interaction

3.6.1 Theory of Selective Photothermolysis

3.6.1.1 Basic Principles

For many years, electromagnetic radiation (EMR) from lasers, lamps, and other sources (including microwave ones) has been used to treat a variety of medical conditions in ophthalmology, dermatology, urology, otolaryngology, and other specialties. For example, in dermatology EMR sources have been used to perform a wide variety of procedures including hair removal, treatment of various pigmented lesions, removal of unwanted veins, tattoo removal, and skin resurfacing. For all these treatments, a natural or artificial chromophore presented in the body is heated by absorption of either monochromatic or broadband EMR. Typical natural (endogenous) chromophores include water, melanin, hemoglobin, protein, lipid, etc. Exogenous chromophores can include dyes, ink, carbon particles, etc. For example, heating of a chromophore may result directly in the destruction of a tattoo or a pigmented lesion. In these cases, the treated target for destruction and the chromophore occupy the same area. These cases are well-described by the theory of selective photothermolysis (SP) [55].

The SP theory provides a quantitative description of optical treatments such as those mentioned earlier. The aim of SP is to provide a permanent thermal damage of targeted structures with the surrounding tissue held intact. The SP theory is based on three general principals:

1. *Wavelength of EMR has to be selected to provide maximum contrast of absorption of the target vs. surrounding tissue and other competitive targets.*

For example, for hair follicle, the SP target chromophore is the melanin in the hair shaft and hair matrix. Competitive chromophores are blood, water, and lipid. Competitive target is also epidermal melanin. In the first approximation, the optimum wavelength can be selected based on the analysis of spectra of absorption of competitive chromophores, as showed in Figs. 3.2–3.5. Additional factor for the wavelength selection is the location of the target and competitive target in different depth of the skin. Attenuation of light in skin is wavelength-dependent (see Fig. 3.8) and should be taken into account for optimum wavelength selection. For example, hair bulb has hair matrix located in subcutaneous fat at a depth of 2–5 mm, and stem cells are located at a depth of about 1 mm.

2. *Pulsewidth of EMR has to be selected to provide maximal contrast of heating of the target versus surrounding tissue.*

To satisfy this criterion, the EMR pulsewidth τ must be small compared to the thermal relaxation time (TRT) of the whole target. The TRT is the time of cooling of the target with decreasing temperature of the target in $e = 2.7$ times after fast adiabatic heating. TRT $\approx d^2/Fk$, where d is the size of the target in mm or cm, k is the coefficient of thermal diffusion of tissue ($k \approx 0.1$ mm^2/s = 0.001 cm^2/s for dermis) and F is the geometrical factor. $F = 8$, 16, and 24 for planar, cylindrical, and spherical target, respectively. Actually, if the condition $\tau <<$ TRT is met, the heat generated within the target due to EMR absorption does not flow out of the structure until it becomes fully damaged (coagulated, injured). This approach provides both selective damage and minimum light energy deposition.

3. *Fluence of the pulse has to be sufficient enough to provide coagulation or ablation of the target.*

The theory of SP is good for the first estimation of treatment parameters. But this theory has several limitations which correspond with the simplification of biological target. Real organ as blood vessel, hair follicle, and others have a complex distribution of chromophores and thermo-sensitive targets. For example, selection of the wavelength on the peak of spectrum of absorption of target can lead to nonuniform heating of the target. The absorption coefficient of blood with 75% oxygenation at 577 nm (peak of oxyhemoglobin absorption) is $\mu_a = 30$ mm^{-1}. The penetration depth of light in blood vessels can be estimated as $h = 1/\mu_a \sim 0.03$ mm = 30 μm. For plexus vessels with diameter 7–30 μm, light will provide the uniform heating of the vessels. But for large vessels with a diameter of 100–200 μm, we can expect overheating just portion of a vessel and not complete coagulation of another portion. It can be a reason for creation of the so-named purpura effect. The usage of pulsewidths shorter than TRT of the target becomes inapplicable when the target absorption is nonuniform over its area and a part of the target exhibits weak or no absorption, but the other part exhibits a significant absorption. If this is the case, the weakly absorbing part of the target has to be damaged by heat diffusion from the highly pigmented/strongly absorbing one (hereafter called the *heater* or *absorber*). An example of such a target is the hair follicle [56]. The highly absorbing areas consist of melanin-bearing structures, (i.e., the hair shaft and the matrix cells). The other follicular tissues including the stem cells do not contain an appreciable amount of any chromophore that absorbs in the red/NIR. These targets (tissues) can be damaged by heat diffusion from the hair shaft or the matrix cells to the surrounding follicular tissues. Another example is the treatment of telangiectasia or leg veins with a wavelength near the maximum hemoglobin absorption. Permanent closure of a vascular malformation or a vein probably requires coagulation of the vascular or vein wall [57,58]. In this case, coagulation of the wall requires heat diffusion from blood to wall. This consideration motivated Altshuler et al. to develop the theory of extended selective photothermolysis (ESP) [59].

3.6.1.2 Extended Theory of Selective Photothermolysis

Thermal damage of target with separation between part of the target and the pigmented area requires deposition of sufficient heat energy into the absorbing area, and good heat exchange between this area and the targeted external structures. Heat deposition depends on the absorption coefficient of the absorber and the EMR power density. Heat exchange depends in turn on the distance between the heater and the outermost part of the target and on the heat transmission coefficient between the absorber and the intervening tissue. However, at a sufficiently high temperature both the heater absorption coefficient and the heat transmission coefficient from the heater to the other targeted tissues may become lower due to phase transitions and destructive processes such as bleaching, melting, boiling, and bubble formation. This results in inefficient use of EMR energy for phase transition processes within the absorber and the intervening tissue. To prevent these undesirable effects, the heater peak temperature has to be limited to a prescribed value, T_{1max}, called hereafter the temperature of heater absorption loss. The temperature of heater absorption loss T_{1max} of most endogenous chromophores (e.g., melanin, hemoglobin, and water) exceeds 100^0C. Simultaneously, to ensure permanent damage of the whole target, the temperature should exceed a second prescribed value, the *damage temperature* T_2, throughout the target area. This temperature is lower than the temperature of heater absorption loss T_{1max}. More precisely, the damage temperature, T_2, is the temperature at which irreversible thermal damage of the target occurs. We suggest that the basic damage mechanism in soft tissue is the denaturation of proteins. As an alternative to the rigorous Arrhenius rate process integral [see Eq. (3.17)] [60], the dynamic denaturation process may be described approximately by using the simpler damage-temperature concept. For human-skin collagen, T_2 is about 65–75°C, and for cells T_2 it is about 60°C if the exposure duration is several tens of milliseconds [60]. Furthermore, the tissue temperature should not exceed the water boiling point, to prevent formation of vapor bubbles that could insulate the absorbing area from the surrounding tissue. Strictly speaking, vapor bubbles can transfer heat from the absorbing area to the target, but this process is unpredictable because vapor bubbles can move in different directions.

To meet the temperature limitations described earlier, the EMR power must be limited, (but there must be sufficient to heat the target up to the damage temperature) and, therefore, the EMR pulse has to be made sufficiently long to deliver enough energy. Other important criterion is selection of wavelengths of light. It is necessary to provide relatively uniform heat deposition in pigmented area during EMR pulse and at same time to absorb maximum EMR energy in the pigmented area. To meet this criterion, optical density of the pigmented area has to be in the range $1 < \mu_a d_1 < 3$. We define the *thermal damage time* (TDT) of the target to be the time required for irreversible target damage with the sparing of the surrounding tissue. For a nonuniformly absorbing target structure, the TDT is the time when the outermost part of the target reaches the target damage temperature T_2 through heat diffusion from the heater. Because the heat-diffusion front becomes blurred when propagating into the tissue, part of the heat energy will leave the target. Therefore, the heated area will be larger than the damaged area. However, we demonstrate later here that the target damage can still be selective, even though the TDT is many times as long as the TRT of the whole target. Apparently, the *optimum EMR pulsewidth*, τ_0, should be shorter than or equal to the TDT. So, in contrast to the standard theory of selective photothermolysis, nonuniformly pigmented structures have to be treated by long EMR pulses: the pulsewidth must typically be longer than the target TRT. In addition, the EMR power should be limited to prevent reduction of the heater absorption by bleaching, vaporization, etc.

Figure 3.16 shows the general structure of a *target* that is thermally denatured by heat diffusion from a *heater*. The heater includes an endogenous or exogenous chromophore (pigment) with a high photon absorption coefficient. The target exhibits weak EMR absorption. The distance between the heater and the target is *d*. Photons from the EMR source are absorbed by the heater. As discussed in the Section 3.6.1.1, the EMR power density has to be adjusted so that during the time of treatment the heater temperature T_1 does not exceed the temperature T_{1max} where the absorption coefficient may begin to drop. The heat propagates from the heater to the target due to either thermal diffusion or two other possible mechanisms. These mechanisms are hot ablation products emanating from the heater, or steam. In this chapter, we will concentrate exclusively on the thermal diffusion mechanism. The process of target thermal damage is completed when its temperature reaches T_2, while the temperature of the surrounding tissue remains below its damage temperature. In general, it is not possible to damage the target without damaging the tissue between the target and the heater. For this target type, precise selective damage of the target is impossible. But

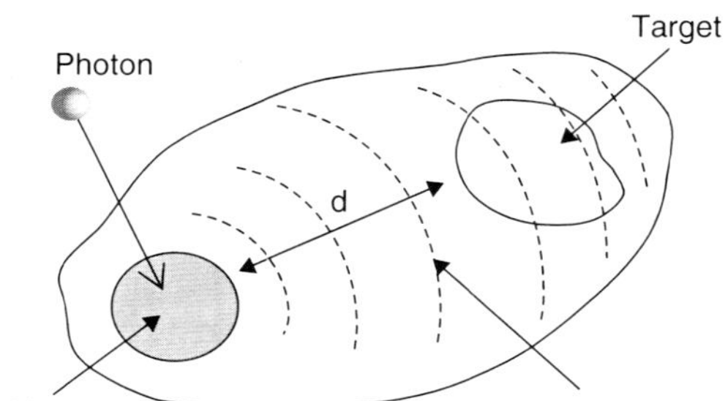

Figure 3.16 The general structure of a target that is thermally denatured by heat diffusion from a heater; the heater includes an endogenous or exogenous chromophore (pigment) with a high photon absorption coefficient; The target exhibits weak EMR absorption; the distance between the heater and the target is *d*.

if the treated target is a layer or a shell of tissue roughly symmetrical with respect to the heater, the target damage can be made precisely selective. This means that the damage zone covers the whole target without extending beyond it. After the end of the EMR pulse, the target temperature at the outermost point goes on growing until it reaches a maximum. The time delay between the end of the EMR pulse, $t = \tau_0$, and the moment when the temperature of the outermost point is reached, the maximum temperature T_2 is denoted by ε and thus TDT = $\tau_0 + \varepsilon$. Therefore, the EMR pulsewidth is equal to or shorter than the TDT. This delay ε is roughly equal to the propagation time of the heat front from the heater to the target $\varepsilon \approx d^2/k$; ε is therefore close to the target TRT. As we shall see subsequently, TDT is significantly longer than the TRT in most cases. So, for selective treatment the pulsewidth should be approximately equal to the TDT: $\tau_0 \cong$ TDT. The TDT depends in turn on T_1, T_2, and the target size. For a better understanding of this dependence we first look more closely at heat diffusion from the heater to the target. Specifically, we will be focused on a target comprising a heavily pigmented long cylinder of diameter d_1 and a surrounding treated area of diameter d_2 (Fig. 3.17). This simple geometry can be used to model thermal damage of hair follicle by hair-shaft heating or blood vessel destruction. We consider two heating modes. The first mode utilizes a *rectangular EMR pulse* (Fig. 3.18a) and the second one utilizes a *flattop temperature pulse* (Fig. 3.18d). In the case of the rectangular EMR pulse, the heater temperature grows during the EMR pulse and reaches T_1 at the end of the pulse (Fig. 3.18b). In the case of the flattop temperature pulse, the temperature of the heater is constant during the EMR pulse, which requires a special pulse shape (Fig.3.18c). For both heating modes, the heater temperature is below the temperature of heater absorption loss $T_{1\max}$ so the absorption coefficient of the heater does not change.

The sequence of thermal profiles during the heating process is depicted in Fig. 3.17. The input parameters for modeling are d_1 = 70 μm, d_2 = 210 μm. The temperature of heater absorption loss T_1 is 100°C (the boiling point of water). The damage temperature is 65°C,

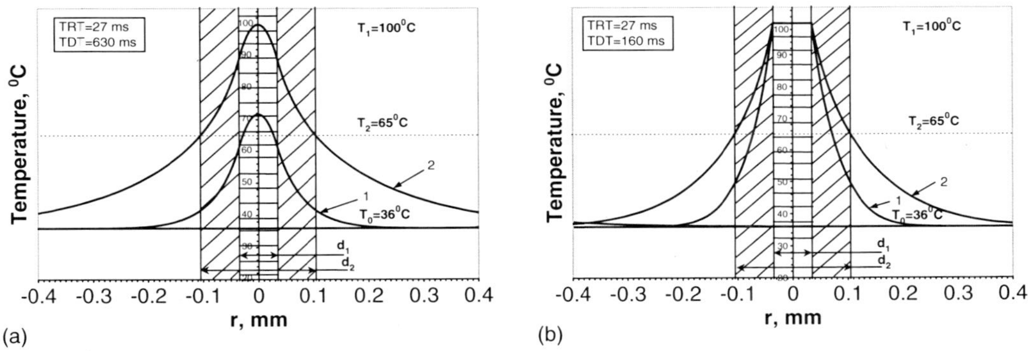

Figure 3.17 Temperature distribution in the tissue with cylindrical absorber with diameter 0.07 mm and target with diameter 0.21 mm [59]: (a) shows the temperature distribution for a rectangular EMR pulse at two points in time: bottom curve at t = TRT = 27.5 ms and top curve at t = TDT = 1.6 s; (b) shows the temperature distribution for a flattop temperature pulse at two points in time: bottom curve at t = TRT = 27.5 ms and top curve at t = TDT = 0.36 s. Maximum temperature of the absorber is T_1 = 100°C. Damage temperature is T_2 = 65°C. Initial temperature is T_0 = 37°C. The absorption of the tissue surrounding the heater was neglected.

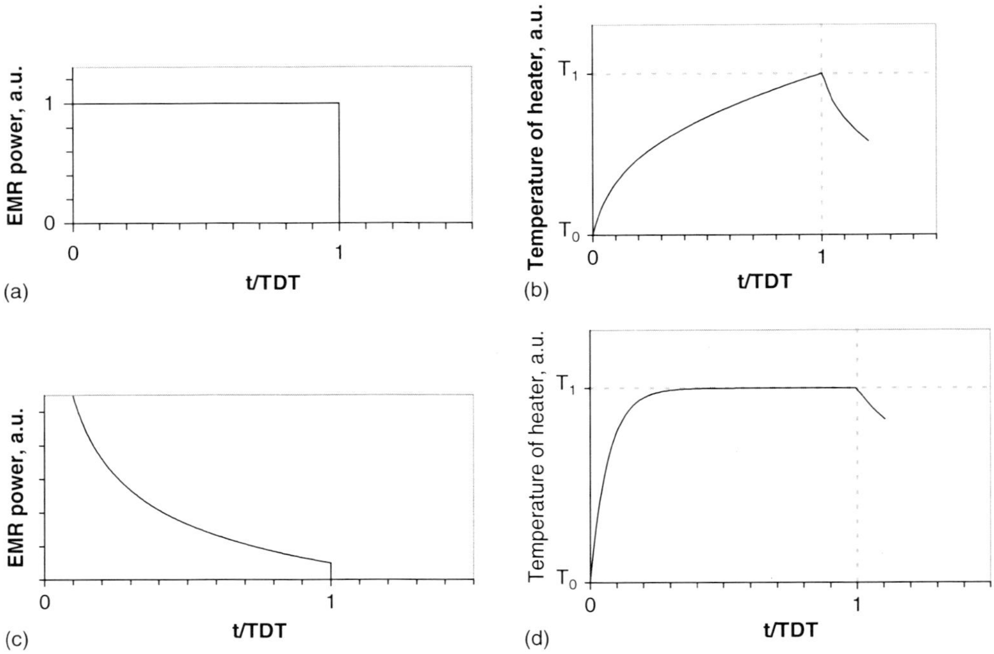

Figure 3.18 Time dependence of electromagnetic radiation (EMR) power and temperature of the heater (absorber) [59] Two basic cases are shown: (a) rectangular EMR pulse; (b) shows heater temperature as a function of time that is produced by a rectangular EMR pulse; (c) shows EMR power as function of time that produces a flattop temperature pulse at the heater and (d) flattop temperature pulse.

which is the average protein denaturation temperature in the 10–1000 ms pulsewidth range. We assume that the EMR absorption is confined to the heater and that the thermal properties of the heater, the target, and the surrounding tissue are the same. Heat diffusion from the heater takes place simultaneously with the growth of the heater temperature due to light absorption. This process is described well by the heat conduction equation.

Figure 3.17a and b shows the temperature profiles in the cylindrical target and the surrounding tissue at different instants of time for the rectangular EMR pulse and the flattop temperature pulse, respectively. Curve 1 in both figures is the temperature profile at the time equal to the thermal relaxation time of the whole target. The latter time is TRT = $d_2^2/16k$. In our case the TRT = 27 ms. We can see that at the time instant t = TRT the boundary temperature of the target is still significantly below the damage temperature. Curve 2 in both figures is the temperature profile at the moment when the boundary temperature of the target reaches the damage temperature T_2. At this moment, the whole target is damaged but the surrounding tissue is still intact. It is this time that has been defined above as the thermal damage time. In our case TDT = 0.63 s for the rectangular EMR pulse and TDT = 0.16 s for the flattop temperature pulse. Based on this example, we can arrive at two main conclusions. First, the ratio TDT/TRT is about 23 and 6 for the rectangular EMR pulse and the flattop temperature pulse modes, respectively. Therefore, in both modes the pulsewidth τ = TDT is significantly longer than the TRT of the entire target. Second, at the time instant when t = TDT, the heated area is significantly larger than the damaged target.

These observations present a striking contrast to the classical case of selective photothermolysis. This distinction is a result of the spatial separation of the heavily pigmented and treated areas. Actually, in contrast to the classical case, the basic damage mechanism is now heat diffusion rather than direct heating by EMR absorption. The heat diffusion front is not sharp and, therefore, heat is spreading outside the damaged area; however, the damage is still rather selective.

We have described the difference in treatment modality between uniformly and nonuniformly pigmented targets. The new extended theory of selective thermal damage of nonuniformly pigmented structures in biological tissue postulates the following:

1. The EMR wavelength should be chosen to provide sufficient contrast between the absorption coefficient of the pigmented area and that of the tissue surrounding the target and provide optical density of pigmented area in the range 1– 3.
2. The EMR power should be limited to prevent absorption loss in the pigmented area, but it must be sufficient to achieve a heater temperature higher than the target damage temperature.
3. The pulsewidth should be made shorter than or equal to the thermal damage time (TDT), which can be significantly longer than the thermal relaxation time of the target.

3.6.2 Treatment Parameters and Applications

3.6.2.1 Treatment Parameters for Planar, Cylindrical, and Spherical Targets

Heat diffusion is strongly dependent on the heater and target geometry. In this section, we discuss three basic geometries: planar, cylindrical, and spherical (Fig. 3.19). In all cases, we

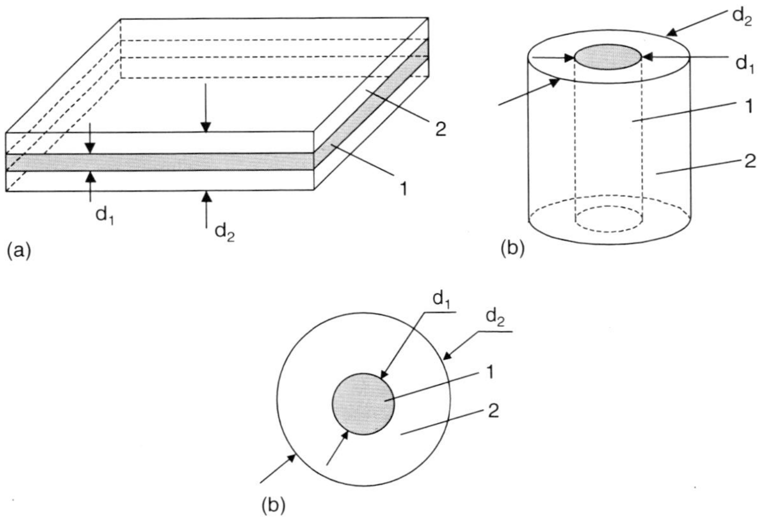

Figure 3.19 Three types of target with different geometry [59]: (a) planar, (b) cylindrical, (c) spherical. 1 is the heater (absorber), 2 is the target, d_1 is the size of the heater, d_2 is the size of the target.

assume a heater with size d_1 located in the center of a target with size d_2. We define the ratio $x = d_2/d_1$ to be *the geometrical factor of the target*. As before, T_1 is the maximum temperature of the pigmented area, and T_2 is the target damage temperature ($T_1 > T_2$). As we show in the Appendix, the target thermal damage time can be expressed by the following formula:

$$\text{TDT} = \text{TRT} \times r(x, \Delta),$$

where $r(x,\Delta)$ is a function of the geometrical factor x and *temperature factor* Δ defined as $\Delta = (T_2 - T_0)/(T_1 - T_0)$, T_0 is the target and heater temperature before irradiation. Normally T_0 is the body temperature and is equal to 373C. TDT is proportional to the TRT. In the Appendix, we present formulas for the TDT of planar, cylindrical, and spherical targets.

Figure 3.20 shows the ratio $r(x,\Delta)$ = TDT:TRT as a function of geometrical factor x for two heating modes: rectangular EMR pulse and flattop temperature pulse. The calculation parameters were $T_1 = 100°C$, $T_2 = 65°C$, $T_0 = 37°C$ ($\Delta = 0.52$).

We emphasize that in the framework of our analytic theory, the ratio $r(x,\Delta)$ does not depend on the size of the entire target and the tissue thermal properties. Several important conclusions follow from Fig. 3.20.

First, the ratio TDT/TRT is an increasing function of geometrical factor x.

Second, the actual value of this ratio is very different for plane, cylindrical, and spherical targets. For a plane target, the TDT is several times higher than the TRT. The TDT exhibits appreciable growth with increasing the target dimensionality. It is implied herein that the planar, cylindrical, and spherical targets are one-, two-, and three-dimensional, respectively.

Third, for the same TDT/TRT, the relative size of the damaged zone x is smallest for the spherical target. Next in this order is the cylindrical target. The plane target exhibits the largest damage area. The relative size x of the damaged zone around the heater decreases when increasing the target dimensionality. The latter two conclusions are intuitively apparent. Actually, conductive heating of a weakly absorbing tissue should proceed more effectively for a low-dimensional target. This "dimensionality" concept is a useful target parameter. It is also applicable to nonsymmetrical targets. The temperature profile is sharper and better localized for the spherical heater compared to the cylindrical one, and it is better for the cylindrical than the planar heater. For the classical case of selective photothermolysis, the target geometry is not important because thermal damage is confined to the same area as the EMR absorption and direct heating. In our case, thermal damage due to heat diffusion is confined to an area that is distinct from the heater. The dependence of heat diffusion on heater geometry is very strong.

Fourth, the ratio TDT/TRT depends strongly on heating mode. The rectangular EMR pulse mode (Fig. 3.18a) represents the gentlest heating mode because the heater temperature reaches maximum T_1 at the end of the pulse (Fig. 3.18b). The ratio TDT/TRT is maximum for this mode. The flattop temperature pulse mode (Fig. 3.18d) represents the most aggressive heating mode because the heater temperature reaches a maximum just after the beginning of the pulse and the maximum heater temperature takes place during the EMR pulse. The ratio TDT/TRT is a minimum for the flattop temperature pulse mode. As mentioned earlier, the flattop temperature pulse mode can be realized by using an EMR pulse with a special temporal profile. The initial power density should be very high to raise the heater temperature abruptly (for a time interval of the order of or shorter than the TRT). After the maximum temperature is reached, the power density should undergo a steep fall

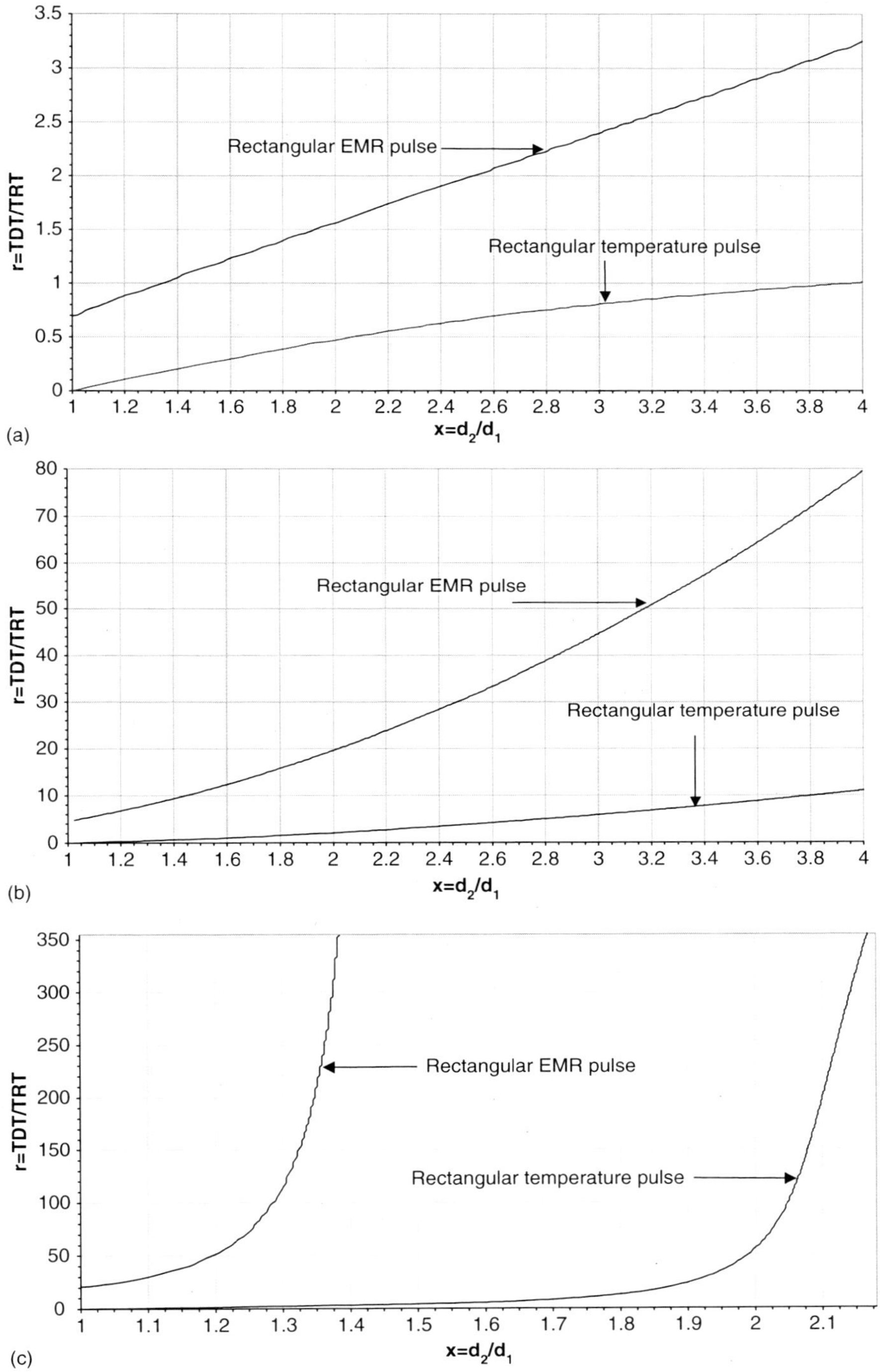

Figure 3.20 The ratio r of TDT and TRT as function of ratio x of size of the target d_2 and size of the heater d_1 for the planar (a), cylindrical (b), and spherical (c) targets [59]. Ratio r = TDT:TRT is given for two heating modes: rectangular EMR pulse and flattop temperature pulse.

to prevent overheating. Then, to maintain the heater temperature at the prescribed level (Fig. 3.18d), the power density should fade gradually to compensate the heat flow out of the heater. The pulse power should be precisely adjusted to keep the heater temperature below the temperature of heater absorption loss. The power depends on heater absorption and size and the EMR attenuation in tissue (see Appendix). In reality, it is probably difficult to exactly create these two modes. Thus, the real value of TDT can be between these two extreme cases.

The ratio TDT/TRT depends on the temperature factor $\Delta = (T_2 - T_0)/(T_1 - T_0)$. Table 3.2 shows the influence of initial target temperature T_0 and maximum heater temperature T_1. All calculations were done for a cylindrical target with the same size as Fig. 3.18 for the rectangular EMR pulse. The TDT increases by a factor of 2.6 by precooling from 37°C to 27°C and decreases by a factor of 2.3 by preheating to 45°C. The fluence should be changed at the same time. If the heater temperature can reach a high value without losing absorption, the TDT can be significantly reduced. The ratio TDT/TRT is about 1.5–2 (TDT = (1.5–2)·TRT) for the case when the heater temperature is 200–250°C. In biological tissue, such a high temperature can be expected in melanin in the hair shaft or in an exogenous chromophore such as carbon. But we must remember that the thermal diffusivity can drop in the tissue surrounding the heater due to water vaporization. So this case is very difficult to predict.

As we have shown here, heat diffusion from the heater is very different for different target geometries. The heater temperature should depend on the heater geometry. Figure 3.21 shows the heater center temperature as a function of pulsewidth for a rectangular EMR pulse with the same power. Spherical, cylindrical, and planar heaters have similar sizes d_1, thermal properties, and EMR absorption coefficients. The thermal relaxation time of the heater τ_r depends on geometry, and the ratio is 1:2:3 for spherical, cylindrical, and planar heaters, respectively. If the pulsewidth τ is significantly shorter than the thermal relaxation time of the heater τ_r ($\tau << \tau_r$) the temperature rise of all the heaters exhibit the same elevation of temperature. However, as shown in Fig. 3.21, the temperature behavior of heaters with different geometries is very different for pulsewidth equal to or longer than τ_r. A steady-state heater temperature for a rectangular EMR pulse is possible only for a spherical heater. For cylindrical and especially for planar heaters, the temperature is continuously rising when the pulsewidth is increasing (the power density should be constant, the energy density should be proportional to the pulsewidth). This is because 3D heat diffusion from the spherical heater (in contrast with 2D and 1D heat diffusion from the cylindrical and

Table 3.2 Thermal Damage Time as a Function of the Temperature Factor [59]

Initial Temperature T_0 (°C)	Temperature of Heater T_1 (°C)	Temperature of Damage T_2 (°C)	Temperature Factor Δ	TDT/TRT
37	100	65	0.44	21
27(precooling)	100	65	0.52	53
45(preheating)	100	65	0.36	9.1
37	200	65	0.17	2.0
37	240	65	0.14	1.5

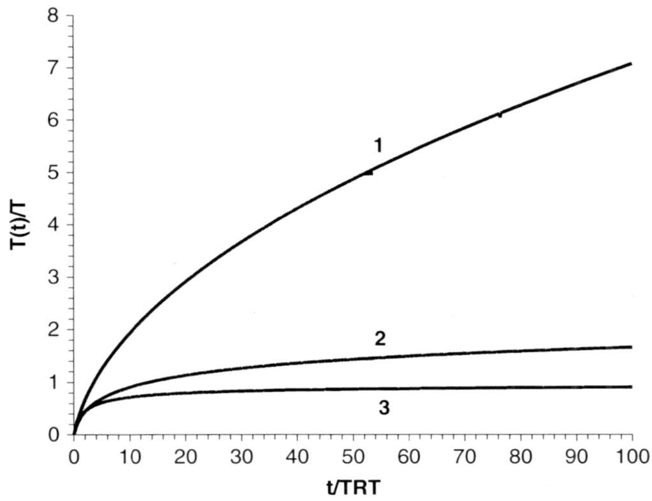

Figure 3.21 Increase of heater temperature $T(t)$ against time for the planar (1), cylindrical (2), and spherical (3) heaters of the same size on exposure to the rectangular EMP pulse. The time is normalized to the temperature relaxation time of the same heater. The temperature is normalized to the steady-state temperature of the spherical heater (notice, there is no steady-state temperature for the planar and cylindrical heaters).

planar heaters, respectively) can compensate constant heating from the rectangular EMR pulse. This phenomenon is very important for uniform targets when the heater and target are the same. For example, the temperature of epidermis (planar target) for a rectangular optical pulse at a wavelength strongly absorbed by melanin will continuously rise during a long pulse ($\tau >> \tau_r$). However, for a spherical target such as the hair bulb matrix, the temperature will stabilize at a steady-state level. To produce a constant heater temperature (flattop temperature pulse as shown in Fig. 3.18d, the EMR pulse shape must be special (Fig. 3.18c), with a strong peak in the beginning and decaying amplitude.

3.6.2.2 Applications of the Extended Theory of Selective Photothermolysis

Photoepilation. Photoepilation utilizes light to cause thermal or mechanical damage of hair follicles. To achieve hair growth delay, it is sufficient to either damage matrix cells of anagen hair follicles or coagulate blood vessels of the papilla, or possibly destroy part of the outer root sheath [56]. For permanent hair-follicle damage, in accordance with current knowledge, it is necessary to damage stem cells that are located in the bulge area at the interface of the outer root sheath and the connective tissue sheath [61]. One can also irreversibly damage a hair follicle at the level of the dermis by replacing it with connective tissue.

The matrix cells produce the hair shaft. The matrix cells contain melanosomes that produce hair melanin. The concentration of melanin in the matrix cells is significantly higher than in the hair shaft. Melanin is distributed uniformly and densely in the matrix cells. So for a pulsewidth longer than the TRT of individual melanosomes (1 μs), the matrix cells act as a uniformly pigmented target. This is a typical example of a target where standard selective photothermolysis theory is applicable. For selective and effective treatment, the energy and pulsewidth have to be significantly shorter than the TRT of the matrix cells. The matrix cells form a dome-shaped structure with the smallest cells close to half of the hair shaft diameter d_h. So the TRT of the hair matrix can be estimated as the TRT of a layer with thickness $d_h/2$, that is, $d_h^2/32k$. The pulsewidth values for effective treatment are presented in Table 3.3.

Another method of halting hair shaft growth is to coagulate blood vessels in the papilla. The loop of blood vessels in the papilla is located in the center of the matrix cell dome. Blood absorption is significantly lower than melanin absorption in the neighboring matrix cells because of the small vessel size. So, the most effective method of papilla blood vessel coagulation is to utilize heat diffusion from the matrix cells that absorb light. This is a typical case where the extended theory of selective photothermolysis can be applied. The highly pigmented heater (the matrix cells) and the lightly pigmented target (the blood vessels) are separated by a distance of about $d_h/2$. We calculated the TDT of the hair papilla blood vessels using our thermal diffusion model, assuming the maximum heater temperature to be $T_1 = 100°C$ and the blood vessel coagulation temperature to be $T_2 = 65°C$. The TDT value for different hair sizes and heating modes are presented in Table 3.3. We can see a significant difference in optimum treatment pulsewidth for: (1) the matrix cells by direct light absorption and (2) the papilla blood vessel loop by heat diffusion.

Let us now consider damage of hair stem cells. They are located in the basal cell layer of the outer root sheath of the lower isthmus. The stem cells do not have any pigment that can effectively absorb light in the therapeutic window (600–1200 nm), which is also the best wavelength range for photoepilation [56]. However, the stem cells can be damaged by heat diffusion from the melanin-rich hair shaft or an artificial chromophore inside the inner root sheath. This case is very well-described by the cylindrical model that we considered in some detail previously. The optimum treatment pulsewidth of the stem cells is close to the TDT of the follicle structure. If we assume that the maximum hair-shaft temperature is

Table 3.3 Optimum Pulsewidth for Hair-Follicle Treatment [59]

Hair Type/ Diameter	TRT of Hair Shaft (ms)	TRT of Hair Follicle (ms)	Halting Hair Shaft Growth			Permanent Hair-Follicle Destruction	
			TRT of Hair Matrix (ms)	TDT of Papilla Blood Vessel (ms)		TDT of Stem Cell (ms)	
				Flattop Temperature Pulse	Rectangular Light Pulse	Flattop Temperature Pulse	Rectangular Light Pulse
Fine/ 30 μm	0.6	5.4	<0.3	0.5	1.5	30	115
Medium coarse/ 70 μm	3	27	<2	2.7	8.5	170	610
Large coarse/ 120 μm	9.6	87	<5	8	21	510	1800

$T_1 = 100°C$ and the damage temperature of the stem cells is $T_1 = 65°C$, we can calculate the TDT as shown in Table 3.A2 (approximate formulas give us slightly lower values than the exact modeling we presented in Fig. 3.20). Calculation results for a follicle with a ratio of follicle diameter to hair-shaft diameter of $x = 3$ are presented in Table 3.3.

To estimate power density P and fluence F for the rectangular light pulse case, we can use formulas from Table 3.A2 (see Appendix) that we can simplify for the follicle case as:

$$P = \frac{32}{\mu_a q} \cdot \frac{x^2}{\text{TRT}} \cdot \frac{T_1 - T_0}{\ln\left(1.4x^2 \cdot \dfrac{\text{TDT}}{\text{TRT}}\right)} \quad [\text{W/cm}^2],$$

$$F = P \cdot \text{TDT} \; [\text{J/cm}^2],$$

where μ_a (cm^{-1}) is the hair-shaft absorption coefficient, and q is the ratio of the radiance at the target location to the input power density (attenuation factor). For our calculation, we use typical treatment conditions: wavelength is 800 nm, dark hair with $\mu_a = 100\ cm^{-1}$, $x = 3$, $q = 1$ (low isthmus level of follicle), $T_1 = 100°C$, $T_2 = 65°C$, and $T_0 = 36°C$. Using the formulas shown here, we calculate the following values: power density is 560 W/cm^2 for 30 μm fine hair, 100 W/cm^2 for 70 μm terminal hair and 35 W/cm^2 for 120 μm coarse hair. For these hair with equal melanin concentration and different diameters, the fluence necessary to damage the stem cells is the same. The fluence value is 40 J/cm^2. The pulsewidth value appears in Table 3.3 as the TDT for a rectangular light pulse. We can see from Table 3.3 that selective and complete hair-follicle damage can be achieved over a very broad range of pulsewidths. For example, for a hair shaft diameter of approximately $d_1 = 70$ μm and a hair follicle diameter of $d_2 = 210$ μm, this range is 170–610 ms, which is significantly longer than the TRT (27 ms). If the energy of the ablation products is not too high, most of them release their energy inside the hair follicle and heat up the follicle structure. So, all the absorbed energy was utilized to heat the follicle structure (including IRS and ORS). For a pulsewidth significantly shorter than 30 ms with the same fluence, we can expect more hair-shaft ablation and escape of the hair-shaft ablation products from the hair follicle. In this case, the damaged volume of the hair follicle should decrease with decreasing pulsewidth and, for very short pulses, it should be limited to the hair shaft.

Photosclerotherapy. Photosclerotherapy produces thermal or mechanical damage of the vessel structure due to light absorption by blood [62]. The blood hemoglobin exhibits selective light absorption over a wide wavelength range. Optimum vessel closure can be achieved by denaturation of the endothelium that is in direct contact with blood. This case is well-described by the standard theory of selective photothermolysis [55]. Other authors have suggested that permanent vessel closure requires denaturation of the vessel wall structure [57,58]. These structures do not contain any strongly absorbing chromophores and can be damaged by heat diffusion from blood. Therefore, we will apply the extended theory of selective photothermolysis to estimate treatment parameters for different vein sizes. As a first approximation, the vein can be modeled as a cylinder. However, this is true just for limited constrained cases depending on the light wavelength and vein size. The cylindrical model is valid if the light penetration depth in the blood exceeds the vessel internal diameter D. In this case, the blood is heated uniformly. If scattering of the blood is lower than absorption, the light penetration depth is roughly equal to the inverse blood absorption

Table 3.4 Optimum Pulsewidth for Treatment of Spider Veins [59]

Diameter of Vein (mm)	Wall Thickness (mm)	TRT of Blood Volume (ms)	TRT of Vein (ms)	TDT of Vein and Fluence F							
				λ = 577 nm				λ = 1060 nm*			
				Flattop Temperature Pulse	Flattop Light Pulse			Flattop Temperature Pulse		Flattop Light Pulse	
				TDT (ms)	F (J/cm^2)	TDT (ms)	F (J/cm^2)	TDT (ms)	F (J/cm^2)	TDT (ms)	F (J/cm^2)
0.1	0.035	0.5	5	40	10	130	20	40	590	140	1090
0.25	0.08	4	35	150	6	515	15	150	335	610	615
0.5	0.12	35	135	215	4	740	6	215	115	1200	200
1	0.15	260	540	240	4	670	5	240	50	2400	90

*We can predict that TDT and F will be of the same order for another important wavelength (λ = 800 nm) because the absorption coefficients for vein blood at λ = 800 nm and λ = 1060 nm are similar.

coefficient $1/\mu_a$. Thus the cylindrical model can be used when $D < 1/\mu_a$. In this case, the heater diameter is $d_1 = D$ and the target diameter is $d_2 = D + 2w$, where w is the vessel-wall thickness. If $D >>; 1/\mu_a$, the heated zone is a cylindrical blood layer, in contact with the vessel wall. The heated-zone thickness is approximately equal to the penetration depth $1/\mu_a$. This case can be roughly described by the planar model with heater thickness $d_1 = 1/\mu_a$ and target thickness $d_2 = 1/\mu_a + 2w$. Analytic evaluation of the TDT using the expressions of Table 3.A2 is possible in the rectangular light-pulse case, when the whole blood volume is uniformly irradiated. If this is the case, then the cylindrical target model is applicable. In the opposite limiting case of very strong blood absorption, a thin layer of blood adjacent to the vessel wall is only irradiated. In the latter case one may apply the planar model. Here, the analytic and numeric results show rather good agreement. In other cases both the TDT and the input flux were evaluated numerically. The maximal heater temperature is $T_1 = 100°C$ (limited by blood coagulation and vaporization), vessel wall denaturation temperature is $T_2 = 65°C$, and the initial body temperature is $T_0 = 37°C$. The calculations were performed for two wavelengths λ: $\lambda = 577$ nm (maximum hemoglobin absorption), where $1/\mu_a = 43$ μm; and moderate hemoglobin absorption $\mu = 1060$ nm, where $1/\mu_a = 1400$ μm [63,64]. To estimate the power density and fluence for a rectangular light pulse, we can use formulas from Table 3.A2 and direct modeling for a flattop temperature pulse. Calculation results for different types of spider veins are presented in Table 3.4.

As we see from Table 3.4, the TDT of the entire vein wall structure is shorter for the 577 nm wavelength that coincides with the strong hemoglobin absorption peak. The TDT can be very long, typically for large veins. The perfusion factor can also be important, but is not discussed in the present study. Blocking of the blood flow can be important for optimum treatment with such a super long pulse. The large veins need significantly lower fluence, power density, and longer pulsewidth than small veins. This dependence is even stronger for light with low absorption in hemoglobin. The water absorption of surrounding tissue can be important for long treatment wavelengths ($\lambda > 750$ nm) due to parallel nonselective heating by water absorption that can increase the tissue temperature T_0. As we showed earlier, the TDT will be shorter in this case, and the fluence will be lower.

The parameters for spider-vein treatment suggested by the new theory are very different compared to typical clinical parameters used for photosclerotherapy, and should be clinically proven.

Following from Tables 3.3 and 3.4, the treatment time (pulsewidth) for hair follicles and spider veins is on the order of several hundreds of milliseconds. As was shown in ref. [65], parallel cooling of the epidermis (simultaneous heating by light absorption and heat removal by heat diffusion into the skin and cooling agent) is very effective in this pulsewidth range. This cooling mode is important for epidermal protection and makes the procedure more effective, because high fluence can be delivered through the epidermis.

Appendix: Determination of Amplitude and Duration of Rectangular EMR Pulses

The appendix summarizes the EMR pulse parameters for the treatment of the basic targets exhibiting a high degree of symmetry, that is, the planar, cylindrical, and spherical ones.

The notations and the basic parameters of the problem are explained in Table 3.A1 [59].

Table 3.A1

Variable	Dimensionality	Name	Assumptions and Relations
k	$cm^2\ s^{-1}$	Thermal diffusivity	Assumed to be the same all over the target
ρ	$g\ cm^{-3}$	Density	Assumed to be the same all over the target
c	J/(g K)	Specific heat	Assumed to be the same all over the target
μ_a	cm^{-1}	Tissue absorption coefficient	Assumed to be zero outside the heater
q	a.u.	The ratio of radiance to the input power density	
d_1	cm	Thickness or diameter of the heater	
d_2	cm	Thickness or diameter of the target	$d_2 > d_1$
d_3	cm	Mean spacing between the targets	$d_3 > d_2$
T_0	°C	Initial temperature of both the target and the surrounding tissue	$T_0 = 37°C$
T_{1max}	°C	Temperature of heater absorption loss	$T_{1max} = 100–250°C$
T_1	°C	Maximum temperature of the heater(absorber)	$T_2 < T_1 \le T_{1max}$
T_2	°C	Temperature of irreversible damage of the tissue	$T_2 = 70°C$
Δ	a.u.	Temperature factor, temperature ratio	$\Delta \equiv \frac{T_2 - T_0}{T_1 - T_0} < 1$
x	a.u.	Geometrical factor, diameter ratio	$x \equiv d_2/d_1 > 1$

Our analysis was based on the heat conduction equation. We have found approximate analytic solutions of TDT and input power density P.

The final expressions for the important variables in question are outlined in simplified form in Table 3.A2 [59]. It is implied that the thermal constants, that is, the density, the thermal diffusivity, and the thermal conductivity do not vary significantly within the target and the surrounding tissues.

The present discussion is restricted to rectangular EMR pulses only. Our goal herewith is to determine the pulsewidth and the pulse amplitude. This may be performed in the following order:

1. Based on the target geometry, the size of the pigmented area (heater) d_1, size of the target d_2, distance between targets d_3, the geometrical factor $x = d_2/d_1$ and density factor can be determined.
2. Based on the thermal properties of the target and chromophore, the temperature of pigmented area absorption loss T_{1max}, temperature of target damage

Table 3.A2

Quantity	Notation [Dimensionality]	Approximate Expression for a Particular Target Geometry		
		Planar	Cylindrical	Spherical
Thermal relaxation time of the heater	τ_r [s]	$\tau_r = \frac{d_1^2}{8 \cdot k}$	$\tau_r = \frac{d_1^2}{16 \cdot k}$	$\tau_r = \frac{d_1^2}{24 \cdot k}$
Thermal relaxation time of the target	TRT [s]	$\mathrm{TRT} = \frac{d_2^2}{8 \cdot k} = x^2 \cdot \tau_r$	$\mathrm{TRT} = \frac{d_2^2}{16 \cdot k} = x^2 \cdot \tau_r$	$\mathrm{TRT} = \frac{d_2^2}{24 \cdot k} = x^2 \cdot \tau_r$
Thermal damage time	TDT [s]	$\mathrm{TDT} = \frac{\mathrm{TRT}}{2 \cdot x^2} \cdot \left[\left(\frac{D-\Delta}{1-\Delta}\right)^2 - 1\right]$ $D = \exp(-x^2) + 1.8 \cdot x \cdot \mathrm{erf}(x)$	$\mathrm{TDT} = \frac{\mathrm{TRT}}{x^2} \cdot \exp\left(\frac{D - 0.3 \cdot \Delta}{1-\Delta}\right)$ $D = 0.6 + 2 \cdot \ln(x) - Ei(-1.4 \cdot x^2)$	$\mathrm{TDT} = \begin{cases} 0.9 \cdot \frac{\mathrm{TRT}}{x_t^2}\left[\left(\frac{1-\Delta}{D-\Delta}\right)^2 - 1\right] \\ D - \Delta > 0 \\ \infty, \quad D - \Delta \le 0 \end{cases}$ $D = 0.7 \cdot \frac{\mathrm{erf}(1.3 \cdot x)}{x}$
Input power density	P [W/cm^2]	$P = \frac{\rho \cdot c}{\mu_a q} \cdot \frac{1.1 \cdot x^2}{\mathrm{TRT}} \cdot \frac{T_1 - T_0}{\sqrt{1 + 2.1 \cdot x^2 \cdot \frac{\mathrm{TDT}}{\mathrm{TRT}} - 1}}$	$P = \frac{\rho \cdot c}{\mu_a q} \cdot \frac{x^2}{\mathrm{TRT}} \cdot \frac{T_1 - T_0}{\ln\left(1 + 1.4 \cdot x^2 \cdot \frac{\mathrm{TDT}}{\mathrm{TRT}}\right)}$	$P = \frac{\rho \cdot c}{\mu q} \cdot \frac{0.3 \cdot x^2}{\mathrm{TRT}} \cdot \frac{T_1 - T_0}{1 - \frac{1}{\sqrt{1 + 1.2 \cdot x^2 \cdot \frac{\mathrm{TDT}}{\mathrm{TRT}}}}}$
Input fluence	F [J/cm^2]	$F = P \cdot \mathrm{TDT}$	$F = P \cdot \mathrm{TDT}$	$F = P \cdot \mathrm{TDT}$

T_2, initial tissue temperature T_0, the temperature factor $\Delta = (T_2 - T_0) / (T_1 - T_0)$ can be determined.

3. Using these parameters and formulas, we can use the formulas in Table 3.A2, rows 1–3 to determine τ_r, TRT and finally TDT. As explained earlier, the TDT is approximately equal to the optimum duration of the EMR pulse τ_0.
4. Based upon an estimate of the EMR attenuation factor at the depth of target q and the absorption coefficient of pigmented area μ_a and the heater temperature T_1, the power density on the skin P using can be determined by using the formulas from Table 3.A2 row 4. The power density should be limited in order not to bleach the pigmented area, but it should be significantly high to rich the temperature of target damage T_2.
5. Treatment fluence is given by $F = P \cdot \text{TDT}$ or $F = P \cdot \tau_0$.

In contrast to both the planar and cylindrical targets, the TDT for the spherical target may be evaluated to infinity (see Table 3.A2, row 3, rightmost column). This means that one cannot ensure the safety of the heater in an attempt to damage the whole target. After a part of the target becomes damaged, the heater temperature reaches the crucial value T_1. This gives rise to phase transitions, bleaching, bubble formation, and other nonlinear processes lying outside the scope of this chapter. Therefore, our simple theory provides the means to describe thermal damage of sufficiently small spherical targets only. More precisely, for a given value of Δ the diameter ratio x must not exceed the value obtained from the equation $D = 0$, where variable D is a function of x determined by the last expression of Table 3.A2, row 3, rightmost column.

References

1. V.V. Tuchin, *Tissue Optics. Light Scattering Methods and Instrumentation for Medical Diagnosis*, PM 166, SPIE Press, Bellingham, WA, 2007.
2. V.V. Tuchin, "Lasers and Fiber Optics in Biomedicine," *Laser Physics* **3**(4), 767–820; **3**(5), 925–950, 1993.
3. OGI Optics Courses. http://omlc.ogi.edu/classroom/.
4. G. Vargas, E.K. Chan, J.K. Barton, H.G. Rylander III, and A.J. Welch, "Use of an Agent to Reduce Scattering in Skin," *Laser. Surg. Med.* **24**, 133–141, 1999.
5. R.M.P. Doornbos, R. Lang, M.C. Aalders, F.W. Cross, and H.J.C.M. Sterenborg, "The Determination of *In Vivo* Human Tissue Optical Properties and Absolute Chromophore Concentrations Using Spatially Resolved Steady-State Diffuse Reflectance Spectroscopy," *Phys. Med. Biol.* **44**, 967–981, 1999.
6. R.R. Anderson and J.A. Parrish, "Optical Properties of Human Skin," in *The Science of Photomedicine*, J.D. Regan and J.A. Parrish (eds.), Plenum Press, New York, 1982, pp. 147–194.
7. Yu.P. Sinichkin, N. Kollias, G. Zonios, S.R. Utz, and V.V. Tuchin, "Reflectance and Fluorescence Spectroscopy of Human Skin *In Vivo*" (Chapter 13), in *Handbook of Optical Biomedical Diagnostics*, V.V. Tuchin (ed.), SPIE Press, Bellingham, WA, 2002, pp. 725–785.
8. N. Kollias and A.N. Baqer, "Absorption Mechanisms of Human Melanin in the Visible, 400-720 nm," *J. Invest. Dermatol.* **89**, 384–388, 1987.
9. W.-F. Cheong, S.A. Prahl, and A.J. Welch, "A Review of the Optical Properties of Biological Tissues," *IEEE J. Quantum Electr.* **26**(12), 2166–2185 1990; updated by W.-F. Cheong, further additions by L. Wang and S.L. Jacques, August 6, 1993.
10. M.J.C. van Gemert, S.L. Jacques, H.J.C.M. Sterenborg, and W.M. Star, "Skin Optics," *IEEE Tranc. Biomed. Eng.* **36**(12), 1146–1154, 1989.

11. S.L. Jacques, C.A. Alter, and S.A. Prahl, "Angular Dependence of the He-Ne Laser Light Scattering by Human Dermis," *Lasers Life Sci.* **1**, 309–333, 1987.
12. Pierre Agache and Philippe Humbert (eds.), *Measuring the Skin*, Springer, Berlin, 2004.
13. F.O. Libnau, O.M. Kvalheim, A.A. Christy, and J. Toft, "Spectra of Water in the Near- and Mid-infrared Region," *Vibrational Spectroscopy* **7**, 243–254, 1994.
14. T.L. Troy, D.L. Page, and E.M. Sevick-Muraca, "Optical Properties of Normal and Diseased Breast Tissues: Prognosis for Optical Mammography," *J. Biomed. Opt.* **1**(3), 342–355, 1996.
15. E.K. Chan, B. Sorg, D. Protsenko, M. O'Neil, M. Motamedi, and A.J. Welch, "Effects of Compression on Soft Tissue Optical Properties," *IEEE J. Select. Tops Quant. Electr.* **2**(4), 943–950, 1996.
16. A.N. Bashkatov, E.A. Genina, V.I. Kochubey, and V.V. Tuchin, "Optical Properties of Human Skin, Subcutaneous and Mucous Tissues in the Wavelength Range from 400 to 2000 nm," *J. Phys. D: Appl. Phys.* **38**, 2543–2555, 2005.
17. E. Salomatina, B. Jiang, J. Novak, and A.N. Yaroslavsky, "Optical Properties of Normal and Cancerous Human Skin in the Visible and Near Infrared Spectral Range," *J. Biomed. Opt.* **11**(6), 064026-1-9, 2006.
18. C.R. Simpson, M. Kohl, M. Essenpreis, and M. Cope, "Near-Infrared Optical Properties of *Ex Vivo* Human Skin and Subcutaneos Tissues Measured Using the Monte Carlo Inversion Technique," *Phys. Med. Biol.* **43**, 2465–2478, 1998.
19. T.L. Troy and S.N. Thennadil, "Optical Properties of Human Skin in the Near Infrared Wavelength Range of 1000 to 2200 nm," *J. Biomed. Opt.* **6**, 167–176, 2001.
20. R. Graaff, A.C.M. Dassel, M.H. Koelink, et al. "Optical Properties of Human Dermis *In Vitro* and *In Vivo*," *Appl. Opt.* **32**, 435–447, 1993.
21. A. Kienle, L. Lilge, and M.S. Patterson, "Investigation of Multi-Layered Tissue with *In Vivo* Reflectance Measurements," *Proc. SPIE* **2326**, 212–214, 1994.
22. S. Nickell, M. Hermann, M. Essenpreis, T.J. Farrell, U. Krämer, and M.S. Patterson, "Anisotropy of Light Propagation in Human Skin," *Phys. Med. Biol.* **45**, 2873–2886, 2000.
23. O. Khalil, S.-J. Yeh, M.G. Lowery, X. Wu, C.F. Hanna, S. Kantor, T.-W. Jeng, J.S. Kanger, R.A. Bolt, and F.F. de Mul, "Temperature Modulation of the Visible and Near Infrared Absorption and Scattering Coefficients of Human Skin," *J. Biomed. Opt.* **8**(2), 191–205, 2003.
24. J.M. Shmitt, A. Knüttel, and R.F. Bonnar, "Measurement of Optical Properties of Biological Tissues by Low-Coherence Reflectometry," *Appl. Opt.* **32**, 6032–6042, 1993.
25. A. Knüttel and M. Boehlau-Godau, "Spatially Confined and Temporally Resolved Refractive Index and Scattering Evaluation in Human Skin Performed with Optical Coherence Tomography," *J. Biomed. Opt.* **5**, 83–92, 2000.
26. A. Knüttel, S. Bonev, and W. Knaak, "New Method for Evaluation of *In Vivo* Scattering and Refractive Index Properties Obtained with Optical Coherence Tomography," *J. Biomed. Opt.* **9**, 265–273, 2004.
27. S. Tauber, R. Baumgartner, K. Schorn, and W. Beyer, "Lightdosimetric Quantitative Analysis of the Human Petrous Bone: Experimental Study for Laser Irradiation of the Cochlea," *Laser Sur. Med.* **28**, 18–26, 2001.
28. F.A. Duck, *Physical Properties of Tissue: A Comprehensive Reference Book*, Academic Press, London, 1990.
29. A.N. Bashkatov, *Controlling of Optical Properties of Tissues at Action by Osmotically Active Immersion Liquids*, Cand. Science Thesis, Saratov State University, Saratov, 2002.
30. A.N. Bashkatov, E.A. Genina, V.I. Kochubey, M.M. Stolnitz, T.A. Bashkatova, O.V. Novikova, A.Yu. Peshkova, and V.V. Tuchin, "Optical Properties of Melanin in the Skin and Skin-Like Phantoms," *Proc. SPIE* **4162**, 219–226, 2000.
31. R.R. Anderson, "Polarized Light Examination and Photography of the Skin," *Arch. Dermatol.* **127**, 1000–1005, 1991.
32. N. Kollias, "Polarized Light Photography of Human Skin," in *Bioengineering of the Skin: Skin Surface Imaging and Analysis*, K.-P. Wilhelm, P. Elsner, E. Berardesca, and H.I. Maibach (eds.), CRC Press, Boca Raton, FL, 1997, pp. 95–106.
33. S.L. Jacques, J.C. Ramella-Roman, and K. Lee, "Imaging Skin Pathology with Polarized Light," *J. Biomed. Opt.* **7**(3), 329–340, 2002.

34. V.V. Tuchin, L. Wang, and D. . Zimnyakov, *Optical Polarization in Biomedical Applications*, Springer-Verlag, Berlin, 2006.
35. D.A. Zimnyakov, Yu.P. Sinichkin and V.V. Tuchin, "Polarization Reflectance Spectroscopy of Biological Tissues: Diagnostic Applications," *Radiophysics and Quantum Electronics* **47**(10/11), 860–875, 2004.
36. H.J.C.M. Sterenborg, M. Motamedi, R.F. Wagner, S. Thomsen, and S.L. Jacques, "*In Vivo* Fluorescence Spectroscopy for the Diagnosis of Skin Diseases," *Proc. SPIE* **2324**, 32–38, 1994.
37. R.R. Anderson, "*In Vivo* Fluorescence of Human Skin," *Arch. Dermatol.* **125**, 999–1000, 1989.
38. Yu.P. Sinichkin, S.R. Utz, A.H. Mavlutov, and H.A. Pilipenko, "*In Vivo* Fluorescence Spectroscopy of the Human Skin: Experiments and Models," *J. Biomed. Opt.* **3**, 201–211, 1998.
39. H. Zeng, C. MacAulay, D.I. McLean, and B. Palcic, "Spectroscopic and Microscopic Characteristics of Human Skin Autofluorescence Emission," *Photochem. Photobiol.* **61**(6), 639–645, 1995.
40. R. Graaff, J.G. Aarnoudse, J.R. Zijp, P.M.A. Sloot, F.F.M. de Mul, J. Greve, and M.H. Koelink, "Reduced Light Scattering Properties for Mixtures of Spherical Particles: A Simple Approximation Derived from Mie Calculations," *Appl. Opt.* **31**, 1370–1376, 1992.
41. V.V. Tuchin, *Optical Clearing of Tissues and Blood*, PM 154, SPIE Press, Bellingham, WA, 2005.
42. H. Schaefer and T.E. Redelmeier, *Skin Barrier: Principles of Percutaneous Absorption*, Karger, Basel, 1996.
43. V.V. Tuchin, "Optical Clearing of Tissue and Blood Using Immersion Method," *J. Phys. D: Appl. Phys.* **38**, 2497–2518, 2005.
44. V.V. Tuchin, G.B. Altshuler, A.A. Gavrilova, A.B. Pravdin, D. Tabatadze, J. Childs, and I.V. Yaroslavsky, "Optical Clearing of Skin Using Flashlamp-Induced Enhancement of Epidermal Permeability," *Lasers Surg. Med.* **38**, 824–836, 2006.
45. V.V. Tuchin, A.N. Bashkatov, E.A. Genina, Yu.P. Sinichkin, and N.A. Lakodina, "*In Vivo* Investigation of the Immersion-Liquid-Induced Human Skin Clearing Dynamics," *Tech. Phys. Lett.* **27**(6), 489–490, 2001.
46. L.S. Dolin, F.I. Feldchtein, G.V. Gelikonov, V.M. Gelikonov, N.D. Gladkova, R.R. Iksanov, V.A. Kamensky, R.V. Kuranov, A.M. Sergeev, N.M. Shakhova, and I.V. Turchin, "Fundamentals of OCT and Clinical Applications of Endoscopic OCT" (Chapter 17), in *Coherent-Domain Optical Methods: Biomedical Diagnostics, Environmental and Material Science*, Vol. 2, V.V. Tuchin (ed.), Kluwer Academic Publishers, Boston, 2004, pp. 211–270.
47. N. Guzelsu, J.F. Federici, H.C. Lim, H.R. Chauhdry, A.B. Ritter, and T. Findley, "Measurement of Skin Strech via Light Reflection," *J. Biomed. Opt.* **8**, 80–86, 2003.
48. G. Müller and A. Roggan (eds.), *Laser-Induced Interstitial Thermotherapy*, SPIE Press, Bellingham, WA, 1995.
49. A.J. Welch and van M.J.C. Gemert (eds.), *Optical-Thermal Response of Laser Irradiated Tissue*, Plenum Press, New York, 1995.
50. H. Niemz, *Laser-Tissue Interactions. Fundamentals and Applications*, Springer, Berlin, 1996.
51. C.H.G. Wright, S.F. Barrett, and A.J. Welch, "Laser-Tissue Interaction," in *Lasers in Medicine*, D.R. Vij and K. Mahesh (eds.), Kluwer, Boston, 2002.
52. V.V. Tuchin, "Light-Tissue Interactions," in *Biomedical Photonics Handbook*, Tuan Vo-Dinh (ed.), CRC Press, Boca Raton, FL, 2003, pp. 3-1–3-26.
53. A.A. Oraevsky, S.J. Jacques, and F.K. Tittel, "Measurement of Tissue Optical Properties by Time-Resolved Detection of Laser-Induced Transient Stress," *Appl. Opt.* **36**(1), 402–415, 1997.
54. A.A. Karabutov and A.A. Oraevsky, "Time-Resolved Detection of Optoacoustic Profiles for Measurement of Optical Energy Distribution in Tissues" (Chapter 10), in *Handbook of Optical Biomedical Diagnostics*, PM107, V.V. Tuchin (ed.), SPIE Press, Bellingham, WA, 2002, pp. 585–674.
55. R.R. Anderson and J. Parrish, "Selective Photothermolysis: Precise Microsurgery by Selective Absorption of the Pulsed Radiation," *Science* **220**, 524–526, 1983.
56. C.C. Dierickx, M.C. Grossman, W.A. Farinelli, and R.R. Anderson, "Permanent Hair Removal by Normal-Mode Ruby Laser," *Arch. Dermatol.* **134**, 837–842, 1998.

57. M.J.C. Van Gemert, A.J. Welch, and A.P. Amin, "Is There an Optimal Laser Treatment for Port Wine Stains?" *Lasers Surg. Med.* **6**, 76–83, 1986.
58. R.A. Neumann, R.M. Knobler, H. Leonhartsberger, and W. Gebhart, "Comparative Histochemistry of Port-Wine Stains after Copper Vapor Laser (578 nm) and Argon Laser Treatment," *J. Invest. Dermatol.* **99**(2), 160–167, 1992.
59. G.B. Altshuler, R.R. Anderson, D. Manstein, H.H. Zenzie, and M.Z. Smirnov, "Extended Theory of Selective Photothermolysis," *Lasers Surg. Med.* **29**(5), 416–432, 2001.
60. F.C. Henriques and A.R. Moritz, "Studies of Thermal Injury. I. The Relative Importance of Time and Surface Temperature in the Causation of Cutaneous Burns," *Amer. J. Pathol.* **23**, 695–720, 1947.
61. T. Sun, G. Cotsarelis, and RM Lavker, "Hair Follicle Stem Cells: The Bulge Activation Hypothesis," *J. Invest. Dermatol.* **96**(Suppl. 5), 77S–78S, 1991.
62. J.S. Dover and K.A. Arndt,. "New Approaches to the Treatment of Vascular Lesions," *Lasers Surg. Med.* **26**, 158–163, 2000.
63. A. Roggan, M. Friebel, K. Doershel, A. Hahn, and G. Mueller, "Optical Properties of Circulating Human Blood in the Wavelength Range 400–2500 nm. *J. Biomed. Opt.* **4**, 36–46, 1999.
64. A.N. Yaroslavsky, I.V. Yaroslavsky, T. Goldbach, and H.J. Schwarzmaier, "Optical Properties of Blood in the Near-Infrared Spectral Range," *Proc. SPIE* **2678**, 314–324, 1996.
65. H.H. Zenzie, G.B. Altshuler, M.Z. Smirnov, and R.R. Anderson, "Evaluation of Cooling Methods for Laser Dermatology," *Lasers Surg. Med.* **26**, 130–144, 2000.

4

Select Laser and Pulsed Light Systems for Cosmetic Dermatology

Paul Wiener and Dale Wiener

Palomar Medical Technologies, Inc., Burlington, MA, USA

Gurpreet S. Ahluwalia (ed.), *Cosmetic Applications of Laser and Light-Based Systems*, 125–142,

4.1 Introduction

The implementation of lasers in the cosmetic arena dates back almost to the beginning of lasers themselves. In 1960, Dr. Theodore Maimen demonstrated the first working of a ruby laser. Within a few years, Dr. Leon Goldman, a dermatologist at the University of Cincinnati, demonstrated some early applications using a ruby laser on the skin [1]. The desire of dermatologists and other physicians to treat various conditions, stimulated more research and experimentation.

For the most part, lasers used in the areas of scientific research were modified or reengineered for use in the medical arena. Most notably, the argon (Ar) laser and the carbon dioxide (CO_2) laser became the first lasers used for cosmetic procedures. The CO_2 laser primarily ablates tissue, so for dermatologic procedures, most applications were used limited to "removing lumps, and bumps or warts." Argon laser, with wavelengths of 488 nm (blue light) and 514 nm (green light) was the first laser to offer some selectivity for a specific chromophore (blood). Green light offered some reasonable absorption characteristics for blood, so doctors found that they could coagulate blood vessels quite effectively. However, because argon laser was a continuous wave laser, the heating of the blood vessels was usually excessive. For skilled users, argon laser was very effective, but for the majority of users and for the majority of applications, the argon laser required too much of technical know-how to be used on most patients. Other continuous wave or quasi-continuous wave lasers, (such as copper-vapor lasers and dye lasers) produced wavelengths that were more selective for chromophores such as blood (oxyhemoglobin) and pigmentation (melanin). For most part, they were more effective than the older argon technology, but they still lacked optimization regarding the target chromophores.

In the mid-1980s, Drs. Parrish and Anderson at Massachusetts General Hospital developed the theory of selective photothermolysis, which spurred the development of a class of medical devices that increased the selectivity of lasers and pulsed light sources for specific target chromophores [2]. Working within the confines of this theory, physicians, researchers, and corporate R&D staff were able to develop lasers and light sources that not only included wavelength selectivity, but also specific pulse durations and appropriate energies. In other words, laser manufacturers now designed systems that were better optimized to safely treat specific dermatologic conditions with the most appropriate lasers and light sources. From this point onward, lasers and light sources moved into a realm of safer and more effective systems, which provided increased patient satisfaction. Additional modifications of this theory have provided a broader basis for the development of other applications that had not even been thought of 10–20 years ago; so since the mid-1980s, the potential applications have broadened significantly to cover an extensive list of skin conditions.

This chapter will encompass the development and implementation of light-based technology in the cosmetic field. In this fast-moving market, continued research and development has yielded innovations in technology that have basically created new treatment opportunities that were unavailable only five years ago. Nearly every step of development has been spurred off from another older principle or technology. This chapter categorizes the discussion into four sections: coherent light-based systems, multiple wavelength intense pulsed light-based systems, nonablative and ablative fractional skin resurfacing, and fractional skin tightening. By tracing the path that technology has taken, we can understand better where it is at present, and where the research and development will take us in the future.

4.2 Coherent Light-Based Systems

4.2.1 Pulsed Dye Lasers

4.2.1.1 Vascular Lesions

With the advances in theory and technology, new devices evolved for conditions that had typically been resistant to old technology. As systems began to utilize increasingly selective wavelengths, and improved safety measures, it became possible to selectively target a specific chromophore with minimal damage to the surrounding tissue. This was and is the key to the treatment of smaller facial vessels, which range from <0.1 to 1 mm. In this case, an optimal wavelength and pulse duration are combined with safety measures to target and treat these small vessels. This brought the pulsed dye laser (PDL) into the spotlight of the laser market.

When developing light-based technology, in order to treat facial vessels (telangiectasias), there are three key factors; pulse duration, fluence, and wavelength. The target chromophore in these treatments is oxyhemoglobin, which has an absorption peak around 577 nm (Fig. 4.1), so the PDL uses filters to emit light at and around this wavelength. The light is sent into the vessel, selectively absorbed by the oxyhemoglobin, converted to heat, and then diffused to the endothelial lining of the vessel. The endothelium is composed

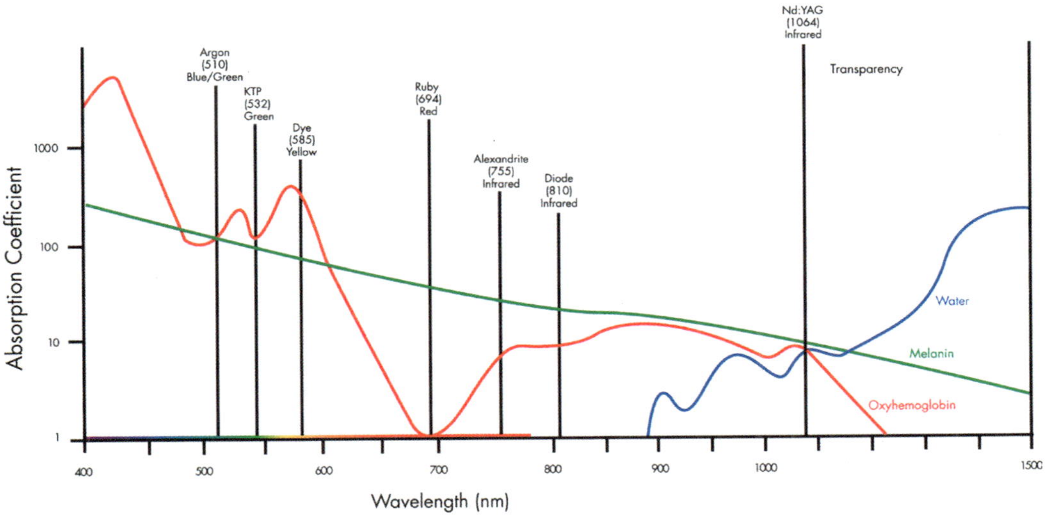

Figure 4.1 Absorption spectrum of major skin chromophores—melanin, oxyhemoglobin, and water—against wavelength emissions.

primarily of proteins, so thermal denaturation occurs, causing the vessel to collapse. The theories behind this mechanism of treatment are called the theory of selective photothermolysis, and the extended theory of selective photothermolysis. Literally, photothermolysis means light-based thermal change; the change in this case is the rise in temperature and the subsequent denaturation of the interior vessel linings. The PDL was able to produce optimal parameters to produce this effect.

For wavelength choice, the key principle, as stated earlier, is oxyhemoglobin absorption. This type of absorption has three peaks: 418, 542, and 577 nm. While the highest absorption coefficient is available at 418 nm, there is a very high level of melanin absorption and very short wavelength for sufficient penetration. The presence of a competing chromophore causes energy absorption in areas other than the target, decreasing the amount of energy delivered to the target, and thus decreasing treatment results and increasing the risk of side effects. The 577 nm peak represents the best of both worlds; a high absorption coefficient with a sufficient amount of energy penetration. Most leading PDLs are filtered between 585 and 600 nm, and obtain good results. They are used to treat all forms of facial telangiectasia. (Figure 4.2, showing treatment with Palomar Medical Technology, Inc.'s Lux G pulsed light hand piece, demonstrates the type of vessels being discussed.)

Correct pulse duration depends mainly on the thermal relaxation time of the target, which is an estimation of how long it takes for heat to dissipate from the target. The thermal relaxation time can be used to allow an operator to precisely tune his or her pulse duration to each treatment. For most treatments of small facial vessels, the correct pulse duration with PDL will be between 0.45 and 40 ms, depending on vessel diameter. This pulse duration will be coupled with fluence high enough to cause sufficient injury to the targeted vessel. Using short pulse duration and a high fluence creates a high peak power that is necessary for the treatment of small vessels. The fluence must be high enough to raise the oxyhemoglobin inside the vessel to above 70°C, which is the temperature required for the endothelial denaturation, causing vessel collapse. In large clusters of lesions such as port wine stains, multiple treatments are necessary; it is not uncommon to see lightening of the lesion even after 20 treatments and purpura is a common side effect. Purpura is a dark-red/purple colored reaction posttreatment that comes from aggressively treating smaller vessels. These

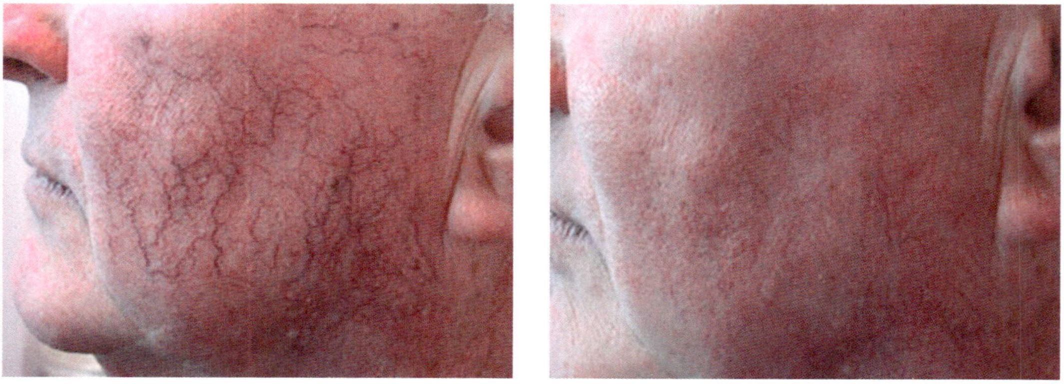

Figure 4.2 Facial vessels before (left) and after (right) three treatments with the Palomar Lux G hand piece.

vessels explode and the blood shows up in this type of reaction, not unlike that of a common hematoma (bruise). Purpura will occur when high levels of fluence are used at short pulse durations, or too much pulse overlapping occurs. This purpura is not uncommon in PDL treatments, and usually resolves itself with topical ointments and bandages in 7–10 days.

4.2.2 Potassium Titanyl Phosphate Lasers

4.2.2.1 Vascular Lesions

In the quest for better vessel treatments, the past twenty years has brought us nearly twenty separate types of light-based technology, all with varying degrees of success. In the case of treating facial vessels and telangiectasias, there has been no universally consistent system. Gold standards in the past few years have typically been PDLs (Candela V-Beam), pulsed light systems with contact cooling and light filters in the 500 nm range (Palomar Starlux with Lux G hand piece), and to a lesser degree, potassium titanyl phosphate (KTP) lasers. Although the KTP laser has decreased in popularity in the past five years, several companies still make very nice systems that have produced some remarkable results, and are still widely used in top practices worldwide (Laserscope Aura/Gemini).

When dealing in the 500–600 nm wavelength range, several factors come into play. One is the high level of oxyhemoglobin absorption, which is our main goal in the treatment of any blood vessel, regardless of size or location. The absorption levels are highest in the lower end of the spectrum. Another factor to consider is competing chromophores, such as melanin. At these wavelengths, there is a high melanin absorption coefficient, so any energy that is directed toward the oxyhemoglobin in the vessels will be diluted by the melanin that lies above it. For that reason, most original vascular lasers (pulsed dye) are up at 585–600 nm, where the melanin absorption coefficient is lowest, but there is still a good amount of blood absorption. The KTP wavelength (at 532 nm) was thought to be too shallow a penetration, as a result of the high melanin absorption coefficient mentioned earlier.

A KTP laser is an Nd:YAG (neodymium:yttrium aluminum garnet) laser at a fundamental wavelength of 1,064 nm which is frequency doubled to half the wavelength or 532 nm by passing the beam through a KTP crystal. This setup, putting the system at 532 nm, created an extremely efficient pigmentation removal laser, and also a system that could attack the more superficial facial vessels. As the principles of light show us, the absorption coefficients at this wavelength for competing chromophores (melanin) are too high to obtain any real significant penetration. Certain early studies done by Van Gemart [3], however, show significant thermal damage at 1.4 mm, which almost contradicts physics because of the high level of melanin absorption at that wavelength. The reasons and mechanisms for this penetration are still debatable.

During treatments of facial vessels and telangiectasias with the KTP laser systems, vessel size is really the greatest deciding factor in treatment technique. The difference in vessel size dictates the pulse duration and the fluence (energy) that is used for the treatment. For larger vessels, longer pulse durations and higher fluences are typically used, while smaller vessels are treated by shorter pulse durations. The smaller vessels <0.3 mm (Fig. 4.3) are generally treated using the 4–6 mm spot sizes, and anywhere from 9–13 J/cm^2

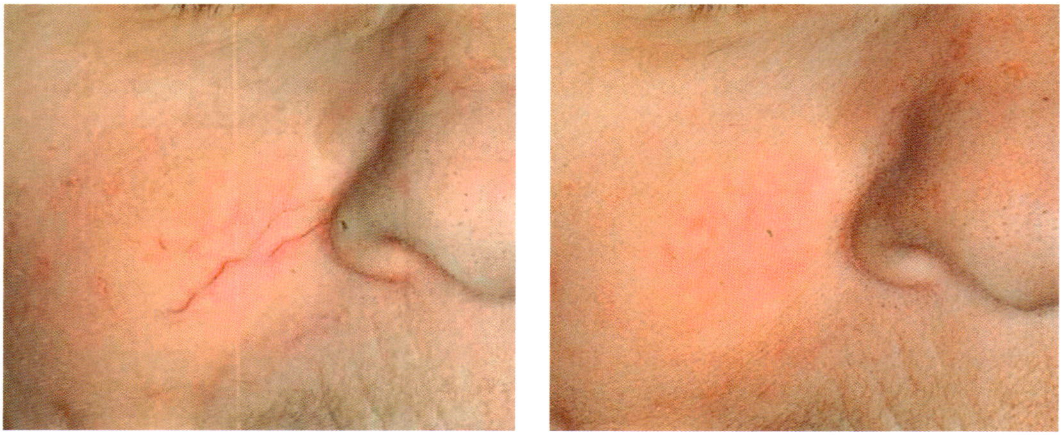

Figure 4.3 Treatment of smaller vessels (<0.3 mm) with the Palomar Lux G hand piece.

with a pulse duration of 10–30 ms. Larger vessels >0.4 mm often respond better to the pulse durations in the 30–50 ms range, as it takes longer for the energy to evenly distribute itself around the vessel lining. For these types of individual distinct vessels, the tracing technique is used, where the vessel is traced with slightly overlapping pulses. For general redness and erythema associated with rosacea, the 10 mm spot size is used to easily treat the whole area of vessels.

4.2.2.2 Pigmented Lesions

As discussed earlier, before Van Gemart proved the depth capabilities of the 532 nm KTP laser, the general attitude was that the high absorption coefficients of the competing chromophores such as melanin would not allow the laser to penetrate to a sufficient depth to treat vascular conditions. Although this was disproved, the fact remains that there is a large amount of melanin absorption at the 532 nm wavelength. This creates an ideal condition for the treatment of lentigines, freckles, and macular seborrheic keratosis on skin types I–IV. (Figure 4.4, showing treatment with Palomar Lux G hand piece, demonstrates the type of lesion being discussed.)

Treatments are performed using a 2–4 mm spot size, a pulse duration of 10–15 ms, and a fluence of 15–20 J/cm^2. The immediate endpoint is a darkening of the lesion, which may occur over the following 10 minutes. In many cases, especially with lighter pigment, a second pass may be necessary. The lesion will develop a micro-crusting type of reaction, and will come to the surface of the skin and basically exfoliate off the epidermis. This will occur over a period of 5–10 days after treatment.

Complications can arise when using KTP lasers in basically the same manner as any other laser system. Undercooling, overtreatment, and misplacement of hand piece aperture can all cause minor side effects. When improper contact is made with the skin, the cool tips cannot do their job, and a burn may follow. Too much overlap can cause blistering and burning as a result of too much thermal damage in one place. Side effects are more common when treating

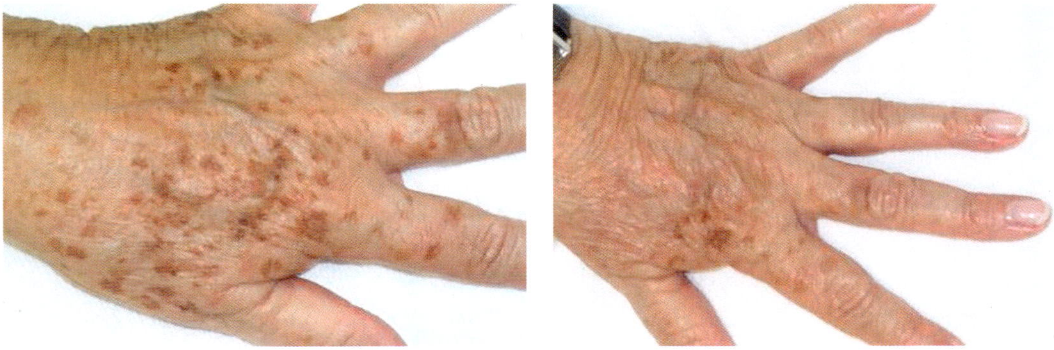

Before After

Figure 4.4 Improvement in pigmented lesions seen after (right) a single treatment with the Palomar Lux G hand piece. Photos courtesy of Palomar Medical Technologies, Inc.

patients in the IV–VI skin-type range, as a result of the high melanin absorption level at 532 nm, so generally only skin-type I–IV are safely treated. Most issues can be resolved with topical steroids, ice, and proper postcare.

Intense pulse light (pulsed light) is another very effective method of treating unwanted epidermal pigment. This technology is discussed in detail later in this chapter.

4.2.3 Long Pulsed Nd:YAG Lasers

4.2.3.1 Vascular Application

When treating vascular issues with light-based technology, the depth of the vessel plays a major role in the decision of which technology to use. Superficial facial vessels are generally smaller in size, and closer to the surface of the skin, and as a result, are easily treated by highly absorbed wavelengths in the 480–670 nm range. When deeper, larger vessels, usually in the leg area, are the target, those shorter wavelengths mentioned earlier cannot achieve deep enough penetration. This is due to the amount of competing chromophores at those wavelengths, namely melanin. The key in treating these deeper vessels is to utilize a wavelength that has some selectivity for oxyhemoglobin, but lacks the competing chromophores present at the shorter wavelengths. Long pulsed Nd:YAG lasers at 1,064 nm are the perfect fit.

The Nd:YAG laser is ideal for treating larger, deeper vessels up to 3–4 mm. This wavelength combines some selectivity for absorption in oxyhemoglobin, while not having the high level of competing chromophores present at the lower 400–700 nm wavelengths. This gives the YAG laser the ability to penetrate deep enough into the skin to treat deeper vessels that previously required painful injections known as sclerotherapy. This allows operators to treat deep vessels with a laser treatment that leaves the patient with a relatively minor urticarial histamine type response that dissipates over the two weeks posttreatment.

Although many patients seek treatment of facial vessels, the YAG laser is not the optimal tool for this procedure. Although effective on some of the larger vessels, the YAG laser presents a new problem not encountered in the lower wavelengths, water absorption.

At 1,064 nm, there is an elevated level of water absorption, and because the skin in the nasal area has such a high water content, the high absorption coefficient will pull the laser deeper into the skin than desired. This elevated light absorption, combined with the lack of competing chromophores may result in a burn, a change in pigmentation, and in the worst cases, a permanent, depressed scar. These facial vessels can be much more efficiently and safely treated using systems in the lower wavelength range (pulsed dye or pulsed light with contact cooling).

The characteristics that make the YAG inefficient and dangerous for treating facial telangiectasias actually make it ideal for treating leg veins. Leg vessels by nature are usually larger and deeper than facial vessels, so a system is needed that can penetrate deep enough to selectively target such a vessel. The YAG utilizes a spot size that corresponds to the vessel size, and is then moved along the vessel, firing every few millimeters depending on the vessel size. As with other vascular treatments, pulse duration and fluence are adjusted according to the size as well, the larger the vessel, the longer the pulse duration will be, and higher the fluence. One pass should be sufficient (Fig. 4.5), but in many of the more extreme cases, a second pass may be necessary. Never stack the pulses on top of one another, as this is a sure way to cause a complication. The endpoint should be an urticarial histamine-type (raised red cat scratch) response, and although some vessels may close at the time of treatment, not seeing the closure does not always warrant a second pass. Vessels up to 4 mm can be treated with the laser alone, but in the case of larger vessels, proper treatment requires a combination of YAG and injection therapy (sclerotherapy).

Immediately after treatment, cooling is essential to reduce the inflammation and the risk of thermal injury, and a topical steroid may be used to resolve any blistering issues. Compression stockings may be worn for 1–2 weeks posttreatment, and some light walking and exercise is essential to resolve the majority of these lower extremity vascular issues. In case of blood pooling, lancing with an 18-gauge needle is sufficient to remove excess blood, and using a 500–700 nm pulsed light hand piece with contact cooling, such as the Palomar Starlux with the Lux G hand piece, will help to resolve most hemosiderin staining issues. Treatments should be every 6–8 weeks, and most cases should be resolved in three to six treatment sessions.

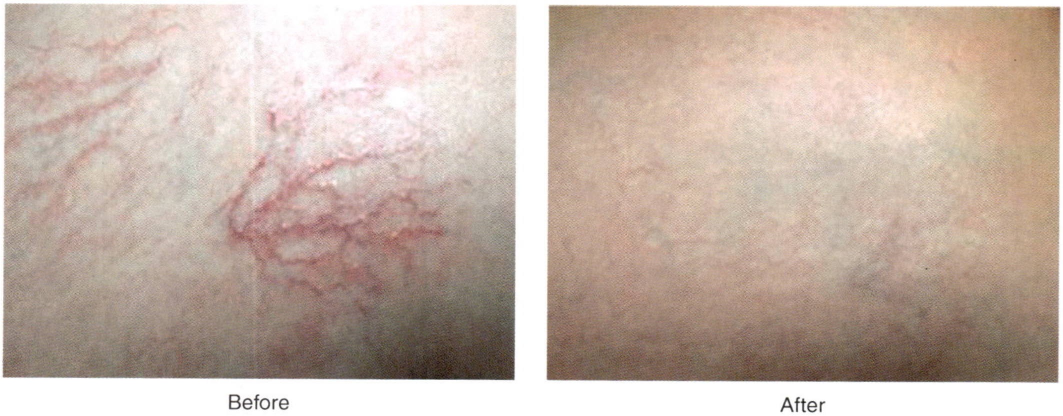

Figure 4.5 Treatment of larger leg vessels (>0.4 mm) with Palomar Lux 1064 YAG.

4.3 Multiple Wavelength Intense Pulsed Light-Based Systems

4.3.1 Overview

Until this point, our discussion has been focused upon lasers. These have been single wavelength, coherent beams of light, which are used to treat various conditions. Although many of these systems are extremely effective in their respective functionalities, the fact that they all emit a single wavelength of light limits their versatility. Having one wavelength, they all have one level of absorption, and therefore have only limited purpose. The introduction of the intense pulsed light system (know as "pulsed light" technology) has been facilitated by the theory of bands of multiple wavelengths being isolated and utilized for the treatment of a wide range of conditions. Through the use of a Xenon flashlamp and selective dichroic light filters, a pulsed light hand piece can be tailored to the proper wavelengths possessing the optimal absorption for the condition that is being treated.

4.3.2 Principles

As seen in Fig. 4.6, a higher blood and epidermal melanin absorption is in the 400–700 nm range, with a drop in the 700–900 nm range (deeper melanin selective wavelengths), and then an increase in blood absorption in the 900–1,200 nm range. This means that by using the 400–600 nm range, we can create a spectrum that will effectively treat vascular and epidermal pigmented lesions The longer spectrum (700–1,200 nm) will be helpful in targeting deeper vascular lesions and epdirmal pigments. The utilization of this bimodal filtration allows a system to utilize the wavelengths that it needs, while blocking out the wavelengths in the middle of its band, which simply adds unwanted heat in the skin. Palomar Medical Technologies, Inc.'s Starlux Lux G hand piece was designed to take

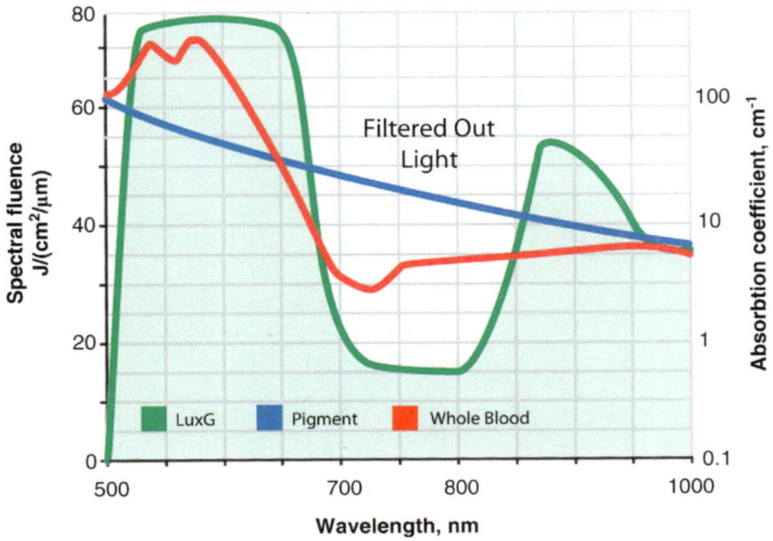

Figure 4.6 Chart demonstrating Palomar's patented double cutoff light filter for the optimization of spectrum relative to oxyhemoglobin absorption.

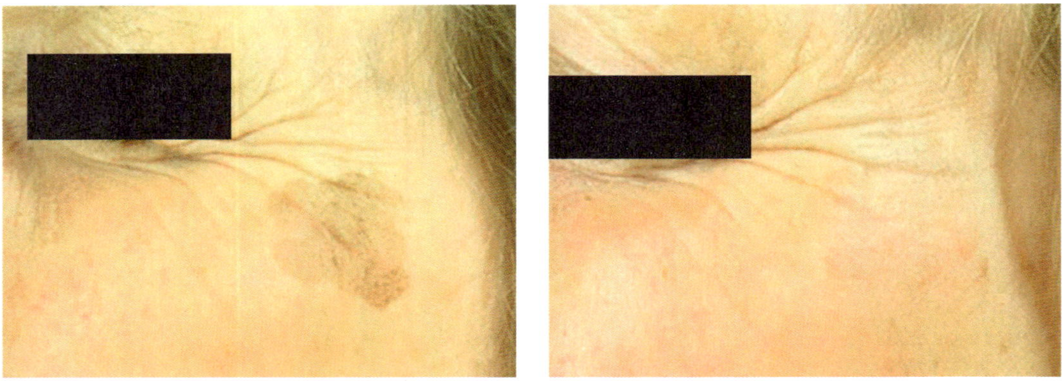

Before After

Figure 4.7 Benign epidermal pigmentation treated with the Palomar Lux G hand piece. Photos courtesy of Haneef Alibhai, MD.

advantage of the optimal peaks of absorption in oxyhemoglobin, which also correspond to high absorption in epidermal pigment.

Most pulsed light Fig. 4.7 systems use a type of slide-in filters that are not extremely precise in their wavelength cutoff. This type of filter is called a "reflective filter" which uses dichroic coatings on the slide in filter glass. Dichroic filters are angle dependent and wavelength dependent filters, which means that light bounces off the coating at various angles. These types of filters are very accurate when used with lasers, because lasers are monochromatic, but these types of filters are less accurate when used with pulsed light systems, because by definition they are multi-wavelength light sources. Therefore, all pulsed light systems that employ slide in filters must use this technology. Palomar's unique, patented dual wavelength filtration system uses both dichroic filters and absorption filters. The dual filtration system provides a more precise specification on the short wavelength end of the spectrum and permits Palomar to produce the unique wavelength outputs of the Lux G and Lux V hand pieces. By definition, if a filter absorbs light to select a wavelength range, the light is converted to heat and the filter must be cooled. Therefore, slide-in reflective filters, such as those used by Sciton and Lumenis are not cooled, and thus can not be absorption filters.

4.3.3 Technology

By the law of physics, a noncoherent flash of light can only travel a fraction of the distance than that of a coherent beam like a 1,064 nm YAG laser, and subsequently cannot penetrate nearly as deep into the skin. For this reason, the pulsed light systems in the market today have not been optimized for treating deep vessels such as leg veins, or treating the deeper dermal lesions such as Neavus of Ota. They are however ideal for treating most sun-based pigmentation, facial vascular issues, and most types of hair removal. As with any light, there is a large amount of scatter off the skin when the pulse flashes. A process developed and patented by Palomar Medical and Massachusetts General Hospital, called photon recycling, uses a series of mirrors to capture the scattered light and recycle it back into the target [4]. This allows for maximum energy to reach the target in a single pulse.

An essential part of any pulsed light treatment is the use of proper cooling. Using proper materials and the proper technology, a system can employ all the benefits of multiband wavelengths and dual filtration, as well as parallel, contact cooling during the actual flash [5–8]. By using a water-cooling system that has two stage capabilities, a system can remain cold even after several hundred shots. A sapphire tip is necessary for proper heat dissipation. Many cheaper systems will use quartz, which insulates the heat in the skin, increasing the opportunity for side effects. A pulsed light system can range from $4,000–$120,000 depending on the materials used in its construction, and it is extremely important for consumers to understand that they are not even close to being equal technologies. The cheaper the product, the poorer the quality of the materials and technology that is being used.

Although sun-based pigmentation is a relatively easy treatment, in order to deal with vascular issues, fighting acne, and unwanted hair, a proper combination of dual filtration, two-stage contact cooling, photon recycling, and a sapphire tip is necessary to safely and effectively treat the condition.

4.3.4 Photofacial Applications

A key development in the cosmetic light-based technology arena over the past eight years has been the photofacial, which has become one of the most popular treatments on the market. This "light-based" facial is aimed at the treatment of vascular issues and epidermal pigmentation (see Figs. 4.8 and 4.9, respectively), which improves the overall appearance of the skin. The pulsed light system, by nature of incoherent light, uses the majority of its energy up in the very superficial layers of the skin, making it ideal for this type of treatment

The treatment is normally performed with a pulsed light hand piece in the 500–1,200 nm wavelength range, making it ideal for the treatment of both vascular and pigmented lesions. Using the Palomar Lux G hand piece on the Starlux system, the operator utilizes the double cutoff filter (500–670 nm, 870–1,200 nm) to efficiently treat both targets without depositing unnecessary heat in the skin from the 700–900 nm range. The typical treatment involves the treatment of pigmentation in the forehead and outer cheek area using a 10–20 ms pulse

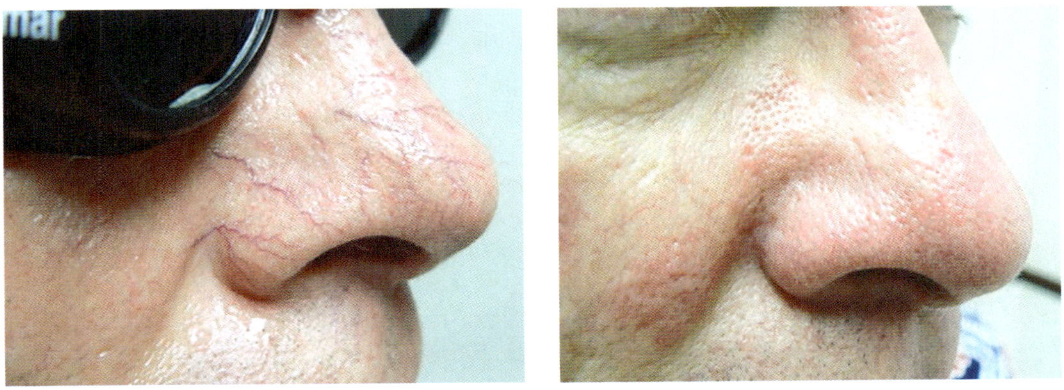

Before After

Figure 4.8 Larger facial vessels in and around the alar groove treated with the Palomar Lux G hand piece.

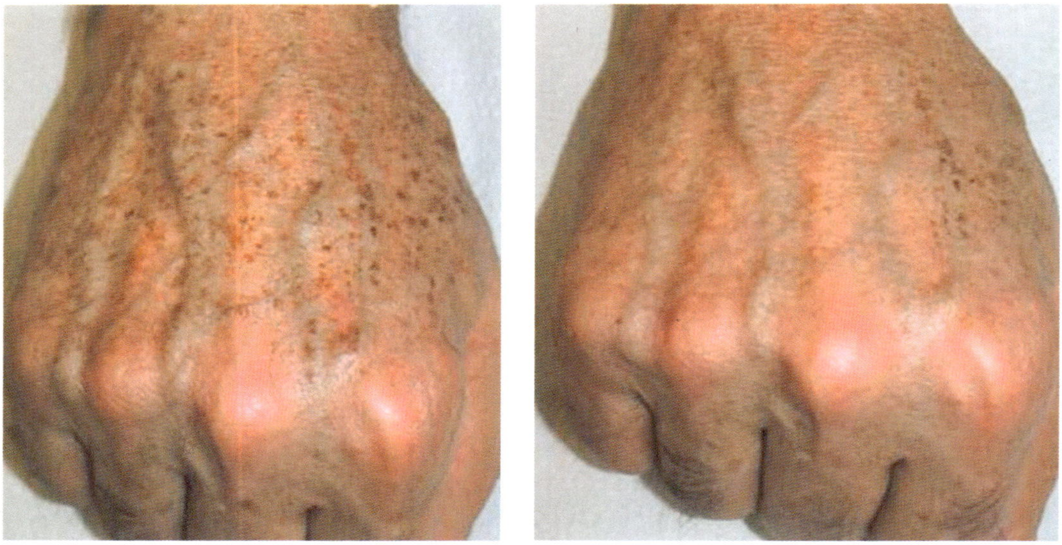

Figure 4.9 Benign epidermal pigmentation treated with the Palomar Lux G hand piece.

duration, followed by the treatment of smaller vessels on the ridge of the cheekbone below the eyes as well as the chin, which is mostly done with a 10 ms pulse duration. The final part of the treatment is usually targeting the large vessels in the alar groove (Fig. 4.8), which usually requires two pulses, one at 100 ms pulse duration, and one at 20 or 10 ms, all with high fluence. The skin in this area, being very sebaceous, is quite forgiving to this type of light, and therefore can be treated more aggressively than any other area on the body. The desired result for the pigmentation is a slight darkening, enough to indicate an exfoliation of the lesion 7–15 days later, depending upon the body area treated (Fig. 4.9). For the vessels, the desired endpoint is the vasoconstriction or closing of the vessels.

4.3.5 Treatments

Treatments with pulsed light systems involve direct contact to the skin, and can involve multiple passes when the treatment warrants it. One of the unique features of a pulsed light system is the extremely large range of pulse duration options that are available. On some of the top of the line systems, an operator can choose between 5 and 100 ms, depending on what condition they are treating. The cases involving larger vessels in the Alar area are usually the ones that require a multiple pass approach. For vessels with larger diameters (Fig. 4.10), a multi pass approach, working from longer pulse duration to shorter pulse duration, may be the best choice. For example, a 0.7 mm vessel may require a 100 ms pass at 50–70 J/cm^2, followed by a 20 ms at 30–50 J/cm^2, using correct contact to the skin, so as to utilize the contact cooling sapphire. For smaller vessels, a 20 ms pass followed by a 10 ms pass may be necessary. The idea is to "chase the vessel size" as the vessel is injured by light, and the diameter is reduced. Although the larger vessels are treated by a long pulse with high fluence, oftentimes, a small amount of blood flow still remains. The shorter pulse follow-up will create a higher peak power, which is necessary to treat the smaller amount

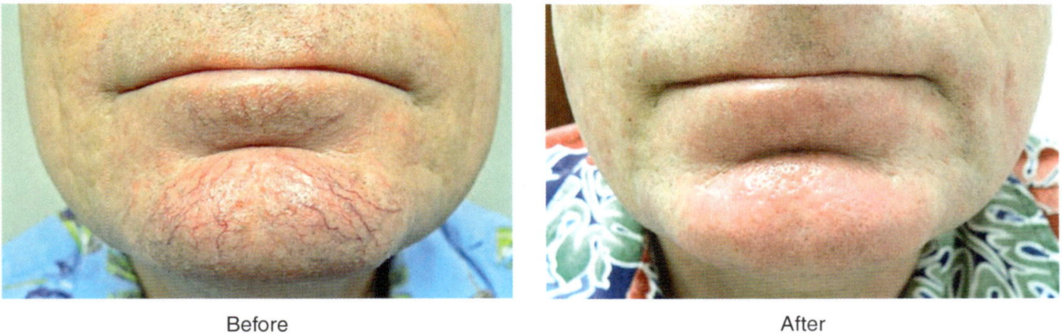

Figure 4.10 Larger facial vessels treated with the Palomar Lux G hand piece. Photos courtesy of Michael Sinclair, MD.

of oxyhemoglobin left in vessel. The same type of multiple pass, variable pulse duration treatment approach is also used in the fighting acne, and in some resistant hair removal cases. While even the best pulsed light systems are limited by their depth of penetration, they are now regarded as an alternate new gold standard for the treatment of all superficial blood vessels, sun-based pigmentation, hair removal, and also helpful in fighting acne. It is this type of versatility that makes the right pulsed light system an important part of any practice.

4.4 Nonablative Fractional Skin Resurfacing

4.4.1 Overview

Using the nonablative fractional delivery of 1,540–1,550 nm light, an operator is able to treat conditions such as scar tissue or melasma and resurface the skin with very little discomfort or downtime. By creating spatially confined columns of coagulated tissue, the skin undergoes a natural healing process involving collagen production and tissue replacement (Fig. 4.11).

Fractional technology, invented and developed jointly by Palomar Medical Technologies, Inc., and Wellman Labs at MGH in Boston (US Patent 6,997,923 B2 exclusively licensed to Palomar Medical Technologies, Inc.), allows for controlled thermal energy to be delivered in a fractional manner, resulting in the surrounding tissue remaining relatively

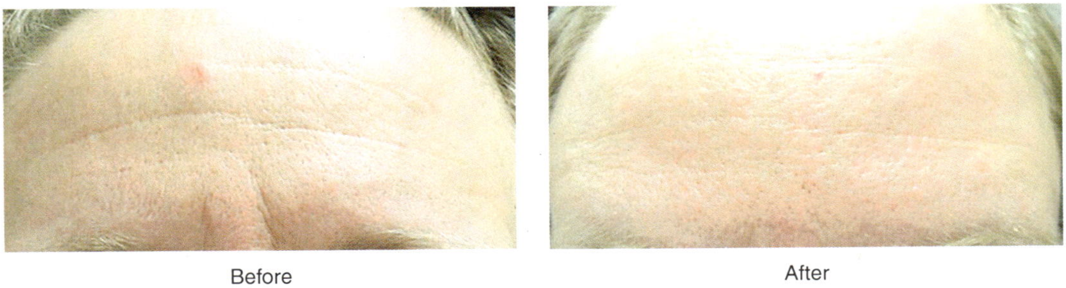

Figure 4.11 Skin resurfacing after two treatments with the Palomar Lux 1540 hand piece. Photos courtesy of Michael Sinclair, MD.

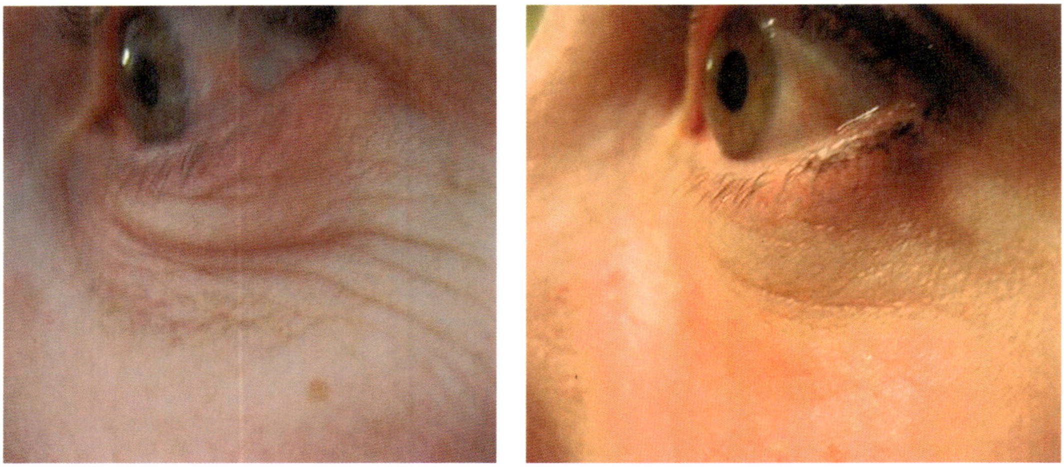

Figure 4.12 Skin resurfacing 4 weeks after three treatments with the Palomar Lux 1540 hand piece.

unchanged. This creates less downtime and less risk of complications. In addition to skin resurfacing, this new method of energy delivery has been applied, with great success, to the treatment of acne scarring, surgical/laceration scars, stretch marks, and melasma.

4.4.2 Principles

The principle on which this type of technology is based requires the use of nonablative wavelengths of light to selectively coagulate tissue in the epidermis and dermis. The light is focused by sending it through micro-lens arrays to create multiple beams of light or by focusing one beam and using a scanner to create hundreds of coagulation columns ranging from 300 μm deep to 1.4 mm in depth, depending on the energy per microbeam. This reaction spurs the body's natural healing response to remove and replace the coagulated tissue, stimulating fibroblast collagen production, and subsequently resurfacing the tissue (Fig. 4.12). This is different than in the past when ablative erbium:YAG and CO_2 lasers were used to totally remove the top layer of the epidermis and a portion of the dermis, often called full-surface ablation as opposed to fractional ablation discussed later. Although the results were, and still are very good, this type of skin ablation creates a large amount of down time, and an increased risk of complications. These include but are not limited to, infection, hypertrophic, and atrophic scarring, and permanent hypopigmentation.

By using different lasing materials and "doping methods", fundamental wavelengths can be shifted for different uses. In the case of nonablative skin resurfacing, the goal of researchers was to have the ability to create the same type of healing effect as seen in earlier ablative treatments, but without the significant downtime or increased risk of complications. Wavelengths in the 1,320–1,580 nm have been tried at this point, with varying success. The key in this case is to have a wavelength that has a good penetration into the skin for depth, and has a low collagen denaturation threshold (level of thermal damage at which the tissue will coagulate). This combination can be found in the 1,530–1,590 nm range, which is where most of the more successful systems in the market have set their beam wavelengths.

4.4.3 Technology

As with any type of new technology, there are a number of systems to choose from, ranging from \$40,000–\$130,000, depending on the technology. Palomar Medical Technologies, Inc.'s Lux 1,540 nm erbium:glass laser and the Reliant 1,550 nm erbium "doped" diode fiber Fraxel systems are currently the two top systems in the market, both for the materials used in their construction, and the large amount of clinical and technical research that went into their development.

The Palomar system is based on lasing light from an erbium rod through a glass fiber, creating the desired 1,540 nm wavelength necessary to coagulate the tissue at an efficient rate with enough depth for some clinical effect to take place. Using stamping technique, the Palomar system creates its coagulation columns uniformly, and with very little exposure (10 ms) to the light. This creates minimal discomfort, and a quick treatment for smaller areas, such as acne scarring on the cheek (Fig. 4.13), or skin rejuvenation (Fig. 4.14). Topical anesthetic is not typically used with this fractional nonablative system in a majority of treatments, as any discomfort is managed with posttreatment ice application alone. Density and depth are the two keys to the successful treatment of most of these conditions, so energy per microbeam is usually in the 40–70 mJ range, and multiple passes are performed

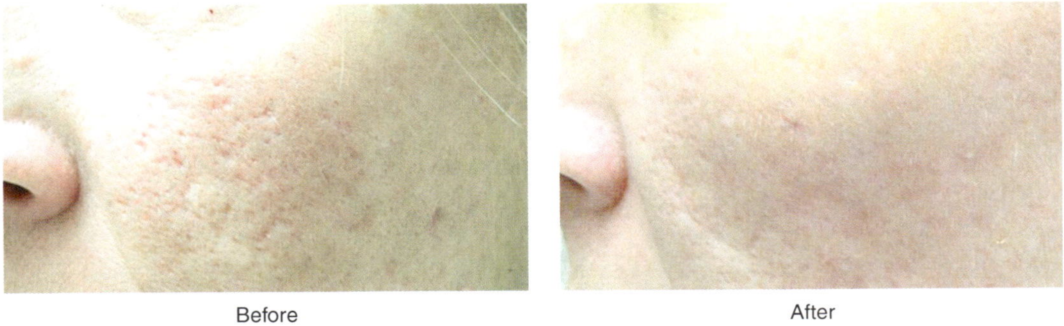

Figure 4.13 Acne scarring treated with Palomar Lux 1540 hand piece.

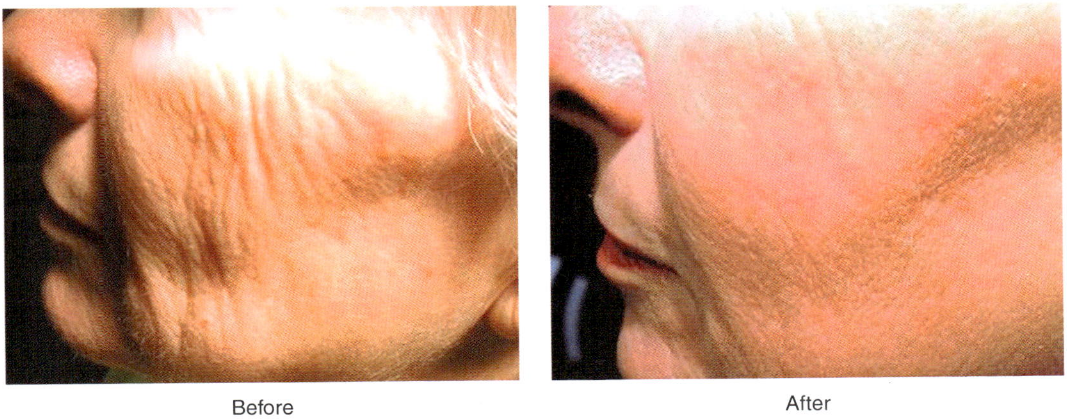

Figure 4.14 Skin resurfacing 8 weeks after treatment with the Palomar Lux 1540 hand piece. Photos courtesy of Dwight Scarborough, MD.

to a specific area before moving on to others. This system uses a stamping technique and this, in part, helps control discomfort.

The Reliant Fraxel system utilizes an erbium doped diode fiber to achieve the desired 1,550 nm wavelength. It uses a scanning device to paint the pulses of energy onto the skin. This process creates more heat in the skin, ultimately creating pain. This creates an increase in epidermal temperature relative to that of a device that uses a stamping technique, and consequently, topical anesthetics are needed for all treatments with devices that use scanners. This also may lead to an increased occurrence of postinflammatory hyper pigmentation (PIH), which is realistically the one potential side effect that can occur with these types of treatments. This creates an added expense of a disposable, as well as added time and space taken up by the patient in the office. Regardless of the pain with the Fraxel system, it has shown very good and consistent results, and along with the Palomar, continues to be one of the most successful systems in offices worldwide.

4.4.4 Clinical Applications

Acne scarring is one of the conditions that has been most successfully treated with this type of nonablative fractional technology worldwide. Using high energies and creating enough density, most acne scarring (including ice pick scarring in Asian skin) can be treated to the point of patient satisfaction, as the collagen collapse is not a dynamic problem. Surgical scars and striae (stretch marks) are also responding extremely well to these treatments, as even the hypopigmented scar tissue has responded with increased melanin production and decreased fibrosis. Melasma has been successfully managed with the Lux 1540. Inadvertent UV exposure can promote rapid regression of this condition.

As nothing in the nonablative technology market has lived up to results observed with conventional ablative resurfacing for wrinkles, the industry needed to find a way not only to achieve these results but also to reduce downtime and side effects associated with this procedure. Recent research by Palomar (fractional erbium:YAG 2,940 nm) and Reliant (fractional CO_2) has brought about the fractional delivery of ablative wavelengths, removing the entire epidermis and dermis in the same ablative way as before, but in a fractional delivery system which does so only with columns of ablated tissue. This creates the same type of wound to the skin as conventional full-surface ablative techniques, and as a result, the same type of healing response, but with a benefit of leaving unaffected tissue surrounding these micro-ablated columns. The result is less downtime with faster healing, resulting in fewer complications compared to conventional full-surface ablative resurfacing. Clinically, the results are much closer to what was observed with conventional resurfacing.

4.5 Fractional Skin Tightening

4.5.1 Overview/Principles

The same principles utilized in fractional skin resurfacing at roughly a millimeter of depth are applied to the condition of lax and loose skin at 2–4 mm of depth. Using wavelengths in the 850–1,350 nm, the Palomar Lux Deep IR hand piece uses a halogen lamp to create a pulse duration ranging from 5–10 seconds, depending on desired depth. While there are several skin tightening systems in the market (Cutera Titan, Alma Accent, Thermage), Deep IR is the only one to combine IR wavelengths with a fractional delivery system, making it possible to penetrate deeper with more energy as well as less discomfort and side effects. This technology is aimed at the thermal treatment of deep dermis to help tighten loose skin in the jowling areas, abdomen, arms, legs, and so on.

4.5.2 Technology

The Deep IR system is run off a halogen lamp filtered at 850–1,350 nm. This creates a wavelength spectrum that will penetrate to a proper depth and then be absorbed by the water content and converted to heat. The system combines long pulse durations (seconds) and fractional delivery to achieve this penetration.

In order to allow high amounts of deeply penetrating wavelengths of light to target water in the mid-deep dermis, extensive cooling is required to protect the upper portion of the skin. The Deep IR runs on the Starlux 500 platform that utilizes advanced two-staged cooling. In addition to cooling, another important safety feature unique to this light-based technology is a series of electronic field sensors (E-Field Sensor) which indicates a proper contact to the skin. Once proper contact is established, the system begins a precooling stage (0–5 seconds), a firing stage (5–10 seconds), followed by a one second post-cooling phase. If at any time, the contact to the skin is lost, the firing phase is stopped, but the cooling is continued. The patient feels a slight warmth toward the end of the firing phase of the treatment.

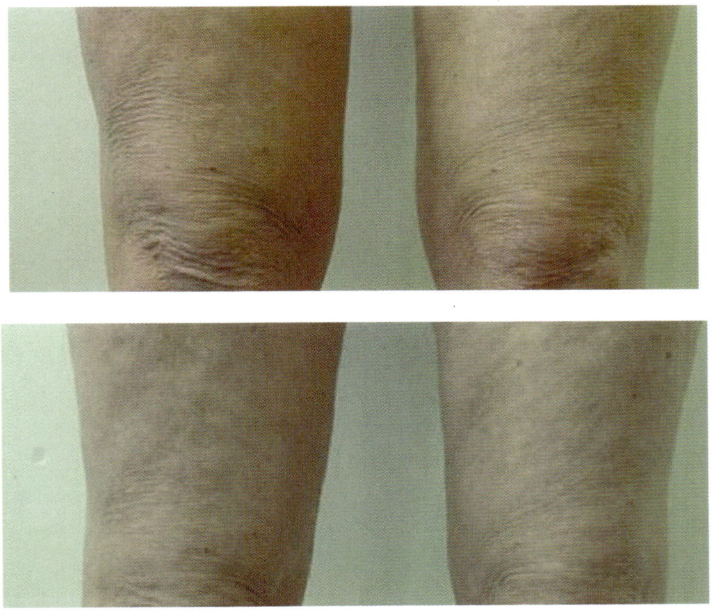

Before

1 Mo FU

88J/cm^2, 5s, 1 Tx

Figure 4.15 Skin tightening on thighs using the Palomar Lux Deep IR hand piece.

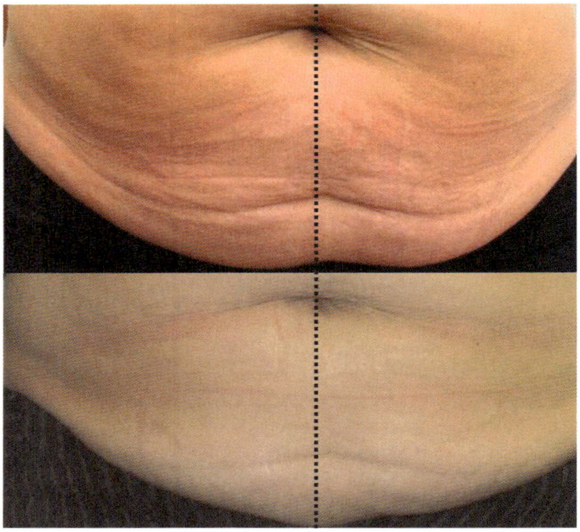

Figure 4.16 Skin tightening on abdomen using the Palomar Lux Deep IR hand piece. Photos courtesy of K. Katri, MD.

4.5.3 Clinical Applications

Fractional-skin tightening is aimed at the loose or crepey skin present on the face, arms, legs (Fig. 4.15), abdomen (Fig. 4.16), and neck. This is a good combination treatment with the fractional skin resurfacing technology and a pulsed light photo facial system, giving a patient deep tightening, shallow epidermal/dermal resurfacing, and unwanted pigmentation and vascular removal. Deep IR protocol should consist of 3–5 treatments spaced one month apart. There usually is some noticeable short-term improvement in the treatment area but the ultimate results will be observed overtime since the treatment effect is ongoing for several months after the treatments are completed.

References

1. Goldman L, Rockwell Jr., RJ. Laser systems and their applications in medicine and biology. *Adv. Biomed. Eng Med. Phys.* 1968; **1:** 317–382.
2. Anderson RR, Parrish JA. Selective photothermolysis: precise microsurgery by selective absorption of pulsed radiation. *Science* 1983; **220**: 524–527.
3. Gemart V. Modeling laser treatment or port wine stains. In: Tan OT, ed., Management and Treatment of Benign Cutaneous Vascular Lesions. Philadelphia: Lea & Febiger, 1992.
4. Anderson RR. Radiation-delivery device, US Patent 5,824,023, 1998. (This patent refers to the recycling and refocusing of scattered light back into the target.)
5. Gregory B. Altshuler, et. al., Method and apparatus for photothermal treatment of tissue at depth, US patent 7,351,252, April 2008.
6. Gregory B. Altshuler, et. al., Cooling system for a photo cosmetic device, US patent 7,204,832, April 2007.
7. Gregory B. Altshuler, et. al., Method and apparatus for controlling the temperature of a surface, US patent 6,648,904, November 2003.
8. Gregory B. Altshuler, et. al., Method and apparatus for dermatology treatment, US patent, 6,273,884, August 2001.

PART 2
HAIR MANAGEMENT BY LIGHT-BASED TECHNOLOGIES

5

Hair Removal Using Light-Based Systems

David J. Goldberg

Skin Laser & Surgery Specialists of New York & New Jersey, and Department of Dermatology, Mount Sinai School of Medicine, New York, NY, USA

5.1 Introduction

Over the years, numerous laser and light-based devices have been developed to effectively remove unwanted hair. These innovations have been based on applying principles of laser physics to selectively targeted hair follicles. As our understanding of hair biology has grown and laser technology has advanced, various different light-based sources have become increasingly effective and efficient in removing undesirable hair. This review will focus on the most popular laser devices commonly used today, which include the alexandrite, diode, and 1064 Nd:YAG lasers. In addition, intense pulsed light (IPL) devices for hair removal are now growing in popularity, and will also be reviewed.

Gurpreet S. Ahluwalia (ed.), *Cosmetic Applications of Laser and Light-Based Systems*, 145–156,

Successful light-based hair removal is predicated on an understanding of hair biology. All human hair show various stages of hair growth [1]. The anagen or growth phase is variable in duration (up to six years) and leads to the catagen or regression phase, which is usually constant at around three weeks. The telogen or resting phase follows just prior to the resumption of the anagen phase and lasts approximately three months. At any given time, the majority of hair follicles (80–85%) are in anagen and the remaining follicles are either in the catagen phase (2%) or the telogen phase (10–15%).

The anagen duration varies greatly depending on age, season, anatomic region, sex, hormonal levels, and certain genetic predisposition. For example, scalp hair are in the anagen cycle from 48–72 months, while thigh and leg hairs are in anagen from 1–6 months. It is these variations that lead to the tremendous disparity in hair cycles reported by various investigators [1–5].

Long-term hair removal requires a laser or light source impact on one or more growth centers of hair. Anecdotal approaches have suggested that the pluripotential stem cells of the bulge, dermal papilla, and hair matrix must be treated in the anagen cycle for effective hair removal [2]. If the damage is not permanent during this cycle, it has been suggested that follicles will move into the telogen stage as they fall out. Thus, all the follicles may become synchronized after the first laser treatment, providing the patient a temporary reduction of approximately three months. The hair follicle will then return to anagen based on the natural hair cycle. However, this time-honored theory of optimal anagen treatment times has been challenged by Dierickx et al., whose widely accepted findings suggest that anagen/telogen cycling does not have the significant impact on laser induced response as was earlier thought [7]. In accordance with this view, most clinicians have found that attempts to correlate hair removal efficacy with growth cycles to be fruitless.

Another view has emerged since a recent study by Orringer et al. examined the effects of laser hair removal on the immunohistochemical properties of hair follicles using both a 1064 nm Nd:YAG laser and an 800 nm diode laser [8]. After a single laser treatment, they found that the immunostaining properties of the follicle, including the bulge region, remained mostly unchanged. They concluded that laser hair removal may occur by a functional alteration of follicular stem cells, rather than the commonly viewed theory of light-source-based hair destruction.

Currently, there is no agreement on a definition for treatment-induced "permanent" hair loss. In addition, there are no studies evaluating the long-term durability of laser hair removal. Permanence, defined as an absolute lack of hair in a treated area for the lifetime of the patient, may be an unrealistic goal. Most researchers agree with Dierickx et al., who have proposed to define "permanent" hair loss as a significant reduction in the number of terminal hair after a given treatment, that is stable for a period longer than the complete growth cycle of hair follicles at any given body site [6]. If no hair regrows after this time period, it can be assumed that the growth centers have no capacity to recover from injury, and are not simply in telogen.

It is now accepted that almost any laser can induce at least a temporary hair loss. Fluences as low as 5 J/cm^2 can induce this effect, which tends to last 1–3 months. The mechanism of action appears to be an induction of catagen and telogen. Permanent hair reduction, occurring at higher fluences is seen in 80% of individuals, seen with a variety of light-based systems, and is fluence-dependent.

5.2 Laser and Light-Based Devices

5.2.1 Alexandrite Laser (Fig. 5.1)

A multitude of lasers and light sources have been shown to be effective for laser hair removal. The first effective light-based systems were 694 nm ruby lasers. Although highly effective for the removal of pigmented hairs on light-skinned individuals, such systems are now rarely used. Today most light-based hair removal is undertaken with 755 nm alexandrite, 800 or 810 nm diode, and 1064 nm Nd:YAG lasers as well as a variety of nonlaser intense pulsed light devices.

Finkel et al. were among the first group to evaluate the efficacy of the alexandrite laser in removing unwanted hair [9]. They treated 126 patients with an alexandrite laser over a 15-month period. All subjects were treated with a 2 msec pulse duration with fluences between 20–40 J/cm^2 (average of 25 J/cm^2) using a 7 mm spot size. Cooling of the epidermis was accomplished with a topical cooling gel. The total number of treatments varied between three and five sessions and treatments were spaced between 1 and 2.5 months. The average hair count taken before a second treatment was 65% of the hair, as compared to the numbers present at baseline. However, the numbers varied in different anatomic locations. As would be expected, there was progressive improvement with each laser hair removal session. The average amount of hair present three months after the final treatment was

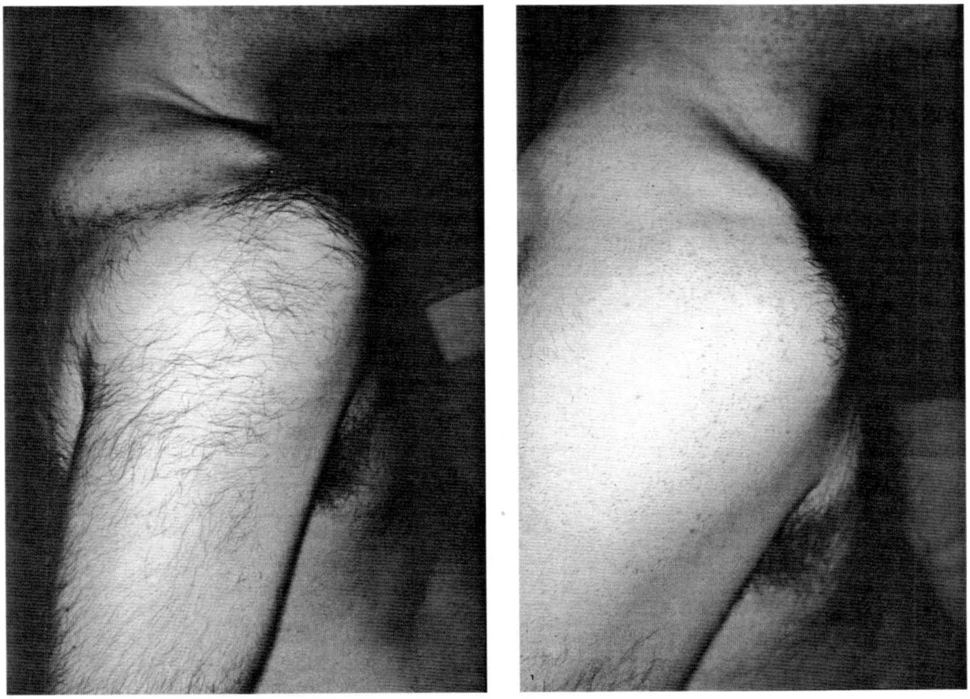

Figure 5.1 Unwanted hair before laser hair removal (left); six months after five alexandrite laser hair removal sessions (right).

markedly less than that seen after the first session. An average of only 12% of hair persisted at the final analysis (88% reduction). The best responding area was the sideburns (95% hair reduction) and the poorest responding area was the abdomen (75% hair reduction).

Narukar et al. evaluated both a 20 and a 5 msec pulse duration alexandrite laser in skin phenotypes IV–V [10]. All individuals were treated with fluences less than 20 J/cm^2. In this study, better results were obtained with the 20 msec pulse duration. Thus, longer pulse duration laser hair removal systems may be more beneficial, and clearly safer, in darker-complexioned individuals.

Nanni et al. also evaluated the hair removal efficacy of different alexandrite laser pulse durations [11]. In their 36 subjects study, they examined hair removal efficacy of a 5, 10, or 20 msec duration pulsed alexandrite laser. The mean fluence used was 18 J/cm^2. An average of 66% hair reduction was recorded at the one-month follow-up, 27% average hair reduction was observed at the three-month follow-up, and only a 4% hair decrease remained at the six-month follow-up visit. No significant differences were seen in hair regrowth rates between the different pulse durations. The authors noted that after the one treatment utilized in this study, there was on an average, no significant reduction in hair growth by the six-month follow-up. The authors concurred that multiple sessions of treatment are required for optimal results.

Jackson et al. reported similar findings in eight hair removal patients treated with Fitzpatrick skin phenotypes III–IV [12]. They used a 5 and 20 msec alexandrite laser and fluences between 14–20 J/cm^2. Both pulse durations led to equal hair removal efficacy. The same authors then evaluated the effect of a 20 msec versus a 40 msec alexandrite laser in a similar patient population. Fifteen subjects were treated with fluences varying between 12–17 J/cm^2. Although clinical response was similar, greater posttreatment pigmentary changes were observed in the 20 msec group compared to the 40 msec group. In addition, the authors noted greater pain when the longer pulse duration system was used.

Rogers et al. evaluated alexandrite laser hair removal in 15 subjects [13]. All were Fitzpatrick skin phenotypes I–III with blond or brown hair. They utilized a fluence of 22 J/cm^2 delivered in 20 msec pulse durations. The authors found that 80% of treated individuals had post-laser erythema, which lasted on average for 2–3 days; 47% showed perifollicular erythema which lasted on average for 90 hours. At two months, 55% of the hair was absent. However, at three months only 19% of the hair was absent. These findings might have been improved if only darker hair had been treated.

Touma and Rohrer evaluated a 3 msec alexandrite laser, used in conjunction with –30°C cryogen spray cooling [14]. They evaluated 21 subjects, 12–15 months after one treatment with average fluences of 33 J/cm^2. The presumed permanent hair reduction was noted to be 30% at this period. Furthermore, the authors noted a 29% reduction in the width of the remaining hair.

Avram et al. also evaluated the same 3 msec alexandrite laser, used in conjunction with a 30°C cryogen spray cooling [15]. Using a variety of fluences, he noted a 40–60% hair reduction after three treatments performed at 4–8 week intervals. It was interesting to note that 15% of treated individuals showed 80% hair reduction after three treatments and 15% of treated individuals showed less than 30% hair reduction after three treatments. This suggests that results can vary from individual to individual and from one anatomic region to the next. The findings are also consistent with anecdotal reports suggesting that there are rare individuals who, for unknown reasons, may not respond to laser hair removal.

Goldberg et al. compared the effect of pulse duration and multiple treatments on alexandrite laser hair removal efficacy in Fitzpatrick skin type I–III patients [16]. Fourteen subjects (3 men and 11 women) between the ages of 19 and 51 years were studied. All subjects had black or brown terminal hair. A pulse duration of 2 msec, and a repetition rate of 5 pulses per second was compared with an alexandrite laser with a pulse duration of 10 msec and a repetition rate of 3 pulses per second. An energy fluence of 25 J/cm^2 and a spot size of 7 mm were constant in both groups. Consecutive treatment and evaluations occurred at two to three month intervals for a total of three treatment visits. Six months after the last treatment, the average percentage of hair reduction was 33.1% for the 2 msec pulse duration and 33.9% for the 10 msec pulse duration alexandrite laser. These were statistically equivalent. It is worth noting that there was a slightly greater, albeit statistically insignificant, loss of thicker hair (such as those seen on the back of men) with the 10 msec alexandrite laser.

In 2001, Eremia at al. reported their results of a cryogen-cooled 3 msec alexandrite laser in 89 patients who received a minimum of three treatment sessions 4–6 weeks apart [17]. Fluences of 30–50 J/cm^2 were used. They noted a mean hair reduction of 74% with best results in patients with Fitzpatrick skin types I–IV [16]. Similarly, Lloyd using a 20 msec alexandrite laser achieved a 78% clearance one year after laser hair treatment [18].

5.2.2 Diode Laser

Diode lasers emit monochromatic laser light at a wavelength of 800 nm or 810nm [19–21]. Dierickx evaluated the effectiveness and safety of an 810 nm pulsed diode laser for the permanent reduction of unwanted hair. He studied 95 subjects composed mainly of Fitzpatrick II–III skin phenotypes and those with brown or black hair. Subjects were treated and examined at baseline, 1, 3, 6, 9, and 12 months after treatment. The authors evaluated one versus two treatments, single versus multiple-pulse treatments of the same area, and the fluence-response relationship. Pulse durations from 5–20 msec and fluences between 15–40 J/cm^2 were utilized in this study.

Treatment results demonstrated two different effects on hair growth: hair growth delay and permanent hair reduction. A measurable growth delay was seen in all patients (100%) at all fluence/pulse width configurations tested; this growth delay was sustained for 1–3 months. Significant fluence-dependent, long-term hair reduction occurred at all fluences in 88% of subjects. Clinically obvious long-term hair reduction usually required ≥30 J/cm^2. After two treatments at 40 J/cm^2, with a 20 msec pulse duration, the average permanent hair reduction at the end of the study (12 months after final treatment) was 46%. Two treatments significantly increased hair reduction as compared to one treatment, with an apparently additive effect. At a fluence of 40 J/cm^2, the initial treatment removed approximately 30% of terminal hairs, and the second treatment given one month later removed an additional 25%. Triple pulsing of the same area did not significantly increase hair reduction over single pulsing, but did increase the side effects. Additional findings included reduction in hair diameter and reduction in color of regrowing hairs (lighter hair).

In another study, Campos et al. treated 38 subjects with 1–4 treatments using an 800 nm diode laser [22]. They utilized fluences of 10–40 J/cm^2 and evaluated results four months after a final treatment. They noted that 59% of subjects had only sparse growth at the end of the study. Better results were seen after multiple sessions and with the use of higher fluences. However, this study was limited by its short final follow-up duration.

In a longer 20-month study, Lou et al. treated 50 Fitzpatrick skin types II–III subjects using an 800 nm diode laser with fluences of 10–40 J/cm^2 and pulse durations varying between 5–30 msec [23]. They noted increasing improvement with increasing numbers of sessions. Hair regrowth plateaued at six months with hair regrowth varying between 47% and 66% at the end of the study.

Other groups have also studied the diode laser for hair removal. Bouzari et al. studied the affect of treatment interval on hair reduction with an 800 nm diode laser [24]. All patients studied, received either two or three treatments. Patients treated at 45-day intervals had a 78.1% mean hair reduction, while those treated at 90-day intervals had only a 28.7% reduction. They concluded that shorter intervals were more successful at eliminating hair. Fiskerstrand et al. compared the efficacy between two different long-pulsed diode lasers in a split-face study [25]. They found the MedioStar® and Lightsheer® systems to be equal in efficacy. A different group studied the effect of various spot sizes on hair removal efficacy with a long-pulsed diode laser and found that a larger spot size (14 mm) was superior to a smaller spot size (8 mm) [26]. Finally, Adrian et al. found the long-pulsed diode laser safe and effective at removing hair in African-American patients [27].

5.2.3 Nd:YAG Laser (Fig. 5.2)

Bencini et al. were the first to evaluate hair removal efficacy with a long-pulsed millisecond Nd: YAG laser [28]. Such a system theoretically combines the pigmentation safety of a near-infrared laser with the photothermal benefits of millisecond pulsed technology. They studied 208 subjects who were treated during an 11-month period at multiple different areas. The vast majority (203 subjects) were Fitzpatrick skin phenotypes II–IV, while five were Fitzpatrick skin phenotype V. Hair colors were divided as follows: 124 subjects having dark hair, 2 with white hair, 78 with blond hair, and 4 with red hair.

The authors used a 3 or 4 mm spot size and fluences between 23 and 56 J/cm^2. In general, lower fluences were utilized for darker or finer hair and higher fluences for lighter or thicker hair. A single treatment resulted in a 20–40% hair loss of the treated area, lasting over 24 weeks. Higher fluences, however, also caused more discomfort. This led the subjects to often choose comfort over greater efficacy in their treatments.

In a similar study, a millisecond Nd: YAG laser was evaluated using 15–30 msec pulse durations and fluences of 50–60 J/cm^2 (S. Kilmer, personal communication). Twenty five subjects with 100 treatment sites were evaluated. Skin phenotypes I–V were evaluated; anatomic sites included the face, arms, legs, axilla, bikini line, and back. Response was assessed three months after a single treatment. The median hair count reduction three months after a single treatment was 32% for treatment parameters of 60 J/cm^2 and 30 msec; 24% for the treatment parameters of 50 J/cm^2 and 15 msec.

Tanzi and Alster used an Nd:YAG laser in 36 patients with types I–VI skin [29]. They performed three laser treatments at 4–6 week intervals using fluences of 30–60 J/cm^2. The pulse durations used were varied based on Fitzpatrick skin type and were as follows: 10 msec for skin types I–II; 20 msec for skin types III–IV; 30 msec for skin types V–VI. Hair counts were taken at one, three, and six months after treatment. Peak hair removal was

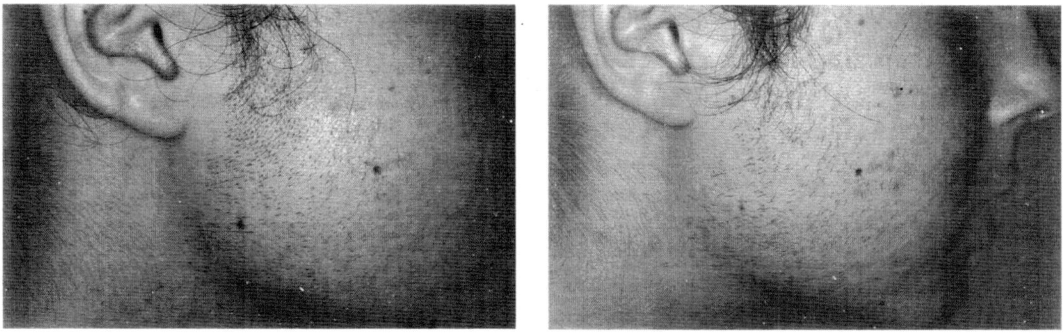

Figure 5.2 Unwanted hair before laser hair removal (left); four months after one Nd:YAG laser hair removal session (right).

seen one month after the final treatment with mean hair reduction of 58–62% seen on facial sites, and mean hair reduction of 66–69% seen on nonfacial sites. At six months, a 41–46% mean reduction was seen on facial sites; a 48–53% reduction was noted on nonfacial sites. This study confirmed that Nd:YAG lasers are safe even in the darkest of skin phenotypes.

Levy et al. treated 29 patients with a 4 msec 1064 nm Nd: YAG laser at fluences of 56–70 J/cm^2 and then evaluated for hair loss at three, six, and nine months after treatment [30]. They reported an average reduction at nine months of 46%. Finally, Raff et al. studied two different long-pulsed Nd:YAG laser systems on 42 patients with various pulse lengths, fluences, and spot sizes[31]. The average hair reduction, 12 months after the last treatment, was 48% with the Lyra XP® and between 30–35% with the Smartepil II®. They concluded that larger spot size and longer pulse duration improve efficacy.

5.3 Intense Pulsed Light

Nonlaser induced selective photothermolysis can also be used for hair removal and has been utilized with a filtered flashlamp intense pulsed light (IPL) sources. The IPL technology emits a broad spectrum of light that is filtered to allow release of varying emitted wavelengths of light. Cutoff fillers are utilized to select either shorter or longer wavelengths. A higher cutoff filter blocks many shorter wavelengths and thus is generally used when treating darker skin types.

So far, only a few investigators have evaluated this modality for hair removal (Table 5.1). Gold et al. published the first significant series of patients treated with IPL [32]. They evaluated hair removal efficacy in 31 subjects, the majority between 30 and 50 years of age. Although a variety of anatomic sites were treated, the most common areas were the neck (27%), lip (22%), and chin (19%). All sites were treated at the same time and evaluated 2, 4, 8, and 12 weeks after treatment. Treatment parameters varied according to the pigmentation of the skin and treated hair. Four cutoff filters were used: 590, 615, 645, and 695 nm. Fluences ranged between 34–55 J/cm^2 and were delivered in sequences of between 2–5 pulses, each pulse varying between of 1.5–3.5 msec in length. The authors reported that after a single treatment, approximately 60% hair removal was noted at 12 weeks. They concluded that IPL was a safe and effective method for long-term hair reduction.

Table 5.1 Summary of Studies Using IPL Sources for Hair Removal

Investigators	Details	Main Findings
Amin and Goldberg [36]	–10 subjects –Compared two different IPL parameters to alexandrite and diode lasers –Two total treatments, spaced monthly	–Hair removal efficacy was found to be identical with the IPL and both lasers –All four laser and light-based devices had ~50% hair reduction at 210 days
Toosi et al. [38]	–232 subjects –Randomized controlled trial –Compared diode laser, Alexandrite laser and IPL –Between three and seven treatments	–No statistically significant difference in efficacy between alexandrite laser (69%), IPL (67%) and diode laser (72%) at 6 months
Hee et al. 2006	–55 Asian subjects –Four treatments spaced 4–6 weeks –Two different IPL parameters utilized	–8 months after last treatment IPL I: 52.8% clearance; IPL II: 83.4% clearance –Efficacy for IPL II was statistically significantly higher than for IPL I
Marayiannis et al. [37]	–389 subjects –Retrospectively compared IPL vs. alexandrite laser –Multiple treatments	–No statistically significant difference in efficacy between alexandrite laser and IPL
Sadick et al. [34]	–34 subjects –Multiple treatments spaced monthly	–Mean hair reduction of 76% after a mean of 3 treatments –In subgroup of 14 patients followed > 12 months: 83% hair reduction (mean 3.9 treatments)
Weiss et al. [33]	–48 subjects –Two total treatments, spaced monthly	–Week 8: 42% hair reduction –6 Months posttreatment: 33% hair reduction
Gold et al. [32]	–31 subjects –Single treatment	–Approximately 60% hair removal was noted at 12 weeks

IPL = intense pulsed light.

Weiss et al. evaluated the efficacy of IPL hair removal in 48 subjects with Fitzpatrick skin types I–V [33]. All patients were treated two times, with a one-month interval between the treatments. Anatomic sites treated included both facial regions and nonfacial regions. At week eight, a 42% hair reduction was noted. At six months, hair reduction was found to be 33%. Sadick et al. also performed hair removal with an IPL device and had successful results [34]. They impressively reported an 83% hair reduction after a mean of 3.9 patients in 14 patients, who were followed up between 12 and 26 months after their last treatment.

Lee et al. studied the photo-epilatory effects of two different wavelengths of the same IPL device in an Asian population [35]. They treated 28 patients with a 600–950 nm filter and 27 patients with a 645–950 nm filter. Longer pulse widths were utilized in the 645–950 nm groups. Four treatments were performed at intervals of 4–6 weeks. Eight weeks after the final treatment, the average clearance was 52.8% (600–950 nm filter) and 83.4% (645–950 nm filter). They concluded that removing 45 nm of the emitted spectra and applying a longer pulse width provided a safer and more effective treatment in this population.

Several groups have compared IPL to laser sources commonly used for hair reduction. Amin et al. compared IPL hair removal with two different sets of parameters to laser hair removal with an alexandrite and diode laser. Hair reduction efficacy was found to be identical with the IPL and both lasers [36]. Another group retrospectively compared the efficacy of long- and short-pulse alexandrite lasers with an intense light source for photo-epilation in 389 patients [37]. Overall, no statistically significant difference in efficacy was found between patients treated with the alexandrite laser and those who received IPL treatment. However, those treated with an alexandrite laser required 6–8 treatments, while those who received IPL treatment required 8–9 treatments. Finally, Toosi et al. also compared IPL hair removal to the alexandrite and diode laser in 232 patients. Six months after the last session, they found no statistical difference between light sources [38].

5.4 Laser/Light-Based Removal of Nonpigmented Hair

A more different problem is the laser and light-based removal of nonpigmented hair. Since laser hair removal of unwanted hair is based on melanin absorption, blonde, and white hair have proved to be very difficult to treat. A variety of methods have been utilized with less impressive results than are seen with the removal of pigmented hair.

Recently, a group in Germany investigated the use of a topically applied liposomal melanin spray in 42 patients prior to diode laser treatment with the hope to provide a chromophore in patients with blonde, white or gray hair [39]. Overall, results were disappointing, and patients in the treatment group showed only a 14% reduction in hair six months after three treatment cycles. The authors concluded that this was not an efficacious or cost-effective treatment.

A different approach to target nonpigmented hair has been utilizing a combined light/bipolar radiofrequency device. Sadick and Laughlin treated 36 adults with white and blonde hair using a combined light/bipolar radiofrequency device [40]. They obtained an average hair reduction of 48%, six months after four treatments. In a similar study, Goldberg et al. evaluated a combined pulsed light bipolar radiofrequency device with and without pretreatment with a topical photosensitizer (5-ALA) [41]. Six months after two treatments, increased efficacy was found in the group who received pretreatment with topical 5-ALA (48% vs. 35% reduction in hair).

5.5 Conclusion

Many different light-based systems can successfully remove unwanted hair. The majority of laser and light-based systems used today appear to be equivalent in efficacy. The decision of which system to use is likely to be based largely on practitioner experience with a particular system and also on patient population treated. (Nd:YAG lasers are preferred by many for darker skin types.)

Although significant advances in light-based hair removal have been made, questions as to appropriate treatment intervals remain. It is clear that a simple understanding of anagen and telogen cycles does not adequately predict what time intervals between laser sessions are optimal.

Finally, home-based light-based hair removal systems are soon to be available throughout the world. These technologies have the potential to replace other temporary home methods for the removal of unwanted hair. It can be expected that home-based devices will deliver lesser energies than seen with the larger office-based units, making them quite safe. However, the lesser-delivered fluences will also make them less effective. How these home-based devices will affect the office-based laser hair removal market is yet to be determined.

References

1. Seago SV, Ebling FIG. The hair cycle on the human thigh and upper arm. *Br J Dermatol* 1985; 113: 9–16.
2. Goldberg DJ. Unwanted hair: evaluation and treatment with lasers and light source technology. *Adv Derm* 1999; 14: 115–39.
3. Cotsarelis G, Sun TT, Lavker RM. Label-retaining cells reside in the bulge area of pilosebaceous unit: implications for follicular stem cells, hair cycle and skin carcinogenesis. *Cell* 1990; 61: 1321–7.
4. Oliver RF. Whisker growth after removal of the dermal papilla and lengths of follicle in the hooded rat. *J Embryol Exp Morphol* 1966; 15: 331–47.
5. Olsen EA. Methods of hair removal. *J Am Acad Dermatol* 1999; 40: 143–55.
6. Dierckx CC, Grossman MC, Farinellli WA, et al. Permanent hair removal by normal mode ruby laser. *Arch Dermatol* 1998; 134: 837–42.
7. Dierickx C, Campos VB, Lin WF, Anderson RR. Influence of hair growth cycle on efficacy of laser hair removal. *Lasers Surg Med* 1999; 24(Suppl. 11): 21.
8. Orringer JS, Hammerberg C, Lowe L, Kang S, Johnson TM, Hamilton T, Voorhees JJ, Fisher GJ. The effects of laser-mediated hair removal on immunohistochemical staining properties of hair follicles. *J Am Acad Dermatol* 2006; 55(3): 402–7.
9. Finkel B, Eliezri YD, Waldman A, et al. Pulsed alexandrite laser technology for noninvasive hair removal. *J Clinical Laser Med Surg* 1997; 15: 225–9.
10. Narukar V, Miller HM, Seltzer R. The safety and efficacy of the long pulse alexandrite laser for hair removal in various skin types. *Lasers Surg Med* 1998; 18(Suppl. 10): 38.
11. Nanni C, Alster TS. Long-pulsed alexandrite laser-assisted hair removal at 5, 10, and 20 millisecond pulse durations. *Lasers Surg Med* 1999; 24: 332–7.
12. Jackson BA, Junkins-Hopkins J. Effect of pulsewidth variation on hair removal in ethnic skin. American Society for Dermatologic Surgery Meeting, May 1999, Miami Beach, FL.
13. Rogers CJ, Glaser DA, Siegfried EC, Glaser DA. Hair removal using topical suspension-assisted Q-switched Nd:YAG and long-pulsed alexandrite lasers: A comparative study. *Derm Surg* 2000; 22: 322–7.
14. Touma DJ, Rohrer TE. The 3 msec long pulse alexandrite laser for hair removal. American Society for Dermatologic Surgery Meeting, May 1999, Miami Beach, FL.
15. Avram M. Alexandrite laser hair removal. 8th International Symposium on Cosmetic Laser Surgery, March 1999, New Orleans, LA.

16. Goldberg DJ, Akhami R. An evaluation comparing short and long pulsed durations using the alexandrite laser for noninvasive hair removal. *Lasers Surg Med* 1999; 25: 223–8,.
17. Eremia S, Li CY, Umar SH, Newman N. Laser hair removal: Long-term hair results with a 755 nm alexandrite laser. *Dermatol Surg* 2001; 27: 920–4.
18. Lloyd JR, Mirkov M. Long term evaluation of a long pulsed alexandrite laser for the removal of bikini hair at shortened treatment intervals. *Dermatol Surg* 2000; 26: 633–7.
19. Dierickx CC, Grossman MC, Farinelli BS, et al. Hair removal by pulsed infrared diode laser. *Lasers Surg Med* 1998;10(Suppl.): 42.
20. Dierickx CC, Grossman MC, Farinelli BS, et al. Comparison between a long pulsed ruby laser and a pulsed, infrared laser system for hair removal. *Lasers Surg Med* 1998; 10(Suppl.): 42.
21. Grossman MC, Dierickx CC, Quintana A, et al. Comparison of various lasers for hair removal. *Lasers Surg Med* 1998; 10(Suppl.): 42.
22. Campos VB, Dierickx CC, Farinelli, W, Lin T, et al. Hair removal with an 800 nm pulsed diode laser. *J Am Acad Derm* 2000; 43: 442–7.
23. Lou WW, Quintana AT, Geronemus RG, Grossman MC. Prospective study of hair reduction by diode laser (800 nm) with long-term follow-up. *Dermatol Surg* 2000; 26: 428–33.
24. Bouzari N, Tabatabai H, Abbasi Z, Firooz A, Dowlati Y. Hair removal using an 800-nm diode laser: comparison at different treatment intervals of 45, 60, and 90 days. *Int J Dermatol* 2005; 44(1): 50–3.
25. Fiskerstrand EJ, Svaasand LO, Nelson JS. Hair removal with long pulsed diode lasers: A comparison between two systems with different pulse structures. *Lasers Surg Med* 2003; 32(5): 399–404.
26. Baumler W, Scherer K, Abels C, Neff S, Landthaler M, Szeimies RM. The effect of different spot sizes on the efficacy of hair removal using a long-pulsed diode laser. *Dermatol Surg* 2002; 28(2): 118–21.
27. Adrian RM, Shay KP. 800 nanometer diode laser hair removal in African American patients: A clinical and histologic study. *J Cutan Laser Ther* 2000; 2(4): 183–90.
28. Bencini PL, Luci A, Galimberti M, et al. Long-term epilation with long-pulsed neodymium:YAG laser. *Dermatol Surg* 1999; 25: 175–8.
29. Tanzi EL, Alster TS. Long-pulsed 1064-nm Nd:YAG laser-assisted hair removal in all skin types. *Dermatol Surg* 2004; 30: 13–17.
30. Levy J-C, Trelles MA, de Ramecourt M. Epilation with a long-pulse 1064nm Nd:YAG laser in facial hirsuitism. *J Cosmetic & Laser Ther* 2001; 3:175–9.
31. Raff K, Landthaler M, Hohenleutner U. Optimizing treatment parameters for hair removal using long-pulsed Nd:YAG-lasers. *Lasers Med Sci* 2004; 18(4): 219–22.
32. Gold MH, Bell MW, Foster TD, et al. Long term epilation using the EpiLight broad band, intense pulsed light hair removal system. *Dermatol Surg* 1997; 23: 909–13.
33. Weiss RA, Weiss MA, Marwaha S, et al. Hair removal with a non-coherent filtered flashlamp intense pulsed light source. *Lasers Med Surg* 1999; 24: 128–32.
34. Sadick NS, Weiss RA, Shea CR, Nagel H, Nicholson J, Prieto VG. Long-term photoepilation using a broad-spectrum intense pulsed light source. *Arch Dermatol* 2000; 136(11): 1336–40.
35. Lee JH, Huh CH, Yoon HJ, Cho KH, Chung JH. Photoepilation results of axillary hair in dark-skinned patients by IPL: A comparison between different wavelength and pulse width. *Dermatol Surg* 2006; 32: 234–40.
36. Amin S, Goldberg DJ. Clinical comparison of four hair removal lasers and light sources. *J Cosmetic & Laser Ther* 2006; 8: 65–8.
37. Marayiannis KB, Vlachos SP, Savva MP, Kontoes PP. Efficacy of long- and short pulse alexandrite lasers compared with an intense pulsed light source for epilation: A study on 532 sites in 389 patients. *J Cosmet Laser Ther* 2003; 5(3–4): 140–5.
38. Toosi P, Sadighha A, Sharifian A, Razavi GM. A comparison study of the efficacy and side effects of different light sources in hair removal. *Lasers Med Sci* 2006; 21(1): 1–4.
39. Sand M, Bechara FG, Sand D, Altmeyer P, Hoffmann K. A randomized, controlled, double-blind study evaluating melanin-encapsulated liposomes as a chromophore for laser hair removal of blond, white, and gray hair. *Ann Plast Surg* 2007; 58(5): 551–4.

40. Sadick NS, Laughlin SA. Effective epilation of white and blond hair using combined radiofrequency and optical energy. *J Cosmet Laser Ther* 2004;6: 27–31.
41. Goldberg DJ, Marmur ES, Hussain M. Treatment of terminal and vellus non-pigmented hairs with an optical/bipolar radiofrequency source – with and without pre-treatment using topical aminolevulinic acid. *J. Cosmetic & Laser Ther* 2005; 7 :25–8.

6

Removal of Unwanted Facial Hair

Pete Styczynski[1], John Oblong[1], and Gurpreet S. Ahluwalia[2]

[1]The Procter and Gamble Company, Cincinnati, OH, USA
[2]The Gillette Company, a wholly owned subsidiary of The P&G Company, Needham, MA, USA

Gurpreet S. Ahluwalia (ed.), *Cosmetic Applications of Laser and Light-Based Systems*, 157–179,

6.1 Introduction

Facial hair represents a significant psychological burden for many women around the world. Although excessive facial hair is often associated with disease states such as polycystic ovaries and hormonal imbalance, the range of facial hair from a vellus to terminal state is commonly found among women with no underlying hormonal etiology. Facial hair in women is perceived as unfeminine and unhealthy, and is associated with aging (P&G internal data). In addition, women are often not only bothered by the presence of facial hair but also frustrated with the removal process [1,2]. There are strong emotive responses as a result of facial hair that include lack of self-confidence, shamefulness, and being uncomfortable in social and intimate situations [3]. For these reasons, among others, the management and removal of female facial hair represents an unmet global consumer need. This chapter will focus on the overall contribution of facial hair to the perception of beauty, the biology of facial hair, and the current methods of removal with emphasis on laser technologies, as well as some future trends for removal of unwanted facial hair.

6.2 Overall Contribution of Facial Hair to the Perception of Beauty

While it has often been stated that beauty is in the eye of the beholder, for women there are not many more troubling conditions from a beauty perspective than the presence of facial hair. Attractiveness is largely driven by facial beauty, and is independent of ethnic extraction [4]. Several studies have demonstrated that the presence of facial hair on women is associated with a compromised quality of life. A report by Loo and Lanigan [5] suggested that excessive facial hair was similar to eczema and psoriasis and exceeded acne in terms of a dermatology life quality index score, indicating a strong negative impact. One of the most universal features desired by women, in particular, is flawless skin. From an evolutionary perspective, facial beauty is associated with good health and genetic makeup and, as such, promotes reproduction. The presence of hair on a woman's face, along with the skin texture, age, and facial shape, affects the perception of beauty, health, and reproductive capacity [4]. In addition, a negative self-perception arising from facial hair causes women to feel "abnormal" and "unfeminine". While these notions of beauty are largely reinforced in the popular media, they are based on an evolutionary bias.

The psychosocial affect of facial hair was most recently highlighted in a report by Lipton et al. [3] where 88 women in the United Kingdom were surveyed about their hair removal practices, and the emotional impact of facial hair. In this study, the impact of facial hair on the psychological and social well-being of women was explored. This work revealed that more than 80% of the women surveyed often or almost always put a "lot of effort into facial hair removal". Moreover, a similar overwhelming percentage is often or almost always "frustrated" by the process. Interestingly, Asian and mixed-race women were more likely to perceive the extent of their facial hair as being more severe than Caucasian women. The emotional impact of facial hair described in the work by Lipton et al. [3] indicated that the majority of the women were "bothered" and "worried" about their facial hair to the extent that it had a negative impact on their self-confidence, resulting in them being uncomfortable in social situations. In addition, feelings of shame and even anger were related to the presence of facial hair.

Consistent with what Lipton reported, internal data from P&G consumer insight studies on US females revealed that 65% of respondents (out of 423 subjects) reported having fine facial hair. In about a third of respondents, facial hair was found along the jawline, and another third indicated the presence of facial hair on the upper lip region. Fifty percent of respondents claimed to be bothered by fine facial hair, and 57% felt that fine facial hair makes them feel less feminine. In addition, 40% said facial hair makes them feel less confident. Not only does the presence of facial hair contribute to feelings associated with a lack of femininity, but the actual process of removing facial hair can also represent a source of such feelings. Some example comments from a nonpublished P&G consumer study include:

- I wish it was more feminine/pleasurable
- The experience of removing facial hair is not beautiful!
- Facial hair gets in the way of my beauty
- When I look in the mirror, I want to see me!
- I want others to see me for who I am, not for the facial hair

The study also suggested that removal of fine vellus hair on the cheek area and coarse upper lip hair rank as being among the top 10 global beauty needs. In addition, the study highlights a particularly low satisfaction with current hair removal methods for facial hair. While many women are interested in preventing or reducing the presence of facial hair, a P&G survey of 1098 women in the United States revealed that only about 4% are using products to treat facial hair, highlighting the gap in the ability to meet the needs of the consumer in this area.

A report by Housman et al. [6] noted the demographic differences with respect to the reporting frequency of excessive facial hair. Women in the 30-to-59-year age group were more likely to report excessive facial hair than women aged in the 18–29 and over- 60 age groups. An increase in facial hair as a function of age is often associated with a shift in the estrogen/androgen balance and, therefore, it was surprising to learn that the over-60 age group was less likely to report excessive facial hair. As pointed out by Housmann et. al., this may be a result of the life stage such as less likely to be pursuing career and relationships, in contrast to the 30–59 years age group. Overall, the study suggested that excessive facial hair is a common problem for many women, and has a negative impact on their quality of life.

While it has been well-documented that facial hair is common and bothersome to many women, the psychological impact of facial hair—especially in today's hypermedia world—cannot be overstated. In an article by Toerien et al. [2] entitled "Body Hair Removal: The 'Mundane' Production of Normative Femininity," the authors highlight the association between being hair-free and youthfulness, and attractiveness. Their findings were based on a study of 678 women residing in the United Kingdom with 86% between the ages of 17 and 50. Of this sample, 41% reported having depilated facial hair at some point, and a significant relationship was determined between the age factor and whether facial hair was ever removed, with 39% of the subjects in the 21–30 years age group reporting to have removed facial hair versus 54% of those in the 31–50 years age group, and 64% in the 51 years and older group. Consistent with similar studies, these data suggest that facial hair represents a problem for a large percentage of the female population which increases with age as assessed by the necessity for removal.

6.3 Survey of Facial Hair Removal/Management Methods

There are several means of removing facial hair that are currently practiced with shaving, plucking, waxing, and depilation being the main methods. While each method has certain advantageous attributes, the duration and acceptability vary and all these methods possess significant shortcomings in the context of a beauty regimen. Figures 6.1 and 6.2 show the distribution of hair removal method and the frequency of removal, respectively, on the upper lip as determined from a P&G Ipsos Insight Study with a basis size of 423 subjects. Consistent with what has been reported elsewhere in the literature, the major methods

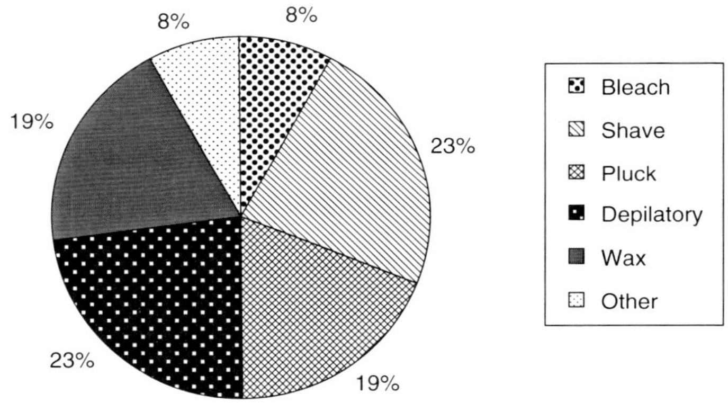

Figure 6.1 Distribution in commonly employed methods of hair removal from the upper region lip by women.

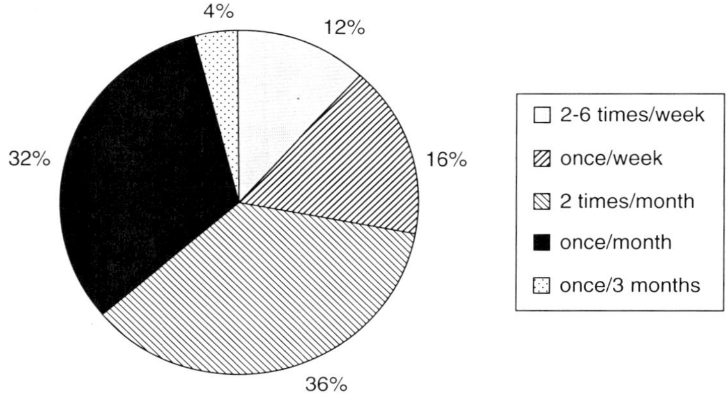

Figure 6.2 Frequency of hair removal from the upper-lip region by women.

of hair removal include shaving, plucking, depilation, and waxing. Eighty-four percent of the women surveyed reported removing hair at least once a month with 16% removing hair weekly.

6.3.1 Shaving

Shaving represents a fast and inexpensive method of facial hair removal with minimal side effects. Although there is no evidence to suggest that shaving influences the density of the hair, there is a common perception that hair regrowth after shaving is thicker [7]. This perception may be the result of the blunt ends that are produced by shaving. Irritation and razor bumps, as well as nicks and cuts can also arise from shaving. While this method of facial hair removal is quick and convenient, its duration is only a day or two and, moreover, there is a strong association of shaving the face and loss of femininity.

6.3.2 Depilation

Another common approach to facial hair removal as evidenced by the recent growth of products is depilation. Depilatories (typically thioglycolates in an alkaline formulation) remove the hair by chemically dissolving the disulfide bonds found between cysteine residues that make up the keratinized hair shaft [7]. This approach can be efficacious on an acute scale and longer lasting than shaving, since hair is removed to depths slightly below the surface of the skin. However, anywhere from 10–25% of individuals will have side effects that include skin irritation, chemical dermatitis, and even allergic dermatitis. In addition, a major drawback of depilatories is the unpleasant odor produced by hydrogen disulfide gas that is released as a by-product from the thiol chemistry during use.

6.3.3 Plucking/Waxing

Removal of hair by plucking is often the first method employed, and can be performed on a spot-treatment basis with tweezers, or over a larger area using an electronic device. With this technique, hair is removed at the root as the entire shaft and bulb are removed from the skin. Although the effect typically lasts for several weeks, this method is not practical for larger surface areas with a large number of hair, and there can be pain associated with the process. Another method of hair removal that is commonly used is waxing. Warm wax is applied to the skin and then removed, pulling hair out from the root. This procedure is used in the facial area, torso, and legs and lasts for several weeks, similar to plucking. Side effects common to this procedure include skin irritation, redness, as well as pain.

6.3.4 Electrolysis

Electrolysis was considered to be the first method available to obtain permanent hair removal. A fine-wired probe is placed into the hair follicle and a weak electric current is applied in an attempt to destroy the hair bulb region [7]. This method is time consuming and the rate of success is largely dependent upon the skill and technique of the electrologist.

In addition, this method is inconvenient, as multiple sessions are often required. In addition, there is potential for skin irritation and redness, as well as pain and scarring.

6.3.5 Drug Therapy

It is well-known that androgens have a potent influence on hair growth and the overall physical characteristics of hair [8,9]. For example, at the onset of puberty vellus hairs are converted to terminal hair in various body regions such as the axilla, genital area, legs, chest, and face—the latter two occurring preferentially in males. The contribution of androgens to hair growth is paradoxical, in that they have the opposite affect on the crown of the scalp in certain individuals—leading to male pattern baldness. While there have been studies which show differences in the sensitivity of hair follicles to androgens, it is not clear if the paradoxical response of hair follicles to androgens is mediated locally at the level of the hair follicle, or is the result of systemic androgens, or perhaps more likely, a combination of both.

An anti-androgen approach to treat unwanted facial hair that has been characterized as hirsutism is predicated on the blockade of the androgen receptor using spironolactone (SL). This agent has been successfully used to reduce the growth of androgen-dependent hair, but often requires high doses, as its affinity for the androgen receptor is more than an order of magnitude less than that of dihydrotestosterone (DHT), the potent form of testosterone [8,9]. There are side effects associated with SL therapy that range from milder conditions of gastritis and dry skin [8] to more severe ones such as polyuria and hypotension. In addition, SL is contraindicated in patients with renal insufficiency, hyperkalemia, and pregnancy. While the use of anti-androgens may be suitable for people with severe hirsutism, they are not useful for the control of facial hair that is not androgen-dependent, such as vellus hair.

A similar approach to controlling facial hair includes the use of 5α-reductase inhibitors. This enzyme reduces testosterone to the more potent dihydrotestosterone (DHT). Elevated 5α-reductase activity has been reported to be associated with idiopathic hirsutism [8] and, therefore, therapeutic strategies which block 5α-reductase seem plausible. Finasteride, an inhibitor of Type II 5α-reductase has been successfully developed by Merck & Co. (Proscar) as a systemic treatment for male-pattern baldness. In an attempt to capitalize on the paradoxical effects of androgens on hair growth, finasteride was evaluated as a treatment for hirsutism, but was shown to be minimally effective when compared with anti-androgens [8].

The only topical treatment approved by the FDA for the treatment of unwanted hair growth is eflornithine-HCl, which is marketed as Vaniqa (13.9% eflornithine). The molecule's mechanism of action involves the irreversible inhibition of ornithine decarboxylase, a rate-limiting enzyme required for the synthesis of polyamines [10]. While originally developed as a novel cancer therapeutic, cancer clinical trials with difluoromethylornithine (eflornithine) showed low efficacy, but was found to be well-tolerated among patients [11]. Increased ODC and DNA synthesis was shown in murine hair follicles [12], and related to the onset of the anagen hair growth cycle stage [13]. These insights gave rise to the notion of hair growth inhibition by polyamine analogs and Shander and Ahluwalia [10] identified the utility of eflornithine (difluoromethylornithine) as a hair-growth inhibitor, which was subsequently brought to the market place through collaboration with Bristol-Myers Squibb.

6.3.6 Laser Treatment

Unlike the methods of hair removal discussed earlier, laser hair removal has the capacity to destroy the cells in the hair follicle bulb and provide a longer lasting, and in some cases, permanent effect. The potential for eliminating hair growth on the face is especially important, as the process of managing facial hair has strong negative connotations, and implies masculinity.

Laser hair removal is based on the ability of melanin in the cells of the hair follicle to convert the energy emitted by the laser into heat and, thereby, give rise to the local destruction of the hair follicle. Since eumelanin acts as the primary chromophore in the hair follicle and high levels are critical to the efficacy of the laser, variability of the response is largely dependent upon the degree of pigmentation of the hair follicle. Dark hair is known to respond well, whereas lightly pigmented and gray hair is much less responsive. Along these same lines, the potential for skin adverse effects as a result of laser treatment is also dependent upon the level of skin pigmentation. Thus, individuals with darker skin are more susceptible to burns than lighter skinned patients and, therefore, the selection of laser settings with proper pulse width and duration are critical to avoid these negative side effects. A more detailed discussion of laser-mediated hair growth management is included in Section 6.5.

6.4 Facial Hair Biology

6.4.1 Hair Follicle Structure, Growth Characteristics, and Regulation

Human body hair can be defined as three major types: lanugo, vellus, and terminal hair. Lanugo hair is present on infants and is lost shortly after birth. In contrast, vellus and terminal hair remain as the major hair type on the human body. There are clear physical distinctions between these hair types. Vellus hairs are finer hairs, nonpigmented, and range in size from 0.01 cm thick to < 2 cm in length and are present on all body parts with the exception of plantar skin (palms and soles). In contrast, terminal hairs are coarser, pigmented, and average in thickness, and 0.05 cm and >2 cm in length. In adults, terminal hair is found on the scalp, limbs, axilla, and male face [14]. The latter three regions represent areas from which consumers remove hair on a nearly daily basis.

The hair follicle is a dynamic skin structure that is the product of numerous cumulative biological processes that permit constrained rapid cell proliferation similar to that found in the epithelial lining in the intestinal system, and which at the same time gives rise to the highly differentiated keratinized hair fiber. The growth of hair is cyclical and includes an active growing stage (anagen) followed by a transitional stage (catagen), and a resting stage (telogen). At any given time, about 90% of terminal facial hair is in anagen in males, which may last up to several years. In contrast, the majority of vellus hair on the face is in telogen. In the upper lip and chin regions of women, terminal hair may be androgen-sensitive similar to the male beard and contribute to the hirsute phenotype, whereas dark vellus type hairs on the cheek may be a function of genetic predisposition, and independent of androgens. Table 6.1 compares hair growth rates and diameters across body sites (P&G internal data).

Anatomically, the hair follicle has been segmented into four major regions: the infundibulum, the sebaceous gland region, the isthmus, and the bulb [14]. The infundibulum is

Table 6.1 Growth Rate and Diameter of Hair Follicles from Various Body Sites

Body Site	Average Growth Rate (mm/day)	Average Diameter Range (μm)
Vellus	0.03	14–25
Upper lip	0.05	16–53
Female lower leg	0.17	40–100
Male beard	0.35	40–150
Scalp	0.35	40–150

contiguous with the surface of the skin and extends to the insertion site of the sebaceous duct. This region of the hair follicle is where the naked hair fiber emerges from the follicle as well as the path of sebum secretion from the sebaceous gland to the surface of the skin. The sebaceous gland region contains the lobular sebaceous gland full of cells filled with lipid to be secreted into the infundibulum. The isthmus is the middle region of the follicle from the sebaceous gland to the bulge area that is near the insertion point for the arrector pili muscle. It is in this region that the inner root sheath disappears. The highly metabolic bulb region includes the epithelial and germinative matrix cells that envelope the mesenchyme-derived dermal papilla cells. In the anagen phase of the hair cycle, the rapidly proliferating matrix cells are in close physical contact with the dermal papilla. This contact is lost as the hair cycle progresses through catagen and into the resting telogen stage. The hair follicle bulb also includes melanocytes which provide melanin for incorporation into the hair fiber. Coupled with the rapid proliferation rate of the hair follicle germinative cells are keratinization processes that give rise to the highly differentiated hair fiber and the inner and outer root sheaths. The size of the hair fiber has been suggested to be a function of the volume of the dermal papilla compartment [15].

The hair growth cycle is characterized by marked changes in metabolic activity as well as morphology, and includes an active growth phase (anagen) in which the hair bulb surrounds the dermal papilla and has a high mitotic index. In humans, anagen hair follicles descend to the level of the hypodermis in such a way that their bulb regions are highly vascularized and immersed in fat—potentially to support the high energy requirements required by the growing follicle. Following a period of active growth that varies by body site and age—among other factors—the follicle moves through a transitional stage (catagen) where the follicle regresses in a way that the dermal papilla and matrix cells become separated and disengaged. The process is characterized by a high level of proteolytic and apoptotic activity. Telogen is the resting phase of the hair cycle, and represents a quiescent hair follicle showing little or no cell proliferation or metabolic activity. It remains unclear what controls the cycle, although numerous studies have elegantly demonstrated the contribution of a variety of cytokines in its regulation [16–18].

6.4.2 Regulation of Facial Hair Growth

Both males and females have hair follicles present on the face. At puberty hair follicles on the cheeks, chin, and upper lip regions of the male face transition from vellus to terminal hair, as characterized by an increase in hair follicle diameter and pigmentation. This transition is driven by systemic androgen activity as well as by the local conversion of testosterone to dihydrotestosterone (DHT) as catalyzed by 5α-reductase, and results in the prolongation of anagen in the majority of male facial hair follicles in these regions. The sensitivity of hair follicles to androgens is variable within an individual and may be independent of systemic androgen levels. The intracrine effect of androgens is highlighted by the fact that the conversion of testosterone to DHT is catalyzed by 5α-reductase in the skin and hair follicle. At least two isoforms of 5α-reductase are localized in the skin with the acidic, Type II isoform thought to be the primary form in beard hair follicles [19]. The actions of androgens on hair growth are extensively reviewed elsewhere [20].

Clinical hirsutism has been defined by Ferriman and Gallwey based on the density of terminal hair at eleven different body sites that included the upper lip and chin as well as the areas on the torso and extremities. While the presence of terminal facial hair in women may result from a genetic predisposition based on ethnic extraction, an altered hormonal balance such as in polycystic ovary syndrome (PCOS) or from undefined causes as in the case of idiopathic hirsutism [8], it is clear that the subjective assessment of facial hair has created a dynamic range for which methods of removal and management are desired. Thus, it is important to view facial hair beyond the clinical definition of hirsutism to include both terminal and vellus hair to the extent that it represents a concern, and impacts self-perception.

6.4.3 Facial Hair Phenotypes and Regional Differences

Female facial hair can present itself as a coarse terminal hair in a male-type pattern in the upper lip and chin regions (hirsutism), or as a finer vellus-type on the chin and cheeks. The characteristics of vellus-type hair in terms of its density and pigmentation are more commonly associated with genetic extraction but, nonetheless, are largely unwanted. It has been suggested that the size of the dermal papilla compartment is correlated with the size (diameter) of the hair follicle [21]. More recently, Elliot et al. [15] reported that the dermal papilla volume from hair follicles derived from terminal male facial hair follicles was nearly 40 times greater than that of female facial vellus follicles. Moreover, the increase in dermal papilla volume was associated with a 17-fold increase in the number of cells and a 2.4-fold increase in the volume of each cell. These data suggest that the transition of a follicle from vellus to terminal hair phenotype or vice versa is, at least in part, related to the number of cells and/or the average extracellular matrix volume in the dermal papilla compartment.

The majority of clinical evaluations and products research to date has been focused upon identifying technologies that impact terminal hairs. This hair is pigmented to varying shades and is what consumers shave or manage via other means on a daily to weekly basis. Among females, body sites may include underarms, lower legs, bikini areas, as well as upper lip and chin. In contrast, vellus hair is present on all parts of the body with the exception of plantar skin and is somewhat less noticeable, since it is generally nonpigmented, and does not grow as aggressively, nor as long as terminal hair. It is typically referred to as "peach fuzz" and Western consumers are fearful of the myth that if they shave this hair, it will transform into a pigmented terminal hair and will worsen the condition.

It has been suggested that an average of 10% of vellus hair follicles are actively growing (anagen) and 90% are in the resting phase (telogen). However, it has been reported [22] that female cheek vellus hairs are in a 44:56 ratio of anagen to telogen, in contrast to 90:10 ratio. The ratio of 44:56 is similar to what has been determined for female forearm hairs, which are very fine terminal hairs with relatively low pigmentation content (P&G unpublished data).

Published technical literature on vellus hair is very limited. There have been some histological evaluations of vellus hair follicles that found unique attachments of the erector pili muscle to the outer root sheath of the hair, particularly in facial vellus hairs. This difference is morphologically ascribed as a "skirt" structure that is different than that present in terminal hair. In addition, there is a high concentration of CD34+ cells localized around the "skirt", suggesting the presence of bulge cells as found in terminal hair. The physiological role of the skirt, however, remains unclear.

In an attempt to characterize facial vellus hair, we analyzed via scanning through an electron microscope, the hair diameter and cuticle patterning of individual facial vellus hair collected from the middle jawline areas of nine panelists (10 hair per panelist). Two of the nine panelists had jawline hairs that were pigmented. While these hair were more similar to terminal hair in appearance, their overall length on the face suggests vellus like growth properties. Figure 6.3 highlights the hair diameter values calculated and the graph shows diameter values from all nine panelists, including the two with pigmented jawline hairs. The hair diameter ranged from about 12 μm to just over 30 μm with an average of 22 μm for all hair. Although the panel size was small, these preliminary findings suggest that vellus hair diameters are significantly smaller than the typical diameter of terminal hairs present on scalp and body (50–150 μm).

Scanning electron microscopy (SEM) was used to visualize hair fibers from vellus and terminal hair. The SEM images below compare Type I and Type II upper lip hairs to facial vellus and scalp terminal hairs. Type I hair resemble terminal scalp hair, not only in diameter but also in cuticle pattern, whereas Type II hairs appear more like vellus hair regarding these parameters (Fig. 6.4). These insights may serve as a basis for the identification of different technologies or regimens toward the removal of unwanted facial hair.

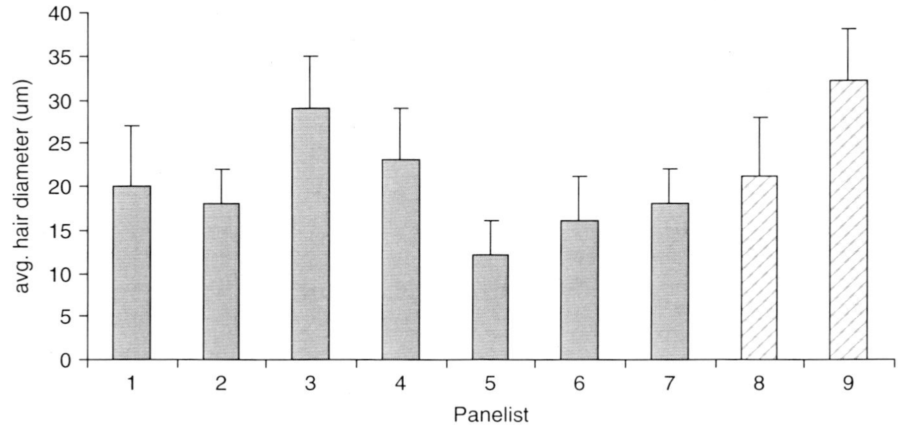

Figure 6.3 Diameter determination from vellus facial hair collected from the jawline. (hash bars represent pigmented hairs)

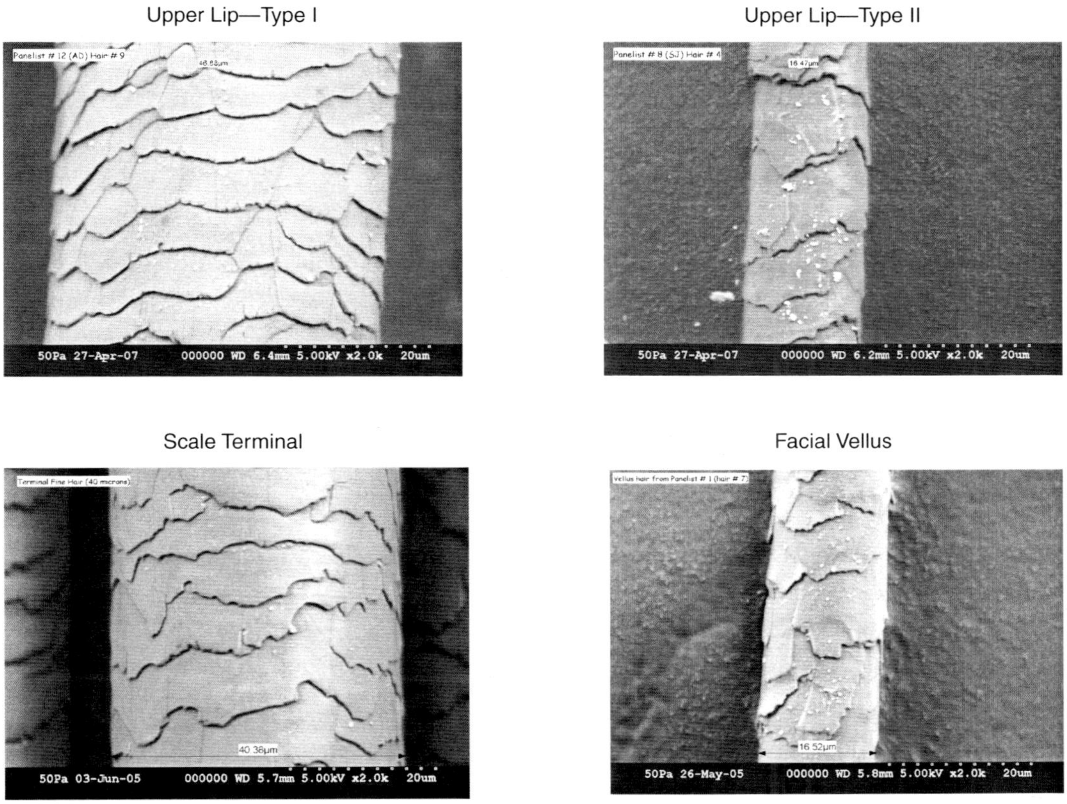

Figure 6.4 A SEM comparison of Type I and Type II upper lip hair fibers with scalp and vellus facial hair.

To assess the efficacy of chronic technologies on vellus hair, a study was conducted at Bower Research Center (Boulder, CO) in 65 women (40 placebos; 25 Vaniqa) with upper lip hair. The effect of Vaniqa™ on Type I and Type II facial hair was evaluated following twice daily application for 12 weeks. Half-face digital images and Hi-Scope images (Fig. 6.5) were captured at baseline, 4, 8, and 12 weeks. The study also included self-assessments and focus groups. Vaniqa™ elicited a 51% (significant at $p < 0.05$) reduction of average growth rate in Type I hair at week 4, which was maintained through weeks 8 and 12 (Fig. 6.6). With Type II facial hair, Vaniqa elicited as much as a 35% reduction in average growth rate at week 4, with a significant reduction at weeks 8 and 12 (Fig. 6.7).

Earlier analysis of human facial vellus hair showed that the average diameters were significantly less than human terminal hair. While terminal hairs range from 40 to 150 μm, vellus hairs were typically 20–27 μm in diameter, with the higher diameters being from pigmented vellus hair. It was noted that the spacing of the cuticle plates on the outside of the hair was larger on vellus hair when compared to terminal hair. To address this issue, we provided STEM visualization of vellus hairs and compared them with images of terminal hair. The number of cuticle layers ranged from 2 to 5 for vellus hairs, whereas for terminal hairs it ranged from 5 to 12 (Fig. 6.8) and the diameter measurements as noted earlier were

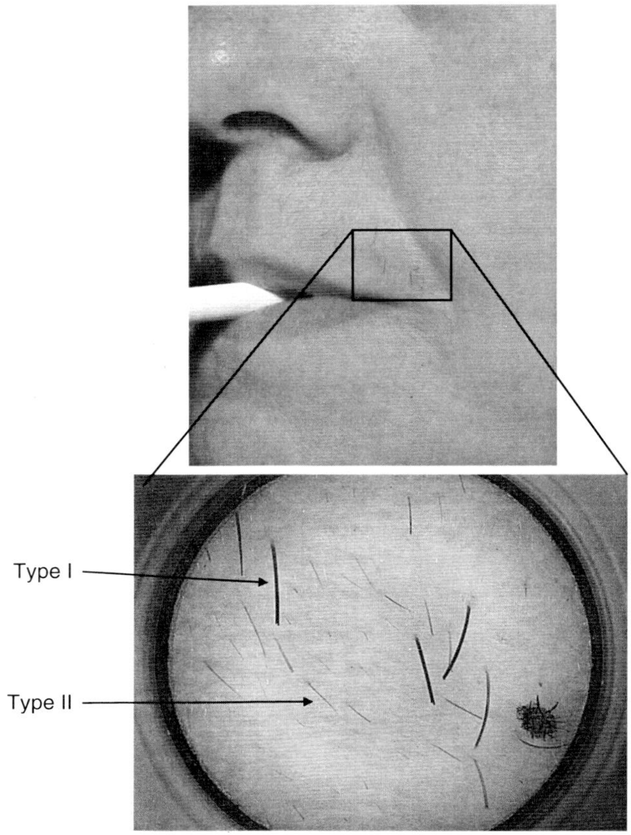

Figure 6.5 Depiction of REAL and Hi-Scope Type I and Type II hair from the upper lip.

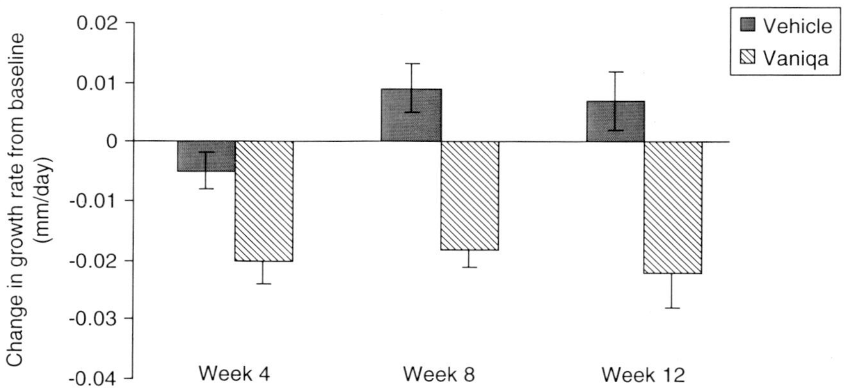

Figure 6.6 Effect of Vaniqa™ on Type I facial hair based on Hi-Scope image analysis.

confirmed. Thus, female facial vellus hair can be viewed as having an overall geometrical and structural differential that is approximately half of that seen with coarser terminal hairs. It was hypothesized that the seemingly weaker nature of vellus hairs could render them susceptible to a depilatory formula with lower thioglycolate levels and pH ranges. This could translate to an effective depilatory that may have a lower irritation profile. This

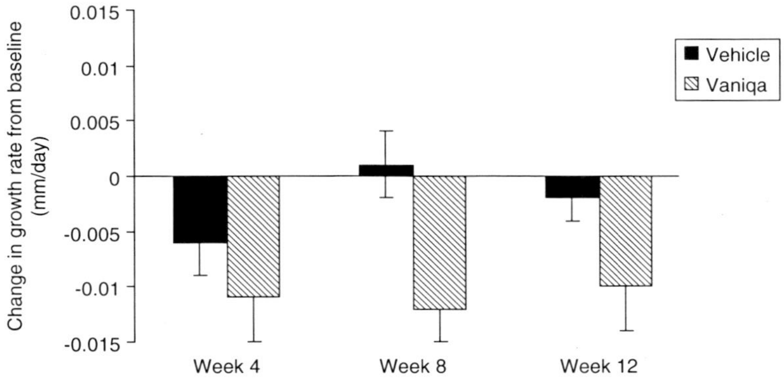

Figure 6.7 Effect of Vaniqa™ on Type II facial hair based on Hi-Scope image analysis

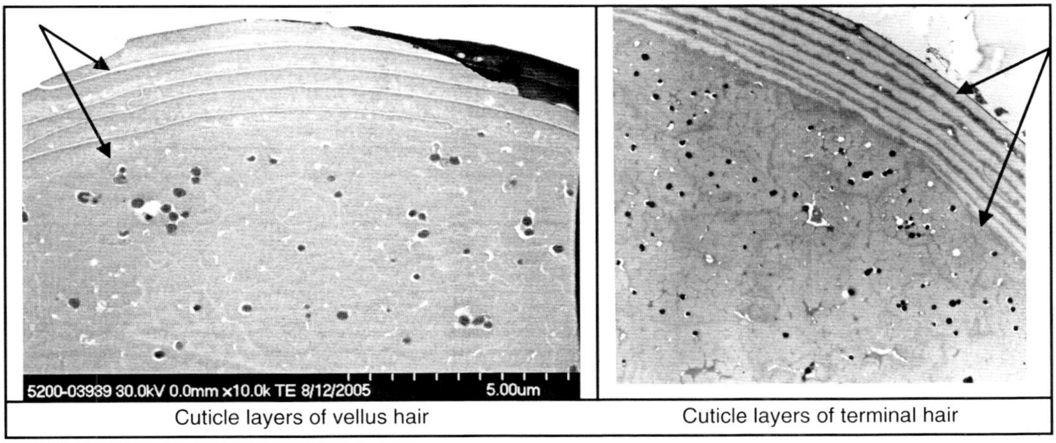

Figure 6.8 STEM of a partial cross section from a vellus hair and terminal hair with arrows depicting the cuticle layers.

may also hold true for pigmented vellus hair since the diameters are still significantly less than those of terminal hair.

While not a common occurrence, vellus and terminal hair are able to transition from one state to another based on dramatic changes as hormonal alterations, severe dietary alterations, laser therapy, or as a drug-induced side effect. Historically, it has been viewed that vellus hair on the body exists as a fairly dormant follicular unit. The cycle has been described as very slow with only 10% of hair being in an active growing state at any given time, with the remaining being in the telogen resting phase. In addition, it was thought that the hair represents a slower growth population and is not androgen-dependent like terminal hairs. However, efforts by Blume et al. [22] more closely examined these population distributions between vellus and terminal hairs and differences between male and females. Several key observations include:

- Greatest density of vellus hair was on forehead and cheeks, followed by back, shoulder, and chest
- Forty-six to forty-nine percent of vellus hair on cheeks and forehead were growing

- Vellus hair showed a 4–10-fold slower growth rate than terminal hair.
- Overall hair growth was not affected by sex or age, even though the maximum length of vellus hair decreased as a function of aging.
- No correlation between sebum excretion rates and vellus hair growth was found.

In summary, the findings from this study set the stage for establishing a model for both identifying key technical drivers for consumer noticeability of unwanted facial hair, and establishing the success criteria for identifying acute and chronic intervention technologies. This will allow for calculation of what percentage inhibition of hair growth must occur from a topical in order to significantly shift the timing for appearance of vellus hair that signals a "need to manage" point.

6.5 Contribution of Laser-Based Technologies to Facial Hair Removal

6.5.1 Facial Hair—A Key Area of Concern for Women

Among body areas for which laser hair removal is indicated, facial hair removal is most popular. An automated phone survey of over 5000 women conducted by the Gillette group in the Boston area (Needham, MA) had 145 respondents who indicated having received professional laser or light-based treatments for hair removal. This represents about 3% market penetration for laser hair removal procedure. Reported estimates for the US market range from 1 to 2%. Of the 145 respondents, 54% had received laser/light-based treatments on the face as at least one of the treatment sites (Fig. 6.9). A recent study of women living with facial hair [3] showed a high level of emotional distress and psychological morbidity

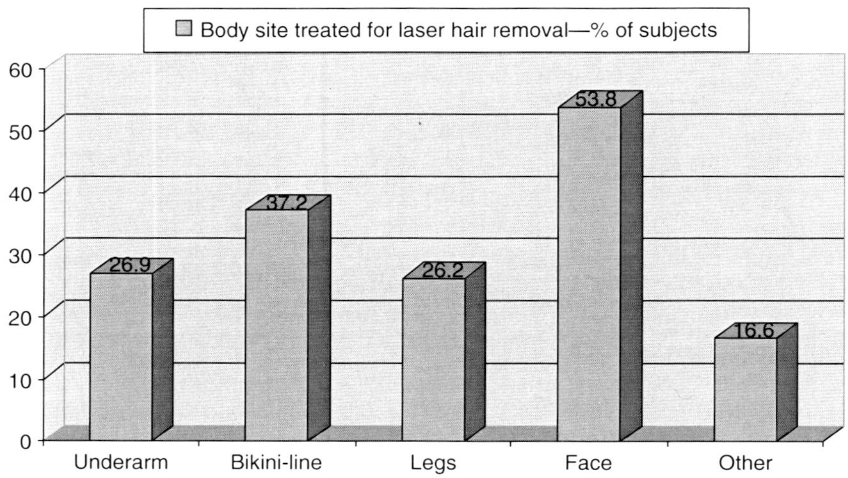

Figure 6.9 Percentage of respondents (of 145 people surveyed) treated for laser hair removal across various sites.

in this population, so it is not surprising that facial hair removal remains one of the key areas for which women seek treatment. Women feel overwhelmed with the effort they need to invest to control the growth of facial hair, resulting in their low self-esteem.

6.5.2 Types of Professional Hair Removal Lasers

Based on the principle of selective photothermolysis [23], several lasers have been developed to selectively affect the hair follicle. Eumelanin in the hair follicle acts as a chromophore that can absorb the laser radiation of a select wavelength, converting it into thermal energy. Light-based systems for hair removal are either laser-based, a single wavelength coherent beam of light, or are intense pulsed light (IPL) systems based on noncoherent beam of light composed of multiple wavelengths. Four types of lasers, differentiated by the wavelength they produce, are used for hair removal: ruby (694 nm), alexandrite (755 nm), diode (800 nm) and Nd:YAG (1064 nm). There are several IPL systems available with varied range of wavelengths. By appropriately selecting the laser wavelength, pulse duration, and fluence energy, critical follicle components responsible for hair growth, that is, follicle matrix, papilla, and stem cell population can be thermally injured while sparing the surrounding skin tissue. Other factors that can affect the efficacy outcome are hair-growth cycle, the proportion of growing or anagen hair at the time of treatment, the amount of melanin in hair follicle, and the depth of the hair follicle in skin. Generally, the longer wavelengths penetrate deeper into skin than the shorter wavelengths [24]. In addition to the characteristic wavelength, the laser aperture size plays a role in determining the depth of penetration of laser energy. The larger spot size allows for deeper penetration of light into skin.

6.5.3 Clinical Efficacy for Facial Hair Removal

6.5.3.1 Efficacy Determination

The clinical efficacy of laser hair removal systems is determined by the amount of hair remaining after completing a course of laser treatments. For facial hair, treatment regimens generally range from four to eight treatments given at a 3–6-week interval. Even though dramatic and stable reduction can be achieved, complete hair eradication by laser is rarely possible [25–28]. The hair remaining after treatment and/or any new hair growth tends to be softer, thinner, and less pigmented, which adds to the aesthetic benefits, and thereby results in a high level of subject satisfaction [26,27,29]. To obtain objective data, especially for comparative purposes, the amount of hair is determined by using a manual hair-count method. As hair density is rarely uniform on the whole treatment area, the counts must be performed on the same site before and after treatments to get a valid difference. Natural skin land marks or implanted micro-dot tattoos can be used to get back to the same site during the course of treatment. In conjunction with the hair count, the rate of hair growth after treatment can be used to access efficacy [30]. These objective methods provide a conservative measure of effectiveness, as laser treatment often results in altering the character of the regrowing hair, which is not captured in the hair count or the growth rate difference, but rather adds to the benefits perceived by the subjects. Some researchers count only the terminal hair whose diameter and color is similar to the hair present before the treatment. This is considered a good representation of efficacy, since the sparser, finer, and the lightly pigmented hair remaining generally do not bother the subjects. In addition to these objective measures, the patient satisfaction of the outcome can be used to gauge the treatment benefits [26,27,29]. Though patient satisfaction is an important measure, it is also highly subjective and dependent on the patient's expectations of the results and the actual results achieved.

6.5.3.2 Clinical Efficacy Data for Facial Hair

McDaniel et al. [31] reported long-term efficacy of alexandrite laser treatment on the upper lip, leg, back, and bikini area in 22 patients with Fitzpatrick skin types I–III. The hair removal efficacy varied with the site, pulse duration, and the number of treatments administered. A spot size of 10 mm, and a fluence of 20 J/cm^2, was compared at pulse durations of 5, 10, and 20 msec. A single pulse of 10 msec duration at 20 J/cm^2 was found to be most effective. Hair reduction at six months post single dose laser treatment was 40, 56, 50, and 15% for the upper lip, leg, back, and bikini area, respectively. The hair reduction on the upper lip increased from 40 to 54% when a second treatment was administered at 8 weeks after the initial treatment.

Women suffering from PCOS generally have an increased level of androgen hormone that results in terminal hair growth on the face, somewhat similar to the male beard, referred to as facial hirsutism. A randomized controlled trial was conducted in 88 PCOS patients using an alexandrite laser at a pulse width of 20 msec, and spot size of 12.5 mm [32]. The study was split into the high fluence (23.6 J/cm^2) treatment group and the low-fluence (4.8 J/cm^2) sham control group. Each group received treatments for six months at 4–6 week intervals. The treatment group reported a significant reduction in the severity of facial hair, time spent on hair removal, as well as reduction in depression and anxiety. The sham control group showed minimal hair reduction benefit.

In another study [5], the quality of life measures were studied in 45 hirsute patients receiving laser hair removal treatments with long-pulsed ruby, diode, or long-pulsed alexandrite lasers. The patients completed dermatology life quality index (DLQI) questionnaires at various time intervals during the six-month study. A major improvement in DLQI scores was observed at one to two months after treatment, but longer-term benefit was not observed. There was a high level of patient satisfaction (71%) and willingness to undergo further treatments (78%), despite the fact that by 6 months after treatment, facial hair growth had returned to the pretreatment levels in almost all patients (97%). In this study, there was no evidence of permanent hair reduction on face.

A prospective study of 38 subjects (21 women, 17 men) was conducted by Campos et al. using an 800 nm diode laser [33]. Face (37%) and back (21%) were the two most common areas treated. The fluence level used was in the range of 20–40J/cm^2 for skin types I–IV, and from 10–20 J/cm^2 for the darker skin types V and VI. The mean number of laser treatments administered was 2.7 (range, 1–4). The clinical outcome data showed the face to be more responsive than the back; however the difference was not statistically significant. A long-term stable reduction in hair (permanent reduction) was seen in a majority of the patients at a mean fluence of 33.4 J/cm^2 receiving a mean 2.8 treatments. Multiple treatments had an additive effect on efficacy.

The studies by Campos [33] and others [34–37] have shown the efficacy benefit of using higher fluence levels in achieving a stable long-term reduction (permanent) of hair. This can be more safely achieved in skin types I–IV, whereas in darker skin types (Fitzpatrick V/VI), there is a risk of causing serious dermal injury at the higher fluence levels. It is generally accepted that at a set fluence, the dermal adverse events are proportional to skin pigmentation, especially for skin types IV–VI. The risk of dermal injury can be reduced by providing adequate cooling to the epidermis during treatment, and by using longer wavelengths. Levy [34] treated 29 women with facial hirsutism representing all Fitzpatrick skin types using a long-pulse (4 msec) Nd:YAG (1064 nm) laser at fluence range of 56–70 J/cm^2 and a spot size of 3 mm. Objective hair reduction at 3, 6, and 9 months after 3 monthly administered treatments was 43, 36, and 46%, respectively. This stable long-term reduction was similar to the results obtained by Sommer [38] who treated 43 patients with Fitzpatrick skin type I–IV for hirsutism using a near maximal tolerated dose of the ruby laser. The mean fluence exposure for this group was 48 J/cm^2. Upper lip, chin, and neck were the three areas most often treated. There is a good literature agreement on greater dermal safety of Nd:YAG compared to other lower wavelength lasers for the darker skin types V/VI [34,39,40].

6.5.3.3 Treatment of Lighter Hair Color

Although the laser energy has been shown to be an effective means of reducing hair growth on the face, its efficacy is limited by the requirement for the heat-absorbing chromophore, melanin. Thus, facial hair that is not darkly pigmented is less susceptible than darker hair, resulting in uneven efficacy within an individual and variability across the population. For hair follicles with very low pigmentation such as in blonde, gray, or extremely light hair, the laser system is generally ineffective. Two approaches have been explored to treat lightly-pigmented or unpigmented hair; one is related to targeting preformed melanin to hair follicles using liposomal preparations, and the other is related to combining a melanin responsive Laser/IPL with a photo-thermal energy that does not require melanin as chromophore, such as a radiofrequency (RF) source.

Sadick et al. studied the combination of IPL/RF technology in 40 patients with varied hair colors [41]. There were at least 11 subjects with blonde, red, and white color hair in the study. The face was the most common site, treated with ten enrolled subjects. Four treatments were administered over a period of 9–12 months at 8–12 week intervals. The fluence applied for IPL ranged from 15–26 J/cm^2 depending on the patient's skin and hair color, and the RF energy range was 10–20 J/cm^2 depending on the treatment site. The average clearance recorded at 18 months was 80–85% for brown and black hair and 40, 60, and 60% for white, red, and blonde hair. The average clearance on the face was 65%. The treatments were well-tolerated by the patients, with no significant dermal adverse events. The study provides evidence that unpigmented or lighter hair can be effectively treated with the IPL/RF combination.

Although there are several literature reports on the targeted delivery of macromolecules to hair follicle, its utility in man is yet to be proven. Melanin is a large polymer molecule which is simply too large to penetrate the skin using conventional topical formulation. Attempts have been made to enhance melanin penetration by encapsulating it in a liposomal preparation. To enhance the lasers' efficacy for gray and blonde hair, not only does melanin need to penetrate the skin surface, it also needs to be delivered in sufficient amounts to the base of the hair follicle, which has been challenging to achieve. A recent study by Sand et al. [42] looked at the effect of topical preparation of encapsulated melanin on hair growth inhibition of white/gray hair by laser (three cycles of 800 nm diode laser at 28–40 J/cm). Hair counts were taken eight weeks after each laser treatment and six months after the last treatment. A statistically significant 14% reduction was observed with the melanin group versus 10% with placebo. These findings were not considered to be clinically significant.

6.5.4 Adverse Effects

6.5.4.1 Dermal Safety

Based on the hair removal mechanism of various lasers, the severity and frequency of dermal side effects are directly related to the fluence dose of the laser energy and the skin pigmentation levels. For safe and effective hair removal treatment, the treating clinician generally adjusts the laser parameters for each subject based on the hair and skin color. In addition, the contact surface material (typically sapphire) of the hand piece, skin cooling, and the clinician experience all play a part in providing a safe and effective treatment. Generally, the amount of laser energy (fluence) delivered is selected based on the patient's Fitzpatrick skin type, or the skin pigmentation level. Care is taken to avoid excessive heating of the epidermal surface. An actively cooled sapphire tip may be used to provide heat conduction from epidermis before, during, and after each laser pulse. For treatment, the hand piece is firmly placed on the skin for about 0.5 s to cool the surface (epidermis can be cooled in 0.2 s). By effectively protecting the epidermis, cooling allows higher laser energy to be delivered to the deeper follicle target. Gentle compression of the hand piece further brings the target follicle closer to the skin surface and is thought to result in a greater efficacy of the treatment. Compression can also blanch the underlying blood vessels, thus minimizing laser absorption by the competing chromophore oxyhemoglobin providing greater epidermal safety. The ruby, alexandrite, and diode laser systems are roughly equivalent in their efficacy and dermal safety profile, with a slight preference for ruby when treating lighter skin and darker hair, and for Nd:YAG when treating darker skin. Nd:YAG is slightly less efficient because of its lower melanin absorption index but is much safer on the darker skin types V and VI.

6.5.4.2 Eye Safety

One of the safety concerns related to facial hair treatment is the use of laser near the eye or the periorbital area. A direct laser light exposure to open eye carries the risk of serious injury to retina, including permanent blindness. The laser irradiance level in most epilation procedures can easily exceed the maximal permissible levels (MPE) for retinal injury in as short as few millisecond exposure time. Therefore, it is highly recommended that the patients, the treating clinicians, and others in the treatment room wear eye safety goggles that are appropriate for the intensity and the wavelength of the laser being used. Though the risk for retinal injury can be effectively managed with proper eye protection, the risk remains for the pigmented eye structures such as iris during treatments near the eye where the laser light can reach the interior of the eye indirectly by scatter and diffusion through the periocular tissue. The light that enters the bony orbit can get absorbed by the darkly pigmented iris. A case report on damage to iris structure after laser eyelid epilation without the use of protective eye shield was reported by Herbold [43]. A case of diode laser-induced iris atrophy was reported by Brilakis [44]. In this study, a similar hair removal procedure was performed on the eyelid, again without proper eye protection [44]. Pham performed laser removal of eyelashes on five subjects using metal eye shields placed behind the eyelids and found no evidence of eye damage assessed immediately and 3–6 months after treatment [45].

6.5.4.3 Paradoxical Hair Growth

The laser hair removal treatment in some patients has been found to result in a terminal growth of hair in untreated areas in close proximity to the treated ones [46–50]. In most cases, this paradoxical hair growth occurs at a site adjacent to the treatment area that has high vellus growth, and is relatively free of terminal hair. Face is the most common site where the terminal hair growth stimulation has been noted. It is considered a rare side effect of laser treatment, and the most susceptible population is of Mediterranean descent with darker skin types.

In a retrospective analysis of 750 patients of Mediterranean ancestry, with 4374 subjects administered laser treatments over a 7-year period, terminal hair induction was noted in 30 (4%) subjects [51]. In 28 of the 30 cases, the terminal hair growth was on the face. There was no relation as to which laser system or energy fluence was used. The majority of the subjects who experienced hair stimulation were of skin types III or IV.

In another retrospective study of 543 patients treated over five years in a dermatology clinic in Spain, the incidence rate for hair growth stimulation was much higher at 10.5% [52]. The treatments were administered on the face: cheek, chin, and neck. Increased terminal hair growth was noted both within the treatment area and in areas adjacent to the treated site. Majority of the subjects in this study were Type III (68%), although Type II (13%) and Type IV (19%) were also present. There was no correlation between the hair growth stimulation and the skin type. Researchers in this study believe that the hair stimulation was related to subtherapeutic dose of the laser reaching the follicles bordering the treatment area. To this end, the researchers treated additional 200 subjects over a two-year period with concomitant use of cold packs placed on areas surrounding the treatment site, and have not had a single incidence of hair stimulation since.

Unlike the studies from Greece [51] and Spain [52], a retrospective study from Canada involving a similar patient group size of 489 the incidence rate for terminal hair stimulation was found to be only 0.6% (3/489 patients) [53]. All three subjects had Type IV skin. The study concluded that though hypertrichosis is real, it is a rare occurrence in the Canadian population studied.

In a study of 49 patients with facial hirsutism with polycystic ovarian syndrome, 5 of the 49 subjects treated with an IPL source showed terminal hair stimulation [46]. All subjects had Type III skin and the study was conducted in Spain. This 10% incidence rate was similar to the larger study from Spain by Willey [52]. Though paradoxical hair growth has mostly been seen on face, a case study after diode laser treatment on a man's back was reported by Bernstein [54].

Isolated cases of localized hypertrichosis have been noted under conditions that result in dermal trauma, stress, or inflammation. Increased localized hair growth has been reported after fracture cast application [55], bug bites [56], site surrounding a burn [57], or scar [58], and site of chronic inflammation [59,60]. An excellent review on different forms and causes of hypertrichosis has been written by Wendelin [59]. Since the laser-induced paradoxical hair growth has mostly been associated with the Mediterranean population, it is not clear what the underlying cause is for this stimulation.

6.6 Future Trends—Synergy with Cosmetic Technologies

Lasers represent a major advance in the control of unwanted hair growth. However, the constraints around pigmentation and acceptable energy levels for "in-home" use may constrain the potential of lasers for widespread commercial success. Thus, the use of chemical technologies in combination with energy-emitting devices may represent an opportunity for meeting the threshold consumer needs and overcoming pigment and energy limitations. A recent report by Hamzavi et al. [61] evaluated the combination of eflornithine cream and laser on facial hair removal. In this double-blinded study, women received laser hair removal on the upper lip region. The right- or left upper lip received eflornithine cream or placebo control in a randomized fashion. Laser treatments were conducted up to six times every four weeks, using a long pulse alexandrite laser (10–40 msec pulse duration) at fluences of 7–40 J/cm^2. Using both subjective assessments by investigators and subjects as well as objective hair counts, laser in combination with eflornithine produced greater efficacy than laser treatment alone. In addition, the perceptible onset of efficacy was earlier in the combination group. This study suggests that the combination of laser hair removal with chemical approaches to hair growth control may provide an optimal approach to the management of facial hair.

Another study combined the use of radiofrequency-derived energy with topical aminolevulinic acid (ALA) [62]. This study compared six months of pulsed light bipolar radiofrequency with or without pretreatment with ALA. Unwanted facial hair was treated twice at four- and six-week intervals with half of the radiofrequency treatment area pretreated with topical ALA. This study reported the removal of terminal white hair from 35% with the radiofrequency device alone to 48% when ALA was used in combination. These data suggest that topically applied chemical technologies can augment the hair removal efficacy of energy-emitting devices.

Similarly, Sand et al. [42] reported the results of a study that investigated the use of melanin-encapsulated liposomes with laser therapy for the removal of blond, white, and gray hair. In this study, liposomal melanin was applied prior to laser treatment. Although the combination of the melanin and laser were more effective than treatment with the laser alone, the overall clinically observed reduction in hair growth was too weak to warrant further development of the liposomal melanin formulation.

Together these studies suggest that the combination of laser hair removal and topical hair growth reduction may provide a most favorable regimen for managing facial hair. Achieving optimal efficacy, onset of activity, and the duration of response will have the best opportunity to not only satisfy the consumer, but to improve their lives.

References

1. Jackson, J., Caro, J.J., Caro, G., Garfield, F., Huber, F. Zhou, W., Lin, C., Shander, D., Schrode, K. and the Eflornithine Study Group. The effect of eflornithine 13.9% cream on bother and discomfort due to hirsutism. *Int. J. Dermatol.* 2007; 46: 976–981.

2. Toerien, M., Wilkinson, S. and Choi, P. Body hair removal: the "mundane" production of normal feminity. *Sex Roles* 2005; 52: 399–406.
3. Lipton, M.G., Sherr, L., Elford, J., Rustin, M.H.A. and Clayton, W.J. Women living with facial hair: the psychological and behavioral burden. *J. Psychosomatic Res.* 2006; 61: 161–168.
4. Fink, B. and Neave, N. The biology of facial beauty. *Int. J. Cos. Sci.* 2005; 27: 317–325.
5. Loo, W.J. and Lanigan, S.W. Laser treatment improves quality of life of hirsute females. *Clin. Exp. Dermatol.* 2002; 27: 439–441.
6. Housman, T.S., Derrow, A.E., Snively, B.M., Lahiry, S., Rapp, S.R., Hawes, D.F., Grummer, S.E., Feldman, S.R., Feischer, A.B. and McMichael, A.J. Women with excessive facial hair: a statistical evaluation and review of impact on quality of life. *Cos. Dermatol.* 2004; 17: 157–165.
7. Shenenberger, D.W. and Utecht, L.M. Removal of unwanted facial hair. *Am. Fam. Phys.* 2002; 66: 1907–1911.
8. Azziz, R., Carmina, E. and Sawaya, M. Idiopathic hirsutism. *Endocrine Rev.* 2000; 21: 347–362.
9. Azziz, R. The evaluation and management of hirsutism. *Clinical Gynecologic Series: An Expert's View* 2003; 101: 995–1007.
10. Shander, D. and Ahluwalia, G.S. Management of unwanted facial hair by topical application of eflornithine. In: *Cosmeceuticals and Active Cosmetics*, Elsner, P. and Maibach, H.I., eds. Taylor & Francis: Boca Raton, FL, 2005, pp. 489–510.
11. Marton, L.J. and Pegg, A.E. Polyamines as targets for therapeutic intervention. *Ann. Rev. Pharmacol. Toxicol.* 1995; 35: 55–91.
12. Probst, E. and Krebs, A. Ornithine decarboxylase activity in relation to DNA synthesis in mouse interfollicular epidermis and hair follicles. *Biochem. Biophys. Acta.* 1975; 407: 147–157.
13. Nancarrow, M.J., Nesci, A., Hynd, P.I. and Powell, B.C. Dynamic expression of ornithine decarboxylase in hair growth. *Mech. Dev.* 1999; 84(1–2): 161–164.
14. Abell, E. Embryology and anatomy of the hair follicle. In: *Disorders of Hair Growth: Diagnosis and Treatment*, Olsen, E., ed. McGraw-Hill: New York, NY, 1994.
15. Elliot, K., Stephenson, T.J. and Messenger, A. Differences in hair follicle dermal papilla volume are due to extracellular matrix volume and cell number: implications for the control of hair follicle size and androgen responses. *J. Invest. Dermatol.* 1999; 113: 873–877.
16. Boivin, W.A., Jiang, H., Utting, O.B. and Hunt, D.W.C. Influence of interleukin-1a on androgen receptor expression and cytokine secretion by cultured human dermal papilla cells. *Exp. Dermatol.* 2006; 15: 784–793.
17. Hoffman, R., Eichler, W., Huth, A., Wenzel, E. and Happle, R. Cytokines and growth factors influence hair growth in vitro. Possible implications for the pathogenesis and treatment of alopecia areata. *Arch. Dermatol. Res.* 1996; 288: 153–156.
18. Foitzik, K., Lindner, G., Meuller-Roever, S., Maurer, M., Botchkareva, N., Botchkarev, V., Handjiski, B., Metz, M., Hibino, T., Soma, T., Paolo Dotto, G. and Raus, R. Control of murine hair follicle regression (catagen) by TGF-β1 in vivo. *FASEB J.* 2000; 14: 752–760.
19. Thigpen, A.E., Silver, R.I., Guileyardo, J.M., Casey, M.L., McConnell, J.D. and Russell, D.W. Tissue distribution and ontegeny of steroid 5α-reductase isozyme expression. *J. Clinical Invest.* 1993; 92: 903–910.
20. Randall, V.A. Androgens and human hair growth. *Clinical Endocrinol.* 1994; 40: 439–457.
21. Van Scott, E.J. and Ekel, T.M. Genomic relationships between the matrix of the hair bulb and its dermal papilla in normal and alopecic scalp. *J. Invest. Dermatol.* 1958; 31: 281–287.
22. Blume, U., Ferracin, J., Verschoore, M., Czernielewski, J.M. and Schaefer, H. *Brit. J. Dermatol.* 1991; 124: 21–28.
23. Anderson, R.R. and Parrish, J.A. Selective photothermolysis: precise microsurgery by selective absorption of pulsed radiation. *Science* 1983; 220: 524–527.
24. Boulnois, J.L. Photophysical processes in recent medical laser development: a review. *Laser Med. Sci.* 1986; 1: 47–66.
25. Haedersdal, M. and Wulf, H.C. Evidence-based review of hair removal using lasers and light sources. *J. European Acad. Derm. Venereolo.* 2006; 20: 9–20.
26. Bjerring, P., Zacharia, H., Lybecker, H. and Clement, M. Evaluation of the free-running ruby laser for hair removal—a retrospective study. *Acta. Derm. Venereol. (Stockh)* 1998; 78: 48–51.

27. Solomon, M.P. Hair removal using the long-pulsed ruby laser. *Ann. Plast. Surg.* 1998; 41: 1–6.
28. Connolly, C.S. and Paolini, L. Study reveals successful removal of unwanted hair with LPIR laser. *Cosmet. Dermatol.* 1997; 10: 38–40.
29. Preston, P.W. and Lanigan, S.W. Patient satisfaction with laser hair removal. *J. Cosmet. Dermatol.* 2004; 17: 68–72.
30. Liew, S.H., Grobbelaar, A.O., Gault, D.T. and Linge, C. Objective measurement of rate of hair growth following laser hair removal. *Cosmet. Dermatol.* 1999; 12: 18–20.
31. McDaniel, D.H., Lord, J., Ash, K., Newman, J. and Zukowski, M. Laser hair removal: a review and report on the use of the long-pulsed alexandrite laser fro hair reduction of the upper lip, leg, back and bikini region. *Dermatol. Surg.* 1999; 25: 425–430.
32. Clayton, W.J., Lipton, M., Elford, J., Rustin, M. and Sherr, L. A randomized controlled trial of laser treatment among hirsute women with polycystic ovary syndrome. *Br. J. Dermatol.* 2005; 152: 986–992.
33. Campos, V.B., Dierickx, C.C., Farinelli, W.A., Lin, T.D., Manuskiatti, W. and Anderson, R.R. Hair removal with an 800-nm pulsed diode laser. *J. Am. Acad. Dermatol.* 2000; 43: 442–447.
34. Levy, J.L., Trelles, M.A. and Ramecourt, A.D. Epilation with a long-pulse 1064 nm Nd:YAG laser in facial hirsutism. *J. Cosmet. Laser Ther.* 2001; 3: 175–179.
35. Dierickx, C.C., Grossman, M.C., Farinelli, B.S., et al. Hair removal by pulsed infrared diode laser. *Laser Surg. Med.* 1998; 22(Suppl. 10): 42.
36. Dierickx, C.C., Grossman, M.C., Farinelli, B.S., et al. Comparison between a long pulsed ruby laser and a pulsed infrared laser system for hair removal. *Laser Surg. Med.* 1998; 22(Suppl. 10): 42.
37. Grossman, M.C., Dierickx, C.C., Quintana, A., et al. Removal of excess body hair with an 800 nm pulsed diode laser. *Laser Surg. Med.* 1998; 22(Suppl. 10): 42.
38. Sommer, S., Render, C. and Sheehan-Dare, R. Facial hirsutism treated with the normal-mode ruby laser: results of a 12-month follow up study. *J. Am. Acad. Dermatol.* 1999; 41: 974–979.
39. Kilmer, S.L. Laser hair removal with the long pulse 1064 nm Coolglide™ laser system. *Laser Surg Med.* 2000; 26(Suppl. 12): 84.
40. Bencini, P.L., Luci, A., Galimberti, M. and Ferranti, G. Long-term epilation with long-pulsed Nd:YAG laser. *Dermatol. Surg.* 1999; 25: 1–4.
41. Sadick, N.S. and Shaoul, J. Hair removal using a combination of conducted radiofrequency and optical energies—an 18-month follow-up. *J. Cosmet. Laser Ther.* 2004; 6: 21–26.
42. Sand, M., Bechara, F.G., Sand, D., Altmeyer, P. and Hoffman, K. A randomized, controlled, double-blinded study evaluating melanin-encapsulated liposomes as a chromophore for laser hair removal of blond, white and gray hair. *Ann. Plas. Surg.* 2007; 58: 551–554.
43. Herbold, T.M., Busse, H. and Uhlig, C.E. Bilateral cataract and corectopia after laser eyelid epilation. *Ophthalmology* 2005; 112: 1634–1635.
44. Brilakis, H.S. and Holland, E.J. Diode-laser-induced cataract and iris atrophy as a complication of eyelid hair removal. *Am. J. Ophthalmol.* 2004; 137: 762–763.
45. Pham, R.T.H., Tzekov, R.T., Biesman, B.S. and Marmor, M.F. Retinal evaluation after 810 nm Dioderm laser removal of eyelashes. *Dermatol. Surg.* 2002; 28: 836–840.
46. Moreno-Arias, G., Castelo-Branco, C. and Ferrando, J. Paradoxical effect after IPL photoepilation. *Dermatol. Surg.* 2002; 28: 1013–1016.
47. Moreno-Arias, G., Castelo-Branco, C. and Ferrando, J. Side-effects after IPL photoepilation. *Dermatol. Surg.* 2002; 28: 1031–1034.
48. Marayiannis, K.B., Vlachos, S.P., Savva, M.P. and Kontoes, P.P. Efficacy of long- and short-pulse alexandrite lasers compared with an intense pulsed light source for epilation: a study of 532 sites in 389 patients. *J. Cosmet. Laser Ther.* 2003; 5: 140–145.
49. Vlachos, S.P. and Kontoes, P.P. Development of terminal hair following skin lesion treatments with an intense pulsed light source. *Aesthetic Plast. Surg.* 2002; 26: 303–307.
50. Bouzari, N., Tabatabai, H., Abbasi, Z., Firooz, A. and Dowlati, Y. Laser hair removal: comparison of long-pulse Nd:YAG, long-pulsed alexandrite, and long-pulsed diode lasers. *Dermatol. Surg.* 2004; 30: 498–502.

51. Kontoes, P.P., Vlachos, S.P., Konstantinos, M., Anastasia, L. and Myrto, S. Hair induction after laser-assisted hair removal and its treatment. *J. Am. Acad. Dermatol.* 2006; 54: 64–67.
52. Willey, A., Torrontegui, J., Azpiazu J. and Landa, N. Hair stimulation following laser and intense pulsed light photo-epilation: review of 543 cases and ways to manage it. *Laser Surg. Med.* 2007; 39: 297–301.
53. Alajlan, A., Shapiro, J., Rivers, J.K., MacDonald, N., Wiggin, J. and Lui, H. Paradoxical hypertrichosis after laser epilation. *J. Am. Acad. Dermatol.* 2005; 53: 85–88.
54. Bernstein, E.F. Hair growth induced by diode laser treatment. *Dermatol. Surg.* 2005; 31: 584–586.
55. Leung, A. and Kiefer, G.N., Localized acquired hypertrichosis associated with fracture and cast application. *J. Nat. Med. Assoc.* 1989; 81: 65–76.
56. Tisocco, L.A., Del Campo, D.V., Bennin, B. and Barsky, S. Acquired localized hypertrichosis. *Arch. Dermatol.* 1981; 117: 127–128.
57. Shafir, R. and Tsur, H. Local hirsutism at the periphery of burned skin. *Br. J. Plast. Surg.* 1979; 32: 93.
58. Gupta, S., Gupta S., Kanwar, A.J. and Kumar, B. Hypertrichosis surrounding scar of knee replacement surgery. *J. Am. Acad. Dermatol.* 2003; 50: 802–803.
59. Wendelin, D.S., Pope, D.N. and Mallory, S.B. Hypertrichosis. *J. Am. Acad. Dermatol.* 2003; 48: 161–179.
60. Olsen, E.A. Hypertrichosis. In: *Disorders of Hair Growth*, 2nd ed. McGraw Hill: New York, NY, 2003, pp. 401–430.
61. Hamzavi, I., Tan, E., Shapiro, J. and Lui, H. A randomized bilateral vehicle-controlled study of eflornithine cream combined with laser treatment versus laser treatment alone for facial hirsutism in women. *J. Am. Acad. Dermatol.* 2007; 57: 54–59.
62. Goldberg, D.J., Marmur, E.S. and Hussain, M. Treatment of terminal and vellus non-pigmented hairs with an optical/bipolar radiofrequency energy source—with and without pre-treatment using topical aminolevulinic acid. *J. Cosmet. Laser Ther.* 2005; 7: 25–28.

7

Synergy of Light and Radiofrequency Energy for Hair Removal

Neil S. Sadick [1] ***and Rita V. Patel*** [2]

[1] *Department of Dermatology, Weill Medical College of Cornell University, New York, NY, USA*

[2] *Department of Dermatology, University of Miami Miller School of Medicine, Miami, FL, USA*

7.1 Introduction

The process of unwanted hair removal has made technological strides over the past decade to the point where permanent results can be accomplished by irreversibly damaging hair follicles via the use of lasers [1]. While less expensive methods of hair removal such as plucking, waxing, and chemical dermabrasion are still popular, there has been a strong consumer movement to longer-lasting techniques [2–5]. The search for a more enduring method with compelling results across all hair colors and skin types remains the main challenge facing the cosmetic dermatologist today. However, with the advent of nonablative devices, which have been successfully engineered to treat various aesthetic demands such

Gurpreet S. Ahluwalia (ed.), *Cosmetic Applications of Laser and Light-Based Systems*, 181–194,

as wrinkles and displeasing pigmented lesions [6], hair removal using lasers and intense pulsed light (IPL) have begun to yield promising results and are set to become a safe, efficient, and reliable method for long-term hair removal [7–12].

Hair follicles have three main phases of growth. Anagen is the period of active growth, catagen is the involutionary stage, and telogen is the resting period. The anagen phase represents the best targeted stage for treatment, as the degree of melanin in the follicle is maximal, allowing for greater energy absorption with subsequent heat production that damages the hair follicle [14]. Melanin, the main target chromophore of light energy, is located in follicles of the terminal hair that extend deep into adipose tissue lying 2–7 mm below the skin surface, depending on the body site [13,14].

Removal of hair by light can be accomplished, in theory, by three mechanisms: photothermal destruction through local heating, photomechanical destruction through the generation of shock waves, or photochemical destruction through the creation of toxic mediators such as singlet oxygen or free radicals [15]. Laser hair removal is accomplished though follicular unit destruction [16]. The ability to remove hair without damaging the surrounding skin is based on selective photothermolysis: the theory that at a particular wavelength, pulse duration, and fluence, thermal injury is confined to the target, containing a light-absorbing molecule, known as a chromophore [17].

Photoepilatory devices are guarded under the principle of thermokinetic selection by which, target structures of large volume (i.e., hair) are not able to transmit absorbed energy to secondary structures, versus smaller-volume structures of the same chromophore (i.e., epidermis) [18]. Therefore, an appropriate pulse length of laser or light energy must be chosen that will selectively heat the target structure and the neighboring desired targets, for example the hair papillae, the germinative cell layer, and bulge areas of the hair follicle [7]. The duration of impulse must be selected to be above the thermal relaxation time of the epidermis, and below the thermal relaxation time of the targeted cell, the hair follicle [19].

The average thermal relaxation time of the epidermis is 3–10 ms, whereas the hair follicle thermal relaxation time varies from 80 to 100 ms [19]. In addition, hair follicle depth, diameter, density, and cyclical phase variations differ depending on the anatomic site of the body; for instance, 85% of the hair on the scalp is in the anagen phase [19]. The wavelength, fluence, depth of penetration, pulse duration, and spot size are variables characteristic to each laser system, and must be taken into account when choosing the appropriate nonablative modality for permanent hair removal [19].

Permanent hair reduction associated with long-term photoepilation requires light-induced interaction with the primary bulge and secondary matrix germinative regions of the pilosebaceous unit, which include areas from the bulge down to the matrix [7]. It has been emphasized that both the hair follicle and the bulge area are important for hair growth, as studies have shown that the mere destruction of the bulge area results in only temporary hair reduction [20,21]. Pantrichodestruction or damage to the entire germinative area of the follicle must therefore occur for permanent removal [19]. Partial germinative zone injury will lead to the formation of dystrophic hair that have the potential to regrow, whereas nongerminative zone injury will lead to an exogen shedding of hair that will regrow as a nonaffected terminal hair during the subsequent anagen follicular cycle (Table 7.1) [7,19,22].

Table 7.1 Possible Effects of Photoepilation

Action	Effect
Chromophore not targeted (i.e., light hair telogen)	No effect
Hair shaft destroyed	Exogen shedding regrowth
Partial germinative area destruction	Regrowth with dystrophic hair
Total germinative area destruction	Long term (possibly permanent) removal

7.2 Principle of Selective Photothermolysis: Combination of RF and Light Energies

A number of lasers systems are now available for hair removal (Table 7.2), as the absorption spectrum of melanin within hair follicles extends from the ultraviolet to the near-infrared wavelengths [1,15,23,53]. The most simple classification system divides the lasers according to wavelength: short, intermediate, or long. The laser surgeon must choose a short wavelength technology (500–800 nm) for light skin phenotypes with light brown/blond hair of thin diameter, whereas longer wavelengths (800–1200 nm) would be used for darker skin phenotypic individuals who have coarse dark brown/black hair [9].

Most of the methods intended for long-term hair removal use optical energy delivered to the tissue and absorbed mostly by the hair shaft, while the epidermis and surrounding tissue have minimal absorption [24]. There are two kinds of light sources that provide sufficient energy to achieve thermal destruction of hair: lasers and intense pulsed light. Laser light energy is monochromatic, and can be selectively absorbed by the hair shaft at the skin depth of a few millimeters. Intense pulsed light is composed of a broad spectrum white light of output wavelengths (500–1200 nm) that can be tuned to achieve better results by filtering out some parts of the spectrum generated by the light source in accordance with skin pigmentation [25–27].

However, various limitations on the laser and IPL modalities curtail their utilization capacity. The effective removal of unwanted hair using optical energy has been essentially limited to black, dark, and medium tones of brown hair [28]. Light energy must penetrate

Table 7.2 Indications for Short, Intermediate, and Long Wavelength Laser Hair Removal

Wavelength	Skin/Hair Type
Short	Fitzpatrick skin types I–III
Ruby (694 nm)	Thin hair shaft
Alexandrite (755 nm)	Blond to light brown hair
Intermediate	Fitzpatrick skin types II–V
Diode (800–900 nm)	Intermediate hair shaft
	Light brown to dark brown hair
Long	Fitzpatrick skin types IV–VI
Nd:YAG (1064 nm)	Intermediate to coarse hair shaft
	Medium brown to black hair

through the epidermis first to reach the hair follicle, thereby creating a major barrier for penetration. The benefits of high energy fluences are countered by the risks of damaging the epidermis. One limitation is the treatment of darker skin types. Even though advances in technology and development of longer wavelength lasers such as the 1064 nm Nd:YAG have enhanced the ability to treat darker skin types, the high concentration of endogenous melanin within the epidermis of these individuals increases the risk of crusting, blistering, and dyschromia [29–31]. An additional drawback of these technologies is the inability to treat light pigmented hair, such as red, blond, gray, and white, which do not contain high concentrations of melanin [8].

Light-based technologies have become restricted by the limitations placed on their methodology. The innovation of radiofrequency energy (RF), however, has brought new life to the optical-based systems, as this energy modality is not sensitive to melanin concentration in the shaft or the epidermis. High-frequency current in the range of 0.3–10 Megahertz, or RF current, produces a pure thermal effect on biological tissue that is dependent on the electrical properties of the tissue [32]. The high efficiency of RF current for tissue heating has made it useful for electrosurgery, and an attractive source of energy for various dermatologic applications [33,34]. The mechanism of tissue heating is based on generating joules of heat by electrical current. The heat generated is described by Joule's law:

$$H = j^2/\sigma,$$

where j is the density of electrical current, and σ is electrical conductivity [35]. Impedance is the value that opposes conductivity. The RF component generates heat from a current of ions that acts according to the physical principle of impedance, that is, electrical current will always follow the path of least resistance [33,36]. For example, blood has a very high electrical conductivity; therefore, it has low impedance. Bone has a very low electrical conductivity or high impedance. Electrical current will always follow the path of highest conductivity (lower impedance); therefore, it does not penetrate bone, but will flow around it. Impedance is also directly, but inversely, correlated with heat, as shown by the equation mentioned earlier [20]. Higher temperatures produce lower impedance and therefore direct the flow of current [37].

Electrical conductivity depends on the frequency of electrical current, type of tissue, and its temperature. The distribution of electrical current depends on the geometry of the electrodes. Typically, two configurations are used in medicine, either monopolar or bipolar. The major difference is how the RF current is controlled and directed at the target. However, there is no difference in the ultimate effect at the same RF fluence [20].

The bipolar system is most commonly used in conjunction with light energy during laser hair removal, examples which include the Aurora and Polaris systems (Syneron Medical Ltd., Yokneam, Israel). A bipolar system passes an electrical current between two electrodes at a fixed distance. Both electrodes are applied to the treated area, and electrical current propagation is limited by the area between electrodes. The behavior of electrical current in a bipolar system is depicted in Fig. 7.1. The penetration of electrical current can be estimated as half the distance between the electrodes, for instance if the electrodes are placed 8mm apart, the penetration depth is about 4mm [20].

The technology presented in this chapter utilizes a new approach combining either IPL or diode laser with conductive RF modalities simultaneously applied to tissues (Table 7.3) [37].

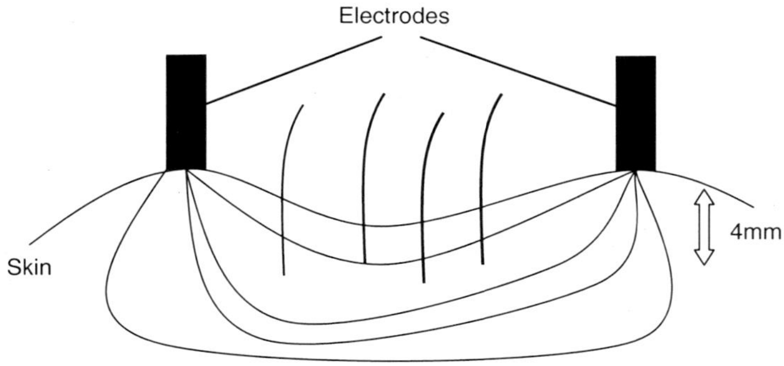

Figure 7.1 Schematic representation of the flow of electrical current through the epidermis using a bipolar system. The geometry of the radiofrequency (RF) electrodes is designed to deliver the RF current at a depth of 4 mm, which can target the deepest hair follicles in all anatomic locations.

Its advantages in initial clinical trials show results in areas where purely light-based systems have not shown efficacy; that is in the treatment of light hair and dark skin phenotypes [20]. Both forms of energy are pulsed and delivered to the tissues with a hand-held device. In the case of the Aurora, the light source is a high-power xenon-lamp that is filtered to transmit the wavelength range of 680–980 nm. The Polaris, on the other hand, manipulates an 810 nm diode laser pulse. The conducted RF electrical energy is bipolar, and can generate energy up to 50J/cm [3,38]. The bipolar RF generator consists of flashlamp pulsed light delivered through a contact sapphire light guide, and the bipolar RF energy is delivered through electrodes embedded in the system applicator when brought into contact with the skin surface [39]. The device also includes an active dermal monitoring system that measures changes in the skin impedance, which is adjustable by the user to provide an integrated safety mechanism (impedance safety limit) to prevent overheating of the dermis. A thermoelectric cooling hand piece provides contact cooling at a temperature of approximately 5°C before, during, and after energy delivery [27].

Table 7.3 Comparison of Combined Optical and RF Hair Removal Technologies: System Specifications

System	Clinical Applications	Laser Type	Light Fluence (J/cm^2)	RF Energy (J/cm^3)	Pulse Repetition Rate (pps)	Treated Area (mm)	Cooling on Skin Surface (°C)
Aurora	Light hair removal	Intense pulsed light (680–980 nm)	10–45	5–25	0.7	12 x 25	5–20
Polaris	Thicker, darker hair removal	High power 810 nm diode	Up to 50	Up to 100	Up to 2	8 x 12	5

RF: radiofrequency.

The synergistic effect of combining RF and optical energy allows for lower optical fluences to be utilized creating a safe treatment for darker skin types [34]. The treatment efficacy is not compromised due to the addition of conducted bipolar RF energy that selectively heats the hair follicle. The conducted RF selectivity mechanism is not affected by the absorption of melanin in the skin or hair shaft [31,45,46].

The combined modality system operates by using pulsed light energy to heat the hair shaft through a selective absorption of energy by the melanin in the hair shaft. Heat then dissipated to the follicle damages the hair shaft (Fig. 7.2a). The conducted RF energy selectivity mechanism is different than light, and the chromophore-independence of RF is perhaps its greatest advantage in allowing for treatment of any hair follicle, regardless of color [20,40]. The RF field for the bipolar system is controlled by impedance properties of the tissue; the current will always flow to the area of minimal impedance between the electrodes. There are two major factors that control the tissue impedance. The material of the hair shaft is mostly keratin which is not conductive; therefore the RF energy will go around the follicle directly (Fig. 7.2b). Second, as the temperature increases, the impedance of the tissue decreases. The light energy creates a preheated area inside the tissue, and directs the RF field into this preheated area versus the surrounding tissue of lower impendence. While the pulse durations are usually initiated at the same time, the RF pulse is set at a longer duration versus that of the optical energy to preheat the target and increase RF selectivity [21,27]. Using the thermal damage time over the thermal relaxation time of the hair follicle allows for longer pulses to extend the zone of thermal damage from melanin-laden structures in the central portion of the follicle to the outermost layers of the root and connective tissue sheath [41]. This heat profile created is uniform across the targeted hair follicle and shaft giving excellent results with minimal risk to the surrounding tissue (Fig. 7.2c) [42].

Based on the mechanisms described earlier, the optimal method utilized is a near-simultaneous application of the optical and bipolar RF energies with a precooling of the epidermis [27]. The steps are summarized as follows:

1. Hydrate and cool the epidermis.
2. Apply optical energy to selectively heat the target and bipolar RF energy to the heated target. The applied energy should be at the level at which the temperature of the epidermis does not exceed the target temperature.
3. Discontinue optical pulse and continue RF pulse for additional selective heating of the target [10,15,27,43,44].

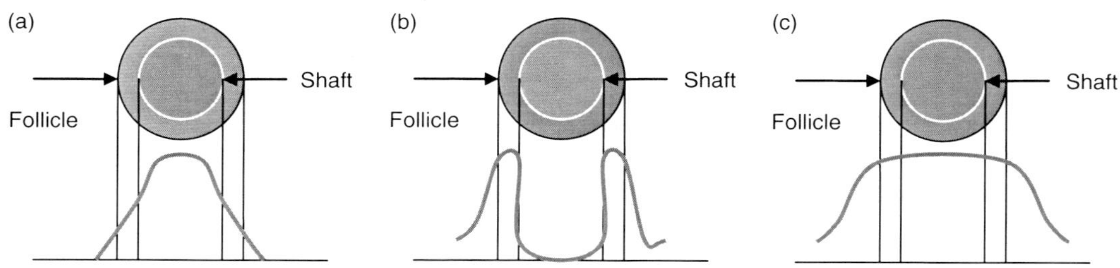

Figure 7.2 Temperature profile of hair structure: (a) shaft is heated by light energy; (b) follicle is heated by RF energy; (c) follicle and shaft heated by combined optical and RF energies.

Table 7.4 Summary of Clinical Studies with Combined Optical and RF Hair Removal Systems

Study	Application/Study Group	System/Treatment Settings	Results/Findings
Del Gliglio and Shaoul [24]	N = 60 Skin types II–V Different hair colors 2–3 treatments	Aurora Optical: 10–28 J/cm^2 RF: 10–20 J/cm^3	At 6 months, hair clearance ranged from 69% to 84% Most effective in axillary region 52% hair clearance for white or blond hair
Sadick and Shaoul [38]	N = 20 Skin types II–V Different hair colors 3–5 treatments	Aurora Optical: 15–30 J/cm^2 RF: 10–20 J/cm^3	80–85% clearance for black and brown hair 40–60% clearance for white, blond, or red hair
Sadick and Laughlin [45]	N = 36 women Skin types I–V Blond or white facial hair 2–5 treatments	Aurora Optical: 15–30 J/cm^{2}RF: 10–20 J/cm^3	44–52% clearance for white or blond hair
Laughlin [46]	N = 10 Skin types V–VI 1 treatment	Aurora Optical: 16–20 J/cm^2 for Type V, 14–17 J/cm^2 for Type VI RF: 18 J/cm^3 for type V,20 J/cm^3 for type VI	50% of subjects obtained hair clearance >35% No adverse reactions
Goldberg et al. [50]	N = 15 women Skin types II–IV White and fine, nonpigmented facial hair Half-face pretreated with 20% solution 5-ALA 2 treatments	Aurora Optical: 24–30 J/cm^2 RF: 20 J/cm^3	With pretreatment with topical 5-ALA, 48% clearance of terminal white hair No long-term pigmentary changes
Yaghmai et al. [53]	N = 69 Skin types I–VI Different hair colors 1 treatment	Aurora Optical: 14–30 J/cm^2 RF: 10–20 J/cm^3	At 3 months, 46% hair clearance Most effective on the arms
Schroeter et al. [56]	N = 17 Skin types I–II Blond hair only Avg. of 8.5 treatments/pt	Aurora Optical: 23.2 J/cm^2 RF: 18.6 J/cm^3	58% blond hair clearance Correlation found between hair removal and number of treatments No correlations between hair clearance and age
Schulze et al. [57]	N = 17 Skin types I–IV 1–4 treatments	Polaris Optical: 35–50 J/cm^2 RF: 10–50 J/cm^3	50% hair clearance after avg. of 1.9 treatments Greater hair reduction with thicker, darker hair Pain (occasionally severe) was proportional to hair pigmentation, density, and thickness
Sadick, Mulholland & Shaoul [42]	N = 45 Skin types II–VI Different hair colors 3 treatments	Polaris Optical: 30–42 J/cm^2 RF: 30–40 J/cm^3	At 6 months, hair clearance ranged from 65 to 70% Most effective on the bikini line

RF: radiofrequency; 5-ALA: topical aminolevulinic acid.

7.3 Review of Clinical Studies with Combined Optical and RF Hair Removal Systems

With the advent of combined optical and RF technology, various clinical trials have been conducted to elucidate the potential of the system's application on different hair types and skin colors by varying the treatment settings, as well as the number of treatments administered (Table 7.4). Del Giglio and Shaoul [24] conducted a recent multicenter study using 60 subjects with Fitzpatrick skin types ranging from II–V and varying hair colors for treatment. In the study, optical energy ranged from 15 to 28 J/cm^2, and RF energy ranged from 10 to 20 J/cm^3. All subjects received three treatments, six to eight weeks apart. Hair counts were performed prior to treatment, and three months after the last treatment. Maximum hair reduction was observed at two to eight weeks. At three months, hair clearance ranged from 64 to 84%, depending on the anatomic site. In addition, 12 subjects with white, blond hair displayed an average clearance of 52% at six months. Treatment was most effective in the axillary region. In most patients, high RF energy (15–20 J/cm^3) was used, and the results indicate that efficacy is determined by the level of RF energy, and not optical energy.

A study which consisted of 40 adult subjects with Fitzpatrick skin types II–V and various hair colors conducted by Sadick and Shaoul [38] showed maximum hair reduction occurring from six to eight weeks following treatments, and progressive decrease of hair density was observed following each subsequent treatment. Subjects received four treatments at 8- and 12-week intervals, over a period of 9–12 months. Depending on the skin and hair phenotypes, light energy ranged from 15 to 30 J/cm^2. Higher optical energy was used in lighter skin phenotypes and hair color. The RF current ranged from 10 to 20 J/cm^3, depending on the anatomic site, with higher RF energy used in facial areas as compared to the lower body regions. The results were monitored until six months after the last treatment, at which time the average clearance of 75% was observed at all body locations, with the best results (85% clearance) seen in the axillae and legs. As expected, darker phenotypes provided greater hair-removal efficiency.

Sadick and Laughlin [45] examined the long-term photoepilatory effect on blond and white hair using 36 women with skin phenotypes ranging from I–V. The chin and upper lip were treated with four sessions over a 9- to 12-month period. The level of RF energy was 20 J/cm^3, while optical fluences varied from 24 to 30 J/cm^2. Maximum reduction in hair counts was observed at six to eight weeks after each treatment, with an average clearance of blond and white hair of 48% at six months follow-up after the last treatment (Fig. 7.3a and b). Side effects were minimal, with 8% of patients reporting transient hyperpigmentation that did not require therapy, and 14% having mild erythema which resolved in 24 hours. While the slightly higher photoepilatory efficiency for blond hair (52% clearance) versus that of white hair (44% clearance) shows the promise of this combined modality system as a favorable alternative to treat this previously difficult-to-treat patient management subgroup, it should be noted that pure laser sources utilized for the removal of darker hair phenotypes are still more advantageous clinically.

Laughlin [46] conducted a study using ten patients, seven of who were East Indian patients with Fitzpatrick skin type V and three African–American patients with Fitzpatrick skin type VI. The RF energy was set at 18 and 20 J/cm^3 for skin types V and VI, respectively. Optical energy was set from 16 to 20 J/cm^2 for skin type V and from 14 to 17 J/cm^2 for skin type VI, respectively. Serial photography and clinical examination were used to

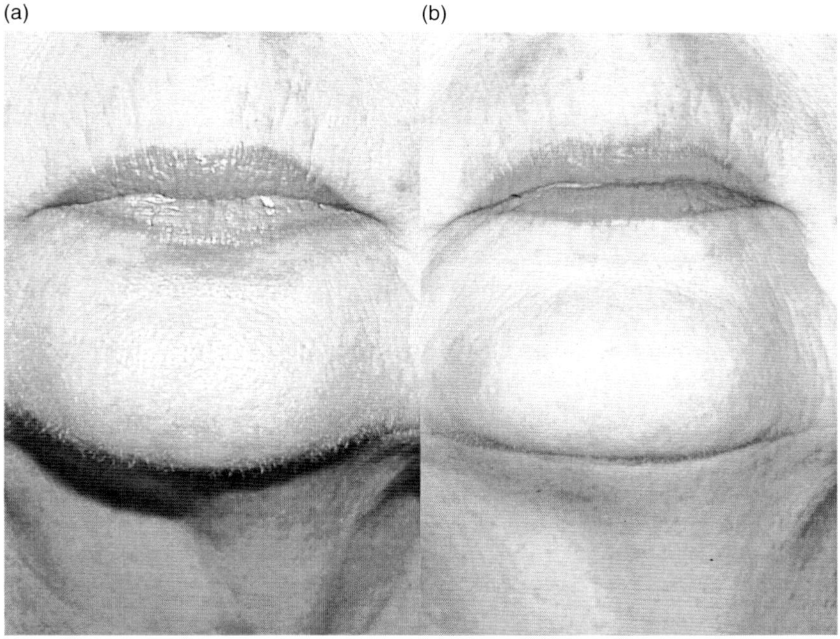

Figure 7.3 (a) Before combined intense pulsed light (IPL)/radiofrequency (RF) white hair removal; (b) after combined IPL/RF white hair removal (four treatments; month 18 (6 months after last treatment)); optical energy = 26 J/cm^2, RF energy = 20 J/cm^3.

evaluate the subjects at one to three days, two weeks, one month, and four to seven months after the final treatment to determine hair loss and adverse effects, notably dyschromia and scarring. Two blinded observers working independently carried out hair counts. The results showed that 50% of subjects obtained a hair loss of more than 35%. The mean hair loss for the entire group was 30.20%, with a range of 13–75.4%. None of the study participants developed any blistering within the first 72 hours of treatment, and the absence of early epidermal injury differentiates this method of treatment from those methods using pure optical energy, where blistering can occur [17,47]. Lasers and IPLs are associated with infrequent complications; the greatest risk being associated with the treatment of darker skin types [32,48]. The adverse events including hyperpigmentation, hypopigmentation, blistering, and crusting, are often associated with the treatment of skin types IV and above [49]. However, epilation with combined RF and optical energy in this study of patients with the darkest of skin types was associated with a zero rate of dyschromia, which suggests that this method of treatment could eliminate the expected risks of photoepilation in skin type V and VI by providing for a better therapeutic margin of safety for the treatment of pigmented skin.

Goldberg et al. [50] studied 15 subjects with nonpigmented facial hair and Fitzpatrick skin types II–IV. Ten of the subjects were clinically determined to have white terminal hair, and the remaining five were noted to have fine, nonpigmented vellus hair. The level of RF energy was set at a constant 20 J/cm^3 for all study subjects, while the chosen optical fluences varied between 24 and 30 J/cm^2 and were delivered in a short pulse profile. Hair counts were taken from standardized digital photos obtained before and six months after final treatments, where baseline hair count was determined by two nontreating physicians.

Half of the face was pretreated with a 20% solution of aminolevulinic acid, a light absorbing photosensitizer shown to promote photoepilation, one hour prior to treatment with the optical bipolar RF device [51]. Topical photosensitizers including 5-amino levulinic acid (ALA) have been previously studied for hair removal [12]. ALA is the first product in the hemesynthesis cascade, which, when present in excess, is converted to protoporphyrin II, a potent photosensitizer. Twenty percent of ALA is absorbed by hair follicles and subsequently converted to protoporphyrin IX in a period of several hours. Subsequent exposure of skin to light energy activates protoporphyrin IX, leading to the formation of singlet oxygen, which damages follicular cell membranes [12,52]. In this study, hair loss of 35% was reported at six months after the first and only treatment. When pretreatment with topical ALA was provided, the average hair removal of terminal white hair was found to be 48%. Nonpigmented hair can be successfully removed with a combined optical bipolar RF source, and these results are further improved with the preapplication of a topical photoabsorbing agent. White hair was also found to contain melanin within the follicular structure, albeit not in the actual follicle itself. This melanin is sufficient for absorbing the optical energy delivered by the combined optical RF device.

A multicenter study involving 69 patients with skin types I–VI were evaluated at 1, 7, 30, and 90 days after a one-treatment session [53]. The optical energy component was delivered with a pulse duration of 25 ms and energy fluence ranging from 14 to 20 J/cm^2, while the RF energy was delivered with a pulse duration of 200 ms with energy density ranging from 10 to 20 J/cm^3. At 90 days after a single treatment, the mean hair count was reduced from baseline by an average of 47%, with best results achieved when treating the arm (65%), followed by the axilla (49%) and legs (44%). The percentage of hair reduction was statistically significant for all three hair colors, with a mean hair count reduction of 43% for black hair, 49% for brown hair, and 35% for blond hair. The use of longer pulse durations permits thorough thermal injury to the entire follicular unit, resulting in more permanent hair removal while producing less thermal damage, as is seen in this study [34,54,55].

Shroeter et al. [56] recruited 17 patients with blond hair; seven of who had skin type I and 10 patients with skin type II. The mean optical energy used per patient was 23.3 J/cm^2 and the mean RF was 18.6 J/cm^3. A mean hair reduction of 57.4% was obtained after an average of 8.5 treatments. A clear trend between hair removal and number of treatments was established in this study, with better results depending on the increasing number of treatments.

Schulze et al. [57] treated facial hair in 17 subjects with Fitzpatrick skin types I–IV. The optical fluences ranged between 35 and 50 J/cm^2, and pulse duration of 100 ms. Radiofrequency energy densities ranged from 10 to 50 J/cm^3. Treatment areas received between one and four treatments over a six-month span, with four to six week intervals between treatments. There was a reported mean hair reduction of 50% after an average of 1.9 treatments at a mean follow-up period of 2.6 months. There was a trend toward greater hair reduction with thicker and darker hair, although in two cases there was a marked reduction in thinner hair. Pain was proportional to hair pigmentation, density, and thickness, and occasionally was severe.

A multicenter study conducted by Sadick, Mullholland, and Shaoul [42] recruited 45 patients with Fitzpatrick skin types II–VI and with various hair colors. Treating a variety of body sites such as the legs, bikini line, axilla, and back, the laser energy density range used varied from 30 to 42 J/cm^2, while the RF energy range was 30–40 J/cm^3. Maximum reduction in hair was observed from two weeks to two months after a single treatment (Fig. 7.4a and b). Maximum average clearance was seen in the bikini line (78%) followed

(a) (b)

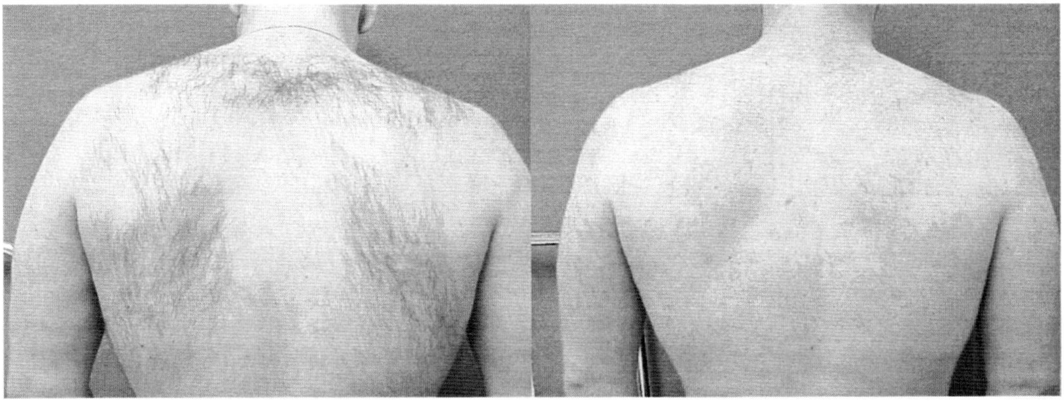

Figure 7.4 (a) Before combined 800 nm diode/radiofrequency (RF) hair removal; (b) after combined 800 nm diode/RF hair removal (three treatments; month 9 (6 months after last treatment)); optical energy = 36 J/cm^2, RF energy = 36 J/cm^3

closely by the legs (75%), axilla (72%), and back (65%). The treatments were well-tolerated by the subjects when concomitantly employing a forced air-cooling device over the use of topical anesthesia.**Figure 7.4** (a) Before combined 800 nm diode/radiofrequency (RF) hair removal; (b) after combined 800 nm diode/RF hair removal (three treatments; month 9 (6 months after last treatment)); optical energy = 36 J/cm^2, RF energy = 36 J/cm^3.

7.4 Conclusion

Hair removal using lasers and light-based methodologies are limited by the chromophore dependence on melanin located in the hair shaft, creating a small window of efficacy and safety for the treatment of light hair hues and darker skin phenotypes. More recently, this dependence on the melanin chromophore has been eliminated by using an alternative source of energy, RF, and synthesizing its use with current light and laser modalities. The combination of optical and RF has proven to be a safe and effective method for the permanent removal of unwanted hair, and has shown promising effects in both those individuals with darker skin types as well as in those with blond or white hair, while also creating a high safety profile with minimal patient discomfort. Further investigation with more long-term studies and comparison trials are needed to further elucidate the integrated RF and optical energy technology, especially in the earlier refractory group of photoepilatory individuals.

References

1. Grossman MC, Dierick C, Farinella W, et al. Damage to hair follicles by normal mode ruby laser pulses. J Am Acad Dermatol 1996; 35(8): 889–94.

2. Lynfield YL, Macwilliams P. Shaving and hair growth. J Invest Dermatol 1970; 55: 170–2.
3. Richard RN, Uy M, Meharg G. Temporary hair removal in patients with hirsutism: A clinical study. Cutis 1990; 45: 1999–2002.
4. Wagner RF Jr. Physical methods for the management of hirsutism. Cutis 1990; 45: 319–26.
5. Natow AJ. Chemical removal of hair. Cutis 1986; 38: 91–2.
6. Alster TS, Lupton JP. Lasers in dermatology: An overview of types and indications. Am J Clin Dermatol 2001; 2: 291–303.
7. Dierickx CC, Grossman MC, Farinelli WA, et al. Permanent hair removal by normal-mode ruby laser. Arch Dermatol 1998; 134: 837–42.
8. Williams R, Havoonjian H, Isagholian K, et al. A clinical study of hair removal using the long-pulsed ruby laser. Dermatol Surg 1998; 24: 837–42.
9. Grossman M, Dierickx C, Quintana A, et al. Removal of excess body hair with an 800 nm pulsed diode laser. Lasers Surg Med 1998; 22(Suppl. 10): 42.
10. Weiss RA, Weiss MA, Marwaha S, et al. Hair removal with a non-coherent filtered flashlamp intense pulsed light source. Lasers Surg Med 1999; 24: 128–32.
11. McDaniel DH, Lord J, Ash K, et al. Laser hair removal: a review and report on the use of long-pulsed alexandrite laser for hair reduction of the upper lip, leg, back, and bikini region. Dermatol Surg 1999; 25: 425–30.
12. Nanni CA, Alter TS. Long-pulsed alexandrite laser-assisted hair removal at 5, 10, and 20 millisecond pulse durations. Lasers Surg Med 1999; 24: 332–7.
13. Ross EV, Ladin Z, Kreindel M, et al. Theoretical considerations in laser hair removal. Dematol Clin 1999; 17: 333–55.
14. Goldberg DJ. Unwanted hair evaluation and treatment with lasers and light pulse technology. Adv Dermal 1999; 14: 115–37.
15. Dierickx, C. Laser assisted hair removal: state-of-the-art. Dermatol Ther 2000; 13: 80–9.
16. Wanner M. Laser hair removal. Dermatol Ther 2005; 18: 209–16.
17. Anderson RR, Parrish JA. Selective photothermolysis: a precise microsurgery by selective absorption of pulsed radiation. Science 1983; 220: 524–7.
18. Sadick NS. Laser and flashlamp photoepilation: A critical review of modern concepts bridging basic science and clinical applications. J Aesthetic Dermatol Cosmetic Surg 1999; 1: 95–101.
19. Sadick, NS. Laser hair removal. Facial Plast Surg Clin North Am 2004; 12: 191–200.
20. Liew SH, Grobbelaar AO, Gault DT, et al. The effect of ruby laser on ex vivo hair follicles: Clinical implications. Ann Plast Surg 1999; 42: 249–54.
21. Cotsarelis G, Sun TT, Lavker RM. Label-retaining cells side in the bulge area of pilosebaceous unit: Implications for follicular stem cells, hair cycle, and skin carcinogenesis. Cell 1990; 61: 1329–37.
22. Hashizume H. Tokura Y, Takigawa M, et al. Hair follicle development expression of heat shock proteins in hair follicle epithelium. Int J Dermatol 1997; 36: 587–92.
23. Sadick NS, Weiss RA, Shea CR, et al. Long term photoepilation using broad spectrum intense pulse light sources. Arch Dermatol 2000; 136: 1336–40.
24. Del Giglio A, Shaoul J. Hair removal using a combination of electrical and optical energies—multiple treatments clinical study six months follow up. http://www.syneron.com/Solutions/Clinical_Results/Clinical_Papers/Hair_Removal.html, accessed 3/2/07.
25. Lask G, Eckhouse M, Slakine A, et al. The role of laser and intense light sources in photoepilation: comparative evaluation. Cutaneous Laser Therapy 1999; 1: 3–13.
26. Gold MH, Bell MW, Foster TD, et al. Long term epilation using the EpiLight broad band, intense pulsed light hair removal system. Dermatol Surg 1997; 23: 909–13.
27. Shroeter, C. Hair removal with the PhotoDerm VL as an intense light source: A histological study. Proceedings of the 18th Annual Meeting of the ASLMS, San Diego, CA, April 1998.
28. Ort RJ, Dierickx C. Laser hair removal. Semin Cut Med Surg 2002; 21: 129–44.
29. Nanni CA, Alster TS. Laser-assisted hair removal: side effects of Q-switched Nd:YAG, long-pulsed ruby, and alexandrite lasers. J Am Acad Dematol 1999; 8: 165–71.
30. Lorenz S, Brunnberg S. Lanthaler M, et al. Hair removal with the long-pulsed Nd:YAG laser: a prospective study with one year follow up. Lasers Surg Med 2002; 30: 127–34.

31. Bencini PL, Luci A, Galimberti M, et al. Long term epilation with long pulsed Nd:YAG laser. Dermatol Surg 1999; 25: 175–8.
32. Sadick NS, Makino Y. Selective electro-thermolysis in aesthetic medicine: A review. Lasers Surg Med 2004; 34: 91–7.
33. Carruthers A. Radiofrequency resurfacing: Technique and clinical review. Facial Plast Surg Clin North Am 2001; 9: 311–19.
34. Tatso JP, Ash SA. Current uses of radiofrequency in arthroscopic knee surgery. Am J Knee Surg 1999; 12: 186–91.
35. Gabriel S, Lau RW, Gabriel C. The dielectric properties of biological tissues: III. Parametric models for the dielectric spectrum of tissues. Phys Med Biol 1996; 41: 2271–93.
36. Duck FA. Physical Properties of Tissue. Academic Press: New York, 1990.
37. Anvari B, Tanenbaum BS, Milner TE, et al. Selective cooling of biological tissues: application for thermally mediated therapeutic procedures. Phys Med Biol 1995; 40: 241–52.
38. Sadick NS, Shaoul J. Hair removal using a combination of conducted radiofrequency and optical energies—an 18-month follow-up. J Cosmet Laser Ther 2004; 6: 21–6.
39. Sadick, NS. Combination radiofrequency and light energies: Electro-optical synergy technology in esthetic medicine. Dermatol Surg 2005; 31: 1211–17.
40. Waldman A, Kriendle M. New technology in aesthetic medicine: ELOS™ electro optical synergy. J Cosmet Laser Ther 2003; 5: 204–7.
41. Rogachefsky AS, Silapunt S, Goldberg DJ. Evaluation of a new super-long-pulsed 810 nm diode laser for removal of unwanted hair: The concept of thermal damage time. Dermatol Surg 2002; 28: 410–14.
42. Sadick NS, Mulholland SR, Shaoul J. Combination of 810 nm high power diode laser with conducted bi-polar RF energy for hair removal. http://www.syneron.com/Solutions/Clinical_Results/Clinical_Papers/Hair_Removal.html, accessed 3/2/07.
43. Kreindel M, Waldman A. Electro-Optical Synergy (ELOS) technology for aesthetic medicine: Advantages and limitations of various forms of electromagnetic energy for safe and effect hair removal. http://www.syneron.com/Solutions/Clinical_Results/Clinical_Papers/Hair_Removal.html, accessed 3/2/07.
44. Lask G, Elman M, Slakine M, et al. Laser-assisted hair removal by selective photothermolysis: Preliminary results. Dermatol Surg 1997; 23: 737–9.
45. Sadick NS, Laughlin SA. Effective epilation of white and blond hair using combination radiofrequency and optical energy. J Cosmet Laser Ther 2004; 6: 27–31.
46. Laughlin SA. Epilation in dark skin (types V and VI) with integrated radio-frequency and optical energy [data on file]. Syneron Medical Ltd. 2002: Tokneam (Israel).
47. Alster TS, Bryan H, William CM. Long-pulsed Nd:YAG laser assisted hair removal in pigmented skin. Arch Dermatol 2001; 137: 885–9.
48. Nanni CA, Alster TS. Complications of laser-assisted hair removal: side effects of Q-switched Nd:YAG, long-pulsed ruby, and alexandrite lasers. J Am Acad Dematol 1999; 41: 165–71.
49. Stratigos AJ, Dover J, Arndt KA. Laser therapy. In: Bolognia JL, Jorizzo JL, Rapini RP, editors, Dermatology. Mosby: New York, 2003: p. 2170.
50. Goldberg DJ, Marmur ES, Hussain M. Treatment of terminal and vellus non-pigmented hairs with an optical bipolar radiofrequency energy source- with and without pre-treatment using topical aminolevulinic acid. J Cosmet Laser Ther 2005; 7: 25–8.
51. Grossman MC, Wimberely J, Dwyer P, et al. Photodynamic therapy for hirsutism. Lasers Surg Med 1995; 17(Suppl. 7): 205.
52. Dierickx CC, Grossman MC. Laser hair removal. In: Dover JS, Goldberg DJ, editors, Laser and Lights, Vol. 2. Elsevier Saunders: Philadelphia, 2005, pp. 61–76.
53. Yaghmai D, Garden JM, Bakus AD, et al. Hair removal using a combination radio-frequency and intense pulsed light source. J Cosmet Laser Ther 2004; 6: 201–7.
54. Manstein D, Dierickx CC, Koh W, et al. Effects of very long pulsed on human hair follicles [abstract]. Lasers Surg Med 2000; 26(Suppl. 12): 85.
55. Altschuler GB, Anderson RR, Manstein D, et al. Extended theory of selective photothermolysis. Lasers Surg Med 2001; 29: 416–32.

56. Schroeter CA, Sharma S, Mbonu NC, et al. Blond hair removal using ELOS systems. J Cosmet Laser Ther 2006; 8: 82–6.
57. Schulze RA, Harrison B, Ross VE. Successful hair reduction with 810 nm diode laser coupled with bipolar radiofrequency [data on file]. Naval Medical Center 2006: San Diego, CA.

8

Hair Removal in Darker Skin Types Using Light-Based Devices

James Henry

The Procter and Gamble Company, Cincinnati, OH, USA

Gurpreet S. Ahluwalia (ed.), *Cosmetic Applications of Laser and Light-Based Systems*, 195–215,

8.1 Introduction

Laser hair removal allows a physician or trained professional to treat large areas of the body quickly, with long-lasting or permanent reduction in hair growth [1]. These reductions in growing hair are caused by the interaction of the laser with the hair melanin. While initially developed for the ideal contrast of dark hair against a fair skin background, an increasing demand is foreseen to treat all individuals regardless of base skin color, particularly in the United States, where the population is becoming more ethnically diverse [2,3]. Significant hair growth reduction has been reported in people with darker skin types; however, the interaction of the laser with skin melanin must be taken into account to prevent long-lasting side effects [3–5]. Permanent changes in pigmentation and skin texture, focal atrophy, and scarring are some of the adverse effects that have been reported with improper laser use [6]. As the understanding of the variations in response in people of color to the laser grows, these side effects associated with laser hair reduction can be reduced [7,8]. This chapter will discuss melanogenesis in skin and hair follicle, how melanin granules or melanosomes are formed, some basic thermal principles describing how hair follicles respond to heat, the relevance of using the Fitzpatrick scale, various cooling systems designed to protect the epidermis, and the efficacy and safety of various lasers and IPL systems for darker skin types.

8.2 Melanin in the Skin

The differences in skin color are the result of genetic background and environmental exposure to the sun [9–11]. Darker pigmented skin evolved in those whose ancestors lived near the equator [12]. It is widely believed that an increase in skin melanin protected the skin from the ultraviolet light exposure in equatorial locations. Lighter skin color may have developed to ensure sufficient vitamin D formation in the epidermis of persons living in northern latitudes [12,13].

Variations in skin color are not due to differences in the melanocyte number, but rather to the size, number, and grouping of melanosomes [14]. In general, dark-skinned subjects have an increased number of large individual melanosomes [15,16]. In these individuals, the melanosomes can be found throughout the epidermis. The melanosomes of light-skinned subjects are predominantly smaller with less melanin, and are found clustered together. In fair-skinned subjects few, if any, melanosomes are found in the upper epidermis. In people of Asian descent, melanosomes are relatively large and found individually, as well as grouped together [17].

Skin color is also affected by exposure to ultraviolet radiation [18–20]. Fitzpatrick developed the most commonly used skin color classification to correlate the amount of tanning or burning in light-skinned individuals in response to UV light [21]. Persons of color were added as three groups, Types IV, V, and VI, to the Fitzpatrick scale. Although widely used today, the Fitzpatrick system is not a good predictor of postinflammatory hyperpigmentation, or keloid scar formation in persons of color [22]. Newer skin-classification systems have been designed to give a better assessment of the risk of these side effects. The Lancer Ethnicity scale combines a person's ancestry as well as skin color to determine their tolerance to cosmetic laser procedures [23]. Specific classifications have been developed for persons of African [24], as well as Hispanic heritage [25].

8.2.1 Melanogenesis in Skin and Hair

Melanogensis occurs by a common biochemical pathway in both the hair follicle and the epidermis [26–28]. Two types of melanin are found in the skin and hair, the brown/black melanin, eumelanin, and the red melanin, pheomelanin [26]. The formation of eumelanin begins with either hydroxylation of intracellular L-phenylalanine or from extracellular tyrosine transported into the cell (Fig. 8.1). L-phenylalanine is hydroxylated to L-tyrosine by the enzyme phenylalanine hydroxylase [29,30]. Phenylalnine hydroxylase requires (6R)-L-erythro 5,6,7,8 tetrahydrobiopterin as a cofactor [29]. 6BH4 may act as an allosteric inhibitor of tyrosinase and its abiogenic isomer, 7BH4, may inhibit PAH [30]. Melanin is also synthesized directly from intracellular L-tyrosine [31,32]. L-tyrosine is hydroxylated to L-3,4-dihydroxyphenylalanine (L-dopa) by tyrosinase or tyrosine hydroxylase isoform I [33,34]. In the eumelanogenic pathway, tyrosinase is the most important enzyme for melanin synthesis [35]. L-dopa can also be formed from the reduction of L-dopaquinone back to L-dopa [32]. This is followed by the oxidation of L-dopa to dopaquinone [35]. Melanogenesis will go through oxido-reduction reactions and intramolecular transformations spontaneously, once L-dopa is formed [28]. The rate of these reactions are determined by local concentrations of hydrogen ions, metal cations, thiols, and the other reducing agents, hydrogen peroxide, and oxygen [35].

Synthesis of pheomelanin starts with the formation of cysteinyldopa from dopaquinone and cysteine [36]. Cysteinyldopa can also be formed from the hydrolysis of glutathionyldopa by glutamyltranspeptidase [37]. Cysteinyldopa is then oxidized to yield 1,4-benzothiazinylalanines. The velocity of post-cysteinyldopa steps of melanogenesis is increased by peroxidase and tyrosinase, which are involved in the transformation of benzothiazinylalanines. The main regulatory mechanism switch from eu- to pheomelanogenesis employs dopaquinone as a key molecule controlling the activity of glutathione reductase. Pheomelanogenesis is also blocked at high tyrosinase activity and high eumelanogenesis rate [36,37].

Both eumelanogenic and pheomelanogenic melanosomes can coexist in the same human cell [38], but not within the same melanosome [39]. Four processes have been used to explain the transfer of melanin granules from the melanosome to keratinocytes of the hair or skin: phagocytosis of the tips of dendrites containing melanosomes, internalization by the keratinocytes, the fusion of plasma membranes, and the transfer of melanosomes to the keratinocytes [40,41].

8.3 Hair Biology

The hair bulb, isthmus, and infundibulum are the three units that make up the vertical sections of the hair follicle [42–44]. The hair bulb begins at the base of the hair follicle and continues to the insertion of the arrector pili muscle. This portion of the hair is 3–7 mm below the surface of the skin in the dermis, and is comprised of the matrix cells, melanocytes, and the dermal papilla [45]. The matrix cells form the outer root sheath, the three layers of the inner root sheet, and the hair shaft [46]. In the mature anagen hair follicle, melanocytes are located in the basal layer of the infundibulum and surrounding the upper dermal papilla [47,48]. To be effective, the laser energy must reach the highly concentrated population of melanin- (color) producing cells in the matrix area [49]. The response to the

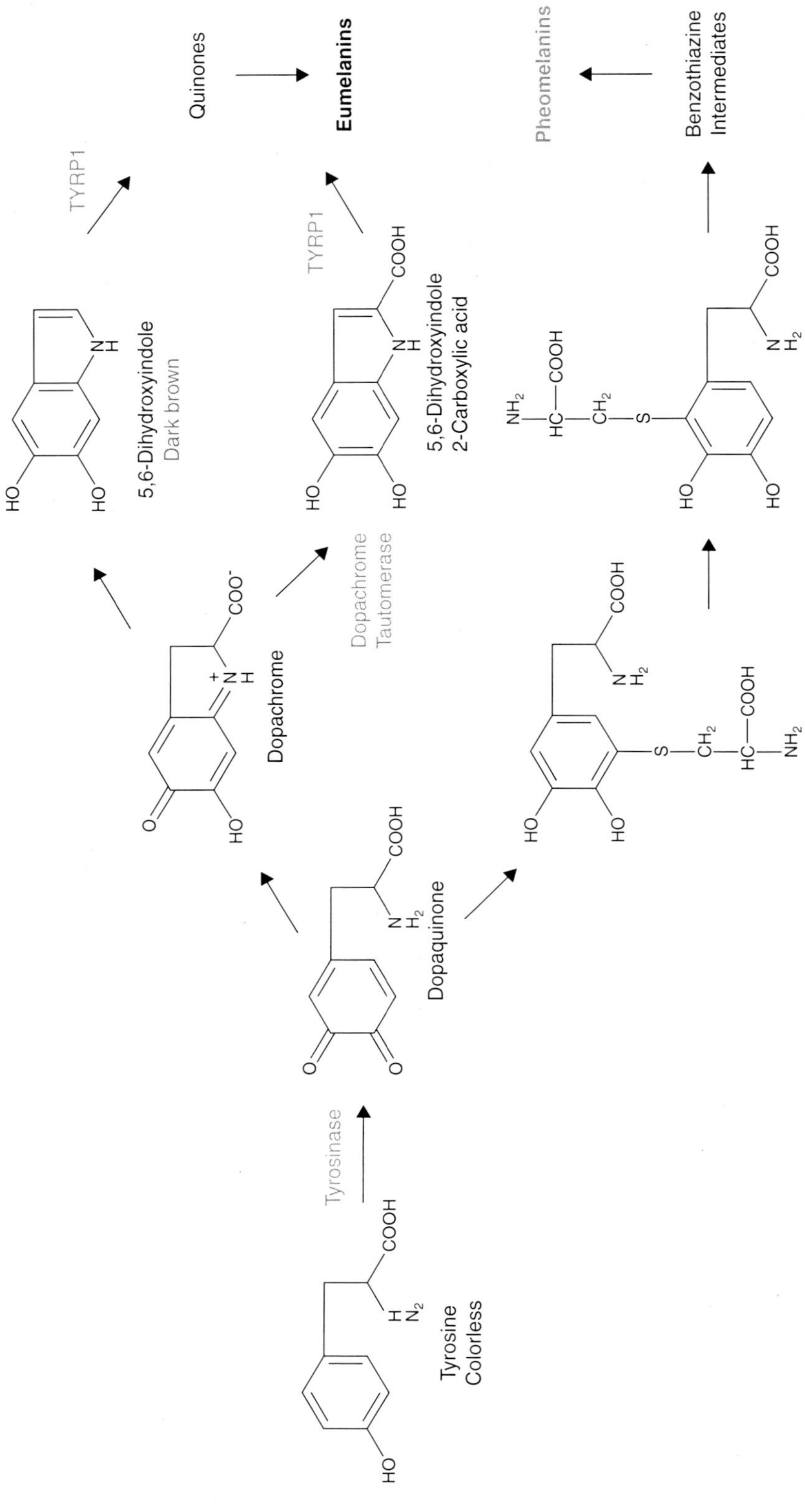

Figure 8.1 The melanin chemical pathway. Richard A. King, MD, PhD, http://albinism.med.umn.edu/newfacts.htm#other

laser will depend on the hair color, with large numbers of electron-dense melanosomes in black-hair follicles being the most responsive [50,51]. Brown-hair bulb melanosomes are somewhat smaller, and may not generate as much heat as black hair, while those of blonde hair have very little melanin and will not produce much heat compared to brown or black hair. In red hair, the melanosomes contain the red pigment phelomelanin, which has the poorest absorption of laser energy.

The dermal papilla provides the factors responsible for controlling the growth of the hair follicle, as well as the nourishment to the rapidly proliferating cells found in the matrix [45]. The close proximity of the melanocytes to the matrix and dermal papilla cells means that enough heat can be transferred to these cells to disrupt their normal function [52,53]. Above the hair bulb is the isthmus, which starts at the insertion of the arecctor pili muscles and continues to the entrance of the sebaceous duct [46]. The isthmus encompasses the bulge region with the pleuripotent stem cells needed to regenerate the hair follicle as it enters the anagen phase of the growth cycle. Cotsarelis et al. have identified a population of slow-growing stem cells that are located at the attachment of the arrector pili muscle (the bulge region) [54]. It is now believed that the stem cells in this region are activated by signals from the dermal papilla during late telogen or early anagen. The stem cells form the more differentiated cells of the matrix and return to their normal noncycling state by mid-anagen. Many believe that the stem cells in this region have to be damaged by the laser treatment for the treatment effects to be permanent hair-growth inhibition [55].

The mammalian hair follicle has three phases in its growth cycle [56]. The growth phase (anagen) is characterized by the rapid growth of the matrix cells. The amount of time a hair is in the anagen phase varies widely, depending on the body site (from years on the scalp, to weeks on the arm). Melanin synthesis in the hair follicle is closely correlated to the growth phase of the hair follicle [57]. After the anagen phase, the hair enters into a short-lived regressive phase (catagen) in which the lower portions of the follicle including the melanocytes undergo apoptosis [58]. The lower portion of the follicle shrinks and forms a thin epithelial cord, which retracts upward to the infundibulum. During catagen, the follicle reduces by about two-third of its original length and the dermal papilla ends up at the level of the arrector pili muscle [46]. The catagen phase lasts about three weeks [45]. The final phase of the hair cycle is the telogen or resting phase [58]. This phase can last from weeks to a few months, dependent upon the body site location. Upon receiving a host of signals, the hair will re-enter anagen with a large increase in epithelial cell division that reforms the matrix region [45]. Some authors have proposed that early anagen phase is the best time to treat the hair with the laser, since the hair bulb is close to the skin surface and melanin synthesis has begun [59].

8.3.1 Some Basic Thermal Principles Describing the Response of Hair Follicles to Heat

To achieve long-lasting or permanent hair reduction, the heat must diffuse, not only to the bulb region, but also to the stem cells of the outer root sheath and the bulb region [60]. To ensure maximum safety and efficacy, the wavelength, pulse duration, and fluence of the laser must be selected in such a way that the thermal injury is confined to the hair follicle [61,62]. The removal of unwanted hair by lasers or other high-energy light sources is based on the process of photothermolysis [55]. To ensure specificity that is necessary for this treatment, melanin of the hair follicle has been selected as the target chromophore [52,53]. The absorption spectrum of melanin is between 250–1200 nm [63]. The reduction in the absorption of the light by melanin with increasing wavelength is offset by the greater depth of penetration of light at higher wavelengths [56]. In the range of wavelengths used for laser hair removal, 650–1064 nm, a balance between specificity and penetration has been achieved [52]. To ensure sufficient thermal destruction and confinement to the hair follicle, the pulse duration should be equal to the thermal relaxation time of the hair shaft (10–100 ms) [55]. Using a pulse width in this range has worked well for subjects who are fair-skinned with dark hair, because the heat generated in the relatively fewer number of melanin granules in the epidermal layer of the skin is relatively low and can easily be removed by cooling the stratum corneum surface [64–67]. However, the absorption of the light within the pigmented epidermis of individuals with higher melanin content has required the modification of the laser parameters to minimize unwanted side effects due to the greater absorption and diffusion into heat in the epidermis [64]. This is achieved in darker skinned subjects by depositing the energy more slowly at longer pulse duration and/or higher wavelengths [5]. The longer duration of energy delivery allows the smaller granules of the epidermis to lose more of this heat [65]. To take full advantage of the adjustments in the laser-treatment parameters for dark-skinned individuals, it is imperative to efficiently cool the skin [69,70].

The FDA has approved the long-pulsed diode and the Nd:YAG systems for permanent hair reduction in those individuals with darker skin [8,55]. While the diode laser may be more effective for hair reduction, the Nd:YAG may have fewer negative side effects on the skin [5]. Pulse durations of 400 ms or longer with the Nd:YAG laser have been used to safely treat individuals with darker skin types [64].

8.4 Side Effects of Laser Hair Removal by Skin Type

When treating people with dark-colored skin, the increase in the density of the melanosmes and sensitivity of the skin to inflammation should be taken into account [65]. Blistering, changes in pigmentation, scabbing, thrombophlebitis, and scar formation are some of the side effects found during removal of unwanted hair with laser [3]. Posttreatment increases in skin pigmentation in subjects with darker skin may be explained by stimulation of melanocytes. Hyperpigmentation, while having a higher incidence rate, had a median duration of 28 days [3,4]. Hypopigmentation could result from redistribution of the melanin in the keratinoyctes, suppression of melanogeniesis, or destruction of melanocytes. Liew et al. have demonstrated a decrease in epidermal melanocyte tyrosinase activity after laser treatment [69]. It has been reported that hypopigmentation induced in laser-assisted hair removal had a median duration of 120 days. The challenge of treating darker skin types is due to the increase in the number and severity of these side effects [2–6].

8.4.1 Cooling the Epidermis

To reduce the damage to the upper layers of the skin, surface cooling is directed at the dermo-epidermal junction where the highest amounts of epidermal melanin reside [68]. The heat generated by the laser treatment must be exchanged between the cooling device and skin stratum corneum [69]. Four types of cooling have been developed to cool the skin during laser treatment: Clear gel (usually chilled), contact cooling (through a window cooled by circulating water), cryogen spray (immediately before/after the laser pulse), and air-cooling [71] (Table 8.1). Topically applied gels are the least expensive cooling method [68]. The cooled gels have a smaller temperature gradient than active cooling devices, so only a small amount of heat can be removed from the skin before the temperature gradient decays, and heat transfer ceases. A sapphire window with 2–6°C circulating water is also used to protect the skin from overheating [72]. These types of systems can drop the temperature of the skin's basal layer by 20°C with a 0.5 s exposure. Evaporative cooling using cryogen sprays has also been used [73,74]. With this method, one is able to create larger temperature gradients between the dermis and epidermis, which allows for greater protection of the epidermis, and reduced risk of inadvertent dermal cooling. For example, the cryogen spray, which is applied for 10–50 ms, followed by the delivery of the laser pulse within 5–10 ms has been used for skin protection. The epidermis is cooled to –10°C for a short period of time (<100 ms), and is limited to cooling to a depth of about 200 um. The cryogen spray relies on the atomization of the spray for uniform dispersal of the droplets; any irregularities in the droplet size may lead to variable localization of the cooling [74]. Although the heat extraction is up to twenty times more efficient than conductive devices, the lack of optical coupling at the skin surface may increase backscatter. In a humid environment, condensation may occur, impeding the subsequent laser pulse.

8.5 Laser Hair Reduction

While the amount of hair reduction varies depending on the treatment parameters and body site, most reports have proved that multiple treatments are necessary to achieve the best results [75–80]. As a rule, 6–10 laser treatments are required during the first year to achieve long-term results [59]. With most laser systems, a single treatment can reduce hair counts by 10–40%, three treatments by 30–70% and after repeated treatments, as much as 90%. These results can be maintained via posttreatment follow-up for as long as 12 months [60]. However, most published studies on laser hair removal are uncontrolled, small-based (less than 50 subjects), and have used a variety of treatment protocols, equipment, skin types, and hair colors [62]. None of the presently utilized lasers has been proven to destroy hair permanently, and long-term results are still lacking [62]. It has been reported that the hair that does grow back will be smaller, lighter, and less noticeable [52]. The goal is to use the highest fluence tolerated by the subject without causing unwanted side effects. In areas

Table 8.1 Comparison of Skin-Cooling Methods [53]

Cooling System	Projected Tolerable Fluence Increase for Type II Skin for 700 nm
Gel at RT	Add 5 J/cm^2
Water and sapphire at 5C	Add 10–15 J/cm^2
Cryogen spray (–30°C, 30 ms burst)	Add 10–15 J/cm^2
Water and glass at 5°C	Add 5–7 J/cm^2

with a high percent of anagen follicles (i.e., face), treatments should be four weeks apart whereas other areas such as legs and back can be treated at 8–12 week intervals. Subjects with Fitzpatrick skin type I–III and dark hair are the best candidates, and respond equally well to the different lasers [78]. The following choices of lasers are recommended based on hair and skin color:

1. Light, thin hair and Fitzpatrick I–II have the choice of the Ruby or Alexndrite laser,
2. Brown hair, medium thickness, and Fitzpatrick skin phenotype II–IV can use the ruby, alexandrite, or IPL
3. Black, coarse hair, and Fitzpatrick skin IV–VI - diode or Nd:YAG (1).

8.5.1 Treatment of Subjects of Color

The most successful treatment of darker skin types has been reported when careful consideration of the epidermal tissue response to the treatment has been taken into account [3,4]. The use of test spots to select the best fluence and pulse duration for each individual has been reported to be helpful in reducing side effects. If test spots are used, it is recommended to choose a nonobvious location that matches the skin color, sun exposure, and hair density of the area to be treated [81]. Then use two to four fluences starting with a lower fluency setting and longer pulse duration, and gradually increasing the fluences and shortening the pulse durations. It may take up to 48 hours for the full cutaneous response to develop, before the most appropriate treatment can be chosen. In regions with higher hair density, lower fluences have been found to prevent pooling of the heat. However, others have claimed a poor correlation between the area tested with test spots, and the area actually being treated [82]. It has been suggested that the lack of correlation between test spots and treatment area is due to pretreatment sun exposure.

In many reports of laser hair removal with lighter skin color, the best results are often seen in those patients who develop perifollicular edema. However, in patients with darker skin, if perifolliclar erythema and edema lasts for longer than a few minutes or hours, this may be a sign that the fluence will need to be lowered to prevent epidermal thermal damage [5]. An increase in pain levels has also been used to monitor epidermal damage [61].

The safest hair removal systems to treat Fitzpatrick types, IV–VI are the long-pulsed Nd:YAG (1064 nm) and the long-pulsed (≥100 ms) diode laser (810 nm) [83]. The use of aggressive cooling is important with the treatment of skin types VI with the diode laser [84]. In darker-skinned individuals selected for laser treatment, reports of side effects are found to have been minimized, if individuals with a history of keloids or hypertropic scarring are treated cautiously [85].

The highest satisfaction of all individuals undergoing laser treatment has been reported for those who have a clear understanding of the outcome of the long-term reduction in hair afforded by this treatment [86]. The lower fluence required for safe treatment of patients with darker skin types may mean that more treatments are required to reach the desired outcome [87, 108]. It is believed that treatments every four to eight weeks will insure the highest results. In treating subjects with darker skin, it is best to avoid crusting, blistering, or scabbing after treatment [75]. In the event that these side effects do occur, topical antibiotic or corticosteroids have been used [5].

8.5.2 Ruby Laser

Ruby lasers generate light at a wavelength of 694 nm, which is absorbed by the melanin in the skin more than some of the longer wavelengths, making it safer for treating individuals with dark hair and light skin. This wavelength of light does not penetrate as deep as some other lasers, making it better suited for treatment of areas with relatively shallow hair follicles (upper lip and chin) [88]. Normal mode ruby lasers produce pulses in the micro- to millisecond range and deliver high-energy fluence ranging from 10 to 40 J/cm^2. However, there are a few reports on the use of this laser on individuals with darker skin types. Lanigan has reported that in skin types IV–VI, the overall occurrence of side effects is three times higher with the ruby laser (30%) compared to the Nd:YAG (9%) (65). Compos et al. found pigmentary changes in five of six patients treated with this laser [88].

8.5.3 Alexandrite Laser

The long-pulse alexandrite laser has a similar mode of action as the ruby laser, but the former is thought to have several advantages [89]. The use of the 755 nm wavelength allows for deeper penetration into the dermis. However, there is a 20% reduction in the absorption of the laser light at this wavelength compared to that seen at 694 nm [60,90]. The longer pulse width of 20 ms is closer to the thermal relaxation time of the hair follicle (40–100 ms) which may more selectively heat the hair follicle [91]. The longer pulse width allows for better cooling of the epidermis, since it is much longer than the relaxation time of the melanin in this portion of the skin (3–10 ms) [52].

Garcia et al. in a study of subjects with skin types IV–VI pretested the treatment area and used posttreatment topical corticosteroids to reduce any posttreatment inflammation [68]. With these precautions, almost 3% of the subjects had side effects. A study comparing long- and short-pulsed alexandrite photoepilation found that erythema, edema, crusting, and pigmentation changes were more noticeable with the long pulse alexandrite lasers [92]. In addition, induction of hair growth was seen in 3% of those subjects treated with the LP-Alex on the face and neck sites [93]. Studies that have been done on Asian subjects have found satisfactory reduction in hair density. Side effects in these reports have ranged from 1 to 3%, and have included crusting, folliculites, and pigment changes. These side effects are reported to be minimal and transient in this subset of subjects with darker skin [92]. Moreno-Arias et al. report long-lasting hypopigmentation in one subject with Type III skin after a single treatment. The author reports that this may be due to long-lasting melanocyte suppression in this subject [94]. A multicenter prospective study found that mild and short-lived acneform reactions occurred in 6% of subjects undergoing laser hair removal. The acneform lesions are most likely to occur in young, dark-skinned individuals treated with Nd:YAG laser. History of PCOS, number of prior treatments, use of aloe vera cooling gel, and the sex of the patient were not correlated with the development of these lesions (Table 8.2) [94].

8.5.4 Diode

The diode laser emits laser energy in the 800 nm wavelength range and the absorption of the laser light by melanin is similar to the alexandrite laser [56]. A recent study has reported the use of the diode laser to treat persons with skin types V and VI skin types using two

Table 8.2 Hair Removal in Different Skin Types with the Alexandrite Laser

Ref.	Fluence	Pluse Width	Spot Size	No. of Patients	Area Teated	No. of Treat-ments	Treatment Intervals	% Hair Reduction	Follow-up	Skin Type	Hair Color	Pain	Eryth-ema	Blis-ters/ Scabs	Hypo-pigmen-tation	Hyper-pigmen-tation
116	15–19 J/cm^2	10–27 ms	12 mm	13	Back	1	NA	NA	1 mon	I-III	dark	NA	NA	NA	NA	NA
117	16 J/cm^2	NA	12/16 mm	11	Axilla	3	6 wk	52%	6 mon	II-IV	NA	NA	NA	NA	NA	NA
82	16–24 J/cm^2	NA	NA	136	Axillae, legs, face	1–3	4 wk	44–77%	3, 9 mon	III–V	NA	NA	NA	8%	yes	yes
118	31–40 J/cm^2	3 ms	10/15 mm	88	Face/trunk	3–4	4–6 wk	71–79%	12 mon	I–IV	NA	NA	NA	NA	10%	2%
90	15–25 J/cm^2	NA	NA	170	Axillla, leg	2-5	NA	60–90%	NA	IV–V	NA	NA	NA	NA	NA	NA
119	18–22 J/cm^2	20 ms	10 mm	133	Face/trunk	3	6–8 wk	72%	3 mon	II–III	NA	NA	NA	33%	26%	40%
120	30–50 J/cm^2	3 ms	10 mm	24	Trunk	1	NA	43–50%	13–18 mon	II–III	dark	NA	NA	NA	4%	8%
121	30 J/cm^2	20 ms	10–15 mm	NA	Face	8	18 mon	75%	3 mon		NA	NA	NA	17%	no	no
122	20 J/cm^2	5, 10, 20 ms	10 mm	NA	Face, leg, back, bikini	1–2	2 mon	15–56%	6 mon	I–III	NA	NA	NA	NA	NA	1.50%
123	14–22 J/cm^2	5, 20 ms	10 mm	400	Face/ trunk	5–7	1 mon	70%	12 mon	III–IV	NA	NA	NA	NA	NA	NA
124	18 J/cm^2	5, 10, 20 ms	10 mm	36	Face/ trunk	1	NA	4–66%	1, 3, 6 mon		NA	NA	NA	NA	yes	NA
125	20–25 J/cm^2	2 ms	10 mm	NA	Trunk	3–6	NA	75–90%	3 mon	I–III	NA	NA	NA	NA	NA	NA
126	20 J/cm^2	20 ms	10 mm	18	Bikini	5	3 wk	78%	12 mon	I–III	black	NA	NA	NA	14%	NA
127	25 J/cm^2	2–5 ms	7 mm	14	Face/ trunk	3	2–3 mon	34%	6 mon	I–III	dark	NA	NA	NA	NA	NA
92	17–39 J/cm^2	30–40 ms	7 mm	319	Face/ trunk	6–8	3 mon	60%	NA	II–V	dark	NA	NA	9%	3%	3%
128	15–25 J/cm^2	10–20 ms	8/10 mm	29	Face/ trunk	5	1 mon	66%	18 mon	III–V	NA	30%	NA	3%	6%	3%
129	25–40 J/cm^2	2–20 ms	9/10 mm	20	Axilla	1–3	1 mon	17–75%	6 mon	I–IV	dark					

treatment regimens [95]. The first treatment consisted of fluences ranging from 15 to 25 J/cm^2 and a pulse width of 30 ms. Mild crusting and hyperpigmentation were noted with these treatment parameters. Histological studies showed mild epidermal damage and occasional subepidermal separation at the dermal epidermal junction. Full thickness epidermal damage and thermal damage in the dermis were found in samples from subjects receiving higher levels of these fluences. The second laser had a pulse width of 100 ms and a fluence of between 20–30 J/cm^2 [96]. The overall histological evaluations were similar with the longer pulse width with less epidermal damage when similar fluences were compared. Greppi treated eight subjects with skin types V–VI for pseudofolliculites barbe with 10 J/cm^2 with a pulse width of 30 ms [97]. Treatments were given 4–6 weeks apart, and subjects received 7–10 treatments. Hypopigmentation was found in two subjects, and hyperpigmentation in three patients. Both types of pigment changes were resolved within eight months (Table 8.3).

8.5.5 The Nd:YAG Laser

This laser emits light at a wavelength of 1064 nm, and skin penetration of this laser will be the highest of the lasers discussed while the melanin absorption will be the lowest [97]. This low melanin absorption has been used to design lasers that are safer for the skin of subjects with darker skin, or tanned skin [64]. In a study comparing the efficacy and complications of the long-pulsed Nd:YAG Laser and the diode laser, and undertaken in Chinese subjects, Chan and his colleagues found that the long-pulsed Nd:YAG was associated with more pain [98]. They report hypopigmentation as the only long-lasting side effect in their treatment of the underarm and legs that was resolved by 36 weeks. In a study of 11 subjects with Types IV–VI, comparing an IPL system to the Nd:YAG, Goh found that the hair reduction was similar between the two systems [83]. Pain was higher in the Nd:YAG; however, no side effects were seen with the Nd:YAG. Three patients experienced blistering, followed by postinflammatory pigmentation with the IPL system [83]. Galadari compared the alexandrite, Nd:YAG, and the Diode laser in a group of subjects with skin types IV–VI [99]. He reports that the amount of hair reduction did not differ between the three lasers; however the Nd:YAG gave fewer side effects than the other two lasers. Alster et al. treated 20 subjects with skin types IV–VI with dark hair. The authors report significant hair reduction one year after the last treatment (70–90%). The adverse effects in this study included mild to moderate treatment of pain and rare occurrences of transient changes in skin pigment. Histological examination found no epidermal damage tissue with this laser treatment [100]. The Nd:YAG laser has also been used to treat pseudofolliculitis barbae. Both Ross and Weaver report improvement in this condition without long-lasting side effects. [101,102]. Lanigan, in a study of subjects with skin types IV–VI, found that side effects associated with treatment increased with skin type. In this study, there were side effects in 2% of those with type IV, and 20% for types V and VI (Table 8.4) [65].

8.5.6 The IPL Devices

The IPL devices use polychromatic light with a broad wavelength spectrum (590–1200nm) [103–107]. It is difficult to compare efficacies of the IPL devices across studies due to differences in number of treatments, treatment sites, and varying follow-up periods. Fodor et al.

Table 8.3 Hair Removal in Different Skin Types with the Diode Laser

Ref.	Fluence	Pluse Width	Spot Size	No. of Patients	Area Treated	No. of Treat-ments	Intervals	% Hair Reduction	Follow-up	Skin Type	Hair Color	Eryth-ema	Blis-ters/ Scab Crust-ing	Hypo-pigmen-tation	Hyper-pigmen-tation
130	35 J/cm^2	30–45 ms	9–12 mm	29	Lip	3	6–8 wk	49%	6 mon	II–IV	Brown–black	1/3	NA	No	No
131	25–40 J/cm^2	15–30 ms	9 mm	8	Face/neck	2–3	1.5–3 mon	25–78%	5 mon	II–IV	NA	NA	13%	NA	13%
132	25–35 J/cm^2	50 ms	12 mm	24	Face/trunk	3	1 mon	51–74%	1–6 mon	II–III	Brown	NA	NA	NA	NA
133	23–115 J/cm^2	200–1000 ms	10 mm	10	Trunk	2	1 mon	34%	6 mon	II–IV	Dark	62%	NA	10%	23%
134	48 J/cm^2	NA	NA	36	Back/leg	4–6	NA	34–43%	1–3 mon	NA	Dark	NA	NA	NA	NA
135	33 J/cm^2	5–30 ms	9 mm	50	Face/trunk	3	NA	~50	9 mon	II–IV	NA	NA	NA	29%	NA
136	40 J/cm^2	20 ms	9 mm	47	Back/leg	2–6	1 mon	20–53%	1–12 mon	II–III	Dark	NA	NA	15%	44%
137	44/50 J/cm^2	60–80 ms	4 mm	30	Leg	3	4 wks	56–74%	3 mon	NA	NA	NA	NA	NA	NA
97	10 J/cm^2	30 ms	9 mm	8	Face/pfb	7–8	4–6 wks	75–90%					25%	38%	25%

Table 8.4 Hair Removal in Different Skin Types with Nd:Yag Laser

Ref.	Fluence	Pluse Width	Spot Size	No. of Patients	Area Treated	No. of Treat-ments	Treatment Intervals	% Hair Reduction	Follow-up	Skin Type	Hair Color	Pain	Erth-ema	Blis-ters/ Scabs	Hypo-pigmen-tation	Hyper-pigmen-tation
109	40–70 J/cm^2	4 ms	6 mm	480	Trunk	8–10	1–3 mon	35–60%	3 years	I – IV	All	NA	yes	11%	4%	2%
110	40 J/cm^2	50 ms	10 mm	23	Lower leg	5	4 wks	27–71%	1–12 mon	I – IV	All	NA	NA	3%	NA	NA
110	40 J/cm^2	10–20 ms	5–7 mm	20	Lower leg	5	4 wks	0–68%	1–12 mon	I – IV	All	NA	NA	3%	NA	NA
111	40 J/cm^2	4 ms	4 mm	7	Lower leg	5	4 wks	43%	12 mon	I – IV	All	NA	NA	3%	NA	NA
112	25–80 J/cm^2	3.5 ms	4 mm	14	Trunk	1	NA	24–29%	1–3 mon	I – IV	Dark	NA	40%	NA	NA	NA
102	50–100 J/cm^2	50 ms	5 mm	28	Thigh	1	NA	40–50%	12 wks	VI	Dark	Yes	NA	NA	4%	NA
113	65 J/cm^2		3 mm	29	Face	3	4 wks	43%	3–9 mon	I – VI	ND	NA	NA	17%	NA	NA
114	10–55 J/cm^2	23–32 ms	6–8 mm	11	Face/trunk	6	1mon	42%	up to 18 mon	III–V	ND	36%	NA	18%	0%	9%
115	30–135 J/cm^2	35–65ms	5 mm	9	Axilla, trunk, bikini	3	6–8 wks	64%	6–10 mon	II–IV	Dark	NA	NA	NA	NA	NA
101	40–50 J/cm^2	50 ms (3×)	5 mm	8	Face,axilla, leg	3	4 wks	51–90%	12 mon	IV–VI	Dark	Yes	NA	NA	Yes	NA
118	50–100 J/cm^2	50 ms		15		1		29%	3–9 mon		ND	90%			5%	5%

Table 8.5 Hair Removal in Different Skin Types with IPL Systems

Ref.	Fluence	Pluse Width	Spot Size	No. of Patients	Area Treated	Treat-ments	Treatment Intervals	% Hair Reduction	Follow-up	Skin Type	Hair Color	Pain	Ery-thema	Blisters/ Scabs	Hypo-pigmen-tation	Hyper-pigmen-tation
105	25–40 J/cm^2	NA	10 × 45 or 8 × 35 mm	80	Face/trunk	2–6	4 wks	60% good to excellent	NA	III–V	NA	NA	16%	6%	1%	9%
138	25–40 J/cm^2	50–80 ms	10 × 20 mm	156	Face/axilla	3–5	6 wks	70–85%	6 mon	III–V	NA	Yes	NA	NA	NA	NA
107	4–6.5 J/cm^2	35 ms	22 × 55 mm	6	Back/chest	5–7	4 wks	58–71%	3–6 mon	II–III	Dark	Yes	Yes	Yes	NA	NA
139	3–90 J/cm^2	2–25 ms	8 × 35 mm	100	Face	6–8	NA	87%	27 mon	II–III	Dark	6%	6%	NA	NA	6%
140	7.5 J/cm^2	35 ms	22 × 55 mm	10	Face/trunk	5	4 wks	80%	1 mon	III–V	NA	80%	NA	100%	NA	100%
116	6.5 J/cm^2	35 ms	22 × 55 mm	12	Face/trunk	1–3	4 wks	27%	4 mon	I–IV	Red–black	76%	NA	NA	NA	NA
141	18.3 J/cm^2	44.5 ms	NA	11	Bikini	4	4/5 wks	75–80%	4 mon	II–IV	All	NA	NA	NA	NA	NA
92	40–43 J/cm^2	3.8 ms	10 × 45 mm	49	Face	3–9	8 wks	NA	NA	NA	NA	100%	61%	18%	2%	16%

reports that most complications were observed in patients with skin types IV and V [103]. The authors found that blister and transitory hyperpigmentation were common in Type V subjects. They also observed a negative correlation between skin type and subject satisfaction. In a small study of subjects with Type V and VI skin, Johnson and Dovale report no lasting pigmentary changes with 5–7 treatments with an IPL device [104]. Lee et al. found that using a 645–950 nm filter and longer pulse width was more effective in reducing hair density, as well as safer, in Asian patients [106].

8.6 Conclusion

As the demand for laser hair removal continues for individuals regardless of skin color, understanding the choice of treatment parameters to prevent unwanted side effects becomes more important. The use of longer wavelengths, the long-pulsed Nd:YAG (1064 nm), and the long-pulsed (≥100 ms) diode laser (810 nm), with longer pulse widths is becoming the laser treatement of choice for hair removal in subjects with darker skin types [87,108]. The highest satisfaction of all individuals undergoing laser treatment is seen when patients are given a clear understanding of what to expect from the treatment.

References

1. Sadick, N. S. (2004). Laser hair removal. *Facial Plastic Surgery Clinics of North America*, 12(2), 191–200.
2. Tanzi, E. L., & Alster, T. S. (2004). Cutaneous laser surgery in darker skin phototypes. *Cutis*, 73(1), 21–24.
3. Jackson, B. A. (2003). Lasers in ethnic skin: A review. *Journal of the American Academy of Dermatology*, 48(6 suppl.), S134–S138
4. Yee, S. (2005). Laser hair removal in Fitzpatrick type IV to VI patients. *Facial Plastic Surgery*, 21(2),139–144.
5. Battle Jr., E. F., & Hobbs, L. M. (2004). Laser-assisted hair removal for darker skin types. *Dermatologic Therapy*, 17(2),177–183.
6. Lim, S. P. R., & Lanigan, S. W. (2006). A review of the adverse effects of laser hair removal. *Lasers in Medical Science*, 21(3), 121–125.
7. Fanous, N., & Yoskovitch, A. (2002). New classification scheme for laser resurfacing and chemical peels. Modifications for the different ethnic groups. *Facial Plastic Surgery Clinics*, 10(4), 405–413.
8. Hexsel, D., Arellano, I., & Rendon, M. (2006). Ethnic considerations in the treatment of Hispanic and Latin American patients with hyperpigmentation. *British Journal of Dermatology*, 156(Suppl. 1), 7–12.
9. Rawlings, A. V. (2006). Ethnic skin types: Are there differences in skin structure and function? *International Journal of Cosmetic Science*, 28(2), 79–93.
10. Wagner, J. K., Parra, E. J., Norton, H. L., Jovel, C., & Shriver, M. D. (2002). Skin responses to ultraviolet radiation: Effects of constitutive pigmentation, sex, and ancestry. *Pigment Cell Research*, 15(5), 385–390.
11. Leenutaphong V. (1995). Relationship between skin color and cutaneous response to ultraviolet radiation in Thai. *Photodermatology Photoimmunology and Photomedicine*, 11(5–6), 198–203.
12. Taylor, S. (2002). Skin of color: Biology, structure, function, and implications for dermatologic disease. *Journal of the American Academy of Dermatology*, 46(2, Part 3), S41– S62.
13. Quevedo Jr., W. C., Fitzpatrick, T. B., & Jimbow, K. (1985). Human skin color: Origin, variation and significance. *Journal of Human Evolution*, 14(1), 43–56.

14. Alaluf, S., Atkins, D., Barrett, K., Blount, M., Carter, N., & Heath, A. (2002). Ethnic variation in melanin content and composition in photoexposed and photoprotected human skin. *Pigment Cell Research*, 15(2), 112–118.
15. Thong H, Jee S, Sun C , and Boissy RE. (2003). The patterns of melanosome distribution in keratinocytes of human skin as one determining factor of skin colour. *British Journal of Dermatology*, 149(3), 498–505.
16. Goldschmidt, H., & Raymond, J. Z. (1972). Quantitative analysis of skin color from melanin content of superficial skin cells. *Journal of Forensic Sciences*, 17(1), 124–131.
17. Choe, Y. B., Jang, S. J., Jo, S. J., Ahn, K. J., & Youn J. (2006). The difference between the constitutive and facultative skin color does not reflect skin phototype in Asian skin. *Skin Research and Technology*, 12(1), 68–72.
18. Youn, J. I., Oh, J. K., Kim, B. K., Suh, D. H., Chung, J. H., & Oh, S. J., et al. (1997). Relationship between skin phototype and MED in Korean, brown skin. *Photodermatology Photoimmunology and Photomedicine*, 13(5–6), 208–211.
19. Del Bino, S., Sok, J., Bessac, E., & Bernerd, F. (2006). Relationship between skin response to ultraviolet exposure and skin color type. *Pigment Cell Research*, 19(6), 606–614.
20. Quevedo Jr., W. C., Fitzpatrick, T. B., Pathak, M. A., & Jimbow, K. (1975). Role of light in human skin color variation. *American Journal of Physical Anthropology*, 43(3), 393–408.
21. Fitzpatrick, T. B. (1988). The validity and practicality of sun-reactive skin types I through VI. *Archives of Dermatology*, 124(6), 869–871.
22. Fanous, N. (2002). A new patient classification for laser resurfacing and peels: Predicting responses, risks, and results. *Aesthetic Plastic Surgery*, 26(2), 99–104.
23. Lancer, H. A. (1998). Lancer ethnicity scale (LES). *Lasers in Surgery and Medicine*, 22(1), 9.
24. Willis, I., & Earles, R. M. (2005). A new skin classification system relevant to people of African descent. *Cosmetic Dermatology*, 18(3), 209–216.
25. Drosner, M., & Adatto, M. (2005). Photo-epilation: Guidelines for care from the European society for laser dermatology (ESLD). *Journal of Cosmetic and Laser Therapy*, 7(1), 33–38.
26. Lin, J. Y., & Fisher, D. E. (2007). Melanocyte biology and skin pigmentation. *Nature*, 445(7130), 843–850.
27. Slominski, A., Wortsman, J., Plonka, P. M., Schallreuter, K. U., Paus, R., & Tobin, D. J. (2005). Hair follicle pigmentation. *Journal of Investigative Dermatology*, 124(1), 13–21.
28. Schallreuter, K., Slominski, A., Pawelek J. M., Jimbow, K., & Gilchrest, B. A. (1998). What controls melanogenesis? *Experimental Dermatology*, 7(4), 143–150.
29. Knappskog, P. M., Flatmark, T., Aarden, J. M., Haavik, J., & Martínez, A. (1996). Structure/function relationships in human phenylalanine hydroxylase: Effect of terminal deletions on the oligomerization, activation and cooperativity of substrate binding to the enzyme. *European Journal of Biochemistry*, 242(3), 813–821.
30. Tolleson, W. H. (2005). Human melanocyte biology, toxicology, and pathology. *Journal of Environmental Science and Health - Part C Environmental Carcinogenesis and Ecotoxicology Reviews*, 23(2), 105–161.
31. Boissy, R. E. (1988). The melanocyte: Its structure, function, and subpopulations in skin, eyes, and hair. *Dermatologic Clinics*, 6(2), 161–173.
32. Jimbow, K., Fitzpatrick, T. B., & Wick, M. M. (1991). Biochemistry and physiology of melanin pigmentation. *Physiology, Biochemistry, and Molecular Biology of the Skin*, 2, 873–909.
33. Marles, L. K., Peters, E. M., Tobin, D. J., Hibberts, N. A., & Schallreuter, K. U. (2003). Tyrosine hydroxylase isoenzyme I is present in human melanosomes: A possible novel function in pigmentation. *Experimental Dermatology*, 12(1), 61–70.
34. Hearing, V. J., & Tsukamoto, K. (1991). Enzymatic control of pigmentation in mammals. *FASEB Journal*, 5(14), 2902–2909.
35. Sulaimon, S. S., & Kitchell, B. E. (2003). Review article the biology of melanocytes. *Veterinary Dermatology*, 14(2), 57–65.
36. Duval, C., Smit, N. P. M., Kolb, A. M., Régnier, M., Pavel, S., & Schmidt, R. (2002). Keratinocytes control the pheo/eumelanin ratio in cultured normal human melanocytes. *Pigment Cell Research*, 15(6), 440–446.

37. Inazu, M., & Mishima, Y. (1993). Detection of eumelanogenic and pheomelanogenic melanosomes in the same normal human melanocyte. *Journal of Investigative Dermatology*, 100(2 suppl.), 172S–175S.
38. Øyehaug, L., Plahte, E., Våge, D. I., & Omholt, S. W. (2002). The regulatory basis of melanogenic switching. *Journal of Theoretical Biology*, 215(4), 449–468.
39. Riley, P. A. (2003). Melanogenesis and melanoma. *Pigment Cell Research*, 16(5), 548–552.
40. Slominski, A., Tobin, D. J., Shibahara, S., & Wortsman J. (2004). Melanin pigmentation in mammalian skin and its hormonal regulation. *Physiological Reviews*, 84(4), 1155–1228.
41. Scott, G., Leopardi, S., Printup, S., & Madden, B. C. (2002). Filopodia are conduits for melanosome transfer to keratinocytes. *Journal of Cell Science*, 115(7), 1441–1451.
42. Kishi, K., Kubota, Y., Nakajima, H., & Nakajima, T. (2007). Molecules involved in the hair cycle. *Japanese Journal of Plastic and Reconstructive Surgery*, 50(2), 157–161.
43. Abell, E. (1994). Embryology and anatomy of the hair follicle. In: Olsen, E. A., ed., *Disorders of Hair Growth: Diagnosis and Treatment*. New York: McGraw-Hill, pp. 1–19.
44. Fuchs, E. (1998). Beauty is skin deep: The fascinating biology of the epidermis and its appendages. *Harvey Lectures*, 94, 47–77.
45. Stenn, K. S., & Paus, R. (1999). What controls hair follicle cycling? *Experimental Dermatology*, 8(4), 229–233.
46. Kligman, A. M. (1959). The human hair cycle. *Journal of Investigative Dermatolology*, 33, 307–316.
47. Tobin, D. J., Hordinsky, M., & Bernard, B. A. (2005). Hair pigmentation: A research update. *Journal of Investigative Dermatology Symposium Proceedings*, 10(3), 275–279.
48. Commo, S., & Bernard, B. A. (2000). Melanocyte subpopulation turnover during the human hair cycle: An immunohistochemical study. *Pigment Cell Research*, 13(4), 253–259.
49. Anderson, R. R., & Parrish, J. A. (1983). Selective photothermolysis: Precise microsurgery by selective absorption of pulsed radiation. *Science*, 220(4596), 524–527.
50. Steggerda, M., & Seibert, H. C. (1942). Size and shape of head hair from six racial groups. *Journal of Heredity*, 32, 315–318.
51. Vernall, D. G. (1961). Study of the size and shape of hair from four races of men. *American Journal of Physical Anthropology*, 19, 345–350.
52. Dierickx, C. C. (2002). Hair removal by lasers and intense pulsed light sources. *Dermatologic Clinics*, 20(1), 135–146. Retrieved July 14, 2007, from Scopus database.
53. Ross, E. V., Ladin, Z., Kreindel, M., & Dierickx, C. (1999). Theoretical considerations in laser hair removal. *Dermatologic Clinics*, 17(2), 333–355.
54. Cotsarelis, G., Kaur, P., Dhouailly , D., Hengge, U., & Bickenbach, J. R. (1999). Epithelial stem cells in the skin: Definition, markers, localization and functions. *Experimental Dermatology*, 8(1), 80–88.
55. Anderson, R. R. (2000). Lasers in dermatology—A critical update. *Journal of Dermatology*, 27(11), 700–705.
56. Ort, R. J., & Anderson, R. R. (1999). Optical hair removal. *Seminars in Cutaneous Medicine and Surgery*, 18(2), 149–158.
57. Tobin, D. J., Slominski, A., Botchkarev, V., & Paus, R. (1999). The fate of hair follicle melanocytes during the hair growth cycle. *Journal of Investigative Dermatology Symposium Proceedings*, 4(3), 323–332.
58. Paus, R., & Cotsarelis, G. (1999). The biology of hair follicles. *New England Journal of Medicine*, 341(7), 491–497.
59. Dierickx, C. C., Campos, V. B., Lin, D., Farinelli, W., & Anderson, R. R. (1999). Influence of hair growth cycle on efficacy of laser hair removal. *Lasers in Surgery and Medicine*, (Suppl. 11), 21–29.
60. Goldberg, D. J. (2007). Laser- and light-based hair removal: An update. *Expert Review of Medical Devices*, 4(2), 253–260.
61. Lepselter, J., & Elman, M. (2004). Biological and clinical aspects in laser hair removal. *Journal of Dermatological Treatment*, 15(2), 72–83.
62. Haedersdal, M., & GØtzsche, P. C. (2006). Laser and photoepilation for unwanted hair growth. *Cochrane Database of Systematic Reviews*, (4).

63. Lacombe, V. G. (2004). Laser hair removal. *Facial Plastic Surgery*, 20(1), 85–89.
64. Nouri, K., & Rivas, M. P. (2003). Review of laser hair removal in Fitzpatrick skin types IV to VI. *Cosmetic Dermatology*, 16(6), 24–26.
65. Lanigan, S. W. (2003). Incidence of side effects after laser hair removal. *Journal of the American Academy of Dermatology*, 49(5), 882–886.
66. Lim, S. P. R., & Lanigan, S. W. (2006). A review of the adverse effects of laser hair removal. *Lasers in Medical Science*, 21(3), 121–125.
67. Blume-Peytavi, U., Gieler, U., Hoffmann, R., Lavery , S., Shapiro, J., & Chamberlain James, L. (2007). Unwanted facial hair: Affects, effects and solutions. *Dermatology*, 215(2), 139–146.
68. Garcia, C., Alamoudi, H., Nakib, M., & Zimmo S. (2000). Alexandrite laser hair removal is safe for Fitzpatrick skin types IV-VI. *Dermatologic Surgery*, 26(2), 130–134.
69. Liew, S., Grobbelaar, A., Gault, D., Green, C., & Ling, C. (1999). The Effect of ruby laser light on cellular proliferation of epidermal cells. *Annals of Plastic Surgery*, 43 (5), 519–522.
70. Altshuler, G. B., Zenzie, H. H., Erofeev, A. V., Smirnov, M. Z., Anderson, R. R., & Dierickx, C. (1999). Contact cooling of the skin. *Physics in Medicine and Biology*, 44(4), 1003–1023.
71. Klavuhn, K. G., & Green, D. (2002). Importance of cutaneous cooling during photothermal epilation: Theoretical and practical considerations. *Lasers in Surgery and Medicine*, 31(2), 97–110.
72. Zenzie, H. H., Altshuler, G. B., Smirnov, M. Z., & Anderson, R. R. (2000). Evaluation of cooling methods for laser dermatology. *Lasers in Surgery and Medicine*, 26(2), 130–131.
73. Aguilar, G., Majaron, B., Viator, J. A.,, Basinger, B., Karapetian, E., & Svaasand, L. O. et al. (2001). Influence of spraying distance and post-cooling on cryogen spray cooling for dermatologic laser surgery. *Proceedings of SPIE*, 4244, 82–92
74. Majaron, B., Kimel, S., Verkruysse, W., Aguilar, G., Pope, K., & Svaasand, L. O. et al. (2001). Cryogen spray cooling in laser dermatology: Effects of ambient humidity and frost formation. *Lasers in Surgery and Medicine*, 28(5), 469–476.
75. Humohreys, T. (2004). Instrument capsules: Laser hair removal in pigmcented skin. *Skinmed*, 3(4), 220–221.
76. Shapiro, J., & Lui, H. (2005). Treatments for unwanted facial hair. *Skin Therapy Letter*, 10(10), 1–4.
77. Wanner, M. (2005). Laser hair removal. *Dermatologic Therapy*, 18(3), 209–216.
78. Warner, J., Weiner, M., & Gutowski, K. A. (2006). Laser hair removal. *Clinical Obstetrics and Gynecology*, 49(2), 389–400.
79. Woolery-Lloyd, H. (2003). Hair removal techniques: A review. *Cosmetic Dermatology*, 16(6), 45–48, 51.
80. Se, H. L. (2002). Laser hair removal: Guidelines for management. *American Journal of Clinical Dermatology*, 3(2), 107–115.
81. Nanni, C. A., & Alster, T. S. (1998). A practical review of laser-assisted hair removal using the Q-switched Nd:YAG, long-pulsed ruby, and long-pulsed alexandrite lasers. *Dermatologic Surgery*, 24(12), 1399–1405.
82. Hussain, M., Polnikorn, N., & Goldberg, D. J. (2003). Laser-assisted hair removal in Asian skin: Efficacy, complications, and the effect of single versus multiple treatments. *Dermatologic Surgery*, 29(3), 249–254.
83. Goh, C. L. (2003). Comparative study on a single treatment response to long pulse Nd:YAG lasers and intense pulse light therapy for hair removal on skin type IV to VI—is longer wavelengths lasers preferred over shorter wavelengths lights for assisted hair removal. *Journal of Dermatological Treatment*, 14(4), 243–247.
84. Rispler, J. (2003). Laser-assisted hair removal for darkly pigmented skin. *Aesthetic Surgery Journal*, 23(2), 143–144.
85. Breadon, J. Y., & Barnes, C. A. (2007). Comparison of adverse events of laser and light-assisted hair removal systems in skin types IV-VI. *Journal of Drugs in Dermatology*, 6(1), 40–46.
86. Haedersdal, M., & Wulf, H. C. (2006). Evidence-based review of hair removal using lasers and light sources. *Journal of the European Academy of Dermatology and Venereology*, 20(1), 9–20.
87. Lask, G., Eckhouse, S., Slatkine, M., Waldman, A., Kreindel, M., & Gottfried, V. (1999). The role of laser and intense light sources in photo-epilation: A comparative evaluation. *Journal of Cutaneous Laser Therapy*, 1(1), 3–13.

88. Campos, V. B., Dierickx, C. C., Farinelli, W. A., Lin, T.-Y. D., Manuskiatti, W., & Anderson, R. R. (2000). Ruby laser hair removal: Evaluation of long-term efficacy and side effects. *Lasers in Surgery and Medicine*, 26(2), 177–185.
89. Tatlidede, S., Egemen, O., Saltat, A., Turgut , G., Karasoy, A., and& Kuran, I. (2005). Hair removal with the long-pulse alexandrite laser. *Aesthetic Surgery Journal*, 25(2), 138–143.
90. Lu, S. Y., Lee, C. C., & Wu, Y. Y. (2001). Hair removal by long-pulse alexandrite laser in oriental patients. *Annals of Plastic Surgery*, 47(4), 404–411.
91. Weir, V. M., & Woo, T. Y. (1999). Photo-assisted epilation--review and personal observations. *Journal of Cutaneous Laser Therapy*, 1(3), 135–143.
92. Marayiannis, K. B., Vlachos, S. P., Savva, M. P., & Kontoes, P. P. (2003). Efficacy of long- and short pulse alexandrite lasers compared with an intense pulsed light source for epilation: A study on 532 sites in 389 patients. *Journal of Cosmetic and Laser Therapy*, 5(3–4), 140–145.
93. Jin, H. , Wang, J., Jiang, G., Wang, H., Liu, Y., & Zuo, Y., et al. (2006). Effectiveness and safety of long-pulsed alexandrite laser for hair removal in 1 702 patients. *Acta Academiae Medicinae Sinicae*, 28(2), 210–213.
94. Moreno-Arias, G. A., & Camps-Fresneda, A. (2003). Long-lasting hypopigmentation induced by long-pulsed alexandrite laser photo-epilation. *Dermatologic Surgery*, 29(4), 420–422.
95. Carter, J. J., & Lanigan, S. W. (2006). Incidence of acneform reactions after laser hair removal. *Lasers in Medical Science*, 21(2), 82–85.
96. Adrian, R. M., & Shay, K. P. (2000). 800 nanometer diode laser hair removal in African American patients: A clinical and histologic study. *Journal of Cutaneous Laser Therapy*, 2(4), 183–190.
97. Greppi, I. (2001). Diode laser hair removal of the black patient. *Lasers in Surgery and Medicine*, 28(2), 150–155.
98. Chan, H. H., Ying S.-Y., Ho, W.-S., Wong, D. S. Y., & Lam, L. (2001). An in vivo study comparing the efficacy and complications of diode laser and long-pulsed Nd:YAG laser in hair removal in Chinese patients. *Dermatologic Surgery*, 27(11), 950–954.
99. Galadari, I. (2003). Comparative evaluation of different hair removal lasers in skin types IV, V, and VI. *International Journal of Dermatology*, 42(1), 68–70.
100. Alster, T. S., Bryan, H., & Williams, C. M. (2001). Long-pulsed Nd:YAG laser-assisted hair removal in pigmented skin: A clinical and histological evaluation. *Archives of Dermatology*, 137(7), 885–889
101. Ross, E. V., Cooke, L. M., Timko, A. L., Overstreet, K. A., Graham, B. S., & Barnette, D. J. (2002). Treatment of pseudofolliculitis barbae in skin types IV, V, and VI with a long-pulsed neodymium:Yttrium aluminum garnet laser. *Journal of the American Academy of Dermatology*, 47(2), 263–270.
102. Weaver 3rd, S. M., & Sagaral, E. C. (2003). Treatment of pseudofolliculitis barbae using the long-pulse Nd:YAG laser on skin types V and VI. *Dermatologic Surgery*, 29(12), 1187–1191.
103. Fodor, L., Menachem, M., Ramon, Y., Shoshani, O., Rissin, Y., & Eldor, L. et al. (2005). Hair removal using intense pulsed light (epilight): Patient satisfaction, our experience, and literature review. *Annals of Plastic Surgery*, 54(1), 8–14.
104. Johnson, F., & Dovale, M. (1999). Intense pulsed light treatment of hirsutism: Case reports of skin phototypes V and VI. *Journal of Cutaneous Laser Therapy*, 1(4), 233–237.
105. Adatto, M. (2003). Hair removal with a combined light/heat-based photo-epilation system: a 6-month follow-up. *Journal of cosmetic and laser therapy,* 5(3–4), 163–167.
106. Lee, J. H., Huh, C. H., Yoon, H. J., Cho, K. H., & Chung, J. H. (2006). Photoepilation results of axillary hair in dark-skinned patients by IPL: A comparison between different wavelength and pulse width. *Dermatologic Surgery,* 32(2), 234–240
107. Ferraro GA, Perrotta A, Rossano F, and D'Andrea F .(2004). Neodymium:yttrium-aluminum-garnet long impulse laser for the elimination of superfluous hair: Experiences and considerations from 3 years of activity. *Aesthetic Plastic Surgery*, 28(6), 431–434.
108. Raff, K., Landthaler, M., & Hohenleutner, U. (2003). Optimizing treatment parameters for hair removal using long-pulsed Nd:YAG lasers. *Lasers in Medical Science*, 18(4), 219–222.
109. Lorenz, S., Brunnberg, S., Landthaler, M., & Hohenleutner, U. (2002). Hair removal with the long pulse d Nd:YAG laser: A prospective study with one year follow-up. *Lasers in Surgery and Medicine.*, 30(2), 127–134.

110. Fournier, N., Aghajan-Nouri, N., Barneon, G., & Mordon, S. (2000). Hair removal with an Athos Nd:YAG 3.5 ms pulse laser: A 3-month clinical study. *Journal of Cutaneous Laser Therapy* ., 2(3),125–130.
111. Levy, J. L., Trelles, M. A, & Ramecourt, A. D. (2001). Epilation with a long-pulse 1064 nm Nd:YAG laser in facial hirsutism. *Journal of Cosmetic and Laser Therapy*, 3(4), 175–179.
112. Goldberg, D. J., & Samady, J. A. (2000). Evaluation of a long-pulse Q-switched Nd:YAG laser for hair removal. *Dermatologic Surgery*., 26(2), 109–113.
113. Bencini, P. L., Luci, A., Galimberti, M., & Ferranti, G. (1999). Long-term epilation with long-pulsed neodimium:YAG laser. *Dermatologic Surgery*, 25(3),175–178.
114. Goldberg, D. J., & Silapunt, S. (2001). Hair removal with a combined light/heat based photo-epilation system. *Journal of Cutaneous Laser Therapy*, 3(1), 3–7.
115. Goldberg, D. J., & Silapunt, S. (2001). Hair removal using a long-pulsed Nd:YAG Laser: Comparison at fluences of 50, 80,and 100 J/cm. *Dermatologic Surgery*, 27(5), 434–436.
116. Eremia, S., Li, C. Y., Umar, S. H., & Newman, N. (2001). Laser hair removal: Long-term results with a 755 nm alexandrite laser. *Dermatologic Surgery*, 27(11), 920–924; 1-year results, 27(11), 925–929.
117. Freedman, B. M., & Earley, R. V. (2000). Comparing treatment outcomes between physician and nurse treated patients in laser hair removal. *Journal of Cutaneous Laser Therapy*, 2(3), 137–140.
118. Laughlin, S. A., & Dudley, D. K. (2000). Long-term hair removal using a 3-millisecond alexandrite laser. *Journal of Cutaneous Medicine and Surgery*, 4(2), 83–88.
119. Raulin, C., & Greve, B. (2000). Temporary hair loss using the long-pulsed alexandrite laser at 20 milliseconds. *European Journal of Dermatology*, 10(2), 103–106.
120. McDaniel, D. H., Lord, J., Ash, K., Newman, J., & Zukowski, M. (1999). Laser hair removal: A review and report on the use of the long-pulsed alexandrite laser for hair reduction of the upper lip, leg, back, and bikini region. *Dermatologic Surgery*, 25(6), 425–430.
121. Salem, A. M., Shokeir, H., Badawi, A., El-Morsy, I., Azaam, O., & Shaarawy, I. et al. (2003). Effective reduction of fine and coarse hair in patients with skin types III-V by long-pulsed alexandrite laser at 5- and 20-millisecond pulse durations. *Cosmetic Dermatology*, 16(7), 39–42.
122. Nanni, C. A., & Alster, T. S. (1999). Long-pulsed alexandrite laser-assisted hair removal at 5, 10, and 20 millisecond pulse durations. *Lasers in Surgery and Medicine*, 24(5), 332–337.
123. Boss, W. K., Jr., Usal, H., Thompson, R. C., & Fiorillo, M. A. (1999). A comparison of the long-pulse and short-pulse Alexandrite laser hair removal systems. *Annals of Plastic Surgery*, 42(4), 381–384.
124. Finkel, B., Eliezri, Y. D., Waldman, A., Slatkine, M. E. (1997). Pulsed alexandrite laser technology for noninvasive hair removal. *Journal of Clinical Laser Medicine and Surgery*, 15(5), 225–229.
125. Lloyd, J. R., & Mirkov, M. (2000). Long-term evaluation of the long-pulsed alexandrite laser for the removal of bikini hair at shortened treatment intervals. *Dermatologic Surg*, 26(7), 633–637.
126. Goldberg, D. J. (2002). Laser hair removal. *Dermatologic Clinics*, 20(3), 561–567.
127. Bouzari, N., Tabatabai, H., Abbasi, Z., Firooz, A., & Dowlati, Y. (2004). Laser hair removal: Comparison of long-pulsed alxeandrite and long-pulsed diode laser. *Dermatologic Surgery*, 30(4), 498–502.
128. Handrick, C., & Alster, T. S. (2001). Comparison of long-pulsed diode and long-pulsed alexandrite lasers for hair removal: A long-term clinical and histologic study. *Dermatologic Surgery*, 27(7), 622–626.
129. Fiskerstrand, E. J., Svaasand, L. O., Nelson, J., & Stuart, C. S. (2003). Hair removal with long pulsed diode lasers: A comparison between two systems with different pulse structures. *Lasers in Surgery and Medicine*, 32(5), 399–404.
130. Kopera, D. (2003). Hair reduction: 48 months of experience with 800 nm diode laser. *Journal of Cosmetic and Laser Therapy*, 5(3–4) 146–149.
131. Bouzari, N., Tabatabai, H., Abbasi, Z., Firooz, A., & Dowlati, Y. C. S. (2005). Hair removal using an 800-nm diode laser: Comparison at different treatment intervals of 45, 60, and 90 days. *International Journal of Dermatology*, 44, 50–53.

132. Bäumler, W., Scherer, K., Abels, C., Neff, S., Landthaler, M., & Szeimies, R. (2002). The effect of different spot sizes on the efficacy of hair removal using a long-pulsed diode laser. *Dermatologic Surgery*, 28(2), 118–121.
133. Sadick, N. S., & Prieto, V. G. (2003). The use of a new diode laser for hair removal. *Dermatologic Surgery*, 29(1), 30–33.
134. Baugh, W. P., Trafeli, J. P., Barnette, D. J., Jr., & Ross, E. V. (2001). Hair reduction using a scanning 800 nm diode laser. *Dermatologic Surgery*, 27(4), 358–364. Kauvar, A. N. (2000). Treatment of pseudofolliculitis with a pulsed infrared laser. *Archives of Dermatology*, 136(11), 1343–1346.
135. Lou, W. W., Quintana, A. T., Geronemus, R. G., & Grossman, M. C. (2000). Prospective study of hair reduction by diode laser (800 nm) with long-term follow-up. *Dermatologic Surgery*, 26(5), 428–432.
136. Hussain, M., Suwanchinda, A., Charuwichtratana, S., Goldberg, D. (2003). A new long pulsed 940 nm diode laser used for hair removal in Asian skin types. *Journal of Cosmetic and Laser Therapy*, 5(2), 97–100.
137. Bedewi, A. (2004). Hair removal with intense pulsed light. *Lasers in Medical Science*, 19(1), 48–51.
138. Schroeter, C. A., Groenewegen, J. S., Reineke, T., & Neumann, H. (2004) Hair reduction using intense pulsed light source . *Dermatologic Surgery*, 30(2), 168–173.
139. Troilius, A., & Troilius, C. (1999). Hair removal with a second generation broad spectrum intense pulsed light source—A long-term follow-up. *Journal of Cutaneous Laser Therapy*, 1(3), 173–178;
140. Trelles, M. A., Allones, I., Calderhead, R. G., & Velez, M (2003) Hair removal evaluated with a filterless flashlamp-based system: A preliminary study in 10 patients. *Journal of Cosmetic and Laser Therapy*, 5(1), 15–24.
141. Marayiannis, K. B., Vlachos, S. P., Savva, M. P., & Kontoes, P. P. (2003). Efficacy of long- and short pulse alexandrite lasers compared with an intense pulsed light source for epilation: a study on 532 sites in 389 patients. *Journal of Cosmetic and Laser Theropy*, 5(3–4):140–145.

9

Effect of Laser and Light-Based Systems on Hair Follicle Biology

Natalia V. Botchkareva[1,2] *and Gurpreet S. Ahluwalia*[1]

[1]*The Gillette Company, a wholly owned subsidiary of The P&G Company, Needham, MA, USA*
[2]*School of Life Sciences, The University of Bradford, Bradford, UK*

Gurpreet S. Ahluwalia (ed.), *Cosmetic Applications of Laser and Light-Based Systems*, 217–237,

9.1 Introduction

Over the last decade, photoepilation has become one of the most popular method for longer-term management of unwanted hair growth. Based on the principle of selective photothermolysis first described by Anderson and Parrish [1], several different types of the lasers and noncoherent light sources have been developed to selectively exert the effects on hair follicle structure and achieve a long lasting hair removal benefit [2–8]. Under this principle, advantage is taken of the endogenous chromophore melanin concentrated in the pigment-producing melanocytes of the hair matrix, and in the keratinocytes of the hair shaft of anagen hair follicles and its absence in the surrounding dermal tissue thus allowing for a selective targeting of the laser energy. Based on the level of hair follicle melanin and the laser parameters used, the amount of thermal energy released can cause a variable damage to the follicle, resulting in a hair removal ranging from a temporary to a permanent effect [9–12]. Topping et al. demonstrated that the light- induced effects are directly related to an increase in the temperature of the hair follicle [13]. The rise in temperature is dependent both on the quantity and the type of the melanin pigment present. Dark pigmented hair follicles with high eumelanin content are more sensitive to the laser, compared to the blonde, red, and gray hair, which are either nonresponsive or show a minimal effect [14,15].

At the basic biochemical and molecular level, the released thermal energy from laser treatment can either simply kill the fibre producing cells of hair follicle by causing denaturation of cellular proteins and phospholipid membranes or can modify the molecular mechanisms and signaling pathways that control the growth and cycling of the hair follicle. Changes in the hair follicle structure and histopathological alterations after exposure to high fluence laser energy to permanently effect hair growth have been studied by several researcher groups. Ono and Taleshita demonstrated injury to the melanin containing cells, hair shaft, and outer root sheath cells following high energy laser treatment [16]. Though high fluence lasers have the potential to permanently effect hair growth as desired by many consumers, in most part it is difficult to achieve because of the collateral effect on the skin at the high laser energy. Pain, erythema, edema, and blistering are some of the potential dermal adverse events observed during or immediately after the irradiation, additionally, pigmentary changes, scarring, and skin sensitivity can also be developed as a delayed side effect [5]. The severity and frequency of these dermal events is dependent on the skin pigmentation level, and the type of laser and the laser parameters used.

In order to fully understand the effects of laser energy on hair growth and design the next generation laser and light-based systems, or modify the laser parameters on current systems to achieve the consumer-desired high hair removal efficacy and low dermal side effects, it is important to understand how thermal changes in follicle effect the biochemical and molecular mechanisms that regulate its growth and cycling. This report reviews the critical hair structures and hair growth processes that are either involved in converting the laser photons to thermal energy, or are part of the biochemical and molecular systems affected by the thermal energy.

9.2 Hair Follicle Anatomy and Hair Cycle

Understanding hair follicle components and hair cycle is important to understanding the overall sensitivity and reaction of follicles to laser exposure. The hair follicle is a multicellular structure that generates an organ-specific product—a pigmented hair fiber, as a result of tightly coordinated interactions between epithelial, mesenchymal, and neuroectodermal cells of the follicle. The hair follicle can be divided into several anatomical compartments. The upper follicle is permanent, but the lower follicle regenerates with each hair follicle cycle. The infundibulum extends from the skin surface to the sebaceous duct. The isthmus is a middle portion, which extends from the duct of sebaceous gland to the exertion of arrector pilli muscle. The inferior segment or lower follicle part consists of the suprabulbar and the bulbar areas. The lower hair bulb comprises extensively proliferating keratinocytes and pigment-producing melanocytes of the hair matrix, whereas suprabulbar area contains differentiated cells derived from the hair matrix. The suprabulbar and bulbar areas are separated by a line across the widest part of the dermal papilla, and it is known as the Auber line. The dermal papilla is a cluster of mesenchymal cells. The extensive epithelial-mesenhymal interactions occurring in the hair bulb leads to the formation of the final hair follicle product—the hair shaft [17,18].

The hair follicle has a multilayered structure, where the major compartments include: the hair shaft, the inner root sheath, the companion layer, the outer root sheath, and the connective tissue sheath (Fig. 9.1). The hair shaft is subdivided into three layers: the medulla, cortex, and the cuticle. The major structural proteins of the cytoskeleton synthesized in the hair shaft are the keratins and keratin-associated proteins [19]. The medulla

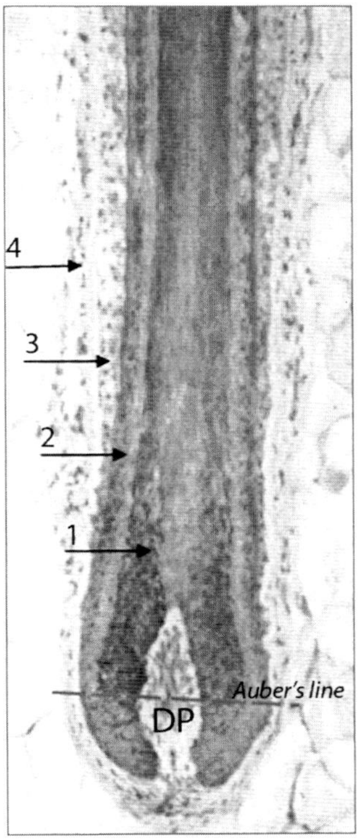

Figure 9.1 Histomorphology of human anagen hair follicle. The major compartments are shown: (1) hair shaft, (2) inner root sheath, (3) outer root sheath, (4) connective tissue sheath. DP: dermal papilla.

is a central part of larger hair. Medullary cells of all types of male and female sexual hair constitutively express the Type I hair keratin hHa7, which is directly regulated by androgens [20]. The hair-shaft cortex is composed of longitudinally arranged fibers. The hair-shaft cuticle covers the hair, and its integrity and properties greatly impact the appearance and character of the hair. It is formed by a layer of scales, which interlock with opposing scales of the inner root sheath that allows the hair shaft and the inner root sheath to move upward in concert. The inner root sheath extends from the base of the bulb to the isthmus. It consists of four layers: companion layer, Henle's layer, Huxley's layer, and the inner root sheath cuticle. The cells of the inner root sheath cuticle partially overlap with the cuticle cells of the hair shaft, which also allows anchoring of the hair shaft to the base of the follicle. Inner root sheath cells express trichohyalin protein that is an intermediate-filament-associated protein, and play a role in the lateral aggregation, alignment, and stabilization of inner root sheath filament bundles [21]. The inner root sheath separates hair shaft from the outer root sheath, which forms the external concentric layer of epithelial cells in the hair follicle. The outer root sheath consists of several layers of cells. The thickness and cellularity of the outer root sheath vary with the level of the follicle: it is single-layered just about the bulb, the cell number is gradually increased in upward, and at the level of the sebaceous gland it becomes multilayered and is structurally similar to the epidermis [22]. The outer root sheath represents a heterogeneous cell population consisting of keratinocytes, stem cell progeny for keratinocytes, neuroendocrine Merkel cells, Langerhans cell, melanocytes precursors, and differentiating melanocytes [23,24].

One of the key components of the hair follicle is the dermal or follicular papilla. The papilla plays a critical role in regulating the hair follicle through various phases of hair cycle and is thought to be responsible for determining the character and pigmentation of the hair shaft. A stereologic study on 235 hair follicles from different sites suggested that the volume of the dermal papilla correlates with the number of matrix cells and the resulting size of the hair shaft [25]. In contrast to the epithelial compartment of the hair follicle, the fibroblasts of the papilla have been reported to show no division, and be highly resistant to apoptosis.

Epithelial stem cells and daughter cells are located in the follicular bulge of the outer root sheath, and can be identify by long-term retaining of BrdU label or by immunodetection of cytokeratins 15 and 19 [26,27]. Undifferentiated melanocytes, which are also located in a bulge area and the outer root sheath, can be visualized by immunostaining for tyrosine-related protein-2 [23].

Hair follicle has the unique property of undergoing cyclic changes of its structure and activity with periods of active hair fiber production (anagen), apoptosis-driven involution (catagen) and relative resting (telogen) phases [17,28]. The follicle regeneration is characterized by dramatic changes in its microanatomy and cellular activity. Cell fate during growth and involution is controlled by local balance of growth regulators that induce either proliferation/differentiation or apoptosis.

Proliferation, differentiation, and apoptosis in hair follicle keratinocytes are controlled by a number of signaling molecules, including the bone morphogenetic protein/ transforming growth factor-beta, epidermal growth factor, fibroblast growth factor, Hedgehog, insulin growth factor, Notch, neurotrophin, tumor necrosis factor, and Wnt families [17,18,28]. The decrease in proliferative activity or activation of apoptosis in the matrix region leads to a hair growth retardation and/or alterations in hair follicle cyclic activity.

In resting hair follicles, signaling exchange between the follicular epithelium and mesenchyme appears to be minimal. Initiation of growth phase (anagen) in telogen hair follicle is accompanied by the activation of a large number of growth stimulatory signaling pathways in both the hair follicle epithelium and mesenchyme. This leads to the activation of the keratinocyte and melanocytes stem cells located in the secondary hair germ, and the formation of hair bulb, in which keratinocytes extensively proliferate, the melanocytes make a pigment resulting in the formation of the hair shaft.

The anagen or growth phase of hair can be divided into six stages. During the early phases of anagen, hair progenitor cells proliferate, envelope growing dermal papilla, and begin to differentiate into the hair shaft and inner root sheath. In mid-anagen, melanocytes located in the hair matrix show pigment-producing activity, and newly formed hair shaft emerges and displaces the telogen hair. In late anagen, the formed hair bulb is located deep in the subcutaneous fat layer, and the new hair shaft emerges from the skin surface [29]. In addition to the hair follicle tissue remodeling, skin innervations and vascular networks also undergo substantial changes with the progression of anagen stage [30,31]. A modulation of angiogenesis in the skin can significantly affect the hair growth rate [32,33].

The hair follicle transformation from active growth to regression (anagen–catagen transition) is characterized by a sudden decline in the dermal papilla secretion of growth factors for hair matrix keratinocytes, leading to the dramatic reduction of their proliferative activity, termination of hair-shaft production, and activation of massive apoptosis in the proximal hair follicle epithelium [28,34,35]. Morphologically and functionally, catagen is divided into eight substages [29]. During catagen, HF compartments involved in hair production are reduced in size that allows them to regenerate in the next hair cycle after receiving the appropriate stimulation (Fig. 9.2). During catagen the dermal papilla is transformed into a cluster

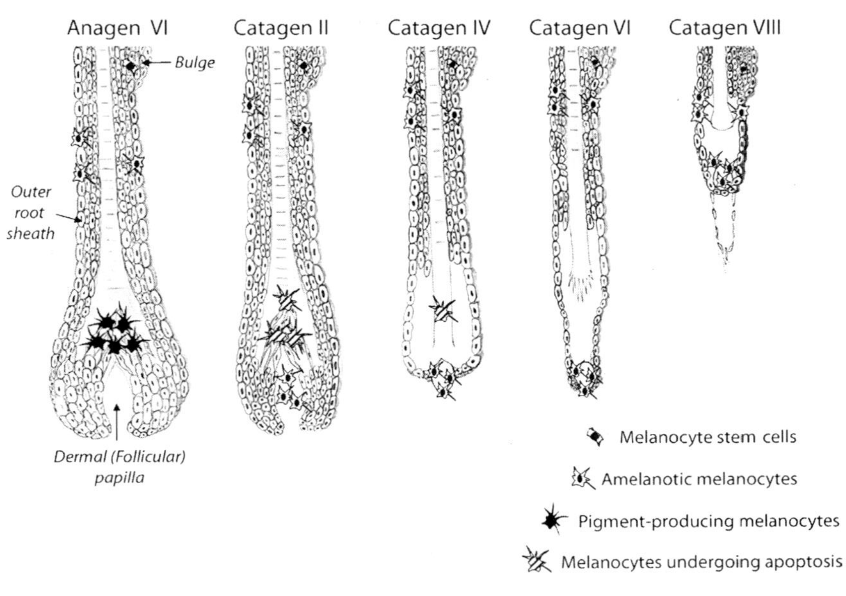

Figure 9.2 The hair follicle transformation from active growth to regression (anagen–catagen transition) and catagen development. Model illustrating changes in distinct subpopulations of the hair follicle melanocytes during catagen. Adopted from Sharov et al. [46].

of quiescent cells that moves from subcutis to the dermis/subcutis border to contact the distal portion of the hair follicle epithelium which forms the secondary hair germ, containing follicular stem cells. The proximal part of the hair shaft contains a keratinized brush-like structure that is surrounded by epithelial cells of the outer root sheath. Anagen-catagen hair follicle transition may be triggered by a variety of stimuli, including signaling via death receptors, and by the withdrawal of growth factors that maintain cell proliferation and differentiation in the anagen hair follicle. The important roles for such growth factors as fibroblast growth factor 5, insulin-like growth factor 1, transforming growth factor beta, neurotrophins, and PTHrp in catagen development have been demonstrated during the last decade [17,18,28,36,37]. Accumulating evidence suggests that apoptosis in every distinct hair follicle compartment is regulated differently. In addition, distinct cell populations in the hair follicle show a differential ability to undergo apoptosis. The majority of the follicular epithelial cells and melanocytes are highly susceptible to apoptosis, while dermal papilla fibroblasts and the populations of keratinocytes and melanocytes selected for survival display a high resistance to apoptosis.

The duration of hair cycle stages varies in different body areas. Normal human scalp hair follicles have the longest anagen phase, which can last up to several years; they display a relatively short catagen phase (1–2 weeks) followed by the telogen phase lasting several months. The majority of the hair follicles on the scalp are in anagen phase (80–85%), whereas the remaining hair follicles (2%) are either in catagen or telogen phases (10–15%). Hair located on other body sites is characterized by longer telogen phases (up to 9 months). Anagen phase of the hair follicle of axilla, arms, legs, and thighs last only 3–4 months. The majority of the hair follicles in these areas are in the telogen phase (50–80%).

9.3 Hair Pigmentation

The factor most important to determining the hair follicle sensitivity to laser treatment is the melanin pigment of the hair follicle. Therefore understanding when, where, and how melanin is formed is important in understanding laser efficiency, as well as in determining the temporary or permanent effect of laser on hair reduction. Hair color is determined by the concentration and type of melanin, which is synthesized in the melanocytes. There are two major types of melanin: the black/brown eumelanin and the reddish/yellow pheomelanin. The presence of different amounts or a combination of eumelanin and pheomelanin result in the variety of hair color. While pheomelanin is the major melanin type in red hair, eumelanin is present in large amounts in brown and black hair. Melanin synthesis in hair follicles is carried out by melanocytes present on the basement membrane surrounding the dermal papilla. The formed melanin is transported to the keratinocytes in the precortical zone that differentiate to form the pigmented hair shaft. Thus, the melanocytes together with hair matrix keratinocytes and dermal papilla fibroblasts form the hair follicle pigmentary unit. Melanogenesis is controlled by several key enzymes that are uniquely expressed in melanocytes [38]. Tyrosinase catalyzes the rate-limiting initial events of melanogenesis, and mutations in tyrosinase gene lead to loss of pigment. Tyrosinase-related proteins (TRP) 1 and TRP2 share 40–45% amino acid identity with tyrosinase and are also important in the melanogenesis process, functioning as downstream enzymes in the melanin biosynthetic pathway [38,39]. The TRP1 appears to be important for eumelanogenesis, as suggested by its lack or defective expression in pheomelanogenic cells [38]. Melanin synthesis occurs in specialized organelles, termed melanosomes, which are transferred from melanocytes to the surrounding keratinocytes upon their maturation [39]. Melanosome structure correlates with the type of melanin produced—eumelanosomes are elliptical and contain fibrillar matrix, whereas the pheomelanosome shape is variable, but predominantly spherical and contain a vesiculoglobular matrix. The size of the melanosomes and their numbers are important in determining pigmentation. In black hair, follicular melanocytes contain the largest number of melanosomes, while in brown hair follicle melanosomes are slightly smaller, and in blonde hair melanosomes are poorly melanized.

Hair is actively pigmented only during the anagen stage of the hair cycle. The hair follicle pigmentary unit cyclically regenerates synchronously with the hair follicle transition through distinct hair cycle stages [40]. The melanogenic activity of the follicular melanocytes is strictly coupled to the anagen stage, decreases during late anagen and early catagen, and ceases during late catagen and telogen (Fig. 9.2) [41]. In anagen hair follicle, the melanocytes may be divided into three distinct subpopulations. The first is located in the hair follicle bulge and represents melanocyte stem cells that repopulate the melanocytes in the new hair bulb formed at the onset of anagen [23,42]. Melanocytes have their own distinct population of stem cells responsible for their regeneration. These stem cells are characterized by the expression of Trp2, Bcl-2, Pax3, with extremely low proliferation rate. Bcl-2 appears to play a key role in the maintenance of melanocyte stem cells, as Bcl-2 knockout mice show progressive hair graying due to the depletion of melanocyte stem cells [43,44]. The second population of follicular melanocytes is located in the hair follicle outer root sheath, expresses TRP2 and relatively weak TRP1, displays proliferative activity during early and mid-anagen, and represents differentiating melanocytes. The third population represents melanogenically-active melanocytes, which are located in the hair matrix above the dermal papilla [23]. These cells proliferate only during mid-anagen, and express a full set of enzymes and other proteins involved in melanin biosynthesis including tyrosinase, Trp1, Trp2 (in mice), and pMel17 (in humans). They actively produce melanin during mid- to late anagen, and transport it to hair-shaft keratinocytes. During catagen, melanocytes progressively disappear from the hair bulb, presumably via apoptosis and/or dedifferentiation [45–47].

The epidermis also contains pigment-producing melanocytes. However, follicular melanocytes show several important differences from the epidermal melanocytes. They are larger, have more extensive dendrites, and contain a greater number and larger size melanosomes, compared to epidermal melanocytes. In addition, the hair bulb melanin unit consists of one melanocyte to five keratinocytes in the hair bulb as a whole, and one to one in the basal layer of the hair bulb matrix, whereas each epidermal melanocyte is associated with about 36 viable keratinocytes. This differentiation is primarily responsible for the selective photothermolysis advantage in favor of hair removal, while sparing the epidermis.

9.4 Principles of Laser Hair Removal and Factors Affecting Treatment Efficacy

The principle of selective photothermolysis that applies to laser hair removal indicates that by using a combination of the appropriate laser wavelength, pulse duration, and fluence targeted to a particular chromophore can selectively affect the target tissue. For the hair follicle, melanin serves as the endogenous chromophore located in the pigment-producing melanocytes of the hair matrix and in the keratinocytes of the hair shaft of anagen hair follicles [48]. Laser light energy is absorbed by melanin and is transformed into heat, which in turn affects the neighbouring cellular structures. Thus, a dark pigmented hair follicles with high eumelanin content generally responds well, whereas the treatment is less effective or even ineffective on grey, blonde, red, or light-brown hair. The effect is further determined by both the sensitivity of cellular component and the rise in local tissue temperature.

Various factors can affect the amount of laser energy reaching the target follicle. A study reported by Liew et al. (1999) on the efficacy of a ruby laser treatment at a fluence of 11 J/cm^2 was compared using variables of patient's age, intracutaneous hair length, epidermal depth, dermal density, skin color, and total melanin content and relative eumelanin content of hair. No correlation was found between the efficacy of treatment and various variables used in the study, except that the patients with higher eumelanin content in their hair had better long-term results. The results indicated eumelanin to be the single most critical factor in ruby laser effectiveness [14].

The fact that the hair follicle pigmentary unit cycles synchronously with the hair cycle stages and the hair is actively pigmented only during the anagen stage of the hair cycle, suggests the sensitivity of the hair follicle to laser treatment to be cycle-dependent. Indeed, the experiments carried out in mice showed that only actively growing pigmented anagen hair follicles were sensitive to hair removal by laser, whereas catagen- and telogen-stage hair follicles were resistant to the treatment [49]. This observation was verified in a human study where it was demonstrated that hair follicles considered being in the early anagen phase with poor melanin showed poor efficacy with a long-pulsed ruby laser at 20 J/cm^2 [50]. In humans, hair show a mosaic or asynchronous pattern of growth with an autonomy for growth and pigmentation residing in each individual hair follicle. Except for scalp hair growth where the majority of the hair follicles reside in anagen stage (80–85%), the other body areas express a large proportion of telogen hair follicles. This suggests that multiple laser treatments would be needed to capture all hair in their anagen phase and a single laser treatment is unlikely to produce the desired hair-removal effects. In contrast to this theory, results from some other studies showed anagen and nonanagen hair follicles to be equally sensitive to laser treatment [4,51]. By light microscopic examination of skin biopsies obtained after a ruby laser treatment at fluence of 11 J/cm^2, Leiw et al. observed that the coagulation damage of hair follicles was not confined to anagen hairs [52]. By using phototrichograms and hair counting in patients six months after the ruby laser treatment, Dierickx et al. (1998) concluded that both anagen and telogen hair had an almost equal probability of loss [51]. Therefore, the question remains as to what are the critical contributing factors that determine the susceptibility of the human hair follicles to the laser treatment, and what role precisely does the hair cycle play in this process. It is possible that sensitive hair follicle components in catagen and telogen hair follicles get affected by temperature changes in the surrounding anagen follicles during treatment.

Other areas of discussion in literature include the role of follicular stem cells in permanent reduction of hair—these cells are responsible for populating the matrix cell population at the start of the hair cycle; the temperature sensitivity of matrix cells—the cell population at the base of hair bulb responsible for the fiber formation; and the role dermal papilla might play in laser-induced changes in hair character, growth rate, and hair cycle transition of hair follicle.

By using select markers, Orringer et al. (2006) determined whether laser treatment could cause alterations in the follicular stem cells of the bulge region. Axillary hair growth was targeted on one side with an 800 nm diode laser and the other side was treated with a 1064 nm Nd:YAG laser. Serial skin samples were obtained at baseline and at various time points after treatment, and used for analysis of cytokeratin 15, cytokeratin 19, and CD34 expression, markers of hair follicle stem cells. Clinically, loss of hair was demonstrated on both sides, and the hair follicle histology was consistent with the thermal injury. However, immunohistochemical markers for the pluripotent stem cells located at the bulge region stained with a similar pattern and intensity for the posttreatment samples as compared to baseline. The results indicated that stem cells are spared from laser-induced damage [53].

On the other hand, a study conducted by Liew and colleagues (1999) showed that stem cells can be affected by laser treatment. The authors treated ex-vivo scalp skin samples with a ruby laser (14 and 20 J/cm^2) and used light microscopy and immunohistology to determine the extent of damage. A monoclonal antibody LP2K against Keratin 19 was used to study the effect on stem cell population. The result indicated that most of the laser-induced changes involved the bulge region, and not the hair bulb [54].

9.5 Mechanism of Laser-Induced Damage and Histopathological Changes in the Hair Follicle

Because laser light energy is transformed into heat upon absorption by melanin, the change in tissue temperature, and the rate and time period of this temperature rise determines the tissue response. In a study using real-time thermal imaging, the temperature change in response to ruby laser was determined to be in a range of 5–10°C with heat dissipation occurring 2 s after exposure [13]. Cellular injury with subsequent inflammation and repair occurs after tissue temperature increases by only 5–10°C [55]. Deactivation of most cellular enzymes begins to occur at a temperature of 40– 45°C, with an initial reversible damage that can become irreversible with sustained exposure. Temperatures above 60°C lead to denaturation of most proteins, whereas a further increase above 70°C leads to denaturation of DNA. Vaporization of tissue water with cell shrinkage, hyperchromasia, membrane rupture, protein denaturation, and collagen hyalinization occurs at temperatures of 60°C–140°C [56]. Though thermal injury is probably the predominant mechanism for hair reduction, a role of photomechanical damage has also been suggested to contribute to the same, whereby the laser-induced sharp temperature gradient between melanosomes and the surrounding structures are thought to lead to a thermal expansion and generation, and propagation of acoustic waves that can mechanically damage the melanosome-laden cells [57].

9.5.1 Early Histopathological Changes Observed after Laser Treatment

In studies using various laser systems at varying fluences, coagulation of cellular proteins in hair follicles was observed after irradiation at high fluence levels. In skin samples biopsied immediately after the normal-mode ruby laser pulses at fluence of 30–60 J/cm^2, selective but heterogeneous follicular damage was detected, consisting of thermal coagulation and asymmetric focal ruptures of the follicular epithelium. The presence of focal ruptures suggested vaporization and steam formation with temperature exceeding 100°C in some areas of the follicle [58]. In another study, ruby laser at a fluence of 11 J/cm^2 caused widespread coagulation and charring of subcutaneous hair shafts, as well as distortion and cracking of the shaft matrix [52]. A study by Kato T et al. (2004), also with the ruby laser showed localized damage in the hair bulb and the bulge region, and dilation of the inner root sheath filled with eosinophilic substances [59]. Exposure of the hair follicles to 11–17 J/cm^2 alexandrite laser resulted in the disappearance of inner root sheath cells and disruption in the cellular polarity of the outer root sheath and dermal papilla [59]. Coagulation of hair follicle components was also observed with alexandrite and diode laser systems at 25–40 J/cm^2 immediately following irradiation, and accompanied by variable amounts of inflammation and pigmentary changes [60]. Similar findings were reported by Ono and Tateshita (2000) who observed histopathological changes in melanin-containing cells, including damaged external root-sheath and swollen dermal papilla after exposure to long-pulsed alexandrite laser at fluence of 25 J/cm^2 [16]. Extensive necrosis of follicular and sebaceous gland epithelium was evident in the specimen obtained 6 h after treatment with Nd-YAG laser at fluence of 23–56 J/cm^2 [61]. The laser-induced damage to follicular cells was most often seen associated with the damaged hair shafts [13]. Collectively, these studies provide evidence in support of thermal-induced injury not only to the melanin rich follicle structures, but also to the surrounding cellular components including epithelial cells of the inner and outer root sheaths, and in some instances, the cells of dermal papilla.

9.5.2 Histopathological Changes Observed at Different Periods of Time after Laser Treatment.

Observations made at extended periods after various laser treatments revealed further progression of histopathological changes in the hair follicle. For instance, seven days after the treatment with ruby laser at 11 J/cm^2, damaged follicular epithelium was observed with increased eosinophilia and pyknotic nuclei, suggesting the development of cell death in the hair follicle. Twenty one days later, most of the damage had extended to the level of the insertion of arrector pilli muscle, and the damaged cells of the follicular bulb appeared shrunken with collapsed outer root sheath [52]. One month later, cystic formation of hair follicles and foreign body giant cells were observed in the skin treated with either ruby (10–18 J/cm^2) or alexandrite (11–17 J/cm^2) lasers. By using the TUNEL technique, the authors detected apoptotic cells in the dermal papilla after ruby laser treatment, and development of apoptosis in the inner root sheath after alexandrite laser treatment [59]. Analysis of the follow-up biopsies after the application of diode laser (25–35 J/cm^2) obtained one and three months after laser treatments revealed that hair follicles contained necrotic keratinocytes, and the histology findings were consistent with catagen phase. In a case study by Sadick, the biopsy showed complete destruction of two hair follicles, evidenced by the presence of two hair shafts lying freely in the dermis, and a reduction in the anagen/telogen ratio to about 1:1 versus the normal 8:2 [62]. A study report by Dierickx on one-year follow-up after ruby laser treatment showed the total number of the hair follicles to be near identical in the control- and laser-treated sites; however, on the laser side there was a reduction in the terminal hair with a proportional increase in the small vellus-like follicles, and the average hair-shaft diameter was substantially decreased compared to the control (68.7 ± 4.2 μm and 22.5 ± 12.2 μm, respectively) [51]. A more extensive damage to hair follicles was detected in biopsies taken three months after treatment with Nd-YAG laser at 23–56 J/cm^2.

A complete disappearance of hair follicles with the occasional presence of arrector pilli muscle and little focal fibrosis was observed [61].

Thus, laser treatment initiates degenerative changes in the hair follicles that may be accompanied by an increase in apoptosis and necrotic events in the follicular epithelium and in some cases in the dermal papilla, which may lead to either hair follicle miniaturization or hair follicle destruction.

9.5.3 Hair Follicle Response to Laser Is Fluence-Dependent

Histological analysis of the hair follicles exposed to the laser suggested that hair follicle response to laser treatment is fluence-dependent, and therefore, hair removal could be achieved through different mechanisms, including modification of hair cycle, hair follicle miniaturization, or hair follicle destruction. If low fluences accelerate anagen-catagen hair follicle transition leading to temporary hair loss, higher fluences could cause hair follicle destruction accompanied by permanent hair reduction [2].

The fluence-dependent hair follicle response was elegantly demonstrated in the experiment carried out on isolated anagen hair follicles using the ruby laser [63]. The authors observed that laser irradiation of hair follicles at fluence of 1.2 J/cm^2 did not result in immediate gross visible effects; however, hair growth rate dramatically decreased $\geq$ 80% and hair follicles exhibited a transition from anagen to catagen during the six-day culture period studied. In contrast, a fluence of 3.8 J/cm^2 resulted in a visible damage to hair follicles observed under light microscopy, and the majority of the hair follicles did not exhibit anagen–catagen transition. Even though the study was performed with isolated hair follicle, and not the intact skin, nonetheless, it clearly demonstrated that low fluences are capable of causing specific changes to hair follicle without resulting in their destruction.

McCoy et al. (1999) evaluated histological changes in axillary hair follicles in response to an exposure to 3-ms pulse of ruby laser at fluence range from 10 to 40 J/cm^2. The results showed fluence-dependent damage to the hair follicle compartments, including the inner root sheath and the hair shaft. The predominance of catagen follicles and late catagen/telogen follicles was seen at one week and at four weeks after a single treatment, respectively, suggesting a premature induction of anagen–catagen transition of hair follicles in response to the treatment [64]. In a study that involved three monthly treatments of axillary area with a long-pulsed alexandrite laser and a long-pulsed diode laser at 25 J/cm^2 or 40 J/cm^2, histological evaluations one and six months after the final treatment revealed follicular miniaturization and fewer number of terminal hairs in all biopsies, irrespective of the laser or the fluence used [60].

Taken together, these studies indicate that laser-mediated hair removal can be achieved through different mechanisms, including modification of hair cycle, hair follicle miniaturization, and hair follicle destruction, and that this follicular response may be related to a threshold increase in temperature within the specific hair follicle.

9.6 Molecular Mechanisms Involved in Hair-Follicle Response to the Laser

The molecular mechanisms involved in laser-mediated hair removal still remain obscure.

It is well known that the response of cells to increased temperature is accompanied by elevated expression of superfamily of proteins termed "heat shock proteins" (HSP). Each subfamily of HSPs is composed of members expressed either constitutively or regulated inducibly. HSPs participate in essential physiological processes, such as regulation of cell cycle, differentiation, and programmed cell death. They are also called "stress proteins" because their synthesis is stimulated by variety of stresses including heat and irradiation. The HSPs induced in response to stress have a substantial role in cell protection against death. The HSPs induced by heat bind to denatured proteins to restore their structure or to bring them to degradation pathways. In addition, stress activated HSPs are able to control several intracellular apoptotic events [65,66]. It has been shown that 815-nm diode laser induces long-lasting expression of 72-kDa heat shock protein in normal rat skin [67].

In experiments carried out in our laboratory we observed that exposure of hair follicles to diode laser irradiation-induced synthesis of HSP in the follicular cells (Fig. 9.3). Shortly after irradiation up-regulated expression of HSP70 was detected in the follicular outer and inner root sheaths, as well as in the dermal papilla. In addition, an appearance of HSP70 expression was detected in the melanogenic area above the dermal papilla, which is in line with the concept that hair follicle melanocytes are the major target for the laser. However, the progressive decrease in HSP70 expression in the hair follicles over time was also seen, suggesting activation of cell death in this area. Thus, our data together with literature reports suggests that enhanced expression of HSP after laser irradiation reflects a physiological response of the hair follicles to irradiation stress, and it might serve as a first protective mechanism against laser-induced cell death. Therefore, we hypothesized that HSP inhibitor application prior to laser treatment may result in decrease of hair follicle resistance to apoptosis. In addition, HSP inhibitors may allow a decreasing the dose of laser irradiation to achieve hair growth inhibitory effects. We tested this hypothesis by cotreatment of the hair follicles with HSP inhibitor KNK437 and diode laser, using the isolated hair follicle ex vivo model. Our data suggest that KNK437 and laser irradiation have a synergistic effect on reduction hair growth in vitro (Fig. 9.3) [68].

Cellular damage caused by the laser could also be accompanied by an increase in p53 expression. p53 is a transcription factor that mediates cellular apoptosis in response to the variety of stresses, including ionizing irradiation, chemotherapy, and exposure to a thermal insults [69–71]. However, in the absence of p53, stressful stimuli cause cell necrosis [72]. Topping et al. observed that a single exposure to 15 J/cm^2 ruby laser caused an appearance of p53 expression in epithelial cells lining the hair follicles with the damaged hair shafts. p53 expression was found to be extended in a radial fashion within the hair follicle. The more severely damaged cells were incapable of expressing p53, most likely because of their death. In hair follicles that reached the highest temperature, p53 expression was also seen in the sebaceous glands [13].

In our laboratory, we explored the effect of laser treatment on p53 expression, catagen induction and growth reduction using isolated human hair follicles. p53 expression was examined four hours after direct exposure of hair bulbs of isolated anagen hair follicles to a diode laser dose that induced catagen formation (2 W for 100 ms) and the dose that reduced hair growth without inducing catagen but causing hair follicle destruction (4 W for 50 ms) (Fig. 9.4). The hair follicles treated with 2W for 100 ms laser showed substantial up-regulation in p53 expression in the outer root sheath, hair matrix, and in the dermal papilla, compared to the untreated hair follicles. In contrast, the hair follicles treated with 4W for 50 ms laser showed a complete absence of p53 immunoreactivity in parts of hair follicle exposed directly to the laser and presumed to reach coagulative temperature levels,

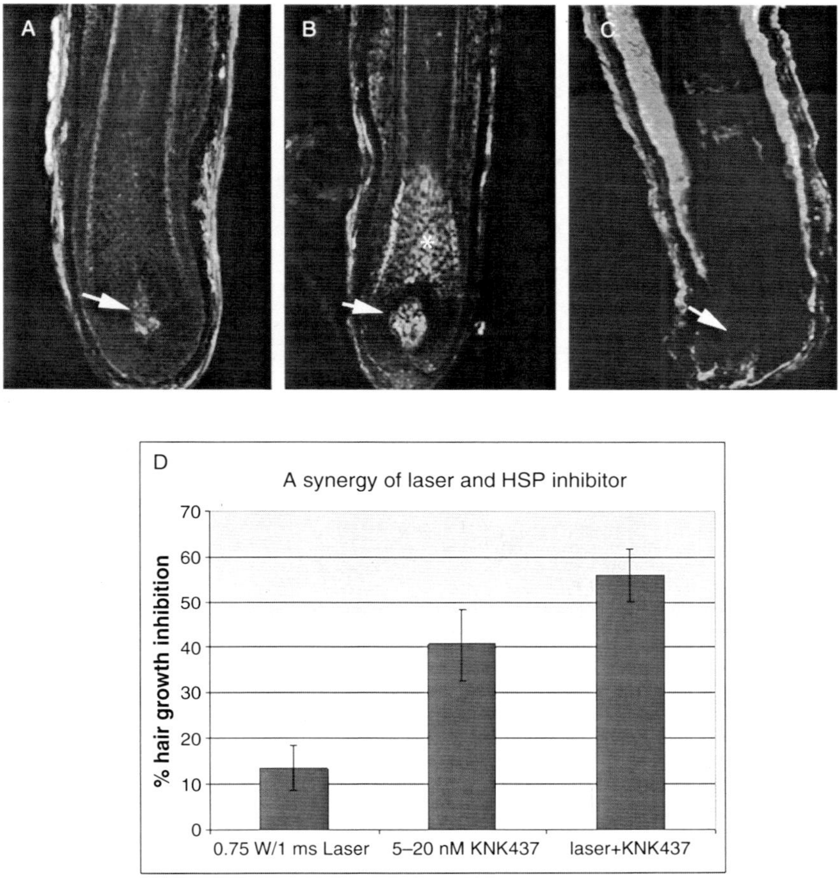

Figure 9.3 Involvement of HSP in response of follicular cells to laser treatment. **A**—in normal anagen hair follicle, HSP70 immunofluorescence (Rhodamine) was detected in the dermal papilla (arrow) and connective tissue sheath; **B**—shortly after irradiation up-regulated expression of HSP70 was detected in the outer and inner root sheaths, as well as in the dermal papilla (arrow), and it appeared in the melanogenic area above the dermal papilla (asterisk); **C**—lack of HSP70 in the dermal papilla (arrow) and its prominent expression in the outer root sheath 24 hours after laser treatment; **D**—substantial hair growth inhibition caused by combination of low laser and KNK437 doses.

while p53 immunoreactivity was still present in the distal parts of the outer and inner root sheaths where temperature had not reached the threshold level for severe damage. This data is in agreement with observations from Topping et al. described earlier [13], and suggests that moderate laser fluences activate pro-apoptotic pathways and p53-dependent cellular changes in the hair follicles, whereas high laser fluences cause severe damage to the hair follicle resulting in its necrotic death.

In addition to p53 protein, another molecule that may regulate cellular apoptotic events is the protein Bcl-2. In our laboratory, we investigated the expression of this anti-apoptotic protein in the hair follicles in response to varying diode laser fluences. Bcl-2 is a mitochondrial-anchored protein, which forms ion channels capable of maintaining membrane homeostasis

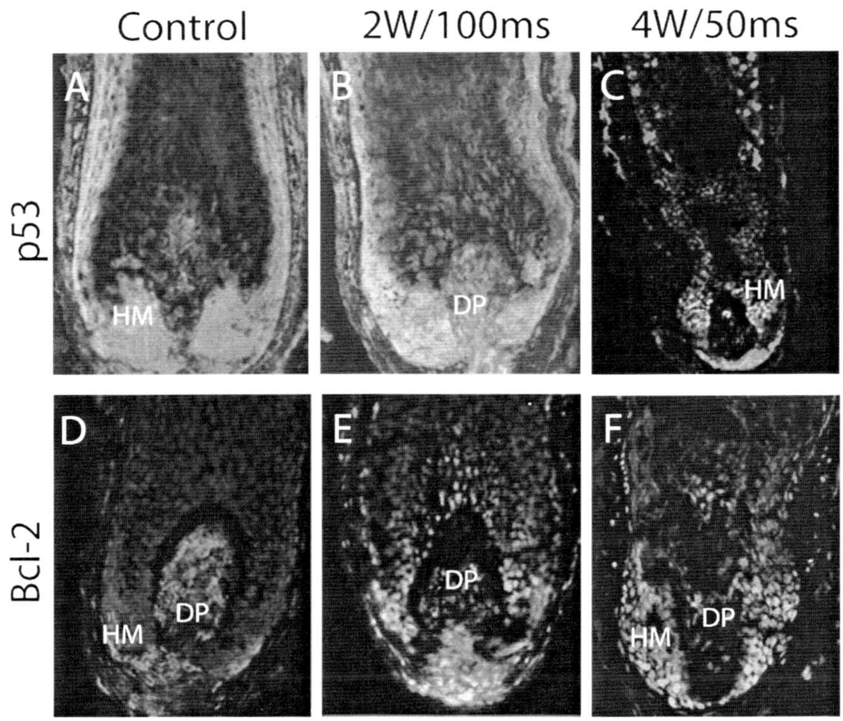

Figure 9.4 Dose-dependent response of hair follicles to different laser treatments. p53 and Bcl-2 expression (Rhodamine) was examined four hours after direct exposure of hair bulbs of hair follicles to different laser doses. **A**—p53 immunoreactivity in the proximal outer root sheath, hair matrix, and dermal papilla of the control anagen hair follicles; **B**—2W/100 ms laser caused substantial up-regulation in p53 immunoreactivity in the outer root sheath, hair matrix, and in the dermal papilla; **C**—absence of p53 immunoreactivity in hair follicle parts exposed to, **D**—Bcl-2 expression in intact dermal papilla and hair matrix; **E**—decrease in Bcl-2 immunoreactivity in the dermal papilla and in the hair matrix cells after 2W/100 ms laser treatment; **F**— a complete absence of Bcl-2 expression in hair follicles exposed to 4 W/50 ms laser. DP: dermal papilla; HM: hair matrix.

and preventing cytochrome *c* release, thus enhancing cell survival [74]. Bcl-2 protects cells against apoptosis via suppression of cysteine proteases called caspases, which are part of the proteolytic caspase cascade that is activated by diverse apoptotic stimuli [75]. It was shown that induction of apoptosis by heat and gamma-radiation in a human lymphoid cell line was accompanied by a decline in Bcl-2 protein levels and was mainly executed in a caspase-dependent pathway [76]. In the hair follicle, Bcl-2 is prominently expressed in the follicular dermal papilla throughout the hair cycle, which demonstrates a high anti-apoptotic potential of this hair follicle compartment [34]. In our studies, we also detected a prominent Bcl-2 immunoreactivity in the dermal papilla and hair matrix keratinocytes in control hair follicles. However, the hair follicles treated with 2W laser at 100 ms pulse showed a significant decrease in Bcl-2 immunoreactivity in the dermal papilla, as well as in the hair matrix cells. A complete absence of Bcl-2 expression was seen in hair follicles treated with 4W laser at 50 ms pulse. The elevated expression of Bcl-2 in a follicular papilla at the low laser power suggests activation of anti-apoptotic response in the hair follicle and the viability of dermal papilla cells, while the absence of Bcl-2 in hair follicles exposure to the higher laser power indicates occurrence of cell death (Fig. 9.4).

Collectively, these findings suggest that hair follicle response to moderate laser fluences is characterized by activation of the apoptotic program, a normal physiological process that hair follicle follows during its cycle transition from the anagen to catagen phase. The apoptosis activation is likely to result in premature hair follicle transition to its resting stage, resulting in the cessation of hair growth and achieving a temporary hair removal. The absence of both pro-apoptotic p53 and anti-apoptotic Bcl-2 protein in the hair follicles after exposure to high laser fluences suggests severe and perhaps irreversible damage to the follicular structure, leading to pathological destruction of the hair follicle and permanent reduction in hair.

9.7 Methods to Assess Laser Effects on Hair Follicle Growth

As detailed earlier, the laser effects on hair follicle are highly dependent on temperature changes in localized compartments within the hair follicle. Various model systems can be used to determine the mechanism of laser-induced changes in the hair follicle, and from these changes, predict the short- and long-term effect of a laser treatment on hair- reduction efficiency.

Most of the studies on histopathological changes in response to laser treatment have been performed using the punch-skin biopsies taken from volunteers after the laser treatment. This approach has some advantages and certain disadvantages. One of the disadvantages is that punch biopsy produces a limited number of hair follicles per sample, and that only a small number of samples can be practically obtained because of the invasiveness of this procedure. Thus, data analysis is performed using a very limited sample size. There is an additional technical challenge, as vertically sectioned skin samples rarely go through the entire length of the same hair follicle. The horizontal sections used in most studies make it difficult to evaluate comparative responses of various components within the hair follicle to a laser treatment. However, the advantage of this approach is that skin samples can be collected at defined periods of time after the laser treatment while the hair follicle is in its natural environment, and therefore allows evaluation of not only long term effects of laser, but also assessment of the changes in the skin structure surrounding the hair follicles.

To determine the acute response of follicles and skin, an alternative approach is the use of ex-vivo skin. The samples of skin containing pigmented hair follicles can be obtained from patients undergoing cosmetic surgery (face-lift or brow-lift procedures). The model has been successfully used by several research groups and in our own laboratory to assess the immediate damage in hair follicles and epidermal melanocytes after laser treatment [13, 54, 68].

The hair-bearing skin are used for obtaining anagen hair follicles and growing them under in-vitro organ culture conditions. It has been demonstrated that dissected mature human anagen follicles can be successfully grown in culture with similar in-vivo hair fiber growth rates [77]. Human scalp/facial follicles are microdissected free of dermal/subcutis tissue, and placed free-floating in a serum-free culture medium. These follicles continue to grow for a short period of time (10–21 days) in culture, and are useful for studying the anagen phase as well as the onset of catagen. This model system has been widely used to investigate hair follicle biology [78]. The major application has been to investigate the possible role of growth factors in controlling hair follicle growth and differentiation. The advantage of hair follicle organ culture is that it is allows for the direct assessment of hair shaft elongation in response to the treatment of interest. The model is good for making assessment of morphological, biochemical, and molecular changes in individual hair follicle compartments at various time periods after the initiation of treatment. The model is also suitable to study anagen—catagen transition following a treatment. Because of the large number of hair follicles that can be obtained from a small piece of skin sample, it provides a better statistical analysis compared to the biopsy samples.

In our laboratory, we have used a combination of ex-vivo and in-vitro models to study the effect of lasers on hair follicle. For these studies, ex-vivo human skin is exposed to varied laser doses, and then the hair follicles are dissected and placed in a culture medium for the determination of hair growth rate and cyclic transition. In addition, changes in follicular morphology, assessment of cell proliferation and apoptosis, as well as the expression of genes known to be involved in hair growth regulation can be studied at different time points after laser exposure.

9.7.1 Morphology

To evaluate hair follicle morphology, we use histochemical detection of endogenous alkaline phosphatase activity. Alkaline phosphatase is an excellent marker of dermal papilla cells, and is known to be expressed in dermal papilla throughout the hair cycle. This allows visualization of dermal papilla morphology as a useful morphological marker for studying the hair follicle cycle [29, 79]. The criteria for recognizing the key stages of hair follicle growth and development of the regression (catagen) stage has been reported in literature [28, 29], and it may help to understand and evaluate hair follicle response to laser treatment.

9.7.2 Apoptosis

In order to identify the potential follicular targets for laser action and to define the onset of laser-induced apoptosis, Terminal deoxynucleotidyl Transferase dUTP Nick End Labeling (TUNEL) staining can be performed to localize apoptotic cells in the irradiated hair follicles. TUNEL is the most commonly used method, which identifies apoptotic cells in situ by labeling of apoptosis-induced DNA fragmentation. Because the melanocytes are known to be the primary target for the laser, double immunostaining for simultaneous detection of apoptotic cells by TUNEL and melanocytes immediately after laser exposure could provide information on the depth of laser penetration and effectiveness of the laser for melanocyte damage (Fig. 9.5). The melanocytes can be visualized histochemically by using the antibody against pMel-17 protein that allows identification of both bulb and scattered melanocytes in the outer root sheath at any hair follicle stage [80].

9.7.3 Proliferation

The proliferative activity of the follicular cells, which influences hair shaft production, could be assessed by applying immunohistochemical methods for visualizing the cell cycle

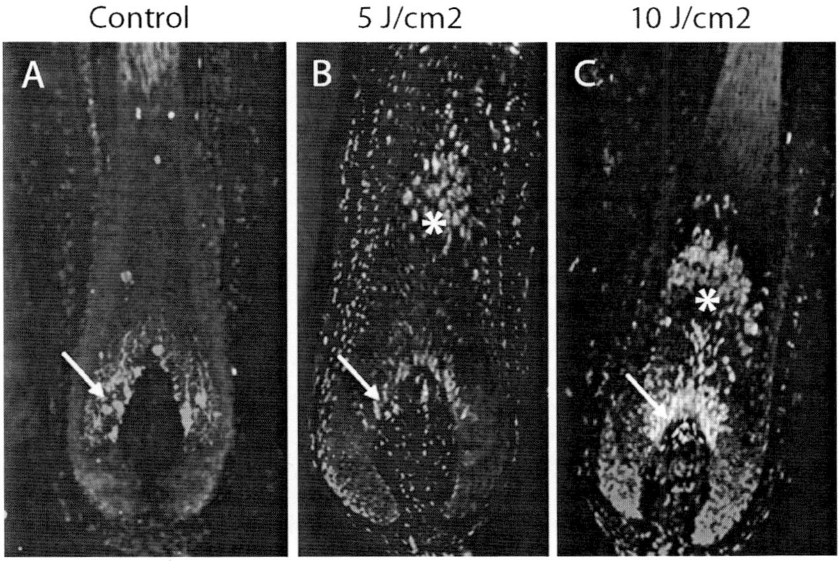

Figure 9.5 Detection of apoptosis in the hair follicle melanocytes after laser treatment by simultaneous visualization of pMel-17 (Rhodamine) and TUNEL staining (FITC). **A**—in anagen hair follicle, pMel-17 immunofluorescence (Rhodamine) was detected in the hair matrix (arrow) and no TUNEL positive cells were detected; **B**—immediately after treatment with 5 J/cm^2 laser, appearance of apoptotic TUNEL positive melanocytes expressing pMel-17 protein were seen in hair follicle melanogenic area above the dermal papilla (arrow), and TUNEL positive cells were detected in the differentiating keratinocytes of the hair shaft containing melanin granules (asterisk); **C**— in hair follicles exposed to 10 J/cm^2 laser, the number of TUNEL-positive cells in the hair matrix and the hair shaft increased (arrow and asterisk, respectively), and appeared in the dermal papilla.

associated antigens. One such antigen that is broadly used for the detection of proliferating cells is Ki-67. The monoclonal antibody against Ki-67 can be used to recognize a proliferation specific antigen expressed in the nuclei during the G1, S, G2, and M phase of growth, but not in the G0 phase [81]. In the hair follicle, Ki-67 immunoreactivity is usually seen in cells of the hair bulb and in distinct cell population of the outer root sheath of the anagen hair follicle, while it is progressively decreased with development of catagen phase. Quantitative analysis of Ki-67 in the hair follicles in response to different laser treatments could provide information on status of proliferative activity of follicular cells, which reflects the ability of the hair follicle to produce the hair shaft.

9.7.4 Hair Follicle Tissue Remodeling

The hair follicle is a multicellular structure, where each epithelial compartment is characterized by the expression of specific differentiation marker, and it can undergo dramatic tissue remodeling as a result of physiological or a pharmacological stimuli. In human hair follicles, the dynamics and tissue remodeling that occur during anagen-catagen transition and catagen progression were very elegantly described by Como and Bernard [82]. Using monoclonal antibodies directed against keratin 14 expressed in the outer root sheath, trichohyalin, transglutaminase I and desmoglein expressed in the inner root sheath, and Ki67 antigen as expression of cellular proliferation, the authors characterized the movement and changes in each of the main hair follicle compartments during the catagen process. The effect of laser on the development of catagen in hair follicles can be studies using these marker proteins.

9.7.5 Cell Viability

The hair follicle cells can undergo either a programmed cell death (apoptosis) whereby they retain the ability to regenerate at the initiation of next growth cycle, or they can undergo a necrotic cell death where they lose their ability to regenerate. One of the markers of cell viability in hair follicle is Bcl-2, which is known to suppress apoptosis in a variety of cell systems [74, 75]. Throughout the hair cycle, Bcl-2 protein is prominently expressed in the follicular dermal papilla [34]. Several studies suggest a positive relationship of hair follicle activity and Bcl-2 levels in the dermal papilla cells. For example, Bcl-2 expression increases in the dermal papilla in response to treatment with the hair growth stimulatory agent minoxidil [83]. Elevated levels of follicle Bcl-2 have also been associated with a decreased sensitivity to chemotherapy-induced alopecia [69]. On the other hand, inhibitory effects of testosterone and 5-alpha-dihydrotestosterone on hair follicle activity are accompanied by a decreased expression of Bcl-2 protein in dermal papilla cells [84]. The dermal papilla Bcl-2 level seems to be an excellent marker for hair follicle viability, and might be useful in studying laser-induced changes in the follicle.

9.8 Conclusions

While high laser fluences and energy levels can cause permanent reduction in hair desired by many consumers, it has the potential for collateral skin damage. The level of efficacy versus dermal side effects is generally dependent on the difference in eumelanin levels between epidermis and the target hair follicle. To understand the mechanism of laser-induced changes in hair growth reduction, whether temporary or permanent, or changes in hair character after treatment, or to determine the treatment regimen for maximal hair reduction benefit, it is important to understand the impact that laser has on key biochemical and molecular targets that regulate hair growth, character, and cycling.

References

1. Anderson RR and Parrish JA, *Selective photothermolysis: precise microsurgery by selective absorption of pulsed radiation.* Science,1983. 220:524–527.
2. Goldberg DJ, *Laser- and light-based hair removal: an update.* Expert Rev Med Devices, 2007. 4:253–260.
3. Lepselter J and M Elman, *Biological and clinical aspects in laser hair removal.* J Dermatol Treat., 2004. 15:72–83.
4. Liew SH, *Laser hair removal: guidelines for management.* Am J Clin Dermatol., 2002. 3:107–115.

5. Nanni CA and Alster TS, *Laser-assisted hair removal: side effects of Q-switched Nd:YAG, long-pulsed ruby, and alexandrite lasers.* J Am Acad Dermatol., 1999. 41:165–171.
6. Sadick NS, *Laser hair removal.* Facial Plast Surg Clin North Am, 2004. 12:191–200.
7. Tanzi EL, Lupton JR, and Alster TS, *Lasers in dermatology: four decades of progress.* J Am Acad Dermatol., 2003. 49:31–34.
8. Warner J, Weiner M, and Gutowski KA, *Laser hair removal.* Clin Obstet Gynecol., 2006. 49:389–400.
9. Solomon MP, *Hair removal using the long-pulsed ruby laser.* Ann. Plast. Surg., 1998. 41:1–6.
10. Haedersdal M and Wulf HC, *Evidence-based review of hair removal using lasers and light sources.* J. European Acad. of Derm. and Venereol., 2006. 20:9–20.
11. Connolly CS and Paolini L, *Study reveals successful removal of unwanted hair with LPIR laser.* Cosmet. Dermatol., 1997. 10:38–40.
12. Bjerring P, et al., *Evaluation of the free-running ruby laser for hair removal—a retrospective study.* Acta. Derm. Venereol.,1998. 78:48–51.
13. Topping A. et al., *The temperatures reached and the damage caused to hair follicles by the normal-mode ruby laser when used for depilation.* Ann Plast Surg., 2000. 44:581–590.
14. Liew, S., et al., *Ruby laser-assisted hair removal success in relation to anatomic factors and melanin content of hair follicles.* Plast Reconstr Surg., 1999. 103:1736–1743.
15. Wimmershoff MB. et al., *Hair removal using a 5-msec long-pulsed ruby laser.* Dermatol Surg., 2000. 26:205–210.
16. Ono I and Tateshita , *Histopathological changes in the hair follicle after irradiation of long-pulse alexandrite laser equipped with a cooling device.* Eur J Dermatol., 2000. 10:373–378.
17. Paus R and Cotsarelis G, *The biology of hair follicles.* N Engl J Med., 1999. 341(7):491–497.
18. Stenn KS and Paus R, *Controls of hair follicle cycling.* Physiol Rev., 2001. 81(1):449–494.
19. Schweizer J. et al., *Hair follicle-specific keratins and their diseases.* Exp Cell Res, 2007. 313:2010–2020.
20. Jave-Suarez LF., et al., *Androgen regulation of the human hair follicle: the type I hair keratin hHa7 is a direct target gene in trichocytes.* J Invest Dermatol., 2004. 122:555–564.
21. O'Guin WM, Sun TT, and Manab M, Manabe, *Interaction of trichohyalin with intermediate filaments: three immunologically defined stages of trichohyalin maturation.* J Invest Dermatol., 1992. 98:24–32.
22. Parakkal PF, *The fine structure of anagen hair follicle of the mouse.* Adv Skin Biol., 1969. 9:441–469.
23. Botchkareva NV., et al., *SCF/c-kit signaling is required for cyclic regeneration of the hair pigmentation unit.* The FASEB J, 2001. 15(3):645–58.
24. Cotsarelis G, Sun TT, and Lavker RM, *Label-retaining cells reside in the bulge area of pilosebaceous unit: implications for follicular stem cells, hair cycle, and skin carcinogenesis.* Cell, 1990. 61:1329–1337.
25. Elliott K, Stephenson TJ, and Messenger A.G, *Differences in hair follicle dermal papilla volume are due to extracellular matrix volume and cell number: implications for the control of hair follicle size and androgen responses* J Invest Dermatol., 1999. 113:873–877.
26. Commo S, O Gaillard, and BernardBA, *The human hair follicle contains two distinct K19 positive compartments in the outer root sheath: a unifying hypothesis for stem cell reservoir?* Differentiation,2000. 66:157–164.
27. Lyle, S, et al., *The C8/144B monoclonal antibody recognizes cytokeratin 15 and defines the location of human hair follicle stem cells.* J Cell Sci , 1998. 111(Pt 21):3179–3188.
28. Botchkareva NV, Ahluwalia G, and Shander D, *Apoptosis in the hair follicle.* J Invest Dermatol., 2006. 126(2):258–264.
29. Muller-Rover, S, et al., *A comprehensive guide for the accurate classification of murine hair follicles in distinct hair cycle stages.* J Invest Dermatol., 2001. 117(1):3–15.
30. Botchkarev, VA, et al., *Hair cycle-dependent plasticity of skin and hair follicle innervation in normal murine skin.* J Comp Neurol., 1997. 386(3):379–395.
31. Mecklenburg, L, et al., *Active hair growth (anagen) is associated with angiogenesis.* J Invest Dermatol., 2000. 114(5):909–16.

32. Ahluwalia G, Styczynski P, and Shander D, *Inhibition of hair growth.* 2000 USA, US6093748.
33. Yano K, Brown , and Detmar,M, *Control of hair growth and follicle size by VEGF-mediated angiogenesis.* J Clin Invest, 2001. 107(4):409–417.
34. Lindner G, et al., *Analysis of apoptosis during hair follicle regression (catagen).* Amer J Pathol., 1997. 151:1601–1617.
35. Straile WZ, Chase HB, and Arsenault C, *Growth and differentiation of hair follicles between activity and quiescence.* J. Exp. Zool., 1961. 148:205–222.
36. Botchkarev, V.A., et al., *Epithelial growth control by neurotrophins: leads and lessons from the hair follicle.* Prog Brain Res., 2004. 146:493–513.
37. Hebert JM., et al., *FGF5 as a regulator of the hair growth cycle: evidence from targeted and spontaneous mutations.* Cell,1994 . 78(6):1017–1025.
38. Hearing VJ, *Biochemical control of melanogenesis and melanosomal organization.* J Invest Dermatol., Symposium Proceedings / the Society For Investigative Dermatology, Inc. [and] European Society For Dermatological Research, 1999. 4(1):24–28.
39. Jimbow, K., et al., *Assembly, target-signaling and intracellular transport of tyrosinase gene family proteins in the initial stage of melanosome biogenesis.* Pigment Cell Res, 2000. 13:222–229.
40. Tobin DJ., et al., *The fate of hair follicle melanocytes during the hair growth cycle.* J. Invest. Dermatol. Symp. Proc, 1999. 4:323–332.
41. Slominski, A., et al., *Melanogenesis during the anagen-catagen-telogen transformation of the murine hair cycle.* J. Invest. Dermatol., 1994. 102: 862–869.
42. Nishimura, E.K., et al., *Dominant role of the niche in melanocyte stem-cell fate determination.* Nature, 2002. 416(6883):854–60.
43. Mak, S.S., et al., *Indispensable role of Bcl2 in the development of the melanocyte stem cell.* Dev Biol., 2006. 291:144–153.
44. Nishimura EK, Granter SR,and Fisher DE, *Mechanisms of hair graying: incomplete melanocyte stem cell maintenance in the niche.* Science, 2005. 307:720–724.
45. Slominski, A., et al., *Hair follicle pigmentation.* J Invest Dermatol., 2005. 124:13–21.
46. Sharov, A., et al., *Changes in different melanocyte populations during hair follicle involution (catagen).* J Invest Dermatol., 2005. 125:1259–1267.
47. Tobin, D.J., et al., *Do hair bulb melanocytes undergo apoptosis during hair follicle regression (catagen)?* J Invest Dermatol., 1998 111:941–947.
48. Polla, L.L., et al., *Melanosomes are a primary target of Q-switched ruby laser irradiation in guinea pig skin.* J Invest Dermatol., 1987 89:281–286.
49. Lin, T.Y., et al., *Hair growth cycle affects hair follicle destruction by ruby laser pulses.* J Invest Dermatol., 1998. 111:107–113.
50. Omi, T., et al., *Histologic effects of ruby laser hair removal in Japanese patients.* Lasers Surg Med.,1999. 25:451–455.
51. Dierickx, C.C., et al., *Permanent hair removal by normal-mode ruby laser.* Arch Dermatol., 1998. 134:837–842.
52. Liew, S., et al., *Ruby laser-assisted hair removal: an ultrastructural evaluation of cutaneous damage.* Br J Plast Surg., 52(8), 1999.636–643.
53. Orringer, J.S., et al., *The effects of laser-mediated hair removal on immunohistochemical staining properties of hair follicles.* J Am Acad Dermatol., 2006. 55:402–407.
54. Liew, S., et al., *The effect of ruby laser light on ex vivo hair follicles: clinical implications.* Ann Plast Surg., 1999. 42:249–254.
55. Thomsen S, *Pathologic analysis of photothermal and photomechanical effects of laser-tissue interactions.* Photochem Photobiol., 1991. 53:825–835.
56. Carroll L and Humphreys TR , *LASER-tissue interactions.* Clin Dermatol., 2006. 24:2–7.
57. Ara, G., et al., *Irradiation of pigmented melanoma cells with high intensity pulsed radiation generates acoustic waves and kills cells.* Lasers Surg Med., 1990. 10:52–59.
58. Grossman, M.C., et al., *Damage to hair follicles by normal-mode ruby laser pulses.* J Am Acad Dermatol., 1996. 35:889–894.
59. Kato, T., et al., *Histological hair removal study by ruby or alexandrite laser with comparative study on the effects of wavelength and fluence.* J Cosmet Laser Ther, 2004. 6:32–37.

60. Handrick C and Alster TS, *Comparison of long-pulsed diode and long-pulsed alexandrite lasers for hair removal: a long-term clinical and histologic study.* Dermatol Surg., 2001. 27:622–626.
61. Bencini, P.L., et al., *Long-term epilation with long-pulsed neodimium:YAG laser.* Dermatol Surg., 1999. 25:175–178.
62. Sadick N and Prieto V, *The use of a new diode laser for hair removal.* Dermatol Surg., 2003. 29:30–33.
63. Shander D, Farinelli WA, and Anderson RR,, *Modification of hair cycle stage, growth rate and structure of hair shaft of cultured hair follicles exposed to normal mode ruby laser irradiation.* J Investig Dermatol. Symp Proc, 1999. 4:351.
64. McCoy S, Evans.A, and James C, *Histological study of hair follicles treated with a 3-msec pulsed ruby laser.* Lasers Surg Med., 1999. 24:142–150.
65. Beere, H.M., et al., *Heat-shock protein 70 inhibits apoptosis by preventing recruitment of procaspase-9 to the Apaf-1 apoptosome.* Nat Cell Biol., 2000 2:469–475.
66. Nadeau SI and Landry J, *Mechanisms of activation and regulation of the heat shock-sensitive signaling pathways.* Adv Exp Med Biol., 2007. 594:100–113.
67. Souil, E., et al., *Treatment with 815-nm diode laser induces long-lasting expression of 72-kDa heat shock protein in normal rat skin.* Br J Dermatol., 2001. 144:260–266.
68. Shander, D., et al., *Reduction of hair growth* 2006USA, 20060134048.
69. Botchkarev, V., et al., *p53 is essential for chemotherapy-induced hair loss.* Cancer Res, 2000. 60:5002–5006.
70. Ohnishi T, *The role of the p53 molecule in cancer therapies with radiation and/or hyperthermia.* J Cancer Res Ther, 2005. 1:147–150.
71. Song S and Lambert PF, *Different responses of epidermal and hair follicular cells to radiation correlate with distinct patterns of p53 and p21 induction.* Am J Pathol., 1999. 155:1121–1127.
72. Moallem SA and Hales B, *The role of p53 and cell death by apoptosis and necrosis in 4-hydroperoxycyclophosphamide-induced limb malformations.* Development, 1998. 125:3225–3234.
73. Botchkarev V., et al., *p53 Involvement in the control of murine hair follicle regression.* Am J Pathol., 2001. 158:1913–1919.
74. Adams JM and Cory S , *Bcl-2-regulated apoptosis: mechanism and therapeutic potential.* Curr Opin Immunol., 2007. 19:488–496.
75. Earnshaw WC, Martins LM, and Kaufmann SH, *Mammalian caspases: structure, activation, substrates, and functions during apoptosis.* Annu Rev Biochem., 1999. 68:383–424.
76. Nijhuis, E.H., et al., *Induction of apoptosis by heat and gamma-radiation in a human lymphoid cell line; role of mitochondrial changes and caspase activation.* Int J Hyperthermia, 2006. 22:687–698.
77. Philpott MP, Green MR, and Kealey T, *Human hair growth in vitro.* J Cell Sci., 1990. 97:463–471.
78. Philpott MP, Sanders,DA, and KealeyT, *Whole hair follicle culture.* Dermatol Clin, 1996. 14:595–607.
79. Handjiski, B.K., et al., *Alkaline phosphatase activity and localization during the murine hair cycle.* Br J Dermatol., 1994. 131(3):303–310.
80. Commo S and Bernard BA, *Melanocyte subpopulation turnover during the human hair cycle: an immunohistochemical study.* Pigment Cell Res., 2000. 13:253–259.
81. Gerdes, J., et al., *Immunobiochemical and molecular biologic characterization of the cell proliferation-associated nuclear antigen that is defined by monoclonal antibody Ki-67.* Am J Pathol., 1991. 138:867–873.
82. Commo S and Bernard BA, *Immunohistochemical analysis of tissue remodelling during the anagen-catagen transition of the human hair follicle.* Br J Dermatol., 1997. 137:31–38.
83. Han, J.H., et al., *Effect of minoxidil on proliferation and apoptosis in dermal papilla cells of human hair follicle.* J Dermatol Sci., 2004. 34:91–98.
84. Winiarska A. et al., *Effect of 5alpha-dihydrotestosterone and testosterone on apoptosis in human dermal papilla cells.* Skin Pharmacol Physiol., 2006. 19:311–321

10

Management of Unwanted Hair

Gurpreet S. Ahluwalia

The Gillette Company, a wholly owned subsidiary of The P&G Company, Needham, MA, USA

Gurpreet S. Ahluwalia (ed.), *Cosmetic Applications of Laser and Light-Based Systems*, 239–252,

10.1 Introduction

The main function of mammalian hair is to provide environmental protection. However, this function has now largely been lost in humans, in whom hair is retained or removed from various parts of the body essentially for cosmetic reasons. Though both men and women remove hair, it is the appearance of hair on a woman's body that is perceived as unnatural. Women feel that hair does not belong to their body, except for the scalp, and constantly seek means to rid themselves of this unwanted hair. In Western culture, the hair-free body is the norm for women [1,2]. The issue of unwanted hair is further magnified, and takes on a whole new significance, for women who suffer from the presence of terminal hair on their face. The psychological aspects of having facial hair is extremely devastating to women, who will go to great lengths to hide the condition, as it is a condition accompanied by a significant psychosocial burden [3]. Hirsute patients treated in an endocrinologist's practice are often the most grateful of all patients, who are satisfied with rather small improvements [4].

Generally, women rely on shaving, depilatory creams, bleaching, waxing, and plucking to remove unwanted hair. Lack of femininity, irritation, and nonperformance are some of the disadvantages that prevent these cosmetic methods of hair removal from being totally acceptable in this population. Moreover, these approaches provide only a short-lived or temporary effect. Even with these disadvantages, it has been estimated that women spend over $2.0 billion a year on cosmetic remedies for this problem. Among emerging technologies, laser-assisted hair removal has received much attention. The technology introduced about ten years ago has seen a tremendous growth, with the current worldwide revenue of about $2.8 billion (Medical Insight publication). Most of this revenue, however, represents the fee to the service providers paid by the consumer. Table 10.1 lists various hair-removal options available to the women.

Hair removal can be broadly put into two categories, depilation and epilation. The two terms are quite often used interchangeably in the scientific and the patent literature. Stedman's medical dictionary probably comes closest to defining them. Epilate is defined as "to extract a hair, to remove the hair from a part by forcible extraction...". Depilate is defined as "to remove hair by any means." An important distinction is that in epilation, the complete hair shaft is removed from its roots, whereas, in depilation superfluous hair is removed and the hair-root is left undamaged.

10.2 Physical Methods

10.2.1 Shaving

Shaving is by far the most popular and convenient method of hair removal. In the United States, more that 90% of women shave their legs and axillae, and about 50% shave their bikini line (Gillette survey). Shaving any area of the face is the least popular method of hair removal in the US population, probably because of the fear of increase in hair growth. This has been proven to be a myth, as shaving does not change the thickness or the growth rate of human hair [5,6]. Side effects from shaving are minimal, however, prickly stubble and a rapid regrowth of hair that results after shaving leaves an undesirable tactile feel and 'shadow' of dark hair under the skin which is particularly noticeable on the face. The popular safety razors are designed to remove hair just below the skin surface by gently lifting and cutting each hair without nicks and cuts, and resulting in a smooth and hair-free skin.

Table 10.1 **Available Options for Hair Removal**

Method	Mechanism	Perceived Efficacy	Best for	Trade-off
Wet Shaving	Removes hair at skin surface	1–3 days	Convenient, easy, and inexpensive	Short-lived results, prickly stubble
Epilation and Waxing	Pulls hair from root	Approx 3 weeks	Longer lasting	Painful, skin effects, minimum hair length required for grabbing
Depilatories	Breaks cystine double bonds and dissolves superfluous hair	5–10 days; 3× longer than shaving	Stubble free softer regrowth, smoother skin	Odor, messy, skin irritation
Cosmeceuticals	Unknown	Minor/moderate, adjunct to other methods	Prolonging hair-free period, delayed regrowth	Does not remove or reduce hair growth
Drug actives	Selective biochemical target in hair follicle	Moderate—adjunct to other methods	Hair density and growth-rate reduction	Expensive, drug effect, does not remove hair
Professional high-energy Laser/IPL	Melanin targeting–follicle destruction	3–6 months (Permanent?)	Professional use	Expense, inconvenient, dermal side effects
Emerging Low-energy Laser/IPL	Hair cycle change/anagen arrest	Est. 2–6 weeks (temporary)	Home use	Periodic treatments

10.2.2 Epilation

Waxing is a time-tested method of hair removal that has seen a renewed popularity. While depilatories remove hair at the skin's surface, "epilatories", such as tweezers and waxes, pluck hair from below the surface. Waxing and tweezing may be more painful than using a depilatory, but the results are longer lasting. Because the hair is plucked at the root, new growth is not visible for several weeks after treatment. Simply stated, waxing is an efficient way of plucking hair from a larger surface. When performed properly, results can last from 4–8 weeks, depending on the site and the individual's hair- growth rate [5,7]. Like shaving, epilation generally does not effect the growth rate of hair, however, duration of the anagen cycle may get effected [8,9].

A large variety of waxing products is available for professional use. Epilatory waxes are also available over the counter for home use. They contain combinations of waxes, such as paraffin and beeswax, oils or fats, and a resin that makes the wax adhere to the skin. There are "hot" and "cold" waxes. For hot waxing, a thin layer of heated wax is applied to the skin in the direction of the hair growth. The best hot waxes melt at just above the body temperature, reducing the danger of burning. The hair becomes embedded in the wax as it cools and hardens. The wax is then pulled off quickly in the opposite direction of the hair growth, taking the uprooted hair with it. Cold waxes work similarly. Strips precoated with wax are pressed on the skin in the direction of the hair growth, and pulled off in the opposite direction. The strips come in different sizes for use on the eyebrows, upper lip, chin, and bikini area. It is recommended that hair be at least 1/8" long before waxing. Labeling of over-the-counter waxes cautions that these products should not be used by people with diabetes and circulatory problems, who are particularly susceptible to infection. Side effects include folliculitis, pseudofolliculitis, postinflammatory hyperpigmentation, and scarring [5,10].

A typical epilatory preparation contains rosin and beeswax, modified by the addition of mineral or vegetable oil and/or other waxes. The preparations may also include such agents as camphor for its cooling effects, a local anesthetic, and/or an antibacterial compound. Since waxing stings the skin, many waxers use pain-reducing gels (usually 4% lidocaine) that are specially formulated to penetrate intact skin. The pain-reducing gels are applied 30–60 min before waxing.

Various mechanical epilation devices are also available that can quickly remove a relatively large body of hair. These devices use rows of tweezers that are designed to firmly grab and pull multiple hair from their root. Hair must be of sufficient length for the device to engage. Skin irritation, folliculitis and in-grown hair are some side effects that have been noted [10].

Another epilation method that is more common in the Middle East and India than other parts of the world is Threading or Khite [11]. Threading is generally performed by a cosmetologist who uses twists of a cotton thread to pull out rows of hair. This method is generally used on the face, especially to remove upper lip hair and shape the eyebrows. Like other epilation method, the results can last several weeks. Skin irritation and folliculitis and secondary pigmentary changes are some of the side effects that have been noted.

10.3 Chemical Methods

10.3.1 Depilatory Creams

Depilatories act like a chemical razor that dissolves the hair fiber, causing it to separate easily from the skin surface with results lasting up to two weeks [5,17]. Two forms of depilatories, chemical and enzymatic, are described in the literature and are commercially available. However, only the chemical-based depilatories are effective and has majority of the market share (>95%). Chemical depilation can be achieved by agents such as alkali-metal sulfites and sulfides, amines, and mercaptans to cause the hair to lose its tensile strength and deteriorate. The use of sulfides as depilatories goes back to the1880s [12]. Though these depilatories are most efficacious, they have a high odor and their use is linked to skin toxicity. Depilatory products containing barium sulfide (Magic Shave) have been specially formulated for use by African–American men who have pseudofolliculitis barbae (PFB), and are unable to shave. On the other hand, the sulfite depilatories have low irritancy potential, do not have the strong odor of sulfides, but are also slow-acting and inefficient [13,14].

Patents covering the use of mercaptans for use in depilatories were first issued in the late 1930s [13,14]. The use of nonpolar aliphatic mercaptans (methyl, butyl, and benzyl mercaptans) is limited to the leather industry because of their strong smell [12–14]. The odor of the polar mercaptans is not as strong, and can be more easily masked. The sulfhydryls that fall into this class are thioglycolic acid, thiolactic acid, beta-aminoethyl mercaptan, thioglycerol, beta-mercapto-alkanesulfonic acids, and dithiothreitol [13,14]. These compounds react more slowly with the hair but have less odor. Another advantage of the mercaptans is that they are safer and can be used on the face [15].

Currently, the most commonly used chemical depilatories are mercaptans, particularly salts of thioglycolic acid. Cosmetically elegant alkaline creams containing thioglycolates were first patented in 1940s for human use (by Nair), and they remain the standard chemical depilatories used today. The active ingredient in nearly all commercially successful chemical depilatories is calcium thioglycolate, which is most often used in conjunction with calcium hydroxide. Thioglycolate depilatories work by hydrolyzing (reducing) disulfide bonds [17]. Hair strength is a function of the disulfide bonds between cysteine molecules. Cysteine forms 15% of the keratin protein in hair fiber and about 1–2% of the keratin in skin. This differential provides for the preferential hydrolysis of hair keratin over skin. There is a rapid reaction between alkaline thioglycolate and hair keratin, that is pH-dependent and reversible [12,16].

The thioglycolate-based compounds have a low systemic toxicity and are stable at the concentrations (2.5–4%) at which they are effective [12,15]. The depilatory preparations are able to produce optimum effect in 5–15 min, depending on the pH of the preparation. The pH must be at least 10, with quickest depilation occurring at a pH of about 12.5. In general, preparations designed for facial use are milder (lower pH and/or low thioglycolate concentration) than those intended for use on the limbs. The thioglycolates are claimed to be safe at concentrations of up to 15%, if used infrequently. Depilatories can work well, but they can also cause serious skin irritation and even chemical burns, and possibly scarring, if the formula is too strong, or is left on for too long [15].

Recent developments in depilatory technologies include a better masking of the mercaptan odor, hands-free application technology, addition of agents such as antiinflammatory agent, antiirritants, emollient oils, skin-soothing agents, vitamins, hair regrowth inhibitors, and other such agents that provide skin benefits. Because of the high pH of the preparations and a strong 'reducing' ability of mercaptan, compatibility with additives also needs to be addressed. Attempts have been made to accelerate the depilatory action of thioglycolates by including reagents that swell the hair fibers. Sodium metasilicate and a blend of N-vinyl lactam, esters of dicarboxylic acid, melamine or dicyanamide are reported to enhance the action of thioglycolate. Urea, imidazolidinone, guanindines and surfactants have also been used for this purpose [15,16]. Others have the thioglycolate mixed with reagents such as mercaptocarboxylic acid, dimethylisosorbide, and long-chain alkylamines for enhancement in activity. Methods to increase the speed and efficacy of depilatories by providing heat to the treated area have also been reported. Work has therefore continued toward developing an ideal depilatory that is highly efficacious, rapid acting, safe for frequent and long-term use, nonirritating, odorless, and can be mass marketed. Though literature is filled with such claims, a product with these attributes is yet to be marketed.

10.3.2 Enzyme Depilatories

While there are number of patents covering the use of enzymes as cosmetic depilatory, technology has not advanced to a commercially viable product. Enzyme action in depilation is complex and reportedly varies with the type of enzyme. Some proteolytic enzymes, such as papain and trypsin solubilize hair to some extent, while others act on non-keratin portions of the skin and hair follicle to affect loosening and removal of hair [13]. Enzymes such as that produced by *Streptomyces fradiae* attack keratin directly by breaking the disulfide bonds of the hair [17,18]. The enzymes with direct action on hair fiber have been the main focus of attention for cosmetic depilatories. Enzymatic depilation has several advantages over the thioglycolate approach, including, odor-free and skin-friendly preparations (pH of 4.5–8 compared to 10+ for thiols). The major drawback, however, is poor efficacy and potential for allergic reaction. Products containing protease enzyme preparations (fruit enzymes) are commercially available, mostly for delaying hair regrowth after depilation, or are present for claim purposes only.

10.3.3 Cosmeceuticals for Hair Reduction

The marketplace is filled with creams, lotions, and sprays with implied claims for hair-growth reduction. From the regulatory perspective, any product with explicit claims of hair-growth reduction or inhibition falls under the 'drug' category, requiring substantial demonstration of efficacy and safety, and approval from the FDA for marketing. The cosmeceutical products, for example, hair minimization, bypass the regulatory requirements by carefully worded claims that imply effects on hair growth. These products are similar to the OTC antiwrinkle creams that claim to reduce the appearance of fine lines and wrinkles.

There is no convincing scientific evidence in the form of published clinical data in a peer- reviewed journal that would supports the hair-reduction claims made by most OTC products in this category. In fact, these products often claim that repeated use makes hair feel softer, finer, and less noticeable, thereby reducing the frequency with which one needs to shave. Examples of marketed products include Jergens Naturally Smooth Shave Minimizing Moisturizer, Curel Shave Minimizing Moisturizer, Biore Beyond Smooth Daily Facial Moisturizer, Kalo Hair Inhibitor, Epil-Stop, St. Ives Smooth Legs Shave Minimizing Moisturizer For Dry Skin, Suave Advanced Smoothing Lotion Shave Minimizing Formula, King of Shaves Women Vanish Hair Minimizing Spritz, and Aveeno Positively Smooth Moisturizing Lotion. A number of topical herbal preparations can also be found on the internet and in infomercials on television, often with misleading and false information on the effectiveness of the marketed products.

10.3.4 Pharmaceuticals (Rx) for Hair-Growth Control

10.3.4.1 Hormonal Treatments

Women suffering from clinically significant hirsutism, male-pattern hair growth in women [19], generally seek treatments beyond simple cosmetic measures or cosmeceutical products to manage their hair growth. A study conducted by McKnight [20] in a random population of 400 European women asked the question how many felt they have excess amount of unwanted body hair. Nine percent of the surveyed women felt that they were particularly 'hairy'. The investigator assessment classified this nine percent to be 'hirsute'. The medical literature estimates that hirsutism occurs in approximately 5% of the population [21,22]. In women, the condition is caused by increased levels of androgens (male hormones) and/or increased hair follicle sensitivity to androgens, and only rarely is it accompanied by any serious underlying medical problem [23–27].

It is generally agreed by both dermatologists and endocrinologists that successful management of the hirsutism condition requires a combination of both medical (anti-hormonal, Rx creams) and cosmetic procedures. Two kinds of products are available, a topical Rx cream Vaniqa that inhibits the rate of hair growth by targeting a hair follicle enzyme ornithine decarboxylase [28], and anti-androgens that work by reducing the androgen-dependent hair growth [29]. For the treatment of clinically hirsute women, drug treatments include systemic use of steroidal and nonsteroidal anti-androgens, including spironolactone, flutamide, cyproterone acetate, finasteride and cimetidine. In the United States, spironolactone (Aldactone) is the most widely used anti-androgen for this indication [30–35]. Spironolactone interferes with the formation of androgens and androgen receptor binding [26]. Its use in hirsutism is suggested for women with normal ovulatory cycles and normal testosterone levels [30,33]. Cyproterone acetate and flutamide, inhibit the biological activity of androgens by blocking the binding of the androgen to its receptor. The overall efficacy of these drugs is similar to sprionolactone [32,36]. Cyproterone acetate (Dianette) is especially recommended for women who exhibit increased testosterone levels or polycystic ovarian syndrome [37–39]; however, the drug is not available in the United States market. Flutamide (Eulexin), is a potent anti-androgen and a receptor-binding agent, and shows efficacy that is similar to spironolactone [40]. Earlier studies used higher doses of the drug, up to 250 mg, that resulted in severe toxicities; however, the drug has now been shown to be effective at much lower doses, that is, 62.5 mg with fewer side effects [41,43]. Potential for liver toxicity is still the major issue with this drug [40,44–46]. The most recent drug to be tried for this indication is a 5-alpha reductase inhibitor, finasteride (Proscar). Originally developed for benign prostatic hypertrpohy in men, finasteride inhibits the formation of active reduced metabolites from testosterone. The results from controlled clinical studies with this drug demonstrated clinically significant efficacy with relatively fewer side effects [40, 47]. The treatment did not show any efficacious advantage over sprinolactone [31,33,40]. All four of these oral medications, because of their anti-androgenic activity, have some systemic side effects. The most significant side effect reported is hepatotoxicity [48,44–46]. Overall, the efficacy of anti-androgen treatment is rather limited, they do not provide a "cure" for this condition, and have not been approved by the FDA as treatments for hirsutism.

10.3.4.2 Vaniqa (Eflornithine), a Topical Drug (Rx) for Unwanted Facial Hair

The attraction of treatment with eflornithine, active in Vaniqa, is that it can effect hair growth when applied topically, and it provides major safety advantages over antiandrogens. Another significant advantage is the increased and earlier onset of efficacy, within 6–12 weeks, compared to a 6–12 month lag for the anti-androgens. A randomized, double-blind controlled study with eflornithine (15% dose) in hirsute women resulted in a 48% reduction in the hair-growth rate based on objective determinations [49,50]. This objective reduction in hair length was also accompanied by perceptible, as well as clinically meaningful improvements noted by the clinical investigator and subjects. Unlike the side effects associated with hormonal anti-androgen therapy, eflornithine treatment was found to be extremely safe, with only minor dermal side effects [51]. In addition, eflornithine enabled the women to more easily mask their condition which, by their account, promoted improvements in their self-confidence [49,52].

10.4 Energy-Dependent Processes

10.4.1 Electro-epilation

For years electrolysis has been a method of choice for women seeking permanent hair removal. There are three types of electro-epilation methods in use: galvanic electrolysis, thermolysis, and a combination or the blend method. The use of galvanic electrolysis can be traced back to 1875 when Charles Michel, an ophthalmologist used this technique to treat trichiasis (ingrown eyelashes) [53]. The procedure involves inserting a thin needle along the hair shaft down to hair follicle region and applying a pulse of low magnitude direct current (DC) to destroy the hair producing follicle tissue [54–56]. The galvanic current results in the formation of sodium hydroxide from water and sodium chloride in the hair follicle cells, resulting in a permanent destruction of the follicular structure [57]. Hair removal by the thermolysis method involves sending a high-frequency alternating current (AC) through the needle, which heats up the water molecules and destroys the follicle. Some electrologists claim to have obtained better results using the blend method, and experts recommend this method of hair removal over galvanic to thermolysis [58–60].

The effectiveness of electrolysis procedures is highly dependent on the operators' expertise in accurately positioning the needle in the follicle bulb. The procedure is time consuming, expensive, and painful as each follicle needs to be individually destroyed [61]. Potential dermal side effects include papules, inflammation, scarring, and skin pigmentory changes [54,55,62].

10.4.2 Laser and Light-Based Systems

Despite its limitations, electrolysis for many years was the most popular method of long-term or permanent hair removal. The introduction of laser-based hair removal procedures in the 1990s has now replaced most of the electrolysis market. This method of hair removal is based on the concept developed by Anderson and Parrish in 1983, commonly known as 'selective photothermolysis' [63]. The principle behind this concept is preferential absorption of certain wavelengths of light by the hair follicle chromophore melanin, resulting in a significant damage to the hair-fiber producing structures with minimal effect on the surrounding tissues [64]. Melanin in hair follicles is produced by, and is concentrated in melanocytes which are present in the vicinity of the hair matrix and dermal papilla cells that regulate hair growth and cycling. By causing significant thermal damage to these cell types, the hair-growth processes can be interrupted or permanently halted, depending on the extent of damage caused. It is proposed that in order to achieve permanent hair removal, it may be necessary to also damage the stem cell population located in the 'bulge' region [65–67].

There are two types of light sources that are used for photo-epilation procedures: laser and intense pulsed light (IPL). Both operate in the red or near-infrared wavelength region of the light spectrum, the principle difference being that the laser sources have a coherent beam of a single wavelength of light, whereas the IPL devices are based on a slice of light spectrum, typically between 590–1200 nm, and the beam is noncoherent. The near infrared wavelengths allow for selective absorption by the hair follicle melanin and combined with deep dermal penetration and careful selection of the pulse durations, the thermal damage can be confined to the hair follicle. There are four major types of hair removal lasers defined by the wavelength they produce: Ruby (694 nm) [67,68], Alexandrite (755 nm) [69], Diode (about 810 nm) [70], and Nd:YAG (1064 nm) [71]. In general, the Ruby laser is most suited for treatments of light skin and dark hair, whereas, Nd:YAG is mostly used for darker skin tones. An optimum combination of laser/light source, wavelength, pulse duration, fluence (energy), and epidermal cooling is selected to effect hair growth on an individual, based on the person's skin and hair color and the site of treatment [72–77]. Effective permanent hair reduction can be achieved without significant dermal adverse effects [67]. However, aggressive use of laser to achieve higher efficacy or using inappropriate laser parameters especially on darker skin tones, can result in significant skin effects ranging from edema, crusting, burning, scarring, to longer-lasting pigmentary changes [78,79]. Laser hair removal is described in detail in later chapters of this book.

10.4.3 Photodynamic Therapy for Hair Removal

The photodynamic therapy (PDT) is based on the principle of light interaction with a phosensitizing molecule resulting in the formation of chemically reactive species in the target tissue. However, the phosensitizer must preferentially accumulate in the target tissue to avoid collateral damage. For topical PDT, aminolevulonic acid (5-ALA) was the first molecule demonstrated to have practical usefulness [80]. Topically applied ALA is metabolized by viable cells into protoporphyrin IX (PpIX), a molecule extremely sensitive to light around 415 nm. Upon excitation to light, PpIX produces reactive oxygen species and free radicals that cause localized tissue damage [81]. The advantage of ALA for hair reduction is that the topically applied ALA gets metabolized to PpIX by the rapidly proliferating hair follicle cells at a rate that is much greater than the epidermal cells. Moreover, since the metabolic activation is not dependent on hair color, it makes ALA an ideal molecule for the PDT of gray and blonde hair, which otherwise exhibit poor efficacy to the laser hair-removal treatments. Grossman et. al. [82] studied topically applied ALA in a hirsute subject, and demonstrated good efficacy. However, dermal phototoxicity is still a significant safety concern for the use of ALA or any other phosensitizer molecule for hair-removal purposes.

10.5 Biochemical Target-Based Hair-Growth Reduction

10.5.1 Patented Technologies on Hair-Growth Regulation

The hair follicle represents a complex structure that includes the undifferentiated, rapidly proliferating cells of the matrix region surrounding the dermal papilla, as well as the highly differentiated or keratinized hair fiber that emerges from the follicle. The matrix cells give rise to the hair shaft proper, as well as to the inner root sheath layers. The dermal papillas are enveloped by the matrix region, and are thought to be a primary regulator of the hair cycle as well as hair phenotype. Although the dermal papilla has been implicated in hair-growth regulation, no definite data has clearly implicated a single factor as a signal for controlling hair growth. This is due to the likely scenario where multiple and redundant factors are involved with hair cycle and hair-growth regulation. The structural and functional heterogeneity that exists within this skin appendage is interesting as it contains some of the most rapidly proliferating cells in the body, as well as the most differentiated. These properties, along with its ability to renew itself, provide unique characteristics that can be selectively exploited to control hair growth.

Pioneering work in the area of biochemical control processes for hair-growth regulation has been performed by researchers from the Gillette Company (now P&G). The researchers used a rational biochemical and pharmacological approach that started with the identification of key hair follicle growth biochemical pathways a based on biochemical/histological and molecular biology studies; next they identified and selected chemical molecules that could specifically alter the activity of an identified target pathway, and have the potential to penetrate skin stratum cornium after their topical application in dermatologically acceptable formulations. It was then demonstrated in in-vitro human hair follicle and/or hamster flank-organ models that the chemical agent is able to effectively inhibit hair growth in a dose-dependent manner parallel with the alteration of the target pathway activity in the follicle. The results of these investigations led to a series of patents in this field. A select list of awarded US patents on hair-growth control along with the metabolic pathway and select hair-growth inhibitors is shown in Table 10.2.

Table 10.2 Biochemical Target-Based Hair-Growth Reduction—Select Patented Technologies for Hair-Growth Reduction by the Gillette Company (now P&G)

Hair-Growth Target	Inhibitor/Active[1]	Pat No.	Inventor(s)
Androgen receptor	17-α allyltestosterone	4,885,289	Breuer, Kaszynski, Shander.....
ODC inhibitor	DFMO, eflornithine	4,720,489	Shander
S-adenosyl methionine decarb.	MGBG, MAOEA	5,132,293	Shander, Harrington, Ahluwalia
γ-glutamyl transpeptidase	Anthglutin	5,096,911	Ahluwalia, Shander
Adenylosuccinate synthetase	L-alanosine	5,095,007	Ahluwalia
Transglutaminase	Isoxazole derivatives	5,143,925	Shander, Funkhouser
L-asparagine synthetase	Ethacrynic acid	5,444,090	Ahluwalia
Sulfhydryl compound	N acetyl cysteine,	5,411,991	Shander, Ahluwalia, Grosso
Cyclooxygenase	NSAIDs, Indomethacin	6,248,751	Ahluwalia, Shander
Lipoxygenase	NDGA, Quercetin	6,239,170	Ahluwalia, Shander
Topical composition; VANIQA	eflornithine	5,648,394	Boxall, Amery, Ahluwalia
Nitric oxide synthetase	Arginine derivatives	5,468,476	Ahluwalia, Shander, Henry
Ornithine amino transferase	5 fluoro methylornithine	5,474,763	Shander, Funkhouser
Cysteine pathway enzymes[2]	O-succinyl serine	5,455,234	Ahluwalia, Shander
Protein kinase C	H-7, Glycyrrhetinic acid	5,554,608	Ahluwalia, Shander, Styczynski
Green tea polyphenols-catechins	EGCG	5,674,477	Ahluwalia
Angiogenesis suppressors[2]	Aurintricarboxylic acid	6,093,748	Ahluwalia, Styczynski, Shander
Arginase	Aminoisobutyric acid	5,728,736	Shander, Henry, Ahluwalia
Energy metabolism[2]	Phloretin, Quinaldic acid	5,652,273	Henry, Ahluwalia, Shander

(Continued)

Table 10.2 Biochemical Target-Based Hair-Growth Reduction—Select Patented Technologies for Hair-Growth Reduction by the Gillette Company (now P&G) (Continued)

Hair-Growth Target	Inhibitor/Active[1]	Pat No.	Inventor(s)
Hypusine biosynthesis pathway[2]	diaminooctane	6,060,471	Styczynski, Ahluwalia, Shander
Glycosaminoglycans and Glycoprotein[2]	Diethylcarbamazine	5,908,867	Henry, Ahluwalia, Shander
Matrix metalloproteinase[2]	Minocycline	5,962,466	Styczynski, Ahluwalia, Shander
Cholesterol synthesis	Statins	5,840,752	Henry, Ahluwalia, Shander
Topoisomerase	Etoposide, novobiocin	6,037,326	Styczynski, Ahluwalia
Androgen conjugation[2]	Ethoxyquin, BHA, BHT	5,958,946	Styczynski, Ahluwalia
Alkaline phosphatase	orthovanadate	6,020,006	Styczynski, Ahluwalia
Protein tyrosine kinase	Tryphostin	6,121,269	Henry, Ahluwalia
Ceramide metabolism	PDMP, PPMP	6,235,737	Styczynski, Ahluwalia
DFMO active enantiomer	L-DFMO	6,743,822	Styczynski, Ahluwalia, Shander
Fatty acid mechanism[2]	Methyl Palmoxirate	7,160,921	Hwang, Henry, Ahluwalia, Shander

[1]Inhibitor/active: The patent discloses several inhibitors or active molecules. Only select compounds are listed here.
[2]Multiple target enzymes are identified and claimed under the broad target pathway.

References

1. Toerien M and Wilkinson S. Gender and body hair: constructing the feminine women. *Women Stud Int Forum* 2003; 26(4): 333–344.
2. Basow SA and Braman AC. Women and body hair – social perceptions and attitudes. *Psychol Women Q* 1998; 22 (4): 637–645.
3. Lipton MG, Sherr L, Elford J, Rustin MHA, and Clayton WJ. Women living with facial hair: the psychological and behavioral burden. *J Psychosomatic Res.* 2006; 61: 161–168.
4. Rittmaster RS. Medical treatment of androgen dependent hirsutism. *J Clin Endocrinol Metb.* 1995; 80 (9):2559–2563.
5. Olsen EA. Methods of hair removal. *J Am Acad Dermatol.* 1999; 40: 143 - 147
6. Lynfield YL, MacWilliams P. Shaving and Hair Growth. *J Invest Dermatol.* 1970; 55:170–172.
7. Hale PA and Ebling FJ. The effects of epilation and hormones on the activity of rat hair follicles. *J Exp Zool.* 1975; 191: 49–62.
8. Johnson E and Ebling FJ. The effect of plucking hairs during different phases of the follicular cycle. *J Embryol Exp Morphol.* 1964; 12: 465 – 474.
9. Richards RN, Uy M, and Meharg G. Temporary hair removal in patients with hirsutism: a clinical study. *Cutis* 1990; 45: 199–202.
10. Wright RC. Traumatic folliculitis of the legs: a persistent case associated with use of home epilating device. *J Am Acad Dermatol.* 1992; 27: 771–772.
11. Abdel-Gawad MM, Abdel-Hamid IA, and Wagner RF Jr. Khite: a non-Western technique for temporary hair removal. *Int J Dermatol.* 1997; 36(3): 217.
12. Harry RG. *Harry's Cosmetology*, 7th ed., Chemical Publishing, New York, 1982, pp. 142–155.
13. Barry RH. Depilatories. In Balsam MS and Sagarin E, eds., *Cosmetics Science and Technology*, Vol. 2. Wiley Intersciences, New York, 1972, pp. 39–72.

14. Rieger, M and Brechner S. Depilatories. In De Navarre, ed., *The Chemistry and Manufacture of Cosmetics*, Vol. IV, 2nd ed., Continental Press, FL, 1975, pp. 1229–1373.
15. Draelos Z. Hair removal. *Cosmetic in Dermatology*, 1978; pp. 99–103.
16. Natow AJ. Chemical removal of hair. *Cutis*, 1986; 38: 91–92.
17. Kunert J and Stránsky Z. Thiosulfate production from cystine by the keratinolytic prokaryote *Streptomyces fradiae*. *Arch Microbiol.* 1988; 150: 600–601.
18. Kunert, J. Biochemical mechanism of keratin degradation by the actinomycete *Streptomyces fradiae* and the fungus *Microsporum gypseum*: A comparison. *J. Basic Microbiol.* 1989;29(9), 597–604.
19. Redmond GP, and Bergfeld WF. Diagnostic approach to androgen disorders in women: acne, hirsutism and alopecia. *Cleve Clin J Med.* 1990; 57:423–427.
20. McKnight E. The prevalence of hirsutism in young women. *Lancet* 1964; 1:410–413.
21. Clayton RN, Ogden V, Hodgkinson J, et al., How common are polycystic ovaries in normal women and what is their significance for the fertility of the population? *Clin Endocrinol (Oxf.)* 1992; 37:127–134.
22. Hatch R, Rosenfield RL, Kim MH, and Tredway D. Hirsutism: implications, etiology, and management. *Am J. Obstet Gynecol.* 1981; 140: 815–830.
23. Rittmaster RS. Clinical relevance of testosterone and dihydrotestosterone metabolism in women. *Am J Med.* 1995; 98(1A): 17S–21S
24. Givens JR and Kurtz BR. Hirsutism, virilization, and androgen excess. In Hurst JW, Ambrose SS, et al., eds., *Medicine for the Practicing Physician*, 3rd ed. Boston: Butterworth-Heinemann, 1992, pp. 568–571.
25. Rittmaster RS. Medical treatment of androgen-dependent hirsutism. Clinical Review 73. *J Clin Endocrinol Metab.* 1995; 80(9): 2559–2563.
26. Ehrmann DA and Rosenfield RL. An endocrinological approach to the patient with hirsutism. *J Clin Endocrinol Metab.* 1990; 71: 1–4.
27. Dawber RPR, Ebling FJG, and Wojnarowska FT. Disorder of hair. In Champion RH, Burton JL, Ebling FJG, eds. *Textbook in Dermatology*. Oxford: Blackwell Scientific, 1992; 2533–2571.
28. Balfour JA and McClellan K. Topical eflornithine. *Am J Clin Dermatol.* 2001; 2: 197–202.
29. Conn JJ and Jacobs HS. The clinical management of hirsutism. *Eur J Endocrinol.* 1997; 136: 339–348.
30. Spritzer PM, Lisboa KO, Mattiello S, and Lhullier F. Spironolactone as a single agent for long-term therapy of hirsute patients. *Clin Endocrinol (Oxf.)* 2000; 52(5): 587–594.
31. Erenus M, Yucelten D, Durmusoglu F, and Gurbuz O. Comparison of finasteride versus spironolactone in the treatment of idiopathic hirsutism. *Fertil Steril.* 1997; 68(6): 1000–1003.
32. Erenus M, Yucelten D, Gurbuz O, Durmusoglu F, and Pekin S. Comparison of spironolactone-oral contraceptive versus cyproterone acetate-estrogen regimens in the treatment of hirsutism. *Fertil Steril.* 1996; 66(2): 216–219.
33. Wong IL, Morris RS, Chang L, Spahn MA, Stanczyk FZ, and Lobo RA. A prospective randomized trial comparing finasteride to spironolactone in the treatment of hirsute women. *J Clin Endocrinol Metab.* 1995; 80(1): 233–238.
34. Barth JH, Cherry CA, Wojnarowska F, and Dawber RP. Spironolactone is an effective and well tolerated systemic antiandrogen therapy for Hirsute women. *J Clin Endocrinol Metab.* 1989; 68(5): 966–970.
35. Shapiro G and Evron S. A novel use of spironolactone: treatment of hirsutism. *J Clin Endocrinol Metab.*; 1980; 51(3): 429–432.
36. Hammerstein J and Cupceancu B. The treatment of hissutism with cyproterone acetate. *Ger Med Mon.* 1969; 14(12): 599–602.
37. Gokmen O, Senoz S, Gulekli B, and Isik AZ. Comparison of four different treatment regimes in hirsutism related to polycystic ovary syndrome. *Gynecol Endocrinol.* 1996; 10(4): 249–255.
38. Belisle S and Love EJ. Clinical efficacy and safety of cyproterone acetate in severe hirsutism: results of a multicentered Canadian study. *Fertil Steril.*; 1986; 46: 1015–1020.
39. Falsetti L and Galbignani E. Long-term treatment with the combination ethinylestradiol and cyproterone acetate in polycystic ovary syndrome. *Contraception* 1990; 42: 611–619.

40. Moghetti P, Tosi F, Tosti A, Negri C, Misciali C, Perrone F, Caputo M, Muggeo M, and Castello R. Comparison of spironolactone, flutamide, and finasteride efficacy in the treatment of hirsutism: a randomized, double blind, placebo-controlled trial. *J Clin Endocrinol Metab.* 2000; 85(1): 89–94.
41. Muderris II, Bayram F, and Guven M. Treatment of hirsutism with lowest-dose flutamide (62.5 mg/day). *Gynecol Endocrinol.* 2000; 14(1): 38–41.
42. Muderris II, Bayram F, and Guven M. A prospective, randomized trial comparing flutamide (250 mg/d) and finasteride (5 mg/d) in the treatment of hirsutism. *Fertil Steril.* 2000 May; 73(5): 984–987.
43. Marugo M, Bernasconi D, Meozzi M, Del Monte P, Zino V, Primarolo P, and Badaracco B. The use of flutamide in the management of hirsutism. J *Endocrinol Invest.* 1994; 17(3): 195–199.
44. Andrade RJ, Lucena MI, Fernandez MC, Suarez F, Montero JL, Fraga E, and Hidalgo F. Fulminant liver failure associated with flutamide therapy for hirsutism. *Lancet* 1999; 353: 983.
45. Wysowski DK and Fourcroy JL. Flutamide hepatotoxicity. *J Urol.* 1996; 155(1): 209–212.
46. Wallace C, Lalor EA, and Chik CL. Hepatotoxicity complicating flutamide treatment of hirsutism. *Ann Intern Med.* 1993; 119(11): 1150.
47. Faloia E, Filipponi S, Mancini V, Di Marco S, and Mantero F. Effect of finasteride in idiopathic hirsutism. *J Endocrinol Invest.* 1998; 21(10): 694–698.
48. Committee on Safety of Medicine/Medicine Control Agency. Hepatic reactions with cyproterone acetate (Cyprostat, Androcur). *Current Problems in Pharmacovigilance* 1995; 21: 1.
49. Shander D, Ahluwalia GS, and Morton JP. Management of unwanted facial hair by topical application of eflornithine. In Elsner P, Maibach HI eds., *Cosmeceuticals and Active Cosmetics*, 2nd ed. Taylor & Francis, 2005, pp. 489–510.
50. Schrode KS, Huber F, Staszak H, Altman DJ, Shander D, Ahluwalia GS, and Morton J. Randomized, double-blind, vehicle controlled safety and efficacy evaluation of eflornithine 15% cream in the treatment of women with excessive facial hair (Abstract). *57th Ann Meeting Am Acad Dermatol*, San Francisco CA. 1999.
51. Hickman JG, Huber F, and Palmisano M. Human dermal safety studies with eflornithine HCL 13.9% cream (VANIQA), a novel treatment for excessive facial hair. *Curr Med Res Opin.* 2001; 16 (4): 235–244.
52. Jackson JD, Shander D, Huber F, Schrode KS, and Mathes BM. The evaluation of quality of life in two studies of women treated with topical eflornithine HCL 13.9% cream for unwanted facial hair. (abstract). *59th Ann Meeting Am Acad Dermatol*, 2001; 259.
53. Michel CE. Trichiasis and distichiasis with an improved method for radical treatment. *St. Louis Clinical Record* 1875 Oct; 2:145–148.
54. Wagner RF Jr, Tomich JM, and Grande DJ. Electrolysis and thermolysis for permanent hair removal. *J Am Acad Dermatol.* 1985; 12(3): 441–449.
55. Wagner RF Jr. Medical and technical issues in office electrolysis and thermolysis. *J Dermatol Surg Oncol.*; 1993; 19(6): 575–577.
56. Hobbs ER, Ratz JL, and James B. Electrosurgical epilation. *Dermatol Clin.* 1987; 5(2): 437–444.
57. Richards RN and Meharg GE. Electrolysis: observations from 13 years and 140,000 hours of experience. *J Am Acad Dermatol.* 1995; 33(4): 662–666.
58. Hinkel AR and Lind RW. *Electrolysis, Thermolysis and the Blend: The Principles and Practice of Permanent Hair Removal.* Los Angeles, CA: Arroway Publishers, 1968. ISBN 0-9600284-1-2.
59. Richards RN and Meharg GE. *Cosmetic and Medical Electrolysis and Temporary Hair Removal: A Practice Manual and Reference Guide.* Toronto: Medric Ltd., 1991, pp. 37–40. ISBN: 0-9694746-0-1.
60. Bono M. *Real World Electrology: The Blend Method.* Santa Barbara, CA: Tortoise Press, 1994. ISBN: 0-9642682-0-5.
61. Wagner RF Jr, Flores CA, and Argo LF. A double-blind placebo controlled study of a 5% lidocaine/prilocaine cream (EMLA) for topical anesthesia during thermolysis. *J Dermatol Surg Oncol.* 1994; 20: 148–150.
62. Petrozzi JW. Verrucae planae spread by electrolysis. *Cutis* 1980; 1: 85.

63. Anderson RR and Parrish JA. Selective photothermolysis: precise microsurgery by selective absorption of pulsed radiation. *Science* 1983; 220(4596): 524–527.
64. Grossman MC, Dierickx CC, Farinelli W, Flotte T, and Anderson RR. Damage to hair follicle by normal mode ruby laser pulses. *J Am Acad Dermatol.* 1996; 35 (6): 889–894.
65. Sun T-T, Costsarelis G, and Lavker RN. Hair follicular stem cells: the bulge-activation hypothesis. *J Invest Dermatol.* 1991; 96(Suppl. 5): 775–785.
66. Lavker RM, Miller S, Wilson C et al. Hair follicle stem cells: their location, role in hair cycle, and involvement in skin tumor formation. *J Invest Dermatol.* 1993; 101(suppl.): S16–S26.
67. Dierickx CC, Grossman MC, Farinelli WA, and Anderson RR. Permanenet hair removal by normal-mode ruby laser. *Arch Dermatol.* 1998; 134: 837–842.
68. Sommer S, Render C, Burd R, and Sheehan D R. Ruby laser treatment for hirsutism: clinical response and patient tolerance. *Br J Dermatol.* 1998; 138: 1009–1014.
69. Nanni CA and Alster TS. Long-pulsed alexandrite laser assisted hair removal at 5, 10 and 20 millisecond pulse durations. *Lasers Surg Med.* 1999; 24: 332–337.
70. Loo WW, Quintana AT, Geronemus RG, and Grossman MC. Prospective study of hair reduction by diode laser (800nm) with long-term follow-up. *Dermatol Surg.* 2000; 26: 428–432.
71. Bencini PL, Luci A, Galimberti M, and Ferranti G. Long-term epilation with long-pulsed neodymium YAG laser. *Dermatol Surg.* 1999; 25: 175–178.
72. Lask G, Elman M, Noren P, Lee P, and Nowfar-Rad M. Hair removal with the epitouch™ ruby laser – a multicenter study (abstract). *Lasers Surg Med.* 1997(Suppl. 9): 32.
73. Nanni CA and Alster TS. Optimizing treatment parameters for hair removal using topical carbon based solution and 1064 nm Q-switched neodymium: YAG laser energy. *Arch Dermatol.* 1997; 133: 1546–1549.
74. Goldberg DJ. Various mechanisms of laser hair removal. *Cosmet Dermatol.* 1997; 10(8): 36–38.
75. Gold MH, Bell MW, Foster TD, and Street S. Long-term epilation using the epilight broad band, intense pulsed light hair removal system. *Dermatol Surg.* 1997; 23: 909–913.
76. Dierickx CC, Grossman MC, Farinelli WA, Manuskiatti W, and Anderson RR. Long-pulsed ruby laser hair removal: comparison between two pulse widths (0.3 and 3 msec) [abstract]. *Laser Surg Med.* 1997(Suppl. 9): 36.
77. Finkel B, Eliezri D, Waldman A, and Slatkine M. Pulsed alexandrite laser technology for noninvasive hair removal. *J Clin Laser Med Surg.* 1997; 15: 225–229
78. Aghassi D, Carpo B, Eng K, and Grevelink JM. Complications of aesthetic laser surgery. *Ann Plast Surg.* 1999; 43: 560–569.
79. Moreno-Arias GA, Castelo-Branco C and Ferrando J. Side-effects after IPL photoepilation. *Dermatol Surg.* 2002; 28: 1131–1134.
80. Kennedy JC, Puttier RH and Pross DC. Photodynamic therapy with endogenous protoporphyrin IX: basic principles and present clinical experience. *J Photochem Photobiol.* 1990; 6:143–148.
81. Touma D and Gilchrest B. Topical Photodynamic Therapy: A New Tool in Cosmetic Dermatology. Seminars in Cutaneous Medicine and Surgery 2003; 22: 124–130.
82. Grossman M, Wimberly J, Dwyer P, et al. Photodynamic therapy for hirsutism. *Lasers Surg Med.* 1995; 17(Suppl. 7): 44.

PART 3
LIGHT-BASED SYSTEMS FOR IMPROVING SKIN APPEARANCE

11

Skin Rejuvenation Using Fractional Photothermolysis: Efficacy and Safety

***Brian Zelickson*[1,2] *and Susan Walgrave*[2]**

[1]*Associate Professor of Dermatology, University of Minnesota, Minneapolis, MN, USA*
[2]*Zel Skin and Laser Specialists, Edina, MN, USA*

Gurpreet S. Ahluwalia (ed.), *Cosmetic Applications of Laser and Light-Based Systems*, 255–270,

11.1 Introduction

Over the past hundred years, the US population has more than tripled, owing to the increased life expectancy (average 77 years) and declining mortality rate. Specifically, the population of those aged 65 and older is expected to increase rapidly starting in 2011, as the first of the baby boom generation reach this age [1]. With the majority of the population being female and living longer, the demand for cosmetic procedures is growing tremendously.

Skin rejuvenation is a term used to define a procedure that can reduce the signs of aging and photodamage. Traditionally, ablative lasers, such as the carbon dioxide (CO_2) and erbium:yttrium-aluminum-garnet (Er:YAG), have been the most successful lasers for resurfacing the skin by improving texture, wrinkles, and pigmentation. These devices accomplish this by ablating the epidermis, and potentially the upper portions of the dermis, inducing a controlled wound and a subsequent healing response [2]. This improvement, however, is associated with prolonged downtime, including erythema, which can last from weeks to months, and carries with it significant risk of complications, including infection, pigmentary changes, and scarring.

Nonablative devices have since been developed to minimize the risk of side effects by sparing the epidermis while targeting structures within the dermis. Although patients typically experience minimal recovery time, the overall textural improvement has been mild and unpredictable, despite histological evidence of collagen remodeling [3–6].

To overcome the shortcomings of both the procedures, fractional photothermolysis was developed in an attempt to achieve a greater efficacy than nonablative procedures, without the downtime and side effects associated with ablative resurfacing. The concept of fractional photothermolysis was first developed in 2001 by Dr R. Rox Anderson of Massachusetts General Hospital. In 2003, Huzaira et al. [7,8] tested this theory with a 1500 nm laser to assess whether the thermal effects could be spatially confined within human tissues. Indeed, the laser created multiple foci of thermal injury that were approximately 50 to 150 μm in diameter and 0 to 550 μm in depth, while sparing surrounding tissues. After further studies by Manstein et al. [9], Reliant Technologies Inc. introduced this technology with the Fraxel® SR 750 laser system in 2004, which led the way for a new era of resurfacing devices.

11.2 Fractional Photothermolysis Defined

To date, there is no agreed-upon definition for fractional resurfacing. For the purpose of this discussion, we will define this as the destruction or removal of a fraction of the skin, including the full thickness of the epidermis and portions of the dermis, where the depth of the injury is greater than the width and ratios of treated to nontreated tissue >10% and <90%. The principle behind fractional photothermolysis is the formation of isolated and microscopic thermal wounds that are surrounded by zones of spared, viable tissue in a geometrical pattern that does not depend on chromophore distribution [9]. Fractional photothermolysis is distinct from selective photothermolysis, which was first described in the early 1980s. With selective photothermolysis, certain wavelengths are chosen that are specifically absorbed by target structures. The laser energy is then converted into heat, destroying these target tissues, while leaving the surrounding tissues undisturbed.

At mid-infrared wavelengths, initially chosen for nonablative fractional photothermolysis, the target chromophore is water, which is uniformly distributed throughout the tissue.

Therefore, damage is confined by using an optically focused beam, which allows microscopic columns of thermal damage to be created in the epidermis and dermis. By keeping the beam tightly focused, high local irradiance is achieved while minimizing the dispersion of energy to surrounding tissues to avoid bulk thermal damage.

Both selective photothermolysis and fractional photothermolysis produce an adjustable three-dimensional microscopic thermal injury with the width and depth of the lesion dependent upon the parameters selected [9]. In fractional photothermolysis, using an infrared wavelength, these microscopic tissue injuries are also referred to as microscopic thermal zones or MTZs. The MTZs are columns of coagulated tissue that extend from the epidermis down to the mid-dermis with sections of noncoagulated viable tissue between the columns. Spared interlesional tissue with fractional treatment of the epidermis appears to stimulate prompt re-epithelialization of damaged tissue [10,11]. With pan-surface ablative procedures, re-epithelialization is prolonged because there is no viable epidermis remaining to participate in the wound-healing process [10,12]. In contrast, complete protection of the entire epidermis in traditional nonablative resurfacing precludes rapid epidermal turnover, reducing the efficacy of the treatment [10]. Fractional photothermolysis affords an excellent compromise by treating a part of the epidermis while allowing spared portions to contribute to the healing process. The stratum corneum also remains intact, following nonablative fractional treatment. This allows the skin to maintain its barrier function to defend against microbial infection [10,11] and minimizes the risk of other potential side effects, such as oozing and crusting. Therefore, the skin also appears visually intact to the human eye due to the preserved stratum corneum and undetectable microscopic injuries [13].

Since the introduction of the first infrared wavelength fractional systems, there have been a series of other "fractional" systems using a variety of parameters. In light of this, it is important to further define fractional resurfacing in terms of the extent of tissue injury, which is dependent upon multiple parameters of the fractional device.

11.3 Fractional Treatment Parameters to Consider

The parameters that influence the extent of fractional injury include wavelength, fluence, spot size, surface density of the microscopic thermal injuries, delivery method of the microbeams (i.e. scanned, stamped), and cooling capabilities. Other important procedural considerations include the number of treatments, interval between treatments, number of passes, interval between passes, and the need for topical applications and anesthetics [14].

Wavelengths between 1400–1600 nm, in the near to mid-infrared range, are ideal for coagulative fractional resurfacing because they can penetrate deep within the dermis [15]. Due to their high water absorption, lasers with wavelengths of 2940 and10600 nm are used for ablative fractional resurfacing. These devices can remove or ablate a column of tissue with a small rim of coagulated tissue surrounding it. This type of injury can lead to a slightly longer healing time than mid-infrared lasers, but studies suggest a greater clinical response.

It is important to note that the wavelength is not the only parameter that is important in determining the width and depth of the tissue injury. Fluence levels of each microbeam also affect the depth of penetration, as well as the nature of the beam itself and how it is delivered. There are fractional lasers that produce a single, focused beam that is pulsed in temporal succession as it scans across the surface (Fig. 11.1a), or lasers that produce multiple, focused, micro-beams emitted simultaneously in a stamping fashion (Fig. 11.1b). The fractional

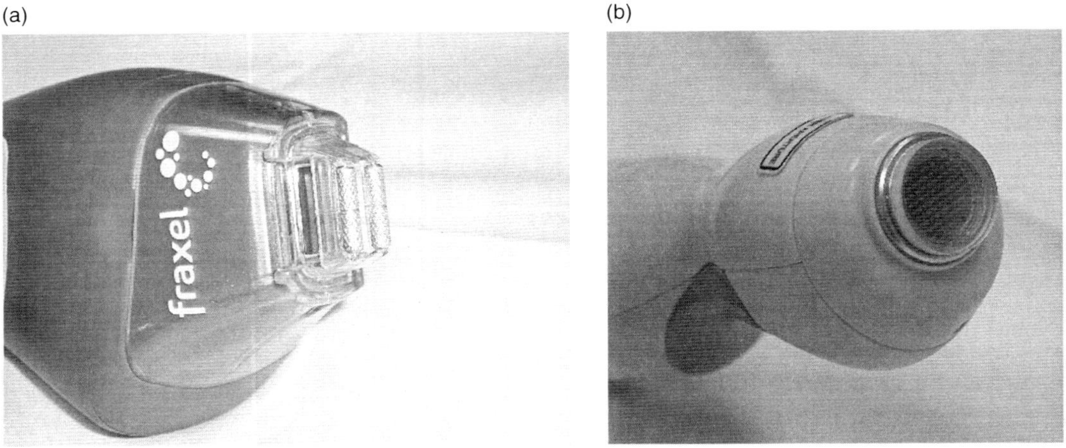

Figure 11.1 (a) Fraxel re:store scanning handpiece. (b) Lux1540™ Fractional stamping handpiece.

damage pattern and surface coverage area is therefore defined by the scanning or stamping characteristics of each device [14].

Consideration should also be given to the diameter of the thermal injury columns, which should be kept small enough to allow for easy migration of keratinocytes and other biological cell mediators from the surrounding tissues [10,16]. These dimensions have not been specifically defined, but studies have found that the distance between microscopic zone centers should be greater than 125 μm to avoid bulk tissue damage and significant side effects [9].

Cooling may also affect the dimensions of the microscopic columns. Integrated contact cooling devices or an external cooling device, such as the SynerCool™ (Syneron Medical, LTD) or Zimmer Cryo 5® (Zimmer Medizin Systems), are used concurrently during treatment, and have been shown to increase patient comfort to a great extent [17]. However, cooling may also significantly decrease the width of the thermal damage zones with less impact on depth [18,19]. When the skin temperature is decreased from body temperature to 20°C, there is a corresponding 40% decrease in the MTZ area at parameters tested (1550 nm, 10 mJ, 250 MTZ/cm^2) [19]. How this affects the clinical outcome is yet to be determined.

Certain parameters, such as the number of passes completed, depend on the density of the microscopic thermal injuries per pass. With this consideration in mind, there should be sufficient passes to allow for approximately 10–90% of the treatment surface area to be covered. Therefore, patients may need more than one treatment to cover most of the skin surface. The time between successive passes should be > 20 s to allow for excess heat to dissipate, and minimize bulk thermal injury [20].

It is also important for individual treatments to be spaced far enough apart to allow for the resolution of clinical swelling and flaking. For most mid-infrared fractional devices, this time is approximately 2–4 weeks.

11.4 Biological Effects of Fractional Photothermolysis

Interesting histological changes ensue after mid-infrared fractional photothermolysis. Columns of epidermal and dermal cell necrosis are seen immediately after treatment, with preservation of the stratum corneum (Fig. 11.2). Each of the microscopic columns of thermal injury is surrounded by a heat-shock zone that releases cell mediators to signal the wound-healing cascade [13]. Specifically, heat-shock protein (hsp) 70 expression is increased, most prominently within the epidermis in areas that underlie necrotic debris, and in dermal tissues that surround the MTZs [2]. Hsp 70 causes up-regulation of transforming growth factor (TGF)-beta which increases collagen production, thereby stimulating dermal remodeling [2,9,10]. Evidence of increased dermal collagen III production is seen after one week [2].

Within an hour of treatment, keratinocytes begin to move to the deep and lateral margins of the epidermal wound [11,13]. By 12 hours, viable cells surround the necrotic debris and begin to form a plug containing this microscopic epidermal necrotic debris (also known as MENDs) [13]. This compact material ranges from 50 to 200 μm in diameter [9,10], and has been found to contain both melanin and elastin [10].

By 24 hours, MENDs are found within the epidermis above each area of the dermal injury with intact stratum corneum [2,13]. Stem cells located in the basal layer appear to be temporarily activated and begin to replace the epidermal tissues [10,21]. At this time, the

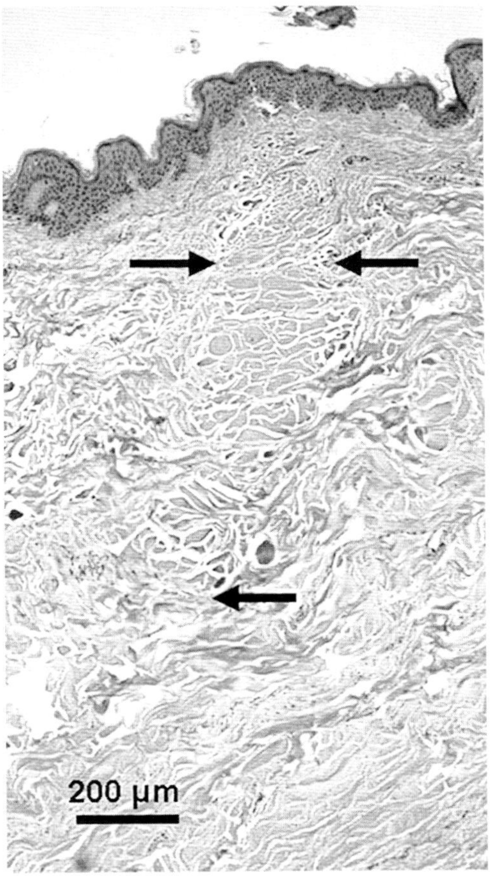

Figure 11.2 H&E stain depicting a microthermal zone (black arrows) immediately after nonablative fractional treatment at 40 mJ (Fraxel re:store™).

continuity of the epidermal basal cell layer is also restored. By 48–72 hours posttreatment, the epidermis has re-epithelialized, with partial restoration of the basement membrane [2].

By seven days there is a complete epidermal regeneration, and exfoliation of coagulated MENDs also starts to occur. This corresponds to a bronze color that patients develop, which disappears with subsequent desquamation.

Three months after treatment, thermal damage columns are completely resolved and replaced with new collagen. Increased undulating rete ridge patterns are also observed [2,9].

Recently, ablative fractional resurfacing has been introduced to overcome extensive epidermal and dermal thermal damage associated with traditional ablative devices. Similar to mid-infrared fractional photothermolysis, microscopic zones are created, but instead are composed of ablated, coagulated tissue (Fig. 11.3). This results in changes that are slightly different histologically. Initial studies examined far-infrared (10600 nm) ablative fractional treatment of forearms in subjects with pulse energies ranging from 5–40 mJ with a single

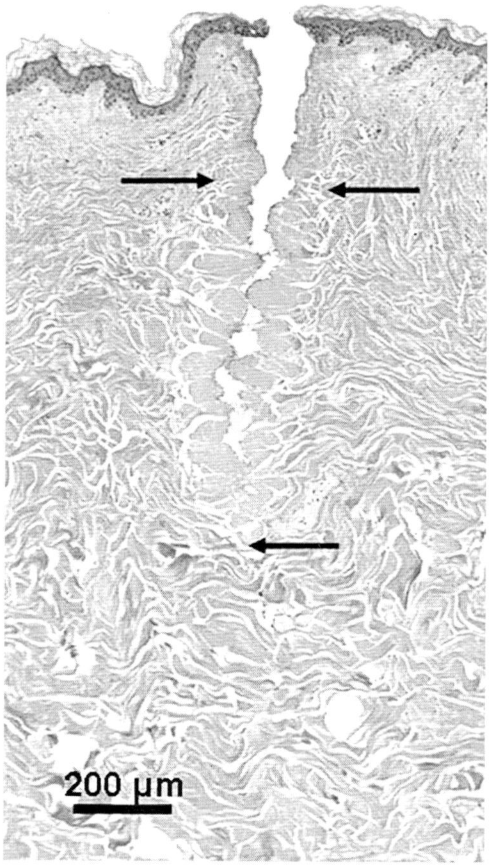

Figure 11.3 H&E stain showing a micro-lesion immediately posttreatment after ablative fractional resurfacing (Fraxel re:pair™) at 40 mJ. The ablative zone is surrounded by a zone of coagulation.

pass. Spot densities of 400 MTZ/cm^2, which created an interlesional distance of approximately 500 μm, were used for pulse energies of 5–30 mJ. Densities of 100 MTZ/cm^2 were used for 40 mJ [22].

Histological examination showed an immediate ablation of the epidermis and dermis after treatment. Ablative zones, lined by a thin layer of eschar, ranged from 71 to 121 μm in width, and 210 to 659 μm in depth for pulse energies of 5–30 mJ. The total lesion size (including the surrounding thin, coagulation zone) ranged from 138 to 270 μm in width, and 298 to 993 μm in depth. These lesions ended in a tapered manner, unlike traditional CO_2 which result in broad, rectangular-shaped thermal coagulation zones parallel to the skin surface. Tapered lesions are beneficial because they allow for deeper dermal ablation while maintaining interlesional tissue viability [22].

Greater increases in depth than width were generally seen with increasing energies. Therefore, the depth-to-width ratio increased with increasing pulse energies [22], unlike nonablative fractional resurfacing which maintains a relatively constant depth-to-width ratio of 5 [20]. This may allow for deeper removal of unwanted dermal material through the transepidermal elimination pathway [11].

By 48 hours, the ablative zone was completely replaced by invaginating epidermal cells. The basement membrane remained partially disrupted, but was completely restored by day 7. Upregulation of hsp 72 was also seen. Microscopic epidermal necrotic debris were also found in the stratum corneum, and were exfoliated by 7 days posttreatment. Increased numbers of spindle cells that were likely to be consistent with fibroblast activity and ongoing dermal remodeling, were also noted this time [22].

By one month, the epidermal invagination had considerably regressed, and this space was replaced by newly synthesized collagen. The coagulation zone surrounding the ablative zone was also diminished, but was still present. Collagen in both of these zones appeared haphazard with abundant spindle cells still present [22].

Three months posttreatment, scattered areas in the dermis mildly resembled residual lesions. Hsp 72 activity decreased significantly while hsp 47 expression increased, consistent with ongoing collagen synthesis and dermal remodeling [22].

Most of the histological studies reported to date have been completed with Fraxel lasers (Reliant Technologies, Inc.). Further studies examining the biological effects with all commercial devices available are warranted.

11.5 Therapeutic Uses and Clinical Efficacy

Several studies have shown fractional photothermolysis to be effective in treating photodamaged skin and rhytids [9,23]. Early studies appear promising for a myriad of other dermatologic conditions, including melasma, scarring, poikiloderma of Civatte, and rejuvenation of ethnic skin [16,24–30].

11.5.1 Photodamage

Long-term clinical improvement of facial and nonfacial photodamaged skin has been seen up to nine months after treatment with the 1550 nm Fraxel laser. Wanner et al. [23] examined fifty patients (skin types I–III) who underwent three treatments (8 mJ, 2000 MTZ/cm^2

for facial areas; 8 mJ, 1,500–2000 MTZ/cm^2 for nonfacial areas) three to four weeks apart. Nine months after treatment, 51–75% improvement in photodamage was observed in 73 and 55% of facial and nonfacial treated skin, respectively. Transient erythema and edema were seen in the majority of patients; however, no protracted pigmentary changes or scarring were observed [23].

In Asian patients with photodamaged skin, pigmentary problems are often more of a concern than rhytids [25,31]. Postinflammatory hyperpigmentation (PIH) is a common complication in these and other dark-skinned patients who attempt laser resurfacing [25,32–34]. Fractional resurfacing, however, can be effective in Asian patients when appropriate parameters are used, and caution is exercised to prevent complications such as hyperpigmentation. Initial studies have shown that there is a lower incidence of PIH when lower microthermal zone densities are used [24,25]. Chan et al. [25] found that Asian patients who received a high-energy, low-density treatment with the Fraxel (average fluence 16.3 mJ, total density 1000 MTZ/cm^2) had a lower prevalence of PIH than those who received a low-energy, high-density treatment (average fluence 8.2 mJ, total density 2000 MTZ/cm^2). Similarly, Kono et al. [24] found that the use of higher densities (even with lower fluences) was associated with an increased risk of developing hyperpigmentation. Patients also experienced more pain, erythema, and swelling when higher densities and increased fluences were used. Overall, the clinical efficacy and patient satisfaction were significantly higher with high-fluence, low-density treatments.

11.5.2 Scarring

Scarring is a common concern for many patients, whether due to surgery, trauma, acne, or burn injuries. Scars can be abnormal in texture, color, and can have a loss of surface dermatoglyphics. Early reports have shown more than 75% clinical improvement in erythema, induration, and texture after a single treatment with the Fraxel 1550 nm. At mid-infrared wavelengths, deeper blood vessels are able to be accessed while simultaneously inducing dermal remodeling.[27] Microvasculature destruction occurs due to the high water content of blood, and direct thermal trauma [26].

Fractional photothermolysis also has shown some success in treating hypopigmented scars. An initial pilot study [28] looked at seven patients with hypopigmented scarring, mainly due to inflammatory acne. Six out of seven patients saw marked (51–75%) clinical improvement after two to four treatments (pulse energies of 7–20 mJ, total densities of 1,000–2,500 MTZ/cm^2). Another study [30] saw 25–50% clinical improvement in 91% of patients with mild to moderate atrophic facial scars after a single treatment, while 87% of patients receiving three treatments saw at least a 51–75% improvement.

Fractional resurfacing may help improve hypopigmentation by causing normal melanocytes from the surrounding tissues to migrate and repopulate the newly resurfaced tissue to blend and minimize the appearance of the scar. Atrophy is likely to improve through dermal collagen remodeling that occurs after treatment [28].

11.5.3 Melasma

Melasma (also known as cholasma or mask of pregnancy) is a relatively common condition primarily affecting women, although 10% of cases also occur in men [29]. It is characterized by brown maculae that often occur symmetrically on the face; however, other sun-exposed areas may also be affected.

An initial case report [16] of a 31-year-old Caucasian woman with resistant melasma found marked improvement after two treatments with the Fraxel 1550 nm laser (6–8 mJ, total density 2000 MTZ/cm^2). She continued to show improvement at her six-month follow-up.

Another pilot study [29] examined 10 female patients with melasma (skin types III–V) who received four to six Fraxel 1550 nm treatments (fluence 6–12 mJ, total density 2000–3500 MTZ/cm^2). Sixty percent of these patients achieved 75–100% clearance. Interestingly, one of these patients was Hispanic, while other Hispanic subjects tended to respond less favorably. Another patient (Asian, Type IV skin) had 100% clearance without any incidence of hyper- or hypopigmentation.

Overall, patients with melasma should be treated at monthly or greater intervals using lower fluence levels and treatment densities to minimize the risk of postinflammatory change. Typically, multiple treatments are needed. It is also important to remind patients that melasma can be recurrent, especially when the causative melanocytes and hormonal profile are still present [18].

11.5.4 Other Therapeutic Uses

Fractional photothermolysis may also be beneficial in the treatment of poikiloderma of Civatte. Poikiloderma is a condition characterized by atrophy, hyper- and hypopigmentation, and dilation of blood vessels (telangiectasia) on the neck and chest [26,35,36]. It is most commonly seen in middle-aged and fair-skinned women with a history of sun exposure. Early case reports have noted improvement in erythema, dyschromia, and texture after treatment with a 1550 nm fractionated laser [26].

11.6 Pretreatment Considerations

Prior to treatment, patients should meet with their clinician to discuss their medical history. For patients with a history of cold sores (oral herpes simplex), valacyclovir or acyclovir should be prescribed starting on or before the day of treatment. Medications, such as Accutane® should also be stopped 6–12 months prior to treatment, and topical retinoids and exfoliating products (i.e., glycolic or lactic acids) should be stopped 2 weeks prior to treatment [13]. Caution should be exercised in treating patients who have a history of impaired wound healing, keloids, are immunocompromised, have darker skin types (IV–VI), or are currently tanned.

For pain management during treatment, a topical, lipid-based anesthetic with lidocaine, such as 2.5% lidocaine/2.5% prilocaine cream (EMLA®) or 23% lidocaine/7% tetracaine ointment, is commonly applied approximately one hour prior to treatment. Care should be taken to prevent topical anesthetic from contacting the eyes, as incidents of corneal abrasions have been reported [13].

When performing ablative fractional treatments or when more aggressive treatment settings are used, nerve blocks (infraorbital and mental) are often administered using 4% articaine HCl with epinephrine or other dental anesthetic. Oral medications, such as anxiolytics and analgesics, may also be used. Anxiolytics also aid in preventing lidocaine toxicity [37]. Despite these measures however, mild to moderate pain is often experienced.

Eye protection is also required during treatment. For treatments off the face, protective goggles alone are sufficient. Metallic ocular shields or other protective eye shields should be worn for treatments on the face. When treating directly over the eyes, proper intraocular shields must be placed.

11.7 Posttreatment Considerations

Immediately after treatment, erythema and edema are common, and may persist for two to three days, or slightly longer with ablative treatments, and when more aggressive settings are used. Edema is typically most prominent in the periorbital area and usually lasts less than 1 week in most patients [18]. With ablative fractional resurfacing, punctate bleeding and serous oozing are also commonly present, but typically resolve within 24 hours. Petechiae and/or purpura are also seen in a minority of patients. To help minimize these effects, patients are encouraged to use ice packs and sleep with their head elevated for one to two days after treatment.

Patients may apply moisturizer, sunscreen, and makeup after nonablative fractional treatments since the stratum corneum remains intact, but these products should be avoided for several days after ablative fractional treatments. A few days later, their skin may have a tanned or bronzed appearance once the erythema resolves. This improves as the skin exfoliates over the next several weeks. Xerosis, and occasionally some associated pruritis are also commonly experienced for several days afterward, usually resolving around the same time that the erythema and edema dissipate [38].

A limited number of patients may experience an acneiform eruption that often occurs in the perioral area. This may be treated with topical acne medications or oral antibiotics if persistent or bothersome. Finally, patients should be strongly encouraged to avoid direct sun exposure for one to two months after treatment, to help minimize any abnormal pigmentation and ensure long-standing clinical improvement.

11.8 Devices Currently Available

See Table 11.1 for a list of devices and their corresponding characteristics.

11.9 Treatment Complications and Management

The most common side effects, such as erythema, edema, pruritis, flakiness, petechiae/purpura, and acneiform eruptions are transient and self-limiting. Oral or topical steroids may be given for persistent erythema, edema, and pruritis. However, these symptoms can generally be managed with adequate moisturization, and antihistamines if necessary.

More serious complications can arise, however. There have been reports of a 10–12% incidence of PIH following fractional photothermolysis [18]. As discussed earlier, this is most common in patients with a history of PIH or melasma, and in patients of darker skin types (IV–VI). Treating with lower densities seems to minimize the disruption of the dermal–epidermal junction, which in turn decreases the risk of developing postinflammatory

Table 11.1 Fractional Devices

Device	Manufacturer	Wavelength (nm)	Medium	Beam Delivery Method	Microbeam Spot Size	Fluence Level Capabilities	Dimensions of Thermal Injury (Width/Depth)
Fraxel re:store™ (see Fig. 11. 4a,b)	Reliant Technologies, Inc.	1550	Erbium-doped	Scanned	140 μm	Up to 70 mJ	50–200 μm/ 400–1300 μm
Lux1540™ Fractional (see Fig. 11. 5a,b)	Palomar Medical, Inc.	1540	Erbium:glass	Stamped	75–100 μm	5-100 mJ	50–350 μm/ ≤1000 μm
Fraxel re:pair™ [39]	Reliant Technologies, Inc.	10600	CO_2	Scanned	130 μm	Up to 70 mJ	200–400 μm/ 400–1800 μm*
ActiveFX™ [40]	Lumenis, Inc.	10600	CO_2	Scanned	1300 μm	5–225 mJ	>1000 μm/ 75–100 μm**
Pixel™ [41]	Alma Lasers Ltd.	2940	Erbium:YAG	Stamped	50 μm	Up to 28 mJ/pixel (7 × 7 matrix), 17 mJ/pixel (9 × 9 matrix)	50 μm/ 50–150 μm***
ProFractional™ [42]	Sciton, Inc.	2940	Erbium:YAG	Scanned	250 μm	6–375 J/cm^2	250 μm/ 25–1500 μm***

(Continued)

Table 11.1 Fractional Devices (*Continued*)

Device	Manufacturer	Wavelength (nm)	Medium	Beam Delivery Method	Microbeam Spot Size	Fluence Level Capabilities	Dimensions of Thermal Injury (Width/ Depth)
Affirm™	Cynosure, Inc.	1440 + 1320 + (560–950)	Nd:YAG + xenon pulsed light	Stamped	150 μm	Up to 8 J/cm² (1440 nm), 14 J/cm² (1320 nm), 20 J/cm² (pulsed light)	150 μm/ 300 μm (1440 nm)– 2000 μm (1320 nm)
Mixto SX	Lasering USA (manuf)	10600 (wavelength)	CO2 (medium)	scanned (beam)	300 um (spot size)	fluence level capabilities and dimensions of thermal injury not available at time of press	
eCO2	Lutronic Corp. (manuf)	10600 (wavelength)	CO2 (medium)	scanned (beam)	120 um (spot size)	2-240 mJ (fluence)	up to 2.5 mm (dimensions)

*Includes surrounding coagulation zone

**Note that at these dimensions, this may not meet our current definition of fractional since the width of the thermal injury is > 500 μm, and, depending on the parameters selected, may not penetrate into the dermis (Fig. 11.6)

***Note that at this depth, this may not meet our current definition of fractional since the thermal injury might not penetrate into the dermis, depending on the parameters selected

(a) (b)

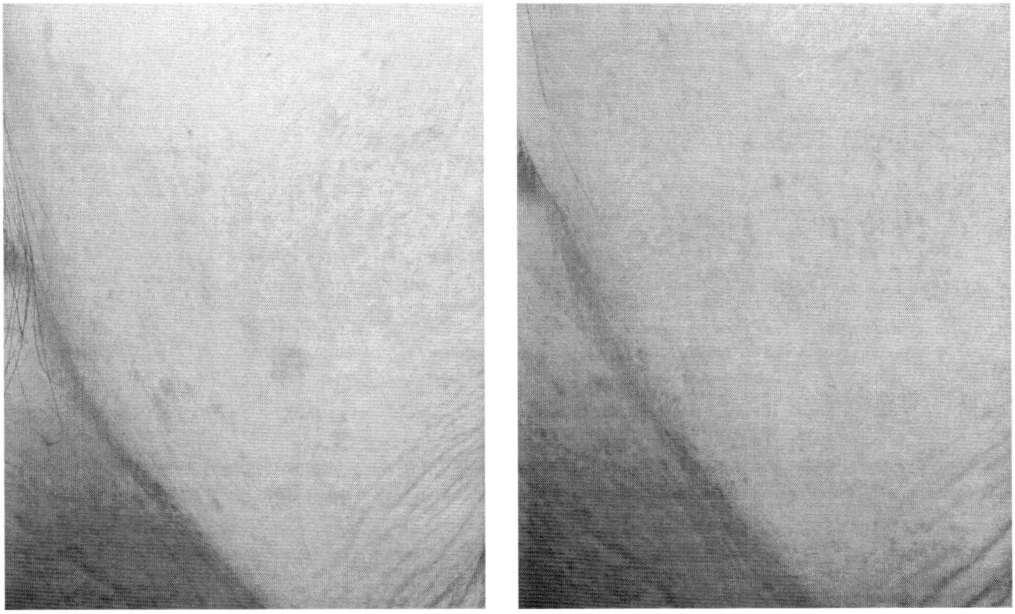

Figure 11.4 (a) Photodamaged skin prior to treatment with the Fraxel re:store. (b) Improvement in pigmentation 1 month after three Fraxel treatments.

(a) (b)

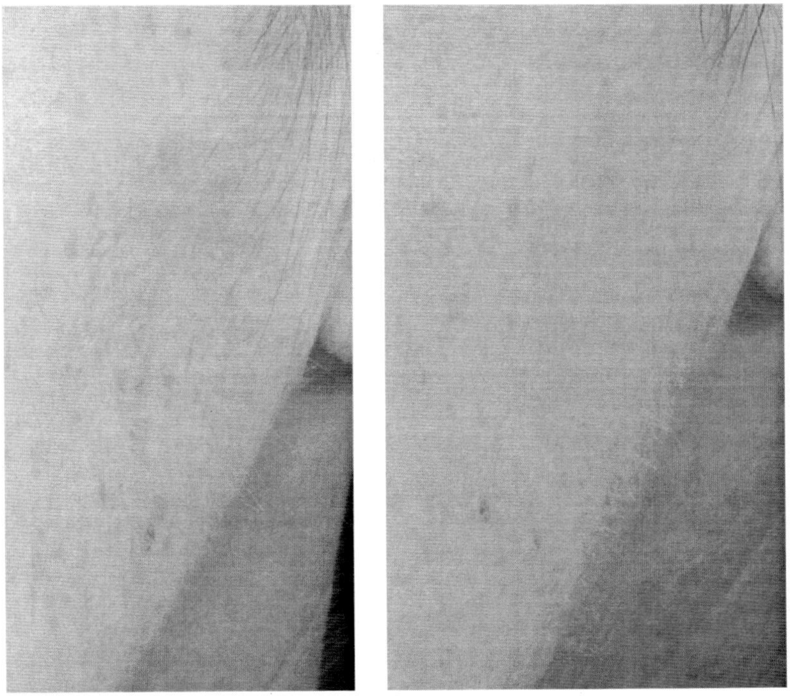

Figure 11.5 (a) Photodamaged skin prior to treatment with the Lux1540™. (b) Improvement in pigmentation 1 month after three Lux1540™ treatments.

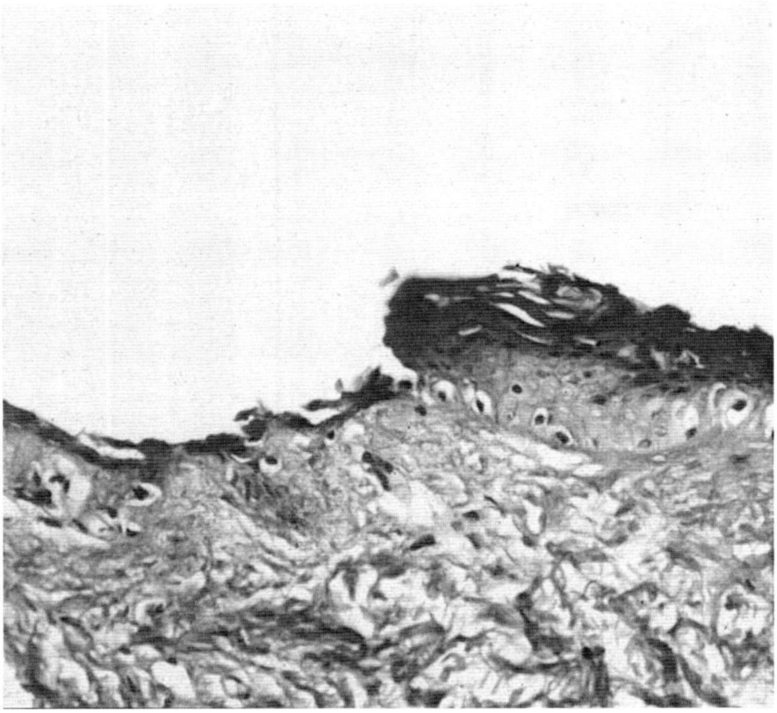

Figure 11.6 Ablation of the epidermis immediately after ActiveFX™ treatment.

hyperpigmentation [24]. Utilizing lower-energy levels and densities, as well as topical hydroquinone pre- and posttreatment, may help limit the incidence of PIH to less than 5% [43].

There have also been case reports of systemic lidocaine toxicity following topical application of lidocaine cream or gel, [37]. CNS toxicity can be seen at plasma lidocaine levels as low as 1–5 μg/ml with clinical symptoms of tinnitus, dysgeusia, light-headedness, nausea, and diplopia. At higher levels, nystagmus, slurred speech, hallucinations, muscle tremors, seizures, and eventually coma and respiratory arrest can ensue [44].

Lidocaine is primarily metabolized through the hepatic system. Therefore, its use should be limited in patients who have compromised liver function, or in those who are taking medications that may inhibit liver enzymes responsible for its metabolism. Other factors to consider in determining toxicity are duration of application, whether it is under occlusion, the total amount of surface area covered, BMI of the patient, and if there is any disruption in the stratum corneum. Management of toxicity includes maintenance of airway and respiration and administration of benzodiazepines [37].

11.10 Conclusions

Skin rejuvenation using the concept of fractional photothermolysis is a safe and effective treatment modality. Creating microscopic lesions in the epidermis and dermis can lead to expulsion of necrotic debris and dermal collagen remodeling that correlates with improvement in dyschromia, wrinkles, texture, and scarring, with minimal downtime and side effects. Many fractional lasers may also be used for nonfacial resurfacing of the neck, chest, and hands, unlike traditional ablative devices. As technologies advance and biological mechanisms unfold, this promises to be useful for an array of dermatologic conditions.

References

1. Hobbs F, and Stoops N. Demographic trends in the 20th century. U.S. Government Printing Office, Washington, D.C., 2002.
2. Laubach HJ, Tannous Z, Anderson RR, and Manstein D. Skin responses to fractional photothermolysis. *Lasers Surg Med.*; 2006;38:142–149.
3. Trelles MA, Allones I, and Velez M. Non-ablative facial skin photorejuvenation with an intense pulsed light system and adjunctive epidermal care. *Lasers Med Sci.*; 2003;18:104–111.
4. Goldberg D, Tan M, Dale SM, and Gordon M. Nonablative dermal remodeling with a 585-nm, 350-microsec, flashlamp pulsed dye laser: clinical and ultrastructural analysis. *Dermatol Surg.*; 2003;29:161–163; discussion 163–164.
5. Levy JL, Trelles M, Lagarde JM, Borrel MT, and Mordon S. Treatment of wrinkles with the nonablative 1,320-nm Nd:YAG laser. *Ann Plast Surg.*; 2001;47: 482–488.
6. Fournier N, Dahan S, Barneon G, Diridollou S, Lagarde JM, Gall Y, and Mordon S. Nonablative remodeling: clinical, histologic, ultrasound imaging, and profilometric evaluation of a 1540 nm Er:glass laser. *Dermatol Surg.*; 2001; 27:799–806.
7. Huzaira M, Anderson RR, Sink K, and Manstein D. Intradermal focusing of near-infrared optical pulses: a new approach for non-ablative laser therapy. *Lasers Surg Med.*; 2003; 15(suppl):66.
8. Khan MH, Sink RK, Manstein D, Eimerl D, and Anderson RR. Intradermally focused infrared laser pulses: thermal effects at defined tissue depths. *Lasers Surg Med.*; 2005; 36:270–280.
9. Manstein D, Herron GS, Sink RK, Tanner H, and RR Anderson. Fractional photothermolysis: a new concept for cutaneous remodeling using microscopic patterns of thermal injury. *Lasers Surg Med.*; 2004; 34:426–438.
10. Hantash BM and Mahmood MB. Fractional photothermolysis: a novel aesthetic laser surgery modality. *Dermatol Surg.*; 2007; 33:525–534.
11. Hantash BM, Bedi VP, Sudireddy V, Struck SK, Herron GS, and Chan KF. Laser-induced transepidermal elimination of dermal content by fractional photothermolysis. *J Biomed Opt.*; 2006;11: 041115:1–9.
12. Shook BA and Hruza GJ. Periorbital ablative and non-ablative resurfacing. *Facial Plast Surg Clin N Am.*; 2005; 13:571–582.
13. Chiu RJ and Kridel RWH. Fractionated photothermolysis: the Fraxel 1550-nm glass fiber laser treatment. *Facial Plast Surg Clin N Am.*; 2007;15: 229–237.
14. Childs J, Zelickson BD, et al. Lattice of optical islets: theory and experiment. Manuscript in preparation, 2007.
15. Kaufman J and Narurkar V. Fractional resurfacing. Available at http://www.modernmedicine.com. Accessed June 22, 2007.
16. Tannous ZS and Astner S. Utilizing fractional resurfacing in the treatment of therapy-resistant melasma. *J Cosmet Laser Ther.*; 2005;7:39–43.
17. Fisher GH, Kim KH, Bernstein LJ, et al. Concurrent use of a handheld forced cold air device minimizes patient discomfort during fractional photothermolysis. *Dermatol Surg.*; 2005; 31:1242–1244.
18. Rahman Z, Alam M, and Dover JS. Fractional laser treatment for pigmentation and texture improvement. *Skin Therapy Letter*; 2006; 11:7–11.
19. Laubach HJ, Chan HH, Rius F, Anderson RR, and Manstein D. Effects of skin temperature on lesion size in fractional photothermolysis. *Laser Surg Med.*; 2007; 39:14–18.
20. Bedi VP, Chan KF, Sink RK, et al. The effects of pulse energy variations on the dimensions of microscopic thermal treatment zones in nonablative fractional resurfacing. *Laser Surg Med.*; 2007; 39:145–155.

21. Geronemus RG. Fractional photothermolysis: current and future applications. *Lasers Surg Med.*; 2006; 38:169–176.
22. Hantash BM, Bedi VP, Kapadia B, et al. In vivo histological evaluation of a novel ablative fractional resurfacing device. *Laser Surg Med.*; 2007;39:96–107.
23. Wanner M, Tanzi EL, and Alster TS. Fractional photothermolysis: treatment of facial and nonfacial cutaneous photodamage with a 1,550-nm erbium-doped fiber laser. *Dermatol Surg.*; 2007; 33:23–28.
24. Kono T, Chan HH, Groff WF, et al. Prospective direct comparison study of fractional resurfacing using different fluences and densities for skin rejuvenation in Asians. *Lasers Surg Med.*; 2007; 39:311–314.
25. Chan HHL, Manstein D, Yu CS, Shek S, Kono T, and Wei WI. The prevalence and risk factors of post-inflammatory hyperpigmentation after fractional resurfacing in Asians. *Lasers Surg Med.*; 2007; 39:381–385.
26. Behroozan DS, Goldberg LH, Glaich AS, Dai T, and Friedman PM. Fractional photothermolysis for treatment of poikiloderma of Civatte. *Dermatol Surg.*; 2006; 32:298–301.
27. Behroozan DS, Goldberg LH, Dai T, Geronemus RG, and Friedman PM. Fractional photothermolysis for the treatment of surgical scars: a case report. *J of Cosmet Laser Ther.*; 2006;8:35–38.
28. Glaich AS, Rahman Z, Goldberg LH, and Friedman PM. Fractional resurfacing for the treatment of hypopigmented scars: a pilot study. *Dermatol Surg.*; 2007; 33:289–294.
29. Rokhsar CK and Fitzpatrick RE. The treatment of melasma with fractional photothermolysis: a pilot study. *Dermatol Surg.*; 2005; 31:1645–1650.
30. Alster TS, Tanzi EL, and Lazarus M. The use of fractional laser photothermolysis for the treatment of atrophic scars. *Dermatol Surg.*; 2007; 33:295–299.
31. Chung JH, Lee SH, Youn CS et al. Cutaneous photodamage in Koreans: influence of sex, sun exposure, smoking, and skin color. *Arch Dermatol.*; 2001;137:1043–1051.
32. Nanni CA and Alster TS. Complications of carbon dioxide laser resurfacing. An evaluation of 500 patients. *Dermatol Surg.*; 1998; 24:315–320.
33. Chan HH, Fung WK, Ying SY, and Kono T. An in vivo trial comparing the use of different types of 532 nm Nd:YAG lasers in the treatment of facial lentigines in Oriental patients. *Dermatol Surg.*; 2000; 26:743–749.
34. Chua SH, Ang P, Khoo LS, and Goh CL. Nonablative 1450-nm diode laser in the treatment of facial atrophic acne scars in type IV to V Asian skin: a prospective clinical study. *Dermatol Surg.*; 2004; 30:1287–1291.
35. Graham R. What is poikiloderma of Civatte? *Practitioner*; 1989;233:1210.
36. Geronemus R. Poikiloderma of Civatte. *Arch Dermatol.*; 1990;126:547–548.
37. Marra DE, Yip D, Fincher EF, and Moy RL. Systemic toxicity from topically applied lidocaine in conjunction with fractional photothermoloysis. *Arch Dermatol.*; 2006;142:1024–1026.
38. Fisher GH and Geronemus RG. Short-term side effects of fractional photothermolysis. *Dermatol Surg.*; 2005; 31:1245–1249.
39. Hantash BM, Bedi VP, Chan KF, and Zachary CB. Ex vivo histological characterization of a novel ablative fractional resurfacing device. *Laser Surg Med.*; 2007; 39:87–95.
40. Lumenis Aesthetic. UltraPulse Encore for ActiveFX & MaxFX treatments. 2007. Available at http://www.aesthetic.lumenis.com/wt/page/ultrapulse. Accessed July 16, 2007.
41. Keller GS. Fractional ablative skin resurfacing with the pixel laser. Alma Lasers 2006. Available at http://www.almalasers.com/cms/mycms/AlmaLasers/white_papers/WP_Pixel_Keller_August_17_2006.pdf. Accessed July 16, 2007.
42. Sciton, Inc. ProFractional. 2007. Available at http://www.sciton.com. Accessed July 16, 2007.
43. Burns AJ, ed. Fractional resurfacing in plastic surgery. Medical Insight, Inc. Aliso Viejo, CA: Medical Insight, Inc; 2005. Available at http://www.miinews.com/stage/pdf/Fraxel_CME_1005.pdf. Accessed July 16, 2007.
44. Auletta MJ and Grekin RC. Local anesthesia for dermatologic surgery. New York, NY: Churchill Livingstone Inc; 1991.

12

LED Low-Level Light Photomodulation for Reversal of Photoaging

Robert A. Weiss[1], *Roy G. Geronemus*[2], *and David H. McDaniel*[3]

[1]*Maryland Laser Skin & Vein Institute, Hunt Valley, MD, USA*
[2]*New York University Medical Center, New York, NY, USA*
[3]*Eastern Virginia Medical School, Virginia Beach, VA, USA*

12.1 Introduction

12.1.1 Photomodulation

Photorejuvenation is a greatly sought after treatment for restoring photodamaged skin, but the vocabulary used to describe it is often varied and confusing. Photorejuvenation refers to a process which utilizes light energy sources to reverse or structurally repair sun-induced changes over time. This skin degeneration, or photoaging, is compounded by environmental damage to the skin from smoking, pollutants, and other insults, causing free radical

Gurpreet S. Ahluwalia (ed.), *Cosmetic Applications of Laser and Light-Based Systems*, 271–280,

formation. Nonablative photorejuvenation refers to the controlled use of thermal energy to accomplish skin rejuvenation without disturbance of the overlying epidermis. Nonablative modalities include primarily intense pulsed light (IPL), but other visible wavelengths are used as well, including pulsed dye laser (PDL) and 532 nm green light (KTP laser) [1]. Various infrared wavelengths with deeper penetration are used for remodeling dermal collagen in all skin types, regardless of pigmentation, and these wavelengths include 1064, 1320, 1450, and 1540 nm [2,3]. All these devices entail thermal injury, either by heating the dermis to stimulate fibroblast proliferation, or by heating blood vessels for photocoagulation [4–6]

A radical change in this concept is the theory of photomodulation. Light emitting diode (LED) photomodulation is a novel approach to photoaging and remains the only category of nonthermal light treatments designed to regulate the activity of cells rather than invoke thermal wound-healing mechanisms[7,8]. This incurs far less risk for patients than other light modalities. The first written report on using photomodulation for facial wrinkles was by McDaniel and his group in 2002 [9].

Photomodulation evolved from the use of LED and low-energy light therapy (LILT) use for stimulating the growth of plant cells [10]. The notion that cell activity can be up- or down-regulated by low-energy light had been entertained in the past, but consistent or impressive results had been lacking [11,12]. Some promise had been shown with wound healing for oral mucositis [12]. Wavelengths examined earlier included a 670 nm LED array [12], a 660 nm array [13], and higher infrared wavelengths [14]. Fluence and duration of exposure were varied in these studies, with high energy required for modest results [12].

To investigate LED light for modulating skin properties, a model of fibroblast culture was utilized in conjunction with clinical testing. Particular packets of energy with specific wavelengths combined with using a very specific propriety pulse sequencing "code" were found to up-regulate Collagen I synthesis in fibroblast culture using RT-PCR to measure collagen I [9]. The up-regulation of fibroblast collagen synthesis correlated with the clinical observation of increased dermal collagen on treated human skin biopsies [15]. Both in the fibroblast and clinical model, collagen synthesis was accompanied by the reduction of matrix metalloproteinases (MMP), in particular, MMP-1 or collagenases being greatly reduced with exposure to 590 nm/870 nm low-energy light. This novel concept of using very low energy and narrow-band light with specific pulse code sequences and durations was termed LED photomodulation [9]. The device which utilizes pulsed code sequences of LED light to induce photomodulation is Gentlewaves® (LightBioScience, LLC, Virginia Beach, VA).

12.2 Clinical Applications

12.2.1 Photorejuvenation

LED photomodulation can be used both alone, and in combination with a variety of common nonablative rejuvenation procedures in an office setting. Several antiinflammatory and wound-healing applications have surfaced as well, and these are discussed in the subsequent sections. Treatments are delivered using the Gentlewaves® yellow/IR light LED photomodulation unit (LightBioScience, Virginia Beach, VA) with a full-face panel device. Energy density is set at 0.10 Joules/cm^2. One hundred pulses are delivered with pulse duration of 250 ms and an off-interval of 100 ms. Treatment time is approximately 35 s.

We have treated over 4000 patients over the last 4 years. Of these treatments, 15% were LED photomodulation alone, and 85% were concomitant with a thermal-based photorejuvenation procedure. Using specific pulsing sequence parameters, which are the basis for the "code" of LED photomodulation, the original multicenter clinical trial was conducted with 90 patients receiving a series of 8 LED treatments over 4 weeks [16–19]. This study showed very favorable results, with over 90% of patients improving by at least one Fitzpatrick photoaging category, and 65% of the patients demonstrating global improvement in facial texture, fine lines, background erythema, and pigmentation. Results peaked at four to six months following completion of a series of eight treatments [19]. Another study, but this one retrospective, conducted in Baltimore using the same Gentlewaves® 590 nm/870 nm LED array demonstrated similar results (Fig. 12.1). In addition, these results were confirmed by digital microscopy [20].

Most recently, Gentlewaves® 590 nm LED array was used in an independent clinical laboratory, and these data were confirmed. An additional clinical trial involving 65 subjects used cast impressions of lateral canthus wrinkles (Crow's feet). These replicas, illuminated by reproducible shadows of light from each wrinkle, were analyzed with the aid of commercially available image-analyzing software (Quantirides, Monaderm, Monaco). This analysis showed a significant reduction of the number of wrinkles occurring from the second to the fourth month after treatment accompanied by a significant reduction of the length of wrinkles at five months posttreatment. Self-assessment by subjects showed a significant improvement of wrinkles, skin textural softness, and skin glow (data on file at L'Oreal, Paris, France).

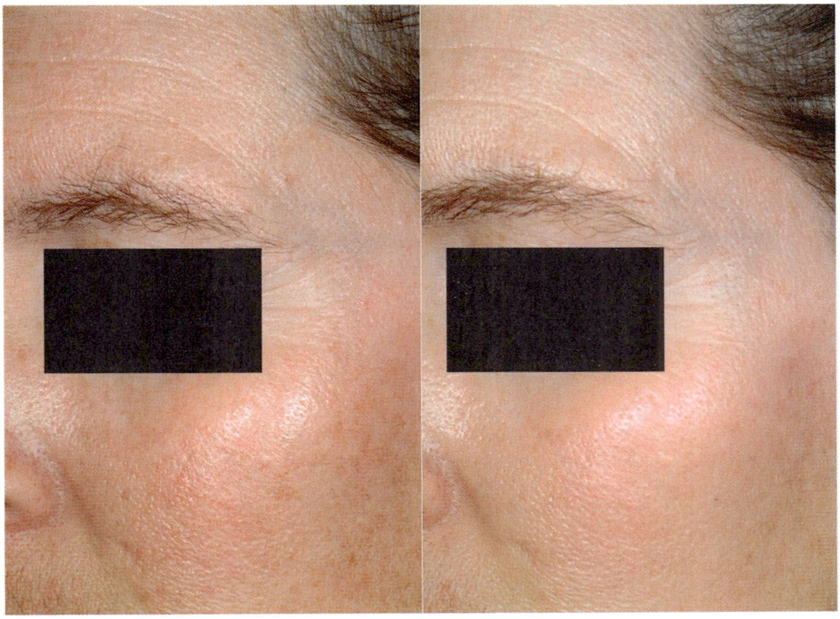

Figure 12.1 Smoothing of the skin seen after 8 treatments over 4 weeks of Gentlewaves® photomodulation (Light BioScience, LLC, Virginia Beach, VA) The after-image illustrates 8 weeks after baseline. Reduction in wrinkles, pigmentation, and improvement of texture are noted.

Others have confirmed that additional wavelengths of LED light, using red and infrared wavelengths, may be effective for improvement in skin texture. Although these treatments were much longer in duration, 36 patients receiving 9 treatments over a 5-week period, showed improvement in skin softness [21]. Each treatment was given in continuous mode (no pulsing) with a treatment time of 20 min using 633 nm and 830 nm as an LED array (Omnilux™, Phototherapeutics, Lake Forest, CA). Another recent report using this system involved 31 subjects with facial rhytids, who received 9 light therapy treatments using combined wavelengths of 633 nm and 830 nm. Fluences were relatively high, utilizing 126 J/cm^2 for 633 nm and 66 J/cm^2 for 830 nm. Improvements to the skin surface were reported at weeks 9 and 12 by profilometry performed on periorbital casts. Results showed that 52% of subjects showed a 25–50% improvement in photoaging scores [22].

In contrast, the clinical effects of patients who receive Gentlewaves® LED photomodulation alone without concomitant treatment report a softening of skin texture, and reduction of roughness and fine lines, which range from significant reduction to subtle but real changes in the "creamy" texture of their complexion. The US FDA first cleared LED devices to be used in the reduction of peri-ocular wrinkles in 2005. Gentlewaves™ was the initial device approved (Fig. 12.2), and then Omnilux™ followed as a 510(k) substantial equivalence approval.

12.3 Antiinflammatory Effects

Over the course of our multiyear experience with photomodulation, we have observed the reduction of erythema from a variety of causes. Reduction of erythema may be induced from wide-ranging skin injuries, including but not limited to thermal laser treatments, UV burns, radiation therapy, and blunt trauma. In addition, anectodal experience with a series

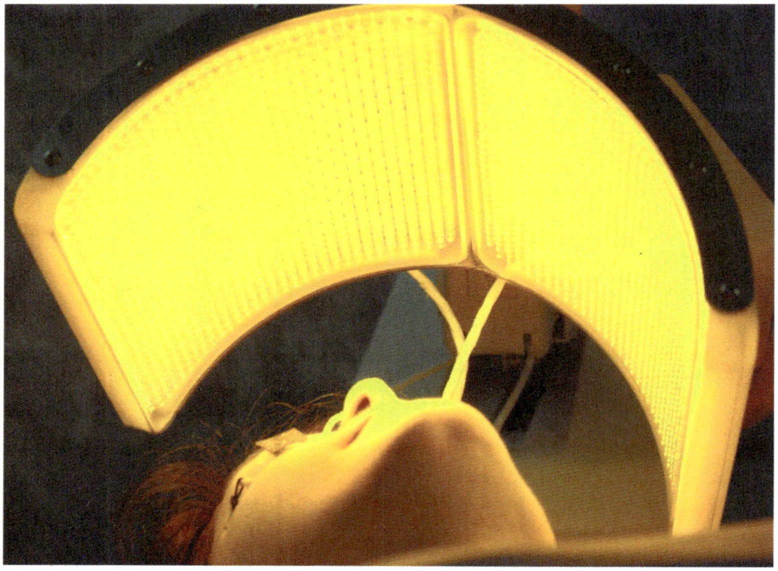

Figure 12.2 GentleWaves® LED Photomodulation® device. Largest unit for medical professionals, and covers the entire face.

of Gentlewaves® LED photomodulation treatments for atopic eczema or to reduce bruising and/or second degree burns has been encouraging. Treatment of atopic eczema in patients withdrawn from all topical medications has led to a rapid resolution within three to four treatments over one to two weeks (Fig. 12.3).

Use of LED photomodulation in combination with other laser modalities results in more effective clinical results, as well as faster resolution of erythema. We believe that the faster resolution of erythema from numerous causes is a result of the antiinflammatory effects of LED photomodulation. The mechanism has not yet been elucidated, although down-regulation by photomodulation of a number of inflammatory mediators from cells such as lymphocytes or macrophages is suspected. Studies on human skin fibroblasts and clinical biopsies have shown reduction in IL-1B1 and IL-6 [23].

A recent study looked at whether LED photomodulation therapy could accelerate resolution of post-intense pulsed light (IPL) erythema [24]. Fifteen subjects were randomized to receive LED treatment to one side of the face immediately following a single IPL treatment for photodamage. Results showed mean erythema scores on the first visit were statistically significantly lower on the LED-treated side. This led the authors to conclude that LED photomodulation treatment accelerates the resolution of erythema and reduces posttreatment discomfort following IPL treatment [24]. This study confirms our observations.

A landmark study on radiation dermatitis examined whether LED photomodulation can alter and improve the outcome of intensity-modulated radiation treatments (IMRT) on

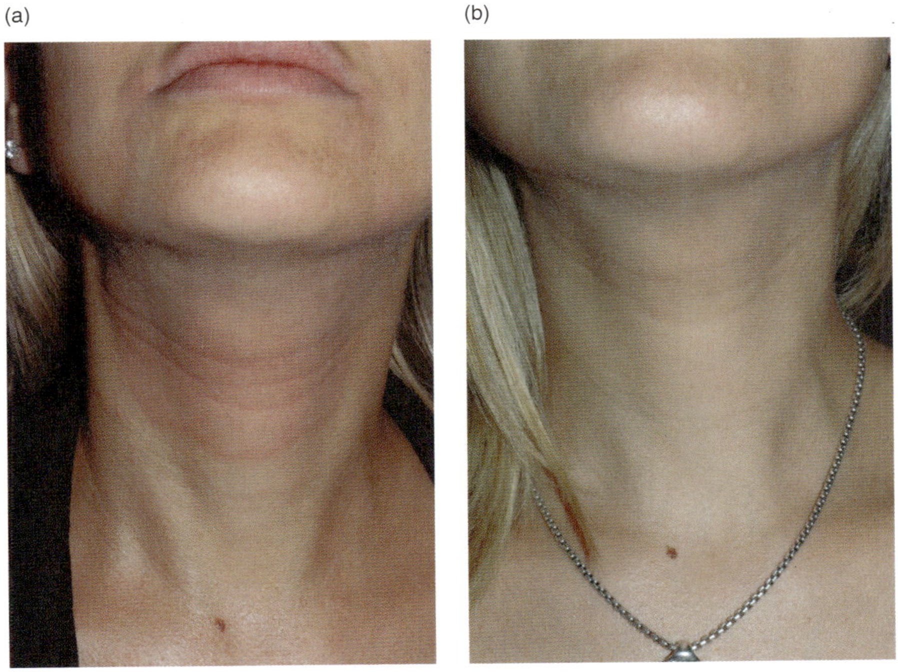

Figure 12.3 (a) Before shows flare of eczema following withdrawal of all therapy. (b) Atopic eczema after three treatments with Gentlewave® LED photomodulation. The after-image shows effects of reduction of inflammation by LED photomodulation within 10 days.

overlying breast skin. Nineteen patients with breast cancer were treated with Gentlewaves® LED photomodulation immediately after every session of radiation. Treatments were for post-lumpectomy patients who received a full course of IMRT [25]. Skin reactions were monitored weekly, using National Cancer Institute (NCI) criteria for grading. Age-matched controls (n = 28) received IMRT without LED photomodulation. The results of this study showed that LED-treatment had a significantly positive effect. Of LED patients, 18 (94.7%) had Grade 0 or 1 reaction and only (5.3%) had Grade 2 reaction. Among controls, 4 (14.3%) had a Grade 1 reaction and 24 (85.7%) had a Grade 2 or Grade 3 reaction. On the non-LED treated group, 67.9% had to interrupt treatment due to side effects of skin breakdown with moist reactions, but only 5% of the LED-treated group had interrupted treatment. The authors concluded that not only did Gentlewaves® LED photomodulation treatments delivered immediately after each IMRT reduce the incidence of adverse NCI criteria skin reactions, but also allowed the full course of treatment and resulted in a final smoother skin texture with improved skin elasticity postradiation treatment.

Additional data indicates an antiinflammatory effect for LED photomodulation following UV-induced erythema. Using a solar simulator, findings indicate a photoprotective effect when delivered after UV radiation [23]. This concept is a rescue from UV damage, even after inadvertant UV radiation has occurred. We have observed a noticeable reduction in UV erythema when LED photomodulation is supplied within hours after UV exposure. The use of 590 nm/870 nm Gentlewaves® LED photomodulation produced significant down-regulation of dermal matrix degrading enzymes which were stimulated by the UV exposure [23]. In addition, a pilot study with precise CO2 laser epidermal destruction has shown promise by using this device for accelerated wound healing.

Parallel to wound healing, the use of photomodulation has been extended to a protective or preventative effect following several types of toxic injury. Experiments using LED light to protect the retina against the toxic actions of methanol-derived formic acid in a rodent model of methanol toxicity have been successful. In a recent study, LED treatment protected the retina from the histopathologic changes induced by methanol LILT on mitochondrial oxidative metabolism in vitro, and retinal protection in vivo [26]. They also suggest that photomodulation may enhance recovery from retinal injury and other ocular diseases, in which mitochondrial dysfunction is postulated to play a role.

Our group has earlier reported on the treatment of human retinal pigment epithelial (RPE) cells in vitro using Gentlewaves® LED photomodulation produced by acute injury from blue light wavelengths [27]. The results showed reduction of cell death from 94% down to between 10–20% (as measured at 24 h). Another in vitro test on human RPE cells showed a 7-fold reduction in VEGF expression at 24 h post-LED exposure using LED photomodulation at 590 nm/870 nm delivered at 0.1 J/cm^2 [27]. Incidentally, several cases of retinal degeneration have been improved by treatments with Gentlewaves® photomodulation (personal communication, David Vasily, MD). Further clinical studies on wound healing, cell rescue from injury, and antiinflammatory applications are actively continuing.

12.4 Photodynamic Therapy

The LED red light (630 nm) has been used for several years in combination with a sensitizer (levulinic acid) for photodynamic therapy (PDT) [28]. When exposed to light with the proper wavelength, the sensitizer produces an activated oxygen species, singlet oxygen, that oxidizes the plasma membrane of targeted cells. Due to a lower metabolic rate, there is less sensitizer in the adjacent normal tissue, thus less of reaction. One of the absorption peaks of the metabolic product of levulinic acid, protoporphyrin, absorbs strongly at 630 nm red. A red LED panel emitting at 630 nm (Omnilux PDT™, Phototherapeutics, Lake Forest, CA) has been used for this purpose in Europe and Asia [29]. We have also used the full panel 590 nm LED array for facilitating PDT. This therapy is delivered by the application of levulinic acid (Levulan™ DUSA, Wilmington, MA) for 45 min and exposure to continuous (nonpulsed) 590 nm/870 nm LED for 15 min for a cumulative dose of over 70 J/cm^2. Our results have shown reduction of actinic damage, including improvement of skin texture and reduction of actinic keratoses [30].

12.5 Mechanism of Action

The primary means for photomodulated upregulation of cell activity for collagen synthesis by LED is the activation of energy switching mechanisms in mitochrondria, the energy source for cellular activity. Cytochrome molecules are believed to be responsible for the light absorption in mitochrondria. Cytochromes are synthesized from protoporphyrin IX and absorb wavelengths of light from 562 nm to 600 nm. It is believed that LED light absorption causes conformational changes in antenna molecules within the mitochrondrial membrane. Proton translocation initiates a pump which ultimately leads to energy for conversion of ADP to ATP. This essentially recharges the "cell battery" and provides more energy for cellular activity.

Others have confirmed that mitochrondial ATP availability can influence cellular growth and reproduction, with lack of mitochrondrial ATP associated with oxidative stress [31]. Cellular aging may be associated with decreased mitochrondial DNA activity [32]. Earlier work has also demonstrated rapid ATP production within mitochrondia of cultured fibroblasts exposed to 590 nm/870 nm yellow/IR LED light only with the proper pulsing sequence [8,33]. New ATP production occurs rapidly after LED photomodulation, triggering subsequent metabolic activity of fibroblasts [18] . There also appear to be receptor-like mechanisms, which result in the modulation of the expression of gene activity producing up- or down-regulation of gene activity, as well as a wide-ranging cell signaling pathway actions. The choice of photomodulation parameters plays a vital role in determining the overall pattern of gene up- or down- regulation. In our experience, the use of LED yellow/IR light without proper pulsing sequence leads to minimal or no consequences on mitochrondrial ATP production.

12.6 Conclusions

LED arrays using LILT for photomodulation are useful for collagen stimulation, textural smoothing, and reduction of inflammation. Pilot wound-healing studies show slightly accelerated wound resolution. Cellular rescue from UV damage and other toxic insults has been shown in small studies. Our combined multiyear experience and clinical observations confirm that combinations of thermal nonablative photorejuvenation and nonthermal LED photomodulation have a synergistic effect. LED photomodulation is delivered immediately, subsequent to the thermal-based treatment for its antiinflammatory effects, which may reduce the thermally induced erythema and edema of nonablative treatments. Delivery of LED light immediately pre- and post-thermal injury appears to potentiate the effect, as observed from our clinical experience.

A significant study with age and radiation-matched controls for radiation dermatitis indicates that there is a powerful potential to further utilize the specific antiinflammatory and cell-rescue properties of LED photomodulation. Radiation-treated patients may have not only reduced side effects, but also smoother skin in the treated areas over a long term. Preliminary data from DNA microarray analysis of the entire human genome of certain skin cell lines after LED photomodulation, and also after UV injury and subsequent LED therapy are currently being analyzed and they support a versatile role for LED photomodulation in enhancing cellular energy production, as well as diverse effects on gene expression. LED photomodulation appears to negate some of the negative aspects of UV exposure. Many clinical and basic science-research pathways await further exploration for this novel nonthermal, low-risk technology.

References

1. Weiss RA, Weiss MA, Beasley KL, and Munavalli G. Our approach to non-ablative treatment of photoaging. *Lasers Surg Med.* 2005 Jul; 37(1): 2–8.
2. Munavalli GS, Weiss RA, and Halder RM. Photoaging and nonablative photorejuvenation in ethnic skin. *Dermatol Surg.* 2005 Sep; 31(9 Pt. 2): 1250–1260.
3. Weiss RA, McDaniel DH, and Geronemus RG. Review of nonablative photorejuvenation: reversal of the aging effects of the sun and environmental damage using laser and light sources. *Semin Cutan Med Surg.* 2003 Jun; 22(2): 93–106.
4. Weiss RA, Gold M, Bene N, Biron JA, Munavalli G, Weiss M, et al. Prospective clinical evaluation of 1440-nm laser delivered by microarray for treatment of photoaging and scars. *J Drugs Dermatol.* 2006 Sep; 5(8): 740–744.
5. Weiss RA, Goldman MP, and Weiss MA. Treatment of poikiloderma of Civatte with an intense pulsed light source. *Dermatol Surg.* 2000 Sep; 26(9): 823–837.
6. Fatemi A, Weiss MA, and Weiss RA. Short-term histologic effects of nonablative resurfacing: Results with a dynamically cooled millisecond-domain 1320 nm Nd:YAG laser. *Dermatol Surg.* 2002 Feb; 28(2): 172–176.
7. Weiss RA, McDaniel DH, and Geronemus RG. Review of nonablative photorejuvenation: reversal of the aging effects of the sun and environmental damage using laser and light sources. *Semin Cutan Med Surg.* 2003 Jun; 22(2): 93–106.
8. McDaniel DH, Weiss RA, Geronemus R, Ginn L, and Newman J. Light-tissue interactions. I: Photothermolysis vs photomodulation laboratory findings. *Lasers Surg Med.* 2002; 32(Suppl. 14): 25.2002.
9. McDaniel DH, Weiss RA, Geronemus R, Ginn L, and Newman J. Light-tissue interactions. II: Photothermolysis vs photomodulation clinical applications. *Lasers Surg Med.* 2002; 32(Suppl. 14): 25.
10. Whelan HT, Smits RL, Jr., Buchman EV, Whelan NT, Turner SG, Margolis DA, et al. Effect of NASA light-emitting diode irradiation on wound healing. *J Clin Laser Med Surg.* 2001 Dec; 19(6): 305–314.
11. Whelan HT, Buchmann EV, Dhokalia A, Kane MP, Whelan NT, Wong-Riley MT, et al. Effect of NASA light-emitting diode irradiation on molecular changes for wound healing in diabetic mice. *J Clin Laser Med Surg.* 2003 Apr; 21(2): 67–74.

12. Whelan HT, Connelly JF, Hodgson BD, Barbeau L, Post AC, Bullard G, et al. NASA light-emitting diodes for the prevention of oral mucositis in pediatric bone marrow transplant patients. *J Clin Laser Med Surg.* 2002 Dec; 20(6): 319–324.
13. Walker MD, Rumpf S, Baxter GD, Hirst DG, and Lowe AS. Effect of low-intensity laser irradiation (660 nm) on a radiation-impaired wound-healing model in murine skin. *Lasers Surg Med.* 2000; 26(1): 41–47.
14. Lowe AS, Walker MD, O'Byrne M, Baxter GD, and Hirst DG. Effect of low intensity monochromatic light therapy (890 nm) on a radiation-impaired, wound-healing model in murine skin. *Lasers Surg Med.* 1998; 23(5): 291–298.
15. Weiss RA, McDaniel DH, Geronemus RG, Weiss MA, Beasley KL, Munavalli GM, et al. Clinical experience with light-emitting diode (LED) photomodulation. *Dermatol Surg.* 2005 Sep; 31(9 Pt. 2): 1199–1205.
16. Weiss RA, McDaniel DH, Geronemus RG, and Weiss MA. Clinical trial of a novel non-thermal LED array for reversal of photoaging: clinical, histologic, and surface profilometric results. *Lasers Surg Med.* 2005 Feb; 36(2): 85–91.
17. McDaniel DH, Newman J, Geronemus R, Weiss RA, and Weiss MA. Non-ablative non-thermal LED photomodulation - A multicenter clinical photoaging trial. *Lasers Surg Med.* 2003; 32(Suppl. 15): 22
18. Geronemus R, Weiss RA, Weiss MA, McDaniel DH, and Newman J. Non-ablative LED photomodulation- Light activated fibroblast stimulation clinical trial. *Lasers Surg Med.* 2003; 32(Suppl. 15): 22.
19. Weiss RA, McDaniel DH, Geronemus R, Weiss MA, Newman J. Non-ablative, non-thermal light emitting diode (LED) phototherapy of photoaged skin. *Lasers Surg Med.* 2004; 34(Suppl. 16): 31.
20. Weiss RA, Weiss MA, Geronemus RG,and McDaniel DH. A novel non-thermal non-ablative full panel led photomodulation device for reversal of photoaging: digital microscopic and clinical results in various skin types. *J Drugs Dermatol.* 2004; 3(6): 605–610.
21. Goldberg DJ, Amin S, Russell BA, Phelps R, Kellett N, and Reilly LA. Combined 633-nm and 830-nm led treatment of photoaging skin. *J Drugs Dermatol.* 2006 Sep; 5(8): 748–753.
22. Russell BA, Kellett N, and Reilly LR. A study to determine the efficacy of combination LED light therapy (633 nm and 830 nm) in facial skin rejuvenation. J *Cosmet Laser Ther.* 2005 Dec; 7(3–4): 196–200.
23. Weiss RA, McDaniel DH, Geronemus RG, and Weiss MA. Clinical trial of a novel non-thermal LED array for reversal of photoaging: clinical, histologic, and surface profilometric results. *Lasers Surg Med.* 2005 Feb; 36(2): 85–91.
24. Khoury JG and Goldman MP. Use of light-emitting diode photomodulation to reduce erythema and discomfort after intense pulsed light treatment of photodamage. *J Cosmet Dermatol.* In press 2008.
25. DeLand MM, Weiss RA, McDaniel DH, and Geronemus RG. Treatment of radiation-induced dermatitis with light-emitting diode (LED) photomodulation. *Lasers Surg Med.* 2007 Feb; 39(2): 164–168.
26. Eells JT, Henry MM, Summerfelt P, Wong-Riley MT, Buchmann EV, Kane M, et al. Therapeutic photobiomodulation for methanol-induced retinal toxicity. *Proc Natl Acad Sci USA* 2003 Mar 18; 100(6): 3439–3444.
27. McDaniel DH, Weiss RA, Geronemus R, and Weiss MA. LED photomodulation "'reverses" acute retinal injury. Annual Meeting of the American Society for Laser Medicine and Surgery, Boston, MA, April 6, 2006.
28. Tarstedt M, Rosdahl I, Berne B, Svanberg K, and Wennberg AM. A randomized multicenter study to compare two treatment regimens of topical methyl aminolevulinate (Metvix)-PDT in actinic keratosis of the face and scalp. *Acta Derm Venereol.* 2005; 85(5): 424–428.
29. Chen HM, Yu CH, Tu PC, Yeh CY, Tsai T, and Chiang CP. Successful treatment of oral verrucous hyperplasia and oral leukoplakia with topical 5-aminolevulinic acid-mediated photodynamic therapy. *Lasers Surg Med.* 2005; Aug; 37(2): 114–122.

30. Weiss RA, McDaniel DH, Geronemus RG, Weiss MA, Beasley KL, Munavalli GM, et al. Clinical experience with light-emitting diode (LED) photomodulation. *Dermatol Surg.* 2005; Sep; 31(9 Pt. 2): 1199–1205.
31. Zhang X, Wu XQ, Lu S, Guo YL, and Ma X. Deficit of mitochondria-derived ATP during oxidative stress impairs mouse MII oocyte spindles. *Cell Res.* 2006 Oct; 16(10): 841–850.
32. Sorensen M, Sanz A, Gomez J, Pamplona R, Portero-Otin M, Gredilla R, et al. Effects of fasting on oxidative stress in rat liver mitochondria. *Free Radic Res.* 2006 Apr; 40(4): 339–347.
33. Weiss RA, Weiss MA, McDaniel DH, Newman J, and Geronemus R. Comparison of non-ablative fibroblast photoactivation with and without application of topical cosmeceutical agents. *Lasers Surg Med.* 2003; 32(Suppl. 15): 23.

13

Global Total Nonsurgical Rejuvenation: Lasers and Light-Based Systems in Combination with Dermal Fillers and Botulinum Toxins

Vic A. Narurkar

Bay Area Laser Institute, San Francisco, CA, and Department of Dermatology, University of California Davis School of Medicine, Sacremento, CA, USA

13.1 Introduction

The last decade has witnessed an unparalleled growth in the demand for nonsurgical procedures. This trend is a result of numerous advances in devices, fillers, and botulinum

Gurpreet S. Ahluwalia (ed.), *Cosmetic Applications of Laser and Light-Based Systems*, 281–289,

toxins, in addition to an aging population. It is becoming increasingly evident that, while each modality, when used in monotherapy, may produce satisfactory clinical outcomes, combination therapies with multiple modalities are often necessary for optimal patient satisfaction. Devices are best for addressing anomalies of the facial canvas, dermal fillers are best for addressing facial volume loss, and neurotoxins are best for addressing dynamic facial lines of expression. There is some evidence that a synergy probably exists between these modalities, even at an ultrastructural level. This chapter will review combination therapies with lasers and light, dermal fillers, and botulinum toxins.

13.2 Lasers and Light-Based Systems

Table 13.1 summarizes the mechanisms of lasers and light-based systems. The central dogma to these therapies is predicated on the theory of selective photothermolysis (SP), where in theory any target can be selectively destroyed if an optimal thermal relaxation time of the target is matched with the optimal chromophore. Biologically, the active chromophores are melanin, hemoglobin, and water. As devices have matured, it has become increasingly evident that there are other mechanisms besides SP. These include refinements of SP such as fractional photothermolysis, photopneumatic therapy, and photodynamic therapy.

When SP was first introduced, single wavelength lasers were the only modality of accomplishing these clinical effects. More recently, broadband light sources (often referred to as pulsed light) which use a flashlamp with selective filters can replicate and often surpass clinical results that are only attainable with single wavelength lasers. With the development of sophisticated cooling systems to protect the epidermis, the use of selective filters and photon recycling, pulsed light systems can replicate and often surpass many laser applications.

13.3 Ablative Laser Resurfacing

Skin rejuvenation can be accomplished by ablative, nonabrasive, and fractional modes of injury. In the 1990s, ablative laser resurfacing was introduced with the advent of pulsed 10,600 nm carbon dioxide and pulsed 2940 nm erbium-YAG lasers. While ablative

Table 13.1 Summary of Devices in Lasers and Light-Based Therapy

Device	Mechanism
Pulsed light; pulsed dye lasers; Q switched lasers	Selective photothermolysis with melanin and hemoglobin chromophores
Photopneumatic devices	Pneumatic therapy with vacuum and broadband light for melanin and hemoglobin chromophores
Pulsed light; pulsed dye lasers; blue and red light sources	Photodynamic therapy to activate endogenous chromophores
Nonablative and ablative fractional lasers	Fractional photothermolysis with chromophore of water

resurfacing produced excellent clinical outcomes, it lost popularity due to prolonged recovery times, persistent erythema, risks of hypopigmentation, and limitation to lighter skin types and facial areas. Even in a relatively fair skin, ablative laser resurfacing often produced an unnatural sheen to the skin, which was most evident at lines of demarcation such as that between the face and neck. For rhytids and laxity, ablative laser resurfacing produced very impressive results. However, ablative laser resurfacing did not address deep meilolabial folds, volume loss in the lips, and central facial volume loss. Moreover, while perioral rhytids showed very impressive results , these rhytids often necessitated adjuvant therapies. Botulinum toxin A to the perioral areas and superficial placed fillers such as collagens and hyalurons addressed the perioral area; medium- to deep-placed fillers such as hyaluronic acids and calcium hydroxyapatite can address the meilolabial folds, and volume enhancing and collagen stimulating fillers such as polylactic acid and calcium hydroxyapatite can address mid- facial volume loss.

13.4 Nonablative Skin Resurfacing

The undesired prolonged recovery and risks of ablative laser resurfacing led to the development of nonablative laser resurfacing with a myriad of lasers in the infra-red region. While safety was generally accomplished with these modes, the clinical results were, at best, modest. The premise of nonablative devices is "inside-out" resurfacing, where the epidermis remains intact and collagen remodeling occurs from selective dermal heating of water. While histologic and ultrastructural images of this technique were impressive, the modest clinicalimprovement and the variability in results led to the lack of continuation of this modality for clinical use. Simultaneously, the late 1990s witnessed the development of photorejuvenation, whereby lasers and light sources were utilized to treat facial canvas dyschromias and vascular anomalies. The terms "photofacial" and "photorejuvenation" were coined to explicate this process, which involved the selective photothermolysis of ecstatic facial telangiectasias and benign pigmented lesions such as lentigines. The nonselective thermal transfer also produced some dermal collagen remodeling. The clinical results were quite impressive for pigments and vessels, but not very impressive for true rhytids. Both infra-red laser nonablative laser resurfacing and visible laser and light source photorejuvenation absolutely require the use of botulinum toxin A for dynamic rhytids, and dermal fillers for static rhytids and volume replacement.

13.5 Photodynamic Therapy

Enhancement of visible laser and light-based photorejuvenation can be accomplished with photodynamic therapy (PDT). Photodynamic therapy was originally introduced with overnight incubation of 5 amino-levulanic acid, followed by activation with a 420 nm blue light for the treatment of actinic keratosis. This approach did not gain momentum due to significant discomfort and recovery time. Photodynamic therapy has undergone a dermatologic renaissance with the use of a short contact incubation of 5 amino-levulanic acid of 1–3 hours, followed by activation by a variety of lasers and light sources including vascular lasers (532 nm, 585 nm, 595 nm), pulsed light sources, blue light, and combination devices.

Fewer photofacial treatments are necessary and more advanced photodamage can be treated. There is some evidence that there is more consistent improvement of skin texture, and possibly fine rhytids with this approach (Fig. 13.1).

13.6 Photopneumatic Therapy

Another modification of visible light sources is the advent of photopneumatic therapy. Photopneumatic therapy takes a radical approach by manipulating optics of the skin instead of device characteristics by applying simultaneous vacuum suction to narrower band broadband light sources using blue and green photons. Suction stretches the skin, allowing the blue and green photons to have deeper dermal penetration, thereby enhancing photon delivery to dermal targets. Lower fluencies are necessary, thereby reducing the pain seen in traditional laser and light-based technologies. Simultaneous pore cleansing with vacuum produces a "porofacial" effect with improvements in acneiform lesions and pore-size reduction. A recent addition to photopneumatic therapy is the addition of topical agents using a modified tip to enhance dermal delivery of topical agents. This approach may address rhytids in addition to dyschromias (Fig. 13.2).

13.7 Nonablative and Ablative Fractional Resurfacing

The risks of ablative resurfacing and the limitations of nonablative resurfacing and photorejuvenation led to the development of fractional resurfacing, which is now the preferred

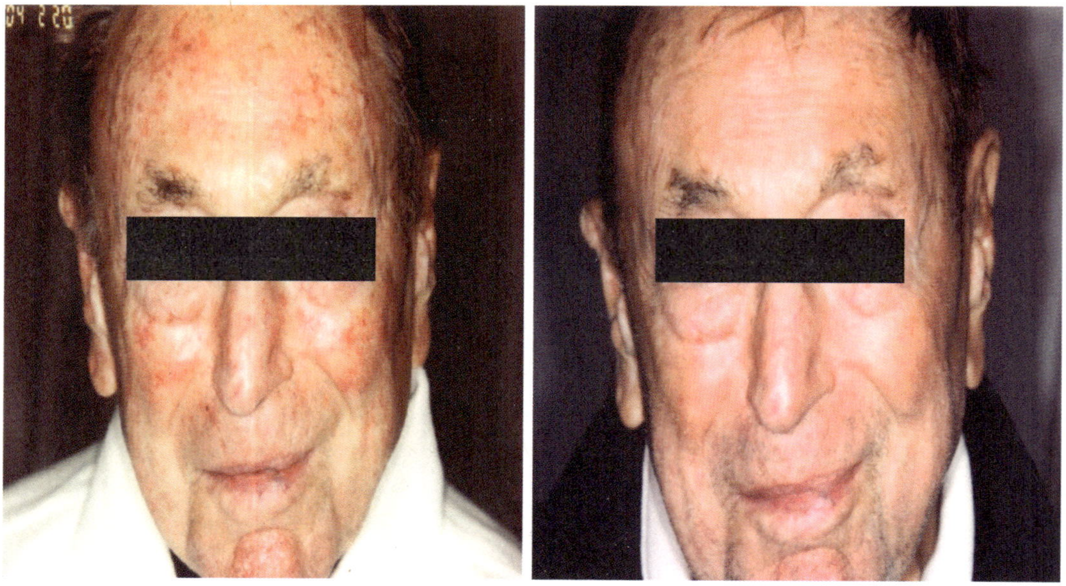

Before After

Figure 13.1 Pre- and post-photodynamic therapy with pulsed light.

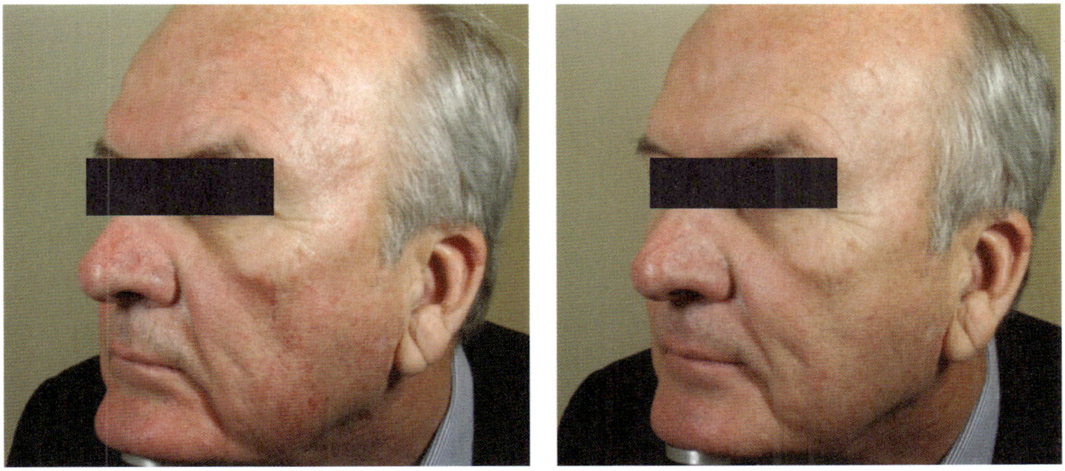

Before After 1 treatment

Figure 13.2 Pre- and post-photopnematic therapy with topical delivery system.

mode of resurfacing. Fractional resurfacing can be divided into true nonablative fractional resurfacing and true ablative fractional resurfacing, with the former having the longest duration of clinical experience at the time of publication. True fractional nonablative resurfacing requires preservation of the stratum corneum, creation of microthermal zones of injury, and extrusion of epidermal contents. Mid- infra-red wavelengths such as 1550 nm were employed as they showed an excellent affinity for water, and little collateral competitions for chromophores. "Pseudo" nonablative fractional resurfacing employed existing ablative and nonablative wavelengths, and did not really address the issues of bulk heating, which was the predominating risk factor in both traditional ablative and nonablative modes. True nonablative laser resurfacing creates columns of microthermal injury, the depths and widths of which can be adjusted to reflect the clinical entity being treated. For example, deep scars and rhytids necessitate deeper dermal penetration, while superficial pigmentary anomalies require superficial dermal penetration.

True nonablative laser resurfacing is now the preferred modality of skin resurfacing, and can be utilized to treat both facial and nonfacial areas. Mild to moderate rhytids show consistent improvement, unlike near infra-red nonablative devices and visible laser and light photorejuvenation devices. Indications for true nonablative resurfacing include periorbital and perioral rhytids, facial and nonfacial rhytids, facial and nonfacial photodamage, melasma, acne scars, and surgical scars. Three to five treatments are usually necessary for optimal results (Fig. 13.3).

The most recent introduction to skin resurfacing is the concept of deep dermal ablative fractional resurfacing. While photorejuvenation and true nonablative fractional resurfacing yield very impressive results, there are subsets of patients with significantly more advanced photoaging where traditional ablative laser resurfacing remains the only viable option. In addition, several patients desire single treatment modalities with reduced recovery period. Ablative fractional resurfacing with deep dermal ablation employs traditional ablative wavelengths of light (2940 and 10,600 nm). Initial results show impressive

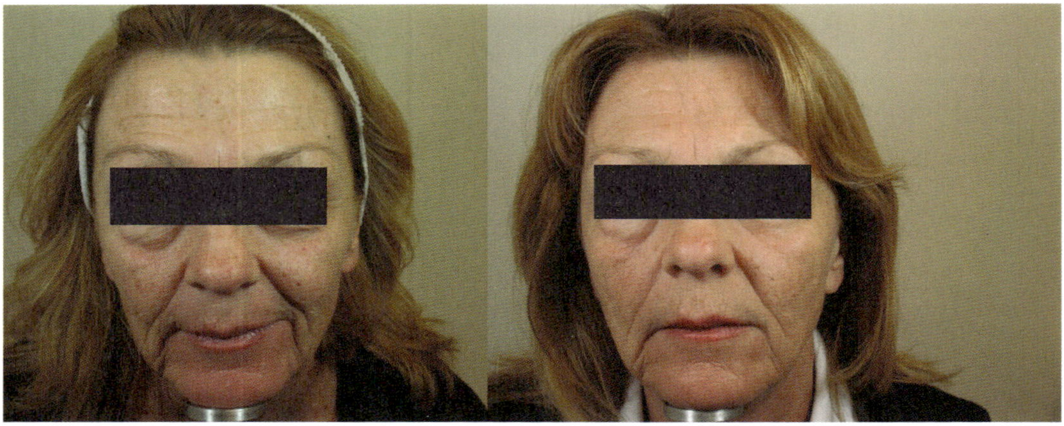

Before After

Figure 13.3 Pre- and post-3 fraxel laser treatments for photoaging.

outcomes on rhytids and laxity. While published data is very limited at this point, early observations with true deep dermal ablative fractional resurfacing do not show risks of hypopigmentation, which was reported to be as high as 20% in traditional ablative resurfacing. Long-term follow-ups and studies are necessary to confirm these early observations.

As with traditional ablative resurfacing, for optimal correction of rhytids, adjuvant therapy with botulinum toxins and dermal fillers is indicated.

13.8 Botulinum Toxins

Botulinum toxins are essential in global facial rejuvenation for relaxation of muscles of facial expression. The most widely studied botulinum toxin is Botulinum toxin A, which is primarily employed in the upper one-third portion of the face to address glabellar, forehead, and periocular rhytids. The lower one-third portion of the face is also gaining momentum with the treatment of the depressor angulii oriis and orbiculars oriis for meilolabial droop and perioral rhytis, respectively. Certain areas such as deep glabellar grooves, meilolabial folds with droop, and perioral rhytids are best addressed in combination with dermal fillers. With many laser and light-source procedures, multiple treatments are necessary for optimal outcomes. The addition of botulinum toxin A on the same day of procedures such as photofacials and fractional resurfacing produces an immediate effect, allowing for greater patient satisfaction (Fig. 13.4). Morever, the timing and synergy for combining botulinum toxin A and photofacial and fractional laser procedures are ideal, as when the patient has completed the series of treatments over a span of three to five months, the effects of botulinum toxin A have worn off. There are a few studies that suggest a synergy between botulinum toxin A and photofacials. Split-face studies show an "enhanced" photofacial effect on patients treated with simultaneous intense pulsed light and botulinum toxin A, suggesting a synergy between the two modalities.

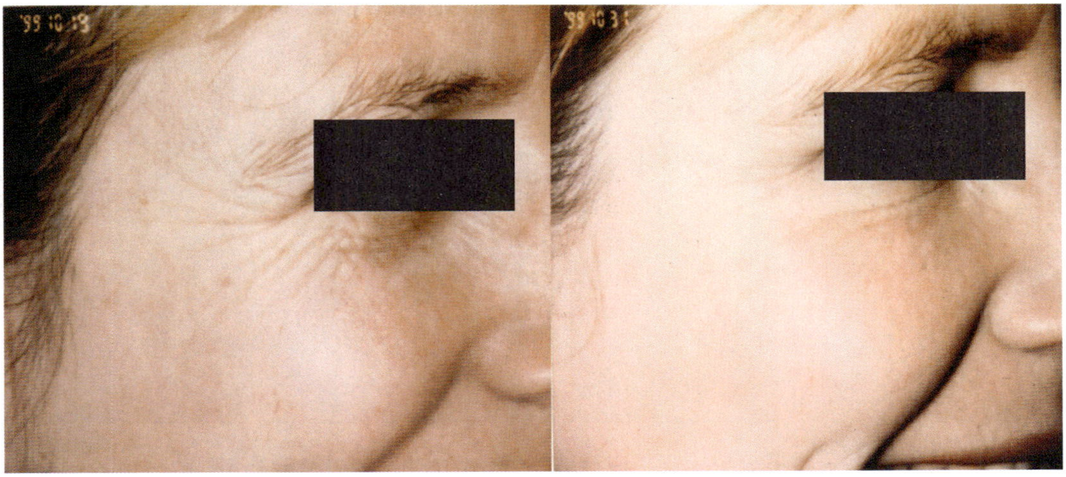

Before After

Figure 13.4 Pre- and post-pulsed light and botulinum toxin A.

13.9 Dermal Fillers

With devices "resurfacing" the facial canvas and botulinum toxin A "relaxing" the muscles of facial expression, dermal fillers address the issue of loss of volume and complete the triad of procedural nonsurgical facial restoration. During the last five years, a myriad of dermal fillers have been approved by the FDA and the way fillers are now utilized are dramatically different from the way they were used when the collagens were first introduced. Dermal fillers can be divided into temporary, mixed, and permanent. For many years, fillers were used to "fill in lines". The trend in the twenty-first century is to address global volume loss, as opposed to "individual" rhytids. Regional volumetric restoration is the best way to utilize dermal fillers. Moreover, fillers can be used in combination with each other, based on the characteristics of the volume loss and the nature of rhytids, once again emphasizing the concept of combination therapy. With certain areas, combination of fillers and botulinum toxin are essential—such as deep glabellar rhytids, droopy meilolabial folds, and perioral rhytids. It is safe to perform photofacials and fractional laser resurfacing over fillers. However, if a patient is considering multiple modalities, it is better to complete the light and laser procedures first, and introduce dermal fillers after the completion of device-based therapies. There is anecdotal evidence that the longevity of fillers may be enhanced after light and laser-based therapies, as these modalities induce neocollagenesis. Studies are underway to quantify these observations. In addition to the face, the hands are becoming an ever popular area for dermal fillers. Combination therapies for hand rejuvenation include using laser/light for lentigines, fractional lasers for resurfacing, and dermal fillers such as calcium hydroxyapatite and polylactic acid for volume loss (Fig. 13.5).

13.10 Conclusions

Total global nonsurgical restoration employs the three "R"s—resurfacing, relaxation, and refilling. Resurfacing employs nonablative, fractional nonablative, fractional ablative,

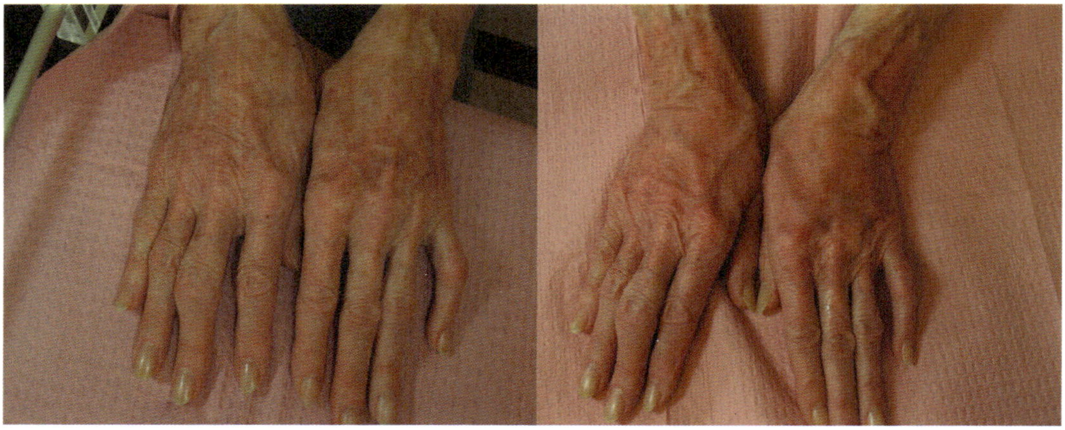

Figure 13.5 Pre- and post-combination with fraxel and radisse filler.

and ablative modalities. The nature of the device to be employed for resurfacing depends on the extent of anomalies of the facial canvas. Relaxation employs botulinum toxin A on both the upper and lower face. Refilling employs dermal fillers, sometimes used in combination with each other. While monotherapy of all these modalities show impressive clinical outcomes, the best patient satisfaction is achieved when these modalities are used in combination, as they each address anomalies in a complementary fashion. The trend in device-based rejuvenation is to employ devices which produce clinically significant results with reduction in recovery and side effects. Hence, photorejuvenation and fractional resurfacing predominate devices for the facial canvas. The trend in dermal fillers is to enhance longevity and address global volume loss, often requiring different classes of dermal fillers used in combination to accomplish these goals. The trend in botulinum toxins complement these therapies by addressing dynamic rhytids in a global fashion. Therefore, with the combination of devices, fillers, and botulinum toxins, total nonsurgical restoration can be accomplished.

Suggested Reading

1. Anderson RR, and Parrish JA. Selective photothermolysis : precise microsurgery by selective absorption of pulsed radiation. *Science*. 1983 Apr 29; 220(4596): 524–7.
2. Galeckas KJ, Collins M, Ross EV, and Uebelhoer N. Split-face treatment of facial dyshcromia: pulsed dye laser with a compression handpiece versus intense pulsed light. *Dermatol Surg*. 2008 May; 34(5): 672–80.
3. Uebelhoer NS, Bogle MA, Stewart B, Arndt KA, and Dover JS. A split-face comparison of pulsed 532 nm KTP laser and 595 nm pulsed dye laser in the treatment of facial telangiectasias and diffuse telangiectatic erythema. *Dermatol Surg*. 2007 Apr; 33(4): 441–8.
4. Dover JS, Bhatia AC, Stewart B, and Arndt KA. Topical 5-aminolevulanic acid combined with intense pulsed light in the treatment of photoaging. *Arch Dermatol*. 2005. Oct; 141(10): 1247–52.
5. DeHoratius DM and Dover JS. Nonablative tissue remodeling and photorejuvenation. *Clin Dermatol*. 2007. Sep–Oct; 25(5): 474–9, Review.

6. Weiss RA, Weiss MA, Beasley KL, and Munavalli G. Our approach to non-ablative treatment of photoaging. *Lasers Surg Med.* 2005 Jul; 37(1): 2–8.
7. Nikolau VA, Stratigos AJ, and Dover JS. Nonablative skin rejuvenation. *J Cosmet Dermatol.* 2005 Dec; 4(4): 301–7.
8. Narurkar VA. Skin rejuvenation with microthermal fractional photothermolysis. *Dermatol Ther.* 2007 Mar; 20(Suppl. 1): S10–3, Review.
9. Narurkar VA. Lasers, light sources and radiofrequency devices for skin rejuvenation. *Semin Cutan Med Surg.* 2006 Sep; 25(3): 145–50.
10. Tannous Z. Fractional resurfacing. *Clin Dermatol.* 2007 Sep–Oct; 25(5): 480–6, Review.
11. Collawn SS. Fraxel skin resurfacing. *Ann Plast Surg.* 2007 Mar; 58(3): 237–40.
12. Shamban AT, Enokibiri M, Narurkar V, and Wilson D. Photopneumatic technology for the treatment of acne vulgaris. *J Drugs Dermatol.* 2008 Feb; 7(2): 139–45.
13. Carruthers JDA, Weiss R, Narurkar V, and Flynn T. Intense pulsed light and botulinum toxin A for the aging face. *Cosmetic Dermatology.* 2003, Vol 16,S5, 2–16.
14. Beer J and Waibel J. Botulinum toxin A enhances the outcome of fractional resurfacing of the cheek. *J Drugs Dermatol.* 2007 Nov; 6(11): 1151–2.
15. Bruce S. Complementary effects of topical antiaging treatments in conjunction with aesthetic procedures. *J Drugs Dermatol.* 2008 Feb; 7(2 suppl.): S23–7.

14

Skin Rejuvenation Using Microdermabrasion

Mary P. Lupo

Department of Dermatology, Tulane Medical School, and Lupo Center for Aesthetic and General Dermatology, New Orleans, LA, USA

14.1 Introduction

Microdermabrasion is a popular noninvasive office procedure that is sought after by a public that is forever in a quest to improve their appearance with little time off from their day-to-day activities. The major trend of change in aesthetic medicine over the past 20 years has been toward minimally invasive, and minimal downtime procedures. Microdermabrasion is one of the procedures that has fueled that change. Extraordinarily safe, with high patient satisfaction levels, it has become the modern-day version of the facial that really does something to make the skin look and feel better. While many purists complain that microdermabrasion does not have enough science behind it, it is clear that microscopic and molecular changes can be proven, and that clinical responses, while variable, do occur.

This chapter reviews this trendy procedure as well as published information on the subject, so that readers can decide the value of microdermabrasion in their own aesthetic practice.

Gurpreet S. Ahluwalia (ed.), *Cosmetic Applications of Laser and Light-Based Systems*, 291–300,

14.2 History

Since the FDA approval of injectable bovine collagen in the early 1980s, the possibility of improving one's appearance during a quick office visit captivated the public. Soon after, topical tretinoin was hailed as the first agent that could improve the signs of photoaging [1]. Next, the concept of "lunchtime" peels revolutionized skin resurfacing.

No longer did patients have to endure the extensive wound-healing time required for wire-brush or diamond-fraize dermabrasion, or trichloroacetic acid or phenol peels. Now the patient could do a glycolic peel one to two times per month, and see a gradual improvement of their skin without downtime. The continued growth of nonsurgical rejuvenation can be reasonably expected, given the availability of new and more long-lasting dermal fillers, combined with the success of injectable neurotoxin and the ongoing improvements in laser and light devices, and cosmeceuticals.

The procedure dubbed microdermabrasion was first presented to the aesthetic medical community by the Italians in the 1980s [2,3]. It was released in the United States in 1994.

It still enjoys enormous popularity. Over 993,000 microdermabrasion procedures were performed in the United States in 2006 [4]. The procedure known as microdermabrasion is performed by different types of equipment. The most well-known is a closed loop system under negative pressure, dispensing aluminum oxide crystals out of a hand piece with a vacuum that sucks up the used particles along with skin surface impurities and pilosebaceous surface plugs. Other systems use different particles such as baking soda or sodium chloride, or positive rather than negative pressure.

Suggestions by some of a danger of particulates to the eyes or respiratory system have prompted newer systems that abrade the skin without any particles. One such device employs a gritty paddle without any suction or pressure, but instead utilizes vibration to complement the abrasive action. In one limited study, less treatment-associated erythema was reported and improved tone and texture was demonstrated [5]. Another unique technology nicknamed "wet dermabrasion" uses medical-grade diamonds embedded in the hand piece along with the benefits of the vacuum. The skin is elevated by the vacuum to come into contact with the gritty head, while selected cosmeceutical fluids irrigate the skin. These various fluids are chosen based on patient's cosmetic needs. The solution and exfoliated tissue are then evacuated into a container [6]. The concept of exfoliation to improve penetration during the process of irrigation with a cosmeceutical simultaneously is intriguing.

Whatever the machine type, there is considerable chance of operator variation. Factors such as the amount of pressure, flow and size of particles (or pressure on abrasive paddle, or coarseness of grit on hand piece), number of passes, and dwell-time on skin, all can result in differential clinical and microscopic results. The fundamental action is clear: removal of layers of stratum corneum, and deeper layers of the epidermis that then results in clinical and biological changes.

14.3 Published Microscopic and Molecular Findings

Several papers have been published that evaluated the effects of microdermabrasion microscopically. Improvement of the stratum corneum is a common finding. One study showed it to have smoothed and become more compacted with epidermal hyperplasia [7]. In another study, where histology was performed immediately after treatment (acute) and after six procedures, found that acutely the stratum corneum was thinned with focal compaction and homogenization, but after six sessions, change was not seen in the stratum corneum, but the epidermis had thickened [8]. Freedman also demonstrated the thickening of the epidermis, a normalization of the "basket weave" appearance to the stratum corneum and a flattening and widening of the rete pegs [9]. Improved epidermal atrophy, horny plugs, loss of polarity, and basal cell liquefaction have been reported [7]. Another investigator found thinning of the epidermis, increased orthokeratosis, and decreased rete ridge pattern [10]. Moy described epidermal changes he felt were suggestive of hydration of keratinocytes [5].

One study on skin barrier changes with microdermabrasion showed improved hydration of the newly generated stratum corneum after microdermabrasion, with no change in sebum secretion [11]. Another study of serial microdermabrasion found a significantly significant increase in the ceramide level of the stratum corneum after serial microdermabrasion, and a trend toward improved lipid-barrier function [12]. Interestingly, two studies found, in contrast that there was no alteration of the stratum corneum by their methodology [8,13]. Molecular analysis following a single microdermabrasion failed to show an increased expression of the enzymes acetyl coenzyme carboxylase or 3-hydroxy-3-methylglutaryl coenzyme A (HMG-CoA) reductase suggesting that in the stratum, corneum was unaltered, and that there was no stimulation of epidermal repair [13]. Such divergent results suggest a need for standardization of treatment parameters, but given the variation in equipment and techniques, this may be difficult indeed.

Perhaps even more important to discernable skin rejuvenation would be evidence of dermal effects. Three studies documented an increase in elastic fibers in the dermis [5,8,9]. Another showed improvement of elastosis in the dermis [7]. These types of changes could signal a benefit for not only photoaging, but for striae as well. Changes in dermal collagen have been documented microscopically. Comite showed an increase in organized collagen [14]. An immunohistochemical study found increased staining for Type I collagen in the papillary dermis [5]. Papillary dermal edema has been demonstrated [6,10]. Papillary dermal hyalinization and new collagen fibers were also found in Freedman's study[9]. Return of collagen fibers to a more fibrillar appearance has also been reported [7]. Demonstrable overall increases in dermal collagen have been observed [9,14]. Freedman's results also showed an increase in fibroblasts, appearing larger and more dense in number, especially near dermal capillaries. Interestingly, Tan and associates' histology studies did not show any significant change in dermal elastin or collagen content, again highlighting the variability of technique and results [10].

Improvement in microscopic parameters regarding pigment would also be an important finding for clinical improvement of the skin's appearance. More regular distribution of melanosomes and less melanization of the epidermis was observed in one study [8]. Clinical brightening of irregular pigment could be the result of this biologic effect. Two investigators found vascular ectasia and a dermal perivascular infiltrate [9,10]. The clinical benefits of these microscopic changes would be the "glow" so often attributed to microdermabrasion.

When other factors have been studied, increases in skin temperature and increased blood flow post treatment via skin thermography have been documented [10]. The same paper looked at other mechanical alterations and found a decrease in sebum immediately posttreatment, improved skin elasticity, and improved skin compliance (less stiffness) after 5–6 sessions. Extensive biochemical and immunohistological analyses after a single microdermabrasion showed activation transcription factors activator protein AP-1 and nuclear factor NF-κB, which regulate the expression of many genes involved in inflammation, wound healing, growth, differentiation, and apoptosis [13]. In addition, the same study showed substantial elevations in interleukin IL-1β and tumor necrosis factor TNF-α gene expression, as well as that of three matrix metalloproteinases, MMP-1 (interstitial collagenase) MMP-3 (stromelysin-1), and MMP-9 (gelatinase-B). These findings could suggest that microdermabrasion can facilitate extracellular dermal matrix repair. The corresponding clinical benefit of this finding is obvious.

14.4 Published Clinical Findings

Microdermabrasion is used for many skin disorders. Most dermatologists routinely use it in their practices for photoaging, dyschromia, including postinflammatory hyperpigmentation (PIH) and melasma, noninflammatory acne. Less frequently, it is used for striae distensae and traumatic scars, such as dog bites (Fig.14.1). One of the first published reports in the United States on microdermabrasion was its use and benefit on all types of scars: acne, traumatic, varicella, and burns [15]. Its successful use in photoaging has been published as well [14,16]. Some experts have advocated its use for actinic keratosis, keratosis pilaris, seborrheic keratosis, milia, and hypertrophic scars [8,17]. The published benefit and

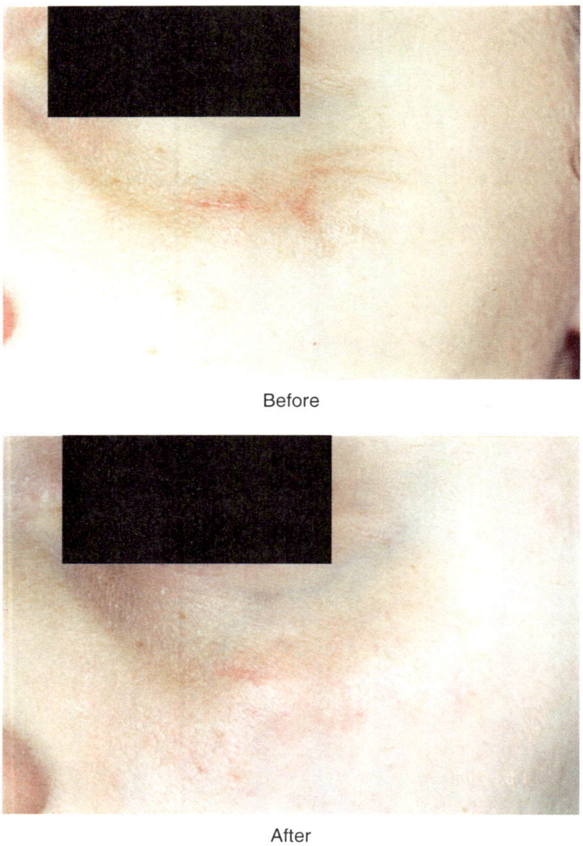

Figure 14.1 Treatment of dog's bite scar with microdermabrasion.

routine use in melasma as well as mottled pigmentation is also well- known [8,18,19]. Improved skin roughness and overall appearance, as well as the improvement of oily skin, dilated pores and fine wrinkles have been reported [7,8]. Diminished pore size, which is a common perception of patients, has been reported in the literature [7,20].

Not all investigators have reported clinical benefits. Shim's study, for example, showed no benefit visible for fine wrinkling or comedonal acne [8]. Another study of questionable scientific method, using no controls and having patients on concurrent retinoids and oral antibiotics, showed improvement in acne with microdermabrasion [21]. What could be an important finding is the improvement of postinflammatory hyperpigmentation seen in this study, although the retinoids that patients were using could exert this benefit as well. The format of this particular experiment makes any pure scientific interpretation difficult.

Dermatologists routinely use combination protocols in their offices for faster and more dramatic clinical results. One interesting study showed that microdermabrasion improved the results of retinoic acid 5% peels, over the results obtained from the peel alone [22].

Not surprising was a report that the retinoid adapalene 0.1% improved the results of microdermabrasion [20]. The more aggressive protocol of combining microdermabrasion and superficial glycolic acid peels has been advocated by some [23]. When an interesting study was performed, comparing the preference of patients between 20% glycolic peels on one side and microdermabrasion on the other, there was a slight preference for the glycolic treatment results [24]. The investigator ratings did not show differences, or even significant improvement in this study. It has been reported that a possible benefit of microdermabrasion over peels from the patient perspective is the short duration of erythema compared to glycolic peel: one day for microdermabrasion, four days for the peel [25].

Microdermabrasion has also been thought to improve the penetration of topical pharmacologic agents as well as cosmeceuticals. One study reported a 20-fold increase in penetration of magnesium ascorbyl phosphate (a hydrophilic pro-drug of ascorbic acid that is stable at neutral pH) into the skin [26]. Other hydrophilic compounds have been better injected into the skin, including 5-fluorouracil, and 5-aminolevulenic acid (ALA) using microdermabrasion [27]. Yet another study found a 5–15-fold higher penetration of ALA after microdermabrasion [28]. Katz and associates have found that microdermabrasion shortens the necessary incubation time of ALA prior to 595nm-pulsed dye laser photodynamic therapy to just 10 min [29]. Wang and associates, however, found no benefit from the addition of microdermabrasion prior to application of 5% lidocaine in a protocol to evaluate the effects on acne of a 1450nm diode laser [30]. In addition, there was no statistically significant benefit to adding the microdermabrasion in respect to the acne improvement itself.

14.5 Complications and Contraindications

One of the main reasons for the popularity of microdermabrasion is its safety. The main risk of any surface abrasion is pigmentary disruption. Lasers, intense pulsed light devices, as well as microdermabrasion are often utilized to treat sun-induced hyperpigmentation, melasma, and PIH, yet practicing physicians are well aware of the risk of inducing PIH from these procedures, especially in darker skin types. Indeed, one of the first published reports on the use of microdermabrasion did report PIH [15]. For this reason, the concurrent use of sunscreens is advisable to all patients. The risk of infection, though minimal, has been cautioned in the literature [31]. As with all equipment, precautions against cross contamination and spread of infection must be exercised. The mechanical friction on the skin could result in histamine release and an urticarial response [32]. Concerns regarding both foreign body granuloma and eye irritations have been voiced [20]. Certainly, a recent trend has been away from the particulate form of microdermabrasion and toward the particle-free technology using abrasive paddles and hand pieces because of these issues. In addition, aerosolization of particles into the lungs is a concern, but one that seems to be avoidable with the proper use of masks by the technicians utilizing the machines on a daily basis. The slight risk of petechiae and purpura from suction should be considered, and prolonged dwell time needs to be avoided. Combining peels and microdermabrasion, especially by inexperienced technicians, can result in complications (Fig. 14.2).

The supervising physician should pay attention to the baseline condition of patients. Active infections such as flat warts, herpes simplex, molluscum contagiosum, or any possible bacterial infection such as staphylococcus aureus are all absolute contraindications to treatment. Some have suggested that rosacea and telangiectasias, as well as active pustular acne are relative contraindications [8]. This is probably good thinking, especially since laser and light treatments give superior results for these diagnoses anyway. All patients on retinoids should be treated with caution.

14.6 Practical Tips

The author has been using microdermabrasion routinely since 1997, starting first with particle abrasion, and then moving to dermal infusion with abrasive hand piece technology five years later. The optimal use for microdermabrasion technology is in a combination protocol using prescription retinoids, cosmeceuticals, and sunscreens to achieve faster improvement of many skin conditions. The best use, in the author's opinion, is for dyschromia of all types: melasma (Fig. 14.3), PIH from acne (Fig.14.4), folliculitis (Fig.14. 5), and

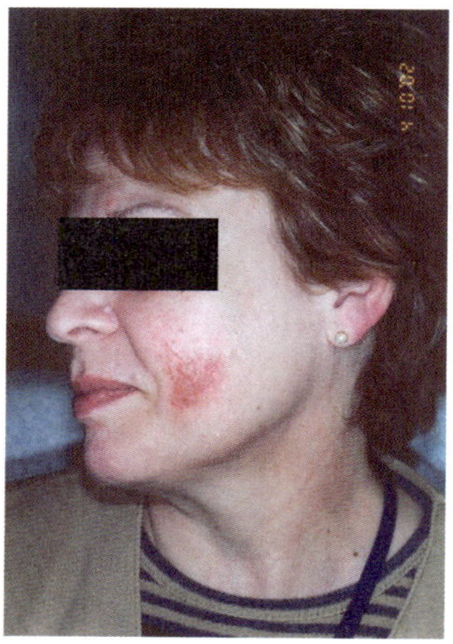

Figure 14.2 Complication from microdermabrasion procedure.

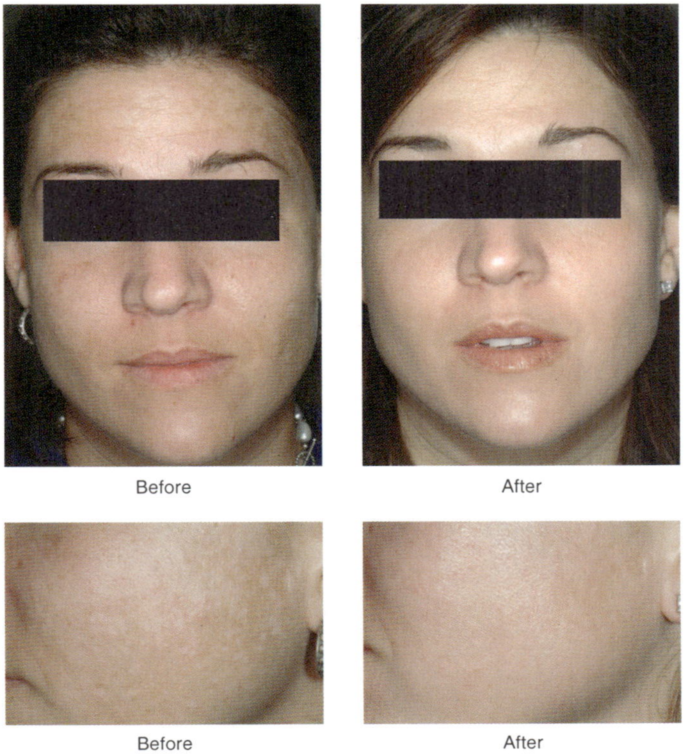

Figure 14.3 Improvement in melasma after microdermabrasion.

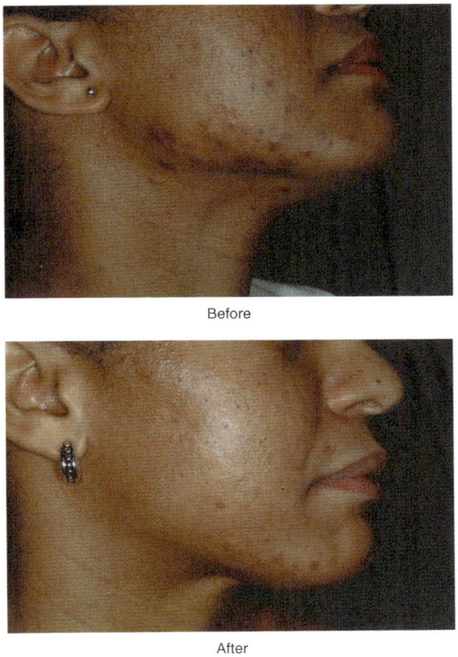

Figure 14.4 Effect of microdermabrasion on acne scaring and discoloration.

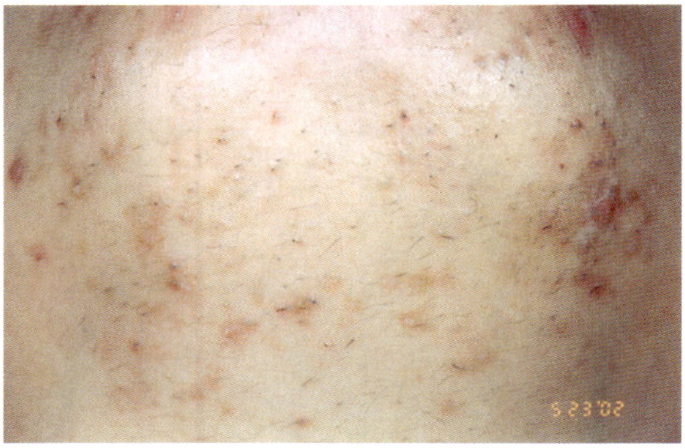

Before

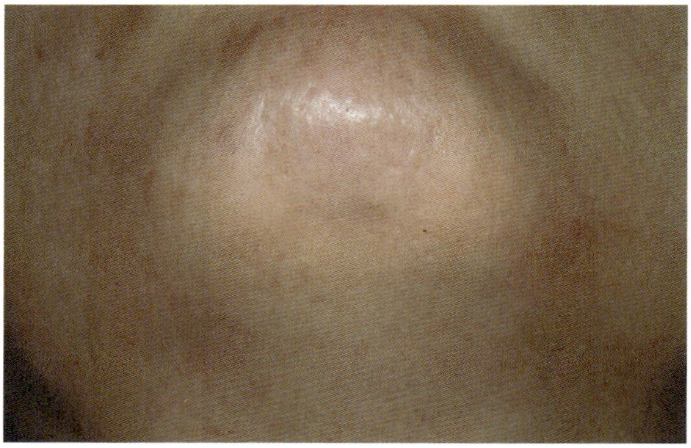

After

Figure 14.5 Folliculitis improvement with microdermabrasion.

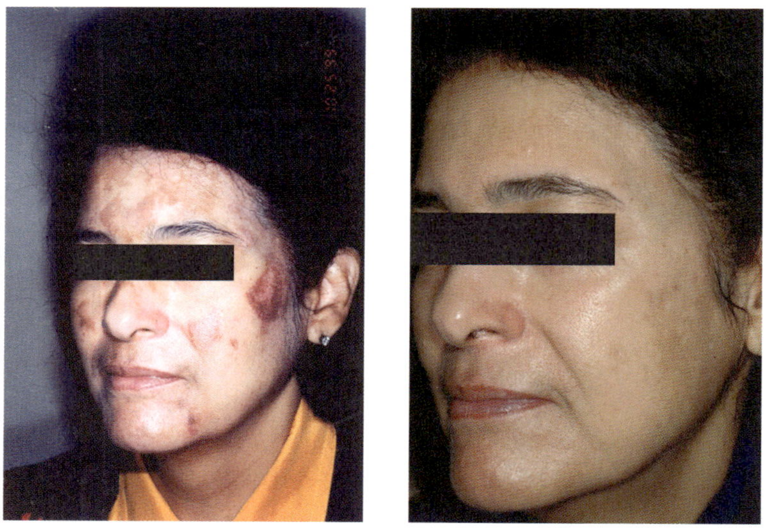

Before After

Figure 14.6 Improvement in postinflammatory hyperpigmentation caused by a TCA peel using microdermabrasion procedure.

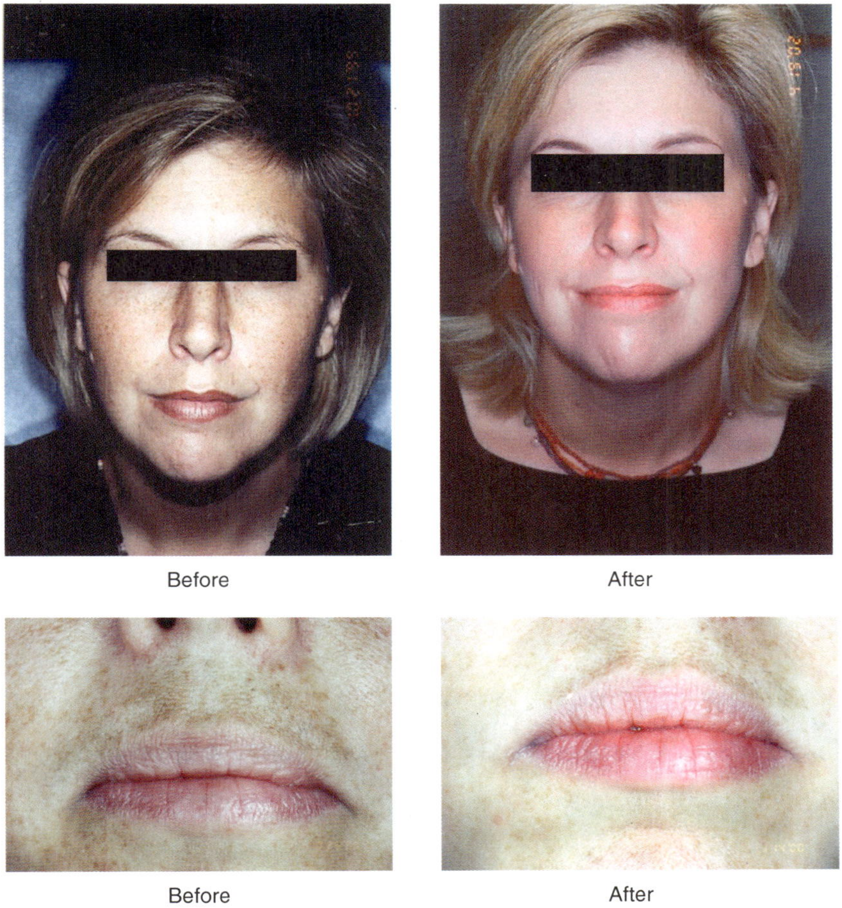

Figure 14.7 Treatment of photoaged skin with microdermabrasion.

PIH from a TCA peel (Fig.14.6), as well as pigment changes as a result of photoaging (Fig. 14.7). In addition, mild traumatic scars improve as shown in Fig.14. 1. This safe and effective complementary therapy should be an effective addition in any aesthetic practice.

References

1. Kligman AM, Grove JL, Hirose R, et al. Topical tretinoin for photoaged skin. *J Am Acad Dermatol.* 1986; 15:836–59.
2. Monteleone G. Microdermabrasion with Aluminum Hydroxide Powder in Scar Camouflage. Third meeting of the Southern Italy Plastic Surgery Association. Benevento, Italy. December 9–10, 1988.
3. Buttatarro F and Frasca N. The Programmed Microdermabrasion: Technique and Directions. Fourth National Congress Italian Association of Surgical Dermatologists. Rome, Italy. June 15–17, 1989.
4. Data of the American Society for Aesthetic Plastic Surgery. Available at http://www.surgery.org/public/consumer/trends/cosmetic_surgery_trends. Accessed June 30, 2007.
5. Kist D, Flor M, and Zelickson B. Vibradermabrasion-New Technique for Superficial Exfoliation. *Cosmetic Dermatol.* 2005; 18: 131–5.

6. Moy LS and Maley C. Skin Management: A Practical Approach. www.plasticsurgeryproducts online.com. January 2007. Accessed July 1, 2007.
7. Hernandez-Perez E, Ibiett EV. Gross and microscopic findings in patients undergoing microdermabrasion for facial rejuvenation. Dermatol Surg 2001; 27:637–40.
8. Shim EK, Barnette D, Hughes K, and Greenway HT. Microdermabrasion: a clinical and histopathologic study. *Dermatol Surg.* 2001; 27:524–30.
9. Freedman BM, Rueda-Pedraza E, and Waddell SP. The epidermal and dermal changes associated with microdermabrasion. *Dermatol Surg.* 2001; 27:1031–4.
10. Tan MH, Spencer JM, Pires LM, Ajmeri J, et al. The evaluation of aluminum oxide crystal microdermabrasion for photodamage. *Dermatol Surg.* 2001; 27:943–9.
11. Rajan P and Grimes P. Skin barrier changes induced by aluminum oxide and sodium chloride microdermabrasion. *Dermatol Surg.* 2002; 28:390–3.
12. Lew BK, Cho Y, and Lee M. Effect of serial microdermabrasion on the ceramide level in the stratum corneum. *Dermatol Surg.* 2006; 32:376–9.
13. Karimipour DJ, Kang S, Johnson TM, Orringer JS, et al. Microdermabrasion: a molecular analysis following a single treatment. *J Am Acad Dermatol.* 2005; 52:215–23.
14. Comite SL, Krishtal A, and Tan MH. Using microdermabrasion to treat sun-induced facial lentigines and photoaging. *Cosmetic Dermatol.* 2003; 16:40–2.
15. Tsai RY, Wang CN, and Chang HL, Aluminum oxide crystal microdermabrasion; a new technique for treating facial scarring. *Dermatol Surg.* 1995; 21:539–42.
16. Coimbra M, Rohrich RJ, Chao J, and Brown SA. *Plast Reconstr Surg.* 2004; 15:113(5):1438–43.
17. Sadick NS and Finn N. A review of microdermabrasion. *Cosmet Dermatol.* 2005; 351–4.
18. Rendon MI and Benitez AL. Use of a triple-combination agent and various procedures for treatment of melasma. *Cosmetic Dermatol.* 2005; 18:495–503.
19. Cook-Bolden F, Nestor M, and Rodriguez M. The use of a triple-drug combination product and procedures for the treatment of hyperpigmentary disorders. *Cosmetic Dermatol.* 2005; 18:589–594.
20. Bhalla M and Thami GP. Microdermabrasion: reappraisal and brief review of literature. *Dermatol Surg.* 2006; 32:809–14.
21. Lloyd JR. the use of microdermabrasion for acne: a pilot study. *Dermatol Surg.* 2001; 27:329–31.
22. Hexsel D, Mazzuco R, Dal'forno T, and Zechmeister D. Microdermabrasion followed by a 5% retinoid acid chemical peel vs. a 5% retinoid acid chemical peel for the treatment of photoaging—a pilot study. *J Cosmetic Dermatol.* 2005; 4:111–6.
23. Briden E, Jacobsen E, and Johnson C. *Cutis* 2007; 79(1 Suppl Combining):13–6.
24. Alam M, Omura NE, Dover JS, and Arndt KA. Glycolic acid peels compared to microdermabrasion: a right-left controlled trial of efficacy and patient satisfaction. *Dermatol Surg.* 2002; 28:475–9.
25. Song JY, Kang HA, Kim MY, et al. Damage and recovery of skin barrier function after glycolic acid peeling and crystal microdermabrasion. *Dermatol Surg.* 2004; 30:390–4.
26. Lee WR, Shen SC, Kuo-Hsien W, et al. Lasers and microdermabrasion enhance and control topical delivery of vitamin C. *J Invest Dermatol.* 2003; 121:1118–25.
27. Lee WR, Tsai RY, Fang CL, et al. Microdermabrasion as a novel tool to enhance drug delivery via the skin; an animal study. *Dermatol Surg.* 2006; 32:1013–22.
28. Fang JY, Lee WR, Shen SC, et al. Enhancement of topical 5-aminolevulenic acid delivery by erbium:YAG laser and microdermabrasion; a comparison with iontophoresis and electroporation. *Br J Dermatol*, 2004; 151:132–40.
29. Katz BE, Truong S, Maiwald DC, et al. Efficacy of microdermabrasion preceding ALA application in reducing the incubation time of ALA in laser PDT. *J Drugs Dermatol.* 2007; 6:140–2.
30. Wang SQ, Counters JT, Flor ME, and Zelickson BD. Treatment of inflammatory facial acne with the 1,450 nm diode laser alone versus microdermabrasion plus the 1,450 nm laser: a randomized, split-face trial. *Dermatol Surg.* 2006; 32:249–55.
31. Shelton RM. Prevention of cross-contamination when using microdermabrasion equipment. *Cutis* 2003; 72:266–8.
32. Farris PK and Rietschel RL. An unusual acute urticarial response following microdermabrasion. *Dermatol Surg.* 2002; 28:606–8.

15

Wrinkles: Cosmetics, Drugs, and Energy-Based Systems

John E. Oblong

The Procter and Gamble Company, Cincinnati, OH, USA

15.1 Introduction

The desire by women throughout the world to be able to restore their facial skin appearance to a more youthful state has been one of the main drivers for the continued growth and

Gurpreet S. Ahluwalia (ed.), *Cosmetic Applications of Laser and Light-Based Systems*, 301–316,

expansion of the antiaging aesthetic market. On the cosmetic side of this multi-billion dollar market, there are currently available to the consumer a literal plethora of products claiming to provide various skin-benefit attributes. While historically the market has been driven primarily by moisturizers, there have been significant technical breakthroughs and alterations in social perceptions that have fueled the robust expansion and growth, both in the cosmetic as well in the professional market (cosmetic surgeons, dermatologists, plastic surgeons). The numerous technological advances have allowed the professional to be able to provide to patients significant transformations with reduced negative side effects and at lower costs. Due to limitations, this chapter will review a few of the options available, with a focus upon technologies that can significantly impact rhytides (fine lines and wrinkles).

15.2 Cosmetics

Cosmetics are products designed to provide an appearance benefit to consumers; these do not provide the same efficacy as prescription drugs per se but fill a meaningful need for consumers and are readily available to the world consumer in their respective marketplaces. As a starting premise, traditionl moisturizers in and of themselves are able to have some modest effect on reducing the appearance of fine lines and wrinkles by hydrating and plumping the skin, but these effects can be transient and are definitely weak in overall robustness. Thus, the consumer who wishes to have a more significant effect on reducing the appearance of their fine lines and wrinkles may seek out more potent technology. In the cosmetic market, there are a limited number of technologies that have been shown to have some degree of impact upon reducing/eliminating the appearance of fine lines and wrinkles. In contrast to the cosmetic products available, the consumer does have the opportunity to seek out more aggressive therapy from professionals, where some of the newer technical developments are significantly impacting the overall market.

15.2.1 Retinoids

Topically used retinoids are comprised of a class of compounds built around the core structure of naturally occurring retinol (vitamin A), which itself is derived from the dietary hydrolysis of β-carotene. Several varying retinoids have been sold as cosmetics and as prescription drugs to treat chronologically aged and photodamaged skin. More relevant to this chapter, nearly all of them when formulated into moisturizers have been shown to consistently improve the appearance of fine lines and wrinkles, both in terms of potency and breadth of response. The commonly used retinoids in the cosmetic market include retinol and esters such as retinyl acetate, retinyl propionate, and retinyl palmitate. Less common but still used is retinal, the oxidized product of retinol. Metabolically, all these retinoids can be converted intracellularly to the true active form, *trans*-retinoic acid (tRA) (Fig. 15.1). This product has been on the market for some time as a topical prescription drug that was originally approved for treating acne (Retin-A), and subsequently approved for treating photodamaged skin, including fine lines and wrinkles (Renova). Since most formulations containing retinoids applied topically can elicit negative effects such as dryness, redness, burning, and irritation [1], there have been extensive research efforts to identify retinoid analogues that have reduced side effects when formulated into moisturizers [2]. While

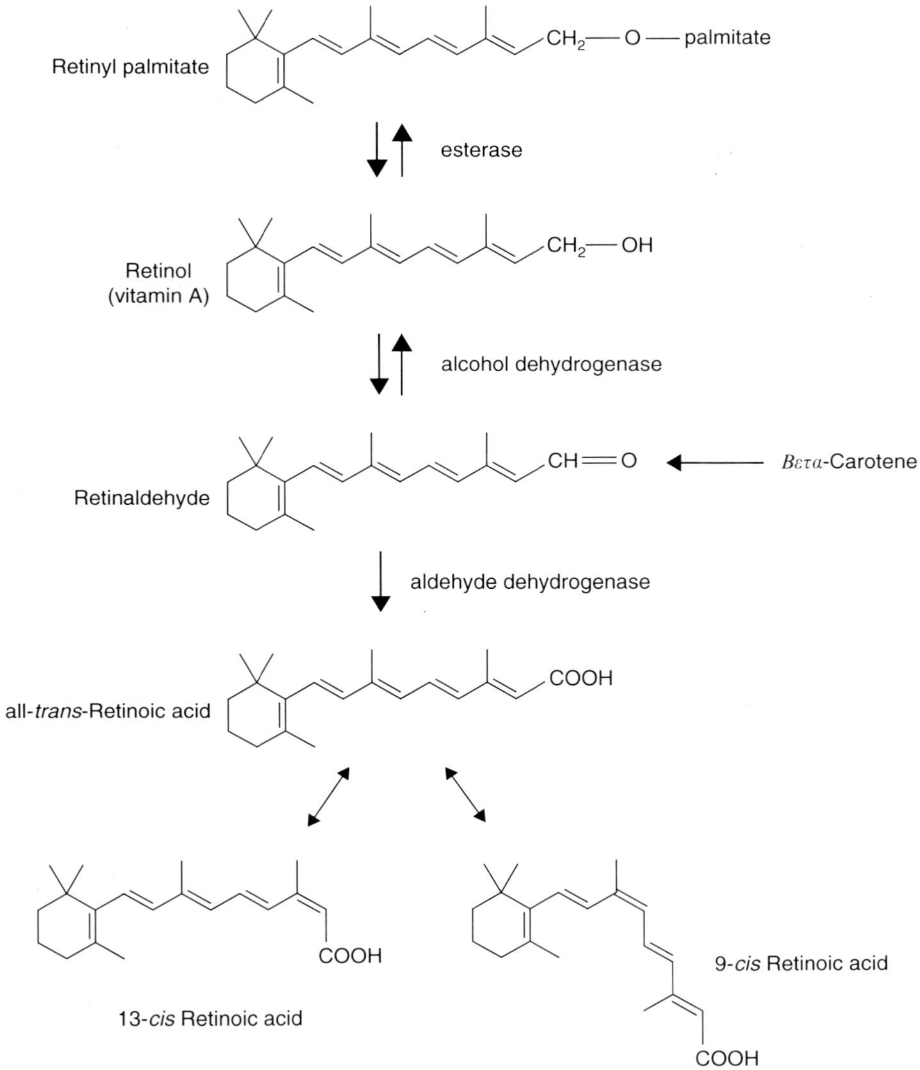

Figure 15.1 Metabolism of retinoids to all-*trans* retinoic acid.

products have been successfully developed and are marketed, it is not apparent that there is a clear reduction in the negative side effects.

Topical retinoids are capable of eliciting robust changes in skin biological systems and this is most evident at the molecular level by changes in gene expression patterns, which ultimately lead to the various biochemical and cellular responses observed [3,4]. The primary point of regulation for gene expression changes occur via binding of tRA (and isoforms) as a ligand to select heterodimer nuclear receptor complexes, which then bind to retinoic acid-response elements in the promoter region of select genes, and turn on gene expression. The receptor complexes are comprised of members from the retinoic acid receptor (RAR) and retinoid X receptor (RXR) family of proteins, both of which are further comprised of α, β, and γ isoforms. The availability of various combinations between the representative isoforms from each family into a heterodimeric complex provides a key regulatory point for regulating diverse gene expression profiles.

While topically delivered retinoids have the ability to elicit profound changes in skin [5,6], the basic premise for these changes can be ascribed to a normalization of skin that has undergone morphological and molecular changes due to aging and environmental insults (primarily UV damage from chronic sun exposure). On a macro level, this includes thickening of the skin to diminish the appearance of fine lines and wrinkles via increased epidermal proliferation and differentiation (net increase in epidermal thickness), increased production of epidermal glycosaminoglycans (GAG), and increased net content of collagen in the dermis (net increase in dermal thickness). From a kinetics response perspective, the significant changes in skin from topical retinoids are observed after continued usage for two months or longer, even though some minor effects can be observed within a few weeks of usage (Fig. 15.2).

It is established that formulations containing topical retinoids besides tRA can improve the appearance of fine lines and wrinkles in chronologically photodamaged skin, particularly retinol [7,8], retinal [9,10], and both retinyl acetate and retinyl propionate (Fig 15.2) [11]. Numerous publications have examined the structural, biochemical, and gene expression changes associated with tRA treatment of photodamaged and aged facial skin [3,12,13].

Two synthetic retinoid analogues, adapalene and tazarotene, have been approved as prescription drugs for the treatment of acne. Of these, only 0.1% tazarotene (Avage) has been approved by the FDA for the treatment of fine lines, wrinkles, and hypo- and hyperpigmentation. The side effects from tazarotene are the typical retinoid-mediated ones but are considered to be more irritating than both adapalene and tretinoin. In addition, it appears that the kinetics of skin response to tazarotene is much quicker than other retinoids, both in efficacy and irritation.

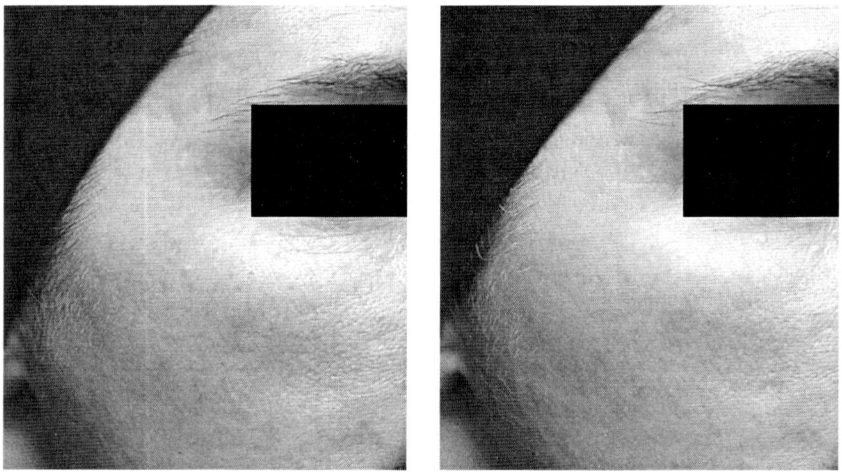

Baseline image Week 12 image

Figure 15.2 Visual impact of topical moisturizers containing 0.3% retinyl propionate upon photodamaged skin.

15.2.2 Niacinamide

In general, vitamins have been used extensively in cosmetic products for some time. One of the relatively newer entrants is niacinamide (nicotinamide) and the amide form niacin (nicotinic acid), water soluble members of the vitamin B complex. Physiologically, the primary biochemical function of dietary niacinamide and niacin is to be used as precursors for the biosynthesis of the key enzyme cofactors nicotinamide adenine dinucleotide (NAD) and the phosphorylated derivative (NADP). These cofactors play a critical role in serving as reducing and oxidizing equivalents for a host of enzymatically catalyzed biochemical reactions.

It has been hypothesized that formulations containing niacinamide can restore the reduced levels of NAD(P)H in keratinocytes and fibroblasts from older, aged, and photo-damaged skin to levels approximating those present in younger-aged skin [14]. This normalization to a homeostatic state is thought to allow cells to function at optimal capabilities, particularly as related to metabolic competence. While precisely how the dinucleotide cofactors might contribute to all these effects has not been elucidated, several specific actions of niacinamide have been described [15–17].

The published literature on topical niacinamide has shown the ability for it to elicit a range of effects [15–17] including reduction in the appearance of fine lines and wrinkles (Fig. 15.3), appearance of evening skin tone, reduction in pigmentation, and improvement in skin barrier. These data were obtained from double-blind, placebo-controlled, left–right randomized studies and was shown to reduce the overall appearance of skin fine lines/wrinkling after 8–12 weeks of treatment.

While it would be anticipated that niacin and esters thereof should be able to mimic niacinamide, there are limitations to their topical usage due to the rapid and acute vasodilation that can be elicited. This onset of redness is a negative side effect for the cosmetic consumer. It is presumed that there is rapid hydrolysis of the esters, yielding free niacin. Since the mid-1990s, there have been several publications reporting the synthesis and biological properties of synthetic ester derivatives that can have varying water solubility properties. The longer chain esters (e.g., myristoyl-nicotinate) are apparently more resistant to

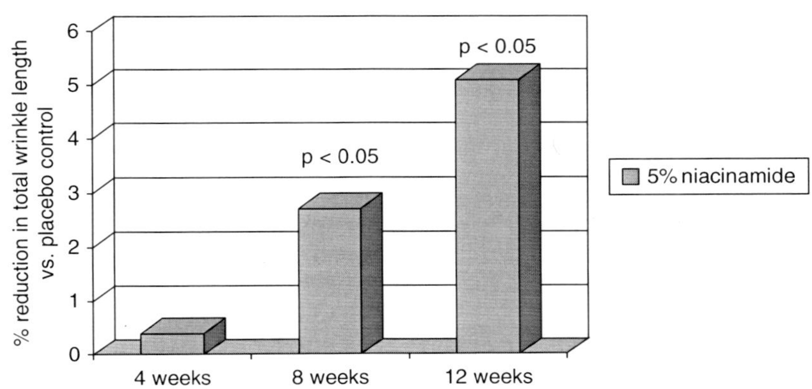

Figure 15.3 Impact of topical moisturizers containing niacinamide on reduction of facial fine lines and wrinkles based on quantitative computer image analysis.

this hydrolysis, and thus appear to be more suitable for topical use. It has been reported that myristoyl-nicotinate can elicit a range of skin benefits at the 1–5% dose range [18].

15.2.3 Ascorbic Acid

One of the most widely used vitamins in cosmetic products is ascorbic acid (vitamin C). The popularity is based partly on the critical physiological role it plays as an antioxidant in various biochemical processes, and is general accepted by the public as providing various health benefits. Relative to skin, its greatest impact is upon collagen synthesis as well as its antioxidant properties. In dermal fibroblasts, it serves as a cofactor for both prolyl hydroxylase and lysyl hydroxylase, key enzymes that posttranslationally hydroxylate proline and lysine residues in Types I and III collagen as part of the posttranslational processing of procollagen, ultimately impacting the integral structural properties of collagen fibrils upon assembly into the final bundled quartenary structure [19,20]. In addition, the antioxidant properties of ascorbic acid make it very attractive to be used topically to help block potential surface damage from oxygen radicals induced by UV and other environmental insults. Since the connection between oxygen radical damage of skin's surface and aging has been well-established [21], the usage of topical antioxidants to combat UV-induced damage has always been an appealing and compelling cosmetic ingredient story. What has not been as well-established however is whether the antioxidants can lead to any significant noticeable change in the aging skin's appearance.

While the usage of topical ascorbic acid is intriguing because of the biochemistry, there are significant limitations which appear to mute some of the anticipated efficacy. These limitations include poor skin penetration, instability due to rapid oxidation, and formulation compatibility with other components in finished products. Attempts to overcome these limitations and thereby increase availability have spurred the development of analogues, which include ascorbyl phosphate (as the magnesium and sodium salts), ascorbyl palmitate, and ascorbyl glucoside.

Several clinical studies have been published discussing the antiskin-aging benefit of ascorbic acid [22]. Relative to fine lines and wrinkles, a stabilized 3% ascorbic acid applied topically was found to be well tolerated by the skin, and it reduced facial wrinkles as determined by skin replica analysis [23]. In another study, 17% vitamin C (10% as ascorbic acid and 7% as tetrahexyldecyl ascorbate) in an anhydrous gel was found to reduce facial photoaging, as determined by dermatologist grading [24]. From a histological assessment of biopsy specimens, there was improvement in net collagen by an increased Grenz zone.

In summary, topical ascorbic acid and the various analogues can be shown to have some modest effects against photodamaged facial skin endpoints, but are not considered robust agents for fine line and wrinkle benefits compared with other technologies available to consumers [25].

15.2.4 Peptides

Peptides represent a newer class of ingredients that have entered the market. In theory, while there are a relatively a large number of peptide variants that could be designed based on amino-acid sequence, a number of residues, and use of derivatives/isomers of respective residues, there have been several peptides that have garnered greater market usage in the cosmetic industry. These include palmitoyl-lysine-threonine-threonine-lysine-serine (pal-KTTKS; Matrixyl®), acetyl-glutamate-glutamate-methionine-glutamine-arginine-arginine (Ac-EEMQRR; Argireline®), and the tripeptide copper binding glycine-histidine-lysine (Cu-GHK). These three particular peptides were designed based on amino-acid sequences present in the proform of Type I collagen, the major form of collagen present in human skin. These peptides are capable of stimulating new collagen synthesis in vitro, presumably based on a wound- healing response to damaged collagen generated from matrix metalloproteinases (MMP) enzymatic activity [26,27]. The copper-bound GHK peptide was shown to stimulate the wound-healing processes in laboratory model systems by increasing the production of dermal matrix components such as collagen and specific matrix remodeling MMPs [28–30]. The ability of these peptides to stimulate collagen synthesis is based in part upon the normal physiological response to a wound-healing event, which includes a degradation of the extracellular matrix followed by a restoration via de novo collagen and GAG synthesis.

Since the skin penetration profile of peptides is not strong, some peptides have a covalent attachment of lipophilic moieties, such as the long-chain palmitoyl group on pal-KTTKS that can dramatically improve their delivery into skin's surface. Supporting this, pal-KTTKS has been shown in both small- and large-base human clinical studies to be capable of improving the appearance of fine lines and wrinkles at the relatively low dose of 3 ppm [31–33].

In contrast to pal-KTTKS, the reported effects of formulations containing other peptides require much higher levels, such as 2% for Cu-GHK and 10% for Ac-EEMQRR. One published study [34] describes increases in skin thickness, hydration, and smoothness from the topical use of a Cu-GHK containing commercial product (peptide dose not indicated) in an open-label study involving 40 subjects. A series of clinical studies of 8–12 weeks duration describing skin improvements such as reduced wrinkling, apparently using topical 2% Cu-GHK, have been presented [35,36]. For Ac-EEMQRR, a conference platform presentation [37] describes 30% reduction in wrinkle depth with 10% of this peptide used topically in a 30-day study.

Another peptide that is currently in products is Ac-EEMQRR, which is mechanistically very different than the collagen derived peptides described earlier. Mechanistically, Ac-EEMQRR is more akin to a mimic of Botox® which functions by inhibiting neurotransmitter release, thereby causing transient muscle relaxation that control the underlying skin around fine lines and wrinkles [37]. In theory, its therapeutic effects should occur in a much shorter time frame after treatment, since Botox itself has more of an immediate acute effect via muscle relaxation.

15.2.5 Kinetin (N_6-Furfuryladenine)

Kinetin is a naturally occurring adenosine derivative found primarily in plants, where it functions as a potent hormone. Its structural basis is built off an adenosine core structure, which can be naturally derivatized to yield various analogues. Kinetin has been used topically in cosmetic products for some time, but its specific mechanism of action has not been elucidated. While it can stimulate plant-cell growth and have antisenescence effects, in human fibroblast cell culture, even very low levels (ppm)of Kinetin can delay the onset of changes associated with cell aging, for example, appearance of lipofuscin, appearance of multinucleate cells, and microtubule disorganization [38]. Clinically, moisturizers containing kinetin have been evaluated at 0.1% levels in three reported studies and has been found to have effects on appearance of fine lines, wrinkles, texture, and hyperpigmentation. While there was no placebo control included, it is not clear how robust the effects were, though it was deemed to be significant based also on dermatologist grading.

15.3 Beyond Cosmetics—Prescription Technologies

Chemical and laser peels administered by professionals represent some of the methods by which consumers as patients can dramatically improve the appearance of facial skin by literally wiping away decades of photodamage. However, these very invasive methods can cause such side effects such as disformation, hyper- and hypopigmentation, and surface appearance of burns, as well as significant downtime during the healing process from the controlled wounding. These negative attributes have helped fuel the development of less invasiveness methodologies that ideally could still deliver significant improvements in the photodamaged skin.

The starting premise for identifying less invasive technologies that are still effective is the fundamental understanding that chemical and laser ablation of the epidermis causes a robust wound repair response that restores the epidermis, but also leads to a significant turnover of the upper layers of the dermis via degradation of older collagen and newly synthesized collagen being deposited [39]. Below is a brief overview of some of the newer technologies available to the professional that can restore the skin to a younger looking state without the significant side effects associated with ablative procedures.

15.4 Dermal Fillers and BOTOX Cosmetic

Some of the best selling and widely known technologies that became available in the mid-1990s to treat photodamaged skin are injectable agents. One of these is an injectable form of the neurotoxin associated with botulism, Botulinum toxin. Commercially known as Botox, this protein-based toxin is injected in the areas around the eyes and forehead to treat fine lines and wrinkles (particularly in the furrow area between the eyes); the effect is a relaxation of the skin due to absence of muscle tension, and thereby relieving the appearance of fine lines and wrinkles. Mechanistically, this occurs by temporarily preventing muscle contractions by blocking the release of acetylcholine over a four-step process that involves internalization of the large subunit (Fig. 15.4). Part of the popularity of this procedure is that the effects are noticed immediately and can last between two and three months, albeit sometimes it can go up to six months in some extreme cases.

Some of the limitations of this procedure include the fact that the effect is localized to the injection area and more distal skin areas with fine lines or texture imperfections are not affected. In addition, there are potential side issues such as relaxation of neighboring tissue, including eyelids, as well as loss of sensation. The procedure is viewed as minimal in its invasiveness, and can be administered outside of the professional's offices.

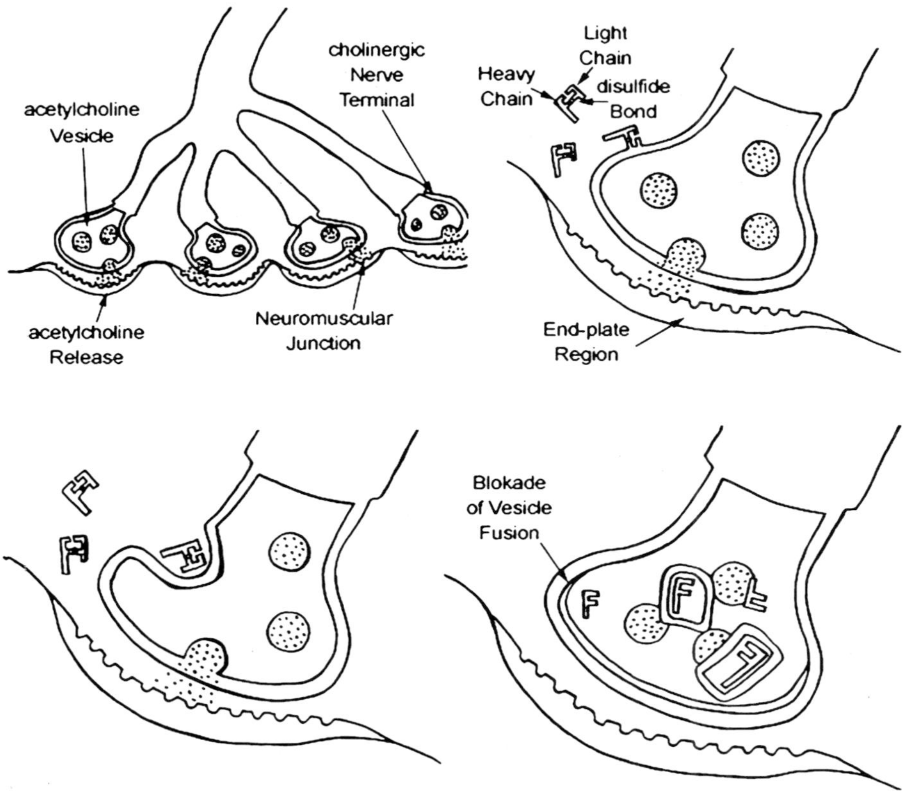

Figure 15.4 Description of the various steps in which Botox reduces neuromuscular activity. Binding of the neurotoxin dichain near the nerve synaptic region allows for internalization and subsequent blocking by the light chain domain to acetylcholine vesicles [66].

Clinically, the results can be dramatic and, appear rapidly after treatment to the patient, with no significant downtime which is an important fact for the patient. In addition to the treatment of glabellar frown lines [40], other benefits have also been reported for other skin-aging symptoms including frown lines [41] and treatment of the neck area for sagging skin [42]. The effects can last up to six months, with some patients noticing the reappearance of wrinkles after two to three months. While a re-treatment will have an immediate effect as before, it is not clear what longer-term side effects can occur from repeated treatments on muscle and skin integrity.

In contrast to the mechanistic action of Botox on temporarily impacting muscle contractions, another injection procedure that is very popular is the usage of dermal fillers to temporarily efface fine lines and wrinkles. In this case, the wrinkled area of the patient's skin is injected with a sterile solution comprised of collagen and hyaluronic acid, or a combination of both. The material is deposited in the dermal area around the injection site and it serves as a physical filler that pushes out from underneath to minimize the appearance of fine lines and wrinkles. Effects can be seen rapidly, but side effects include temporary pain discomfort to the patient, as well as bruising and swelling.

15.5 Nonablative Instruments and Techniques

Over the past ten years, the industry has seen significant growth in developing unique energy-emitting devices that allow the professional to treat patients for fine lines and wrinkles without disrupting the barrier and intact layers the skin. This has been driven both by (1) the need to develop efficacious approaches with reduced negative side effects and long recovery periods for the patient and (2) reapplications from advancements in the aeronautic and military industries for harnessing light as a controlled source of energy. Since ablative techniques cause significant damage to the photodamaged epidermis, the skin responds by undergoing a dramatic wound-healing repair process, yielding a restored epidermis and underlying dermis that is significantly closer in histological and biochemical responsiveness to a younger state. In contrast, nonablative techniques spare the epidermis and underlying dermis by utilizing relatively low fluencies and/or a cooling of the epidermis. As with ablative, nonablative techniques are hypothesized to mechanistically stimulate a wound-healing response via controlled thermal elevation or stimulation of pseudo-chromophores via select wavelengths. While one of the significant advantages of nonablative techniques over ablative is that they allow the professional to treat patients and provide noticeable benefits without the associated significant negatives, and the clinical results from nonablative techniques are not as dramatic as those by the ablative techniques, this has still not precluded the growth and usage of nonablative procedures in the professional's offices. For further information on the market relative to aesthetic devices, excellent summaries and references can be obtained at www.miinews.com.

A classification system for the types of nonablative procedures has been proposed, based partly on the specific endpoints that they impact [43]. Type I has been ascribed to the treatment of pigmentation, and Type II for the treatment of wrinkles and skin tone. In this chapter, the focus will be upon Type II modalities.

15.6 Phototherapy

15.6.1 Laser and Intense Pulsed Light

The usage of light as a form of phototherapy for the treatment of fine lines and wrinkles has seen extensive growth and usage amongst nonablative instrumentation. As noted earlier, the ability of controlled wavelengths of light to impact skin biology and structure is attained by selective stimulation of wound-repair processes in the underlying dermis while sparing the epidermis from any significant damage, particularly compromising the stratum corneum and barrier. The wavelengths and fluencies that have been studied, reported on, and commercialized are fairly broad, and this chapter will focus upon the major ones, including nonablative lasers, intense pulsed light (IPL) and light emitting diodes (LED).

In contrast to usage of lasers for ablative facial skin resurfacing, lasers and IPL for nonablative procedures is based on using longer wavelengths in the mid-infrared range (i.e., 1300–1600 nm) that can penetrate deep into the dermis of skin without compromising the barrier or damaging a precooled epidermis. Some of the lasers that have been studied and used in the field include the 1540 nm erbium, 1450 nm diode, and 1320 Nd:YAG. Numerous published reports have shown that there are significantly noticeable treatment effects upon fine lines and wrinkles from various regimens of treatments. Some of the earliest published work reported findings that using a nonablative laser could lead to significant improvements in fine lines and wrinkles [44], and that long- term treatment can lead to increased efficacy [45]. Amongst an Asian patient base, it was reported that as few as three treatments over a two-month period were found to have satisfactory effects on diminishing fine lines and wrinkles [46]. A direct comparison between long-pulse laser and IPL showed that both were able to positively improve fine lines and wrinkles, but that the long-pulse laser had fewer negative side effects [47]. This is also consistent with IPL having a weaker effect on fine lines and wrinkles compared to long-pulsed lasers. In general, IPL has a more robust effect upon pigmentation endpoints in contrast to structural alterations associated with fine lines and wrinkles. In addition, laser treatments have been reported to last as long as two years, including the fact that treated patients responded better to follow-up treatment than nontreated patients [48]. While some of the side effects from nonablative lasers and IPL include transient erythema and postinflammatory hyperpigmentation, these are significantly less than that from ablative techniques. Overall, it has been reported that patient satisfaction is relatively high from these techniques [49].

15.6.2 Light Emitting Diodes

The usage of LEDs for phototherapy treatment of fine lines and wrinkles is a relatively newer approach employed in the aesthetics market for skin rejuvenation [50]. Due to the comparatively low fluency of energy that is emitted from LEDs, the methodology is sometimes referred to as Low Level Laser (or Light) Therapy (LLLT). Overall, the usage of LED as a means of treating fine lines and wrinkles is considered to be much weaker compared with other energy emitting devices, but has a much lower risk profile of negative side effects.

Mechanistically, it is speculated that the red-light energy emitted from these devices is capable of triggering, via photomodulation, a natural response by the body to the energy, thereby activating such processes as improved cellular metabolism [51], mitochondrial efficiency [52], circulation [53] and net increases in collagen synthesis [54]. This is based in part on in vitro findings on the ability of these wavelengths in the absence of heat to elicit select effects from tissue-cultured cells [55].

Relative to published work ascribing the effects of LEDs against photodamaged and aged facial skin, a clinical study on 93 patients showed significant responses for peri-ocular wrinkles and texture changes to treatment with an LED array at 590 nm [56]. Subsequent work on a larger number of patients suggests an interesting combination of using LEDs in combination with other thermal emitting devices to further enhance the skin rejuvenation effects [56]. Another variable in the usage of visible low fluency light as a biological modulator is usage of a combination of wavelengths concurrently. Russell et al. [57] reported findings on the positive improvements of facial wrinkles by treating patients with the combination of 633 nm and 830 nm. Examining the effects from the combination of 633 and 830 nm at both the clinical and biochemical level, it was reported that the treated patients saw significant improvements in their fine lines and wrinkles and that these correlated with significant changes in the levels of collagen and elastic fibers in treated skin as well as the expression patterns of several cytokines and protein levels of TIMP-1 and 2, supporting the theory that the LED treatments can elicit both molecular and structural changes in the skin [58].

This body of work highlights the potential for identifying more potent combinations of wavelengths and fluency settings that would lead to more robust effects and without the significant side effects associated with more invasive techniques.

15.6.3 Radiofrequency

The usage of radio frequency (RF) as a nonablative approach to treating aged and photodamaged skin has been growing in popularity since its introduction in the mid-1990s. The technique is based on the mechanistic principle that the elevation of temperature inside the dermal layers leads to a transient structural denaturation of collagen fibrils, followed by contraction and tightening of the skin upon cooling. The net effect observed by the patient is an overall firming of the skin, including reduction in fine lines and wrinkles. It should be noted that a temperature elevation to at least 42° C should also elicit a heat shock response from dermal fibroblasts. This is of particular relevance for an impact upon fine lines and wrinkles, since it has been reported that heat shock treatment of dermal fibroblasts elicits a net increase in collagen production. Thus, it cannot be ruled out that the newly synthesized collagen also plays a mechanistic role in the benefits observed from RF treatment.

The level of efficacy attained via RF is relatively modest [59] but measurable changes can occur with as few as two treatments over a one-month period [60]. In addition, there do not appear to be many significant negative side effects [61]. Not surprisingly, this further supports the view that there appears to be a direct correlation with nonablative devices between lowered efficacy and reduced negative side effects.

15.6.4 Fractional Photothermolysis

One of the newer nonablative techniques introduced into the market is fractional photothermolysis (FP). This is a laser-based platform that utilizes an arrayed laser network to cause microscopic treatment zones (MTZ) of heat-induced damage that extends through the epidermis and into the dermis [62]. The level of injury on a macroscale is relatively small and the micro zones of damaged skin is surrounded by larger areas of undamaged skin, which serves as a source of cells and signaling components for the wound-healing repair process at the injured sites. While referred to as a nonablative technique, there is some degree of compromising of the skin's barrier, albeit the laser paths are essentially cauterized by the thermal energy, and the healing occurs quite rapidly. One perspective is to view it as a subfraction of an ablative laser peel, albeit into the dermis.

The general mechanism is essentially a controlled wound-healing response induced by the MTZ. This has been evaluated histologically and shown to include such changes as elevation in hsp70 expression patterns and elevated Type III collagen synthesis [63]. The net gain in collagen as well as other cascades of stimulated wound healing leads to a significant improvement in fine lines and wrinkles.

Clinical results have shown that fractional photothermolysis can lead to significant, and sometimes robust improvements in fine lines and wrinkles [64]. The patient experiences significantly less side effects and downtime as compared to the more ablative laser techniques in laser peels. It has been reported that patients undergoing fractional photothermolysis do undergo transient erythema as well as varying degrees of edema and also dryness and flaking of skin, but overall do not experience anything as significant as side effects from more invasive techniques [65]. As a general rule, patients are advised to be prepared to have at least a day or two of recovery before being able to apply makeup or spend time in direct sunlight.

15.7 Discussion

Today's consumer is constantly seeking products and therapies to help restore a more youthful appearance to their aged and photodamaged facial skin, and the technology development cycle has provided numerous options that span a range of efficacy, costs, and balance of side effects. As a starting point in the decision process on which products or services to choose from, there is generally a direct correlation between efficacy, potency, cost, and negative side effects. The more potent the technology or treatment, the higher the cost and, usually, the higher the risk for negative side effects. While the decision lies with the consumer in terms of how important it is to attain a younger appearance, it is clear that societal pressure to maintain a more youthful appearance has a strong influence. Likewise, there is greater acceptance to be proactive in this pursuit of vanity.

Of the claimed antiaging ingredients that are used cosmetically, some, but not all, can provide measurable degrees of improvement in the appearance of photodamaged facial skin. This is particularly true with continued usage, including in combination with an effective sunscreen. It is difficult to quantitatively compare the magnitude of the effects among the various technologies, but in general they are viewed as being lower than trans-retinoic acid, the current benchmark for topically antiaging ingredients. This gap highlights the technical and business opportunity to identify safe and effective technologies that can be used in the cosmetic market. Considering the enormous diversity of compounds to be found in natural extracts, variations of peptide sequences and combinations, the future possibilities seem limitless for identifying new materials and cosmetic mechanisms to improve the appearance of aging skin.

In contrast, the current methods and procedures used by professionals to treat photodamaged and aged skin can obviously provide nearly an order of magnitude or greater improvement in the aged appearance of facial skin. Significantly, the effects can also appear much more rapidly than that attained with topical cosmetic products. The balance to this higher order efficacy is higher costs, greater chance for negative side effects, and varying amounts of downtime. However, the development cycle will most certainly continue to identify technologies that will improve upon current methods. In addition, the intersection of cosmetic products with professional treatments will theoretically provide the best of both worlds to the consumer as a true cosmetic platform.

References

1. Fluhr, J.W. Vienne, M.P., Lauze, C., Dupuy, G., Gehring, W., and Gloor, M. (1999) Tolerance profile of retinol, retinaldehyde, and retinoic acid under maximized and long-term clinical conditions. *Dermatol.* **199S**: 57–60.
2. Orfanos, C.E., Zouboulis, C.C., Almond-Roesler B., and Geilen, C.C. (1997) Current use and future potential role of retinoids in dermatology. *Drugs.* **53**: 358–388.

3. Goldfarb, M.T., Ellis, C.N., Weiss, J.S., and Voorhees, J.J. (1989) Topical tretinoin therapy: its use in photoaged skin. *J. Am. Acad. Dermatol.* **21**: 645–650.
4. Varani, J., Fisher, G.J., Kang, S., and Voorhees, J.J., (1998) Molecular mechanisms of intrinsic skin aging and retinoid-induced repair and reversal. *J. Invest. Dermatol. Symp. Proc.* **3**: 57–60.
5. Weiss, J. C., Ellis, C.N., Headington, J.T., Tincoff, T., Hamilton, T.A., and Voorhees, J.J. (1988) Topical tretinoin improves photoaged skin: a double-blind vehicle-controlled study, *J. Am. Med. Assoc.*, **259**: 527–532.
6. Davies, P.J.A., Basilion, J.P., and Haake, A.R. Intrinsic biology of retinoids in the skin. In *Physiology, Biochemistry, and Molecular Biology of the Skin*; Goldsmith, L.A. Ed. Oxford University Press: New York, New York, 1997, Vol. 1, 385–409.
7. Kang, S., Duell, E.A., Fisher, G.J., Datta, S.C., Wang, Z.Q., Reddy, A.P., Tayakkol, A., Yi, J.Y., Griffiths, C.E., Elder, J.T., et al. (1995) Application of retinol to human skin in vivo induces epidermal hyperplasia and cellular retinoid binding proteins characteristic of retinoic acid but without measurable retinoic acid levels of irritation. *J. Invest. Dermatol.* **105**: 549–556.
8. Kafi, R., Kwak, H.S., Schumacher, W.E., Cho, S., Hanft, V.N., Hamilton, T.A., King, A.L., Neal, J.D., Varani, J., Fisher, G.J., Voorhees, J.J., and Kang, S. (2007) Improvement of naturally aged skin with vitamin A (retinol). *Arch. Dermatol.* **143**: 606–612.
9. Creidi, P. and Humbert, P. (1999) Clinical use of topical retinaldehyde on photoaged skin. *Dermatol.* **199S**: 49–52.
10. Diridollou, S., Vienne, M.P., Alibert, M., Aquilina, C., Briant, A., Dahan, S., Denis, P., Launais, B., Turlier, V., and Dupuy, P. (1999) Efficacy of topical 0.05% retinaldehyde in skin aging by ultrasound and rheological techniques. *Dermatol.* **199S**: 37–41.
11. Oblong, J.E. and Bissett, D.L. Retinoids (Chapter 6). *Procedures in Cosmetic Dermatology Series: Cosmeceuticals*, : Zoe Draelos, Ed. Elsevier Publishers, 2005.
12. Kligman, A.M., Grove, G.L., Hirose, R., and Leyden, J.J. (1986) Topical tretinoin for photoaged skin. *J Am Acad, Dermatol.* **15**: 836–859.
13. Griffiths, C.E., Finkel, L.J., Tranfaglia, M.G., Hamilton, T.A., Ellis, C.N., and Voorhees, J.J. (1993) An in vivo experimental model for effects of topical retinoic acid in human skin. *Brit. J. Dermatol.* **129**: 389–394.
14. Jongkind, J.F., Verkerk, A., and Poot, M. (1987) Glucose flux through the hexose monophosphate shunt and NADP(H) levels during in vitro ageing of human skin fibroblasts. *Gerontology* **33**: 281–286.
15. Matts, P.J., Oblong, J.E., and Bissett, D.L. (2002) A review of the range of effects of niacinamide in human skin. *Intl. Fed. Soc. Cosmet. Chem. Mag.* **5**: 285–289.
16. Bissett, D.L., Oblong, J.E., and Saud, A. (2003) Topical niacinamide provides skin aging appearance benefits while enhancing barrier function. *J. Clin. Dermatol.* **32S**: 9–18.
17. Bissett, D.L., Miyamoto, K., and Sun, P. (2004) Topical niacinamide reduces yellowing, wrinkling, red blotchiness, and hyperpigmented spots in aging facial skin. *Int. J. Cosmet. Sci.* **26**: 231–238.
18. Jacobson, M.K., Kim, H., and Kim, M. (2002) Modulating NAD-dependent DNA repair and transcription regulated pathways of skin homeostatis: evaluation in human subjects. 60th Annual Meeting of the American Academy of Dermatology, New Orleans, LA, February 22–27.
19. Pinnell, S.R. (1985) Regulation of collagen biosynthesis by ascorbic acid: a review. *Yale J. Biol. Med.* **58**: 553–339.
20. Geesin, J.C., Darr, D., Kaufman, R., Murad, S., and Pinnell, S.R. (1988) Ascorbic acid specifically increases type I and type III pro-collagen messenger RNA levels in human skin fibroblasts. *J. Invest. Dermatol.* **90**: 420–424.
21. Emerit, I. (1992) Free radicals and aging of the skin. *EXS.* **62**: 328–341.
22. Humbert, P.G., Haftek, M., Creidi, P., Lapiere, C., Nusgens, B., Richard, A., Schmitt, D., Rougier, A., and Zahouani, H. (2003) Topical ascorbic acid on photoaged skin: clinical, topographical and ultrastructural evaluation: double-blind study vs. placebo. *Exptl. Dermatol.* **12**: 237–244.
23. Raschke, T., Koop, U., Dusing, H.J., Filbry, A., Sauermann, K., Jaspers, S., Wenck, H., and Wittem, K.P. (2004) Topical activity of ascorbic acid: from in vitro optimization to in vivo efficacy. *Skin Pharmacol. Physiol.* **17**: 200–206.

24. Fitzpatrick, R.E. and Rostan, E.F. (2002) Double-blind, half-face study comparing topical vitamin C and vehicle for rejuvenation of photodamage. *Dermatol. Surg.* **28**: 231–236.
25. Farris, P.K. (2005) Topical vitamin C: a useful agent for treating photoaging and other dermatologic conditions. *Dermatol Surg.* **31**:814–817.
26. Katayama, K., Armendariz-Borunda, J., Raghow, R., Kang, A.H., and Seyer J.M. (1993) A pentapeptide from type procollagen promotes extracellular matrix production. *J. Biol. Chem.* **268**: 9941–9944.
27. Pickart, L. (2003) Copperceuticals and the skin. *Cosmet. Toilet.* **118**: 24–28.
28. Buffoni, F., Pino, R., and Dal Pozzo, A. (1995) Effect of tripeptide-copper complexes on the process of skin wound healing and on cultured fibroblasts. *Arch. Int. Pharmacodyn. Ther.* **330**: 345–360.
29. Simeon, A. Wegrowski, Y. and Bontemps, Y. (2000) Expression of glycosaminoglycans and small proteoglycans in wounds: modulation by the tripeptide-copper complex glycyl-L-histidyl-L-lysine-Cu2+. *J. Invest. Dermatol.* **115**: 962–968.
30. Canapp, S.O. Farese, J.P. and Schultz, G.S. (2003) The effect of topical tripeptide-copper complex on healing of ischemic open wounds. *Vet. Surg.* **32**: 515–523.
31. Lintner, K. Mas-Chamberlin, C. and Mondon, P. (2002) Pentapeptide facilitates matrix regeneration of photoaged skin. *Ann. Dermatol. Venereol.* **129**: 1S401.
32. Mas-Chamberlin, C.; Lintner, K.; and Basset, L. (2002) Relevance of antiwrinkle treatment of a peptide: 4 months clinical double blind study vs excipient. *Ann. Dermatol. Venereol.* **129**:1S456.
33. Robinson, L., Fitzgerald, N., Berge, C., Doughty, D., and Bissett, D. (2002) Pentapeptide offers improvement in human photoaged facial skin. *Ann. Dermatol. Venereol.* **129**: 1S405.
34. Kruger, N.; Fiegert, L.; and Becker, D. (2003) For the treatment of skin aging: trace elements in form of a complex of copper tripeptide. *Cosmet. Med.* **24**: 31–33.
35. Appa, Y., Stephens, T., and Barkovic, S. (2002) A clinical evaluation of a copper-peptide-containing liquid foundation and cream concealer designed for improving skin condition. 60th Annual Meeting of the American Academy of Dermatology, New Orleans, LA, February 22–27.
36. Leyden, J.J., Grove, G., and Barkovic, S. (2002) The effect of tripeptide to copper ratio in two copper peptide creams on photoaged facial skin. 60th Annual Meeting of the American Academy of Dermatology, New Orleans, LA, February 22–27.
37. Blanes-Mira, C., Clemente, J., and Jodas, G. (2003) A synthetic hexapeptide (Argireline) with antiwrinkle activity. Presentation, 37th Annual Conference of the Australian Society of Cosmetic Chemists, Queensland, Australia, March 13–16.
38. Levy, S.B. Kinetin. In *Textbook of Cosmetic Dermatology*, 3rd edition, Baran, R., Maibach, H.I. Eds. Taylor & Francis, Abingdon, Oxon, UK, 2005, 129–132.
39. Kligman, A.M., Baker, T.J., and Gordon, H.L. (1975) Long-term histologic follow-up of phenol face peels. *Plast. Reconstruct. Surg.* **75**: 652–659.
40. Ellis, D.A. and Tan, A.K. (1997) Cosmetic upper-facial rejuvenation with botulinum. *J. Otolaryngol.* **26**: 92–96.
41. Lowe, N.J., Maxwell, A., and Harper, H. (1996) Botulinum A exotoxin for glabellar folds: a double-blind, placebo-controlled study with an electromyographic injection technique. *J. Am. Acad. Dermatol.* **35**: 569–572.
42. Brennan, H.G. and Koch, R.J. (1996) Management of aging neck. *Facial Plast. Surg.* **12**: 241–255
43. Dierickx, C.C. and Anderson, R. R. (2005) Visible light treatment of photoaging. *Dermatol. Ther.* **18**: 191–208.
44. Bernstein, E.F., Ferreira, M., and Anderson, D. (2001) A pilot investigation to subjectively measure treatment effect and side-effect profile of nonablative skin remodeling using a 532 nm, 2 ms pulse-duration laser. *J. Cosmet. Laser Ther.* **3**: 137–141.
45. Brazil, J. and Owens, P. (2003) Long-term clinical results of IPL photorejuvenation. *J Cosmet Laser Ther.* **5**: 168–174.
46. Tay, Y.K., Khoo, B.P., Tan, E., and Kwok, C. (2004) Long pulsed dye laser treatment of facial wrinkles. *J Cosmet Laser Ther.* **6**: 131–135.
47. Goldberg, D.J. and Samady, J.A. (2001) Intense pulsed light and Nd:YAG laser non-ablative treatment of facial rhytids. *Lasers Surg Med.* **28**: 141–144.

48. Fournier, N., Lagarde, J.M., Turlier, V., Courrech, L., and Mordon, S. (2004) A 35-month profilometric and clinical evaluation of non-ablative remodeling using a 1540-nm Er:glass laser. *J Cosmet Laser Ther.* **6**: 126–130.
49. Laury, D. (2003) Intense pulsed light technology and its improvement on skin aging from the patients' perspective using photorejuvenation parameters. *Dermatol. Online J.* **9**: 5.
50. Kelly, K.M., Majaron, B., and Nelson, J.S. (2001) Nonablative laser and light rejuvenation: the newest approach to photodamaged skin. *Arch. Facial Plast. Surg.* **3**: 230–235.
51. Pal, G., Dutta, A., Mitra, K., Grace, M.S., Amat, A., Romanczyk, T.B., Wu, X., Chakrabarti, K., Anders, J., Gorman, E., Waynant, R.W., and Tata, D.B.. (2007) Effect of low intensity laser interaction with human skin fibroblast cells using fiber-optic nano-probes. *J. Photochem. Photobiol. B.* **86**: 252–261.
52. Karu, T.I., Pyatibrat, L.V., and Afanasyeva, N.I.. (2004) A novel mitochondrial signaling pathway activated by visible-to-near infrared radiation. *Photochem. Photobiol.* **80**:366–372.
53. Ihsan, F.R. (2005) Low-level laser therapy accelerates collateral circulation and enhances microcirculation. *Photomed. Laser Surg.* **23**: 289–294.
54. Posten, W., Wrone, D.A., Dover, J.S., Arndt, K.A., Silapunt, S., and Alam, M. (2005) Low-level laser therapy for wound healing: mechanism and efficacy. *Dermatol. Surg.* **31**: 334–340.
55. Kao, B., Kelly, K.M., Majaron, B., and Nelson, J.S. (2003) Novel model for evaluation of epidermal preservation and dermal collagen remodeling following photorejuvenation of human skin. *Lasers Surg. Med.* **32**: 115–119.
56. Weiss, R.A., Weiss, M.A., Geronemus, R.G., and McDaniel, D.H. (2004) A novel non-thermal non-ablative full panel LED photomodulation device for reversal of photoaging: digital microscopic and clinical results in various skin types. *J. Drugs Dermatol.* **3**: 605–610.
57. Russell, B.A., Kellett, N., and Reilly, L.R. (2005) A study to determine the efficacy of combination LED light therapy (633 nm and 830 nm) in facial skin rejuvenation. *J. Cosmet. Laser Ther.* **7**:196–200.
58. Lee, S.Y., Park, K.H., Choi, J.W., Kwon, J.K., Lee, D.R., Shin, M.S., Lee, J.S., You, C.E., and Park, M.Y. (2007) A prospective, randomized, placebo-controlled, double-blinded, and split-face clinical study on LED phototherapy for skin rejuvenation: clinical, profilometric, histologic, ultrastructural, and biochemical evaluations and comparison of three different treatment settings. *J. Photochem. Photobiol. B.* **88**: 51–67.
59. Narins, D.J. and Narins, R.S. (2003) Non-surgical radiofrequency facelift. *J. Drugs Dermatol.* **2**: 495–500.
60. Fritz, M., Counters, J.T., and Zelickson, B.D. (2004) Radiofrequency treatment for middle and lower face laxity. *Arch. Facial Plast. Surg.* **6**: 370–373.
61. Carruthers, A. (2001) Radiofrequency resurfacing: technique and clinical review. *Facial Plast. Surg. Clin. North Am.* **9**: 311–319.
62. Manstein, D., Herron, G.S., Sink, R.K., Tanner, H., and Anderson, R.R. (2004) Fractional photothermolysis: a new concept for cutaneous remodeling using microscopic patterns of thermal injury. *Lasers Surg. Med.* **34**: 426–438.
63. Laubach,. H.J., Tannous, Z., Anderson, R.R., and Manstein, D. (2006) Skin responses to fractional photothermolysis. *Lasers Surg. Med.* **38**: 142–149.
64. Wanner, M., Tanzi, E.L., and Alster, T.S. (2007) Fractional photothermolysis: treatment of facial and nonfacial cutaneous photodamage with a 1,550-nm erbium-doped fiber laser. *Dermatol. Surg.* **33**: 23–28.
65. Fisher, G.H. and Geronemus, R.G. (2005) Short-term side effects of fractional photothermolysis. *Dermatol. Surg.* **31**: 1245–1249.
66. Benedetto, A.V. (1999) The cosmetic use of Botulinum neurotoxin type A. *Int. J. Derm.* **38**: 641–655.

PART 4
TREATMENT OF SKIN AND HAIR DISORDERS USING LIGHT-BASED TECHNOLOGIES

16

Cellulite Reduction: Phototherma Therapy for Cellulite

Jillian Havey[1] and Murad Alam[1,2,3]

[1]Department of Dermatology, Feinberg School of Medicine, Northwestern University, Chicago, IL, USA

[2]Department of Otoloaryngology-Head and Neck Surgery, Feinberg School of Medicine, Northwestern University, Chicago, IL, USA

[3]Department of Surgery, Feinberg School of Medicine, Northwestern University, Chicago, IL, USA

Gurpreet S. Ahluwalia (ed.), *Cosmetic Applications of Laser and Light-Based Systems*, 319–340,

16.1 Cellulite

16.1.1 History of Cellulite

Figure 16.1 Cellulite is a connective tissue disorder that afflicts over 90% of non-Asian women. Source: Peter Paul Rubens (1640). *The Three Graces* (Museo del Prado, Madrid). Permission received from Bruno Dillen - www.artinthepicture.com on 8/8/07.

Cellulite, also known as dermopanniculosis, status protrusus cutis, and adiposis edematosa, afflicts over 90% of non-Asian women. The origin of the word cellulite dates back to 1922, when the French doctors Alquier and Pavot defined the condition as a dystrophy of the mesenchymal tissues characterized by interstitial fluid retention [1]. In 1978, Nurnberger and Muller expounded on Alquier and Pavot's definition when they illustrated that cellulite is caused by papillae adipose, herniations of fat that protrude at the dermo-hypodermal interface from the subcutis through a weakened dermis [2] (Fig. 16.1).

A more recent explanation that is frequently referenced is given by Goldman, who describes cellulite as a normal physiologic state in postadolescent women. Goldman hypothesizes that cellulite is a means to maximize subcutaneous adipose retention, ensuring sufficient caloric availability for pregnancy and lactation [3].

However, Goldman's definition must be clarified. Cellulite, which is mainly located on the lateral aspects of the thighs and buttocks, is thought to primarily exist due to the underlying connective tissue anatomy rather than from excessive adipose tissue. Therefore, cellulite is not synonymous with obesity, which is marked by hypertrophy of adipocytes. Since it can be located in any area of the body that contains subcutaneous adipose tissue, thin and obese women, alike, are inflicted with this condition. While many women may view cellulite as a pathologic condition, there is no morbidity or mortality associated with it. However, if serious cases of cellulite are not adequately treated, this condition can cause pathological tissue alterations such as lipodystrophic and fibrosclerotic degeneration [4] (Fig. 16.2).

More recent advances have focused on the endothelium, which modulates blood-tissue exchanges and maintains microcirculatory homeostasis by balancing fibrinolytic, vasodilatroy, vasoconstrictory, and coagulant factors. The female microcirculation has been distinguished from the male system by the presence of oestrongen receptors in smooth muscle cells and endothelial cells. The distribution of female fat can be explained by the presence of these oestrogen from these adipocyte receptors modulate the lipase activities of the microcirculatory system. The effects of hormones on cellulite formation will be discussed in the next section.

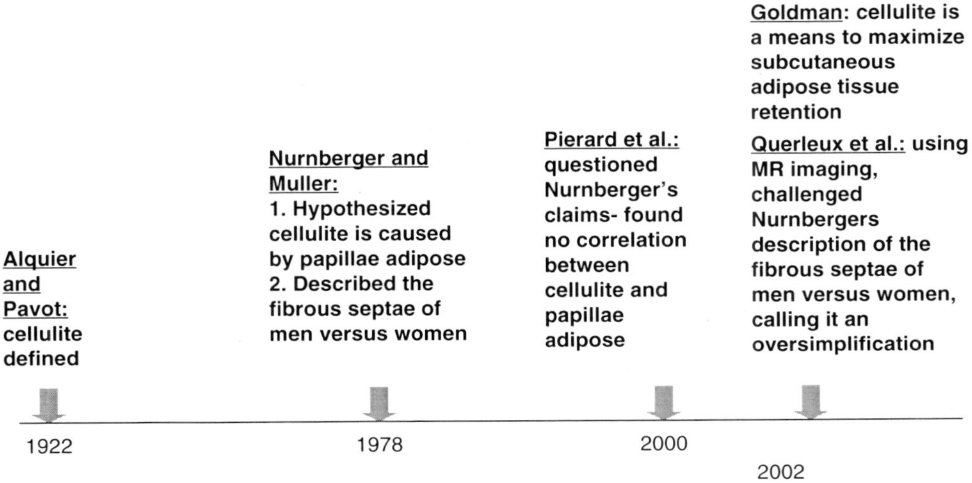

Figure 16.2 An outline of the history of cellulite.

16.1.2 Physiology of Cellulite

Cellulite can be illustrated further by dividing it into two distinct grades: incipient cellulite and full-blown cellulite. Incipient cellulite, which is hardly visible, is characterized by a discrete padded look or 'orange peel' aspect, demonstrable by the 'pinch test' (pinching the skin of a female thigh, for example, where cellulite is commonly manifested). Full blown cellulite, which is extremely noticeable on gross inspection of the skin, is recognized by a lumpy-bumpy and dimpled skin surface. Clinical evidence exists that demonstrates that full-blown cellulite is related to incipient cellulite. A continuum of morphologies exists between the two conditions [5]. In certain cases, striae distensae (stretch marks) are found within the connective tissue strand network [6].

The lumpy-bumpy appearance of a skin surface with cellulite results from a weakening and thinning of the connective tissue network that normally tethers the dermis to deeper skin layers. While some connective tissue strands become enlarged, others become loose, allowing edema and proteoglycan deposits to become part of this complex network. Consequently, the dermal-hypodermal interface is remodeled and the conformation of adipose tissue is altered [7].

16.1.3 Histology of Cellulite

The histology of cellulite was first described by Nurnberger and Muller in 1978, who attributed cellulite formation to sexually dimorphic skin architecture. They hypothesized that cellulite is determined by fatty protrusions through the dermal-hypodermal interface, and reported that these deep adipose indentations were present in the dermis of women, but not of men [2]. Rosenbaum's results substantiated the latter's claims when they discovered that female subjects, both with and without cellulite, exhibited a discontinuous and irregular dermo-hypodermal interface characterized by adipocyte protrusion into the dermis. On the contrary, the connective tissue dermal-adipose tissue border in male subjects was continuous and even [8] (Fig. 16.3). Pierard et al. [9] found no correlation between a clinical evidence of cellulite and papillae adiposae, and thus questioned Nurnberger's claims. Instead, they hypothesized that cellulite results from the stretching of fibrous septae, which in turn, causes the connective tissue support to deteriorate, allowing fat herniation.

Around 1978, Nurnberger and Muller also differentiated between the dermo-hypodermal interfaces of women versus men. They attributed cellulite formation to the sexually dimorphic skin architecture of the fibrous septae, where dermal herniations of subcutaneous fat occur mainly in women due to vertical fascial bands. The fibrous septae of men take on a criss-cross pattern of 45° tilted planes which they claimed, is more resistant to fat herniations. Further, according to these studies, female fat lobules are larger than those in males and are compartmentalized by fibrous septae oriented perpendicular to the dermis. This orientation makes it easier for fat lobules to protrude vertically into the dermis, perpetuating fat herniations and a dimpled cutaneous surface. In contrast, the smaller male fat lobules are separated by obliquely arranged septae, thus preventing herniation [2].

The hypotheses of Nurnberger and Muller have been both supported and refuted since the inception of in vivo imaging methods. Querleux et al. were the first to employ magnetic resonance imaging (MRI) to visualize the 3-dimensional architecture of the fibrous septae [10]. Magnetic resonance imaging evidenced for the first time that women with cellulite had a

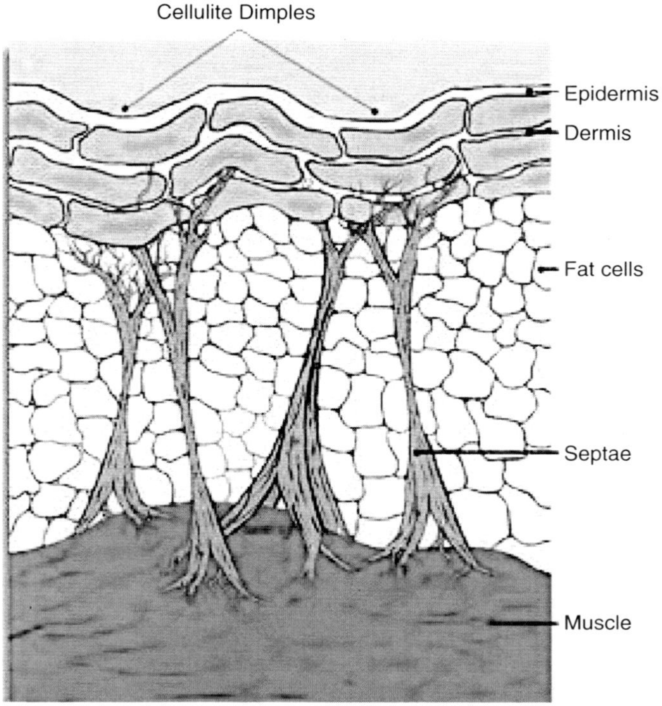

Figure 16.3 The 'orange peel' appearance of a skin surface with cellulite results from a weakening and thinning of the connective tissue network that normally tethers the dermis to deeper skin layers. In women, subcutaneous fat, separated by parallel fibrous septae, extrudes into the dermis and causes dimpling of the dermal surface. Source: www.globalhealthremedies.com.

significantly thicker inner fat layer compared to normal women ($p<0.01$). In addition, the adipose layers of women with cellulite are significantly thicker compared to normal women or men ($p=0.0001$).

The results of MRI have challenged some of Nurnberger's claims. Using MRI on women with cellulite, Querleux et al. also found a smaller percentage of fibrous septae parallel to the skin surface, but a higher percentage of septae perpendicular to the surface [10]. These results are partly in harmony with Nurnberger's hypotheses, but demonstrate that the declarations of the latter (perpendicular pattern in women and criss-cross pattern in men) may be a bit of an oversimplification. MRI gives strong evidence that the directions of the fibrous septae network are more heterogeneous than originally theorized.

16.1.4 Pathogenesis of Cellulite

Although MRI has provided a sophisticated look at the histology of cellulite, the pathogenesis of this condition is still unknown. Contrary to popular belief, physical conditioning and a healthy diet do not prevent the development of cellulite. A variety of topical treatments, massage-based therapies, and surgical and laser techniques have been used to reduce

Table 16.1 Theories of Pathogenesis of Cellulite*

A. Bassas-Grau, et al. (1964)	Edema causes cellulite in the subcutaneous connective matrix of women
B. Querleux et al. (2002)	Laxity of fibrous septae in women leads to fat herniation and cellulite
C. Kligman et al. (1997)	Inflammatory factors and vascular changes cause cellulite

*Several conflicting theories about cellulite pathogenesis.

cellulite, and most have had suboptimal clinical effects. Discovering the best treatment plan for cellulite has been the focus of many scientists and journal publications. However, finding a solution raises even more of a challenge when a consensus as the exact definition and aetiopathology of cellulite has not yet been reached. Surprisingly, scientists are still debating over three conflicting theories that relate to the ethiopathogenesis of cellulite (Table 16.1).

The first of three contradictory theories indicates that excessive hydrophilia of the intercellular matrix causes edema and fibrosclerosis. Bassas-Grau et al. noted that the subcutaneous connective matrix of women with cellulite had an excess of hyperpolymerized acid mucopolysaccharides, which thereby caused edema [11]. Although no other evidence supports the claims of Bassas-Grau, this hypothesis justified the use of topical treatments such as hyaluronidase, which had a therapeutic influence on cellulite treatment for many years.

A second theory focuses on the anatomical conformation of the female subcutaneous tissue as compared to male anatomy. This theory reflects on the works of Querleux, who through MRI confirmed the different orientation of the connective septae bands in women compared to men. According to the data, MRI rules out the possibility of edema occurring in between the adipocytes, but confirms the laxity of the fibrous septae in women, which leads to fat herniation and the lumpy appearance of cellulite [1].

Finally, a third theory attributes the cause of cellulite to inflammatory factors and vascular changes. From biopsies of cellulite, Kligman has claimed that chronic inflammatory cells, such as macrophages and lymphocytes, are diffusely concentrated in the fibrous septae. This phenomena, according to Kligman, attributes the cause of adipolysis and minor inflammation in cellulite patients to these fibrous septae [12].

16.2 Hormonal Influence on Cellulite Development

Cellulite is a connective tissue disorder seen mainly in postadolescent females. In the rare occasions that cellulite is witnessed in men, it is more common in males with androgen-deficient conditions such as hypogonadism, Klinefelter's syndrome, and in patients receiving estrogen-based therapy for prostate cancer [14]. Furthermore, where cellulite is displayed in both obese and thin females, obese males are rarely inflicted with the condition. These findings imply that a hormonal component plays an important role in the etiology of cellulite.

Estrogens and androgens are involved in the formation of cellulite. Estrogen stimulates lipogenesis (lipid synthesis) and inhibits lipolysis (fatty acid oxidation—not synonymous

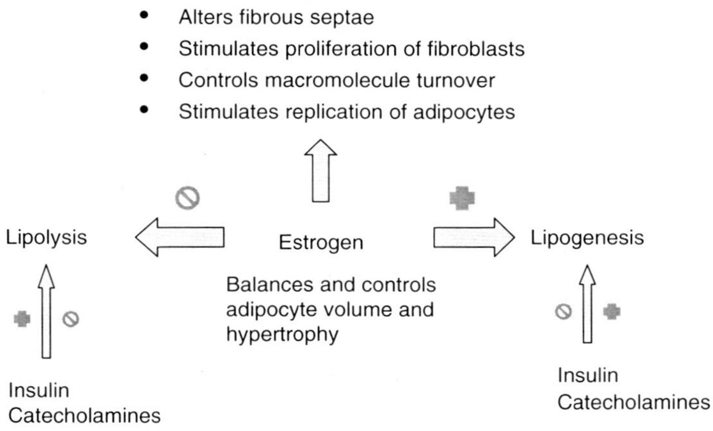

Figure 16.4 The role of estrogen in the onset of cellulite.

with apoptosis), two adipocyte processes that are critical to the metabolic functions of adipocytes that occur throughout the life of the cell. Although the balance of lipogenesis and lipolysis varies according to gender, race, and age, it is important to note that this balance is the ultimate determinant of adipocyte volume and potential adipocyte hypertrophy [15]. Estrogens have a potent influence on the anatomical enlargement of adipocytes in women. The latter may be a reason for the onset of cellulite at puberty, as well as the proliferation of cellulite during pregnancy, estrogen therapy, and menstruation.

The onset of cellulite at menstruation and the exacerbation of cellulite with age may be due to the secretion of matrix metalloproteinases (MMPs) such as collagenase and gelatinase. Endometrial cells must secrete these two enzymes to allow menstrual bleeding to occur. Additionally, endometrial collagenase secretion also causes dermal collagen breakdown. With extended cyclical collagenase release (and with increased age), more and more dermal collagen is torn down, explaining the worsening of cellulite observed with age [5].

The mechanism of cellulite initiation by estrogens has recently been found to be more involved than simply lipogenesis and lipolysis. Estrogen perpetuates cellulite formation by stimulating the proliferation of fibroblasts, controlling macromolecule turnover, and stimulating the replication of adipocytes. Estrogens are also responsible for altering glycoaminoglycans (GAGs) and collagen, which aggravates fibrosclerosis in connective tissue fibrous septae [16].

However, estrogen is not the only perpetrator of cellulite formation. Other hormones, such as insulin and catecholamines stimulate or inhibit lipolysis or lipogenesis, thus altering adipocyte metabolism. Finally, thyroid hormones are known to contribute to lipolysis enhancement within the fatty tissue as well [16] (Fig. 16.4).

16.3 Methods for Cellulite Measurement

The severity of cellulite is ranked according to a universal scale (Table 16.2). While this scale serves as a clinical evaluation of cellulite, there is also a broad spectrum of noninvasive and invasive methods available to assess this condition. A description of these tools follows:

Table 16.2 Cellulite Grading System (I–IV)*

Grade I: No or minimal cellulite-based on observation on gluteal muscle contraction, the 'pinch test', or standing.
Grade II: Irregular skin topography. Cellulite is enhanced by gluteal contraction or pinching. Skin pallor or decreased temperature and sensation often evident.
Grade III: Classic orange peel dimpling, "peau d'orange," at rest on observation. Small subcutaneous nodularities may be present and palpable.
Grade IV: Characteristics of Grade III, as well as more severe puckering and palpable nodules on inspection.

*The Cellulite Grading System is used to classify the severity of cellulite in patients with this condition. More severe forms of cellulite can cause tissue alterations, such as lipodystrophic and fibrosclerotic degeneration.

16.3.1 Simple Observation

Observation with tangential lighting is one of the most common methods used to measure cellulite due to its cost effectiveness, reproducibility, and relative ease. Such observation involves direct or photographic examination and visualization of the skin. Clinicians look for dimpling or puckering of the skin and palpate for subcutaneous nodules. This method is often best achieved in a dark room with tangential lighting where shadows can be created to view subtle dermal surface elevations and depressions [17].

For example, Rao, et al. took high-quality digital photographs with the Fuji S1 Twinflash camera system (Canfield Inc, Fairmont, New Jersey) in order to assess cellulite on the posterior and lateral thighs. Subjects stood against a black backdrop, and the Verilux Happy Lite system was shown at the level of the subject's knees in order to enhance the quality of the image.

16.3.2 Thigh Circumference

Thigh circumference is another common method used to evaluate cellulite due to its cost effectiveness. A flexible ruler is wrapped around the predetermined sites of the subject's thigh. For example, in one study, thigh circumference measurements were taken for the lower and upper thigh at 18 and 26 cm from the superior pole of the patella, respectively [18]. Although this instrument is reproducible and easy, it is not always a direct indication of cellulite. Other factors, such as edema from congestive heart disease, inflammation from external sources, or trauma, can causes changes in thigh circumference [17] (Fig. 16.5).

16.3.3 Weight or Body Mass Index

Body mass index (BMI) is a universal means of estimating body-fat composition based on a population average. However, since cellulite is not synonymous with obesity, BMI is not an effective tool for measuring cellulite (Fig. 16.6).

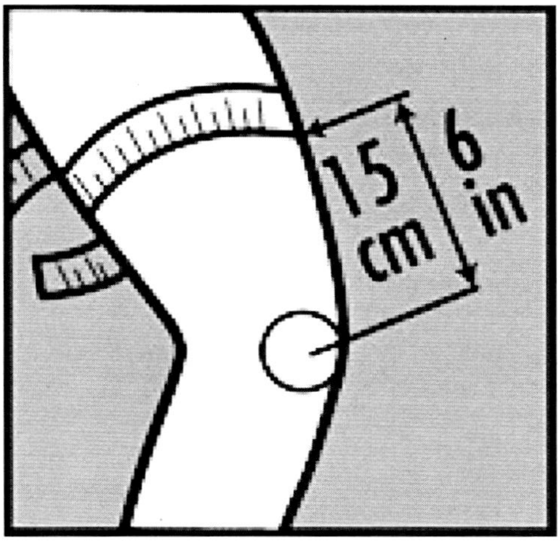

Figure 16.5 Thigh circumference is a method used to evaluate cellulite due to its cost effectiveness. Source: betterbraces.com.

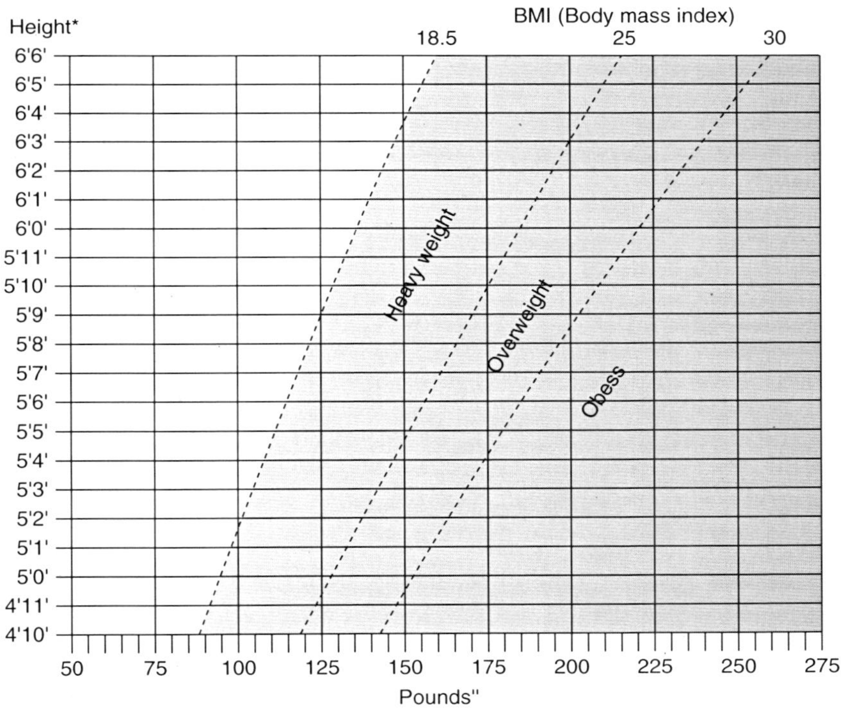

Figure 16.6 BMI is a universal means of estimating body fat composition. However, since cellulite is not synonymous with obesity, BMI is not an effective tool for measuring cellulite. Source: Vince Ferguson, http://sixweeks.com/BMIgraph.html.

16.3.4 Skin Elasticity

The resilience of the dermis can be evaluated with a suction elastometer. By measuring skin tension, clinicians can gauge the amount of intact connective tissue and, consequently, the amount of cellulite present. Although this tool, hypothetically, may be plausible, clinical evidence of its effectiveness has not yet been seen [18] (Fig. 16.7).

16.3.5 Electrical Conductivity

Electrical conductivity measures tissue resistance to electron flow, which is theoretically said to determine percentages of body composition such as water, lean mass, and fat mass. The efficacy of this tool is also under consideration [17].

16.3.6 Deep Skin Biopsy of Cellulitic Areas and Tissue Analysis

This invasive approach determines the histological architecture of cellulite. The problem with this tool is that the histologic results may not correlate with the clinical appearance or outcome.

16.3.7 High Frequency Magnetic Resonance Imaging

This is the most recent approach and a very efficient tool for evaluating cellulite. Magnetic Resonance Imaging has the ability to distinguish between the physiology of different skin

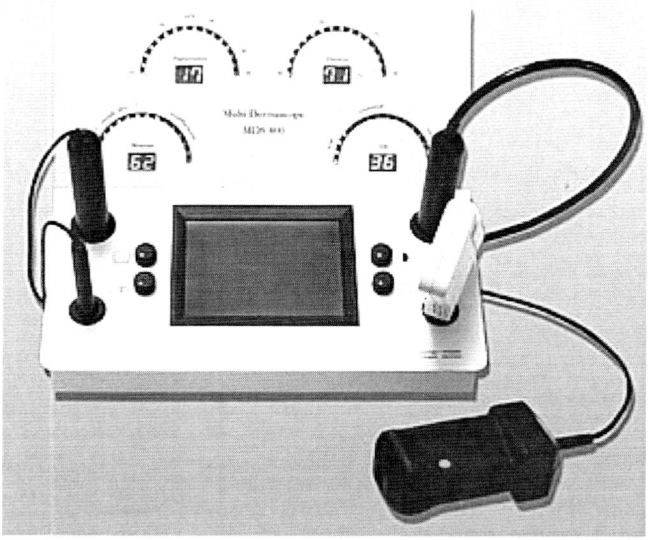

Figure 16.7 Skin elasticity is measured with a skin elasticity machine, which can gauge the relative amount of cellulite present. Source: Courage + Khazaka, Germany, http://www.pastiche.net.nz/technology.html.

layers, specifically the thickness of the skin, as well as the thickness and the volume of the hypodermis. Querleux et al. were the first to utilize MRI to detect Camper's fascia, as well as the superficial and deep hypodermal adipose layers [10]. Additionally, this technology has confirmed that a distinguished characteristic of cellulite is a thicker deep adipose layer.

16.3.8 High Frequency Ultrasonography

Ultrasound is also an accurate tool used to measure dermal thickness. This technology can also detect the presence of deep indentations of adipose tissue in the skin, as demonstrated earlier by histology. In a study by Kligman et al., dermal thickness was measured using a B-scan 20 MHz device [19].

The above forms of measurement vary in their effectiveness, but all have been used in studies that will be mentioned throughout this chapter. Some of these tools can be viewed as subjective, while others serve as purely objective evaluators of cellulite. For instance, simple observation, photographs, and patient and physician questionnaires are a strictly subjective form of measurement. In contrast, MRI, thigh circumference, and ultrasonography are more objective in their nature. The advantages and disadvantages of these two classes of assessment will emerge as the efficacy of various treatment plans for cellulite is scrutinized. However, despite multiple therapeutic modalities for cellulite, there is little scientific evidence that any of the treatments under question are advantageous. Therefore, in order to assess the precision of the data, it is best to examine the means by which cellulite in these studies is being measured. Given that subjective measurement may be more prone to experimental bias and error, the precision of these results may need to be inspected with greater attention than purely objective data. Several researchers agree that the best standardized and objective tools to assess cellulite are MRI and ultrasound [20].

16.4 Photothermal Therapy

There are two novel reported mechanisms that have demonstrated the ability to clinically modify the skin's connective tissue: laser and pulsed light (optical energy) and radiofrequency. While these methods differ primarily in the way they generate heat within the adipose tissue, they both are able to produce temperatures (65–75°C) high enough to cause connective tissue remodeling and reduction [21] (Table16.3).These two new forms of therapy have been combined with other modalities for reducing cellulite and have also been used on their own. Photothermal therapies are considered the future tools of aesthetic dermatology. The most efficacious and commonly used devices are discussed here.

16.4.1 Accent™ Radiofrequency System (Alma Lasers Inc, Ceasaria, Israel; Fort Lauderdale, Fl)

Electric currents have been used in medicine for various purposes for more than a century. Lower frequencies have been used in biostimulation for atrial fibrillation, for example, due to their ability to cause spasms in muscle tissue. Radiofrequency (RF) is defined by

Table 16.3 Techniques to Control Optical Energy and Radiofrequency Devices*

• Ions or dipole molecules (RF)
• Natural biologic chromophore (optical energy)
• Method of electromagnetic energy application
• Electromagnetic power (watts)
• Exposure time (sec)
• Tissue resistance (RF)
• Tissue impedance (RF)

*Laser/pulsed light (optical energy) and radiofrequency devices can thermally modify the connective tissue of the skin [22].

high frequency electric currents (0.3–100 MHz), and is efficacious in heating tissue in electrosurgery.

Since weakened connective tissue and diminished microcirculation play a role in the pathogenesis of cellulite, long-term correction of cellulite may be possible through these two instigators. The Accent™ (Alma Lasers, Ft Launderdale, FL) RF system is a device employed to reduce cellulite via volumetric thermotherapy. The device was approved by the FDA in April 2007 "for use in dermatologic and general surgical procedures for the noninvasive treatment of wrinkles and rhytids using combined treatment with unipolar and bipolar". Although the FDA has approved this device for treatment of wrinkles, it is also marketed toward cellulite reduction.

The mechanism of action of radiofrequency waves is as follows. It has been hypothesized that thermal energy is capable of denaturalizing collagen. Through heating collagen at 65+ degrees Celsius, the protein's bonds break and a transformation from a highly organized crystalline structure to a disorganized gel occurs. While optical energy is dependent on the chromophore concentration of the skin in order to attain thermal tissue destruction, RF is reliant on the electrical properties of tissues [21]. There have been claims that RF is capable of both increasing adipose tissue disruption while noninvasively removing fat deposits in the absence of fat necrosis [21].

The Accent system has Bipolar and Unipolar RF handpieces and a RF generator (40.68 MHz) that improves cellulite through three mechanisms: (1) deep dermal heating, causing neocollagenesis and remodeling; (2) adipocyte apoptosis and disruption of fat cells via thermal effects; (3) lymphatic drainage and increased blood circulation. The Unipolar handpiece penetrates the tissue to a depth of 20 mm, heating the subcutaneous tissue with a RF power of 100–200 watts, and the Bipolar handpiece penetrates the skin superficially to a depth of 2–4 mm with a RF power of 60–100 watts. The Unipolar device also includes a thermo-electric coupling cooler (TEC) [21].

Before surgery, the treatment area is demarcated with 5 cm × 6 cm grids (15 seconds exposure time) or 10 cm × 6 cm grids (30 seconds exposure time), and the skin is lubricated with treatment oil to prevent friction between the applicator and the skin. The system is set at the appropriate watts and time settings, and the TEC is turned on. The technician begins application with horizontal sweeping strokes followed by vertical strokes until the preset time has expired. Epidermal temperature is monitored with a laser thermometer to tract treatment phases. In Phase I (Up-Slope) the patient's skin is monitored until it reaches a temperature of ~410°C, after which Phase II (Maintenance) commences. During Phase II, the time of exposure and the energy level are reduced by 10–15% and three to five passes should be applied. Common side effects include patient perception of pain from the heat and erythema. After treatment, the patient is asked to rest for 10 minutes and is then free to go. The recommended treatment regimen is composed of six treatments, once every two weeks (Fig. 16.8).

There are several advantages to the Accent procedure. It is fast, virtually pain- free, and requires little to no recovery time. Several studies have validated the efficacy of the Accent RF system. In a study by del Pino et al., 26 healthy female patients aged 18–50 years received two treatment sessions 15 days apart, and real-time ultrasound scanning was used to assess the results. The ultrasound scanning device is able to determine the distance between the stratum corneum and the Camper's fascia and from the stratum corneum and the muscle. The results illustrated that 68% of the treated patients experienced a 20% contraction of subcutaneous adipose tissue volume. In addition, patient self-assessment revealed that most women were satisfied with their results, and the women who were the most content were those who had the highest grade of cellulite defects. The authors concluded that the ultrasound was the best tool for assessing the effects of the RF device, because it gave an image of the alterations in the subcutaneous tissues. The results of their study found an increased echodensity of the connective tissue structures, demonstrating that RF, as hypothesized, affects the collagen of the subcutaneous tissues [21].

Overall, the Accent RF system exhibits promising results with minimal side effects. Future clinical trials may validate its efficacy and make its use more widespread.

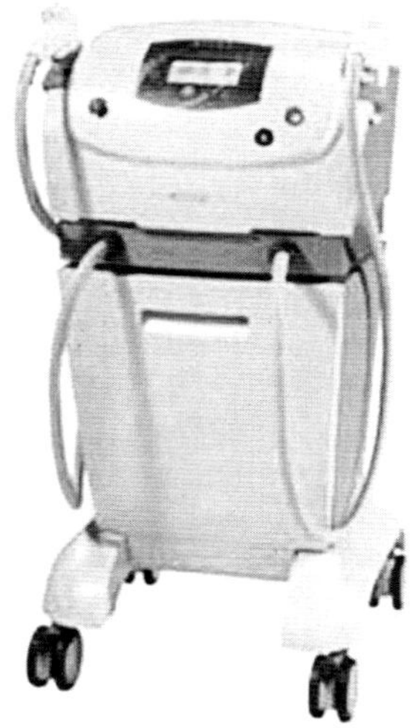

Figure 16.8 The Accent Radiofrequency® system (Alma Lasers, Inc, Ceasuria, Israel; Fort Lauderdale, FL) is a relatively new FDA-approved device employed to reduce cellulite via volumetric thermotherapy using unipolar and bipolar handpieces.

16.4.2 VelaSmooth™ System (Syneron Medical Ltd, Yokneam, Israel)

The next frontier in the non-invasive treatment of cellulite may be lasers. One of these systems, Velasmooth™ (Syneron Medical Ltd, Yokneam, Israel), was approved by the FDA in July 2007 for "temporary reduction of thigh circumference". This system combines RF, 700 nm near-infrared light (IR), and negative tissue massage. The IR aspect of the device reduces the occurrence of adverse effects such as skin pigmentation and scarring, while the RF energy can penetrate deeper layers of the skin (a feature that IR alone does not possess). In addition, the combination of the RF and the IR is thought to create heat that aids in the dissociation of oxygen from oxyHb, thereby increasing the amount of oxygen available in adipose tissue for fat metabolism. The negative tissue massage contributes to the process by aiding in the physical breakdown of fat clusters and in promoting lymphatic drainage [23] (Fig. 16.9).

Velasmooth is recommended for healthy women with a 20–30% BMI. The recommended treatment regimen includes bi-weekly treatments (45–60 minutes each) for a total of 8–12 sessions. With maintenance, the results of Velasmooth are expected to last several weeks to months. However, without maintenance, results have been found to fade after 6–8 weeks.

Prior to treatment with Velasmooth, the designated skin regions are first rinsed with soap and water to wash off any lotion or other debris. The skin is then hydrated with a conductive fluid. The Velasmooth device is programmed to the following settings: IR light of 20 W (700–1500 nm); RF power of 20 W (1 MHz); vacuum suction level of 200 mbar (750 mmHg negative pressure). The affected area is treated by moving the handpiece of the Velasmooth device in a backward and forward motion. The typical protocol calls for 4–6 passes of the device to be delivered to the area over a 30-minute period. Patients have reported a feeling of a mild regional heating sensation during the massage portion of the treatment.

In a study by Sadick and Mulholland, 35 female subjects received 8–16 bi-weekly treatments with the Velasmooth device. Circumferential thigh measurements were taken before treatment and four weeks after the initial treatment, and photographs for review by blinded physicians were also obtained. The results indicated that after 4 weeks of treatment, 70% of all patients experienced a reduction in thigh circumference, and after the full 8 weeks of treatment, all study patients showed a reduction in thigh circumference and cellulite appearance. Physician reviewers found that patients showed an average of 40% improvement in their cellulite, and 90% of the patients would highly recommend the treatment to other women. Finally, a histological assessment of the treated area noted no epithelial or mesenchymal morphologic damage, indicating that Velasmooth does not significantly damage deeper skin structures [24].

Sadick and Magro treated 16 women with cellulite with the Velasmooth system bi-weekly for 6 weeks (12 total treatments). Thigh circumferential measurements of the treated leg and the contralateral control thigh were taken before and after treatment. Results demonstrated that the overall thigh circumference decreased in 71.87% of the treated legs, with a mean decrease of 0.53 cm in the upper thigh. Significant appearance of cellulite was noted [23]. In a similar study by Alster and Tanzi, 18 out of 20 adult women (90%) treated with Velasmooth for cellulite noticed overall clinical improvement. The mean reduction in circumferential thigh measurements was reported to be 0.8 cm [25].

In another clinical trial, the efficacy of a combination of multiple therapies was tested. Kulick evaluated a device that combined RF, infrared energy, and mechanical rollers (ELOS

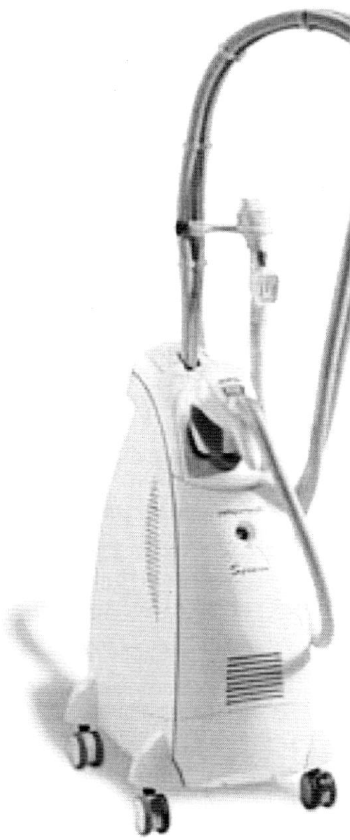

Figure 16.9 The VelaSmooth™ System (Syneron Medical Ltd, Yokneam, Israel; http://www.syneron.com/Solutions/Physicians/Products/velasmooth.html) combines radiofrequency, 700 nm near-infrared light, and negative tissue massage for the reduction of cellulite.

technology). Treatment lasted for 15 minutes and machine settings were set as follows: RF (1 MHz–20W), IR light (700–1500 nm–12.5W) and suction (750 mmHg negative pressure/250 ms on–150 ms off). After treatment, compressive stockings were worn for 48 hours. Patients received two treatments per week for four weeks (eight treatments total). Patients were given follow-up self-assessments at 3 and 6 months, and blinded physicians evaluated photographs at these times as well. Patient self-evaluation revealed 75% perception of improvement at 3 months, and 50% at 6 months. Physician evaluations disclosed a greater than 50% improvement at 3 months and 50% improvement at 6 months [26].

Ongoing clinical trials of the Accent RF system have demonstrated that the device has a high safety and efficacy. Del Pino et al. reported that erythema was a possible short-term side effect that resolved within 15–30 minutes posttreatment. Persistent erythema has been shown to resolve within 24 hours [21,26]. Patient self-perception of transient pain has also been noted during treatment or immediately following treatment. Kulick reported bruising, which resolved within one week post-treatment. After the first and second treatment sessions, it was observed that in subsequent treatment sessions, patients experienced a reduction in the extent of bruising [26].

It is still unknown as to what extent clinical improvement in cellulite via Velasmooth is due to the combination of RF and IR versus mechanical massage. Endermologie, a device exerting a mechanical massage effect, has been shown in past studies to be an effective way of reducing the appearance of cellulite. The studies mentioned earlier demonstrated a reduction in cellulite with the Velasmooth system, which includes a mechanical manipulation component. There are no known studies to date, that have directly compared the efficacy of these two treatments. Perhaps future studies could determine to what extent the massage treatment versus the RF/light energy treatment contributes to cellulite reduction in Velasmooth (Fig. 16.10).

Like other treatment modalities, there are reported side affects of Velasmooth. A small number of the patients from Sadick and Mulholland's study experienced temporary swelling and discomfort, and two patients reported local crusting that resolved after 72 hours. The authors speculated that these side effects may have been due to coupling of electrodes, or improper vacuum contact [24]. Less than half of Sadick and Magro's subjects reported mild redness, discomfort, and mild bruising, edema, erythema, hyperpigmentation, and hypopigmentation within the first two weeks posttreatment. All these symptoms subsided on their own without any additional interventions [23]. Overall, although the clinical studies on the Velasmooth device are still ongoing, the results thus far appear to be efficacious.

16.4.3 TriActive Laser (Cynosure, Chelmsford, MA)

The second major FDA-approved laser technology for reducing cellulite is the TriActive laser (Cynosure, Chelmsford, MA), which combines six near-infrared diode laser energy (808 nm) and mechanical massage with superficial cooling. The FDA approved this device in 2003, stating its intended use is "indicated for minor muscle aches, pain, and spasm, and for improvement in local circulation and reduction in the appearance of cellulite". This system has been shown to produce positive clinical effects on cellulite after a total of 12–16 bi-weekly treatments. The recommended protocol is composed of treatments three times a

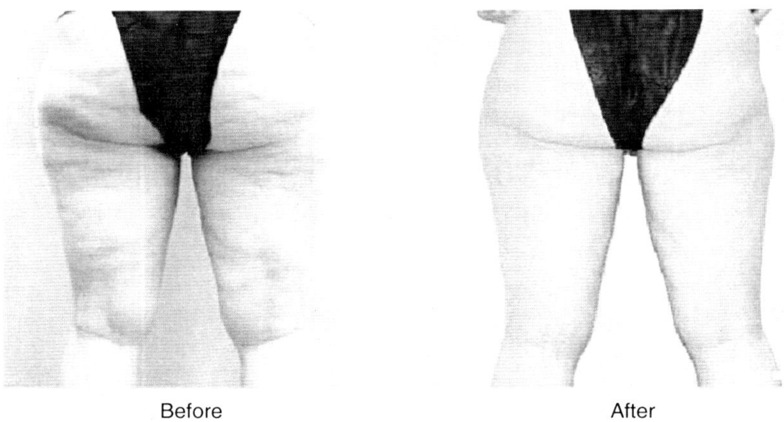

Figure 16.10 Before and after photos of a patient who underwent treatment for cellulite with the VelaSmooth™ System. Photo courtesy of Dr. Boey, Syneron Medical Ltd.

week for two weeks, followed by five weeks of biweekly treatment [14]. This technology was designed with the intention of tightening the skin by increasing lymphatic drainage, increasing superficial blood flow, and stimulating underlying muscles and fascia. With the addition of a laser, TriActive stimulates microcirculation, whereas localized cooling reduces fluid retention (Table 16.4).

TriActive does not have to be administered by a physician because it is a Class I device. Massage therapists are often trained and hired to perform the treatment. The price for treatment is $2000–2500 for a series of 16 treatments. TriActive has been found to improve the results of liposuction and is often used in concordance with this procedure.

The TriActive handpiece is equipped with six 808 nm diode lasers, a cooling face, and a suction port. The suction port for mechanical massage is controlled by the frequency (Hz), which measures the number of aspirations per second, and the duty cycle, which can be manipulated to increase or decrease the intensity of the massage. The function of the cooling face and the diode lasers are outlined in the preceeding paragraphs [27] (Fig. 16.11).

Since TriActive is a fairly new technology, there is little evidence supporting its efficacy, and clinical trials are ongoing. However, in a study by Frew and Katz, 10 female patients with cellulite, ages 18–60 years, were treated with TriActive bi-weekly for a total of sixteen treatments. Half of the affected body areas were treated with TriActive (diode laser, contact cooling, and suction) and the contralateral sides were treated with contact cooling and suction only. Digital photographs reviewed by blinded physicians and patient surveys revealed that 90% of patients reported improvement, and 80% were satisfied and would continue treatment. The only reported side effects to treatment were minimal bruising, which resolved after a week or two [28]. Similarly, Pabby and Goldman administered 10 TriActive treatments over a 5-week period to 11 cellulitic female patients. Blinded evaluation of pre- and posttreatment photos and thigh circumference measurements revealed that all subjects exhibited observable improvement in cellulite and average thigh circumference reduction was 0.83 cm [27]. Like the VelaSmooth system, the long-term efficacy of the TriActive laser is still under review. However, without maintenance, results will diminish within 6–8 weeks (Fig. 16.12).

The VelaSmooth system and the TriActive are the most efficacious treatments for cellulite reduction. Not only are these two treatments relatively new, but, unlike other forms of treatment for cellulite, they both have minimal side effects. In a study by Nootheti, et al.,

Table 16.4 VelaSmooth™ versus TriActive Laser Systems*

VelaSmooth
Infrared light (700–1500 nm)
Radiofrequency (1 MHz)
Vacuum suction massage (750 mmHg)
TriActive
Infrared light (808 nm)
Vacuum suction massage
Superficial cooling

*VelaSmooth and TriActive laser systems differ only in that VelaSmooth incorporates radiofrequency into its protocol, while TriActive incorporates superficial cooling.

Figure 16.11 The TriActive™ handpiece is equipped with six 808 nm diode lasers, a cooling face, and a suction port. Photo courtesy of Cynosure, Inc.

the efficacy of VelaSmooth and TriActive were tested against one another. In a randomized controlled trial, twenty female patients were treated twice a week for six weeks with VelaSmooth on one side of their body and TriActive on the contralateral side. Patient evaluation consisted of blinded review of photographs and circumferential thigh measurements before and after treatment. The circumferential upper thigh results revealed a 28% versus a 30% improvement rate in VelaSmooth verus TriActive, respectively. These differences in treatment efficacy were not significant ($p>0.05$). Blinded review of photographs found that 25% of subjects showed an improvement in cellulite appearance and 55–75% of subjects showed a change in cellulite grade. Again, there was no significant difference in the photo results of the VelaSmooth versus TriActive. While no significant difference was found in the efficacy of these two laser systems, there was a difference in the side effects of the treatments. The bruising incidence and intensity was 30% higher in legs treated with VelaSmooth compared to TriActive. While 7 out of 20 subjects reported bruising in the VelaSmooth leg, only 1 subject reported bruising in the Triactive leg [29].

In conclusion, the safety and efficacy of these methods for reducing cellulite must continue to be evaluated. The efficacy of all of these systems has not yet been perfected, and safety is still a concern. The objective and subjective ways in which clinical trials are evaluating the efficacy of different treatments must be monitored for bias. Future combinations of light energies, radiofrequencies, topical treatments, subcutaneous injections, and/or mechanical massage may provide the foolproof answer for improving cellulite appearance (Table 16.5). The long-term efficacy of all of these treatments is promising with maintenance, but without maintenance, positive results have been shown to dissipate over time. Cellulite has been an aesthetic issue that women have been trying to combat with different

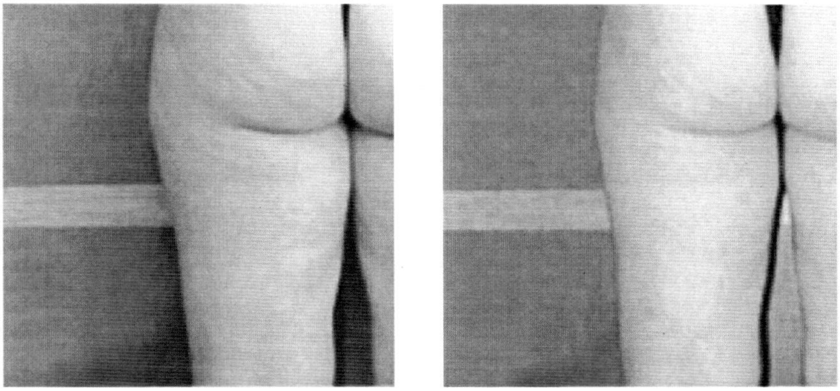

Figure 16.12 Before and after photos of a patient who underwent treatment for cellulite with the TriActive™ System. Photo courtesy: Cynosure, Inc.

Table 16.5 Body Shaping Technologies of the Future That Can Offer Other Alternatives for Cellulite Reduction

Manufacturer	Device	Energy Source	Mechanism of Action	Regulatory Status (Europe and USA)
Alma Lasers Inc	Accent XL	Unipolar and Bi-polar Radiofrequency	Heat generated by tissue resistance to radiofrequency-induced current.	CE cleared. FDA 510k
Syneron Medical Ltd	VelaSmooth	Vacuum coupled Bi-polar Radiofrequency, Infared Light	Tissue manipulation, heat generated by tissue resistance to radiofrequency-induced current, infrared- induced heat for skin tightening	CE cleared FDA cleared
Cynosure, Inc	TriActive	Laser plus vacuum massage	Cellular stimulation plus vacuum massage	FDA 510k
DEKA/ Cyanosure	SmartLipo	1064 nm Nd:YAG laser	Laser assisted destruction of fat cells, coagulation of small blood vessels, collagen shrinkage	CE cleared. FDA cleared
DermaMed International, Inc.	C-Sculpt	LED, cooling, and massage	Cellular stimulation plus vacuum massage.	
General Project	Slim Project	Computerized vacuum assisted massage	Tissue manipulation.	
LipoSonix, Inc	LipoSonix System	Ultrasound	Fat cells disrupted by sound waves.	CE pending FDA pending

(Continued)

Table 16.5 Body Shaping Technologies of the Future That Can Offer Other Alternatives for Cellulite Reduction (Continued)

Manufacturer	Device	Energy Source	Mechanism of Action	Regulatory Status (Europe and USA)
Pollogen Ltd.	Regen	Monopolar plus Bi-polar	Heat generated by tissue resistance to radiofrequency-induced current.	CE cleared
SmoothShapes, Inc.	SmoothShapes100	650 nm and 900 nm Diode Laser	Tissue manipulation by vacuum massage, laser assisted reduction of fat layer.	CE cleared FDA cleared
Syneron Medical Ltd	VelaShape	Bi-Polar RF, Infared Light and Mechanical Massage	Increase in localized metabolism adipose tissue	CE cleared FDA cleared
Thermage Inc.	ThermaCool System with 3.0 cm^2 ThermaTip DC	Monopolar Radiofrequency	Heat generated by tissue resistance to radiofrequency-induced current.	FDA cleared for non-invasive treatment of wrinkles and rhytides; temporary improvement in appearance of cellulite (not for face of shallow tissue areas)
ThermaMedic	ThermaLipo	RF-AMFLI technology	Heat generatd by tissue resistance to radiofrequency-induced current	
Ulthera, Inc.	Ulthera System	Ultrasound	Tissue heated by sound waves, stimulate new tissue production	CE cleared FDA pending.
UltraShape Ltd.	CONTOUR 1 ver2	Ultrasound	Mechanical and sound wave- induced disruption of fat cells	CE cleared

methods for decades. It will be interesting to follow the future innovations of surgeons and dermatologists in the search for the perfect 'cure' to this widespread issue.

References

1. Terranova F, Berardesca E, and Maibach H: Cellulite: Nature and aetiopathogenesis. *International Journal of Cosmetic Science*, 2006; 28:157–167.
2. Nurnburger F and Muller G: So called cellulite: an invented disease. *The Journal of Dermatologic Surgery and Oncology*, 1978; 4:230–234.

3. Goldman MP: Cellulite: A review of current treatments. *Journal of Cosmetic Dermatology*, 2002; 15:17–20.
4. Bacci PA, Allegra C, Albergati F, Brambilla E, Botta G, and Mancini S: Randomized, placebo-controlled double-blind clinical study of the efficacy of a multifunctional plant complex in the treatment of so-called 'cellulite'. *International Journal of Cosmetic Surgery and Aesthetic Dermatology*, 2003; 5:53–68.
5. Draelos ZD: The disease of cellulite. *Journal of Cosmetic Dermatology*, 2005; 4:22–222.
6. Quatresooz P, Xhauflaire-Uhoda E, Pierard-Franchimont C, and Pierard GE: Cellulite histopathology and related mechanobiology. *International Journal of Cosmetic Science*, 2006; 28: 207–210.
7. Pierard GE: Commentary on cellulite: Skin mechanobiology and the waist-to-hip ratio. *Journal of Cosmetic Dermatology*, 2005; 4:151–152.
8. Rosenbaum M PV, Hellmer J, et al.: An explanatory investigation of the morphology and biochemistry of cellulite. *Plastic & Reconstructive Surgery*, 1998; 101:1934–1939.
9. Pierard GE NJ and Pierarf-Franchimont C: Cellulite: From standing fat herniation to hypodermal stretch marks. *American Journal of Dermopathology*, 2000; 22:34–37.
10. Querleux B, Cornillon C, Jolivet O, Bittoun J, Querleux B, Cornillon C, Jolivet O, and Bittoun J: Anatomy and physiology of subcutaneous adipose tissue by in vivo magnetic resonance imaging and spectroscopy: Relationships with sex and presence of cellulite. *Skin Research & Technology*, 2002; 8:118–124.
11. Bassas-Grau E and Bassas-Grau, M.: Consideraciones clinicas etiopatogenicas y terapeuticas sobre la mal nomada 'cellulitis'. *Annals of Medicine*, 1964; 16:2–17.
12. Kligman AM: Cellulite: Facts and fiction. *Journal of Geriatric Dermatology*, 1997; 5:136–139.
13. Richelsen B: Increased alpha-2 but similar beta-adrenergic receptor activities in subcutaneous gluteal adipocytes from females compared with males. *European Journal of Clinical Investigation*, 1986; 16:302–309.
14. Avram MM: Cellulite: A review of its physiology and treatment. *Journal of Cosmetic & Laser Therapy*, 2004; 6:181–185.
15. Rotunda AM, Avram MM, and Avram AS: Cellulite: Is there a role for injectables? *Journal of Cosmetic & Laser Therapy*, 2005; 7:147–154.
16. Rossi AB and Vergnanini AL: Cellulite: A review. *European Academy of Dermatology and Venereology*, 2000; 14:251–262.
17. Rao J, Gold MH, and Goldman MP: A two-center, double-blinded, randomized trial testing the tolerability and efficacy of a novel therapeutic agent for cellulite reduction. *Journal of Cosmetic Dermatology*, 2005; 4:93–102.
18. Rao J, Paabo KE, and Goldman MP: A double-blinded randomized trial testing the tolerability and efficacy of a novel topical agent with and without occlusion for the treatment of cellulite: A study and review of the literature. *Journal of Drugs in Dermatology*: 2004; 3:417–425.
19. Kligman A, Pagnoni A, and Stoudemayer T: Topical retinol improves cellulite. *Journal of Dermatological Treatment*, 1999; 10:119–125.
20. Rawlings AV: Cellulite and its treatment. *International Journal of Cosmetic Science*, 2006; 28: 175–190.
21. Emilia del Pino M, Rosado RH, Azuela A, Graciela Guzman M, Arguelles D, Rodriguez C, and Rosado GM: Effect of controlled volumetric tissue heating with radiofrequency on cellulite and the subcutaneous tissue of the buttocks and thighs. *Journal of Drugs in Dermatology*, 2006; 5:714–722.
22. Brown A and de Almeida O: Novel radiofrequency CRF device for cellulite and body reshaping therapy. Almalasers.com, Accessed 6/26/07.
23. Sadick N, Magro C, Sadick N, and Magro C: A study evaluating the safety and efficacy of the velasmooth system in the treatment of cellulite. *Journal of Cosmetic & Laser Therapy*, 2007; 9:15–20.
24. Sadick NS and Mulholland RS: A prospective clinical study to evaluate the efficacy and safety of cellulite treatment using the combination of optical and RF energies for subcutaneous tissue heating. *Journal of Cosmetic & Laser Therapy*, 2004; 6:187–190.

25. Alster TS and Tanzi EL: Cellulite treatment using a novel combination radiofrequency, infrared light, and mechanical tissue manipulation device. *Journal of Cosmetic & Laser Therapy*, 2005; 7:81–85.
26. Kulick M: Evaluation of the combination of radio frequency, infrared energy, and mechanical rollers with suction to improve skin surface irregularities (cellulite) in a limited treatment area. *Journal of Cosmetic & Laser Therapy*, 2006; 8:185–190.
27. Goldman MP, Bacci PA, Leibaschoff G, Hexsel D, and Angelini F, eds. *Cellulite: Pathophysiology and Treatment*. London, Taylor and Francis Group, 2006, p. 327.
28. Frew K and Katz B: The efficacy of diode laser with contact cooling and suction (triactive system) in the treatment of cellulite. 13th Congress of the European Academy of Dermatology and Venereology. Juva Skin and Laser Center, NY, 2005.
29. Nootheti PK, Magpantay A, Yosowitz G, Calderon S, and Goldman MP: A single center, randomized, comparative, prospective clinical study to determine the efficacy of the velasmooth system versus the triactive system for the treatment of cellulite. *Lasers in Surgery & Medicine*, 2006; 38:908–912.

17

Treatment of Acne: Phototherapy with Blue Light

Voraphol Vejjabhinanta, Anita Singh, and Keyvan Nouri

Department of Dermatology and Cutaneous Surgery, University of Miami, Miller School of Medicine, Miami, Florida, USA

17.1 Introduction

Acne vulgaris is a common skin disease that affects most people at some time or the other during their lives. Its prevalence has been estimated to be about 85–100% in boys aged 16–17 years, and 83–85% in girls of the same age [1,2]. In fact, in the United States, it is estimated that 85–100% of all adolescents will be afflicted with this disease and approximately 25 million adults and 40 million adolescents are affected by this condition [3,4]. Even though it is common in teenagers and early adults, acne can occur in all age groups [5,6]. Twelve percent of women who were at least 25 years old had acne, and this percentage did not diminish until after the age of 44 years [7].

Acne is a multifactorial disorder of pilosebaceous units, and it affects the areas of skin with the greatest density of sebaceous follicles. These areas include the face, neck, chest,

Gurpreet S. Ahluwalia (ed.), *Cosmetic Applications of Laser and Light-Based Systems*, 341–351,

and the back. In some cases, it can occur at some pressure areas, such as the jawline (due to usage of helmet) or buttock area, or it can occur after using some chemical substances such as cosmetics or hair-styling products. In addition, acne can be caused by a variety of factors, including genetics, hormones, mechanical irritation, and chemical products and organisms.

Acne is characterized by noninflammatory, as well as inflammatory lesions. The noninflammatory lesions consist of open and closed comedones. Open comedones (blackheads) are small follicular papules containing a central black keratin plug, formed mainly due to the oxidation of the melanin pigment. Closed comedones (whiteheads) are follicular papules without a visible central plug, due to the keratin plug being trapped deep beneath the epidermal surface. Closed comedones are potential sources of follicular rupture and inflammation. Inflammatory acne is characterized by erythematous papules, nodules, and pustules [8].

The severity of the acne has been classified by the American Academy of Dermatology according to the following specifications: Mild acne is characterized by the presence of comedones, few papules and pustules (generally <10) but no nodules; Moderate acne has several to many papules and pustules (10–40) along with comedones (10–40); Moderately severe acne is characterized as the presence of <40 papules and pustules along with larger and deeper nodular inflamed lesions (up to 5); Severe acne is characterized by the presence of numerous or extensive papules and pustules, as well as many nodular lesions [9]. Although this condition will improve in most patients with time, in some it does not, and these patients have serious long-term effects from acne. These include redness, hyperpigmentation, and permanent scars (atrophic, hypertrophic, and keloids) [10], which may cause psychological problems such as social phobia, lowered self-image, and even depression in some patients [11].

The cost of acne treatment in the United States is estimated to be more than $1 billion per year, with $100 million spent on over-the-counter anti-acne agents [12]. There are many methods for acne treatment. These acne treatments can be classified into three groups: topical medications, oral medications, and surgical/physical agents [13].

The benefits of topical therapy include its easy application and the lower systemic complications or systemic side effects, which is beneficial for some patients or parents concerned about taking systemic agents. However, there are some unwanted reactions such as dryness, irritation, allergic contact dermatitis, or less efficacy when compared with systemic agents.

Oral medications are prescribed to patients with moderate to severe acne. They are more effective than topical medications; however, there is an increased risk of systemic complications/side effects. For example, tetracycline is an effective antibiotic for acne; however, its side effects include gastrointestinal disturbance, photosensitivity and teeth discoloration [14,15]. In addition, isotretionine, which is very effective in recalcitrant acne [16,17], has multiple side effects. These side effects include dry mouth, dry eyes, abnormal serum lipid profile, and its teratogenic effects to the fetus [18,19].

Surgical/physical therapies are another option for acne treatment. These therapies can be used alone, or in combination with topical/oral medications to accelerate the treatment. There are many methods for surgical/physical therapies, including microdermabration, chemical peeling with glycolic acid, salicylic acid, or tricholoacetic acid. Most of these treatments claim to treat hypercornification or follicular obstruction. Side effects include burning, a stinging sensation, irritation; and in severe cases, postinflammatory hyperpigmentation or scarring after procedures [20,21]. However, because of advancement in emulsion technology and drug-delivery systems, many cosmetic companies make these treatments in special over-the-counter formulas, which reduce concentrations and side effects, but remain still effective.

Nevertheless, there are many patients who have a contraindication to these therapies, or do not show improvement in their acne problems, and are not satisfied with the current treatment. These patients need a new modality which is safe, painless, and different from other on-going methods and is more effective. These are the reasons why phototherapy has evolved (Table 17.1).

17.2 Etiology

The etiology of acne is complex and multifactorial. There are four key components that contribute to the development of acne. These include hypercornification of follicular epithelium at the infra-infundibulum with the development of a keratin plug, blocking the outflow of sebum to the skin surface; hyperplasia of the sebaceous glands and hypersecretion of sebum with the onset of puberty, or increased activity due to the stimulation of androgen hormone; lipase-synthesizing bacteria (*Propionibacterium acnes*) colonizing the upper and mid-portion of the hair follicle, converting lipids within sebum to proinflammatory fatty acids; and finally, immune response and induction of inflammation in the follicle associated with the release of cytotoxic and chemotactic factors (Fig. 17.1) [22–25].

Follicular hypercornification or hyperkeratinization is an abnormal proliferation of follicular epithelium. This epithelium sheds and mixes with other materials, such as sebum and bacteria, and collects within the follicular canal, forming a microcomedone. When these materials plug the drainage of the pilosebaceous unit for awhile, debris will accumulate in this canal and enlarge, forming a macrocomedone. This process is believed to be the most important in the pathogenesis of acne. There have been many effective modalities to treat this condition, that is, using keratolytic agents such as salicylic acid, glycolic acid, and retinoic acid to control this condition. However, it takes several weeks to get a good result, and some of them can cause irritation to the skin.

Hyperfuncional sebaceous glands are the second-most important feature in the pathophysiology of acne. Hyperfunctional sebaceous glands lead to increased sebum production in response to stimulation of the androgen hormone. High levels of this hormone, especially during puberty, can produce a greater amount of sebum. A mixture of debris and a greater amount of sebum can plug the follicle and create the appropriate environment for *P. acnes*. Treatment of hyperfunctional sebaceous glands includes isotretinoin and anti-androgens. These are effective medications; however, systemic side effects and teratogenicity have to be considered.

Table 17.1 Reasons Why Phototherapy Developed

- Some patients (or their parents) do not want to take systemic drugs or are reluctant to use oral medications
- Other treatments did not work
- Need new modalities
- Need a treatment that looks professional or high-tech
- Reduce drug resistance

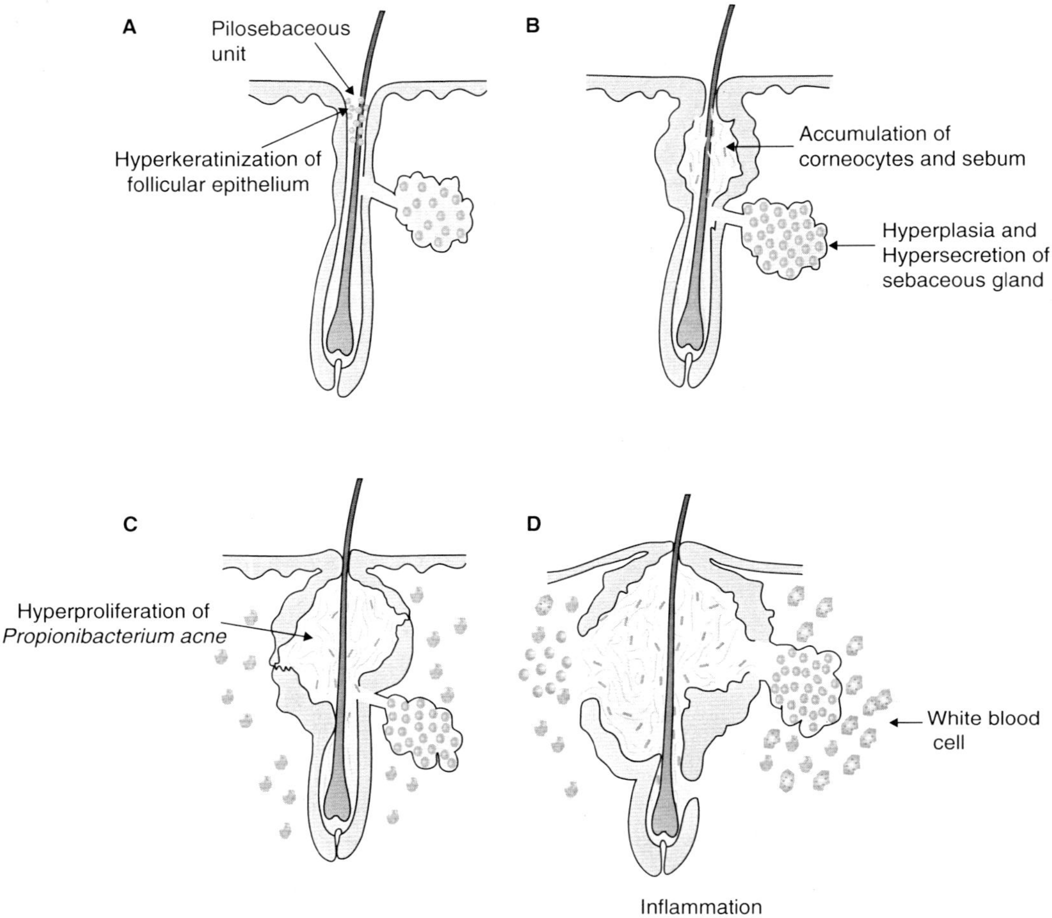

Figure 17.1 Pathogenesis of acne: (A) hyperkeratinization of follicular epithelium; (B) hypersecretion of sebaceous glands; (C) hyperproliferation of *Propionibacterium acnes*; (D) inflammation.

P. acnes is a gram positive, anaerobic bacteria which is usually a part of the normal flora of the skin. *P. acnes* can change triglycerides, which are the main component of sebum, to free fatty acids. These free fatty acids can trigger inflammation. Treatment of *P. acnes* includes both topical and oral antibiotics; however, resistance to these medications is a problem for treatment.

Perifollicular inflammation can occur due to a response to the overgrowth of *P. acnes*, free fatty acid which leaks from the follicle, and the follicular rupture that allows the contents (cell debris, bacteria, and sebum) into the adjacent dermis [26]. Although corticosteroids can control inflammation, systemic forms can depress immunity, and topical forms can cause steroid-induced acne. If we can normalize hypercornification or rid the area of *P. acnes*, we can control the acne.

Acne that is left untreated may cause excoriation, secondary bacterial infections (e.g. gram positive or gram negative bacteria), pigmentary alterations, prolonged erythema, scars (atrophic and hypertrophic/keloids), and psychological problems. Currently, there are many physical modalities for acne treatment. The blue-light system is widely used and accepted as the simple, safe, and painless method for the improvement of acne. Its benefits, indications, and technique will be discussed in this chapter.

17.3 Basic Principles

Propionibacterium acnes in the pilosebaceous unit and follicular inflammation are two of the four main pathogenesis of acne. In addition, if we can decrease the number of sebaceous glands, we can improve acne. Any method which does not disrupt systemic hormonal levels is also beneficial. If we can control these trigger factors, we can reduce the severity of acne, or improve acne.

Phototherapy is a noninvasive, out-patient procedure with a proven efficacy in acne treatment. It involves the activation of photosensitizing agent (endogenous synthesis or stimulating from exogenous agents) by visible light to produce activated oxygen species within a target area, resulting in their destruction [27,28].

17.3.1 Mechanism of Action

As mentioned earlier, *P. acnes*, which colonize the lower infundibulum, can digest triglycerides in the sebum and then release free fatty acids to produce inflammation or release cytokines to stimulate abnormal keratinization and inflammation. *P. acnes* also produces high amount of porphyrin, particularly corpoporphyrin, which is called an endogenous photosensitizer. δ-Aminolevolunic acid (ALA) is another chemical substance which can induce exogenous porphyrin synthesis, particularly uroporphyrin, coproporphylin, and protoporphyrin IX. When these porphyrins are exposed to the light, especially to 381–560 nm wavelength, they can produce singlet oxygen which is a cytotoxic agent (Fig. 17.2) [29].

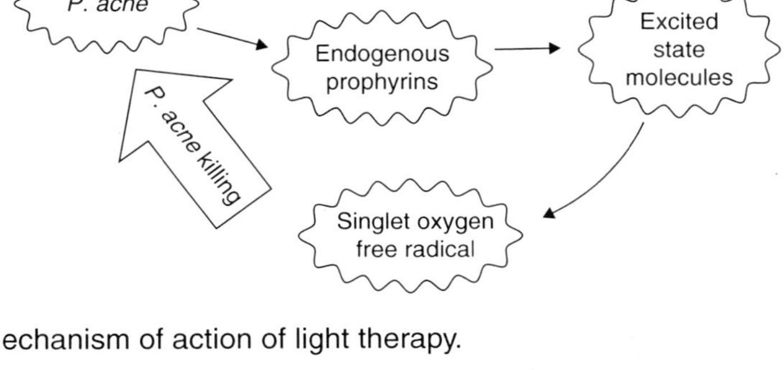

Figure 17.2 Mechanism of action of light therapy.

This agent can destroy *P. acnes*, or in high amounts especially obtained from exogenous ALA application, it can cause damage to sebaceous glands [30].

Porphyrins have a wide range of absorption spectrum from UV to visible light. They have the greatest absorption to wavelengths near 400–420 nm, with their highest absorption peak (called the Soret band) occurring at 415 nm. Further, additional small peaks (Q bands) occur at longer wavelengths between 450 and 700 nm (509, 544, 584 and 634 nm) (Fig. 17.3) [31,32].

Violet/blue light source, with peak wavelengths between 400–420 nm (Soret band), is a popular treatment for acne. It has been very effective, with no reports of mutagenicity, and low side effects when the target is endogenous porphyrin, which is produced by *P. acnes*. Blue light also has antiinflammatory effects for inflamed acne by down-regulating interleukin (IL)-1, a proinflammatory cytokine, which is a chemo-attractant of inflammatory cells, and a stimulant of other inflammatory mediators [33].

17.3.2 Blue Light for Acne

Sunlight exposure is believed to have a beneficial effect on some inflammatory skin conditions. It contains the ultraviolet electromagnetic spectrum, which can kill many organisms, and also has a broad spectrum of electromagnetic radiation. The most important spectrums of solar radiation which reach the surface of the earth are ultraviolet radiation, visible lights, infrared, and radio waves. However, there are many factors that influence the level of energy that reach us, such as area, time of the day, and climate. Artificial light source can help overcome this defect. Many studies show that artificial UV radiation can improve many skin conditions such as psoriasis and vitiligo.

Some patients report improvement of acne after exposure to sunlight [34–36] though some studies state that evidence for improvement of acne due to sunlight is lacking [37]. In addition, UV radiation has complications associated with it, such as erythema, sunburn, pigmentary alterations (prolong hyperpigmentation, melasma), photoaging, and the risk for the development of skin cancer [38].

The basic principle of photomedicine states that when chromophores (any molecule which can absorb some specific electromagnetic spectrum, such as light, and cause a reaction) receive light in the appropriate spectrum, they can produce a reaction which results in

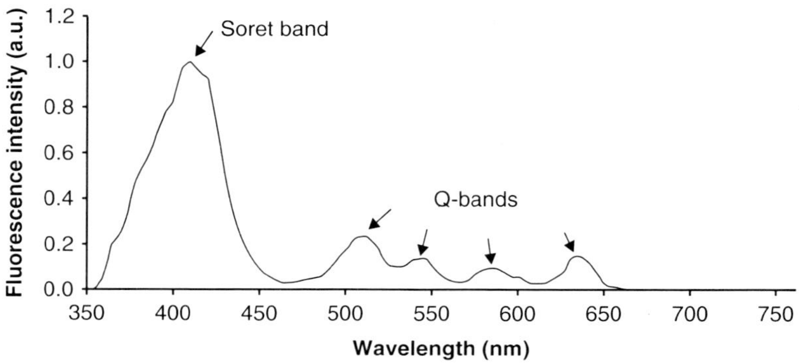

Figure 17.3 Excitation of protoporphyrin IX.

changes within the surrounding chromophore area [39]. *P. acnes*, which are involved in the pathogenesis of acne, can produce endogenous chromophores called porphyrins, mainly coproporphyrin III and protoporphyrin IX (Pp-IX) as part of its normal metabolic processes. Corpoporphyrin III can be seen on the face under 365 nm (UVA1- Wood's lamp) (reddish-orange fluorescences). When these porphyrins (which can react to wavelengths that fall into the ultraviolet to visible light spectrum, but peak at the Soret band of blue light at 400–420 nm) are stimulated by light; this leads to photoexcitation, and the formation of singlet oxygen-free radicals that cause destruction of *P. acnes* [40,41]. Blue light systems are used based on this theory, and many studies prove their efficacy.

17.4 Clinical Studies

In a study done by Elman, 46 patients were treated with blue light (405–420 nm, Clear light™). Of these 46 patients, 80% of them noticed around a 60% improvement of papulo-nodular acne lesions after eight treatments of 8–15 minutes. These patients had prolonged remission of their acne, which was evident in the eight weeks following the end of their treatment. No side effects were noted in any of the patients [42].

In another study done by Tzung, they decided to test the efficacy of blue light by treating only one side of the face, and leaving the other side of the face untreated, for comparison. The treated side was selected randomly, and was treated twice weekly for four consecutive weeks. The other half of the face was left untreated as control. Compared with the non-irradiated side, eight sessions of blue light irradiation was found to be effective in acne treatment [43].

In a study by Gold and colleagues, they compared the efficacy between blue light used for 16 minutes biweekly for 4 weeks to 1% clindamycin used twice daily. Blue light therapy was shown to reduce inflammatory acne vulgaris lesions by an average of 34%, as compared to 14% for topical 1% clindamycin solution [44].

In a study by Tremblay, 45 patients were treated with pure blue light (415 nm and 48 J/cm^2), receiving two treatments of 20 minutes per week for a period of 4–8 weeks. They found that the mean improvement score was 3.14 at 4 weeks, and 2.90 at 8 weeks. In fact, nine patients experienced complete clearing at eight weeks. The treatment was well-tolerated, with 50% of patients reporting being highly satisfied with the treatment. This study suggested that blue light is effective in the treatment of acne vulgaris. There were no reported side effects in this study [45].

Kawada and colleagues conducted an open study on acne patients who were treated twice a week up to five weeks with blue light. They found that the acne lesions were reduced by 64%. Two patients experienced dryness, but none of the patients discontinued treatment due to adverse effects [46].

In a study by Omi and colleagues, they investigated the use of blue light on acne. They recruited a total of 28 adult healthy volunteers with facial acne (mean age 28.1 years, range 16–56 years). These patients were treated with a total of 8 serial biweekly 15-min treatment sessions. This study found that overall, there was a 64.7% improvement in acne lesions. They concluded that blue light is a useful treatment for acne [47].

Morton et al. tried to determine the effect of narrow-band blue light in the reduction of inflammatory and noninflammatory lesions in patients with mild to moderate acne. They performed an open study utilizing a blue light source in 30 patients with mild-to-moderate facial acne. Over a 4-week period, patients received eight 10- or 20-minute light treatments, peak wavelength 409–419 nm at 40 mW/cm^2. This study concluded that eight 10- or 20-minute treatments over 4 weeks with a narrowband blue light was found to be effective in reducing the number of inflamed lesions in subjects with mild to moderate acne. The onset of the effect was observable at week 5. However, the treatment was found to have little effect on the number of comedones [48].

Although theoretically porphyrins should respond well to blue light, it is a shorter wavelength and therefore does not penetrate well into the skin [49]. Longer wavelengths with Q-bands, such as red light, have been combined with blue light in acne therapy. Red light (wavelength 600–650 nm) penetrates deeper into the skin than blue light. In fact, 635 nm light may penetrate through the skin up to 6 mm, compared with 1–2 mm for light at 400–500 nm. Red light has also been shown to be effective in acne treatment by activating porphyrins in the Q band, and decreases inflammation by controlling cytokine release from macrophages [50–54].

Using shorter wavelengths can be more efficient at activation of PpIX; however, we have to be concerned that it may not penetrate deeply enough to produce the wanted effects. Photodynamic therapy is described as the use of cytotoxic oxygen-free radicals generated from photoactivated molecular species to achieve a therapeutic response.

5-ALA is a precursor in the heme biosynthesis pathway of protoporphyrin (Pp)-IX. Normally, Pp-IX levels in the tissue are not synthesized high enough to produce major tissue damage [55]. However, exogenous application of 5-ALA can increase the intracellular level of Pp-IX. Hyperplastic cells are believed to uptake Pp-IX more than the normal cells. This is due to hyperproliferating cells needing more Pp-IX for iron to synthesize additional cells and therefore accumulate Pp-IX.

With the knowledge that 5-ALA can increase the intracellular levels of Pp-IX, it is believed that if we apply 5-ALA to the skin, active cells will accumulate this agent and result in accumulation of Pp-IX. This accumulation of Pp-IX can therefore be used as a target for light therapy, which produces reactive singlet oxygen-free radicals that can cause cellular damage. Pp-IX accumulates in the mitochondria and when it is exposed to light, it causes damage to the mitochondria and cause cytochrome C to leak. This stimulates endonuclease activity and plasma and the nuclear membrane loses integrity [56,57].

Topical photodynamic therapy (PDT) has been beneficial for various skin conditions. The FDA approved PDT using a 5-ALA preparation, Levulan (DUSA, USA), for actinic keratoses, and in Europe PDT with methyl ester of 5-ALA (MAL) (Galderma, France) was approved for actinic keratoses and basal cell carcinomas.

However, the problems associated with photodynamic therapy is that it is expensive to perform, it is time consuming, and it is more painful than blue light alone. Patients have reported erythema, stinging, pruritus, pain, and tightening after this intervention [58].

17.5 Future Direction

Treatment of acne is a challenging process due to its multifarious pathogenesis. For this reason, more studies are needed to explore other possible treatments. For example, additional studies are needed to find more effective ways of enhancing the delivery of ALA to its target, possibly through microdermabration, peeling, or using the ester form of ALA (methyl ALA). The use of antibiotics for *P. acnes* and whether or not we should stop using antibiotics before blue light exposure needs to be explored more thoroughly. In addition, studies for the treatment of periodic acne in women (acne related with menstruation) and ALA with intense Laser (PDL) or Light (Intense blue light or other intense pulsed light systems) are other possible areas that can be investigated.

References

1. Kraning K and Odland GF. Prevalence, morbidity and cost of dermatologic diseases. *J Invest Dermatol.* 1979; 73(Suppl.):395–401.
2. Goodman G. Acne—natural history, facts and myths.*Aust Fam Physician.* 2006 Aug; 35(8):613–6.
3. Del Rosso JQ. Acne in the adolescent patient: interrelationship of psychological impact and therapeutic options. *Today's Ther Trends* 2001; 19:473–84.
4. Elman M, Lebzelter J. Light therapy in the treatment of acne vulgaris. *Dermatol Surg.* 2004; 30:139–46.
5. Williams C and Layton AM. Persistent acne in women: implications for the patient and for therapy. *Am J Clin Dermatol.* 2006; 7(5):281–90.
6. Harrington CI. Post-adolescent acne and marital break-up. *Br J Dermatol.* 1997 Sep; 137(3):478–9.
7. Goulden V, Stables GI, Cunliffe WJ. Prevalence of facial acne in adults. *J Am Acad Dermatol.* 1999; 41:577–80.
8. Cotran RS, Kumar, V, Collins T. "Acne Vulgaris." In: *Robbins and Cotran Pathologic Basis of Disease*, 7th Edition. Saunders: Philadelphia, 2005.
9. Pochi PE, Shalita AR, Strauss JS, et al. Report of the Consensus Conference on Acne Classification. Washington, D.C., March 24 and 25, 1990. J Am Acad Dermatol 1991; 24:495–500.
10. Leyden JJ. Therapy for acne vulgaris. *N Engl J Med* 1997; 336:1156–62.
11. Koo JY and Smith LL. Psychological aspects of acne. *Pediatr Dermatol.* 1991; 8:185–8.
12. *Management of Acne: Summary, Evidence Report/Technology Assessment.* No. 17. Agency for Healthcare Research and Quality: Rockville, MD, March 2001.
13. William JD, et al. "Acne." In: *Andrew's Disease of the Skin: Clinical Dermatology*, 10th Edition. Elsevier: Canada 2006, pp. 231–250.
14. Sanchez AR, et al. Tetracycline and other tetracycline-derivative staining of the teeth and oral cavity. *Int J Dermatol.* 2004; 43:709–715.
15. Nelson R and Parker SR. Doxycycline-induced staining of adult teeth: the first reported case. *Arch Dermatol.* 2006 Aug; 142(8):1081–2.
16. Yan AC. Current concepts in acne management. *Adolesc Med Clin.* 2006 Oct; 17(3):613–37.
17. James WD. Clinical practice: Acne. *N Engl J Med.* 2005 Apr 7; 352(14):1463–72.
18. Nau H. Teratogenicity of isotretinoin revisited: species variation and the role of all-trans-retinoic acid. *J Am Acad Dermatol.* 2001 Nov; 45(5):S183–7.
19. Robertson JA, et al. Accutane cases: a teratogen information service's approach. *Teratology.* 2002 Jul; 66(1):1–2.
20. Camacho FM. Medium-depth and deep chemical peels *J Cosmet Dermatol.* 2005 Jun; 4(2):117–28.
21. Fung JF, et al. Chemical injury to the eye from trichloroacetic Acid. *Dermatol Surg.* 2002 Jul; 28(7):609–10.
22. Leeming JP, Holland KT, Cunliffe WJ. The pathological and ecological significance of microorganisms colonizing acne vulgaris comedones. *J Med Microbiol.* 1985; 20:11–16.
23. Norris JFB, Cunliffe WJ. A historical and immunocytochemical study of early acne lesions. *Br J Dermatol.* 1988;118:651–9.
24. Cunliffe WJ, et al. Comedogenesis: some new aetiological, clinical and therapeutic strategies. *Br J Dermatol.* 2000 Jun; 142(6):1084–91.

25. Zouboulis CC, et al. What is the pathogenesis of acne? *Exp Dermatol.* 2005 Feb; 14(2):143–52.
26. Cunliffe WJ. The sebaceous gland and acne—40 years on. *Dermatology* 1998; 196:9–15.
27. Dougherty TJ, et al. *Photodynamic therapy.* J Natl Cancer Inst. 1998 Jun 17; 90(12):889–905.
28. Ashkenazi H, et al. Eradication of Propionibacterium acnes by its endogenic porphyrins after illumination with high intensity blue light. *FEMS Immunol Med Microbiol.* 2003 Jan 21; 35(1):1724.
29. Arakane K, et al. Singlet oxygen (1 delta g) generation from coproporphyrin in Propionibacterium acnes on irradiation. *Biochem Biophys Res Commun.* 1996 Jun 25; 223(3):578–82.
30. Kim SK, et al. Combination of topical 5-aminolevulinic acid-photodynamic therapy with carbon dioxide laser for sebaceous hyperplasia. *J Am Acad Dermatol.* 2007 Mar; 56(3):523–4.
31. Pottier RH, et al. Non-invasive technique for obtaining fluorescence excitation and emission spectra in vivo. *Photochem Photobiol.* 1986 Nov; 44(5):679–87.
32. Harber LC and Bickers DR. *The Porphyria in Photosensitivity Diseases, Principle of diagnosis and treatment.* BC Decker Inc Publisher, 1989, 241–287.
33. Shnitkind E, et al. Anti-inflammatory properties of narrow-band blue light. *J Drugs Dermatol.* 2006 Jul–Aug; 5(7):605–10.
34. Berg M. Epidemiological studies of the influence of sunlight on the skin. *Photodermatol.* 1989 Apr; 6(2):80–4.
35. Hirsch RJ, Shalita AR. Lasers, light, and acne. *Cutis* 2003; 71:353–4.
36. Sigurdsson V, Knulst AC, van Weelden H. Phototherapy of acne vulgaris with visible light. *Dermatology* 1997; 194:256–60.
37. Magin P, et al. A systematic review of the evidence for "myths and misconceptions" in acne management: diet, face-washing and sunlight. *Fam Pract.* 2005 Feb; 22(1):62–70. Epub January 11, 2005 .
38. van Weelden H, et al. The carcinogenic risk for modern tanning equipment: is UVA safer than UVB. *Arch Dermatol Res* 1988; 280:300–7.
39. Konig K, et al. Fluorescence detection and photodynamic activity of endogenous protoporphyrin in human skin.*Opt Eng* 1992; 31:1470– 4.
40. Manyak MJ. Photodynamic therapy: present concepts and future applications. *Cancer J.* 1990; 3:104–9.
41. Nitzan Y, et al. Induction of endogenic porphyrin production in bacteria and subsequence photoinactivation by various light sources. In: *Photochemotherapy: Photodynamic Therapy and Other Modalities III* (Berg K, Ehrenberg B, Malik Z and Moan J, Eds.). SPIE: Bellingham, WA, pp. 89–94.
42. Elman M, et al. The effective treatment of acne vulgaris by a high-intensity, narrow band 405–420 nm light source. *J Cosmet Laser Ther.* 2003 Jun;5(2):111–7.
43. Tzung TY et al. Blue light phototherapy in the treatment of acne. *Photodermatol Photoimmunol Photomed.* 2004 Oct; 20(5):266–9.
44. Gold MH, et al. A multicenter clinical evaluation of the treatment of mild to moderate inflammatory acne vulgaris of the face with visible blue light in comparison to topical 1% clindamycin antibiotic solution. *J Drugs Dermatol.* 2005 Jan–Feb; 4(1):64–70.
45. Tremblay JF, et al. Light-emitting diode 415 nm in the treatment of inflammatory acne: an open-label, multicentric, pilot investigation. *J Cosmet Laser Ther.* 2006 Apr;8(1):31–3.
46. Kawada A, et al. Acne phototherapy with a high-intensity, enhanced, narrow-band, blue light source: an open study and in vitro investigation. *J Dermatol Sci.* 2002 Nov; 30(2):129–35.
47. Omi T, et al. 420 nm intense continuous light therapy for acne. *J Cosmet Laser Ther.* 2004 Nov;6(3):156–62.
48. Morton CA, et al. An open study to determine the efficacy of blue light in the treatment of mild to moderate acne. *J Dermatolog Treat.* 2005; 16(4):219–23.
49. Kjelstad B and Johnson A. An action spectrum for blue and near ultraviolet inactivation of Propionibacterium acnes; with emphasis on a possible porphyrins photosensitization. *Photochem Photobiol.* 1986; 43:67–70.
50. Papageorgiou P, et al. Phototherapy with blue (415 nm) and red (660 nm) light in the treatment of acne vulgaris. *Br J Dermatol.* 2000; 142:973–8.

51. Cunliffe WJ and Goulden V. Phototherapy and acne vulgaris. *Br J Dermatol.* 2000; 142:853–6.
52. Young S, et al. Macrophage responsiveness to light therapy. *Lasers Surg Med.* 1989; 9:497–505.
53. Lee SY, et al. Blue and red light combination LED phototherapy for acne vulgaris in patients with skin phototype IV. *Lasers Surg Med.* 2007 Feb; 39(2):180–8.
54. Goldberg DJ and Russell BA. Combination blue (415 nm) and red (633 nm) LED phototherapy in the treatment of mild to severe acne vulgaris. *J Cosmet Laser Ther.* 2006 Jun ;8(2):71–5.
55. Kalka K, et al. Photodynamic therapy in dermatology. *J Am Acad Dermatol.* 2000 Mar; 42(3):389–413.
56. Taub AF. Photodynamic therapy in dermatology: history and horizons.*J Drugs Dermatol.* 2004;3:S8–S25
57. Gold MH, Goldman MP. 5-aminolevulinic acid photodynamic therapy: where we have been and where we are going. *Dermatol Surg.* 2004; 30:1077–83.
58. Goldman MP, et al. Comparative benefit of two thermal spring waters after photodynamic therapy procedure. *J Cosmet Dermatol.* 2007 Mar; 6(1):31–5.

18

Treatment of Pseudofolliculitis Barbae

Douglas Shander and Gurpreet S. Ahluwalia

The Gillette Company, a wholly owned subsidiary of The P&G Company, Needham, MA, USA

18.1 Introduction

Shaving bumps, or pseudofolliculitis barbae (PFB), is a condition of the beard area occurring in black men and other people with curly hair. Highly-curved hair can grow back into the skin, causing inflammation and a foreign-body reaction surrounding the site of the ingrown hair. Keloidal scars that look like hard bumps of the beard area and neck can develop over time with continued insults by ingrown hair in the same area. In addition to occurring in African American men, the condition also occurs in Caucasians with curly beard hair, and shaving bumps also develop in the back of shaved necks. Pseudofolliculitis barbae is a serious problem in women with excessive facial hair and the problem is especially challenging in African American men and women, as ingrown hair can cause disfiguring

Gurpreet S. Ahluwalia (ed.), *Cosmetic Applications of Laser and Light-Based Systems*, 353–367,

hyperpigmentation. Shaving bumps also occur in axillae, thighs, and the bikini line, and are often accompanied by uncomfortable itching and irritation [1].

In the uniformed professions including military serviceman, firefighters, and police officers, clean-shaven faces are essential for situations requiring the use of gas masks with a tight seal, especially in combat areas such as Iraq and Afghanistan. Historical formal and informal norms in the military affecting group morale and advancement potential put strong pressures on men to be clean-shaven. Each military branch has their own regulations on shaving, and this has often been a source of tension and friction for African American men with PFB conditions which thwart their ability to maintain a hair- free appearance on their facial skin. Over time, the requirements for shaving have been partially relaxed to accommodate men who cannot tolerate shaving or the repeated use of caustic depilatories, if the presence of hair does not directly impact performance of their duties. Nonetheless, even today the US Marine Corps can summarily discharge servicemen for not complying with shaving requirements even if it does not impact their performance of duties, and male officers in highly visible positions such as recruiting or public affairs are expected to keep clean-shaven faces.

As a result of political and social pressures, much of the clinical research on PFB over the last 35 years has been supported and conducted by the military forces. Studies from the 1970s indicate that PFB incidence is estimated to be between 50% and 80% in African American men [2,3] and a method to assess the severity of PFB was developed by dermatologists practicing in the military [4]. The best method to prevent ingrown hair was use of depilatories, which eradicated ingrown hair above and below the skin but the caustic nature of these chemical depilatories that break disulfide bonds in hair and skin makes this strategy impractical. One outcome of the research of the 1970s led to the conclusion that in men who are able to tolerate it, a daily shaving regimen was the best treatment strategy to minimize PFB. Although counterintuitive, the rationale behind this conclusion was based on informal observations and shaving studies that less frequent shaving regimens increased the opportunity to develop an increased number of ingrown hair. Conversely increased shaving frequency maintained shorter hair lengths and prevented hair shafts from reaching the threshold lengths necessary to reenter skin. Most articles on PFB management recommend shaving frequency to be limited to two or three times a week, and daily shaving to be avoided. Our systematic review of the literature on PFB failed to reveal any informal or controlled clinical research studies supporting the benefits of less frequent shaving over daily shaving. Indeed, the limited number of published studies that have tested daily shaving demonstrate that daily shaving is beneficial over less frequent shaving in men with mild-to-moderate PFB.

Over the last ten years, the military has supported studies on the use of professional hair removal lasers as a treatment for PFB, including studies employing diode and Nd;YAG lasers which were conducted at the San Diego Naval Medical Center, as well as an extensive year-long treatment study using a Nd:YAG laser at the Tripler Army Medical Center in Honolulu Hawaii. Between 2004 and 2006, the US congress has approved 3.39 millions dollars earmarked for a US Army sponsored research program on a long wavelength diode laser being developed by Palomar Medical Technologies in Burlington, MA. Most recently, a million-dollar grant was earmarked for the assessment of a shaving kit at Brooke Army Medical Center at Fort Sam Houston Texas. The kit "Shaver's Choice Skin Therapy Shaving System" is made by SC3 LLC in Madison, Miss and has a mild astringent towel wipe, a shaving gel, and an after-shave skin therapy lotion. In the course of this review, we summarize the results of a variety of treatment strategies and provide conclusions on individual and combination treatment regimens based on ideal shaving regimens, topical hair growth retardants, and a judicious use of lasers to maximize the efficacy and minimize risks of adverse events, and the costs and inconvenience of continuing medical supervision.

In addition to the military support for PFB research, The Gillette Company Corporate R&D Research Group supported substantial fundamental and clinical research studies. The first studies sponsored by Gillette were those conducted by Kligman and Strauss in the 1950s. The Gillette Research Institute (GRI) which was previously located in Maryland also did substantial amount of research on biophysical properties of hair and research on biological control hair follicle proliferation and differentiation processes of hair-shaft formation. In addition, studies on dermatological conditions related to hair- growth processes such as unwanted facial hair and PFB were conducted to exploit practical applications of the more fundamental research efforts.

The GRI researchers conducted fundamental and clinical research studies on PFB for over thirty years between 1968 through 2000, and the culmination of these efforts was identification of optimal shaving frequency regimens confirming and extending earlier findings at the US military facilities. In addition, research at GRI first demonstrated the efficacy and mechanism of action of the hair growth retardant Eflornthine HCl as an effective treatment modality in managing unwanted facial hair in women, as well as providing a viable treatment option for PFB in men.

18.2 Etiology of PFB

In the 1950s, the the seminal studies by Kligman and Strauss [2] established that PFB lesions are indeed caused by ingrown hair which grow out of the follicle and then renter the skin. This propensity of hair to enter skin after exiting the follicle is the result of the curvature of the follicle structure. The curved follicle directs the hair exiting from follicle to come in very close contact with, or be embedded in the grooves of the skin at a short distance from the follicle orifice. Their conclusions were based on skin biopsies from men with PFB, as well as direct visual observations and careful and exacting analysis using a surgical microscope. They reported that the only method for almost complete resolution of PFB was growing a beard, which allowed the hair to erupt from the skin after growing to depths of 2–3mm. It was proposed that the spring force of extended hair was insufficient to hold them in the skin as their length increased in a fashion similar to a coiled rope losing its tension as it length increases. They speculated that the skin texture of men with PFB prevents efficient shaving, as hair embedded in the grooves of skin were not accessible for full engagement of the cutting edge of blades and the uncut hair attained sufficient lengths to reenter the skin. Figure 18.1 is a skin biopsy of a subject with PFB, and demonstrates the propensity of curved hair to renter the skin after exiting the follicle. More recently, epiluminescence dermatoscopy has dramatically reconfirmed the concordance between lesions and ingrown hair, and the demonstration of this hair– skin interaction to patients with PFB to enhance their compliance with treatments and improve treatment success by preventing hair from reentering the skin [6].

A study conducted at the Gillette research institute was designed to characterize the pattern of ingrown hair development in men with PFB. The number of actual ingrown hair in one-square-inch of facial skin in the neck region was quantified in six men with PFB using

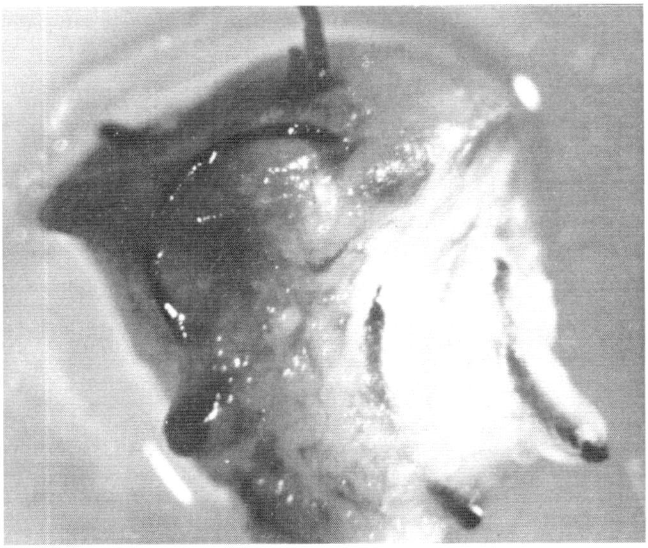

Figure 18.1 Skin biopsy of PFB subject demonstrating rentry of hair into skin and curved hair below skin.

a surgical microscope in order to quantify hair density and the percentage of ingrown hair. In six volunteers, the hair density ranged from 114 to 224 hair per square inch and the percent of ingrown hair ranged from 23 to 50% with an average of 33%. Not all ingrown hair were associated with lesions, but the substantial number of ingrown hair representing potential papule and pustule formation clearly demonstrates the difficulties and challenges in managing PFB. Indeed, it has been noted by dermatologists familiar with PFB that it is a very intractable and recalcitrant condition which can be managed, but not cured. In most men with PFB growing a beard can resolve the problem and informal estimates suggest a population of 10–20% of men who do not respond to growing a beard. To better understand the progression of PFB, we conducted a study on eight men whose PFB condition was resolved by abstaining from shaving for eight weeks. After resolving their PFB, the men began shaving every other day over a 15-day interval. We chronicled and quantified the recurrence of lesions during this period. In all subjects, the level of PFB in terms of total bump counts returned to pre- beard levels within 15 days of resumption of shaving. Figure 18.2 illustrates the dramatic and progressive return of lesions in one subject, whose response was typical of the group.

Poor shaving efficiency has traditionally been ascribed as the root cause of ingrown hair. A shaving study was designed to specifically characterize the quality of shaving efficiency in men with PFB. Images of hair in the circumscribed area of neck regions were captured with a video-microscope. The images were from circumscribed regions of the neck in 10 men with PFB, as well as 19 men not exhibiting PFB. Images were obtained before and after shaving to compare the efficiency of shaving in men with PFB versus men without PFB. Hair was characterized as fully cut at or below skin line, or only partially cut or totally missed after shaving. In men with PFB, only 41% of facial hair was fully cut at or below skin line, compared to over 81% in men without PFB. Men with PFB totally missed an average of 30% of hair compared to only 3% in men without PFB. This study clearly

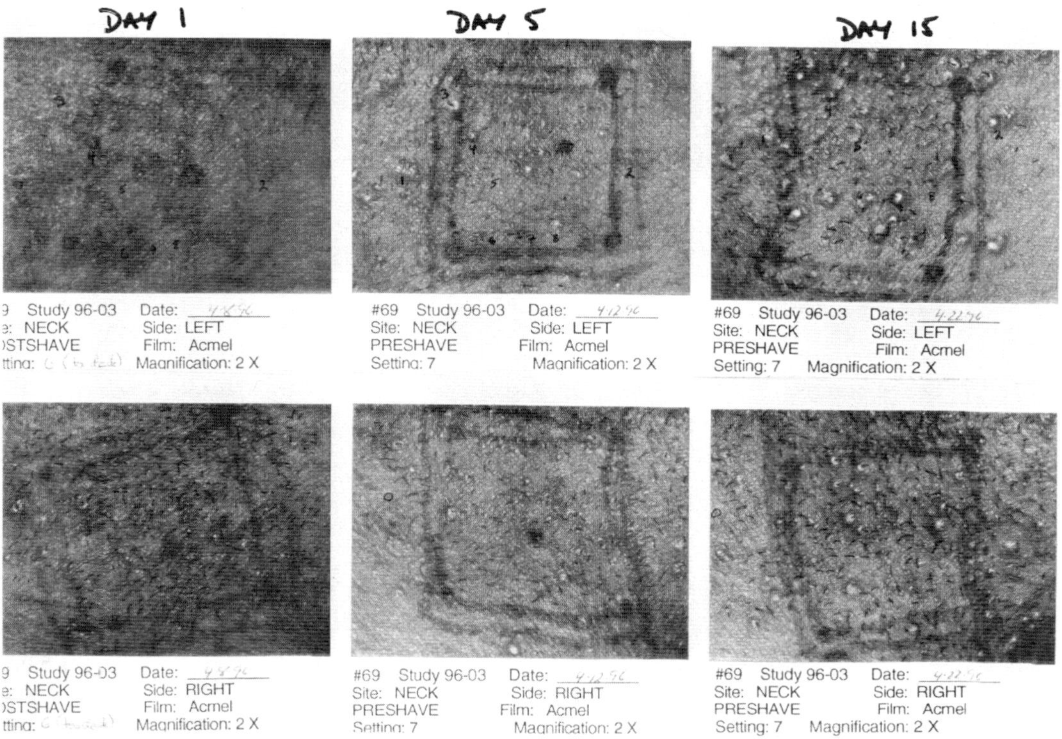

Figure 18.2 Development of shaving bumps over 15 days after resumption of shaving in skin cleared of PFB by eight weeks of shaving abstinence.

confirmed the suggestions of earlier researchers that inefficient shaving was a causative factor promoting PFB. Design and development of more efficient means of mechanical hair removal remains a continuing challenge, as hair is often embedded in the grooves of the highly textured skin in men with PFB.

18.3 Biochemical Factors Involved in Elaboration of Curly Hair

Recent research on the molecular biological factors underlying curly hair has demonstrated that the biological factors controlling the processes of keratin formation and regionalized localization of specific types of keratin that are associated with curly hair are represented in all curly hair fibers independent of ethnic types. The biochemical factor promoting curly hair fibers to become ingrown is the selective accumulation of hHa8 keratin on the concave side of the hair fiber which differentiates sooner and more completely than the cortex on the convex side [7].

Insight into the developmental process contributing to asymmetry in cortex demonstrated that delayed differentiation of the inner and outer root sheath on the convex side of the beard fibers is the result of a sustained zone of highly proliferating bulb matrix cells on the convex side of curly hair shafts, which fail to differentiate at the same time as the cells on the concave side. The differential proliferation rates on each side of the follicle results in the asymmetric spatial and temporal differentiation program [8]. One other risk factor associated with the keratin formation process is the K6hf keratin gene affecting proteins synthesized in the companion layer of the hair follicle which is connected to the inner root sheath. The presence of a malformation in keratins in this region, which is involved in molding the hair shaft, may impact the growth and structure of the hair in a manner predisposing the hair shaft to become coiled and curved during its ascent through the skin and onto the skin surface [9].

18.4 Shaving Regimens for Managing PFB

Men with the most severe cases of PFB often cannot tolerate any shaving, but men with less severe PFB who can shave daily benefit form daily shaving. Strauss and Kligman [2] determined that daily shaving was actually the most effective shaving regimen to minimize lesions in most subjects, but could worsen the condition in those most prone to developing ingrown hair. Ideally, shaving on a daily basis is the most preferred, economical, and least invasive stratagems to manage the ingrown hair causing PFB.

Perricone [10] reported results of an investigator-blinded placebo-controlled study that demonstrated daily shaving resulted in a significant decrease in PFB lesions and the combination of daily shaving with glycolic acid treatment markedly reduced the number of PFB lesions compared to the shaving on a daily basis without glycolic acid. Daily shaving in both cases improved PFB in terms of lesion reduction. The mechanism of action of glycolic acid in enhancing the benefits of daily shaving was not clarified and speculations include an effect of glycolic acid on modification of skin surface or hair fiber structure to enhance shaving efficiency. Adhering to a daily shaving regimen is a difficult challenge for men with PFB and a critical observation from the studies on glycolic acid treatments was the self-reported claims by panelists that glycolic acid treatment actually facilitated compliance, with adherence to a daily shaving regimen. Indeed, the ability of glycolic acid to support maintenance of a daily regimen may have been the primary benefit of glycolic acid.

Preliminary pilot studies at the Gillette Research Institute on a panel of eleven men in a carefully monitored study which required all subjects to perform weekday shaving at the GRI demonstrated that seven weeks of alternate-day shaving produced a significant and stable 30% reduction in lesions from baseline, after both five and seven weeks of compliance with this regimen. Continued shaving on a daily basis for another five weeks resulted in a 60% reduction from baseline values. The results of a recent controlled- investigator blinded study commissioned by Gillette which was executed at the Wake Forest University Dept. of Dermatology have again demonstrated benefits of daily shaving versus alternate-day shaving. Eight weeks of daily shaving significantly reduced PFB lesions 38% from baseline in men who shaved on a daily basis compared to an 11% reduction of the lesions in men who shaved three times a week (manuscript submitted). The results of these studies reinforce the original observations of Kligman and Strauss, as well as the earlier reported research of dermatologists conducting closely monitored studies in military bases. It is important to emphasize that men with PFB are often reluctant to shave frequently because of pain and discomfort, and it is best to closely monitor the shaving frequency studies to be certain that participants are compliant.

18.5 Treatment of PFB with Topical Eflornithine

In our research at the Gillette Company Research Institute, we conducted a series of Phase I, Phase II, and Phase III studies between 1988 and 2000, evaluating the efficacy of eflornithine HCl in facial-hair retardation. In the first study ever conducted, the use of Eflornithine in the treatment of unwanted facial hair in women established the efficacy of its use as a topical hair retardant. In that pilot study, Shander et al. [11] found that there was a remarkable benefit of the topical eflornithine (10% eflornithine HCl delivered in an aqueous hydro-alcoholic vehicle) on improving PFB in women with facial hair. Dermatological observations were recorded by a dermatologist who noted skin conditions prior to treatment, at 12 and 24 weeks of treatment, and at 12 weeks after the withdrawal of treatment. Nine women suffering from pseudofolliculitis treated themselves twice daily over a 6-month period with an aqueous-based vehicle (68% water, 16% ethanol, 5% propylene glycol, 5% dipropylene glycol, 4% benzoyl alcohol, and 2% propylene carbonate) containing 10% eflornithine. At the end of the treatment, 5 subjects exhibited complete clearance, 4 of whom cleared within 12 weeks. Four additional women showed progressive improvements during the 24 weeks of treatment.

A study was proposed to evaluate if eflornithine could reduce ingrown hair and lesions in men with PFB. The open-label study in men with PFB reported by Shander et al. [12] was designed to evaluate the mechanism of action of 13.9 % eflornithine cream applied twice daily in men with PFB including:

(1) reduction of ingrown hair
(2) changes in hair shaft histology specifically related to the ratio of paracortical and orthocortical cells
(3) quantification of the angle of emergence of hair from the skin which was not ingrown since changing the angle of emergence could enhance the ability of blade to contact the low-lying hair and increase the shave efficiency.

The results of this research indicate that Ingrown low lying hair (IGLLHs) are embedded in the grooves and rugosities of the skin, and are predisposed to enter the skin and provoke inflammatory lesions. A reduction in the IGLLHs in hair has significant clinical relevance as these hair have the propensity to become ingrown and cause a papule. Often these hair are not readily visible and can only be observed using high magnification lenses or by the use of facial silicone replicas portraying the facial landscape. We employed the use of facial replicas to identify the ingrown hair, as well as measure the angle of emergence of noningrown hair from the screen.

Facial surface replicas of 48-hour beard growth were obtained at the beginning and end of the 16-week eflornithine-HCL treatment period, and processed to characterize the changes in the number and percentage of IGLLHs, as well as the changes in the angle of emergence. At the start of the study facial replicas from four [4] facial sites exhibiting PFB including the cheek and neck areas on the left- and right-side were prepared in subjects who had refrained from shaving for 48 hours. The sites included four circular areas (0.7 cm diameter) with PFB lesions in the left and right chin and cheek for a total area of 6.22 cm^2. Using photographs at the beginning of the study to pinpoint specific facial features, combined with triangulation techniques measuring distance of the site from fixed facial landmarks, we were able to accurately return to the same sites at the end of the study for making end-of-study replicas.

In addition to quantifying the number of ingrown hair, the angle of emergence of hair that were not ingrown was measured using an automated image analysis technique, in order to provide objective evidence that the treatment altered the shape of hair and their angle of emergence from the skin. Of the 11 men completing topical treatments, 9 were available for facial replica evaluation. One subject was hospitalized for a condition unrelated to topical treatment, one subject shaved on the morning that his final replica was to be obtained. One subject had severe lesions at the start of the study, and his ingrown hair on his face were so dramatically obscured by lesions that pretreatment hair counts could not be accurately obtained as the decrease in his lesion intensity and clinical severity was actually accompanied by an increase in hair.

In eight subjects, the total number of hair in the pretreatment period did not differ from the number in posttreatment. The results from the facial replicas on eight subjects demonstrated a significant (44%) reduction in IGLLHs at the end of the treatment (Fig. 18.3), as well as a significant increase (36%) in the average angle of emergence (AOE) of beard hair measured by image analysis, which is consistent with the reduced tendency to become ingrown.

Clinical evaluations by the Alexander and Delph method [5] were completed in all men at baseline and end of treatment. These reductions in ingrown hair were clinically significant, as 9 of the 11 men demonstrated clinically significant improvements in PFB, including 5 men with marked reductions in lesions and almost total clearance of inflammation, and 4 men with significant improvements in PFB. Figure 18.4 demonstrates a typical

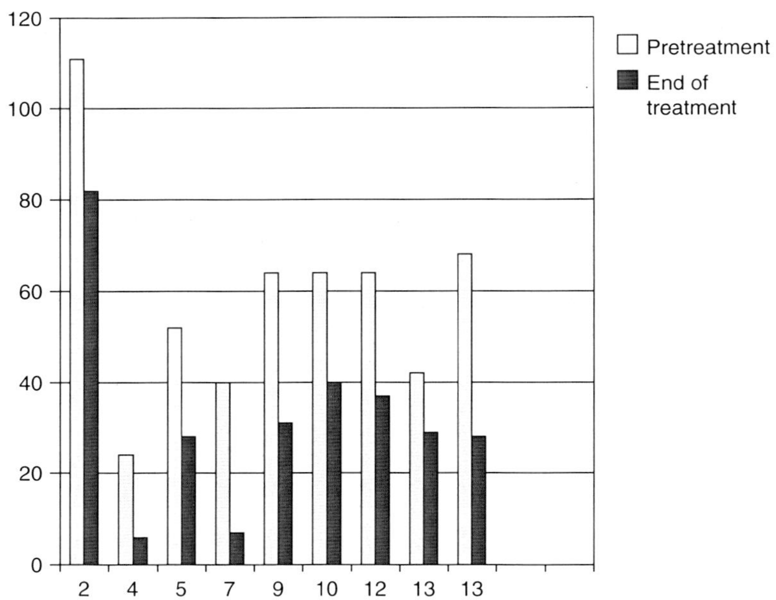

Figure 18.3 Decreased number of ingrown hair following 13.9% eflornithine HCl twice daily treatment for 16 weeks.

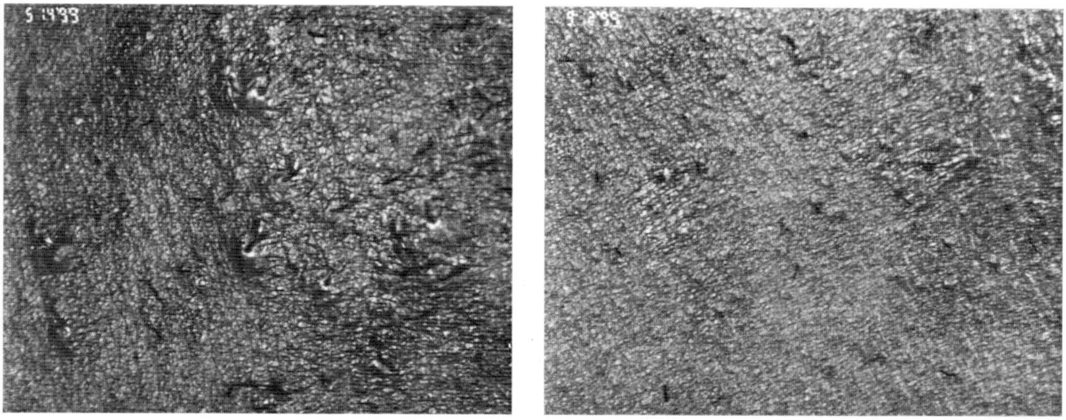

Before treatment End of treatment

Figure 18.4 Marked improvement in PFB after topical Vaniqa treatment.

marked improvement response in which severity is improved from score of 3 to 1 on the PFB severity scale of Alexander and Delph.

Beard-hair samples were also collected at the beginning and end of treatment, and cross-sections were prepared at a thickness of 2 μ, and stained with methylene blue to identify paracortical cells and eosin for the identification of orthocortical cells. The stained sections were then magnified with an Olympus microscope and photographed to provide images for measurements. The cuticle stained an intense dark blue, paracortical cells stained a less intense blue, and orthocortical cells stained a bright red. After the end of treatment, the percentage of eosin-stained orthocortical cells increased, resulting in a striking visual change in cortical staining pattern at the end of treatment. The medulla, when present, stained red as well. The results of this histological analysis are depicted in Fig. 18.5.

For an accurate quantitative image analysis of the paracortical cells, the red staining orthocortical cells and the medulla was subtracted. After extracting out the red level the methylene blue stained areas of paracortical cells area was selected at the best intensity threshold to quantify the area. There was a significant decrease in the percentage of methylene blue-stained paracortical cells from 90 to 76% ($P < 0.001$). Measurements of the cross-sectional area showed that the area of the beard hair sections increased significantly ($p < 0.001$) after treatment with DFMO, while the hair parameters, medulla, cortex, and cuticle, calculated as a percentage of the total cross-section, remained unchanged. Moreover, the average width of the cuticle statistically increased 8% ($p = 0.0022$) from pre- to post-treatment.

We hypothesize that the mechanism of action of Vaniqa in improving PFB involves changes in the ratio of orthocortex to paracortex after administration of eflornithine which modifies the biophysical properties of beard hair. This changes their shape, and the angle of emergence after exiting the follicle orifice thereby reducing the IGLLHs. In addition, the changes in the ratio of orthoccortex to paracortex reflect a morphological modification resulting from decreased cell proliferation, which alters the characteristic asymmetrical pattern of cell differentiation in African American hair follicles (Fig. 18.6).

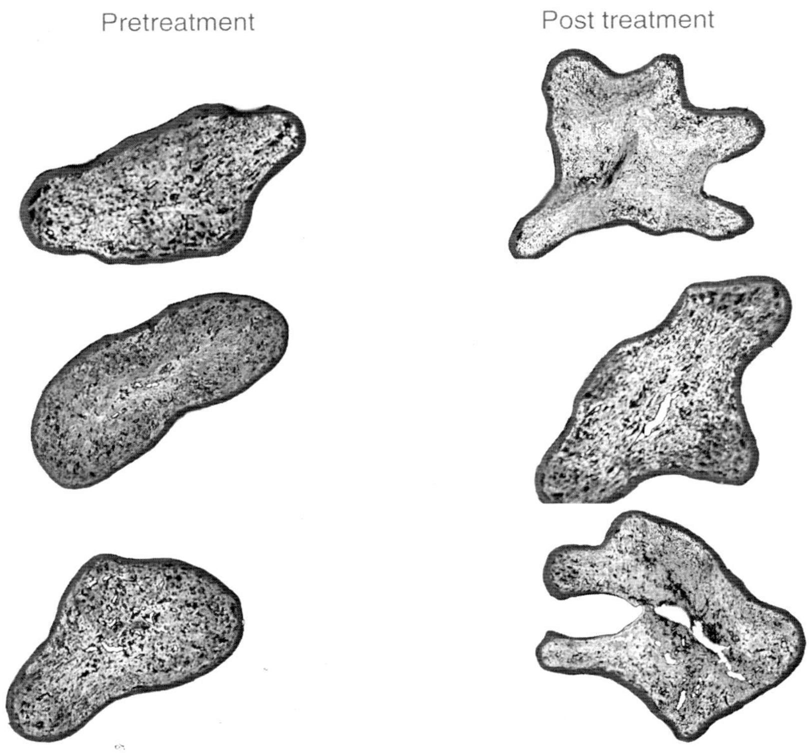

Figure 18.5 Modification of beard hair cortex by topical Vaniqa

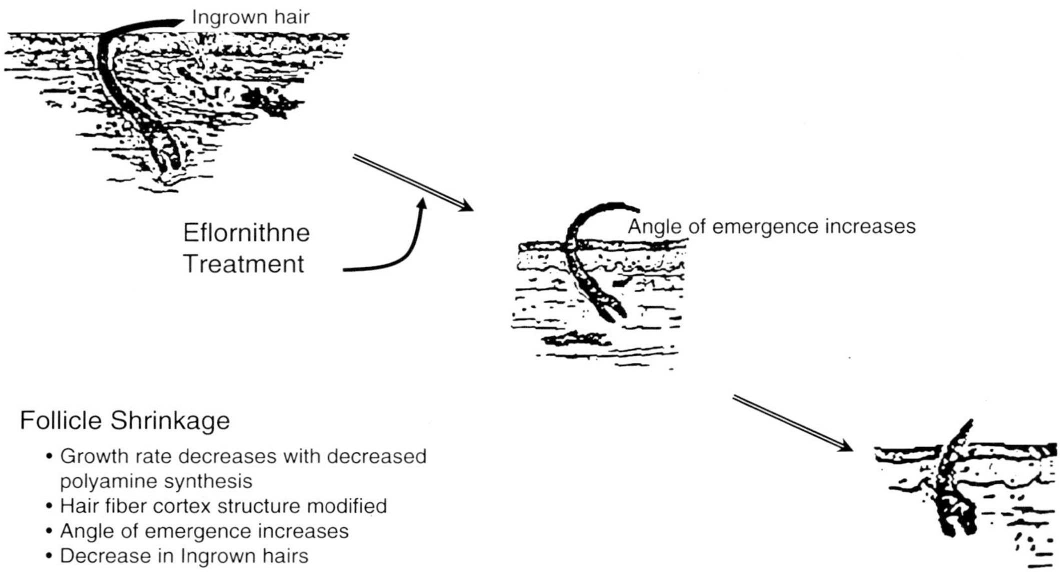

Figure 18.6 Proposed mechanism of action of Vaniqa on PFB

18.6 Laser Treatments for PFB

Laser treatments for PFB including the use of ruby, alexandrite, 800 nm diode and 1080 nm Nd:YAG with and without exogenous chromophores have been evaluated at an increasing rate over the last ten years. The initial rationale for laser treatments for PFB was based on the simple assumption that lasers could achieve a rapid permanent inhibition and/or modification of hair-shaft growth, which would remove the ingrown hair causing PFB. Most laser treatments are designed to inhibit hair growth by acting on endogenous melanin in the hair follicle and hair shaft, which generates heat when irradiated by laser light. The heat is transferred to the follicle targets in proliferative bulb and keratinization zone regions with the objective of destroying or sufficiently damaging the follicle to achieve a severe inhibition or elimination of hair growth, as well as modification of hair shaft formation from damaged follicles, which would result in a weakened shaft incapable of penetrating the skin. The major issues with using endogenous melanin as the target chromophore for inhibition and/or modification of hair-shaft growth are related to the damage of nonfollicle structures such as epidermal melanocytes, resulting in a variety of pigmentation disorders as well as epidermal damage in the skin of African American subjects. Further, repeated laser treatments are necessary to provide the maximal degree of hair-growth elimination or hair modification. Recent evidence-based reviews on the efficacy and safety of hair-growth inhibition by lasers have led to the conclusion that the effects of lasers on hair-growth inhibition or hair-growth retardation require repeated treatments, and there is no evidence for either complete or permanent hair removal [13]. On the other hand, studies on improvement in PFB by laser treatments indicate that complete or permanent hair removal is not necessary to achieve a significant reduction in lesions. Indeed, highly effective improvements in lesion reductions can be achieved with rather modest reductions in hair growth, and the onset of improvements in PFB can be noticed well before maximal hair growth inhibition effects are noted. Adrian and Shay [14] assessed laser treatments for PFB using the 800 nm diode laser, and most patients noted a reduction in PFB symptoms within two to four weeks of an initial laser treatment, regardless of overall hair reduction success [14]. Another study employed the 800 nm diode laser at low pulse settings (10J/cm2 with pulse width of long duration 30 ms) and patients had significant decrease in ingrown hair after just one session in which hair growth was only diminished by 10–25 %; after 7–10 treatments, the PFB condition was fully resolved [15]. A study by Rogers and Glaser [16] employed the combination of a Q- switched 1080 nm Nd:YAG laser in combination with a topical carbon suspension which penetrated into the follicle orifice surrounding the hair follicle. When the carbon suspension acting as the laser chromophore is irradiated by light, the heat generated is transferred to the follicle structures that support the growth and differentiation of hair shaft resulting in modification of hair growth, and therefore a decreased opportunity for hair to enter the skin and cause inflammatory lesions. The major benefit of this strategy was to minimize the risk for collateral epidermal damage, which, in darker skin is highly pigmented and sensitive to heat transfer. Results proved encouraging as significant reductions in PFB were achieved with benefits persisting at least two months after the two treatments.

In summary, the initial studies on treatment of PFB with lasers are encouraging, as the energy threshold requirements to achieve substantial benefits for PFB were less than required for hair growth inhibition, and this indicated that the treatment of PFB with lasers could be managed with less dermal risks than those associated with hair-growth management. Nonetheless, maintenance of PFB improvements by the laser requires repeated treatments which pose problems for subjects with Type V and Type VI skin as the high melanin content in epidermis poses risks of pigmentation changes. The biggest challenge with laser-mediated hair management is not efficacy, but rather issues related to safety, and since the fluence threshold for resolving PFB appears to be less than that required for the inhibition hair growth, this presents an opportunity for PFB treatment.

The recommendations of Battle and Hobbs [17,18] for safer hair removal by the use of laser in darker skin types apply to laser treatment of PFB. Their recommendations are based on the use of long pulse durations to allow more opportunity to cool the skin, as heat in the epidermis is more easily dissipated by concurrent cooling when the heat is generated at a slow rate with longer pulse lengths. Both longer pulsed 800nm diode and 1064 nm Nd:YAG lasers deliver slower heating, and the suggested pulse lengths with the 800nm diode are greater than 100ms and for the Nd:YAG, the pulse lengths should exceed 30ms. A particular advantage for Nd:YAG is the longer wavelength, which enables epidermal sparring as result of bypassing the epidermal layer and focusing more heat generation in the lower dermal regions of the follicle where cell proliferation and early hair-shaft differentiation is initiated. With the diode laser, pulse widths above 400ms should be avoided, as too long a pulse width increases the opportunity for overheating due to excessive thermal transfer in regions of high follicle density, as well as preventing retrograde damage to epidermis as a result of thermal transfer from dermis to epidermis with excessively long pulse widths. The Nd:YAG is always the treatment of choice in subjects with Type VI skin. With either laser system, lower fluences are recommended for facial treatments as increased hair density in the beard area further increases the risks of thermal transfer between follicles.

The 1064 nm Nd:YAG laser theoretically provides the safest wavelength for treatments in dark skin Types IV-VI. Galadari [19] compared hair-removal efficacy and the safety of alexandrite, diode, and Nd;YAG lasers in these skin types.Results demonstrated similar degrees of hair-removal efficacy between the devices, but the fewest incidence of side effects was with the Nd:YAG laser, providing the conclusion that Nd:YAG allows higher fluences without epidermal damage and maximal benefits in targeting the zones related to hair inhibition in the dermis. In addition to maximizing safety, longer pulse durations maximize efficacy according to the concept of thermal damage time [20], which suggests that increased pulse duration maximizes the diffusion of heat generated in the melanized regions of the hair shaft and bulb to the surrounding follicle structures supporting hair-growth cycling and hair-shaft differentiation. Overall, the results of studies evaluating consecutive treatments with the long-pulsed Nd:YAG laser treatment indicate that it represents a safe and effective option for reducing papule formation in patients with PFB barbae, with results lasting up to two to three months after completion of successive treatments in patients with skin Types IV–VI [21–23]. On the other hand, the use of modified superlong pulse 810 nm diode lasers provided long-lasting lesion reductions after successive treatments, but its use is restricted to patients with Type IV and V skin as patients with Type VI skin experienced unacceptable side effects [24].

A prototype 980 nm laser for self-treatment has been tested for PFB, and the results reported at the 25th annual meeting of the American Society for Laser Medicine and Surgery Orlando 2005 are encouraging [25]. The device provided approximately 3–10 J/cm^2. Subjects were treated over a 2-week period receiving 10 treatments. There was a 69%

average reduction in the number of lesions, the range being 48–80% at the end of the treatment period. The majority of subjects indicated improvement in shaving bumps, and said there was greater ease of shaving after three to five treatments. The device was designed to retard hair growth and facilitate shaving, and results confirm the viability of a hair-growth retardation strategy in treating PFB.

18.7 Conclusions

A variety of treatment options exist for managing PFB, and a judicious use of a combination of treatment options may offer the best opportunity for controlling the formation of ingrown hair that result in shaving bumps and inflammation. The magnitude of improvement resulting from compliance with a daily shaving regimen to provide a very safe, effective, economical, and convenient method to manage PFB can, in some subjects, equal that of professional lasers. The use of appropriate topical treatments such as glycolic acid may enhance the ability to comply with daily shaving, as well alter superficial skin topography to minimize the occurrence of ingrown hair and maximize shaving efficiency. Topically applied hair-growth retardants such as Eflornithine HCl have been demonstrated to slow hair growth, modify the morphology of the hair follicle and the hair shafts in a manner which can reduce ingrown hair, and also improve compliance with a daily-shave regimen. These types of conservative approaches which can be used on a sustained basis to maintain therapeutic benefits without the risk of side effects represent an excellent option. The drawback of these approaches is the time it takes to notice benefits which could be as long as two to three months.

The results of clinical studies on the use of professional hair removal lasers to manage PFB is impressive in demonstrating rapid resolution of the condition. However, no data is available on the safety of repeated long-term use of these laser treatments necessary to sustain efficacy. Professional hair-removal laser treatments are expensive, and the convenience and safety of other options which can provide a similar degree of efficacy makes those appealing options as first course of treatments. Recent studies on the combination therapies of laser and Vaniqa in hair removal [26,27] suggest this combination strategy for use in PFB. A recent patent application by Shander et al. [28] using a validated laboratory model for hair-growth inhibition, which is highly predictive of clinical efficacy, demonstrated the use of topical treatments to enhance and promote the effectiveness of low fluence photo-thermal treatments for hair-growth inhibition which are minimally effective when used alone. Overall chemical thermal synergies may represent the most appealing future treatment options that can exploit the effects of light- mediated thermal damage to hair follicles along with specific actions of hair growth retardants, altering growth and differentiation of hair shafts. One can envision periodic boosters with lower fluence laser treatments along with continued treatment of hair retardants to safely and effectively prevent the ingrown hair responsible for PFB. Such treatments may not require complete elimination of hair growth, but rather sufficient modification of the processes of cell proliferation and differentiation to prevent formation of curly hair. This would also allow subjects to comfortably maintain a daily shaving regimen conducive to preventing ingrown hair.

A laser hair-removal device approved for consumer self use which is being introduced for sale in the United States in 2008 [29] is not approved for facial-hair removal, and its use in African American subjects is contraindicated. But it is possible that a device with reduced fluence appropriate for hair retardation and hair-shaft modification may be extremely useful as an adjuvant treatment in combination with topically active hair retardants. The 980 nm diode or related diode lasers with wavelength > 980 nm could be considered as part of the laser/chemistry combination treatment for self-administration. The results of safety and efficacy in reports using IPL in combination with RF to reduce the risks of exposing dark skin to high optical energy fluences of lasers are encouraging [30]. A consumer-use device incorporating the aforementioned electro-optical synergy technology (ELOS) is presently being developed by Syneron Inc. for a variety of dermatological conditions, and may be appropriate for safer treatment of darker skin types.

Although there have been recent claims that laser treatment can cure PFB, there is a lack of well-designed short- and long-term studies with prospectively defined treatment parameters documenting short-term and long-term effects, as well as addressing the critical issue of the adverse effects of repeated laser treatments on dark skin. In addition, none of the published work were designed to adequately control hair-removal regimens before, during, and after treatment, and factors such as frequency of shaving can have a large impact on the overall number of lesions during the course of treatment. Indeed, the benefits of laser treatment may be to enable daily shaving, which then adds a synergistic benefit to the outcome of the laser usage. There are also deficiencies in developing the appropriate inclusion and exclusion criteria related to the severity and previous history of PFB in subjects. The laser methods for PFB treatment require continued treatments to sustain benefits of lesion reduction, and the risks of permanent pigmentation or other skin disorders from multiple treatments may outweigh any benefits for PFB. The ideal strategy would be to employ a convenient inexpensive treatment regimen which do not impose any risks or the need of supervision by medical staff. Such an option may be provided by low fluence self-treatment devices which may effectively reduce the number of lesions, and enable daily shaving to maintain a well-groomed appearance. The success of such devices may be augmented by topical treatments and appropriate shaving regimens.

References

1. Perry PK, Cook-Bolden FE, Rahman Z, Jones E, and Taylor SC. Defining psuedofolliculitis barbae in 2001: A Review of the literature and current trends pages *J Am Acad Dermatol.* 2002; 46(2):113–19
2. Kligman AM and Strauss JS. Pseudofolliculitis of the beard. *AMA Arch Derm.* 1956 Nov; 74(5):533–42
3. Brauner GJ and Flandermeyer KL. Pseudofolliculitis barbae. Medical consequences of interracial friction in the US Army *Int J Dermatol.* 1977 Jul–Aug;16(6):520–5.
4. Braunder GJ and Flandermeyer KL. Pseudofolliculitis barbae. 2. Treatment. *Cutis.* 1979 Jan;23(1):61–4
5. Alexander AM and Delph WI. Pseudofolliculitis barbae in the military. A medical, administrative and social problem. *J Natl Med Assoc.* 1974 Nov;66(6):459–64, 479.
6. Chuh A, Zawar V. Epiluminescence dermatoscopy enhanced patient compliance and achieved treatment success in pseudofolliculitis barbae. *Australas J Dermatol.* 2006 Feb;47(1):60–2.
7. Thibaut S, Barbarat P, Leroy F, and Bernard BA. Human hair keratin network and curvature *Int J Dermatol.* 2007 Oct;46 Suppl 1:7–10.
8. Thibaut S, Gaillard O, Bouhanna P, Cannell DW, and Bernard BA. Human hair shape is programmed from the bulb *Br J Dermatol.* 2005 Apr;152(4):632–8.
9. Winter H, Schissel D, Parry DA, Smith TA, Liovic M, Birgitte Lane E, Edler L, Langbein L, Jave-Suarez LF, Rogers MA, Wilde J, Peters G, and Schweizer J. An unusual Ala12Thr

polymorphism in the 1A alpha-helical segment of the companion layer-specific keratin K6hf: evidence for a risk factor in the etiology of the common hair disorder pseudofolliculitis barbae *J Invest Dermatol.* 2004 Mar;122(3): 652–7.
10. Perricone NV. Treatment of pseudofolliculitis barbae with topical glycolic acid: a report of two studies. *Cutis.* 1993 Oct;52(4):232–5.
11. Shander D, Harrington FE, and Whitmore MC. Treatment of acne and pseudofolliculitis barbae. US Patent 5,328,686, 1994.
12. Shander D, Ahluwalia GS, Funkhouser, MG, and Coble D. Reduction of ingrown hairs and modification of beard hair morphology in African American men with pseudofolliculitis barbae (PFB) treated with topical application of 13.9% eflornithine HCl (Vaniqa) [Abstract 637]. *J Invest Dermatol.* 2005 May;124(A107), 66th Annual Meeting SID, St. Louis.
13. Haedersdal M and Wulf HC Evidence–based review of hair removal using lasers and light sources. *J Eur Acad Dermatol Venereol.* 2006;20:9–20.
14. Adrian RM and Shay KP. 800 nanometer diode laser hair removal in African American patients: a clinical histologic study *J Cut Laser Ther.* 2000;2:183–90.
15. Greppi I. Diode laser hair removal of the black patient. *Lasers Surg Med.* 2001;28(2):150–5.
16. Rogers CJ and Glaser DA. Treatment of pseudofolliculitis barbae using the Q-switched Nd:YAG laser with topical carbon suspension. *Dermatol Surg.* 2000 Aug;26(8):737–42
17. Battle EF Jr and Hobbs LM. Laser-assisted hair removal for darker skin types. *Dermatol Ther.* 2004;17(2):177–83
18. Battle EF Jr, Hobbs LM. Laser therapy on darker ethnic skin. *Dermatol Clin.* 2003 Oct; 21(4):713–23.
19. Galadari I Comparative evaluation of different hair removal lasers in skin types IV,V, VI *Int J Dermatol*, 2003;42:68–70
20. Rogachefsky AS, Silapunt S, and Goldberg DJ. Evaluation of a new super-long-pulsed 810 nm diode laser for the removal of unwanted hair: the concept of thermal damage time. *Dermatol Surg.* 2002 May;28(5):410–14.
21. Ross EV, Cooke LM, Overstreet KA, Buttolph GD, and Blair MA. Treatment of pseudofolliculitis barbae in very dark skin with a long pulse Nd:YAG laser. *J Natl Med Assoc.* 2002 Oct;94(10):888–93.
22. Ross EV, Cooke LM, Timko AL, Overstreet KA, Graham BS, and Barnette DJ. Treatment of pseudofolliculitis barbae in skin types IV, V, and VI with a long-pulsed neodymium:yttrium aluminum garnet laser. *J Am Acad Dermatol.* 2002 Aug;47(2):263–70.
23. Weaver SM 3rd, Sagaral EC. Treatment of pseudofolliculitis barbae using the long-pulse Nd:YAG laser on skin types V and VI.*Dermatol Surg.* 2003 Dec;29(12):1187–91.
24. Smith EP, Winstanley D, and Ross EV. Modified superlong pulse 810 nm diode laser in the treatment of pseudofolliculitis barbae in skin types V and VI. *Dermatol Surg.* 2005 Mar;31(3):297–301.
25. Ross E.V. Lasers in Surgery and Medicine, 2005 (abstract)
26. Hamzavi I, Tan E, Shapiro J, and Lui H. A randomized bilateral vehicle-controlled study of eflornithine cream combined with laser treatment versus laser treatment alone for facial hirsutism in women. *J Am Acad Dermatol.* 2007 Jul;57(1):54–9.
27. Smith SR, Piacquadio DJ, Beger B, and Littler C. Eflornithine cream combined with laser therapy in the management of unwanted facial hair growth in women: a randomized trial. *Dermatol Surg.* 2006 Oct;32(10):1237–43.
28. Shander D, Henry JP, Wu H, Botchkareva N, Ahluwalia GS, and Trumbore MW. Reduction of hair growth. United States Patent Application 20060134048.
29. Wheeland RG. Simulated consumer use of a battery-powered, hand-held, portable diode laser (810 nm) for hair removal: A safety, efficacy and ease-of-use study. *Lasers Surg Med.* 2007 Jul;39(6):476–93.
30. Yaghmai D, Garden JM, Bakus AD, Spenceri EA, Hruza GJ, and Kilmer SL. Hair removal using a combination radio-frequency and intense pulsed light source. *J Cosmet. Laser Ther.* 2004 Dec;6(4):201–7.

19

Light-Based Systems to Promote Wound Healing

Serge Mordon

INSERM & Lille University Hospital, Lille, France

19.1 Introduction

Wound healing is a complex and dynamic process of restoring cellular structures and tissue layers. The human adult wound-healing process can be divided into 3 distinct phases: the inflammatory phase, the proliferative phase, and the remodeling phase. Within these three broad phases is a complex and coordinated series of events that includes chemotaxis, phagocytosis, neocollagenesis, collagen degradation, and collagen remodeling. In addition, angiogenesis, epithelization, and the production of new glycosaminoglycans (GAGs) and proteoglycans are vital to the wound-healing milieu (Fig. 19.1). The culmination of these

Gurpreet S. Ahluwalia (ed.), *Cosmetic Applications of Laser and Light-Based Systems*, 369–380,

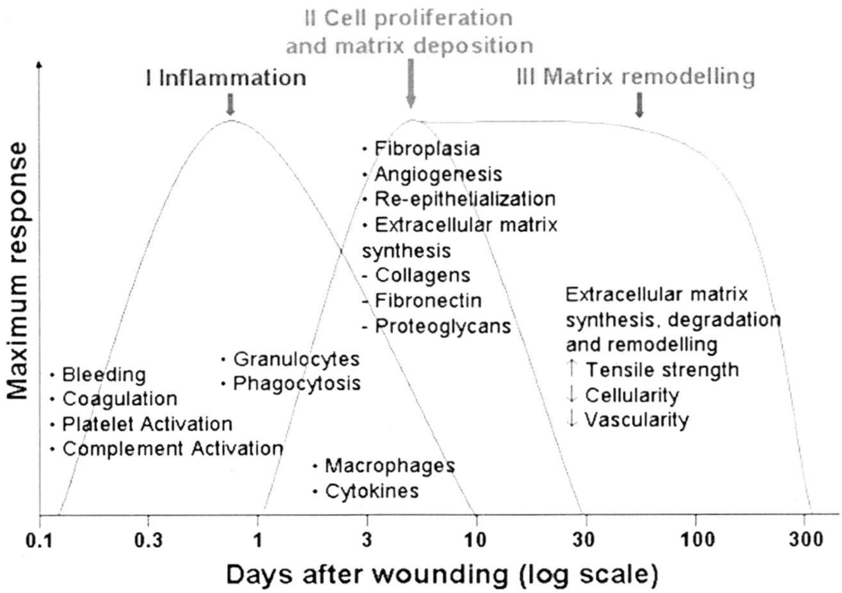

Figure 19.1 The three phases of wound healing from [36].

biological processes results in the replacement of normal skin structures with fibroblastic mediated scar tissue [1].

This process can go awry: (i) it can produce an exuberance of fibroblastic proliferation with a resultant hypertrophic scar, which by definition is confined to the wound site; (ii) further exuberance can result in keloid formation where scar production extends beyond the area of the original insult; (iii) conversely, insufficient healing can result in atrophic scar formation; and (iv) the normal reparative process can be interrupted, leading to a non-healing chronic wound.

If the treatment of acute wounds is well-defined (but it could be still improved), there is still much debate over which treatment modality to use for the treatment of nonhealing chronic wounds. Finally, the wound-healing process has been used for skin rejuvenation. However, this specific application will not be covered in this chapter.

Since the complex wound-healing process is influenced by so many factors, we always must recall the words of Ambroise Paré (1510–1590): "I dressed the wound; GOD healed it."

As we enter the twentieth century, light systems are now proposed to promote wound healing. This chapter summarizes several studies performed to date with different light systems in order to demonstrate if light-based systems could (or could not) play a role to promote wound healing.

19.2 Low-Level Laser Therapy (LLLT)

Low-level lasers have been proposed as early as 1967 by Pr. Endre Mester, in Semmelweis University Budapest, Hungary. Forty years later, there is still much debate about the role LLLT plays in wound healing.

19.2.1 Photomodulation

Photomodulation (a term very often proposed by the authors of papers on LLLT, but for which a clear definition is missing) by light in the red to near infrared (630–1000 nm) is supposed to be the key mechanism to accelerate wound healing. Photomodulation would augment recovery pathways promoting cellular viability, and restoring cellular function following injury. Since photomodulation involves the absorption of a specific wavelength of light by the photoaceptor molecule, two questions remain open: (i) which are the photoacceptors? (ii) which are the action spectra?

It has been postulated that the mechanism of photomodulation at the cellular level is based on the absorption of monochromatic visible and NIR radiation by components of the cellular respiratory chain. Absorption and promotion of electronically excited states cause changes in redox properties of these molecules, and acceleration of electron transfer (primary reactions). Primary reactions in mitochondria of eukaryotic cells are supposed to be followed by a cascade of secondary reactions (photosignal transduction and amplification chain or cellular signaling) occurring in cell cytoplasm, membrane, and nucleus [2]. Cytochrome c oxidase would be a key photoacceptor of light in the far-red to near-IR spectral range. Photostimulation would induce a cascade of signaling events initiated by the initial absorption of light by cytochrome oxidase. These signaling events may include the activation of immediate early genes, transcription factors, cytochrome oxidase subunit gene expression, and a host of other enzymes and pathways related to increased oxidative metabolism [3,4]. It has also been suggested that activation of the respiratory chain by irradiation would increase production of superoxide anions [5]. However, depending on the light dose some of these mechanisms can prevail significantly. Experiments with *E.coli* provided evidence that, at different light doses, different mechanisms were responsible (Fig. 19.2): a photochemical one at low doses, and a thermal one at higher doses [6]. In the event of

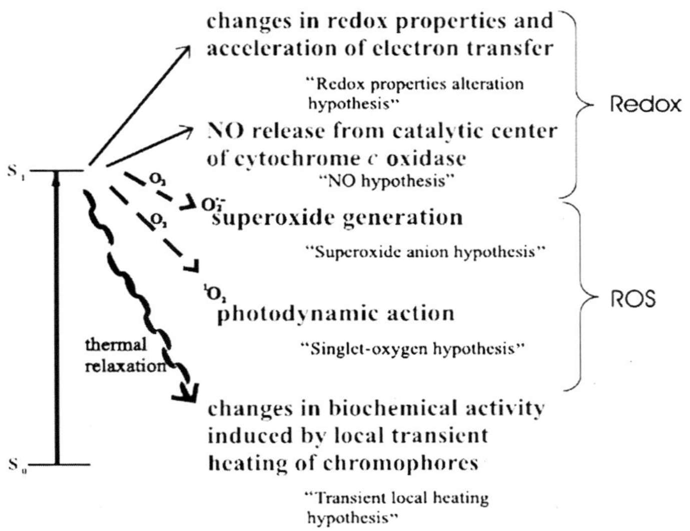

Figure 19.2 Possible primary reactions in photoacceptor molecules after promotion of excited electronic states. ROS = reactive oxygen species (from [7]).

the photoaccepetors being located in the mitochondria, Fig. 19.2 suggests three regulation pathways.

The first one is the control of the photoacceptor over the level of intracellular ATP. It is a well-known fact that even small changes in the ATP level can significantly alter cellular metabolism. The second and third regulation pathways are mediated through the cellular redox state. This may involve redox-sensitive transcription factors (NF-kB and AP-1 in Fig. 19.3), or cellular signaling homeostatic cascades from cytoplasm via the cell membrane to nucleus [7].

In the context of wound healing, only a few studies were initiated to determine the optimal action spectra (or the optimal wavelength). For example, Reedy performed a wound-healing evaluation in diabetic rats using two different wavelengths (He_Ne laser: 632.8 nm) and (Ga-As laser diode: 904 nm) using similar parameters : 7 mW–1 J/cm². Although the results indicated that both the He-Ne and Ga-As lasers enhanced the repair of healing-impaired wounds in diabetic rats compared to the controls, the magnitude of the effects differed considerably between the two lasers. The findings from the biomechanical and biochemical analysis of healed diabetic wounds demonstrated that the He-Ne laser was superior to the Ga-As laser in promoting wound repair. Further, the He-Ne laser produced greater healing effects than the Ga-As laser, with the same energy density. The differences between the He-Ne and Ga-As lasers in promoting wound repair in diabetic rats are attributed to their photochemical interaction with the cells. Evidence suggests that the absorption of light emitted by He-Ne laser at 632.8 nm initiates with the components of respiratory chain, whereas radiation emitted by the Ga-As laser at 904 nm begins at the membrane level, that is, during the cascade of molecular events that leads to photochemical response of the tissue [8].

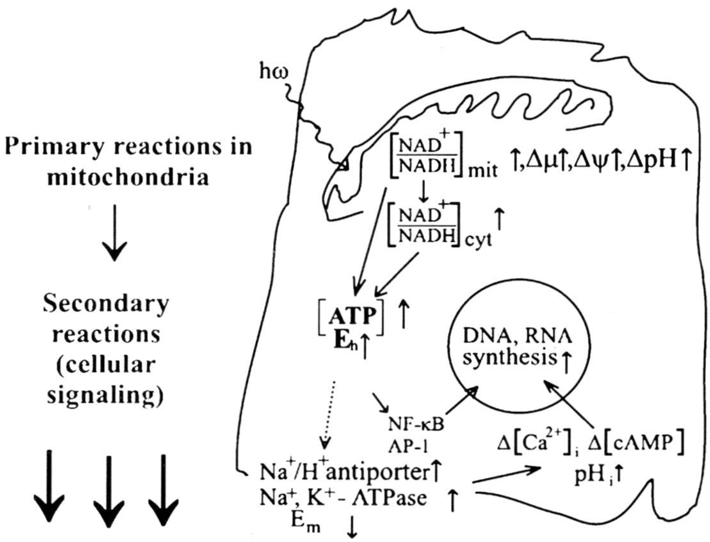

Figure 19.3 Scheme of cellular signaling cascades (secondary reactions) occurring in a mammalian cell after primary reactions in the mitochondria. E_h↑ = shift of the cellular redox potential to more oxidized direction; the arrows ↑ and ↓ indicate increase or decrease of the respective values; [ ] indicate the intracellular concentration of the respective chemicals (from [7]).

19.2.2 Experimental Studies

Lucas et al. have performed a systematic review of cell studies and animal experiment with LLLT on wound healing. Manuscripts were identified by searching Medline, Embase, and SPIE (the International Society for Optical Engineering). It was assessed whether the studies showed a beneficial effect of active treatment or not. The magnitude of the effect was expressed in standardized mean difference. In-depth analyses were performed on (1) studies in which inflicted wounds on animals were irradiated and evaluated; (2) studies with primary outcome measures on dimensions with direct reference to wound healing (ranging from acceleration of wound closure to epithelialization, but excluding surrogate dimensions regarding wound healing; in this case: tensile strength); (3) animal studies with 'true controls'; (4) studies in which animals functioned as their 'own controls' and (5) studies with the highest methodological quality score. The 36 included studies contained 49 outcome parameters of which 30 reported a positive effect of laser irradiation and 19 did not. Eleven studies presented exact data about the effect of active treatment and controls. The pooled effect (SMD) over 22 outcome measures of these studies was –1.05 (95% Cl: –1.67 to –0.43) in favor of LLLT. Methodological quality of the studies was poor. In-depth analysis of studies showed no significant pooled effect size in studies with highest methodological quality scores [9].

Medrado et al. investigated the effects of LLLT on the participation of myofibroblasts in the wound-healing process. Cutaneous wounds were inflicted on the back of 72 Wistar rats (punch-skin removal of 50 mm²). The wounds of two groups of animals were treated immediately after surgery with an AlGaInP diode laser (670 nm–9 mW) at a fluence of 4 J/cm² (exposure time: 31 seconds—Group 1) or 8 J/cm² (exposure time: 31 seconds—Group 2), while a third group consisted of untreated control animals (Group 3). On day 1, 2, 3, 5, 7, and 14 following surgery and laser treatment, fragments of skin were analyzed by histology using conventional sating, immunochemistry, and electron microscopy. Statistically significant differences in the areas of laser-treated and untreated cutaneous ulcers were observed as early as 24 hours after surgery. After 72 hours, the low level laser-treated ulcers exhibited the greatest difference when compared to untreated ulcers. This tendency for treated ulcers to be smaller than untreated ulcers was observed until the seventh day. By the fourteenth day, cutaneous wound in animals were completely healed. These gross changes were positively correlated with the microscopic findings. In treated animals, the extent of edema and the number of inflammatory cells were reduced ($p < 0.05$), but the amount of collagen and elastic fibers appeared slightly increased. The group that received 4 J/cm² laser treatment exhibited significantly more action/desmin-marked cells in correlation with its more marked vascular proliferation than the group treated at 8 J/cm². An enhanced proliferation of fibroblasts and myofibroblats was also observed.

In this study, LLLT reduced the inflammatory reaction, induced increased collagen deposition, and a greater proliferation of myofibroblasts in these experimental cutaneous wounds. However, as clearly stated by the authors, an apparent paradox was noted at the end of this study: LLLT induced a series of morphological changes, presumably favorable to the resolution of wound healing, but did not shorten healing time [10].

In another study performed by Gal et al, the purpose was to evaluate, from the histological point of view, the effect of diode laser irradiation on skin wound healing in Sprague-Dawley rats. Two parallel full-thickness skin incisions were performed on the back of each rat (n = 49) and immediately sutured. After surgery, one wound of each rat was exposed to laser irradiation (continuous mode, 670 nm, AlGaInP diode laser, irradiance: 25 mW/cm²). Each section was irradiated for 8 minutes daily to achieve the total daily dose 30 J/cm²), whereas the parallel wound was not irradiated, and served as control. Both wounds were removed 24, 48, 72, 96, 120, 144, and 168 hours after surgery and routinely fixed and embedded in paraffin sections, stained with hematoxylin and eosin, van Gieson, periodic acid Schiff + periodic acid Schiff diastase, Mallory's phosphotungstic hematoxylin, and azur and eosin, and histopathologically evaluated. As compared to nonirradiated control wounds, laser stimulation shortened the inflammatory phase,and also accelerated the proliferative and maturation phase, and positively stimulated the regeneration of injured epidermis and the reparation of injured striated muscle. LLLT at 670 nm, used at 25 mW/cm² –30 J/cm², positively influenced all phases of rat-skin wound healing [11].

When comparing these two experimental studies, performed with the same AlGaInP diode laser (670 nm), wound healing was improved, but at 4 J/cm² in the first one and 30 J/cm² in the second one. Since it is generally accepted that the biological effect of LLLT depends on three major parameters: wavelength, irradiance (or power density), and fluence (or dose), such a discrepancy between the dose is difficult to interpret. The selection of wavelengths and treatment parameters needs to be rationalized.

19.2.3 Clinical Studies

A recent report of the Agency for Healthcare Research and Quality (AHRQ) has evaluated eleven clinical studies on LLLT for wound healing [12]. Among them, nine were rated poor in quality, while one was rated fair, and only was rated to be of good quality. This higher-quality study did not show a higher probability of complete healing at six weeks with the addition of laser treatment, nor did it show benefit for any of the other reported outcomes [13]. This reports concluded that the available data suggested that the addition of laser therapy did not improve wound healing, as the vast majority of comparisons in these studies did not report any group differences in the relevant outcomes.

The authors have concluded that these studies failed to show unequivocal evidence to substantiate the decision for trials with LLLT in a large number of patients. In fact, there were no differences between the results of these experiments and clinical studies.

In conclusion, future studies should be well-controlled investigations with rational selection of lasers and treatment parameters. In the absence of such studies, the literature does not appear to support widespread use of LLLT in wound healing at this time [14]. The LLLT of wounds may increase certain aspects of healing in the early stages, but not to such a degree as to be clinically undisputable [15].

19.3 Light Emitting Diodes

Light-emitting diodes (LEDs) are much more powerful than the previous generation with quasimonochromatic outputs. These LEDs can offer target specificity to accelerate wound healing. The term "photobiomodulation," is also used since light in the far-red to near-infrared region of the spectrum (630–1000 nm) is supposed to modulate numerous cellular functions [16–19]. Similar to LLLT, experimental results demonstrate that red and near-infrared

LED light treatment stimulates mitochondrial oxidative metabolism in vitro, augments cellular energy production, and accelerates cell and tissue repair in vivo.

19.3.1 Experimental Studies

Due to the recent introduction of powerful LED, there are a very limited number of experimental studies on wound healing. Al-Watban et al. have performed a study to determine the efficacy of polychromatic LED in the enhancement of wound healing in nondiabetic and diabetic rats. A polychromatic LED (a cluster of 25 diodes emitting photons at wavelengths of 510–543, 594–599, 626–639, 640–670, and 842–879 nm) therapy has been evaluated [20]. Although the effect of polychromatic LED therapy in oval full-thickness wound-healing in the diabetic model with the use of 5 and 10 J/cm^2 was promising, further studies to determine optimum dosimetry and efficacy of LEDs were recommended by the authors [21].

19.3.2 Clinical Studies

The number of clinical studies is very limited. Trelles et al., in a small series of ten patients, have used red LED phototherapy (20 minutes, 96 J/cm^2, 633 nm), after blepharoplasty and laser ablative resurfacing. They observed that LED phototherapy cut the time for the resolution of side effects, and the healing time by half to one-third compared with contralateral unirradiated controls. However, Trelles et al. concluded that further studies were warranted with larger populations to confirm these findings [22].

In another study, Trelles et al. have evaluated Er:YAG ablation of plantar verrucae with red LED therapy to assist wound healing. Over 2 years, the author treated 121 plantar warts under local anesthesia in 58 patients with Er:YAG laser ablation, followed by red LED therapy to assist wound healing. (633 nm, 20 minutes, 96 J/cm^2). LED therapy at the same parameters was repeated on postoperative days 2, 6, and 10. To the authors, the Er:YAG laser was ideally suited for precise and speedy ablation of plantar verrucae with minimal thermal damage to surrounding tissue, which, when coupled with visible red LED therapy, had given excellent, accelerated, and pain-free healing in these difficult-to-treat and slow-to-heal lesions, with very low recurrence rates. However, due to the absence of a control group, it is not possible to conclude whether these results could be attributed to the Er:YAG only or to the Er:YAG in combination to the LED therapy [23].

19.4 Lasers

Lasers are now widely used for treating numerous cutaneous lesions, scar revision (hypertrophic and keloid scars), skin resurfacing, skin remodeling, and for fractional photothermolysis (wrinkles removal) [24]. For these treatements, lasers are used to generate heat. The modulation of the effects (volatilization, coagulation, hyperthermia) is obtained by using different wavelengths and laser parameters. The heat source obtained by conversion of light into heat can be very superficial and intense if the laser light is well-absorbed (far-infrared: CO2 or Er:YAG lasers), it can be much deeper and less intense if the laser light is less absorbed by the skin (visible or near- infrared). This heat source will always transfer its energy to surrounding tissues, and whatever the laser used, a 45° C–50° C temperature gradient will be always obtained in the skin (Fig. 19.4). If a wound-healing process exists, it cannot be induced by the dead cells, but only by live cells reacting to this low temperature increase.

The importance of temperature in the wound-healing process has been already recognized as a novel way in which to manipulate the wound-healing environment [25]. The use of heat to treat disease goes back to ancient times. Hippocrates (460–370 BC) wrote "What medicines do not heal, the lance will; what the lance does not, fire will", while Parmenides (510–450 BC) stated "Give me a chance to create fever and I will cure any disease".

The biological effects of far-infrared ray (FIR) on whole organisms remain poorly understood. However, this generated supraphysiologic level of heat is able to induce a heat–shock response, which can be defined as the temporary changes in cellular metabolism. These changes are rapid, transient, and characterized by the production of a small family of proteins termed the heat shock proteins (HSP). In this context, recent experimental studies have clearly demonstrated that HSP 70 which are overexpressed following laser irradiation could play a role with consequently a coordinate expression of other growth factors such as TGF-beta which is known to be a key element in the inflammatory response and the fibrogenic process [24,26,27]. This thermal effect induced by FIR is also known to increase microcirculation [28].

Besides, their use for skin resurfacing, skin remodeling, or rejuvenation and for fractional photothermolysis, lasers are now proposed for surgical scar-healing improvement.

Capon et al. have demonstrated that an 800 nm-diode laser could accelerate wound healing with increased tensile strength (30–58 % greater than in control groups at 7 and 15 days), and could lead to a slender scar if applied almost immediately after conventional skin suture. Histological examination has revealed a much earlier continuous epidermis and dermis, and a reorientation of collagen fiber and elastin network along the skin incision. This observation was particularly interesting, since the most significant difference between normal tissue and scar tissue is due to collagen deposition and alignment during dermal wound healing [29].

In this study, it was confirmed that laser irradiation led to a moderate increase of tissue temperature (<50° C) insufficient to create a thermal damage but high enough to activate HSP 70

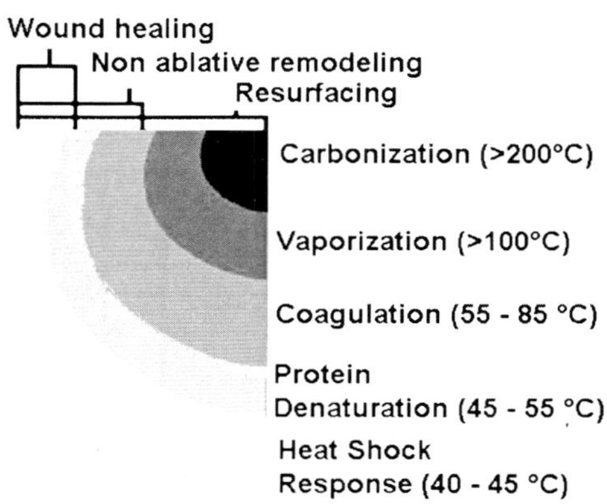

Figure 19.4 Range of thermally dependent interactions from a typical thermal laser.

which was markedly induced in skin structures examined after laser exposure[30,31]. The results observed in this experimental study show that the healing process for the skin occurs with regeneration (in opposition to reparation), a known phenomenon by fetus. Interestingly, some authors have also reported that predominant expression of TGF-B_3 compared with TGF-β_1 and TGF-β_2 induces a "scarless healing" [32,33]. Figures 19.5 and 19.6 summarize the cascade of the wound-healing process and the role of TGF-B1, which could be induced by an elevation of temperature [24].

With a different laser (a 595 nm pulsed dye laser) and a different timing (treatment of surgical scars starting on the day of suture removal), two different studies have shown that the final cosmetic appearance of scar was significantly better for the laser-treated scars when compared to untreated scars [34,35]. In both studies, each scar was divided at the midline into two fields, with half receiving laser-treatment in order to eliminate any bias due to the comparison of different scars. In the laser-treated scar, the fibroblast number was

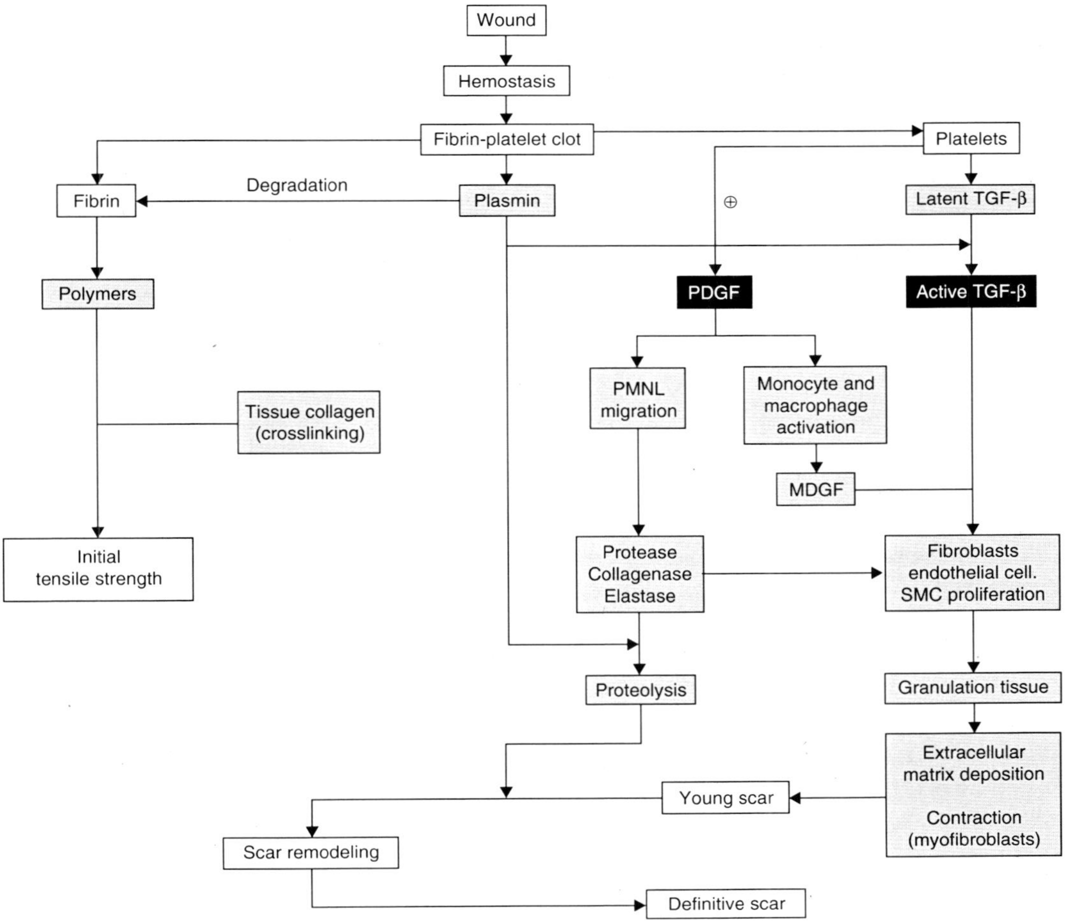

Figure 19.5 Normal wound healing cascade (from [24]). MDGF = macrophage-derived growth factor; PDGF = platelet-derived growth factor; PMNL = polymorphonuclear leukocyte; SMC = smooth muscle cells; TGF-β = transforming growth factor-β.

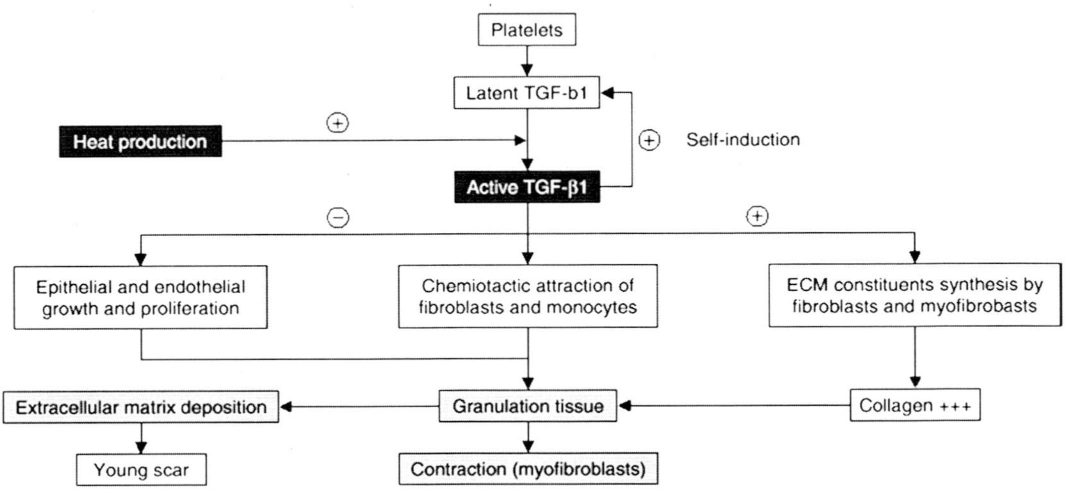

Figure 19.6 role and induction of TGF-β_1 in the wound healing process. TGF-β_1 activation is induced by the heat-shock response [24].

similar to normal skin, the collagen alignment possessed normal multidirectionality, and more elastin fibers were present in the treated sides. Treated halves showed more preservation of normal tissue architecture with more of an elastin tissue network present, whereas the untreated scars had more extensive, visible scarring with decreased elastin tissue networks. To the authors, the 595 nm pulsed dye laser was a safe and effective option to improve the cosmetic appearance of surgical scars in skin types I–IV starting on the day of suture removal.

19.5 Conclusions

Several light-based systems have been proposed to promote wound healing. For LLLT and LED, despite numerous experimental papers, their efficacy in humans needs to be demonstrated in well-designed clinical studies. These systems may increase certain aspects of healing in the early stages, but not to such a degree as to be clinically undisputable. The principle of action of this low level light (photomodulation) is still debated.

Concerning lasers, where the principle of action is based on the generation of low temperatures, numerous studies have demonstrated that lasers play an indisputable role in the wound-healing process, in particular for incisional scars, scar revision, laser-assisted skin closure, laser remodeling, fractional photothermolysis, and laser resurfacing. For these techniques, lasers are used to generate heat. The modulation of the effects (volatilization, coagulation, hyperthermia) is obtained by using different wavelengths and laser parameters. The heat source obtained by conversion of light into heat can be very superficial and intense if the laser light is well-absorbed (far-infrared : CO2 or Er:YAG lasers); it can be much deeper and less intense if the laser light is less absorbed by the skin (visible or near-infrared). This heat source will always transfer its energy to surrounding tissues, and whatever the laser used, a 45–50° C temperature gradient will be always obtained in the skin. If a wound-healing process exists, it cannot be induced by the dead cells, but only by live cells reacting to this low temperature increase. This generated supraphysiologic level of heat is able to induce a heat–shock response, which can be defined as the temporary changes in cellular metabolism.

References

1. Mercandetti M and Cohen A. Wound healing, healing and repair. In eMedicine from WebMD, 2005.
2. Karu T. Molecular mechanism of the therapeutic effect of low-intensity laser radiation. *Lasers Life Sci.* 1988; 2(1):53–74.
3. Karu T. Laser biostimulation: a photobiological phenomenon. *J Photochem Photobiol B.* 1989; 3(4):638–40.
4. Silveira PC, Streck EL, and Pinho RA. Evaluation of mitochondrial respiratory chain activity in wound healing by low-level laser therapy. *J Photochem Photobiol B.* 2007; 86(3):279–82.
5. Karu T, Andreichuk T, and Ryabykh T. Changes in oxidative metabolism of murine spleen following laser and superluminous diode (660–950 nm) irradiation: effects of cellular composition and radiation parameters. *Lasers Surg Med.* 1993; 13(4):453–62.
6. Karu T, Tiphlova O, Esenaliev R, and Letokhov V. Two different mechanisms of low-intensity laser photobiological effects on Escherichia coli. *J Photochem Photobiol B* 1994; 24(3):1556–1.
7. Karu T. Low power laser therapy. In: Vo-Dinh T, editor. *Biomedical Photonics Handbook.* Boca Raton, FL: CRC Press, 2003, pp. 48-1–48-25.
8. Reddy GK. Comparison of the photostimulatory effects of visible He-Ne and infrared Ga-As lasers on healing impaired diabetic rat wounds. *Lasers Surg Med.* 2003; 33(5):344–51.
9. Lucas C, Criens-Poublon LJ, Cockrell CT, and de Haan RJ. Wound healing in cell studies and animal model experiments by Low Level Laser Therapy; were clinical studies justified? a systematic review. *Lasers Med Sci.* 2002; 17(2):110–34.
10. Medrado AR, Pugliese LS, Reis SR, and Andrade ZA. Influence of low level laser therapy on wound healing and its biological action upon myofibroblasts. *Lasers Surg Med.* 2003; 32(3):239–44.
11. Gal P, Vidinsky B, Toporcer T, Mokry M, Mozes S, Longauer F, et al. Histological assessment of the effect of laser irradiation on skin wound healing in rats. *Photomed Laser Surg.* 2006; 24(4):480–8.
12. Samson D, Lefevre F, and Aronson N. Wound-Healing Technologies: Low-Level Laser and Vacuum-Assisted Closure. Rockville, MD: Agency for Healthcare Research and Quality, December 2004.
13. Lucas C, van Gemert MJ, and de Haan RJ. Efficacy of low-level laser therapy in the management of stage III decubitus ulcers: a prospective, observer-blinded multicentre randomised clinical trial. *Lasers Med Sci.* 2003; 18(2):72–7.
14. Posten W, Wrone DA, Dover JS, Arndt KA, Silapunt S, and Alam M. Low-level laser therapy for wound healing: mechanism and efficacy. *Dermatol Surg.* 2005; 31(3):334–40.
15. Surinchak JS, Alago ML, Bellamy RF, Stuck BE, and Belkin M. Effects of low-level energy lasers on the healing of full-thickness skin defects. *Lasers Surg Med.* 1983; 2(3):267–74.
16. Desmet KD, Paz DA, Corry JJ, Eells JT, Wong-Riley MT, Henry MM, et al. Clinical and experimental applications of NIR-LED photobiomodulation. *Photomed Laser Surg.* 2006; 24(2):121–8.
17. Eells JT, Wong-Riley MT, VerHoeve J, Henry M, Buchman EV, Kane MP, et al. Mitochondrial signal transduction in accelerated wound and retinal healing by near-infrared light therapy. *Mitochondrion.* 2004; 4(5–6):559–67.
18. Whelan HT, Buchmann EV, Dhokalia A, Kane MP, Whelan NT, Wong-Riley MT, et al. Effect of NASA light-emitting diode irradiation on molecular changes for wound healing in diabetic mice. *J Clin Laser Med Surg.* 2003; 21(2):67–74.

19. Whelan HT, Smits RL, Jr., Buchman EV, Whelan NT, Turner SG, Margolis DA, et al. Effect of NASA light-emitting diode irradiation on wound healing. *J Clin Laser Med Surg.* 2001; 19(6):305–14.
20. Al-Watban FA and Andres BL. Polychromatic LED therapy in burn healing of non-diabetic and diabetic rats. *J Clin Laser Med Surg.* 2003; 21(5):249–58.
21. Al-Watban FA and Andres BL. Polychromatic LED in oval full-thickness wound healing in non-diabetic and diabetic rats. *Photomed Laser Surg.* 2006; 24(1):10–6.
22. Trelles MA and Allones I. Red light-emitting diode (LED) therapy accelerates wound healing post-blepharoplasty and periocular laser ablative resurfacing. *J Cosmet Laser Ther.* 2006; 8(1):39–42.
23. Trelles MA, Allones I, and Mayo E. Er:YAG laser ablation of plantar verrucae with red LED therapy-assisted healing. *Photomed Laser Surg.* 2006; 24(4):494–8.
24. Capon A and Mordon S. Can thermal lasers promote skin wound healing? *Am J Clin Dermatol.* 2003; 4(1):1–12.
25. Khan AA, Banwell PE, Bakker MC, Gillespie PG, McGrouther DA, and Roberts AH. Topical radiant heating in wound healing: an experimental study in a donor site wound model*. *Int Wound J.* 2004; 1(4):233–40.
26. Toyokawa H, Matsui Y, Uhara J, Tsuchiya H, Teshima S, Nakanishi H, et al. Promotive effects of far-infrared ray on full-thickness skin wound healing in rats. *Exp Biol Med.* (Maywood) 2003; 228(6):724–9.
27. Wagstaff MJ, Shah M, McGrouther DA, and Latchman DS. The heat shock proteins and plastic surgery. *J Plast Reconstr Aesthet Surg.* 2007; in press.
28. Yu SY, Chiu JH, Yang SD, Hsu YC, Lui WY, and Wu CW. Biological effect of far-infrared therapy on increasing skin microcirculation in rats. *Photodermatol Photoimmunol Photomed.* 2006; 22(2):78–86.
29. Dallon J, Sherratt J, Maini P, and Ferguson M. Biological implications of a discrete mathematical model for collagen deposition and alignment in dermal wound repair. *IMA J Math Appl Med Biol.* 2000; 17(4):379–93.
30. Capon A, Souil E, Gauthier B, Sumian C, Bachelet M, Buys B, et al. Laser assisted skin closure (LASC) by using a 815-nm diode-laser system accelerates and improves wound healing. *Lasers Surg Med.* 2001; 28(2):168–75.
31. Souil E, Capon A, Mordon S, Dinh-Xuan AT, Polla BS, and Bachelet M. Treatment with 815-nm diode laser induces long-lasting expression of 72-kDa heat shock protein in normal rat skin. *Br J Dermatol.* 2001; 144(2):260–6.
32. Shah M, Foreman DM, and Ferguson MW. Neutralisation of TGF-beta 1 and TGF-beta 2 or exogenous addition of TGF-beta 3 to cutaneous rat wounds reduces scarring. J *Cell Sci.* 1995; 108 (Pt 3):985–1002.
33. Shah M, Foreman DM, and Ferguson MW. Neutralising antibody to TGF-beta 1,2 reduces cutaneous scarring in adult rodents. *J Cell Sci.* 1994; 107 (Pt 5):1137–57.
34. Nouri K, Jimenez GP, Harrison-Balestra C, and Elgart GW. 585-nm pulsed dye laser in the treatment of surgical scars starting on the suture removal day. *Dermatol Surg.* 2003; 29(1):65–73; discussion 73.
35. Conologue TD and Norwood C. Treatment of surgical scars with the cryogen-cooled 595 nm pulsed dye laser starting on the day of suture removal. *Dermatol Surg.* 2006; 32(1):13–20.
36. Enoch S and Price P. Cellular, molecular and biochemical differences in the pathophysiology of healing between acute wounds, chronic wounds and wounds in the aged. World Wide Wounds, 2004.

PART 5
SYNERGY OF BIOACTIVE MOLECULES WITH LIGHT ENERGY

20

Synergy of Eflornithine Cream with Laser and Light-Based Systems for Hair Management

Gurpreet S. Ahluwalia and Douglas Shander

The Gillette Company, a wholly owned subsidiary of The P&G Company, Needham, MA, USA

20.1 Introduction

When one speaks of beauty in women, it is generally the facial characteristics of women that one is referring to. With this in mind, nothing is more devastating to women, or can impact the

Gurpreet S. Ahluwalia (ed.), *Cosmetic Applications of Laser and Light-Based Systems*, 383–397,

feminine beauty at its core, than the presence of facial hair. It is estimated that over 40 million women in the United States alone suffer from this problem, and at least half of them routinely remove facial hair. A recent report by Lipton et al. [1] showed that women living with facial hair suffer from a high level of emotional distress and psychological morbidity. Women in the study were highly bothered (81%) and self-conscious (70%) about their condition, and felt overwhelmed with the effort that was needed for them to keep their facial hair under control. Another study compared excessive facial hair in women to dermatological conditions, such as psoriasis and eczema, as having a similar dermatology life quality index score [2].

The conventional hair-removal methods that women generally rely on for facial hair include: shaving, depilatory creams, bleaching, waxing, mechanical epilators, and plucking. Each of these methods has certain advantages and shortcomings that influence their acceptability, depending on the desired outcome. While these methods are convenient and low in cost, they have significant limitations, including: the lack of femininity, post-shave stubble, and skin irritation from the shaving method; the offensive sulfur odor and skin irritation from the use of depilatory creams; skin sensitivity and poor performance of bleaching creams used to hide the pigmented terminal hair; skin irritation, in-grown hair, messiness, and pain/discomfort of epilation methods such as waxing, mechanical epilators, and plucking. These are some of the disadvantages associated with the aforementioned methods. Moreover, hair removal by these conventional means is temporary, lasting from 1–2 days after shaving to up to several weeks following epilation (waxing and epilators).

In spite of the wide range of options available, facial hair management remains a significant challenge for women. Depending on the extent of the problem, women may choose just one or multiple methods to control facial hair. The determining factors are the location of the hair: upper lip, chin, cheek or neck; pigmentation: blonde/gray to dark; hair type: terminal, vellus or vellus-like growth; character: coarseness, thickness, density, and the growth rate of hair. Women with significant facial hirsutism, defined as a heavy male pattern terminal hair growth, are often not satisfied with the conventional epilation or depilation methods mentioned earlier and require a combination of both medical and cosmetic treatments to manage their condition. The medical treatment options include an Rx topical cream Vaniqa that is used to inhibit the rate of hair growth, the hormonal treatments that effect the androgen-dependent hair growth and the laser or intense-pulsed light (IPL) treatments to primarily remove the pigmented hair. For treatment of clinically hirsute patients, the systemically administered hormonal drug treatments include steroidal and nonsteroidal anti-androgens namely, spironolactone, flutamide, cyproterone acetate, finasteride, and cimetidine [3–10]. The anti-androgen approach has a sound scientific basis for the androgen-dependent facial hair growth, though it has not been highly successful because of its limited efficacy and safety issues [11–14]. A prolonged treatment time of up to one year is generally required to see good efficacy with the hormonal treatments. The attraction for the eflornithine cream, Vaniqa is that it can be applied topically, similar to a facial moisturizer, and carries no significant safety concerns. The biochemical pharmacology of eflornithine and its use in combination with the laser treatment is described in detail in the following sections.

20.2 Anti-Proliferative Activity of Eflornithine

Eflornithine (α-Difluoromethylornithine; DFMO) is an irreversible inhibitor of ornithine decarboxylase, a critical rate-limiting enzyme in the *de novo* biosynthesis of the polyamines putrescine, spermidine, and spermine [15]. Although the precise function of polyamines in cellular proliferation is not well-understood, they seem to play a fundamental role in the synthesis and/or regulation of DNA, RNA, and protein. The implication that polyamines are somehow involved in cell proliferation stems from the fact that high levels of ornithine decarboxylase (ODC) enzyme and polyamines are found in cancer and other cell types that have a high proliferation rate. The observations that polyamines might play a role in tumor development [16–18], led to an immense amount of work done in late 1970s and all through the 1980s on ODC and polyamine metabolism. The cellular inhibition of ODC by eflornithine causes a marked reduction in putrescine and spermidine, and a variable reduction in spermine depending on the length of treatment and the cell type. A number of excellent reviews have been written on ODC regulation and polyamine metabolism [19–22]. Eflornithine binds the active site of ODC as a substrate, and is then decarboxylated just like the natural substrate ornithine, however, during this catalytic process; a reactive intermediate is formed that covalently attaches itself to the enzyme's active site, causing a permanent deactivation or suicide-inhibition of the enzyme [23,24]. The cellular half-life of ODC of about 30 minutes is one of the shortest of all known enzymes [24]. Thus, in order to achieve a significant antiproliferative effect, the enzyme inhibition must be sustained by a constant inhibitory level of the inhibitor. Among the number of ODC inhibitors that have been synthesized and evaluated, eflornithine by far is the most studied [24].

Exposure of normal as well as a cancerous cell to eflornithine causes a dose-dependent reduction in polyamine levels and results in an anti-proliferative effect on cells. In a comparative study of seven anticancer amino acid analogs using a panel of 60 cultured human tumor cell lines, Ahluwalia et al. found eflornithine to be devoid of the cytotoxic activity that was a common feature for the other six agents evaluated in the study [25]. The antiproliferative effect of eflornithine has therefore been attributed to its cytostatic rather than cytotoxic activity [26]. Probably because of this cytostatic mode of action, eflornithine has not been very effective as a single agent in human anticancer studies [27,28]. However, in the course of anticancer investigations, it was found that eflornithine has a low potential for systemic toxicity at doses that are effective in inhibiting ODC and reducing cellular polyamine levels [25,26,29,30]. Based on this rationale, the therapeutic potential of eflornitine was investigated at low doses (< 1.0 gm/m^2/day) for chemopreventive activity against colon, bladder, and breast cancers [31–33]. Because of a lack of clear therapeutic end point of chemoprevention studies, target tissue polyamine levels, were used as surrogate markers to assess treatment successes [34].

Several active molecules, synthetic and natural, have been investigated by the National Cancer Institute for their potential chemotherapeutic use. These include, nonsteroidal anti-inflammatory drugs (NSAIDs), antioxidents, retinoids and carotenoids, Vitamins (C&E), polyphenols, and eflornithine [32]. Among these, eflornithine has been the most studied and has the most scientific rationale for development as a chemopreventive agent. One of the primary objectives for any chemoprevention study trial is to find a therapeutic drug dose that is either devoid of toxicity or has minor and acceptable level of side effects for chronic and perhaps life-time use. An eflornithine dose of 0.5 gm/m^2/day has been indicated in several long-term dose-seeking studies to be the dose without significant toxicity [35,36]. Ototoxicity, in particular, the low-frequency hearing loss has been shown to be the most sensitive measure of the drug-related toxic effects in human. Based on this toxicity outcome, the observed 'no-effect' dose of eflornithine was determined to be 0.4 gm/m^2/day or 0.74 gm/day for a typical 1.85 m^2 person [37].

20.2.1 Other Uses of Eflornithine

Inhibiting polyamine synthesis is highly toxic to protozoal parasites [38]. Based on this observation, eflornithine was developed for the treatment of meninogoencephalitic stage of *Trypanosoma brucei gambiense* infection (African sleeping sickness). It is available as an Rx intravenous dose under the brand name Ornidyl from Aventis, for treatment of this infection. As a monotherapy, eflornithine has a high cure rate of 94% for Gambian sleeping sickness [39,40], and was found to be the only effective treatment for this disease in cases of melarsoprol relapse [41]. The high intravenous eflornithine dose (400 mg/kg/day or approximately 24 g/day) used for this indication has been associated with systemic adverse effects and hematological toxicities.

Other topical use for which eflornithine has been investigated includes actinic keratoses and pseudofoliculitis barbae (PFB). Topical treatment with eflornithine in a hydrophilic ointment for six months caused a significant reduction in the AK lesions [42]. Use of eflornithine cream as an adjunct to other hair-removal methods is recommended for subjects suffering from the PFB condition [43].

20.3 Effect of Eflornithine on Hair Follicle Growth

In line with the observations that highly proliferative cells tend to have a high polyamine requirement to support growth, it was found that hair follicle, which is one of the most proliferative organ in the body, expresses a high level of ODC activity [44]. Hair growth in humans is a cyclic process with periods of growth (anagen), transition (catagen), and rest (telogen). Using immunocytochemical analysis, Sundberg reported changes in the ODC expression in relation to the hair-follicle growth cycle [45]. In telogen, the ODC expression was detected only in a small group of outer root sheath cells near the 'bulge' region, whereas in anagen ODC expression was found in the whole length of the follicle with particularly high levels in the hyperproliferative matrix region. A similar observation was made by Nancarrow et al. [46], who found high levels of ODC in the anagen growth phase and a very low to nearly undetectable ODC level in the catagen and the resting telogen phases. Hynd and Nancarrow observed that among the three major polyamines, spermidine plays the most critical role in hair growth [47]. Their research further demonstrated an important role of polyamines in hair-fiber formation and keratin- gene expression in hair follicles. In contrast to the high level of ODC in growing follicles, its levels in normal epidermis are very low, unless stimulated by a chemical or mechanical agent. Topical application of a tumor promoter, such as phorbol esters causes a rapid and dramatic increase in the epidermal ODC expression [48,49].

The observations that high polyamine levels are probably required for maintaining hair growth, but not the normal epidermal function and turnover, led to a rational investigation of topical ODC inhibitors for controlling hair growth. Among a number of available ODC inhibitors tested, eflornithine was found to be the most effective in reducing follicular ODC, polyamine levels, and the rate of hair growth [50–52]. Eflornithine was also found to be a highly effective hair-growth inhibitor in an established animal model for the androgen-dependent hair growth [52,53]. One of the key issues for any topically applied agent in terms of both its safety and efficacy is the dermal absorption. Preclinical and clinical pharmacokinetic studies showed that topically applied eflornithine is poorly absorbed through the skin.

Being a highly hydrophilic molecule, eflornithine skin penetration remains limited to less than 1% of the applied dose after single or multiple application [53]. In women, the mean percutaneous absorption after the first (single) dose was found to be 0.34%, which reached 0.82% after multiple doses, representing the likely use conditions [54,55]. The absorbed drug was found to be eliminated, essentially unchanged, via the renal route [53]. Based on eflornithine's favorable preclinical and clinical toxicology and the data from human pharmacokinetic studies, the Gillette Company (now P&G) initiated the development of this molecule as a topical treatment for controlling facial hair growth. The Phase-I trial was conducted in moderately hirsute women as an open-label study using a 10% eflornithine dose in a hydro-alcoholic vehicle. Results of the study indicated no significant adverse events, dermal or systemic, from the drug treatment. The preliminary efficacy observations showed a significant improvement in condition as determined by the clinician scoring on visibility of facial hair, and also the subject's own perception of their improved appearance [56]. Based on the favorable safety and efficacy outcome from Phase-I testing, a dose ranging Phase-II study was conducted at 0, 5, 10, and 15%, concentration of eflornithine. monohydrochloride. monohydrate in a cream-based formulation. Gillette patented the eflornithine cream formulation based on its physio-chemical and hair-growth reduction activity that was later trademarked as Vaniqa™ [57]. The objective analyses in the Phase-II trial, included hair-length measurements taken by a video-imaging instrument. The results demonstrated a statistically significant hair growth reduction of 47% in the 15% eflornithine group compared to a nonsignificant 8% reduction in the vehicle-control group. The hair length-reduction in the 5 and 10% eflornithine groups was 26 and 28%, respectively, that was found to be not statistically different from the vehicle-control group. The physician global scoring of improvement in the condition and the subjects' own perceptions of treatment benefits were found to be consistent with the objective hair length measures [53]. The effective 15% eflornithine cream from the Phase-II study was carried forward to the Phase-III trials conducted in partnership with Bristol-Myers Squibb Company.

20.4 Eflornithine Cream Vaniqa™: An Rx Topical Product for Unwanted Hair Growth

Several factors contributed to the successful development of an eflornithine preparation VaniqaTM (Rx product) that received the US regulatory approval in July 2000. These included: 1. a clearly defined mechanism of action of eflornithine for reduction of hair growth; 2. a favorable systemic and dermal safety profile in doses that can illicit a pharmacological response; 3. a long-term use clinical safety data in humans from the chemoprevention studies; 4. a favorable dermal safety from Phase-I and Phase-II clinical trials and a dose dependant facial hair growth reduction demonstrated in a double blinded, randomized, placebo control studies; 5. a poor safety and/or efficacy profile of other available modalities such as hormonal treatments; and 6. a large unmet consumer need. Vaniqa contains 13.9% of eflornithine monohydrochloride in a moisturizing lotion base for topical application [57]. Studies reporting 15% eflornithine use, in fact, has 15% of eflornithine. monohydrochloride. monohydrate or 11.5% eflornithine in the cream vehicle containing 80.84% water, 4.24% glyceryl stearate, 4.09% PEG-100 stearate, 3.05% cetearyl alcohol, 2.5% ceteareth-20, 2.22% mineral oil, 1.67% steryl alcohol, 0.56% dimethicone, and 0.83% of paraben-based preservative cocktail.

20.4.1 Efficacy of Vaniqa

Vaniqa is the first and only product that has been clinically demonstrated to be safe and effective for hair-growth reduction. Its Food & Drug Administration (FDA)-approved indicated use is for the reduction of unwanted facial hair in women. Unlike the oral hormonal treatments, which can only effect androgen-dependent hair growth, Vaniqa works on all growing hair. The pivotal Phase-III clinical studies on Vaniqa to further assess its safety and efficacy in a larger population, and to obtain regulatory approval were jointly sponsored by the Gillette Company (now P&G) and Bristol-Myers Squibb. In two randomized double-blind studies 594 patients with terminal facial hair, self-treated twice daily with Vaniqa (393 patients) or the placebo cream without eflornithine (201 patients) for a period of 24 weeks [58]. This was followed by a no-treatment phase of eight weeks. A 48-hour hair-growth image was captured after shaving the treatment area at baseline, at several visits during the treatment period, and at the end of the follow-up period. The primary efficacy measure was improvement from baseline in the Physician Global Assessment (PGA) score based on the presence of terminal hair and darkening of the facial skin due to the terminal hair. A four-point scoring system was used for efficacy determination. The treatment group statistically showed significant improvement by the eighth week. By the end of the treatment, 35% of the subjects in the Vaniqa group showed marked improvement or better, compared to only 9% in the vehicle control group. Up to 70% of the subjects in the treatment group had at least some improvement. After stopping the Vaniqa treatment, there was a rapid recovery in hair growth. At eight weeks after treatment (follow-up) hair growth returned to near baseline levels.

In addition to the PGA scoring, there were two secondary efficacy measures: A quality of life assessment and an objective hair-growth measure [59,60]. The quality of life assessment is an important measure, as it addresses the issues that are important to the patient. A set of six questions related to the bother and discomfort that subjects with facial hirsutism feel, was administered at various intervals during the study [59]. The improvement in the 'bother' score from baseline was found to be at least twofold greater in the eflornithine treatment group compared to the vehicle control. At the end of the 24-week treatment period, the difference in improvement in the drug group compared to the vehicle control group was highly significant statistically, for each of the six patient-reported outcomes.

To demonstrate the quantitative changes in hair growth, a video-imaging system was used. Spatial hair mass (area in mm^2) and hair growth (length in mm) were determined at various time points during the study [60]. In two separate studies, the Vaniqa treatment group reached a statistically significant difference in hair mass, compared to the placebo control, in as early as two weeks after starting the treatment, and maintained the statistical difference throughout the treatment period in both the studies.

20.4.2 Vaniqa Safety

Dermal safety of 13.9 % eflornithine cream Vaniqa was assessed in women in four open-label, vehicle-control studies performed by Hickman et al. [61]. The studies included contact sensitization study (230 subjects), a cumulative irritation test (30 subjects), a phototoxicity assessment (25 subjects), and a photocontact allergy study (30 subjects). The data showed that eflornithine cream Vaniqa™ does not have contact sensitizing, photocontact allergic or phototoxic properties. However, under the exaggerated conditions of use, the drug was found to cause some skin irritation. Phase-III clinical trials involving 1967 women in blinded vehicle-control studies (594 subjects), and vehicle control and open label studies (1373 subjects), showed that adverse events for most body systems occurred at similar frequency between the drug and the control groups (Rx Vaniqa package insert). The most frequent treatment-related adverse events were dermal effects at the site of application. A high incidence of acne, 21.3% in the Vaniqa group and 21.4% in the vehicle-control group, indicated that the cream vehicle had comedogenic potential. The treatment-related skin stinging was higher in the Vaniqa™ group (7.9%) than the vehicle control group (2.5%).

20.4.3 Efficacy Limitations of Vaniqa

Though eflornithine is an effective and specific inhibitor of ODC, its efficacy is somewhat limited and variable, probably because of its low skin penetration, a poor target tissue accumulation, and the rapid turnover rate of ODC enzyme. The result is a wide range of efficacy from nearly no effect in some subjects to as much as 70% hair reduction in others. Because eflornithine is a highly hydrophilic molecule with no inherent affinity for penetrating the lipid-rich skin structure, various penetration enhancers have been evaluated to increase its penetration. Agents such as cis-fatty acids, fatty alcohols, fatty acid esters, select terpines, and nonionic surfactants have shown good penetration enhancement ability in in vitro human-skin model [62]. Whether this in vitro effect will translate into efficacy benefits in humans is yet to be studied.

A mean facial hair-growth reduction of 47% by Vaniqa [58] implies that women must use adjunctive hair-removal methods to maintain a hair-free look. The conventional hair-removal methods such as shaving, epilation, and depilatory creams only provide an adjunctive-use benefit without a true synergy. Further, this needed adjunctive use does not necessarily reduce the overall effort women must put into managing their facial hair. Laser hair removal with its own limitations, as a standalone modality, has the potential to provide synergy with Vaniqa in managing facial hair growth.

20.5 Laser Hair Removal

Based on the principle of selective photothermolysis first described by Anderson and Parrish [63], several laser and light-based systems have been developed to affect hair growth. Under this principle, advantage is taken of the high concentration melanin pigment at the base of the anagen hair follicle and its absence in the surrounding dermal tissue that can selectively absorb and convert to heat select wavelengths of laser and light energy. Based on the hair-follicle pigment concentration and the laser parameters used, the amount of thermal energy released can cause mild to severe damage to the follicle, resulting in a range of hair reduction from being a temporary to a permanent effect [64–67]. In general, the efficacy of a laser or light-based treatment is proportional to the melanin concentration in the hair follicle and the amount of laser energy used. A darker hair generally responds well, whereas the treatment is less effective or even ineffective on gray, blonde, red or light-brown hair. The presence of melanin in the skin epidermal layer is the determining factor as to how much laser energy (fluence) can be safely used without inducing dermal damage. The face remains one of the key areas for which women seek laser treatment. Because hair growth on the face tends to be less uniform than other body areas, and pigmentation and growth rate can vary significantly depending on the facial site (upper lip, chin, cheek or neck), it presents certain challenges for achieving a consumer satisfactory effect with the laser treatment alone. Laser hair removal, particularly for the face, is described in detail in other chapters in this book (see Goldberg, chapter 5; Styczynski, chapter 6 and Sadick, chapter 7).

20.6 Synergy of Vaniqa and Laser for Hair Management

Among chemical treatments for controlling facial hair growth, Vaniqa remains the only FDA-approved product for this indication, and among the physical/photothermal methods, laser or light-based treatment is the only modality clinically proven for hair reduction. However, both these methods have certain limitations, the critical one being the level of hair reduction effectiveness. Though eflornithine is color blind, that is, works irrespective of skin and hair color, it has demonstrated only a modest level of efficacy, and though laser can be highly effective, its efficacy and safety is highly dependent on the subject's hair and skin color. The two methods also suffer from a wide range in effectiveness, related in part to inherent technology and a broad range of hair character in the population. By combining these two modalities, one can achieve an efficacy synergy that is highly satisfactory to the consumer and the benefit readily perceptible.

Clinical studies have shown that Vaniqa treatment at the recommended twice daily dose can significantly enhance the hair-removal effectiveness of a laser treatment regimen. A randomized, double-blind and vehicle-controlled study was reported by Smith et al. using laser in combination with and without Vaniqa [68]. The study was conducted at two clinical sites in 54 women who had upper lip and chin hair. In this split-face comparison, patients applied Vaniqa on one side of the face and a placebo cream on the other side. Both sides were treated with a laser. The treatments were performed either with the alexandrite (18–30 J/cm^2; 3 ms pulse) or the Nd:YAG system (30–50 J/cm^2; 50–100 ms pulse). The laser parameters were selected based on the patient's tolerability with the goal of administering a maximal tolerable dose. Subjects received two laser treatments, one at Week 2 and the second at Week 10. Subjects applied Vaniqa and the placebo cream twice a day on split-face for 34 week duration of the study. Physician's Global Assessment score was used as the primary efficacy measure. The scoring system was similar to the one used in the Vaniqa phase-III clinical trials that supported its regulatory approval. The PGA scores were used to determine change from baseline, and to assess differences between the Vaniqa and the placebo-cream side. In addition, subject's self-perceptions of their condition between the left and the right side were used to assess Vaniqa benefits. Results from the PGA scores indicated statistical differences favoring the Vaniqa side for both the upper lip and chin hair at most of the four time points evaluated between Week-6 and Week-22. There were no statistical differences for the PGA scores at Week 34 of the study, indicating a significant hair growth recovery in the 24-week period after the last laser treatment. Subjects, on the other hand, perceived significant differences favoring the Vaniqa side both for the upper lip and chin starting as early as Week-2 and lasting throughout the study to Week-34. One reason for the discrepancy could be that the subjects in their assessment took into account both the vellus and the terminal hair, whereas the PGA scoring was done strictly based on the presence of terminal hair. While laser affects primarily the terminal hair, and Vaniqa both terminal and vellus (both to a lesser degree) combining the two modalities seem to provide a consumer-perceptible benefit that is greater in magnitude than either treatment alone. In addition, the enhancement in efficacy for the combination was achieved without any observable increase in the dermal side effects. There were no significant differences in the types, rates, and severity of dermal adverse events between the Vaniqa and the placebo-cream side.

Another Vaniqa and laser combination clinical trial was conducted by Hamzavi et al. [69,70]. The study included 31 subjects, and was designed as a randomized, double-blind, placebo-controlled split-body test similar in design to the Smith study [68]. The subjects treated one side of their upper lip with Vaniqa and the other side with the placebo cream twice a day in a blinded manner, starting at the time of first laser treatment and continuing until two weeks after the last laser treatment. Both sides received treatment with an alexandrite laser. As compared to just two laser treatments in the Smith study [68], this study included up to six laser treatments performed at monthly intervals. The laser parameters of pulse duration (10–40 ms) and fluence (7–40 J/cm^2) were varied depending on the subject's tolerance and skin reaction; however, both sides on a given subject were treated with the same laser parameters to provide a within-subject comparison. Three efficacy measures used in the study—global scoring by the investigator, objective hair count and patient self-assessment of benefits, all showed a significant benefit difference between the laser and Vaniqa combination compared to laser alone. With the combination, a complete or nearly complete hair-removal efficacy was achieved in over 90% (29/31) of the subjects. On the side with the placebo cream and laser, the level of efficacy was seen in only 68% of the subjects (21/31). Neither study [68,69] had a group-comparing efficacy of Vaniqa alone (without laser). In the Vaniqa Phase-III testing, using a similar global scoring system 47% of the subjects showed a marked improvement or better in their facial hair condition [58]. For the laser/Vaniqa combination, in addition to the PGA scoring, the objective hair count and the patient self-assessment measures all showed a statistical difference favoring the Vaniqa side [59]. The study also demonstrated that the laser and Vaniqa combination results in a longer lasting "hair-free" period. There were no significant dermal adverse events or differences in dermal effects between the Vaniqa and the placebo cream sides.

Vaniqa and laser combination has also been evaluated for removal of gray hair, treatment of darker skin (Type V/VI) and for treating the PFB condition. In a case report, a patient with a mixture of gray and black hair showed a dramatic improvement with the combination of laser (diode) and Vaniqa [71]. Laser treatment alone had left a significant amount of gray hair and the combination of laser and Meladine, a pigment enhancer, also proved to be ineffective. In addition to working on gray hair, the laser/Vaniqa combination seems to work well for hair removal on Fitzpatrick skin types IV–VI. The issue with the darker skin is that lower laser fluences need to be used to maintain dermal safety. Combination of laser and Vaniqa has been described by Callender to be the most effective first-line therapy for the treatment of facial hair in skin types IV–VI [72]. The combination provided quicker results with much greater efficacy. In a retrospective study of 74 African American patients with facial hair who were treated with a combination of Vaniqa (twice daily) and Nd:YAG (long-pulsed or Q-switched), improvements were noted in the severity of hirsutism, PFB condition, and hyperpigmentation (PIH).

20.6.1 Combination of Eflornithine Cream with a Low-Fluence Laser Treatment

The use of lasers for hair removal carries a certain level of potential risk related to dermal side effects, and has certain limitations for treating darker skin or lighter hair. These risks and limitations get further amplified when treatments need to administered to one's face. The risk of causing a permanent or long-lasting dermal effect, such as scarring or pigmentary change is highly dependent on the laser dose used, and the subject's skin type. We have investigated the effect of a relatively low-dose diode laser treatment on the upper-lip hair in a split-body test using eflornithine cream (Vaniqa) on one side and the placebo cream on the contralateral side. The study was carried out as a randomized, double-blind, placebo-control test on 16 women with upper-lip hair. An 810 nm diode laser system (SLP 1000, Palomar Medical Technologies, Burlington, MA) was used with parameter settings of 90 W, 50 ms pulse and 12 mm aperture that translated to a low 4 J/cm^2 skin exposure. Subjects were treated with the laser once every week for 8 weeks, and the Vaniqa and placebo cream applications were made twice daily for 12 weeks. There was a three-month follow-up period after the last Vaniqa treatment. Efficacy response to this low laser dose (fluence) was highly variable among subjects. The denser, darker coarser hair responded well to the laser treatment, with or without the concomitant treatment with Vaniqa (Fig. 20.1). On the other hand, sparser, less pigmented, or thinner hair had responses that ranged from a moderate effect to no response at all from the laser treatment alone. Addition of Vaniqa made a dramatic difference in providing hair reduction benefits for subjects who were either poor or nonresponders to the laser treatment. As shown in Fig. 20.2 (left side), the subject treated with the placebo cream and laser combination had no response, even after the eight-weekly

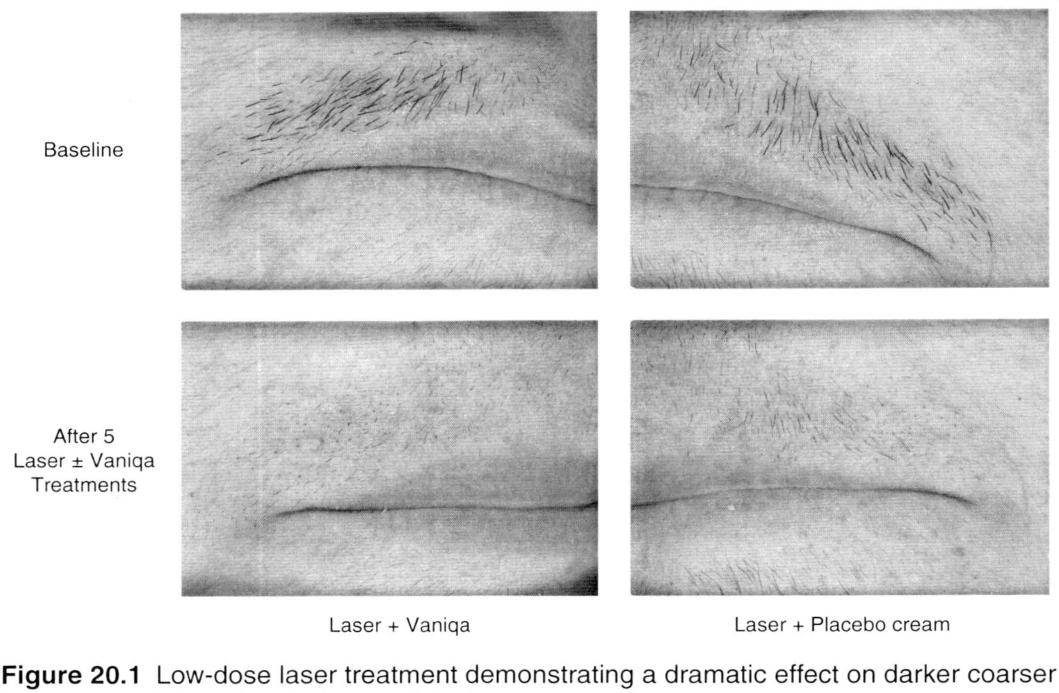

Figure 20.1 Low-dose laser treatment demonstrating a dramatic effect on darker coarser hair with or without the Vaniqa combination.

treatments. The lack of efficacy is probably related to the suboptimal low-fluence used for the thinner, finer hair. However, a combination treatment with Vaniqa and the same laser dose resulted in a complete clearance of terminal hair and a significant reduction in the finer vellus like hair on the contra lateral right side of the subject (Fig. 20.2). For consumers where only a low laser dose can be safely used, the combination can make the difference between an effective treatment and a satisfied consumer, against a complete lack of efficacy and a highly dissatisfied consumer.

A manuscript with the detailed study design and the results is being prepared for a future publication.

20.7 Conclusion

In the past decade, lasers and IPL systems have become increasingly popular for hair removal or permanent hair reduction. However, their market penetration still remains very

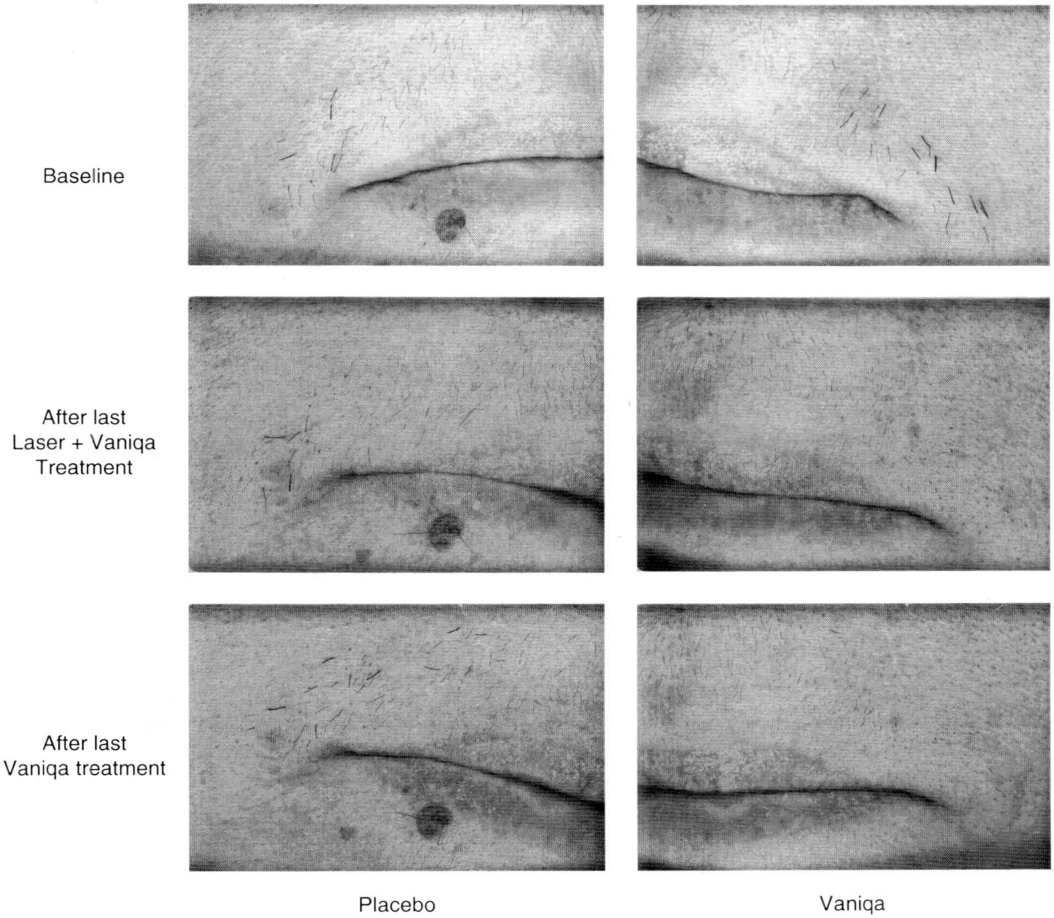

Figure 20.2 Effect of low-level laser on thinner sparser hair growth with and without the Vaniqa combination.

low, at about 1%. There are multiple factors that have contributed to their low acceptability including expense, inconvenience, efficacy limitations, and skin safety risks. The efficacy and safety issues become particularly important when treating the facial areas. Nonuniformity of hair types and pigmentation, and possibly an androgenic hormonal involvement makes the facial hair more resistant to laser treatment. Further, the sensitivity around the beauty aspect of the face prevents one from using laser fluences higher than what a subject can comfortably tolerate, because of the risk of causing dermal damage such as blistering, scarring or permanent pigmentary changes. The use of hair- growth reduction chemical technologies in combination with a laser or IPL system provides an opportunity for enhanced effectiveness. The efficacy synergy demonstrated for the combination of Vaniqa and a laser system represents an opportunity for overcoming limitations of either technologies alone, and meeting the threshold of consumer satisfaction for facial hair removal.

References

1. Lipton MG, Sherr L, Elford J, Rustin MHA, and Clayton WJ. Women living with facial hair: the psychological and behavioral burden. *J. Psychosomatic Res.* 2006; 61: 161–8.
2. Loo WJ and Lanigan SW. Laser treatment improves quality of life of hirsute females. *Clin. Exp. Dermatol.* 2002; 27: 439–41.
3. Spritzer PM, Lisboa KO, Mattiello S, and Lhullier F. Spironolactone as a single agent for long-term therapy of hirsute patients. *Clin. Endocrinol. (Oxf.)* 2000; 52(5): 587–94.
4. Erenus M, Yucelten D, Durmusoglu F, and Gurbuz O. Comparison of finasteride versus spironolactone in the treatment of idiopathic hirsutism. *Fertil. Steril.* 1997; 68(6): 1000–3.
5. Erenus M, Yucelten D, Gurbuz O, Durmusoglu F, and Pekin S. Comparison of spironolactone-oral contraceptive versus cyproterone acetate-estrogen regimens in the treatment of hirsutism. *Fertil Steril.* 1996; 66(2): 216–9.
6. Wong IL, Morris RS, Chang L, Spahn MA, Stanczyk FZ, and Lobo RA. A prospective randomized trial comparing finasteride to spironolactone in the treatment of hirsute women. *J. Clin. Endocrinol. Metab.* 1995; 80(1): 233–8.
7. Barth JH, Cherry CA, Wojnarowska F, Dawber RP. Spironolactone is an effective and well tolerated systemic antiandrogen therapy for Hirsute women. *J. Clin. Endocrinol. Metab.* 1989; 68(5): 966–70.
8. Shapiro G and Evron S. A novel use of spironolactone: treatment of hirsutism. *J. Clin. Endocrinol. Metab.* 1980; 51(3): 429–32.
9. Moghetti P, Tosi F, Tosti A, Negri C, Misciali C, Perrone F, Caputo M, Muggeo M, and Castello R. Comparison of spironolactone, flutamide, and finasteride efficacy in the treatment of hirsutism: a randomized, double blind, placebo-controlled trial. *J. Clin. Endocrinol. Metab.* 2000; 85(1): 89–94.
10. Faloia E, Filipponi S, Mancini V, Di Marco S, and Mantero F. Effect of finasteride in idiopathic hirsutism. *J. Endocrinol. Invest.* 1998; 21(10): 694–8.
11. Andrade RJ, Lucena MI, Fernandez MC, Suarez F, Montero JL, Fraga E, and Hidalgo F. Fulminant liver failure associated with flutamide therapy for hirsutism. *Lancet.* 1999; 353: 983.
12. Wysowski DK and Fourcroy JL. Flutamide hepatotoxicity. *J. Urol.* 1996; 155(1): 209–12.
13. Wallace C, Lalor EA, and Chik CL. Hepatotoxicity complicating flutamide treatment of hirsutism. *Ann. Intern. Med.* 1993; 119(11): 1150.
14. Committee on Safety of Medicine. Hepatic reactions with cyproterone acetate (Cyprostat, Androcur). Med Control Agency. 1995; 21: 1.
15. Pegg, AE, Hibasami H, Matsui I, and Bethell DR. Formation and interconversion of putrescine and spermidine in mammalian cells. *Adv. Enzyme Regul.* 1981; 19: 427–51.

16. Heby O and Russell DH. Changes in polyamine metabolism in tumor cells and host tissues during tumor growth and after treatment with various anticancer agents. In: *Polyamines in normal and neoplastic growth*. 1973. Russell, D.H. (ed.) Raven Press, New York, pp. 221–237.
17. Heby O, Oredsson SM, and Kanje M. Polyamine biosynthetic enzymes as targets in cancer chemotherapy. *Adv. Enzyme Regul.* 1984; 22: 243–264
18. Boutwell RK, O'Brien TG, Verma AK, Weekes RG, Deyoung LM, Ashendel CL, and Astrup EG. The inhibition of ornithine decarboxylase activity and its control in mouse skin epidermis. *Adv. Enzyme Regul.* 1979; 17: 89–112.
19. Pegg AE and McCann PP. Polyamine metabolism and function. *Am. J. Physiol.* 1982; 243: C212–21.
20. Tabor CW and Tabor H, Polyamines. *Annu. Rev. Biochem.* 1984; 53: 749–90.
21. McCann PP, Pegg AE, and Sjoerdsma A. (eds) Inhibition of Polyamine Metabolism: Biological Significance and Basis for New Therapies. 1987; Academic Press, New York.
22. McCann PP and Pegg, AE. Ornithine decarboxylase as an enzyme target for therapy. *Pharmac. Ther.* 1992; 54: 195–215.
23. Metcalf BW, and Bey P. Catalytical irreversible inhibition of mammalian ornithine decarboxylase (E.C.4.1.17) by substrate and product analogs. *J. Am. Chem. Soc.* 1978; 100: 2551–3.
24. Bey P, Danzin C, and Jung M. Inhibition of basic amino acid decarboxylase involved in polyamine biosynthesis. In: Inhibition of Polyamine Metabolism: Biological Significance and Basis for New Therapies. 1987; pp. 1–27. McCann PP, Pegg AE, and Sjoerdsma A. (eds) Academic Press, New York.
25. Ahluwalia GS, Hao Z, Paull K, Stowe E, and Cooney DA. Control of cancer by amino acid analogs. In: *AntiCancer Drugs: Antimetabolite Metabolism and Natural Anti Cancer Agents*. 1994. Powis, G. (ed.) Pergamon Press, New York.
26. Marton LJ, and Pegg AE. Polyamines as targets for therapeutic intervention. *Ann. Rev. Pharmacol. Toxicol.* 1995; 35: 55–91.
27. Dunzendorfer U, and Kroner M. Therapy with inhibitors of polyamine biosynthesis in refractory prostatic carcinoma: an experimental and clinical study. *Onkologie*. 1985; 8: 196–200.
28. Meyskens FL, Kingsley EM, Glatke T. et al. Phase II study of alpha-difluoromethylornithine (DFMO) for the treatment of metastatic melanoma. *Invest New Drugs*. 1986; 4: 257–62.
29. Love R, Carbone P, Verma A. et. al. Randomized phase I chemoprevention dose seeking study of alpha difluoromethylornithine. *J. Natl. Cancer Inst.* 1993; 85: 732–7.
30. Meyskens F, Emerson S, Pelot D, et. al. Dose de-escalation chemoprevention trial of alpha difluoromethylornithine in patients with colon polyps. *J Natl. Cancer Inst.* 1994; 86: 1122–1130.
31. Meyskens FL, and Gerner EW. Development of difluoromethylornithine as a chemoprevention agent for the management of colon cancer. *J. Cellular Biochem* (suppl). 1995; 22: 126–131.
32. Krishnan K, Ruffin MT, and Brenner DE. Chemoprevention of colorectal cancer. *Critical Reviews in Oncology Hematology*. 2000; 33: 199–219.
33. Loprinzi CL, and Messing EM. A prospective clinical trial of difluoromethylornithine (DFMO) in patients with resected superficial bladder cancer. *J. Cellular Biochem. (Suppl.)*. 1992; 161: 153–5.
34. Meyskens FL, Gerner EW, Emerson S, Pelot D, Durbin T, Doyle K, and Lagerberg W. Effect of alpha- difluoromethylornithine on rectal mucosal levels of polyamines in a randomized, double-blinded trial for colon cancer prevention. *J. Natl. Cancer Inst.* 1998; 90: 1212–18.
35. Loprinzi C. et.al. Toxicity Evaluation of Difluoromethylornithine: Doses for Chemoprevention Trials. *Cancer Epidemiol. Biomarkers Prev.*, 1996; 5: 371–4.
36. Love RR, et.al. Randomized phase I chemoprevention dose-seeking study of alpha difluoromethylornithine. *J. Natl. Cancer Inst.* 1993; 85: 732–7.
37. Doyle K.J, et. al. Effects of Difluoromethylornithine Chemoprevention on Audiometry Thresholds and Otoacoustic Emissions. *Arch. Otolaryngol. Head Neck Surg.* 2001; 127: 553–8.
38. Fairlamb AH, Henderson GB, Bacchi CJ, and Cerami A. In vivo effects of Difluoromethylornithine on trypanothione and polyamine levels in bloodstream forms of Trypanosoma brucei. *Mol. Biochem. Parasitol.* 1987; 24: 185–91.

39. Khonde N, Pepin J, andMpia BA. Seven day course of eflornithine for relapsing T.b. gambiense sleeping sickness. *Trans. R. Soc. Trop. Med. Hyg*. 1997; 91: 212–213.
40. Pepin J, Khonde N, Maiso F, et. al. Short-course eflornithine in Gambian sleeping sickness: a multicenter randomized controlled trial. *Bulletin of the World Health Organization*. 2000; 78: 1284–95.
41. Pepin J, and Milord F. The treatment of human African trypanosomiasis. *Adv. Parasitol.* 1994; 33: 1–47.
42. Alberts DS, Dorr RT, Einspahr JG, Aickin M, Saboda, et. al. Chemoprevention of Human Actinic Keratosis by Topical 2-(difluoromethyl)-dl-ornithine. *Cancer Epidemiol. Biomarkers Prev.* 2000; 9: 1281–6.
43. Bridgemen-Shah S. The medical and surgical therapy of pseudofolliculitis barbae. *Dermatologic Therapy*. 2004; 17: 158–63.
44. Probst E and Krebs A. Ornithine decarboxylase acitivity in relation to DNA synthesis in mouse interfollicular epidermis and hair follicles. *Biochem. Biophys. Acta*. 1975; 407: 147–57.
45. Sundberg JP, Erickson AA, Roop DR, and Binder RL. Ornithine decarboxylase expression in cutaneous papillomas in SENCAR mice is associated with altered expression of keratins 1 and 10. *Cancer Res.* 1994; 54: 1344–51.
46. Nancarrow MJ, Nesci A, Hynd PI, and Powell BC. Dynamic expression of ornithine decarboxylase in hair growth. *Mech. Dev.* 1999; 84 (1–2): 161–4.
47. Hynd PI, and Nancarrow MJ, Inhibition of polyamine synthesis alters hair follicle function and fiber composition. *J. Invest. Dermatol.* 1996; 106: 249–53.
48. O'Brien TG, Simsiman RC, and Boutwell RK. Induction of the polyamine biosynthetic enzymes in mouse epidermis and their specificity for tumor promotion *Cancer Res.* 1975; 35: 2426–33.
49. O'Brien TG, Simsiman RC, and Boutwell RK. Induction of the polyamine biosynthetic enzymes in mouse epidermis by tumor -promoting agents. *Cancer Res.* 1975; 35: 1662–70.
50. Shander D, Mudd L, and Usdin V. Inhibition of ornithine decarboxylase (ODC) activity in hamster flank organ: a novel assay for topical screening of antiandrogens. 65th Ann Meeting of Endocrine Soc. 1983; Abstract # 655.
51. Shander D. Hair growth modification with ornithine decarboxylase inhibitors. 1988; US Patent Number 4,720,489.
52. Shander D, Funkhouser MF, and Ahluwalia GS. Pharmacology of hair growth inhibition by topical treatment with eflornithine – HCL monohydrate (DFMO) using flank organ model. 59th Meeting of Am Acad of Dermatol. 2001; Abstract # 228.
53. Shander D, Ahluwalia G, and Morton JP. Management of unwanted facial hair by topical application of eflornithine. 2005; pp. 489–510. In: *Cosmeceuticals and Active Cosmetics*. Elisner P, Maibach HI., (eds). Taylor & Francis, FL.
54. Malhotra, B., Palmisano, M., Schrode, K., Huber, F., Altman, D.J., Ahluwalia, G.S. Percutaneous absorption, pharmacokinetics and dermal safety of eflornithine 15% cream in hirsute women. J. Am Acad. Dermatology (suppl) 1999; vol 40; 57th Annual Meeting (abstract).
55. Malhotra B, Noveck R, Behr D, and Palmisano. M. Percutaneous absorption and pharmacokinetics of eflornithine HCL 13.9% cream in women with unwanted facial hair. *J Clin Pharmacol.* 2001; 41: 972–8.
56. Shenenberger DW and Utecht LM. Removal of unwanted facial hair. *Am Fam Physician.* 2002; 66: 1907–11.
57. Boxall BA, Amery GW, and Ahluwalia GS. Topical composition for inhibiting hair growth. 1997; US Patent Number 5,648,394.
58. Schrode, K. S., Huber, F., Staszak, H., Altman, D. J., Shander, D., Ahluwalia, G.S., Morton, J. Randomized, double-blind, vehicle-controlled safety and efficacy evaluation of eflornithine 15% cream in the treatment of women with excessive facial hair. J. Am Acad. Dermatology (suppl) 1999; vol 40; 57th Annual Meeting (abstract).
59. Jackson., Caro JJ, Caro G, Garfirld F et. al., The effect of eflornithine 13.9% cream on the bother and discomfort due to hirsutism. *Int. J. Dermatol.* 2007; 46: 976–81.
60. 60. Funkhouser, M.F., Shander, D., Schrode, K.S., Huber, F., Staszak, H., Altman, D.J. Use of a video-imaging system to obtain hair measurement data in controlled clinical trial evaluating

the safety and efficacy of eflornithine 15% cream in the treatment of excessive facial hair in women. J. Am Acad. Dermatology (suppl) 1999; vol 40; 57th Annual Meeting (abstract).

61. Hickman, JG, Huber F, and Palmisano M. Human dermal safety studies with eflornithine HCL 13.9% cream (Vaniqa™), a novel treatment for excessive facial hair. *Current Medical Res and Opinion*. 2001; 16: 235–44.
62. Ahluwalia GS, Styczynski P, and Shander D. Reduction of hair growth. 2003. US Patent Application Publication No.: US 2003/0199584 A1.
63. Anderson RR and Parrish JA. Selective photothermolysis: precise microsurgery by selective absorption of pulsed radiation. *Science*. 1983; 220: 524–7.
64. Haedersdal M, and Wulf HC. Evidence-based review of hair removal using lasers and light sources. *J. European Acad. Derm. Venereol*. 2006; 20: 9–20.
65. Bjerring P, Zacharia H, Lybecker H, and Clement M. Evaluation of the free-running ruby laser for hair removal – a retrospective study. *Acta. Derm. Venereol*. (Stockh) 1998; 78: 48–51.
66. Solomon MP. Hair removal using the long-pulsed ruby laser. *Ann. Plast. Surg*. 1998; 41: 1–6.
67. Connolly CS, and Paolini L. Study reveals successful removal of unwanted hair with LPIR laser. *Cosmet. Dermatol*. 1997; 10: 38–40.
68. Smith SR, Piacquadio DJ, Beger B, and Littler C. Eflornithine cream combined with laser therapy in the management of unwanted facial hair growth in women: a randomized trial. *Dermatol Surg*. 2006; 32: 1237–1243.
69. Hamzavi I, Tan E, Shapiro J, and Lui H. A randomized bilateral vehicle-controlled study of eflornithine cream combined with laser treatment versus laser treatment alone for facial hirsutism in women. *J. Am Acad Dermatol*. 2007; 57: 54–59.
70. Hamzavi I, Tan E, Shapiro J, and Lui H. Combined treatment with laser and topical eflornithine is more effective than laser treatment alone for removing unwanted facial hair – a placebo controlled trial. *Lasers in Surgery and Medicine*. 2003; 32 (suppl 15): 32.
71. Ganger LK and Hamzavi IH. Excess salt and pepper hair treated with a combination of laser hair removal and topical eflornithine HCL. *J. Drugs Dermatol*. 2006; 5: 544–545.
72. Callender V and Young C. Combination laser and eflornithine HCL 13.9% cream: A first-line therapy for Fitzpatrick type IV–VI patients with excessive facial hair. *J. Am Acad. Dermatol*. 2005; 52: 209.

21

Photodynamic Therapy for Acne, Rejuvenation, and Hair Removal

Macrene Alexiades-Armenakas

Department of Dermatology, Yale University School of Medicine, New York, NY, USA

Gurpreet S. Ahluwalia (ed.), *Cosmetic Applications of Laser and Light-Based Systems*, 399–414,

21.1 Introduction

Photodynamic therapy (PDT) is a century-old treatment for neoplastic conditions, which most recently has evolved to treat acne and photoaging. PDT is often used for therapeutic challenges; for example, it is currently being tested by the author for the removal of unpigmented hair. An oxygen-dependent reaction between a photosensitizer and light, PDT has employed various photosensitizers and light sources to target distinct cutaneous conditions and tissues [1,2]. Systemic porphyrins, such as hematoporphyrin were the earliest photosensitizers used, but associated with the unwanted side effect of prolonged photosensitivity [1]. Over the past two decades, topical photosensitizers, such as 5-aminolevulinic acid (ALA), and more recently, methylated ALA (MAL) have been developed and have become the most common photosensitizers for dermatologic use. Light wavelengths are chosen according to the porphyrin absorption spectrum: blue wavelengths corresponding to the Soret band, the largest 400 nm peak, and additional wavelengths corresponding to the Q bands in the 500–700 nm range. Broad-band blue light, and red lasers, and light have been studied extensively. The FDA approved indications include topical ALA and blue light since 1999, and topical MAL and red light since 2006. The long-pulsed pulsed dye laser (LP PDL, 595 nm) and intense pulsed light (IPL) are the most recent light sources with favorable results in PDT, with the advantages of greatly minimizing side effects, without compromising efficacy for the treatment of actinic keratoses, photoaging, and acne vulgaris [1,2]. Areas of further research include PDT for the removal of unpigmented blonde or gray hair, a current challenge in dermatology.

21.2 Acne Treatment with PDT

21.2.1 Background

Laser and light treatments for acne have included the FDA-approved blue light and diode 1450 nm laser, both of which act by targeting the sebaceous glands [3]. The disadvantages of these treatments are that they require multiple treatments and are modestly effective, with recurrences being relatively common [3]. The recent strict regulation of isotretinoin, a highly effective acne medication, associated with many untoward effects has reinvigorated the search for acne treatment alternatives, such as PDT. In the 1990s, early studies of red light and lasers following the application of topical ALA for the treatment of acne showed promising efficacy, but with the disadvantage of significant side effects, such as blistering and dyspigmentation [3]. Recently, studies employing topical ALA followed by illumination with either LP, PDL, or IPL have indicated a promising level of safety and efficacy, with little-to-no side effects [4–6]. PDT has become a viable treatment option for acne patients for whom conventional therapies do not work.

21.2.2 Light Treatment Alone

Light alone can activate the PDT reaction by stimulating porphyrins which accumulate within the sebaceous follicle in acne. Porphyrins, mainly coproporphyrin III, are produced by *Propionibacterium acnes*, the anaerobic bacterium that proliferates and causes inflammation in obstructed sebaceous follicles [7]. Endogenous porphyrins serve as photophores to mediate the PDT reaction following exposure to light, particularly blue light [8]. This reaction generates singlet oxygen and mediates bacterial destruction within the sebaceous follicles [8].

Treatment with blue light alone has been mildly to moderately effective in clinical trials, likely due to poor skin penetration by blue light. Blue light once weekly for two weeks in one study showed a 25% improvement in acne severity [9]. Another study of blue light 15 min per day resulted in a 30 and 15% mean reduction for inflammatory and comedonal acne lesions, respectively, after 4 weeks; with a final mean improvement of 63 and 45%, respectively, after 12 weeks [10]. Another study of blue light once weekly for 4 weeks yielded a 43% reduction [11].

While red wavelengths are less effective at porphyrin photoactivation, they do achieve greater penetration depth, thereby targeting the more deeply situated sebaceous follicles. Red light (660 nm) combined with blue light (415 nm) for 15 min daily increased efficacy to a mean lesional reduction of approximately 50% for inflammatory and 25% for comedonal acne after 4 weeks, and 76 and 58% respectively, after 12 weeks [10]. These values resulting from a single treatment with red and blue light combined, were superior to the 30 and 15% mean reduction for inflammatory and comedonal acne lesions, respectively, after 4 weeks; and the final mean improvement of 63 and 45%, respectively, after 12 weeks observed with blue light alone [10]. Recently, red light (635–670 nm) as monotherapy 15 min twice daily for 8 weeks was shown to reduce acne counts (59% decrease) as compared to controls (3% increase) in a split-face design study of 28 patients with acne vulgaris, followed to the 8 week post-treatment interval [12]. Thus, these findings suggest that red light alone demonstrates higher efficacy, as compared to blue light alone in mediating PDT and treating acne.

21.2.3 Systemic ALA with Light

The porphyrins produced by *P. acnes* in the sebaceous follicle are in small quantities, and exogenous ALA has been shown to concentrate in sebaceous units [13]. Intraperitoneal injection of ALA into albino mice was shown to result in the accumulation of protoporphyrin IX (PpIX) in the sebaceous glands of normal skin such that exposure to light of the appropriate wavelength destroyed the sebaceous glands [13]. Experiments aimed at characterizing the intracellular localization of PpIX in sebocytes following ALA application have shown that the fluorescence localizes in cellular membranes and distinctive spots, though efforts at identifying these intracellular spots have been unsuccessful [14]. While these data suggest that systemic ALA is highly effective at targeting sebaceous glands, this systems, approach is hampered by the side effect of prolonged photosensitivity.

21.2.4 Topical ALA and Red Wavelengths

Topical ALA application has also been shown to result in PpIX fluorescence at a higher level within acne lesions than in surrounding normal skin [15]. Subsequent exposure to red light was shown to shrink sebaceous glands [15]. Topical ALA for 3-hour incubation under occlusion followed by red broad-band light (550–700 nm) yielded significant improvement in acne counts, though side effects of blistering, erythema, edema, and dyspigmentation were reported [15]. In another study, topical ALA at 3-hour incubation followed by red diode laser (635 nm) demonstrated similar efficacy and side effects in the treatment of back acne [16]. In the aforementioned study, comparing blue (415 nm) to red (660 nm) light without ALA demonstrated that the longer red wavelengths enhanced efficacy rates, likely due to deeper penetration depth [10]. Similarly, in two studies of topical ALA at 4-hour incubation followed by red (635 nm) laser and polychromatic (600–700 nm) light, significant clearing of acne was observed, however, both light sources were accompanied by significant side effects of crusting and hyperpigmentation [17,18]. While effective, the combination of topical ALA with red wavelengths was associated with side effects of pain, erythema, blistering, crusting and dyspigmentation [19].

21.2.5 Topical ALA and Blue Light

Topical ALA and blue light (peak 417 nm) has yielded more modest efficacy in reported studies, likely due to the shallower penetration depth of these shorter wavelengths. In one study, blue light therapy once weekly for 2 weeks resulted in 25% improvement for light alone, and 32% improvement following ALA application prior to illumination [9]. Similarly, another study demonstrated a 43% acne clearance rate for blue light when it alone was applied once a week for 4 weeks, as opposed to a 60% response when ALA incubation was added prior to illumination for 30–60 min [11]. In another study of 18 patients, short incubation ALA with activation by blue light or intense pulsed light (IPL) demonstrated greater improvement in the IPL group as compared to blue-light group [20]. Topical ALA for a short 1-hour incubation followed by LP PDL as compared to blue light also demonstrated lower effectiveness in the ALA blue-light group [4]. Recently, the phase IIa FDA trials of topical ALA and blue light also failed to demonstrate higher efficacy in the ALA blue-light group as opposed to blue light alone, except when subgroup analysis was performed. The phase IIb trials of topical ALA comparing blue light and LP PDL to controls would be helpful in clarifying these observations.

21.2.6 Topical ALA and LP PDL

Among the Q bands in the PPIX absorption spectrum, a peak at 575 nm is amenable to activation through PDL [21]. PDL (585 nm) following topical 20% ALA was used for the treatment of AK, but with purpura and crusting (see Photoaging) [22]. LP PDL (595 nm) following short incubation ALA was shown to be safe and effective for AK with minimal and rapidly resolving side effects [23]. The advantages of this light source included variable pulse duration in the nonpurpuric range; a longer wavelength with greater penetration depth as compared to blue light; dynamic cooling spray to minimize discomfort; large 10 mm spot size, and great rapidity of treatment with a firing speed of 1 Hz [23]. By targeting hemoglobin, the additional advantage of minimizing erythema within acne lesions or scars is also proffered by this light source.

A reduction in sebaceous hyperplasia was observed following ALA and LP PDL, a finding which was subsequently reproduced [24,25]. On the basis of these findings, topical ALA followed by LP PDL and PDL was assessed in a pilot study of acne patients employing a single treatment with 1-month follow-up [24]. The patient mean percent clearance rates of acne lesions following a single treatment were 69% for the LP PDL and 59% for the PDL [23]. The side effects were minimal, involving mild erythema lasting for 1–2 days.

Topical short incubation (1 hour) ALA followed by LP PDL (595 nm) combined with topical therapy was assessed in a 19-patient study of the treatment of recalcitrant acne of various types and levels of severity [4]. The study patients had failed conventional therapies, including isotretinoin, and exhibited mild-to-severe comedonal, inflammatory, and cystic acne. The mean percent lesional clearance rate per treatment for the LP PDL PDT group was 77%, while all patients including control groups were maintained on topical therapy [4]. Complete clearance was achieved following a mean of 2.9 treatments (range 1–6) and maintained for a mean follow-up interval of 6.4 months (range 1–13). The side effects were minimal and consisted of mild erythema resolving within 1–2 days. Control patients maintained on topical therapy achieved lower rates of clearance: conventional medical therapies achieved 20%, and laser energy alone 32% [4]. The efficacy rate of ALA and LP PDL combined with topical therapy appears to be higher than other light sources used. Prior studies of ALA PDT employing blue and red light or lasers, or IPL demonstrated lesional clearance rates of 32–75% after multiple treatments [5,9,11,15–18], LP PDL PDT combined with topical therapy resulted in a mean lesional clearance rate of 77% per treatment, and was the first PDT regimen to achieve complete clearance for up to 13 months follow-up [4]. LP PDL PDT is performed at monthly intervals, which is more practical than more frequent intervals used in other protocols. A photographic example of a patient with recalcitrant, severe, cystic acne prior to and 14 months following 3 treatments with LP PDL PDT is shown in Fig. 21.1A and B, respectively, with dramatic long-term remission.

In the aforementioned study, LP PDL alone demonstrated efficacy when compared to conventional therapy [4]. The mean clearance rate of LP PDL alone was 32% per treatment

(a) (b)

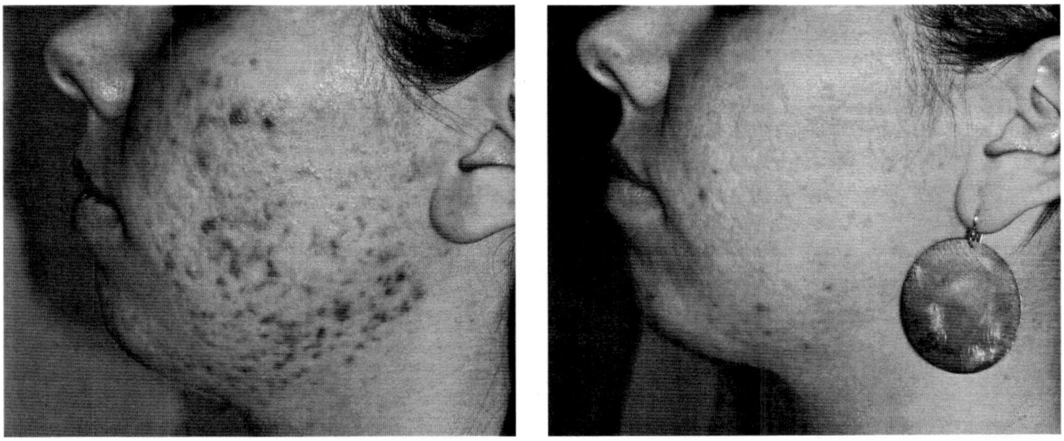

Figure 21.1 Recalcitrant acne responds to LP PDL-mediated PDT. A patient prior to (A) and 14 months following three treatment sessions (B) of topical ALA 1-hour incubation and LP PDL at a fluence of 7.5 J/cm^2, 10 ms pulse duration, 10 mm spot size, and DCD of 30 ms with 30 ms delay. Both the cystic lesions and active erythematous scars improved dramatically following treatment. This patient remained clear to 14 months follow up.

as compared to 20% for the topical control, though the comparison was limited by sample size [4]. This efficacy of LP PDL may be fluence-dependent, as this report employed fluences of 7–7.5 J/cm^2 [4]. An earlier study had suggested efficacy of PDL (585 nm) in treating acne vulgaris [26], though another group was unable to reproduce these findings when low fluences of 3 J/cm^2 and short pulse durations of 350–550 μs were used [27]. It is possible that use of LP PDL at higher fluences and longer pulse durations augments photodynamic activation of porphyrins and photothermal effects on vascular targets, without exceeding the purpura threshold [1]. The mechanisms of acne clearance may also involve the targeting of blood vessels and a resultant antiinflammatory effect as described in the treatment of scars by PDL [28,29]. LP PDL may photoactivate endogenous porphyrins produced by *P. acnes* in the sebaceous follicle, potentially inducing sebaceous gland shrinkage, and decreased bacterial counts as has been shown for blue and red light [1]. Histopathologic evaluation of LP PDL-treated acne lesions will be necessary to elucidate the biological mechanism of the observed clinical findings.

An advantage of the choice of LP PDL as a PDT light source in treating acne is a dramatic improvement in erythematous scars [4]. PDL and LP PDL have been shown to effectively treat active erythematous scars, hypertrophic scars and keloids in particular [28,30,31]. In addition, ALA accumulates in papillary blood vessels and may mediate photodynamic and photothermal injury to blood vessels [32]. PDT has been employed for the treatment of vascular malformations and port wine stains [33]. Lichen sclerosus et atrophicus, a scarring dermatosis with dilated blood vessels in the dermis, was successfully treated by LP PDL-mediated PDT with a 3-year disease- free follow-up [34]. Thus, enhanced resolution of erythematous acne scars may be achieved with ALA and LP PDL. A photographic example of an acne patient whose acne and erythematous acne scars improved from ALA and LP PDL is shown in Fig. 21.2A and B, with acne and scars at baseline and clearing following treatment to 1-year follow-up, respectively.

(a) (b)

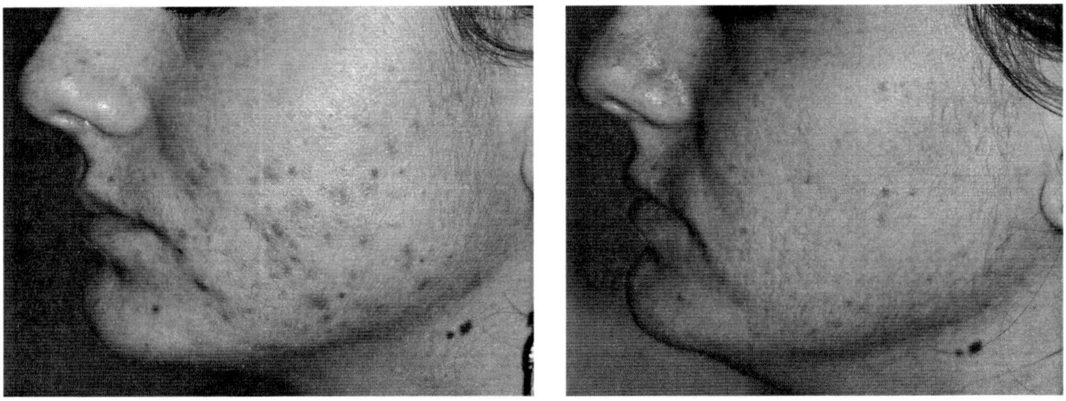

Figure 21.2 Inflammatory acne and erythematous acne scarring treated with ALA and LP PDL. A patient prior to (A) and 1 year following (B) 5 monthly treatments with topical ALA 1-hour incubation and LP PDL at a fluence of 7.5 J/cm^2, 10 ms pulse duration, 10 mm spot size, and DCD of 30 ms with 30 ms delay. This patient has been maintained on topical therapy without recurrence to 2 years' follow up.

21.2.7 Topical ALA and IPL

Topical ALA has also been combined with IPL to treat moderate to severe acne [5]. In one study, IPL (430–1100 nm) following short (1 hour) incubation ALA yielded a response in 12 of 15 patients [5]. Once weekly treatments reduced acne counts by 50.1, 68.5, and 71.8% at the end of the final treatment, with 1-month, and 3-month follow-up without recurrence of treated lesions [5]. In an 18-patient study, ALA with blue light or combination of optical and radiofrequency energy, was evaluated [6]. In that report, ALA was incubated for 15–30 minutes, patients received two to four treatments over a 4–8-week period, and salicylic acid peels [6]. Among the 12 patients who responded, 11 were rated as a 50% response and 5 patients were rated a 75% response [6]. These reports suggest that short incubation ALA with activation by IPL may be an effective, well-tolerated acne treatment and larger, controlled studies are warranted. In addition, due to the photorejuvenative properties of IPL, this may be an approach for the adult acne patient.

21.2.8 Mechanism of PDT in Acne

The data from several published studies indicate that the mechanism of PDT in treating acne lesions is through the targeting of sebaceous gland activity, as is the case for many effective acne treatments. Hormonal activation of the pilosebaceous unit results in hypercornification, sebum production, and proliferation of *P. acnes*. Among the hormones implicated in acne pathogenesis are dihydrotestosterone, dehydroepiandrosterone sulfate and insulin-like growth factor 1, with serum levels correlating with acne lesion counts [35]. Sebocytes have been shown to express androgen receptors, which explains their response to these hormones [36,37]. Exogenous ALA is converted to PpIX which accumulates preferentially in acne lesions, and endogenous porphyrins produced by *P. acnes* also accumulate contributing to the photosensitization of acne lesions, making them amenable to destruction through PDT [7,8,38]. PDT results in decreased sebaceous gland size and vacuolization of sebocytes [8]. This may be the result of direct thermal injury to the sebaceous glands, destruction of *P. acnes*, or manipulation of keratinocyte proliferation in the infundibulum [8]. The level of efficacy of PDT in achieving complete clearance may rely on various factors, including the photosensitizer and light source used, the degree of hormonal stimulation, and the size and level of sebaceous activity at baseline.

21.2.9 Conclusions: PDT for Acne

Topical ALA PDT with short incubation and activation by various light sources including blue and red wavelengths, LP PDL, and IPL is emerging as a safe and effective treatment option for acne patients. In Table 21.1, several published treatment methodologies are presented using topical ALA, and the most effective and commonly used light sources. Although ALA followed by red wavelengths has been reported to be very effective, side effects were an early consideration. Topical ALA combined with blue light has been variably effective in treating acne, and awaits further study. Short incubation topical ALA with activation by LP PDL appears to be both safe and highly effective in treating acne of all types and levels of severity with minimal side effects, and may provide an alternative to isotretinoin in the treatment of recalcitrant acne. Topical ALA and IPL may be an attractive

Table 21.1 Current Topical ALA PDT Methodologies for the Treatment of Acne

Photosensitizer	Incubation (min)	Light Source	Efficacy (%)	Reference
ALA	45–60	Blue light	32–60	[9,11]
ALA	45–60	LP PDL	77	[4]
ALA	60	IPL	50–75	[5,6]

treatment option for adult acne patients, combining photorejuvenation with acne therapy. Topical PDT for the treatment of acne provides a highly needed treatment alternative for recalcitrant acne patients, many of whom are resistant or have acne that recurs following conventional treatments including isotretinoin, and deserves further research to optimize treatment protocols and level of efficacy.

21.3 PDT Photorejuvenation

The application of PDT as a modality for nonablative photorejuvenation is an extension of its efficacy in treating actinic damage and photoaging (for review, see Alexiades-Armenakas) [39]. The application of topical ALA for short incubation times, combined with newer laser and light sources has been shown to be safe and effective for the treatment of actinic keratoses (AK), actinic cheilitis (AC), photodamage, and for photorejuvenation with minimal side effects. Topical ALA has been combined with blue light, LP PDL or IPL. The use of methyl-ALA (MAL) and red light has been developed for AK and potential BCC treatment, though it has not yet been investigated for photorejuvenation.

21.3.1 Topical ALA and Blue Light

Starting in the 1990s, blue light was extensively studied for the treatment of AK due to the superficial localization of these precancerous lesions. The combination of topical ALA for 14–18-hour incubation followed by 1000 seconds of blue light illumination achieved FDA approval for the treatment of AK in 1997 [40]. The Phase III trial, which was randomized, placebo-controlled, and investigator-blinded, demonstrated that 89% of patients achieved greater than 75% clearing of AK at 12 weeks following one to two treatments [41]. Short 1-hour incubation topical ALA followed by blue light has since been demonstrated to be effective in treating diffuse photodamage [42]. The advantages of this approach is the FDA-approved status of the application for AK, which often coexists with photoaging; however the disadvantages include side effects, such as crusting, and up to several weeks of recovery time.

21.3.2 Topical MAL and Red Light

The methylated form of ALA, topical methyl aminolevulinic acid (MAL), has been evaluated largely in conjunction with red light for the treatment of skin neoplasia, and is FDA-approved for the treatment of AK. MAL and red light demonstrated 69 and 89% clearance of AK in two separate studies at a three-month follow-up [43,41]. No published reports have investigated its use for photorejuvenation, though it is possible that it may have potential use for this purpose.

21.3.3 Topical ALA and LP PDL

PDL treatment of photoaged skin stemmed from the clinical and histologic collagen changes observed in PDL-treated hypertrophic scars, striae distensae, and acne scars [45–52]. The 585 nm pulsed dye laser (N-lite) at 350 μs and subpurpuric fluences was the first to be studied for photorejuvenation [53]. An early study reported that a single PDL treatment (585 nm, 4504 s) demonstrated a clinical improvement in 75–90% of mild to moderate wrinkles, and 40% in moderate to severe rhytides [54]. Histologic examination showed increased amounts of normal staining in elastin and collagen fibers in the superficial dermis, with increased cellularity and mucin deposition. The LP PDL at 595 nm was then studied for the treatment of photoaging, with an 18% reported improvement in clinical grading [55]. These findings were attributed to the LP PDL's ability to target facial telangiectasia associated with photodamage [56]. The LP PDL achieved FDA approval for treating photodamage; however, only modest results have been observed, presumably due to predominantly vascular targeting and superficial penetration to the papillary dermis. The application of ALA prior to LP PDL or PDL irradiation thus augments the dermal changes by targeting epidermal photoaging as well. The mechanism of this effect appears to be the activation by the LP PDL at 595 nm of the photosensitizer PpIX which preferentially accumulates in photodamaged cells, resulting in their destruction either by apoptosis or an immune-mediated response [1,2].

Clinical studies have shown that the application of the precursor photosensitizer topical ALA prior to illumination with LP PDL has enhanced the ability of this laser to treat photodamage [23]. Photodynamic therapy mediated by LP PDL is effective in the removal of actinic keratoses, actinic cheilitis, lentigines, fine rhytides, and textural changes due to photodamage [23,57]. In a study of 41 patients with AK, the safety and efficacy of LP PDL (595 nm) was assessed following 3 hours versus 14–18 hours incubation with topical 20% 5-ALA for the treatment of AK [23]. The patient mean percent head AK lesions cleared was approximately 90% at an eight-month follow-up, which was comparable to other treatment modalities such as topical fluorouracil or PDT with blue light [23]. This approach achieved rapid full-face treatment times with minimal discomfort; and minimal posttreatment erythema, which resolved within five days and was the first clinical study to demonstrate that short incubation (3 h) ALA was as effective as long (14–18 h) [23]. It was also significant that no crusting, purpura or dyspigmentation was observed; and erythema was minimal, making it appealing as a cosmetic procedure. PDT employing ALA and LP PDL for full-face treatments of diffuse AK demonstrated improvements in photoaging. Since then, treatment of photoaging with topical ALA and LP PDL has been reported with improvements in texture, fine rhytides, and lentigines [58].

21.3.4 Intense Pulsed Light-Mediated PDT

Among all the light sources, IPL combined with ALA PDT has been the most extensively studied for use in photorejuvenation; this largely stemming from the fact that IPL has independently been shown to rejuvenate skin while spanning wavelengths that activate PPIX. The advantage of IPL (spanning the 400–1200 nm range) is the ability to target both melanin and hemoglobin, thereby improving both dyspigmentation and vascularity. The term "photorejuvenation" was coined to describe the global improvements in photoaging that are observed with the IPL. Filters are placed to exclude shorter wavelengths, thereby selectively targeting various chromophores and typically 5–6 monthly treatments are administered in order to achieve substantial clinical results. IPL alone has yielded modest clinical improvement in rhytides, while pigment and vascular abnormalities of photoaged skin are markedly improved [59]. When studied for rhytide-reduction, histologic evidence of neocollagenesis was observed six months after treatment [60]. Such histologic changes indicative of a dermal remodeling effect, such as an increase in extracellular matrix proteins and neocollagenesis are consistently reported [61]. Patient perception of efficacy is high due to the visible improvements in dyspigmentation and vascularity, which are more easily detectable than mild changes in rhytides, making this device a mainstay in nonablative resurfacing.

The addition of topical ALA prior to IPL has augmented the efficacy observed per treatment, with greater pigmentary, vascular, and rhytide improvement [62–64]. The term photodynamic photorejuvenation has been applied to the use of IPL in the treatment of AK and photodamage [62]. The IPL is an appealing light source for ALA PDT since it spans wavelengths from the blue to the infrared range activating the multiple peaks along the PpIX absorption spectrum. The IPL has been the most rigorously studied light source for the use of PDT in photorejuvenation. A randomized, split-face design clinical study comparing ALA IPL to IPL alone demonstrated greater improvement on the ALA side in erythema, dyspigmentation, and fine rhytides following two monthly treatments [63]. Another IPL following a 1–2-hour incubation of topical ALA resulted in crusting when fluences above a certain threshold were delivered [64]. ALA IPL appears to be more variable in clinical response and side-effect profile, likely due to the variability of different IPL devices in wavelength irradiances. Overall, the response and accompanying side effects from IPL PDT may range from none to marked. The advantages of this light source therefore include its versatility; however, the disadvantage is also its variability in level or response and side effects. Figure 21.3A and B shows a patient with photoaging prior to and following PDT with ALA and IPL, respectively, with improvement in keratoses, erythema, texture, and fine rhytides.

21.3.5 Conclusions: PDT Skin Rejuvenation

For over a century, PDT has been used to treat cutaneous neoplasia, and recently this application has been extended to include photoaging, the cutaneous manifestation of sun-induced aging to the skin. The current protocols in use involve the topical application of ALA for short incubation followed by blue light, LP PDL, or IPL. Among these light sources, the IPL is the best studied and most obvious choice, since it has well-demonstrated efficacy in photorejuvenation which is augmented with antecedent ALA application. Patients with AK and photoaging may be better candidates for topical ALA combined with either blue light or LP PDL, which appear to result in greater AK clearance per treatment as compared to IPL. Finally, with respect to side-effect profile, topical ALA followed by LP PDL or IPL appear to result in the fastest recovery in most cases, the former more

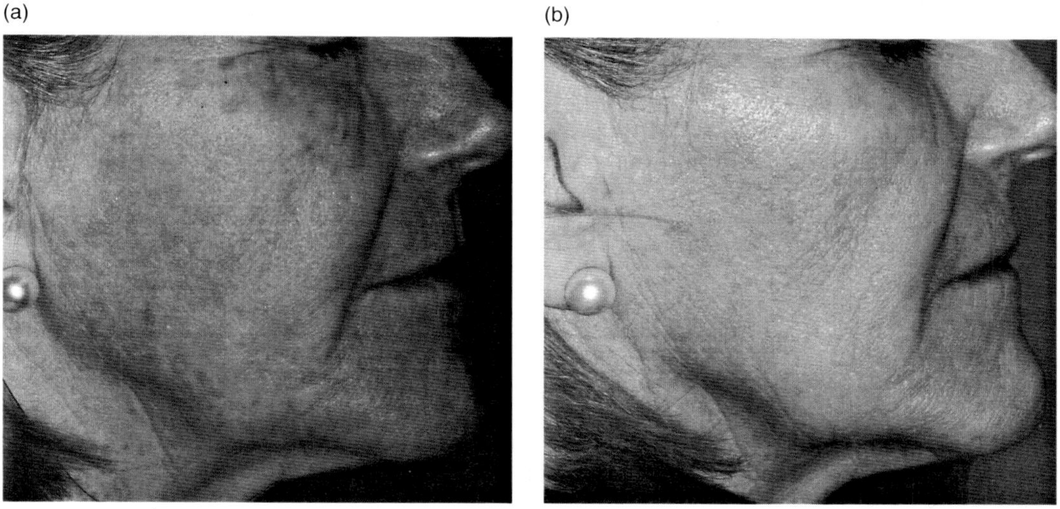

Figure 21.3 Photorejuvenation with ALA and IPL. A patient prior to (A) and following a single treatment (B) with ALA 1-hour incubation and two passes of IPL (Aurora, Syneron) at optical fluence of 16 J/cm^2, radiofrequency fluence of 18 J/cm^2, and long pulse mode. Note the marked reduction in keratoses, erythema, and fine rhytides, with textural improvement.

consistently so, and the latter more unpredictable in the extent of erythema and recovery. Long-term follow-up studies are needed in order to ascertain the long-term safety and proper maintenance treatment intervals in order to maintain the photorejuvenative results.

21.4 PDT for Hair Removal

The application of PDT to the field of hair removal is in its nascent stages of development. As discussed in the acne section earlier, studies have demonstrated uptake of ALA into the pilosebaceous unit. In 1990, Divaris and colleagues demonstrated that intraperitoneal injection of ALA resulted in PpIX fluorescence in sebaceous glands and to a lesser extent in hair follicles of albino mice [13]. Following illumination, a persistent reduction in the number of hair follicles was observed [13]. Identification of PpIX fluorescence was subsequently reported in mice following topical ALA application [65]. Recently, PDT induced damage to not only sebaceous glands, but also to hair follicles in a rat model following application of liposomal ALA and irradiation with a red filtered halogen lamp was demonstrated [66]. Liposomal delivery of topical ALA to intact or depilated rat skin demonstrated PpIX expression in pilosebaceous units, with maximal expression in anagen hair. Inhibition of hair induction after depilation was observed [66].

It is plausible that ALA application topically to depilated skin may facilitate uptake and conversion of ALA into PpIX in hair follicles, making the matrix cells amenable to light destruction. The wavelength would need to be in the red range in order to achieve adequate penetration depth. This would unfortunately be likely to cause the side-effect profile reported extensively for red wavelengths when used with ALA for the treatment of acne; namely, blistering, crusting, and dyspigmentation. It would be interesting to observe whether short incubation ALA and short durations of illumination would be adequate for hair reduction, while ameliorating side effects.

The only published clinical study regarding the use of PDT for hair disorders in humans aimed at treating the hair-loss disease, alopecia areata with topical ALA followed by red light. This study demonstrated no increase in hair growth following treatment, which would be expected, should PDT to hair follicles result in their destruction [67]. Clearly, clinical studies are needed in order to evaluate this potential modality as a mode of hair removal.

The author has conducted an experimental protocol to evaluate topical 5-ALA and IPL (580–980 nm) for the removal of blonde and gray hair. A photographic example of a patient with excessive vellus hair on the face, prior to and following ALA PDT with IPL is shown in Fig. 21.4. Regrowth following treatment increased from an estimated 30% at 1 month (Fig. 21.4A) to 40% at 2 months (Fig. 21.4B), and 50% at 3 months follow-up (Fig. 21.4D).

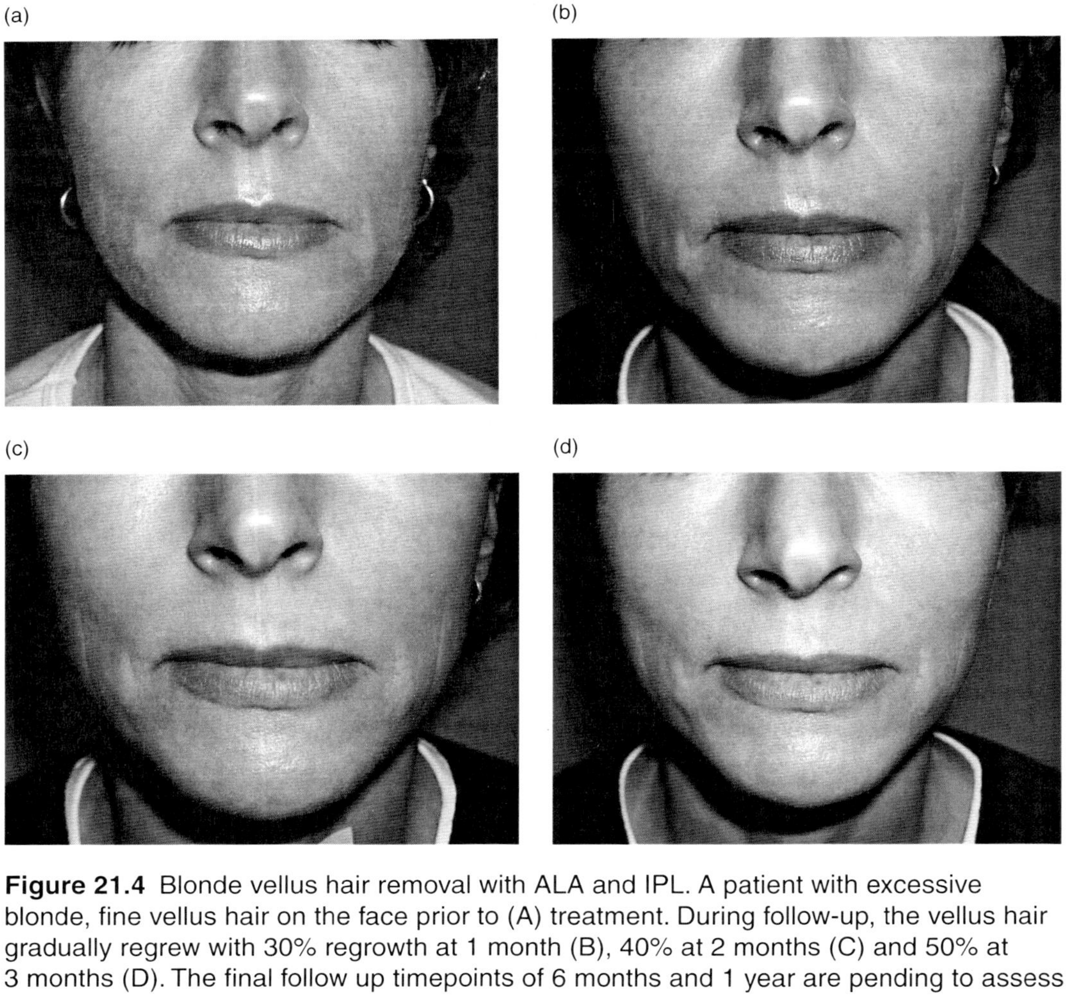

Figure 21.4 Blonde vellus hair removal with ALA and IPL. A patient with excessive blonde, fine vellus hair on the face prior to (A) treatment. During follow-up, the vellus hair gradually regrew with 30% regrowth at 1 month (B), 40% at 2 months (C) and 50% at 3 months (D). The final follow up timepoints of 6 months and 1 year are pending to assess long-term removal.

It will be necessary to follow patients for a long term—6 months and 1 year—in order to make an assessment regarding long-term reduction.

21.4.1 Conclusions: PDT for Hair Removal

Basic science studies indicate that topical ALA results in preferential accumulation of PpIX in hair follicles, with highest expression during the anagen phase. This makes actively growing hair susceptible to light-mediated destruction, as shown in rodent models. Clinical studies are needed in order to determine whether this would be a viable mode of hair removal, provided the protocol is optimized to maintain the efficacy of hair follicle destruction without the risk of significant side effects and complications.

References

1. Alexiades-Armenakas MR. Laser-mediated photodynamic therapy. *Clin Dermatol.* 2006; 24(1): 16–25.
2. Alexiades-Armenakas MR. Aminolevulinic acid photodynamic therapy for actinic keratoses/ actinic cheilitis/acne: vascular lasers. *Dermatol Clin.* 2007; 25: 25–33.
3. Alexiades-Armenakas MR. Treatment of acne with topical PDT. In: Therapy in Cosmetic and Medical Dermatology 2nd Edition (Ed: M.P. Goldman) In: Procedures in Cosmetic Dermatology Series (Ed: Dover JS., Alam M.) Saunders, Philadelphia PA 2007.
4. Alexiades-Armenakas MR. Long pulsed dye laser-mediated photodynamic therapy combined with topical therapy for mild-to-severe comedonal, inflammatory and cystic acne. *J Drugs Dermatol.* 2006; 5(1): 45–55.
5. Gold MH, Bradshaw VL, Boring MM, Bridges TM, Biron JA, and Carter LN. The use of a novel intense pulsed light and heat source and ALA-PDT in the treatment of moderate to severe inflammatory acne vulgaris. *J Drugs Dermatol.* 2004; 3(suppl 6): S15–S19.
6. Taub AF. Photodynamic therapy for the treatment of acne: a pilot study.*J Drugs Dermatol.* 2004; 3(suppl 6): S10–S14.
7. Lee WL, Shalita AR and Poh-Fitzpatrick MB. Comparative studies of porphyrin production in Propionibacterium acnes and Propionibacterium granulosum. *J Bacteriol.* 1978; 133: 811–815.
8. Arakane K, Rya A, Hayashi C et al. Singlet oxygen (1 delta g) generation from coproporphyrin in Propionibacterium acnes on irradiation. *Biochem Biophys Res Commun.*1996; 223: 578–582.
9. Goldman MP and Boyce S. A single-center study of aminolevulinic acid and 417 nm photodynamic therapy in the treatment of moderate to severe acne vulgaris. *J Drugs Dermatol.* 2003; 2: 393–396.
10. Papageorgiou P, Katsambas A, and Chu A. Phototherapy with blue (415 nm) and red (660 nm) light in the treatment of acne vulgaris. *Br J Dermatol.* 2000; 142: 973–978.
11. Gold MH. The utilization of ALA-PDT and a new photoclearing device for the treatment of severe inflammatory acne vulgaris – results of an initial clinical trial. *J Lasers Surg Med.* 2003; 15(suppl): 46.
12. Na JI and Suh DH. Red light phototherapy alone is effective for acne vulgaris: randomized, single-blinded clinical trial. *Derm Surg.* 2007 Oct; 33 (10): 1228–1233.
13. Divaris DX, Kennedy JC, and Pottier RH. Phototoxic damage to sebaceous glands and hair follicles of mice after systemic administration of 5-aminolevulinic acid correlates with localized protoporphyrin IX fluorescence. *Am J Pathol.* 1990; 136 (4): 891–897.
14. Kosaka S, Kawana S, Zouboulis CC, Hasan T, and Ortel B. Targeting of sebocytes by aminolevulinic acid-dependent photosensitization. *Photochem Photobiol.* 2006; 82 (2): 453–7.
15. Hongcharu W, Taylor CR, Chang Y, Aghassi D, Suthamjariya K, and Anderson RR. Topical ALA-photodynamic therapy for the treatment of acne vulgaris. *J Invest Dermatol.* 2000; 115: 183–192.

16. Pollock B, Turner D, Stringer MR, et al. Topical aminolaevulinic acid-photodynamic therapy for the treatment of acne vulgaris: a study of clinical efficacy and mechanism of action. *Br J Dermatol.* 2004; 151: 616–622.
17. Itoh Y, Ninomiya Y, Tajima S, and Ishibashi A. Photodynamic therapy for acne vulgaris with topical 5-aminolevulinic acid. *Arch Dermatol.* 2000; 136: 1093–1095.
18. Itoh Y, Ninomiya Y, Tajima S, and Ishibashi A. Photodynamic therapy of acne vulgaris with topical delta aminolevulinic acid and incoherent light in Japanese patients. *Brit J Dermatol.* 2001; 144: 575–579.
19. Kennedy JC, Marcus SL, and Pottier RH. Photodynamic therapy and photodiagnosis using endogenous photosensitization induced by 5-aminolevulinic acid: mechanisms and clinical results. *J Clin Laser Med Surg.* 1996; 14(5): 289–304.
20. Taub AF. Photodynamic therapy for the treatment of acne: a pilot study.*J Drugs Dermatol.* 2004; 3(Suppl. 6): S10–S14.
21. Pottier RH, Chow YFA, LaPlante JP, Truscott TG, Kennedy JC, and Beiner LA. Non-invasive technique for obtaining fluorescence excitation and emission spectra in vivo. *Photochem Photobiol.* 1986; 44(5): 679–687.
22. Karrer S, Baumler W, Abels C et al. Long-pulse dye laser for photodynamic therapy: investigations in vitro and in vivo. *Lasers Surg Med.* 1999; 25: 51–9.
23. Alexiades-Armenakas, MR and Geronemus, G. Laser-mediated photodynamic therapy of actinic keratoses. *Arch Dermatol.* 2003; Oct; 139 (10): 1313–20.
24. Alexiades-Armenakas MR,, Bernstein L, Chen J, Jacobson L, and Geronemus R. Laser-assisted photodynamic therapy of acne vulgaris and related conditions. *Amer Soc Las Surg. Me*d Abstracts, Anaheim, April 2003.
25. Alster TS and Tanzi EL. Photodynamic therapy with topical aminolevulinic acid and pulsed dye laser irradiation for sebaceous hyperplasia. *J Drugs Dermatol.* 2003; 2(5): 501–4.
26. Seaton ED, Charakida A, Mouser PE, Grace I, Clement RM, and Chu AC. Pulsed-dye laser treatment for inflammatory acne vulgaris: randomized controlled trial. *Lancet.* 2003; 362 (9393): 1342.
27. Orringer JS, Kang S, Hamilton T, et al.Treatment of acne vulgaris with a pulsed dye laser: a randomized controlled trial. *JAMA.* 2004; 291: 2834–2839.
28. Kuo YR, Jeng SF, Wang FS, et al. Flashlamp pulsed dye laser (PDL) suppression of keloid proliferation through down-regulation of TGF-beta1 expression and extracellular matrix expression. *Lasers Surg Med.* 2004; 34 (2): 104–108.
29. Kuo YR, Wu WS, Jeng SF, Huang HC, Yang KD, Sacks JM, and Wang FS. Activation of ERK and p38 kinase mediated keloid fibroblast apoptosis after flashlamp pulsed-dye laser treatment. *Lasers Surg Med.* 2005; 36: 38–42.
30. Alster T. Laser scar revision: comparison study of 585-nm pulsed dye laser with and without intralesional corticosteroids. *Dermatol Surg.* 2003; 29: 25–29.
31. Kono T, Ercocen AR, Nakazawa H, and Nozaki M. Treatment of hypertrophic scars using a long-pulsed dye laser with cryogen-spray cooling. *Ann Plast Surg.* 2005; 54: 487–493.
32. Kelly KM, Kimel S, Smith T, et al. Combined photodynamic and photothermal induced injury enhances damage to in vivo model blood vessels. *Lasers Surg Med.* 2004; 34: 407–413.
33. Evans AV, Robson A, Barlow RJ, and Kurwa HA. Treatment of port wine stains with photodynamic therapy, using pulsed dye laser as a light source, compared with pulsed dye laser alone: a pilot study. *Lasers Surg Med.* 2005; 36: 202–205.
34. Alexiades-Armenakas M.R. Laser-mediated photodynamic therapy of lichen sclerosus. *J Drugs Dermatol.* Nov-Dec 2004; 3(6 Suppl): S25–S27.
35. Cappel M, Mauger D, and Thiboutot D. Correlation between serum levels of insulin-like growth factor 1, dehydroepiandrosterone sulfate, and dihydrotestosterone and acne lesion counts in adult women. *Arch Dermatol.* 2005; 141: 333–338.
36. Zouboulis CC et al. Androgens affect the activity of human sebocytes in a manner dependent on the localization of the sebaceous glands and their effect is antagonized by spironolactone. *Skin Pharmacol.* 1994; 7: 33.

37. Itami S et al. Interaction between dermal papilla cells and follicular epithelial cells in vitro: effect of androgen. *Br J Dermatol.* 1995; 132: 527.
38. Ramstad S, Futsaether CM, and Johnsson A. Porphyrin sensitization and intracellular calcium changes in the prokaryote Propionibacterium acnes. *J Photochem Photobiol B.* 1997; 40: 141–148.
39. Alexiades-Armenakas MR, Dover JS, and Arndt KA. The spectrum of laser resurfacing: non-ablative, fractional and ablative laser resurfacing. *J Amer Acad Dermatol.* March 2008; 58(5):719-37; quiz 738-40.
40. Jeffes EW, McCullough JL, Weinstein GD, Kaplan R, Glazer SD, and Taylor JR. Photodynamic therapy of actinic keratoses with topical aminolevulinic acid hydrochloride and fluorescent blue light. *J Amer Acad Dermatol.* 2001; 45: 96–104.
41. Piacquadio DJ, Chen DM, Farber HF, Fowler JF, Glazer SD, Goodman JJ, Hruza LL, Jeffes EWB, Ling MR, Phillips TJ, Rallis TM, Scher RK, Taylor CR, and Weinstein GD. Photodynamic therapy with aminolevulinic acid topical solution and visible blue light in the treatment of multiple actinic keratoses of the face and scalp: Investigator-blinded, phase 3, multicenter trials. *Arch Dermatol* 2004; 140 (1): 41–6.
42. Touma D and Yaar M, Whitehead S, Konnikov N, Gilchrest BA. A trial of short incubation, broad-area photodynamic therapy for facial actinic keratoses and diffuse photodamage. *Arch Dermatol.* 2004; 140 (1): 33–40.
43. Maloney FJ, Collins P. Ransomized, double-blind, prospective study to compare topical 5-aminolaevulinic acid methylester with topical 5-aminolaevulinic acid photodynamic therapy for extensive scalp actinic keratosis. *Br J Dermatol.* 2007 Jul; 157(1): 87–91.
44. Lehmann P. Methyl aminolaevulinate-photodynamic therapy: a review of clinical trials in the treatment of actinic keratoses and nonmelanoma skin cancer. *Br J Dermatol.* 2007 May; 156(5): 793–801.
45. Alster TS, Kurban AK, Grove GL, et al. Alteration of argon laser-induced scars by the pulsed dye laser. *Lasers Surg Med.* 1993; 13: 368–73.
46. Alster TS. Improvement of erythematous and hypertrophic scars by the 585 nm flashlamp-pumped pulsed dye laser. *Ann Plast Surg.* 1994; 32: 186–90.
47. Alster TS and Williams CM. Improvement of hypertrophic and keloidal median sternotomy scars by the 585 nm flashlamp-pumped pulsed dye laser: a controlled study. *Lancet.* 1995; 345: 1198–200.
48. Kilmer SL and Chotzen VA. Pulse dye laser treatment of old burn scars. *Lasers Surg Med.* 1997; 20(Suppl 9): 34.
49. Alster TS and McMeekin TO. Improvement of facial acne scars by the 585 nm flashlamp-pumped pulsed dye laser. *J Am Acad Dermatol.* 1996; 35: 79–81.
50. McDaniel DH, Ask K, and Zubowski M. Treatment of stretch marks with the 585 nm flash-lamp pumped pulsed dye laser. *Dermatol Surg.* 1996; 22: 332–7.
51. Narurkar V and Haas A. The efficacy of the 585 nm flashlamp-pumped pulsed dye laser on striae distensae at various locations and etiologic factors. *Lasers Surg Med.* 1997; 20(Suppl 9): 35.
52. Wittenberg GP, Fabian BG, Bogomilsky JL, et al. Prospective, single-blind, randomized, controlled study to assess the efficacy of the 585-nm flashlamp-pumped pulsed-dye laser and silicone gel sheeting in hypertrophic scar treatment. *Arch Dermatol.* 1999;135: 1049–55.
53. Zelickson BD, Kilmer SL, Bernstein E, at al. Pulsed dye therapy for sundamaged skin. *Lasers Surg Med.* 1999; 25: 229–36.
54. Bjerring P, Clement m, Heickendroff L, et al. Selective non-ablative wrinkle reduction by laser. *J Cutan Laser Ther* 2000; 2: 9–15.
55. Rostan E, Bowes LE, Iyer S, and Fitzpatrick RE. A double-blind, side-by-side comparison study of low fluence long pulsed dye laser to coolant treatment of wrinkling of the cheeks. *J Cosmet Laser Ther* 2001; 3: 129–36.
56. Goldberg D, Tan M, Dale Sarradet M, and Gordon M. Nonablative dermal remodeling with a 585-nm, 350-microsec, flashlamp pulsed dye laser: clinical and ultrastructural analysis. *Dermatol Surg.* 2003; 29: 161–3; disc. 163–4.

57. Alexiades-Armenakas MR and Geronemus RG. Laser-mediated photodynamic therapy of actinic cheilitis. *J Drugs Dermatol.* 2004; 3 (5): 548–51.
58. Alam M and Dover JS. Treatment of photoaging with topical aminlevulinic acid and light. Skin Therapy Lett. 2004 Dec–2005 Jan; 9 (10): 7–9.
59. Bitter P, Campbell CA, and Goldman M. Nonablative skin rejuvenation using intense pulsed light. *Lasers Surg Med.* 2000; 12: 16.
60. Goldberg D. New collagen formation after dermal remodeling with an intense pulsed light source. *J Cutan Laser Ther.* 2000; 2: 59–61.
61. Zelickson BD and Kist D. Effect of pulsed dye laser and intense pulsed light on the dermal extracellular matrix remodeling. *Lasers Surg Med.* 2000; 12: 17.
62. Ruiz-Rodriguez R, Sanz-Sanchez T, and Cordoba S. Photodynamic photorejuvenation. *Dermatol Surg.* 2002; 28: 742–4.
63. Dover JS, Bhatia AC, Stewart B, Arndt KA. Topical 5-aminolevulinic acid combined with intense pulsed light in the treatment of photoaging. *Arch Dermatol.* 2005; 141 (10): 1247–1254.
64. Hall JA, Keller PJ, and Keller GS. Dose response of combination photorejuvenation using intense pulsed light-activated photodynamic therapy and radiofrequency energy. *Arch Facial Plast Surg.* 2004 Nov-Dec; 6 (6): 374–8.
65. Van der Veen N, de Bruijn HS, Berg RJ, and Star WM. Kinetics and localization of PpIX fluorescence after topical and systemic ALA application observed in skin and skin tumours of UVB-treated mice. *Br J Cancer.* 1996; 73 (7): 925–30.
66. Han I, Jun MS, Kim SK, Kim M, and Kim JC. Expression pattern and intensity of protoporphyrin IX induced by liposomal 5-aminolevulinic acid in rat pilosebaceous unit throughout hair cycle. *Arch Dermatol Res.* 2005; 297 (5): 210–7.
67. Bissonnette R, Shapiro J, Zeng H, McLean DI, and Lui H. Topical photodynamic therapy with 5-aminolevulinic acid does not induce hair regrowth in patients with extensive alopecia areata. *Br J Dermatol.* 2000; 143 (5): 1032–5.

PART 6
REGULATORY AND SAFETY GUIDANCE

22

FDA Regulations for Investigation and Approval of Medical Devices: Laser and Light-Based Systems

Todd J. Banks[1] *and Gurpreet S. Ahluwalia*[2]

[1]*Regulatory Affairs Manager, The Procter & Gamble Company, Cincinnati, OH 45241, USA*
[2]*The Gillette Company, a wholly owned subsidiary of The P&G Company, Needham, MA, USA*

Gurpreet S. Ahluwalia (ed.), *Cosmetic Applications of Laser and Light-Based Systems*, 417–472,

22.1 Introduction

Before a medical device can be introduced into commerce, the manufacturer or the distributor must receive clearance or approval from the Food & Drug Administration (FDA). The FDA provides a clear definition for determining if the device is a 'medical device', which is similar in principle to determining if a chemical agent is a drug or a cosmetic. Generally, if the device is intended for the treatment, prevention, or diagnosis of a disease, or achieves its intended purpose by affecting the structure or any function of the body, it falls under the medical device definition. Medical devices are further classified by the FDA into three classes (I–III) based on their intended use, which also determines the risks and the level of controls needed to ensure safe and effective use of the device. Depending on the device classification, the clearance for marketing is obtained either via premarket notification—510(k), or a premarket approval process (PMA). A simpler and faster 510(k) process is reserved for those devices that can be demonstrated to be substantially equivalent to a legally marketed Class-I or a Class-II medical device. The PMA process is generally followed for the Class-III medical devices.

In the United States, laser and light-based systems used in dermatology, including those indicated for cosmetic application, are regulated as medical devices. The rapid rate of technological advancements and market introduction of these devices has been made possible by the regulatory pathway of 510(k) notification. The process is based on demonstrating to the FDA that the new or modified device is substantially equivalent to a marketed device (predicate) in terms of its intended use, indication for use, and the technology specifications. Depending on the similarities and differences between the predicate and the new device, the FDA may require performance data as a further proof of efficacy and a reasonable assurance of safety. A clinical test is required when the bench tests and *in-vitro* data cannot adequately demonstrate the equivalence to a predicate device. However, before a clinical test can begin an Investigational Device Exemption (IDE), an application must be submitted to evaluating a developmental medical device the FDA which should include an IRB approval on the clinical protocol.

This chapter provides a historical overview of the evolution of FDA medical device regulations and the regulatory approval pathways FDA permits based on the classification of the device. The 510(k) process as it relates to the laser and light-based devices is described in greater detail. In addition, the contents and format of an IDE, a clinical protocol and a 510(k) application are presented.

22.2 History of FDA Regulations for Medical Devices

The US FDA[1] currently regulates the manufacturing and marketing of medical devices in the United States under the Federal Food, Drug and Cosmetic Act[2] (FDCA). It was not until 1938 FDCA that medical devices were subject to any type of federal regulation. However, it was nearly four decades later in 1976 that the device regulations took the next evolutionary step when substantive Medical Device Amendments[3] to the FDCA were passed by Congress, which thereafter required the FDA to establish a comprehensive system of reviewing and approving the marketing of medical devices introduced into interstate commerce. The new law also prohibited, marketing a device until the FDA finds that the device is safe and effective. Despite the increased regulations resulting from the 1976 amendments, the statutory provisions were generally perceived as inadequate. The law was most significantly amended in 1990 by the Safe Medical Devices Act[4] (SMDA); in 1992 by the Medical Device Amendments[5] and yet again in 1997 by the Food and Drug Administration Modernization Act[6] (FDAMA) to further expand the FDA's authority, increase its enforcement powers, and require device manufacturers and others to report adverse device experiences to the FDA (Table 22.1). These changes culminated to establish a comprehensive system of reviewing and approving the marketing of medical devices in the United States. The main changes introduced by these device regulations included the following:

(1) devices would now be classified into three distinct classes based on their perceived risk;
(2) a premarket notification system was introduced to enable the FDA to assess the safety and effectiveness of products prior to marketing; and
(3) a premarket approval system, distinct from the New Drug Premarket Approval requirements for drugs, was introduced for high-risk devices where FDA would review clinical evidence as to the safety and effectiveness of the device before granting approval to the manufacturer to market the item.[7]

22.2.1 Medical Device Amendments 1976

The medical device amendments were written to the Federal Food Drug and Cosmetic Act in May 1976. All new medical devices, including lasers became subject to the FDA premarket clearance. The 1976 amendments established a three-tiered system of medical devices, which categorized the thousands of different types of marketed medical devices into three groups—based on the risk of injury associated with using the device.[8] Relatively risk-free devices, such as tongue depressors, bandages, toothbrushes, and the like, were

Table 22.1 Center for Devices and Radiological Health Milestones[15]

1938	Federal Food, Drug, and Cosmetic (FD&C) Act is enacted. One of the provisions of the new Act, which supersedes the original Food and Drugs Act of 1906, is to extend coverage to devices, making it illegal to sell therapeutic devices that are dangerous or marketed with false claims.
1948	Radiological Health Unit was established by the Bureau of State Services, U S Public Health Service.
1966	Division of Radiological Health was renamed the National Center for Radiological Health.
1968	October 18—Radiation Control for Health and Safety Act of 1968 (Public Law 90-602) was signed by President Johnson. December 20—National Center for Radiological Health becomes the Bureau of Radiological Health as a component of the Environmental Control Administration, Consumer Protection and Environmental Health Service, with a budget of 15.5 million.
1969	October 30—President Nixon issues a message to Congress calling for certain minimum standards and for premarket clearance for certain medical devices. The then Department of Health, Education, and Welfare (HEW) formed The Committee, which became known as the "Cooper Committee". In 1970, the group calls for the inventory and classification of all existing medical devices as the first step toward drafting protective legislation.
1971	May 17—PHS Bureau of Radiological Health transferred to FDA. Its mission: protect against unnecessary human exposure to radiation from electronic products in the home, industry, and healing arts.
1974	February 15—The Bureau of Medical Devices and Diagnostic Products is established.
1976	May 28—Medical Device Amendments to the Food, Drug, and Cosmetic Act of 1938 are enacted, to assure safety and effectiveness of medical devices, including certain diagnostic and laboratory products. The amendments require manufacturers to register with FDA and follow quality-control procedures. Some products must have pre-market approval by FDA; others must meet performance standards before marketing.
1977	May 24—Bureau of Medical Devices and Diagnostic Products is renamed the Bureau of Medical Devices and is reorganized to adopt an organizational structure more suited to implementing the Amendments. Among the changes is the establishment of an Office of Small Manufacturers Assistance, as required by the Amendments. This office will help small manufacturers of medical devices to comply with the law, by providing technical and other nonfinancial assistance.
1978	July 21—Good Manufacturing Practice (GMP) regulations are published in the Federal Register, they become effective December 18, based on the 1976 Medical Device Amendments, the regulations apply to all medical devices and diagnostic products, including those of foreign manufacture intended for US import. In addition to the general controls, more stringent production requirements are imposed on "critical devices", i.e., those that are intended for surgical implantation or for supporting or sustaining life and whose failure can result in a significant injury. December 18—GMP regulations became effective. These "umbrella" regulations set minimum quality assurance requirements for the approximately 4800 companies in the medical device industry.
1980	July 16—The Investigational Device Exempt rule became effective—devices intended solely for investigational use to develop safety and effectiveness data may be exempted from certain requirements of the Act.

(Continued)

Table 22.1 Center for Devices and Radiological Health Milestones[15] (Continued)

1982	First major Congressional oversight hearing on FDA's medical device program and implementation of 1976 Device amendments. October 8—Bureau of Radiological Health and Bureau of Medical Devices merge to become the National Center for Devices and Radiological Health.
1984	March 19—National Center for Devices and Radiological Health is renamed Center for Devices and Radiological Health. September 14—Medical Device Reporting (MDR) regulation published, requiring that manufacturers or importers maintain files when one of their devices may have caused or contributed to a death or serious injury, or when a malfunction had a occurred that could cause a death or serious injury, and to report these to the FDA in a timely manner.
1990	November 28—The Safe Medical Devices Act (SMDA) required medical-device-user facilities to report to the FDA, the manufacturer, or both, whenever they believe there is a probability that a medical device has caused or contributed to a death, illness, or injury. A medical "device user facility" means a hospital, ambulatory surgical facility, nursing home, or outpatient treatment facility that is not a physician's office. The act requires manufacturers to conduct postmarket surveillance on devices that are permanent implants, and whose failure may cause serious health consequences or death, and to establish methods for tracing and locating users depending on such devices. The act authorizes FDA to order device product recalls, to issue "stop use" notices to health professionals and user facilities, and to impose civil penalties (fines) after administrative hearings.
1992	October 27—The Mammography Quality Standards Act is signed into law, requiring all mammography facilities in the United States to be accredited and federally certified as meeting quality standards effective October 1, 1994. After initial certification, facilities must pass annual inspections by federal or state inspectors.
1997	June 1—The Quality System Regulation took effect. FDA Modernization Act mandates the most wide-ranging reforms in agency practices since 1938. Provisions include measures to accelerate review of devices.
2002	October—Medical Device User Fee and Modernization Act (MDUFMA). November 17—Office of InVitro Diagnostic Device Evaluation and Safety (OIVD) formed to promote total product life cycle regulation of medical devices.
2005	August 1—President Bush signs the Medical Device User Fee Stabilization Act of 2005.

placed in Class I, necessitating minimal government supervision and regulation over the manufacture and marketing of these devices. Devices associated with the most significant risks, for example, cardiac pacemakers, prosthetic cardiac valves, and so on, were placed in Class III with comprehensive and stringent regulations relating to the development, testing, manufacturing, and marketing of the devices. Class II devices encompassed those devices representing mild to moderate risks.

The 1976 amendments codified the following:

a. Required to demonstrate safety and effectiveness of medical devices
b. Medical device classification introduced three classes (I, II, III)
    — Class-I: general controls[9] are adequate for assuring safety and effectiveness
    — Class-II: special controls[10] needed
    — Class-III: premarket approval is necessary to demonstrate safety and effectiveness
c. All device manufacturers are required to register with the FDA

22.2.2 Safe Medical Device Amendments 1990

The *Safe Medical Device Amendments* of 1990 was signed into law on November 28, 1990. The objective intent for Congress amending the Food, Drug, and Cosmetic Act was to provide greater assurance regarding the safety and effectiveness of the 1700 types of medical devices that FDA regulates. The new legislation gave FDA additional authority to obtain earlier knowledge of serious device problems, remove defective products from the market more quickly, and track devices from the manufacturer to the consumer.

The law codified the process that permits determination of substantial equivalence to devices already on the market before the 1976 Medical Device Amendments to be marketed without going through a full approval process. In the past, the agency had asked manufacturers for additional data, and now has explicit authority to require manufacturers to submit clinical data to establish that a device is as safe and effective as the device to which it is being compared.

The FDA required manufacturers of products that do not undergo the full-scale approval process to include in their premarket submission, a summary of safety and effectiveness information associated with their devices, or a statement agreeing to make the information available to the public upon request.

The SMDA expanded medical device reporting, already required of manufacturers, to hospitals, nursing homes, and outpatient treatment and diagnostic facilities to include reporting deaths and life-threatening illnesses and injuries attributed to devices.

Under this law, manufacturers of certain permanent life-sustaining or life-supporting devices are required to adopt an effective system of tracking those devices. They would have to maintain records to speed user notification when problems arise.

The 1990 amendments emphasize stronger enforcement authority and allow FDA to order a recall to remove defective products from the market, apply civil penalties for violations of the act, and temporarily suspend premarket approval of products that are found to be hazardous to health.

The law also allowed FDA to use special controls such as guidelines, standards, and postmarket surveillance studies to ensure the safety and effectiveness of devices that need additional controls in order to be safely marketed. Before a device was mass produced, FDA could require manufacturers to conduct design- validation activities to ensure that the device will operate as intended.

In addition, Congress added a humanitarian provision to allow devices to treat or diagnose conditions of diseases affecting fewer than 4000 people to be approved with less effectiveness data than is otherwise required.

22.2.3 The FDA Modernization Act 1997

In 1997, the FDAMA (see US Public Law 105–115, 21 USC 301) expanded the Prescription Drug User Fee Act (PDUFA)[11] policies by codifying a number of practices that had become common at the FDA. The legislation included provisions regarding user access to experimental drugs and medical devices, information on clinical trials, pharmacy compounding, food safety and labeling, and other matters. One provision of the act abolished a prohibition on manufacturers' dissemination of information about unapproved uses of drugs and devices, permitting them to disseminate peer-reviewed journal articles, provided that they commit to file, within a specified time frame, an application to establish the safety and effectiveness of the unapproved use. The statute also added a new provision that required tracking of the status of postmarketing approval studies. This Act provided (a) more focus on devices with greatest risk/benefits, (2) enhanced/early collaboration with industry, (c) allowed certain changes to devices without prior approval and (d) accelerated the time-line for commercializing safe/effective devices.

22.2.4 Medical Device User Fee and Modernization Act (MDUFMA) of 2002

The MDUFMA was enacted largely to improve the efficiency and predictability of the medical device review process. Among other provisions, MDUFMA authorizes the FDA to impose user fees for premarket reviews of certain medical device applications. In exchange, the FDA is expected to meet certain performance goals designed to expedite and improve medical device reviews. The user fees, paid by medical device makers seeking premarket approval, will help advance the device approval process, reduce time to market, and allow consumers earlier access to new treatments and technologies. Furthermore, the FDA has broadened its review process by allowing manufacturers to utilize third parties to conduct their FDA-required inspections. The MDUFMA has four particularly significant features:

- *User fees for premarket reviews* of Premarket Applications, Product Development Protocols, Premarket Reports (a new category of premarket application for reprocessed single-use devices), Biologics License Applications, certain supplements, and 510(k)s.
- *Performance goals* for many types of premarket reviews. These goals become more demanding over time, and include FDA decision goals and cycle goals (cycle goals refer to FDA actions prior to the sponsor's final action on a submission).
- *Establishment inspections may be conducted by accredited persons* (third parties), under carefully prescribed conditions.
- *New regulatory requirements for reprocessed single-use devices*, including a new category of premarket submission, the premarket report.

In an effort to reduce the burden on small businesses, the FDA provides a reduced rate for firms that meet the definition of a small business under FDAAA.[12] The definition of a small business has not changed since 2006, that is, $100 million or less in gross sales and receipts of all affiliates, partners, and parent firms. Small firms with gross sales of $30 million or less would be eligible to have the fee on their first PMA waived. New for FY08, there is a mechanism for firms based outside the United States to qualify for a small business fee reduction.

Fees for Premarket Notification [510(k)s] and Premarket Application

FY 2008 Device Review User Fees (U.S. Dollars)		
Application	**Standard Fee**	**Small Business (≤$100 million in gross receipts or sales) Fee**
510(k)[13]	$3,404	$1,702
Premarket Application[14] (PMA)	$185,000	$46,250

22.3 The FDA Approval Process for Medical Devices

22.3.1 Medical Device Definition

A medical device is defined[16] within the Food Drug & Cosmetic Act as "... an instrument, apparatus, implement, machine, contrivance, implant, *in vitro* reagent, or other similar or related article, including a component part, or accessory which is: recognized in the official National Formulary, or the United States Pharmacopoeia, or any supplement to them, intended for use in the diagnosis of disease or other conditions, or in the cure, mitigation, treatment, or prevention of disease, in man or other animals, or intended to affect the structure or any function of the body of man or other animals, and which does not achieve any of it's primary intended purposes through chemical action within or on the body of man or other animals and which is not dependent upon being metabolized for the achievement of any of its primary intended purposes."

Medical devices distributed in the United Sates are subject to General Controls, premarketing and post marketing regulatory controls.

General Controls include:

1. Establishment Registration by manufacturers, distributors, repackages and relabelers,
2. Medical Device Listing with FDA of devices to be marketed,
3. Manufacturing the devices in accordance with Good Manufacturing Practices,
4. Labeling medical devices in accordance with the labeling regulations, 21 CFR §801 & §809,
5. Medical Device Reporting of adverse events as identified by the user, manufacturer and/or distributor of the medial device.

Premarketing (a.k.a. Special) controls are device- and device-classification specific. Premarketing controls for a medical device may include: clearance to market by 510(k) or approval to market by PreMarket Approval (PMA).

Postmarketing controls include Device Listing, Medical Device Reporting (MDR), Establishment Registration, and Quality System Compliance Inspection.

22.3.2 Device Classification

The FDA[17] has established classifications for approximately 1700 different generic types of devices, and grouped them into 16 medical specialties, referred to as panels. Each of these generic types of devices is assigned to one of the three regulatory classes based on the level of control necessary to assure the safety and effectiveness of the device: Class I, Class II, and Class III. The classifications are assigned by the risk the medical device presents to the user and the level of regulatory control the FDA determines is needed to legally market the device. As the classification level increases, the risks to the user and FDA regulatory control also increase. Accessories to medical devices and devices/products used with a medical device to support the use of the device are considered the same classification as the medical device.

The class to which a device is assigned determines, among other things, the type of premarketing submission/application required for FDA clearance to market. If the device is classified as Class I or II, and if it is not exempt,[18] a 510(k) will be required for marketing. All devices classified as exempt, are subject to the limitations on exemptions. Limitations of device exemptions are covered under 21 CFR §862–§892. For Class III devices, a PMA application will be required unless the device is a preamendment device (on the market prior to the passage of the medical device amendments in 1976, or substantially equivalent to such a device) AND a PMA submission has not been requested by FDA.

Device classification depends on the *intended use*[19] of the device and also upon its *indications for use*. For example, a scalpel's intended use is to cut tissue. A subset of intended use arises when a more specialized indication is added in the device's labeling such as, "for making incisions in the cornea". Indications for use can be found in the device's labeling, but may also be conveyed orally during sale of the product.

Device classification is risk-based, that is, the risk the device poses to the user is a major factor in the class it is assigned. Class I includes devices with the lowest risk and Class III includes those with the greatest risk. As noted earlier, all device classes are subject to General Controls which are the baseline requirements of the Food, Drug and Cosmetic Act.

The *Device Classes* are as follows:

- Class I—General Controls
- Class II—Special Controls
- Class III—Premarket Approval

22.3.2.1 Class I—General Controls

Class I medical devices have the least amount of regulatory control. Class I devices present minimal potential harm to the user. These devices are typically simple in design and manufacture, and have a history of safe use and are only subject to general controls designed to achieve safety and effectiveness through the control of manufacturing, labeling, and related issues,[20] including FDA's Good Manufacturing Practices (GMP).[21] Examples of Class I devices include tongue depressors, arm slings, bandages, manual and electric toothbrushes, examination gloves and hand-held surgical instruments.

General controls include:

1. Establishment Registration[22] (FDA Form 2891[23]) of companies which are required to register under 21 CFR §807.20, such as manufacturers, distributors, repackages, and relabelers. Foreign establishments, however, are not required to register their establishments with FDA.
2. Medical Device Listing[24] (FDA Form 2892[13]) with FDA for devices to be marketed.
3. Manufacturing devices in accordance with GMP[25] in 21CFR §820.
4. Labeling[26] devices in accordance with labeling regulations in 21CFR §801 or §809.
5. Submission of a *premarket notification [510(k)]*, if applicable, before marketing a device.

22.3.2.2 Class II—Special Controls

Class II[27] devices are those for which general controls alone are insufficient to assure safety and effectiveness, and existing methods are available to provide such assurances. In addition to complying with general controls, Class II devices are also subject to special controls. Note, a few Class II devices are exempt from the premarket notification.[28]

Special controls may include special labeling requirements, mandatory performance standards, and postmarket surveillance. Examples of Class II devices include oxygen masks,[29] artificial eyes,[30] and other devices that do not by themselves maintain life, such as powered wheelchairs, infusion pumps, surgical drapes, x-ray machines, and tampons. Laser and light-based systems used in cosmetic dermatology are also Class II medical devices.

22.3.2.3 Class III—Premarket Approval

Class III is the most stringent regulatory category for devices. Class III devices are those for which insufficient information exists to assure safety and effectiveness solely through general or special controls. They are devices that are defined as those used for "supporting or sustaining human life or for a use which is of substantial importance in preventing impairment of human health" or those devices that "present a potentially unreasonable risk of illness or injury."[31] Given that such devices entail the most significant risks, in addition to meeting the aforementioned requirements, they must also be shown to be safe and effective before being marketed. Class III devices are therefore subject to a PMA process by which the FDA reviews clinical evidence as to the safety and effectiveness of the device before granting approval for the device to be marketed or manufactured.[32] Examples of Class III devices which require a premarket approval include replacement heart valves,[33] silicone gel-filled breast implants,[34] artificial knee joints,[35] and extended-wear contact lenses.[36]

Premarket approval is the required process of scientific review to ensure the safety and effectiveness of most Class III devices. However, not all Class III devices require an approved premarket approval application to be marketed. Some Class III devices can be marketed with a premarket notification 510(k), provided they meet the postamendment provision of being introduced to the US market *after* May 28, 1976, and are determined to be substantially equivalent to a preamendment (i.e., introduced to the US market *before* May 28, 1976) Class III device, and for which the regulation calling for the premarket approval application has not been published in the Code of Federal Regulations (21 CFR).

Class III devices which require an approved premarket approval application to be marketed are those:

1. Regulated as new drugs prior to May 28, 1976, also called transitional devices.[37]
2. Devices found not substantially equivalent to devices marketed prior to May 28, 1976.
3. Class III preamendment devices which, by regulation in 21 CFR §814, require a premarket approval application.

22.3.2.4 How To Determine Classification

Most medical devices can be classified by finding the matching description of the device amongst one of the 16 medical specialty "panels" listed in Title 21[38] of the Code of Federal Regulations. For each of the devices classified by the FDA, the CFR gives a general description including the intended use, the class to which the device belongs (i.e., Class I, II, or III), and information about marketing requirements.

It is a point to be noted that manufacturers do have, to a certain degree, the right to contest the classification of a device, and to petition the FDA for changes relative to a device's classification.[39]

To find the classification of a device, as well as to know whether any exemptions may exist, you need to locate the regulation number that is the classification regulation for the device of interest. There are two methods for accomplishing this: go directly to the classification database[40] and search for a part of the device name, or, if you know the medical specialty (device panel) to which the device belongs, go directly to the listing for that panel and identify the device and the corresponding regulation.

Each classification panel in the CFR list those devices classified in that panel. Each classified device has a 7-digit number associated with it, for example, 21 CFR §878.4810[41]—Laser Instrument, Surgical. Once you find your device in the panel's beginning list, go to the section to determine the assigned class. Similarly, in the Classification Database under "laser", you will see several entries for various types of lasers. The three- letter product code, GEX in the database for Laser Surgical Instrument is also the classification identifier which is used on the Medical Device Listing form, FDA-2892.[42]

22.3.3 510(k) Clearance to Market

Most medical devices sold in the United States today are cleared for commercial distribution or marketing by premarket notification (Class II). However, most Class I devices and some Class II devices are exempt[43,44] from the premarket notification and/or good manufacturing practices regulation. The Federal Food, Drug, and Cosmetic Act requires device manufacturers to submit a premarket notification to FDA if they intend to introduce a device into commercial distribution for the first time, or to introduce—or reintroduce—a device that will be significantly changed or modified to the extent that its safety or effectiveness could be affected. Such a change or modification could relate to the design, material, chemical composition, energy source, manufacturing process, or intended use of the device.

The 510(k) submission identifies characteristics of the new or modified medical device as compared to a medical device with similar intended use, currently legally marketed in the United States. The currently legally marketed device is referred to as the "predicate" device.

The information required in a 510(k) submission is defined in 21 CFR §807.87 and includes the following elements:

- Device trade or proprietary name, common or usual name or classification, Class of the device (Class I, II, III)
- Submitter's name and address, contact person, telephone number, and fax number, Representative/Consultant if applicable

- Name and address of manufacturing/packaging/sterilization facilities, Registration number of each manufacturing facility
- Action taken to comply with the requirements of the Special Controls
- Proposed labels, labeling, and advertisements to describe the device, its intended use, and the directions for its use
- 510(k) summary or a 510(k) statement.
- For Class III medical device, a Class III certification and a Class III summary
- Photographs of the device, Engineering drawings of the device
- Identification of the marketed device(s) to which equivalence is claimed including labeling and description of the medical device
- Statement of similarities and/or differences with marketed device(s)
- Data to show consequences and effects of a modified device, performance data (bench, animal, clinical)
- Sterilization information (as applicable)
- Software development, verification, and validation information
- Hardware design and development information
- Information requested in specific guidance documents (as applicable)
- Kit Certification Statement (for a 510(k) submission with kit components only)
- Truthful and Accurate Statement

Additional information may also be included in a 510(k) application. For example, contraindications such as skin conditions (for light-based devices) and situations where the device should not be used; warnings describing the significant safety risks from using the device for intended or unintended use; anticipated adverse effects (undesirable side effects) from the use of the device and precautions such as general situations where the use of the device can result in a hazardous condition during the treatment procedure or can cause harm to the treating clinician, patient, or other bystanders in the room (for laser devices).

Depending on the complexity of the new or modified medical device, the FDA Review of a 510(k) submission takes between 60 and 90+ days. The more complex the changes or comparison required to support the safety and effectiveness of the new or modified medical device, the longer will be the FDA review process.

A Traditional 510(k) submission[45,46] must include the required elements identified in 21 CFR §807.87. CDRH recommends that device sponsors follow the Traditional 510(k) format provided in their guidance document.[47] The 510(k) Screening Checklist[48,49] should be used to assure the 510(k) is complete. It is helpful to attach the 510(k) screening checklist to the submission after the table of contents. It should include page numbers where each of the elements in the 510(k) can be found.

22.3.3.1 Who Is Required To Submit a 510(k)

Neither the FDCA or the 510(k) regulations[50] specify who must apply for a 510(k)—anyone may do so. Instead, they specify the requirements and actions that must occur prior to introducing a device to the US market. The individuals required to file a 510(k) can be categorized as follows:

1. Domestic manufacturers *introducing* a finished device[51] into the US market;
2. Specification developers *introducing* a finished device to the US market;

3. Repackers or relabelers who make labeling changes, or whose operations *significantly affect*[52] the device.
4. Foreign manufacturers/exporters or US representatives of foreign manufacturers/exporters *introducing* a device to the US market.

22.3.3.2 When a 510(k) Is Not Required

The following seven examples illustrate when a 510(k) is *not required.*

1. If an unfinished device is to be sold to another firm for further processing, including components to be used in the assembling of devices by other firms. However, if components are to be sold directly to an end user as replacement parts, a 510(k) is required.
2. If the device is not being marketed or commercially distributed. A 510(k) is not needed to develop, evaluate, or test a device. This includes clinical evaluation. *Note:* if clinical trials are performed with your device, you may be subject to the provisions outlined in the Investigational Device Exemption (IDE) Regulation.[53]
3. If one distributes another firm's domestically manufactured devices, there is no need to submit a 510(k). The device label should clearly communicate who the distributor is.
4. In most cases repackagers or relabelers are not required to submit a 510(k) if the existing labeling or condition of the device is not significantly changed.
5. If the device was legally in commercial distribution before May 28, 1976, a 510(k) submission is not required, unless the device has been modified, or there has been a change in its intended use. These devices are considered "grandfathered".
6. If a foreign made medical device is imported, a 510(k) is not required if:
   - a 510(k) has been submitted by the foreign manufacturer and received FDA marketing clearance, or
   - a 510(k) has been submitted by an importer on behalf of the foreign manufacturer and has received FDA marketing clearance. If one importer submits a 510(k) on behalf of the foreign manufacturer, all other importers of that device, imported from the same foreign manufacturer (the 510(k) Holder) are not required to submit a 510(k) for]that device.
7. If the device is exempted from this requirement by final classification regulation[54] subject to the limitations on exemptions, it means certain Class I or II devices can be marketed for the first time without having to submit a 510(k).

22.3.3.3 When a 510(k) Is Required

A 510(k) is required when:

1. *Introducing a device into commercial distribution (marketing) for the first time.* After May 28, 1976 (effective date of the Medical Device Amendments to the FD&C Act), anyone who wants to sell a device in the United States is

required to make a 510(k) submission at least 90 days prior to offering the device for sale, even though it may have been under development or clinical investigation before that date. If the device was not marketed by your firm before May 28, 1976, a 510(k) is required.

2. *Seeking a different intended use for a device* already in commercial distribution. The 510(k) regulation specifically requires a premarket notification submission for *major* changes in intended use. Intended use is indicated by claims made for a device in labeling or advertising. However, most, if not all, changes in intended use will require a 510(k).
3. There is a change or modification of a device that is already being marketed, *if the modifications could significantly affect*[55] *the safety or effectiveness of the device* or, if the device is to be marketed for a new or different indication.[56] See FDA guidance on determining the significance of a device modification. http://www.fda.gov/cdrh/ode/510kmod.html.

22.3.3.4 Device Modifications

The FDA does not currently accept supplements to amend the submission of a previously cleared 510(k) device. This means that new 510(k) notifications[57] must contain all the needed information. Referencing the earlier submission will not work. The 510(k) must also include supporting data to show that the manufacturer has considered the consequences that the change or new use might have on the safety and effectiveness of the device.[58]

The description of the modified device should include differences from the predicate device that could significantly affect safety and effectiveness. All data from *in vitro*, animal, and human clinical testing, if any, should be included, as well as engineering, bench, and design verification data, and any other information that supports the new indication or the claim that the modified device is as safe and effective as the predicate device.

There are three types of Premarket Notification 510(k)s that may be submitted to FDA: Traditional,[59] Special,[60] and Abbreviated.[61] The Special and Abbreviated 510(k) methods were developed under "the New 510(k) Paradigm"[62] to help streamline the 510(k) review process. The Special 510(k) and Abbreviated 510(k) methods can only be used if certain criteria are met. The Traditional 510(k) method can be used under any circumstances.

There is no Premarket Notification 510(k) "form" to complete. A 510(k) is a document containing information required under 21 CFR §807 Subpart E. All 510(k)s are based on the concept of substantial equivalence (SE) to a legally marketed (predicate) device. All 510(k)s provide a comparison between the device to be marketed and the predicate device or devices.

22.3.3.5 What Is Substantial Equivalence (SE)

Manufacturers should attempt to make a comparison of the new device to its predicate as easy as possible for the FDA reviewer. The 510(k) notification should therefore include discussion of the similarities and differences between the device and its predicate device, and should make use of comparative tables whenever possible (Fig. 22.1). Comparisons might consider such areas as intended use, materials, design, energy used and delivered, anatomical sites, target population, physical safety, and compliance with standards, biocompatibility, and

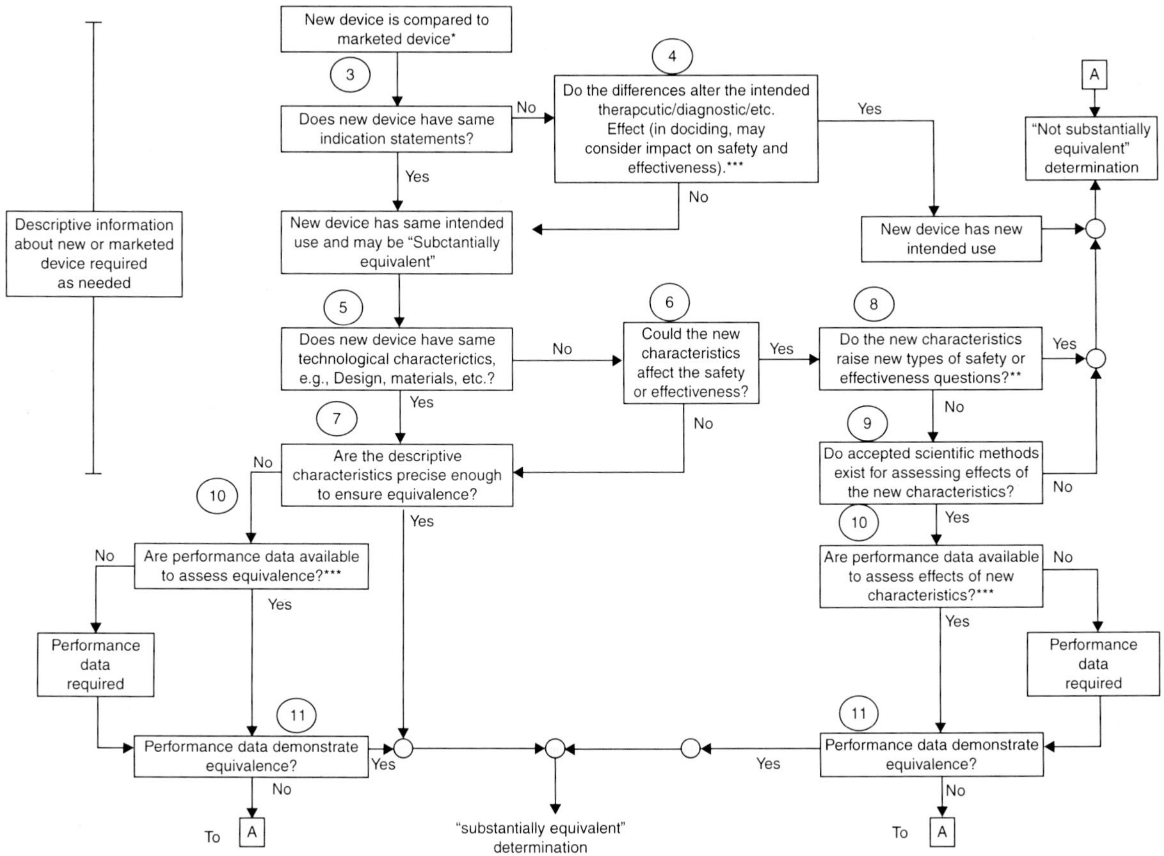

* 510(k) submissions compare new devices to marketed devices. FDA requests additional information if the relationship between marketed and "predicate" (pre-Amendments or reclassified post-Amendments) devices is unclear.

** This decision is normally based on descriptive information alone, but limited testing information is sometimes required.

*** Data may be in the 510(k), other 510(k)s, the Center's classification files, or the literature.

Figure 22.1 Flow chart of 510 (k) "substantial equivalence" decision-making process.

performance. Information used to demonstrate the substantial equivalence of the device to its predicate should be provided in a clear and comprehensible format, making use of tables and graphs where these are helpful to clarify the manufacturer's argument.

Manufacturers should also submit pertinent information about the predicate device, including its labeling, if available. For example, the notification should state whether the predicate is a legally marketed preamendment device or a Class I or Class II postamendment device that has been granted marketing clearance by FDA following the submission of a 510(k). If known, provide the 510(k) document control number (i.e., K followed by 6 digits) for the predicate device. Such 510(k) numbers are available via the Electronic Docket at CDRH.

Unlike a PMA, which requires demonstration of clinically proven safety and effectiveness, the 510(k) requires demonstration of substantial equivalence (SE). SE means that the new device is as safe and effective as the predicate device(s). Significantly, prior to the 1990 amendments, the FDA did not generally require human clinical trials in determining substantial equivalence.[63] However, following the 1990 amendments, the FDA was given express authority to require the submission of performance data, including data from clinical trials, in order to make a substantial equivalence determination.[64]

A device is deemed SE if, in comparison to a predicate device it:

- has the same intended use as the predicate device, *and*
- has the same technological characteristics as the predicate device, *or*
- has different technological characteristics that do not raise new questions of safety and effectiveness, and the sponsor demonstrates that the device is as safe and effective as the legally marketed device.

A claim of substantial equivalence does not mean the new and predicate devices must be identical. Substantial equivalence is established with respect to intended use, design, energy used or delivered, materials, performance, safety, effectiveness, labeling, biocompatibility, standards, and other applicable characteristics. Detailed information on how FDA determines substantial equivalence can be found in the Premarket Notification Review Program (K86-3)[65] blue book memorandum.

Until the applicant receives an order declaring a device SE, they may not proceed to market the device. Once the device is determined to be SE, it can then be marketed in the United States. If FDA determines that a device is *not* SE, the applicant may resubmit another 510(k) with new data, file a reclassification petition,[66] or submit a premarket approval application (PMA). The SE determination is usually made within 90 days and is made based on the information submitted by the applicant.

22.3.3.6 Third-Party Review Program

The Center for Devices and Radiological Health (CDRH) has implemented a Third-Party Review Program.[67] This program provides an option to manufacturers of certain devices of submitting their 510(k) to private parties (Recognized Third Parties) identified by FDA for review, instead of submitting directly to CDRH. By law, FDA must issue a final determination within 30 days after receiving the recommendation of an Accredited Person.

22.3.4 PMA (Premarket Approval)

Premarket approval (PMA) is the FDA process of scientific and regulatory review to evaluate the safety and effectiveness of Class III medical devices. Class III devices are those that support or sustain human life, are of substantial importance in preventing impairment of human health, or which present a potential and unreasonable risk of illness or injury. Due to the level of risk associated with Class III devices, FDA has determined that general and special controls alone are insufficient to assure the safety and effectiveness of class III devices. Therefore, these devices require a PMA application under Section 515 of the FDCA in order to obtain marketing clearance. Please note that some Class III preamendment devices may require a Class III 510(k).

Most Class III medical devices require a PMA. Section 515(c)(1) of the Federal Food, Drug, and Cosmetic Act (FD&C Act) specifies the required contents of a PMA. The PMA application content includes:

- full reports of all information, published or known to, or which should reasonably be known to the applicant, concerning investigations which have been made to show whether or not such device is safe and effective;
- a full statement of the components, ingredients, and properties and of the principle or principles of operation, of such device;
- a full description of the methods used in, and the facilities and controls used for, the manufacture, processing, and, when relevant, packing and installation of such device;
- an identifying reference to any performance standard under Section 514 which would be applicable to any aspect of such device if it were a Class II device, and either adequate information to show that such aspect of such device fully meets such performance standard, or adequate information to justify any deviation from such standards;
- such samples of such device and of components thereof as the Secretary may reasonably require, except that where the submission of such samples is impracticable or unduly burdensome, the requirement of this subparagraph may be met by the submission of complete information concerning the location of one or more such devices readily available for examination and testing;
- specimens of the labeling proposed to be used for such device; and
- such other information relevant to the subject matter of the application as the Secretary, with the concurrence of the appropriate panel under Section 513, may require.

FDA regulations provide 180 days to review the PMA and make a determination. In reality, the review time is normally longer. Before approving or denying a PMA, the appropriate FDA advisory committee may review the PMA at a public meeting and provide FDA with the committee's recommendation on whether FDA should approve the submission. A facility inspection verifying the manufacturing systems present to manufacture the medical device is usually performed prior to FDA PMA approval. After FDA notifies the applicant that the PMA has been approved or denied, a notice is published on the Internet announcing the data on which the decision is based, and providing interested persons an opportunity to petition FDA within 30 days for reconsideration of the decision.

The regulation governing premarket approval is located in Title 21 Code of Federal Regulations (CFR) Part 814, Premarket Approval. A Class III device that fails to meet PMA requirements is considered to be adulterated under Section 501(f) of the FD&C Act, and cannot be marketed.

22.3.4.1 When a PMA Is Required

PMA requirements apply to Class III devices. Device product classifications can be found by searching the Product Classification Database.[68] The database search provides the name of the device, classification, and a link to the Code of Federal Regulations (CFR), if any. The CFR provides the device type name, identification of the device, and classification information.

A regulation number for Class III devices marketed prior to the 1976 Medical Device Amendments is provided in the CFR. The CFR for these Class III devices that require a PMA states that the device is Class III and will provide an effective date of the requirement for PMA. If the regulation in the CFR states that "No effective date has been established of the requirement for premarket approval," a Class III 510(k) should be submitted.

Please note that PMA devices often involve new concepts, and many are not of a type marketed prior to the Medical Device Amendments. Therefore, they do not have a classification regulation in the CFR. In this case, the product classification database will only cite the device type name and product code.

If it is unclear whether the unclassified device requires a PMA, use the three-letter product code to search the PMA database and the Premarket Notification 510(k) database. These databases can be found by clicking on the hypertext links at the top of the product classification database web page. Enter only the three-letter product code in the product code box. If there are 510(k)s cleared by FDA and the new device is substantially equivalent to any of these cleared devices, then the applicant should submit a 510(k).

Further, a new type of device may not be found in the product classification database. If the device is a high- risk device (supports or sustains human life, is of substantial importance in preventing impairment of human health, or presents a potential, unreasonable risk of illness or injury), and has been found to be not substantially equivalent (NSE) to a Class I, II, or III [Class III requiring 510(k)] device, then the device must have an approved PMA before marketing in the United States. Some devices that are found to be not substantially equivalent to a cleared Class I, II, or III (not requiring PMA) device, may be eligible for the *de novo* process as a Class I or Class II device. For additional information on the *de novo* process, see "New Section 513(f) (2)—Evaluation of Automatic Class III Designation: Guidance for Industry and CDRH Staff."[69]

22.3.4.2 Historical Background

PMA requirements apply to Class III preamendment devices, transitional devices, and postamendment devices.

a. Preamendment Devices. A preamendments device is one that was in commercial distribution before May 28, 1976, the date the Medical Device Amendments were signed into law. After the Medical Device Amendments became law, the classification of devices was determined by FDA classification panels. Eventually, all Class III devices will require a PMA. However, preamendment Class III devices require a PMA only after FDA publishes a regulation calling for PMA submissions. The preamendment devices must have a PMA filed for the device by the effective date published in the regulation, in order to continue marketing the device. The CFR will state the date that a PMA is required. Prior to the PMA effective date, the devices must have a cleared Premarket Notification 510(k) prior to marketing. Class III Preamendment devices that require a 510(k) are identified in the CFR as Class III and include the statement "Date premarket approval application (PMA) or notice of completion of product development protocol (PDP) is required. No effective date has been established of the requirement for premarket approval".

b. Postamendment Devices. A postamendment device is one that was first distributed commercially on or after May 28, 1976. Postamendment devices equivalent to preamendment Class III devices are subject to the same requirements as the preamendment devices.

c. Transitional Devices. Transitional devices are devices that were regulated by FDA as new drugs before May 28, 1976. Any Class III device that was approved by a New Drug Application (NDA) is now governed by the PMA regulations. The approval numbers for these devices begin with the letter N. These devices are identified in the CFR as Class III devices, and state that an approval under Section 515 of the Act (PMA) is required as of May 28, 1976 before this device may be commercially distributed. An example of such device is intraocular lenses[70] (21 CFR §886.3600).

22.4 The FDA Classification of Light-Based Medical Devices

- Basis for classification of surgical laser devices
- Laser standards

FDA regulates radiation emitting electronic products. The purpose is to prevent unnecessary exposure to radiation due to the use of these products. There are specific requirements that *apply to all* radiation-emitting electronic products in order to comply with the provisions of the Food, Drug and Cosmetic Act. If the product is also a medical device,[71] the product must also comply with the medical device regulations. Medical lasers have been categorized either in Class II or Class III, depending upon the specific application involved.

22.4.1 Definition of Electronic Product Radiation

Manufacturers and distributors of products meeting the definition of "electronic product radiation" in Section 531 of the FDCA may be subject to certain provisions of the Act including the retention of records and submission of product reports to the FDA, specifically to the Center for Devices and Radiological Health (CDRH). The FDA requirements for these products, record keeping and reporting, are included in the final regulations contained in Title 21 Code of Federal Regulations Parts 1000-1299. According to Section 531 of the FD&C Act:

(1) the term "electronic product radiation" means—
  (A) any ionizing or nonionizing electromagnetic or particulate radiation, or
  (B) any sonic, infrasonic, or ultrasonic wave, which is emitted from an electronic product as the result of the operation of an electronic circuit in such product;

(2) the term "electronic product" means—
   (A) any manufactured or assembled product which, when in operation,
      (i) contains or acts as part of an electronic circuit and
      (ii) emits (or in the absence of effective shielding or other controls would emit) electronic product radiation, or
   (B) any manufactured or assembled article which is intended for use as a component, part, or accessory of a product described in Clause (A) and which, when in operation, emits (or in the absence of effective shielding or other controls would emit) such radiation;
      (i) the term "manufacturer" means any person engaged in the business of manufacturing, assembling, or importing of electronic products.

Most radiation-emitting products are not considered to be medical devices. However, if you make any medical claims, your product is a medical device subject to the provisions of the FDCA for medical devices in addition to the provisions for radiation emitting products.

Examples of electronic products:

Medical: diagnostic x-ray or ultrasound imaging devices, microwave or ultrasound diathermy devices, microwave blood warmers or sterilizers, laser coagulators, ultrasound phacoemulsifiers, x-ray or electron accelerators, sunlamps, ultraviolet dental curing devices;
Nonmedical: microwave ovens, televisions receivers and monitors (video displays), entertainment lasers, industrial x-ray systems, cordless and cellular telephones, industrial RF sealers of plastics and laminates, laser CD players.

22.4.2 Special Issues for Radiation-Emitting Devices

Particularly pertinent for lasers, any device that emits radiation must additionally comply with the Radiation Control for Health and Safety Act passed in 1968,[72] which is also administered by the FDA and which authorizes the development of performance standards and general controls for ionizing radiation products. The Act was designed to protect the public from the dangers of electronic product radiation. Devices that either intentionally emit radiation (such as x-ray equipment) or emit radiation as a consequence of their operation (such as CRTs and television sets) are covered. Further, certain light-emitting products, which emit intense, directed radiation, such as lasers,[73] sunlamps, and ultraviolet lighting[74] are also covered. In addition to specific emissions standards, and to prevent unnecessary exposure to such radiation due to the use of these products,[75] manufacturers and distributors of products meeting the definition of electronic product radiation[76] are required to comply with certain formalities, for example, record keeping, specific labeling requirements, and reporting to the Center for Devices and Radiological Health (CDRH).

22.4.2.1 What Is a Laser?

The term 'laser' is an acronym for '**L**ight **A**mplification by **S**timulated **E**mission of **R**adiation'. Simply stated, a laser tool emits a beam of light which, when focused on a substrate, will thermally affect its target. This is made possible because of the way lasers interact with electrons. There are many different types of medical lasers, some designed to correct vision, remove discolorations, and other imperfections on the skin, whilst others are designed for resurfacing the skin. It is beyond the scope of this chapter to attempt to describe the many different kinds of lasers available.

22.4.2.2 How Does the FDA Regulate Lasers?

It is important to note that the FDA only regulates the sale and marketing of medical devices and does not regulate physicians or nurses in the practice of medicine or in the use of a device. Before a laser can be legally sold in the United States, the company wishing to sell or market the laser must obtain authorization from the FDA. Medical lasers, depending on their application, are usually categorized in Class II or IV[77] and must have premarket approval or premarket clearance from the FDA prior to marketing for any indication. The majority of lasers are cleared through the 510(k) premarket application process. However, there are two minor exceptions to this. Certain unapproved, nonsignificant risk Class III medical devices may be distributed in the United States to individual practitioners who have approval from an Institutional Review Board (IRB) for the investigational clinical use of the device. Alternatively, lasers may be distributed to investigators participating in a study under an IDE approved by the CDRH[78] (although various IDE requirements need to be complied with[79]).

By way of further regulatory rigor, all laser devices distributed for both human and animal treatment in the United States are subject to Mandatory Performance Standards. Laser manufacturers therefore have to meet the federal laser product performance standard and must submit an initial report to CDRH's Office of Compliance prior to distributing the laser.[80] This performance standard specifies the safety features and labeling that all lasers must have, in order to provide adequate safety to users, and includes various technical and service requirements. A laser product manufacturer must certify that each laser model has passed a quality assurance test and complies with the performance standard before introducing the laser into the market. This includes distribution for use during clinical investigations prior to device approval.

The company/manufacturer certifying a laser assumes responsibility for product reporting to CDRH,[81] record-keeping, and notification of defects, noncompliances, and accidental radiation occurrences.[82] However, once the FDA has authorized the commercialization of a laser, a doctor may decide to use that laser for other indications if he/she feels it is in the best interest of a user. The use of an approved device for other than its FDA-approved indication is called off-label use. The FDA does not regulate the practice of medicine.[83] Therefore, the FDA does not have the authority to regulate a doctor's practice and activities. The FDA does, however, regulate the claims manufacturers assert for their devices.

22.4.3 Medical Laser Classification

Each laser product is classified as Class I, IIa, II, IIIa, IIIb, or IV in accordance with definitions established by FDA in paragraphs (b)(5) through (11) of 21 CFR §1040. The product classification is based on the highest accessible emission level(s) of laser radiation to which human access is possible during operation. Lasers are classified into four broad areas, depending on the potential for causing *biological damage*. Laser emission ranges for each class are provided in 21 CFR §1040(d)(2).

The standard establishes the following limits for the Classes:

- *Class I* limits (1040.10(b)(5) and 1040.10(d)(Table I)) apply to devices that have emissions in the ultraviolet, visible, and infrared spectra, and are limits below which biological hazards have not been established. In the visible and near infra-red spectra there are separate Class I limits for radiant energy (power) and integrated radiance (radiance); both limits must be exceeded for the device to move from Class I. These lasers cannot emit laser radiation at known hazard levels. **Class IA**—This is a special designation that applies only to lasers that are "not intended for viewing", such as a supermarket laser scanner. The upper power limit of Class IA is 4.0 mW.
- *Class IIa* limits (1040.10(b)(6) and 1040.10(d)(Table II-A)) apply to products whose visible emission does not exceed Class I limits for emission durations of 1000 seconds or less, and are not intended for viewing. Class IIa limits therefore may not exceed the Class II limits. An example of a Class IIa laser product might be a supermarket scanner.
- *Class II* limits (1040.10(b)(7) and 1040.10(d)(Table II)) apply to products that have emissions in the visible spectrum (400–710 nm) for emission durations in excess of 0.25 seconds, providing that emissions for other durations and/or wavelengths do not exceed the Class I limits. Class II products are considered a hazard for direct long-term ocular exposure. These are low-power visible lasers that emit above Class I levels, but at a radiant power not above 1 mW.
- *Class IIIa* limits (1040.10(b) (8) and 1040.10(d)(Table III-A)) apply to products that have emissions in the visible spectrum and that have beams where the total collectable radiant power does not exceed 5 mw. Class IIIa products include most heliumneon lasers and laser pointers. These are intermediate-power lasers (cw: 1–5 mW), which are hazardous only when viewed directly in the beam.
- *Class IIIb* limits (1040.10(b)(9) and 1040.10(d)(Table III-B)) apply to devices that emit in the ultraviolet, visible, and infrared spectra. Class IIIb products include laser systems ranging from 5 to 500 mw in the visible spectrum. Class IIIb emission levels are ocular hazards for direct exposure throughout the range of the Class, and skin hazards, at the higher levels of the Class. These are moderate-power lasers.
- *Class IV* levels (1040.10(b)(ll)) exceed the limits of Class IIIb and are a hazard for scattered (diffuse) reflection as well as for direct exposure. These are high-power lasers (> 500 mW) which are hazardous to view under any condition (directly or diffusely scattered). Significant controls are required of Class IV laser facilities.

22.4.4 Requirements for Laser Products Performance

The standard specifies performance requirements according to the Class of the laser product and the accessible laser radiation. Note that, where the standard requires a particular performance feature, the feature must be readily identifiable as such on the product. Failure to properly identify required features may lead to difficulties in determining product compliance. The applicability of many requirements depends on whether the product is a laser, or a laser system (1040.10(b) (19),(23)).

A *protective housing* (1040.10(f)(l)) is required for all laser products. The protective housing must prevent human access to laser radiation in excess of the limits of Class I (and collateral radiation in excess of the collateral radiation limits) at all places and times where and when such human access is not necessary in order for the product to accomplish its intended function. The manufacturer must be prepared to justify the necessity of human access to laser radiation greater than Class I limits. If the purpose of the laser system is to generate a laser beam, the justification is self-evident. In other cases, a detailed analysis may be required. Generally, a protective housing must be contiguous. The most common difficulties with protective housings have been human access to laser radiation through cooling vents, or through a poor fit between sections of a protective housing. A protective housing must be sturdy enough to prevent access caused by bending or warping as the product ages.

Safety interlocks (1040.10(f)(2)) may be required on any laser product. They must prevent human access to laser or collateral radiation that exceeds the limits of Class I and Table VI[84] when a protective housing is opened during operation or maintenance, and human access to the interior radiation is not always necessary during such operation or maintenance. (Note that if the housing must be opened during operation and it is necessary to have access to the interior radiation, the level of the interior radiation must be considered when classifying the product, i.e., the classification is determined by the interior level, if it is higher than the exterior level. If the intermittent access to laser radiation occurs only during a maintenance procedure, it does not affect the class of the product.) If access to the interior radiation is sometimes needed, the interlock may be defeatable and the housing must be so labeled. Safety interlocks need not prevent access to interior radiation otherwise accessible only during service.

Safety interlocks to protect from Class IIIb or IV levels must also be redundant or fail-safe; if fail-safe, they must either prevent opening the housing in case they fail, or they must be incapable of failing in a mode that would permit access. Defeatable safety interlocks must provide a visible or audible indication of defeat; further, it must not be possible to close the housing with the interlock remaining defeated. A redundant or fail-safe safety interlock is also required if the failure of a single interlock would allow access to laser radiation in excess of the accessible emission limits of Class II to be emitted directly through the opening created by removal or displacement of the interlocked portion of the protective housing.

A remote interlock connector (1040.10(f)(3)) is required on all Class IIIb and IV laser systems. The purpose of the remote interlock connector is to permit the user to connect a remote barrier interlock, emergency stop switch, or similar device. The circuit must be designed such that, when the terminals of the connector are open, human access to laser radiation is prevented. The electrical potential across the connector terminals must not be greater than 130 volts rms.

A *key control* (1040.10(f)(4)) is required for Class IIIb and IV laser systems in order for the user to prevent unauthorized operation. The key must not be removable in the "on" position.

An *emission indicator* (1040, 10(f)(5)) is required on Class II, IIIa, IIIb, and IV laser systems. The indicator can be visible or audible. On Class IIIb and IV laser systems, the indication must precede emission by a length of time sufficient to allow users and others in the area to recognize that the product has been energized so they can avoid exposure. Depending on the action required and the level of laser radiation involved, the time needed can vary considerably; typical values are in the range of 2–20 seconds. Emission indicators must be duplicated on lasers (heads) and operation controls if they are capable of being separated by greater than 2 meters.

A *beam attenuator* (1040.10(f)(6)) is required on Class II, IIIa, IIIb, and IV laser systems. The beam attenuator is a mechanical or electrical device such as a shutter or attenuator that blocks emission. The beam attenuator blocks bodily access to laser radiation above Class I limits without the need to turn off the laser. The beam attenuator must be available for use at all times during operation. Power switches and key controls do not satisfy the beam attenuator requirement. Manufacturers may apply for approval of alternate means of providing this protection if a beam attenuator is inappropriate to the product.

Operating controls (1040.10(f)(7)) on a Class II, IIIa, IIIb or IV laser product must be located such that it is not necessary for the user to be exposed while manipulating them. Viewing optics, viewports, or display screens (1040.10(f)(8)) may not provide human access to laser or collateral radiation in excess of the limits of Class I and Table VI[85] during operation or maintenance. If the viewing optics employ a shutter or variable attenuator, the shutter or attenuator must be fail-safe; that is, it must be designed such that, upon failure, it is impossible to open the shutter or vary the attenuation. Viewing optics include such devices as viewports, windows, microscopes on welding and drilling devices, and operating microscopes on surgical lasers. Attenuation may be total, or it may be partial as with a filter. Acceptable designs may prevent laser operation until the attenuator has moved into position. Service instructions must include instructions on procedures to avoid hazardous exposure through viewing optics.

A *scanning safeguard* (1040.10(f)(9)) must prevent emission in excess of the limits of the class of the product. For Class IIIb or IV laser products that operate in both scanned and unscanned modes, the scanning safeguard also must prevent emission in excess of the limits of the class of the scanned laser radiation (and whose failure would result in emissions exceeding Class IIIa). Scanned laser radiation is laser radiation that is moved in translation or by changing direction. A scan-failure safeguard must have a reaction time short enough to operate before levels of a higher class are emitted; it is possible to achieve this performance by means of a high inertia scanner in conjunction with an electromechanical shutter.

A *manual reset* (1040.10(f)(10)) is required on Class IV laser systems manufactured after August 20, 1986. It must prevent automatic restart after an interruption due to remote interlock activation or from an interruption for more than 5 seconds due to unexpected loss of main electrical power.

22.5 Clinical Studies with Investigational Laser and Light-Based Systems

22.5.1 Medical Device Studies (Investigational Device Studies)

There are many different types and intended uses of medical devices. Medical devices are implants (i.e., pacemaker), devices to monitor a user (i.e., blood pressure cuff), devices to sustain human life (i.e., ventilator) or devices important in diagnosing, curing, or treating a disease. Other medical devices are, for example, disposable contact lenses or wound dressings. An investigational device is a medical device which is the subject of a clinical trial in order to evaluate the effectiveness and/or safety of the device. Like the pharmaceutical studies, these studies require a research protocol, approval through the IRB ethics committee, and signing an informed consent form. Subjects will usually have to come in more than once for a follow-up visit, just like in pharmaceutical studies.

22.5.2 What Is a Clinical Trial?

A clinical trial (also called clinical research) is a research study using human volunteers designed to determine the safety and effectiveness of a drug, biologic (such as a vaccine), device (such as prosthesis) or other treatment or behavioral intervention. Carefully conducted clinical trials are the fastest and safest way to find treatments that work in people and methods to improve health. *Interventional trials* determine whether experimental treatments or new ways of using known therapies are safe and effective under controlled environments. *Observational trials* address health issues in large groups of people or populations in natural settings.

22.5.3 How Do Clinical Trials Work?

There are different types of clinical trials: treatment, prevention, diagnostic, screening, and quality of life trials—and the trials are conducted in progressive phases (I–IV). To ensure that no one can influence the results of a study, clinical trials employ a range of specialized testing mechanisms intended to prevent bias and provide reliable results:

- *Treatment Trials* test new treatments, new combinations of drugs, or new approaches to surgery or radiation therapy.
- *Prevention Trials* look for better ways to prevent a given disease in people who have never had that disease or to prevent a disease from returning. Preventative approaches include medicines, vitamins, vaccines, minerals, and lifestyle changes.
- *Diagnostic Trials* are conducted to find better tests or procedures for diagnosing a particular disease or condition.
- *Screening Trials* test the best way to detect certain diseases or health conditions.
- *Quality of Life Trials* (or supportive care trials) explore ways to improve comfort and the quality of life for individuals with a chronic illness.
- *Prospective Trials*—Users are identified and then followed over time.
- *Randomized Trials*—Users are grouped by chance into (typically) a treatment group and a control group (also called a placebo group). A control group receives either the current standard treatment or a placebo—an inactive dose or treatment. The results of the control group are then compared with those of the treatment group.
- *Cross-over Trials*—Users receive both the treatment and the placebo at different times, with careful monitoring of their responses to both approaches.
- *Double-blinded Trials*—Neither the user nor the researcher knows if the user is receiving the treatment or the placebo.

In addition, some clinical trials are called *open-label studies*, because both the user and the researcher know that the user is receiving the treatment, and not the placebo.

By federal regulation, every clinical trial in the United States must be approved and monitored by an Institutional Review Board (IRB), an independent committee of physicians, statisticians, community advocates, and others. The IRB is charged with ensuring that all clinical trials within a given medical institution are ethical and that the rights of the participants in those trials are protected.

Clinical trials retain very specific participation guidelines. Establishing and maintaining these guidelines is a critical part of producing meaningful and reliable results. The factors that allow someone to participate in a clinical trial are called "inclusion criteria", while those that disallow someone are called "exclusion criteria". Typical criteria include age, gender, the type and stage of a disease, previous treatment history, and other medical conditions.

22.5.4 What Are the Phases of Clinical Trials?

Clinical trials are conducted in phases. Each phase of a trial has a different purpose and helps scientists to answer specific questions.

While small, early phase trials may be conducted by individuals or small groups of physicians, larger trials are typically conducted by hospitals, pharmaceutical companies, or device manufacturers. If a therapy successfully passes through Phase III trials, the FDA may approve it to be marketed to the public.

Phase	Definition
Phase I Trials	Researchers test a new drug or treatment in a small group of people (20–80) for the first time to evaluate its safety, determine a safe dosage range, and identify side effects.
Phase II Trials	The study drug or treatment is given to a larger group of people (100–300) to see if it is effective, and to further evaluate its safety.
Phase III Trials	The study drug or treatment is given to large groups of people (1000–3000) to confirm its effectiveness, monitor side effects, compare it to commonly used treatments, and collect information that will allow the drug or treatment to be used safely.
Phase IV trials	Postmarketing studies delineate additional information including the drug's risks, benefits, and optimal use.

22.5.5 Research Study

A *research study* is a controlled exploratory study which incorporates a formal research design to test a hypothesis with validated measures. These studies should document that this research can be replicated or generalizable to other settings. The design allows for a statistical test of the specific differences found between the two groups on a validated measure that is relevant to the skills and knowledge intended to be developed.

22.5.6 Pilot Study

A *pilot study* is a preliminary study to the larger study. It is typically limited in the number of subjects than you plan to include in the full study, or you may limit it because your scope is smaller in some other way; for example, the range of types of subjects may be more limited, or the procedures may be more limited. A pilot study can help work out some of the procedural bugs even though you know it is not likely to add anything new or important to your main study. Here are some more reasons to consider a pilot study:

1. It permits preliminary testing of the hypotheses that leads to testing more precise hypotheses in the main study. It may lead to changing some hypotheses, dropping some, or developing new hypotheses.
2. It often provides the researcher with ideas, approaches, and clues you may not have foreseen before conducting the pilot study. Such ideas and clues increase the chances of getting clearer findings in the main study.
3. It permits a thorough check of the planned statistical and analytical procedures, giving you a chance to evaluate their usefulness for the data. You may then be able to make needed alterations in the data collecting methods, and therefore, analyze data in the main study more efficiently.
4. It can greatly reduce the number of unanticipated problems because you have an opportunity to redesign parts of your study to overcome difficulties that the pilot study reveals.
5. It may save a lot of time and money. Unfortunately, many research ideas that seem to show great promise are unproductive when actually carried out. The pilot study almost always provides enough data for the researcher to decide whether to go ahead with the main study.
6. In the pilot study, the researcher may try out a number of alternative measures and then select those that produce the clearest results for the main study.

22.5.7 Pivotal Study

Usually, a Phase III study which presents evidence-based data that the FDA will rely on when deciding whether or not to approve a drug or device. A pivotal study will generally be well-controlled, randomized, statistically powered to adequate size, and whenever possible, double-blind. It represents the optimized drug or device as intended for commercialization, along with the final packaging and labeling to be evaluated in the targeted population.

22.6 Conducting a Clinical Investigation

- Investigational Device Exemption (IDE) - 21CFR Part 812
- IDE content
- Institutional Review Board (IRB)

22.6.1 An Investigational Device Exemption Overview

An Investigational Device Exemption (IDE) allows the investigational device to be used in a clinical study in order to collect safety and effectiveness data required to support a PMA application or a Premarket Notification [510(k)] submission to FDA. Clinical studies are most often conducted to support a PMA. Only a small percentage of 510(k)s require clinical data to support the application. Investigational use also includes clinical evaluation of certain modifications or new intended uses of legally marketed devices. All clinical evaluations of investigational devices, unless exempt, must have an approved IDE before the study is initiated.

Clinical trials using unapproved medical devices on human subjects are performed under an Investigational Device Exemption (IDE). Clinical studies with devices of significant risk must be approved by FDA and by an Institutional Review Board (IRB) prior to initiation of a clinical study. An FDA approval is obtained by submitting an IDE application to FDA.[86]

Studies with devices of nonsignificant risk may not require an IDE, but must be approved by the IRB before the study can begin. The following key elements are inherent in all IDE studies:

- informed consent from all users,
- labeling for investigational use only,
- monitoring of the study, and,
- required records and reports.

An approved IDE permits a device to be shipped lawfully for the purpose of conducting investigations of the device without complying with other requirements of the Food, Drug, and Cosmetic Act that would otherwise apply to devices in commercial distribution. Sponsors need not submit a PMA or Premarket Notification 510(k), register their establishment, or list the device while the device is under investigation. Sponsors of IDEs are also exempt from the Quality System (QS) Regulation except for the requirements for design control. Extensive record-keeping, reporting, and monitoring of the clinical studies is also required.

22.6.1.1 Pre-IDE Meetings

Two types of pre-IDE meetings are possible: meetings in which FDA provides "informal guidance" and meetings where FDA provides "formal guidance" as provided for in Section 201 of the FDA Modernization Act of 1997.

a. "Informal Guidance" Meetings. Sponsors are encouraged to meet with the Office of Device Evaluation (ODE), Reviewing Division before the IDE application is submitted for review so that the reviewing division can provide any advice/guidance which can be used in the development of supporting preclinical data or the investigational plan for incorporation into the IDE application. These meetings may take the form of telephone conference calls, video conferences, or face-to-face discussions. Regardless of the form of the pre-IDE meeting, all meetings should be recorded by the ODE reviewing division and reported on a quarterly basis to ODE senior management. Minutes of the meeting should include the date of the meeting, the attendees, whether material was submitted prior to the meeting for discussion/review by ODE staff, a summary of the discussion, and any recommendations or guidance provided by FDA.

b. "Formal Guidance" Meetings. A sponsor or applicant may submit a written request for a meeting to reach an agreement with FDA regarding FDA's review of an investigational plan (including a clinical protocol). As required by the statute, this meeting should take place no later than 30 days after receipt of the request. The written request should include a detailed description of the device, a detailed description of the proposed conditions of use of the device, a proposed plan (including a clinical protocol) for determining whether there is a reasonable assurance of effectiveness, and, if available, information regarding the expected performance of the device. If an agreement is reached between FDA and the sponsor or applicant regarding the parameters of an investigational plan (including a clinical protocol), the terms of the agreement should be put in writing and made part of the administrative record by FDA.

c. Agreement Meeting. A sponsor or applicant may submit a written request for a meeting to reach an agreement with FDA regarding FDA's review of an investigational plan (including a clinical protocol). The request and summary information should be submitted as a pre-IDE submission and identified as an agreement meeting request. This meeting should take place no later than 30 days after receipt of the request. The written request should include a detailed description of the device, a detailed description of the proposed conditions of use of the device, a proposed plan (including a clinical protocol) for determining whether there is a reasonable assurance of effectiveness, and, if available, information regarding the expected performance of the device. If an agreement is reached between the FDA and the sponsor or applicant regarding the parameters of an investigational plan (including a clinical protocol), the terms of the agreement are put in writing and made part of the administrative record by FDA.

22.6.1.2 Pre-IDE Submissions

Sponsors are encouraged to submit pre-IDE submissions to the ODE reviewing division while the sponsor is preparing a formal IDE submission whenever the sponsor requires informal FDA guidance on troublesome parts of the IDE application, for example, clinical protocol design, preclinical testing proposal, etc. Pre-IDE submissions are logged into the pre-IDE tracking system by the Document Mail Center (DMC). After the document is logged-in, the DMC will jacket the submission in a white folder, attach a tracking sheet, print an acknowledgment letter to the pre-IDE sponsor, and forward the submission and letter to the appropriate reviewing division. The division should verify that the submission belongs to their division and, after signing the acknowledgment letter and placing a copy of it in the pre-IDE, mail the letter to the sponsor.

Upon completion of the review of the pre-IDE submission, the division is responsible for issuing a response to the sponsor in a timely manner, usually within 60 days of receipt. The response may take the form of a letter or comments provided during a meeting or teleconference. If FDA's response is provided via comments during a meeting or a telephone conference call, a memo of the meeting or conference call should be prepared. The division is responsible for ensuring that all memos, reviews, letters, etc. are included in the jacketed file copy for documentation. Upon completion of the review, the document should be returned to the DMC for filing.

22.6.1.3 Approval Process

Investigations covered under the IDE regulation are subject to differing levels of regulatory control, depending on the level of risk. The IDE regulation distinguishes between significant and nonsignificant risk device studies, and the procedures for obtaining approval to begin the study differ accordingly.[87] Some types of studies are also exempt from the IDE regulations.

22.6.1.4 Significant Risk Device

A significant risk device presents a potential for serious risk to the health, safety, or welfare of a subject. Significant risk devices may include implants, devices that support or sustain human life, and devices that are substantially important in diagnosing, curing, mitigating, or treating disease or in preventing impairment to human health, or have the potential to evoke collateral harm. Examples include sutures, cardiac pacemakers, hydrocephalus shunts, and orthopedic implants. The sponsor must demonstrate in the application that there is reason to believe that the risks to human subjects from the proposed investigation are outweighed by the anticipated benefits to subjects and the importance of the knowledge to be gained, that the investigation is scientifically sound, and that there is reason to believe that the device as proposed for use will be effective.

In order to conduct a significant risk device study, a sponsor must:

- submit a complete IDE application[88] to FDA for review and obtain FDA approval of the IDE; submit the investigational plan[89] and report of prior investigations[90] to the IRB at each institution where the investigation is to be conducted for review and approval, and
- select qualified investigators, provide them with all necessary information on the investigational plan and report of prior investigations, and obtain signed investigator agreements from them.

Upon receipt of an IDE application, sponsors are notified in writing of the date that FDA received the original application and the IDE number assigned (receipt of supplements and amendments are not acknowledged). An IDE application is considered approved 30 days after it has been received by FDA, unless FDA otherwise informs the sponsor prior to 30 calendar days from the date of receipt, that the IDE is approved, approved with conditions, or disapproved. In cases of disapproval, a sponsor has the opportunity to respond to the deficiencies and/or to request a regulatory hearing under 21 CFR §16.

Once an IDE application is approved, the following requirements must be met in order to conduct the investigation in compliance with the IDE regulation:

- *Labeling*—The device must be labeled in accordance with the labeling provisions of the IDE regulation[91] and must bear the statement "CAUTION - Investigational Device. Limited by Federal (or United States) law to investigational use".
- *Distribution*—Investigational devices can only be distributed to qualified investigators.[92]

- *Informed Consent*—Each subject must be provided with and sign an informed consent form before being enrolled in the study. The Protection of Human Subjects regulation[93] contains the requirements for obtaining informed consent.
- *Monitoring*—All investigations must be properly monitored[94] to protect the human subjects and assure compliance with approved protocols.
- *Prohibitions*—Commercialization, promotion, and misrepresentation of an investigational device and prolongation of the study are prohibited.[95]
- *Records and Reports*—Sponsors and investigators are required to maintain specified records[96] and make reports to investigators, IRBs, and FDA.

22.6.1.5 Nonsignificant Risk Device

Nonsignificant risk devices are devices that do not pose a significant risk to the human subjects. Examples include most daily-wear contact lenses and lens solutions, ultrasonic dental scalers, and foley catheters. A nonsignificant risk device study requires only an IRB approval prior to the initiation of a clinical study. Sponsors of studies involving nonsignificant risk devices are not required to submit an IDE application to FDA for approval.

Submissions for nonsignificant device investigations are made directly to the IRB of each participating institution. Sponsors should present an explanation to the IRB where the study is undertaken, of why the device does not pose a significant risk. If the IRB disagrees and determines that the device poses a significant risk, the sponsor must report[97] this finding to FDA within five working days. The FDA considers an investigation of a nonsignificant risk device to have an approved IDE when IRB concurs with the nonsignificant risk determination and approves the study.

The sponsor also must comply with the abbreviated IDE requirements:[98]

- *Labeling*—The device must be labeled in accordance with the labeling provisions of the IDE regulation[99] and must bear the statement "CAUTION - Investigational Device. Limited by Federal (or United States) law to investigational use."
- *IRB Approval*—The sponsor must obtain and maintain IRB approval throughout the investigation as a nonsignificant risk device study;
- *Informed Consent*—The sponsor must assure that investigators obtain and document informed consent from each subject according to 21 CFR §50, Protection of Human Subjects, unless documentation is waived by an IRB in accordance with §56.109(c);
- *Monitoring*—All investigations must be properly monitored to protect the human subjects and assure compliance with approved protocols;[100]
- *Records and Reports*—Sponsors are required to maintain specific records[101] and make certain reports as required by the IDE regulation.
- *Investigator Records and Reports*—The sponsor must assure that participating investigators maintain records and make reports as required (see Responsibilities of Investigators); and
- *Prohibitions*—Commercialization, promotion, test marketing, misrepresentation of an investigational device, and prolongation of the study are prohibited.[102]

22.6.1.6 IDE Exempt Investigations

All clinical investigations of devices must have an approved IDE or be exempt from the IDE regulation. Investigations that are exempted from 21 CFR §812 are described in §812.2(c) of the IDE regulation. Studies exempt from the IDE regulation include:

1. a legally marketed device when used in accordance with its labeling;
2. a diagnostic device if it complies with the labeling requirements in §809.10(c) and if the testing:
   a. is noninvasive;
   b. does not require an invasive sampling procedure that presents significant risk;
   c. does not by design or intention introduce energy into a subject; and
   d. is not used as a diagnostic procedure without confirmation by another medically established diagnostic product or procedure;
3. consumer preference testing, testing of a modification, or testing of a combination of devices if the device(s) are legally marketed device(s) [that is, the devices have an approved PMA, cleared Premarket Notification 510(k), or are exempt from 510(k)] AND if the testing is not for the purpose of determining safety or effectiveness and does not put subjects at risk;
4. a device intended solely for veterinary use;
5. a device shipped solely for research with laboratory animals and contains the labeling "CAUTION – Device for investigational use in laboratory animals or other tests that do not involve human subjects". Depending upon the nature of the investigation, those studies which are exempt from the requirements of the IDE regulation may or may not be exempt from the requirements for IRB review and approval under Part 56 and the requirements for obtaining informed consent under Part 50.

22.6.1.7 Who Must Apply for an IDE

The sponsor of the clinical trial is responsible[103] for submitting the IDE application to FDA and obtaining Institutional Review Board (IRB) approval before the study can begin. Foreign companies wanting to conduct a clinical study in the United States MUST have a US sponsor.[104] Under certain circumstances, the clinical investigator may wish to submit an IDE and would, therefore, also act as the sponsor of the study.

22.6.1.8 When To Apply

Study approval must be obtained PRIOR to enrolling users at the study site. Each site must have an approval from the reviewing IRB for that site prior to the beginning the study. For significant risk-device studies, in addition to IRB approvals, the sponsor must also have an approved IDE from FDA prior to beginning the study at any site. The review of applications to FDA and to the IRBs are independent and, therefore, may be submitted simultaneously.

22.6.1.9 FDA Action[105] on IDE Applications

The FDA will notify the sponsor (in writing) of the date it receives an IDE application. The FDA may approve, approve with modification, or disapprove an IDE application. FDA may request additional information about an investigation. The sponsor may provide the requested information or the sponsor may treat such a request as a disapproval of the application and request a hearing in accordance with 21 CFR §16. The clinical investigation may begin after FDA and the IRB approves an IDE for the investigation. An investigation may begin 30 days after FDA receives the IDE application for the investigation of a device if IRB approval has been obtained unless FDA notifies the sponsor that the investigation may not begin.

FDA may disapprove or withdraw approval of an IDE application if FDA finds that:

1. The sponsor has not complied with applicable requirements of the IDE Regulation, any other applicable regulations or statutes, or any condition of approval imposed by an IRB or FDA.
2. The application or a report contains untrue statements or omits required material or information.
3. The sponsor fails to respond to a request for additional information within the time prescribed by FDA.
4. There is reason to believe that the risks to the human subjects are not outweighed by the anticipated benefits to the subjects, or the importance of the knowledge to be gained, that informed consent is inadequate, that the investigation is scientifically unsound, or that the device as used is ineffective.
5. It is unreasonable to begin or to continue the investigation, due to the way in which the device is used or the inadequacy of:
   (i) the report of prior investigations or the investigational plan;
   (ii) the methods, facilities, and controls used for the manufacturing, processing, packaging, storage, and, where appropriate, installation of the device; or
   (iii) the monitoring and review of the investigation.

22.6.1.10 Notice of Disapproval or Withdrawal

If FDA disapproves an IDE application or proposes to withdraw an approval, FDA will notify the sponsor in writing. A disapproval order will contain a complete statement of the reasons for disapproval and will advise the sponsor of the right to request a regulatory hearing under 21 CFR §16. The FDA will provide an opportunity for a hearing before withdrawal of approval, unless FDA determines that there is an unreasonable risk to the public health if testing continues.

22.6.1.11 Promotion of Investigational Devices

Under §812.7, a sponsor, investigator, or any person acting for or on behalf of a sponsor or investigator cannot:

- Promote or test-market an investigational device, until after FDA has approved the device for commercial distribution.

- Commercialize an investigational device by charging the subjects or investigators a higher price than that necessary to recover costs of manufacture, research, development, and handling.
- Unduly prolong an investigation. If data developed by the investigation indicate that premarket approval (PMA) cannot be justified, the sponsor must promptly terminate the investigation.
- Represent that an investigational device is safe or effective.

However, the sponsor may advertise for research subjects to solicit their participation in a study. Appropriate advertising methods include but are not necessarily limited to: newspaper, radio, TV, bulletin boards, posters, and flyers that are intended for prospective subjects. Advertisements should be reviewed and approved by the IRB to assure that it is not unduly coercive and does not promise a certainty of cure beyond what is outlined in the consent and the protocol. No claims should be made, either explicitly or implicitly, that the device is safe or effective for the purposes under investigation, or that the test article is known to be equivalent or superior to any other device. FDA considers direct advertising for study subjects to be the start of the informed consent and subject selection process.

22.6.2 Content of an IDE

An IDE is designed to provide sufficient information which FDA can use to make a determination on whether to approve or disapprove a clinical trial. The following are key components of an IDE for an investigational device with a focus on light-based systems.

22.6.2.1 Investigational Plan (21 CFR §812.25)

a. Purpose of the Clinical Trial. This section describes broad study goals, as well as specific aims of the investigation. A background and rationale for conducting the study is generally helpful for the FDA to understand the broader purpose and determine risk/benefits of the investigation. The background information may include earlier studies conducted with similar devices and technologies, relevant clinical data from literature, and how it applies to the proposed investigation. It is also helpful to provide in this section an overview of the study design that clearly lays out how the study will be conducted and what procedures will be used to determine efficacy and safety.

b. Clinical Protocol. The clinical protocol should provide details on how the subjects will be screened; what the inclusion/exclusion criteria for the study are; at what point the subjects will be enrolled; what evaluations will be conducted before initiating the treatments (baseline evaluation); provide details of the treatments, including procedure, number of treatment visits, and duration of the treatment phase; details for the follow-up phase including, the duration of the follow-up period, number and frequency of visits should be provided.

The study protocol should also define the efficacy and safety measures. Depending on the type of study (research, pilot, or pivotal) the primary and secondary end-points may need to be prospectively defined, that is, what quantitative measure will determine whether the treatment has achieved an effective outcome? Details of the analytical procedures for any objective measures should be provided. The type of subjective measures, for example, subjective questionnaires administered to patients and their usefulness in assessing efficacy and safety should be discussed.

A statistical analysis plan for the data, justification for choosing the sample size, and the randomization scheme should be presented. The details of the statistical plan again will be determined by the development phase of the study. A pivotal study requires a detailed statistical plan, whereas the research-type studies that are primarily designed to gain initial learning that would help modify and set the device parameters may not require a detailed statistical plan.

Risk analysis: This section should clearly identify risk to the patients, and risk management. A detailed description of risks and anticipated adverse events (AEs) should be provided. The risks identified may be theoretically based on the device parameters, mechanism of action, and the known tissue interaction of the emitted energy, or the risks may be based on earlier investigations presented in literature or conducted internally. For example, potential risks for a laser device may include eye injury, pain and discomfort to subjects during treatments, and dermal effects such as erythema, edema, blistering, crusting, and pigmentary changes.

A risk–benefit analysis may be presented, especially if significant AEs are anticipated. For laser and IPL devices, laser parameters and any built-in device features that can affect safety should be described. Procedures for recording and reporting AEs and unanticipated adverse device effects (UADE) should be clearly defined.

c. Device Description. This section provides a detailed description of the device that will be used in treatments. For example, a laser device may consists of a base unit connected to the hand piece that may either directly come in contact with the skin, or may emit laser energy directed at the skin surface for treatment. There may also be a chiller unit that is part of the device components to keep the laser from overheating, or to cool the skin surface during treatments. The laser parameters, including the wavelength, power, pulse width, and frequency, beam-spot size, calculated energy density or fluence on skin should be described. If a secondary skin-cooling method is used, such as contact surface cooling with chilled sapphire tip, cold air spray, cooling gels, or other cooling methods at pre-, post-, and during treatments these should be listed as part of the device description. Additional information may include calibration methods, verification of key laser parameters at certain interval, and cleaning and sanitation procedures.

d. Monitoring Procedures. Sponsor should select qualified individuals to monitor the study. The person or the CRO contracted to monitor the study should be identified. The monitor should be familiar with the investigational device, the clinical protocol, and the informed consent document. It is the monitor's responsibility to ensure that the investigation is conducted in accordance with the approved investigational plan, the investigator's agreement, requirements under the IDE regulations, and any conditions imposed by the IRB or FDA. The monitor also ensures that the reported data are consistent with the recorded source document data and that the rights of the subjects are being protected as per informed consent.

e. Labeling.[106] An investigational device must carry a label identifying it as such: "CAUTION—Investigational Device. Limited by law to investigational use." The label or other labeling shall describe all relevant contraindications, hazards, adverse effects, interfering substances, warnings, and precautions such as "DANGER"; "LASER RADIATION"; "DIRECT EXPOSURE TO BEAM". The label cannot represent that the device is safe or effective for the purposes for which it is being investigated.

f. Consent Material. Before initiating any clinical investigation, informed consent must be obtained. It is a process by which the subjects entering the study must voluntarily confirm their willingness to participate in a particular clinical study, after having been informed all aspects of the study relevant to the subject's decision to participate. A typical consent of document provides information on the principal investigator, the investigational site, the purpose of the study, study details as to the procedures that will be performed before (baseline), during (treatment phase) and after (follow-up) light-based treatments, the time commitment and the number of visits for each phase, risks, and discomforts of the treatments, benefits and alternative treatments, maintenance of confidentiality, and remuneration for participation in the study. The clinical investigator or his or her designee describes the relevant study details to the patients. The informed consent is signed and dated and a copy provided to the subject.

g. IRB Information. The clinical protocol should provide information on the IRB that will be used to obtain approval of the study protocol and the informed consent documents.

22.6.2.2 Other Relevant Information To Be Included in an IDE

a. Prior Investigations (21 CFR §812.27). This section should summarize any prior clinical testing experience by the sponsor with the test device. Additional information could be included from bench, *in vitro* and *in vivo* tests that support the proposed clinical trial. Relevant literature data on safety and effectiveness should also be summarized, including a bibliography of all published material.

b. Responsibility of the Sponsor (21 CFR §812 Subpart C). This section describes the general responsibilities of the study sponsor, including securing the FDA/IRB approvals, selection of qualified investigators and monitors, obtaining signed agreement from the investigator, details on the monitoring procedures, procedures for reporting unanticipated device effects to the agency and IRB, providing periodic reports (annual reports to FDA), procedures on how investigational devices will be controlled, that is, records for shipping and receiving and calibration and maintenance records.

c. IRB Review (21 CFR §812 Subpart D). This section should briefly describe the IRB composition, duties, and function. An IRB is responsible for protecting the rights and welfare of the subjects participating in the study. An IRB approval is required for subject recruitment (any form of advertisement to recruit subjects), consent form, and the clinical protocol. Sponsor is responsible for determining whether the device is a significant or a nonsignificant risk device. The IRB must agree with this determination.

d. Responsibilities of Investigators (21 CFR §812 Subpart E). This section should briefly describe the responsibilities of the investigator in conducting the clinical trial. The regulatory definition of an investigator is "… an individual who actually conducts a clinical investigation, i.e., under whose immediate direction the test article is administered or dispensed to, or used involving a subject, or in the event of an investigation conducted by a team of individuals, is the responsible leader of the team".

The general responsibilities of the investigator include conducting the clinical trial according to the investigational plan, supervising the device use, and protecting the rights, and safety and welfare of the subjects enrolled in the study. The investigator also ensures that appropriate approvals from the FDA and IRB are in place before initiating the study.

e. Records and Reports (21 CFR §812 Subpart G). This section briefly describes what records would be created and maintained.

22.6.3 Institutional Review Boards

Under FDA regulations, an IRB is an appropriately constituted group that has been formally designated to review and monitor biomedical research involving human subjects. In accordance with FDA regulations, an IRB has the authority to approve, require modifications in (to secure approval), or disapprove research. This group review serves an important role in the protection of the rights, safety, and welfare of human research subjects. The purpose of an IRB review is to assure, both in advance and by periodic review, that appropriate steps are taken to protect the rights, safety, and welfare of humans participating as subjects in the research. To accomplish this purpose, IRBs use a group process to review research protocols and related materials (e.g., informed consent documents). The IRB must monitor and review an investigation throughout the clinical study. If an IRB determines that an investigation involves a significant risk device, it must notify the investigator and, if appropriate, the sponsor. The sponsor may not begin the investigation until approved by FDA.

Currently, FDA does not require IRB registration. The institutions where the study is to be conducted should be contacted to determine if they have their own IRB. If the study is conducted at a site that does not have its own IRB, the investigators should be queried to see if they are affiliated with an institution with an IRB that would be willing to act as the IRB for that site in the study. There are also independent/contract IRBs that can be contracted with to act as the IRB for a site. A list of IRBs associated with the Consortium of Independent Review Boards is available from the FDA IDE Staff at 301-594-1190. (Note: FDA does not approve or endorse any IRBs.) In addition, an IRB can be established in accordance with 21 CFR §56.

An IRB must comply with all applicable requirements of the IRB regulation (Part 56) and the IDE regulation (Part 812) in reviewing and approving device investigations involving human testing. The FDA does periodic inspections of the IRB's records and procedures to determine compliance with the regulations.

22.6.3.1 Structure of the IRB

The composition of the IRB is specified in the Department of Health & Human Services (DHHS) regulations.[107] The IRB must be comprised of male and female members from diverse backgrounds who possess the professional competence necessary to review the specific research activities submitted to the Board. In addition, the IRB should reflect an appropriate racial and cultural balance, as well as sensitivity to such issues as community attitudes toward medical research. At least one member of the IRB must be a lay community representative with no formal relationship with the Institution. Lay members of the IRB are not expected to possess the necessary technical expertise to review the scientific aspects of most medical protocols. They should, however, be competent to review the consent process and the consent form.

IRBs must have at least five members. Included in this group should be a least one scientist, one nonscientist, and one individual who has no personal or familial employment by the IRB. Further, membership by at least one attorney and one clergyman is generally recommended. In being certain that appropriate consent is provided, an IRB, by federal regulation, has the following obligations: (1) risks to subjects must be minimized by using procedures that are consistent with sound research design, and that do not necessarily expose subjects to risk and (2) risks to subjects are reasonable in relation to anticipated benefits, if any, to subjects, and the importance of the knowledge that may be expected to result.

22.6.3.2 Responsibilities of the IRB

The National Commission in its report on IRBs[108] articulated the primary purpose of the IRB:

> Investigators should not have sole responsibility for determining whether research involving human subjects fulfills ethical standards. Others who are independent of the research must share this responsibility, because investigators are always in positions of potential conflict by virtue of their concern with the pursuit of knowledge as well as the welfare of the human subjects of their research.

Additional IRB responsibilities specifically imposed by the Federal regulations are: Determination of the acceptability of research projects in terms of institutional commitments and regulations, applicable law, standards of professional conduct and practice, and suspension or termination of research projects not being conducted in accordance with Federal and IRB requirements. Considering the nature of the federally mandated IRB responsibilities, it is obvious that the IRB, of necessity, occupies an important and credible position within the research and administrative structure of the institution it serves.

22.6.3.3 Review Procedures of the IRB

The logistics of IRB review varies from institution to institution. The major principles underlying IRB review, however, do not. First, the IRB must make a decision based on common sense and sound professional judgment as to whether or not the proposed research places the subject "at risk". A subject is considered to be at risk if he is exposed to the possibility of harm, whether physical, psychological, sociological, or other, as a consequence of any activity which goes beyond the application of those established methods necessary to meet his needs. The IRB must consider the fact that certain subject populations (e.g., minors, pregnant women, prisoners, mentally retarded with challenged) may be at greater risk than others.

Certain risks are inherent in life itself, but the IRB is not concerned with the ordinary risks of public or private living. Risk is most obvious in medical and behavioral science research projects involving procedures that may induce a potentially harmful altered physical state or psychological condition. The most obvious examples include surgical procedures, the administration of drugs or radiation, the requirement of strenuous physical exertion, and an intervention that precipitates an emotional disturbance. There is also a wide range of medical, social, and behavioral procedures and projects in which, although there may be no immediate risk, procedures may be introduced which involve discomfort, anxiety, harassment, invasion of privacy, or constitute a threat to the subject's dignity. There are also medical and biomedical projects concerned solely with organs, tissues, body fluids, and other materials obtained in the routine performance of medical services which obviously involve no element of physical risk to the subject, but their use for certain research, training, and service purposes may present psychological, sociological, or legal risks to the subject or authorized representatives. Finally, the risk element should be determined for those studies dependent upon existing information or stored data which have been obtained for quite different purposes but which, when used in a research context, may present risk to the human subject.

If it is judged that the proposed research project will expose a subject to risk, then the IRB must assure itself that (a) the rights and welfare of the subject are adequately protected, (b) the methods used to obtain informed consent are adequate and appropriate, (c) the risks to the subject are outweighed by the potential benefit to him or by the importance of the knowledge to be gained, and (d) the selection of subjects is equitable. The IRB is not responsible for considering scientific merit or methodology unless an alternative experimental design will decrease the potential risk(s) to the subject, and still yield the same potential benefit(s).

Upon completion of the review process and investigator compliance with all modifications recommended by the IRB, an approval to begin the study is issued. Approval is for a maximum period of one year, at which time annual review is required. If a project involves an unusual degree of risk to the subject, approval may be for a period of less than one year as determined by the IRB.

22.6.3.4 Informed Consent—Protection of Human Subjects

No clinical investigator may involve a human being as a subject in research unless the investigator has obtained the legally effective informed consent from the subject. Informed Consent is a written notification to human subjects involved in clinical investigations that provides them with sufficient opportunity to consider whether or not to voluntarily participate in the study. The informed consent document must include all the basic elements of informed consent (outlined later), or it may be a short-form written consent document stating that the elements of informed consent have been presented orally (§50.27). If the short-form method is used, there must be a witness to the oral presentation.

An investigator shall seek such consent only under circumstances that provide the prospective subject or the representative sufficient opportunity to consider whether or not to participate, and minimize the possibility of coercion or undue influence. The information that is given to the subject or the representative shall be in a language understandable to the subject or the representative. No informed consent, whether oral or written, may include any exculpatory language through which the subject or the representative is made to waive or appear to waive any of the subject's legal rights, or releases or appears to release the investigator, the sponsor, the institution, or its agents from liability for negligence. The written consent form must be approved by the Institutional Review Board (IRB) and contain the following basic elements:[109]

1. A statement that the study involves research, an explanation of the purposes of the research, and the expected duration of the subject's participation, a description of the procedures to be followed, and identification of any procedures which are experimental.
2. A description of any reasonably foreseeable risks or discomforts to the subject.
3. A description of any benefits to the subject or to others, which may reasonably be expected from the research.

4. A disclosure of appropriate alternative procedures or courses of treatment, if any, that might be advantageous to the subject.
5. A statement describing the extent, if any, to which confidentiality of records identifying the subject will be maintained, and that notes the possibility that the Food and Drug Administration may inspect the records.
6. For research involving more than minimal risk, an explanation as to whether any compensation and an explanation as to whether any medical treatments are available if injury occurs and, if so, what they consist of, or where further information may be obtained.
7. An explanation of whom to contact for answers to pertinent questions about the research and research subjects' rights, and whom to contact in the event of a research-related injury to the subject.
8. A statement that participation is voluntary, that refusal to participate will involve no penalty or loss of benefits to which the subject is otherwise entitled, and that the subject may discontinue participation at any time without penalty or loss of benefits to which the subject is otherwise entitled.
9. Additional elements of informed consent. When appropriate, one or more of the following elements of information must be provided to each subject:
   a. A statement that a particular treatment or procedure may involve risks to the subject (or to the embryo or fetus, if the subject is or may become pregnant), which are currently unforeseeable.
   b. Anticipated circumstances under which the subject's participation may be terminated by the investigator without regard to the subject's consent.
   c. Any additional costs to the subject that may result from participation in the research.
   d. The consequences of a subject's decision to withdraw from the research and procedures for orderly termination of participation by the subject.
   e. A statement that significant new findings developed during the course of the research which may relate to the subject's willingness to continue participation will be provided to the subject.
   f. The approximate number of subject's involved in the study.

The consent form must be signed by the subject or the subject's legally authorized representative. Each signed consent must be maintained by the clinical investigator and a copy of the informed consent must be provided to the human subject. A combination of oral and written consent may be used. The short-form method of informed consent includes a written summary and a "short form". A written summary is a document of what is to be said to the subject or representative, and must be approved by the IRB. The summary must include all the basic elements of informed consent (discussed earlier). A short form is a document stating that the elements of informed consent have been presented orally to the subject or the subject's legally authorized representative. After the oral presentation is provided, the summary must be signed by the witness and the presenter (investigator or investigator's representative). The short form must be signed by the subject (or the representative) and the witness. A copy of the summary must be provided to the subject (or the representative) in addition to a copy of the short form. The signed documents must be maintained by the clinical investigator.

22.7 510(K) Process for Surgical Laser and Light-Based Devices

- Components of a 510(k) application
- Special requirements for OTC clearance

A brief summary of those requirements outlined under the Radiation Control for Health and Safety Act of 1968 that apply to manufacturers of laser products are presented here. The reader is encouraged to review the regulations in their entirety to foster complete comprehension of the requirements before making any design or procedural decisions.

The following definitions are basic to the regulations:

A *laser* is a device capable of producing or amplifying electromagnetic radiation in the wavelength range from 180 to 1 x 10^6 nanometers[110] by the process of controlled, stimulated emission.

A *laser system*[111] consists of a laser in conjunction with its power supply.

A *laser product*[112] is any device that constitutes, incorporates, or is intended to incorporate a laser or laser system.

A *manufacturer*[113] is any person or organization in the business of making, assembling, or importing laser products. Responsibilities of the laser manufacturer include:

- design and manufacture their products to be in compliance with the standard,
- test their products to assure compliance;,
- certify compliance of their products;
- maintain test and distribution records and a file of correspondence concerning radiation safety, safety complaints, and inquiries;
- use the published reporting guides to submit reports to CDRH, including the Initial and Model Change Reports describing compliance of the product design and testing program and Annual Reports summarizing required records;
- report accidental radiation occurrences (i.e., possible, suspected, and known exposures);
- report any radiation defects or noncompliances; and
- recall (i.e., repair, replace, or refund the purchase price of) defective or noncompliant products.

22.7.1 Components of a 510(k) Application[114]

- Cover letter
- Indication for use
- 510(k) summary
- Truthful and accuracy statement
- Classification summary and certification
- Financial certification or disclosure statement
- Declaration of conformity and summary reports
- Executive summary
- Device description
- Substantial equivalence discussion
- Sterilization and shelf life

- Biocompatibility
- Software validation
- Electrical safety
- Performance testing
- Risk management
- Label comprehension (for OTC clearance)

22.7.2 Content and Format of a Traditional 510(k)

While there is no set formula for presenting these data, there is a general framework in which to follow. That framework is actually an assemblage of major components that, if all are included and the data presented properly, will help ensure that the application is complete. Submitters can make FDA's task easier by using the new cover sheet[115] available from the Center for Devices and Radiological Health (CDRH). Although not required, this cover sheet incorporates a checklist that can help identify the type of submission and whether all pertinent information is included. Other essentials include the following items.

22.7.2.1 Identification

It is important not to overlook the obvious. Proper identification is crucial. Be sure that the applicant's name and street address are included; do not use a post office box. Telephone and fax numbers should be included for the contact person, and if that person is not the applicant, special note of that fact should be made. Be sure the applicant signs the submission and that the date of the application is noted.

Include a table of contents with page numbers indicating the location of the truth-and-accuracy statement; the 510(k) summary or statement; and any attachments, appendices, or illustrations. Be also sure to include the addresses of manufacturing and sterilizing sites.

22.7.2.2 Truth and Accuracy Statement

All 510(k) submitters must include a statement certifying that all information in the application is truthful and accurate, and that no material fact has been omitted. The statement may be in the cover letter, or on a separate page. In either case, the location of this statement should be noted in the table of contents.

22.7.2.3 Device Name

The device name, including both the trade or proprietary name and the classification name, must be included.[116]

22.7.2.4 Registration Number

If applicable, supply the FDA establishment registration number of the owner or operator submitting the premarket notification. Registration is not required in order for a company to submit a 510(k); however, it is a good idea to obtain one. This can be accomplished by filing a Registration Form 2891 with FDA. If this form is already being processed, or if the company intends to register in the future, mention the fact in the notification.

If applicable, include the registration numbers and addresses of each facility used to manufacture the finished devices, including contract sterilizers. The manufacturing process at each facility must be essentially the same, and produce the same device as described in the premarket notification submission.

22.7.2.5 Classification

Include the class of the device, (i.e., Class I, II, or III). If the device has not been classified, the fact needs to be mentioned. If known, include the appropriate classification panel, for example, anesthesiology or orthopedics.[117] If the device does not have a classification panel, state how this was determined. For example, the manufacturer may state that the device was not listed in the classification regulations, nor was it listed in the related FDA publication.

In most cases, an accessory to a classified device takes on the same classification as the parent device. Software and other accessories that accept input from multiple devices usually take on the classification of the parent device with the highest risk.

22.7.2.6 Standards

The submitter should identify any mandatory or voluntary standards met by the device, citing each by paragraph or requirement. Although a device need not meet any particular standard, if substantial equivalence is being claimed to one or more predicate devices that do meet a given standard, then the new device should also meet it. Except for certain radiation-emitting devices, FDA has no mandatory standards, and relies heavily on those accepted by the industry and on its own guidance documents with their standard-like requirements.[118]

22.7.2.7 Labeling[119]

Although manufacturers may submit drafts of their device labeling, including the label on the immediate container of the device, the submission should be representative of the final version. The directions for use should include a specific statement regarding a clinically significant use of the device and any related warnings, contraindications, or limitations (e.g., precautions or adverse effects).[120] The label for a prescription device must bear the following statement: "CAUTION: Federal law restricts this device to sale by or on the order of a __________," the blank to be filled with "physician" or another practitioner who can legally use or order the use of the device.[121] In the process of clearing the 510(k) notification, the FDA may change or limit the labeling content, for example, by limiting its statement of a device's intended uses.

An important component of the 510(k) application is the description of the 'intended use' of the device and the labeled 'indications for use'. The intended use refers to the general functional use of the laser device, such as the effect of the laser radiation on the treatment area. Typically, medical laser devices transfer photon energy to the treatment area, resulting in a local photothermal or a photochemical response. The target site may be the skin surface, or in deeper dermal tissue. The laser parameters (power, pulse width, wavelength, spot-size, fluence) and the target tissue characteristics (body area, pigmentation, vascularity) contribute to the laser–tissue interactions. The tissue response constitutes the intended use of the medical laser devices.

The 'indications for use' is more specific, and refers to a specific disease or a condition for the purposes of its treatment, prevention, mitigation, or diagnoses. New indications for use must be supported by reasonable assurance of safety and efficacy for proposed use of the device, as compared with a predicate device. In the 510(k) application, the terms "intended use" and "indications for use" must be used consistently, and reflect the objective intent of the device and intended outcome as compared with the predicate device. It is important to note that if the new device has an indication of use that substantially differs from the predicate device, and the new indication for use results in altering the intended use of the predicate device, then it may have a new intended use. If the intended use for the new device is not the same as the predicate, or if the new device raises new types of safety or effectiveness questions ,then the device is considered "Not Substantially Equivalent" (NSE) and must be evaluated through the PMA process.

Specific Indication for use:

For laser devices indicated for hair removal, the FDA has allowed the claim of "permanent reduction" of hair, but not "permanent removal". The FDA recognizes that laser treatments do not cause permanent hair removal, but can result in a stable reduction in the number of growing hair. So even though the procedure is generally referred to as "laser hair removal" it cannot be claimed to eliminate all hair in the treatment area. The granted claim for the laser hair removal procedures is: "intended to effect stable, long-term, or permanent reduction" of hair. The permanent or stable reduction is further defined as the reduction in hair density that is maintained over a period of time, representing a full hair-growth cycle, after completing a regimen that may include several treatments. Depending on the body site, the hair cycle may vary from four (axilla and bikini-line) to nine months (legs).

22.7.2.8 Substantial Equivalence Comparison

Manufacturers should attempt to make comparison of the new device to its predicate as easy as possible for the FDA reviewer. The 510(k) notification should therefore include discussion of the similarities and differences between the device and its predicate device, and should make use of comparative tables whenever possible. Comparisons might consider such areas as intended use, materials, design, energy used and delivered, anatomical sites, target population, physical safety, compliance with standards, biocompatibility, and performance. Information used to demonstrate the substantial equivalence of the device to its predicate should be provided in a clear and comprehensible format, making use of tables and graphs where these are helpful to clarify the manufacturer's argument.

For laser and light-based systems, one needs to demonstrate that the predicate and the investigational device has the same intended use, are similar in laser technology and that the investigational device has features that would provide at least a similar level of safety.

Manufacturers should also submit pertinent information about the predicate device, including its labeling if available. For example, the notification should state whether the predicate is a legally marketed preamendment device or a Class I or Class II postamendment device that has been granted marketing clearance by FDA following the submission of a 510(k). If known, provide the 510(k) document control number (i.e., K followed by 6 digits) for the predicate device.

22.7.2.9 Class III Certification and Summary

Before claiming substantial equivalence to a Class III preamendment device, the manufacturer should determine whether a PMA is required for that generic type of device and, thus, for their product. If no PMA is required, the 510(k) submission must still have a special added certification statement.[122,123]

22.7.2.10 Description

The 510(k) notification should include a physical description of the new device, together with an explanation of its intended use, principles of operation, power source, composition, and other information necessary to understand the device. If the notification is for an accessory, it should describe a typical device with which the accessory will be used. All variations of the new device that will be marketed should be listed.

Manufacturers should look at their descriptions from the point of view of the reviewer, in order to spot information that will be necessary to understand the narrative about the device. In many cases, a picture is better than a thousand words. Together with the narrative, it is a good idea to include labeled diagrams, photographs or pictures, engineering drawings, schematics, and any other visual aids. When appropriate, identify all parts of the device—internal and external, assembled, unassembled, and interchangeable—and explain the functions of all significant parts. The device description should include the physical dimensions of the device—for example, its length, width, height, diameter, and weight. Identify any parts intended for single use only.

22.7.2.11 Performance

Performance data are often needed to help demonstrate that the proposed device is as safe and effective as the predicate device. These data may include results from engineering, bench, design verification, human factors, animal, or clinical studies. Tests should be conducted in a manner as similar as possible to how the device will be used during routine application.

- The Bench data in support of efficacy and safety for laser and light-based systems may include laser irradiance measurements, measurements of scattered and reflected light from the treatment area, demonstration of proper functioning of any safety features such as skin-contact sensors, laser beam profile, uniformity of the laser irradiation on the skin surface, maximum permissible exposure (MPE) calculations for eye safety, and laser power degradation on simulated use.

- The nonclinical data may include simulated use on isolated human skin or animal testing to determine safe dose for human use.
- The amount of clinical data required depends on differences in the labeled indication for use, the use conditions, laser parameters and critical device features between the investigational and the predicate device. Generally a well-controlled, randomized, statistically powered study is required to demonstrate the efficacy and safety of the new device. A pilot test may also be required to validate analytical measures of efficacy and safety, the regimen to be used for the pivotal test and to provide justification for the pivotal study sample size.

22.7.2.12 Biocompatibility

Submissions for devices that directly contact the body—for example, gloves and condoms—must include a description of the characteristics of their materials. This description should compare the device to its predicate in sufficient detail to determine biocompatibility, as well as the kind of tests needed to determine biocompatibility. Any material differences between the device to be reviewed and the claimed predicate device must be stated explicitly, but it is just as important to state whether the materials comprising the two devices are identical. Manufacturers need to provide biocompatibility test data for any materials found in the new device that are not present in the predicate device. The data should be in a separate biocompatibility section—a section identified as such—well-organized and complete.

For some devices—such as gloves, condoms, contact lenses, and surgical sutures—the manufacturer should provide an exact identification of all colors in the inks, dyes, markings, radiopaque substances, and other such materials used in the manufacturing process. If the color listed by FDA, is the same as the predicate's color, and there are no apparent concerns, or if the color was included in the general leaching tests, color typically will not be an issue.

22.7.2.13 Software

Applications for computerized devices must follow the appropriate CDRH guidance.[124] Test data must support all performance and safety claims under routine and limited conditions.

Most laser and light-based devices have some element of software control. The FDA published the guidance on May 11, 2005 for software controlled devices: "Guidance for the content of premarket submissions for software contained in medical devices."[125] Components of software documentation may include hazard analysis, software description, and various device components under software control, architecture design chart, software requirements and specification, software design specifications, traceability analysis, and test plan for validation. In addition, it should be demonstrated that should the software fail or have a latent design flaw, the potential injury would not be considered life threatening, and would not result in permanent impairment of a body function.

Software marketed to enhance the performance of a device is regulated as an accessory to that device. If the software is designed to enhance the performance of a group of different devices, it is regulated as an accessory to the device in the group that poses the greatest risk to the user. Any instructions, prompts, or cautions displayed by the system are considered labeling, and must meet the labeling regulations.

22.7.2.14 Sterility

Submissions for devices that are labeled sterile must cite their sterilization method, as well as the method used to validate the sterilization cycle. The notification does not, however, have to include actual validation data. The device's sterility assurance level (SAL) that the manufacturer intends to meet must be included. Note that an SAL of 10^6 is a common industry standard, and that the FDA usually will not accept anything below it. When ethylene oxide (EtO) is to be used to sterilize the device, mention must also be made of the maximum levels of residues of EtO, ethylene chlorohydrin, and ethylene glycol on the device. If radiation sterilization is used, the radiation dose must be stated. The submission should also describe the packaging used to maintain device sterility, but need not include data on packaging integrity.

For devices that contact blood or cerebrospinal fluids, the submission should state whether the device is nonpyrogenic, and describe the method used to make that determination. If the entire device is not labeled sterile or nonpyrogenic, the labeling must clearly identify the parts that are nonpyrogenic and sterile. The relevant guidances on sterility should be consulted.[126,127] Sterilizers intended for use in health-care facilities must meet the appropriate 510(k) requirements in these documents.

Pyrogens. If the device contacts blood or cerebrospinal fluids and will be labeled nonpyrogenic, state the process controls that will be used to control pyrogens. State also what method, such as the *Limulus* amoebocyte lysate (LAL) or the USP rabbit test will be used to determine that each lot is nonpyrogenic.

Sterilization by User. The labeling for devices intended to be sterilized by the user must identify at least one validated method of sterilization. The instructions should be detailed and specific enough for the user to follow and obtain the required SAL. The instructions should also adequately describe any precautions to be followed, such as special cleaning methods required, changes in physical characteristics of the device that may result from reprocessing and resterilization (especially those that may affect the device's safety, effectiveness, or performance), and any limit on the number of times the device can be resterilized and reused without adversely affecting its safety, effectiveness, or performance.

22.7.2.15 Convenience Kits and Trays

If the device is to be marketed as part of a convenience kit, the FDA guidance regarding kits should be consulted.[128] A critical part of that guidance calls for certification by the applicant that the components of the kits have already cleared FDA. If the sterile kit contains finished sterile examination gloves, the notification must contain data to demonstrate that the gloves meet the appropriate American Society for Testing and Materials (ASTM) standards, or their equivalent, and that they pass FDA's 1000-ml water leak test. The gloves must also pass the leak test after undergoing accelerated aging according to one of the protocols described by ASTM, or their equivalent.

If the kit contains sutures, the manufacturer should provide evidence that the sterilant does not contact the sutures during the sterilization process. Some kit assemblers package sutures separately, adding them into the main tray only after it has been processed and sterilized. If sutures are included as part of the kit, the manufacturer cannot change their labeling, packaging, or method of sterilization without prior notification, review, and approval by FDA. Similarly, the supplier of the sutures included in the kit cannot be changed without prior notification, review, and approval by FDA.

If the device kit contains items that are subject to regulation as drugs, a substantially equivalent determination by CDRH will not apply to the drugs in the kit. For information on applicable FDA requirements for marketing drugs in a kit, contact the Center for Drug Evaluation and Research.

22.7.2.16 510(k) Summary or Statement

A premarket notification must include either a summary of the 510(k) safety and effectiveness information upon which the substantial-equivalence determination is based or a statement that this information will be made available by the 510(k) applicant to any person within 30 days of a written request.[129,130] In order to comply with this requirement, manufacturers should familiarize themselves with FDA's exact definition of the terms *summary* and *statement*.[131,132]

Summaries are released by FDA when requested under the Freedom of Information (FOI) Act; statements are used to arrange for this FOI request to be fulfilled by the 510(k) applicant. The decision whether to include a summary or a statement can be changed any time before the substantial equivalence determination is reached, but not after.

510(k) Summaries. If a summary is included, it must be submitted with the 510(k) notification and clearly marked as such in order for FDA to begin its review. There are specific dos and don'ts in putting together a summary. One absolute is that this summary must be complete and correct in order for FDA to complete its review of a 510(k) submission. The FDA will accept summaries and amendments until it issues a determination of substantial equivalence.[133] If a summary has been submitted, requests for copies of it are legally supposed to be furnished by FDA through the FOI process within 30 days after determining that the device is substantially equivalent to another device.

510(k) Statements. If a 510(k) submitter chooses instead to provide a 510(k) statement to satisfy the aforementioned conditions, that statement must be submitted with the notification in order for the FDA to begin its review. The statement should be on a separate letterhead, clearly identified as "510(k) Statement" and signed by the certifier, and it must include specific words beginning with "I certify … ."[134] Written requests by any individual for a copy of the 510(k) must be fulfilled by the statement certifier within 30 days of receipt of the request. Only user identifiers, trade secrets, and confidential commercial information may be purged from the statement. The FDA publishes the names of certifiers on the monthly list of premarket notification submissions for which substantial-equivalence determinations have been made.[135] Those submitting 510(k) applications are not permitted to charge requesters for compiling and disseminating these data.

If a 510(k) submitter fails to comply with the commitment made in the 510(k) statement, the FDA will provide the public with a purged copy of the 510(k) submission.

Noncompliance with the 510(k) statement is a prohibited act; The FDA will use its enforcement powers to obtain compliance.

22.7.3 Requests for Additional Information

After the FDA has accepted a manufacturer's premarket notification, if the agency requests additional information by telephone, fax, or letter, the manufacturer should either submit it within 30 days, or request an extension and state the time needed. When responding, identify the additional information with your company name and 510(k) number. If you submitted a 510(k) summary, include any updates to it. State where the information should be included in your application by referring to the topic, section, or page numbers.

Following this step-by-step submission process will help ensure rapid review of the 510(k) notification. However, submitters must stay abreast with the regulatory process. With the current emphasis on reengineering government, and future concerns about device safety and effectiveness, change will be the only constant.

22.7.4 Special Requirements for Clearance Over-the-Counter (non-prescription) Devices:

Because OTC devices are intended for the mass consumer market, additional studies are required to ensure its safety and effectiveness in the hands of consumer not having the benefits of physician supervision or treatment. Requirements for receiving marketing clearance include:

a. Label comprehension study

   Adequate directions enable a layperson to use an OTC device safely for its intended purposes. For OTC labeling to be clear and truthful (and not false or misleading), it must contain directions, warnings, and information on the intended uses and side effects and be presented in such a manner "as to render the label likely to be understood by ordinary consumers, including individuals with low comprehension ability, as assessed under customary conditions of purchase and use" (21 CFR 330.10 (a)(4)(v)).

b. User manual development

   The FDA requires device manufacturers to supplement their products with printed manuals. The applicable law (21 CFR §801 and §809) considers user manuals to be an extension of a device's labeling that clarifies its proper and safe use. Validation testing is designed to show that lay people can easily understand the usage instructions, and would use the device in a manner that would provide safe and effective treatment.

22.7.5 Specific-Purpose Products

Medical laser products are products that are medical devices manufactured, designed, intended, or promoted for *in vivo* irradiation of the human body for diagnosis, surgery, therapy, cosmetic interventions, or for determining the relative position of the human body.

Class IIIa, IIIb, and IV medical laser products must contain a means for measuring the delivered exposure or treatment level of radiation, accurate within 20%. This requirement is not applicable to Class IIIa aiming devices, except to ophthalmic application. The instruction manual must include a procedure and schedule for recalibration of the measurement system. A modified aperture label is also specified.[136]

22.7.6 Variances and Exemptions

The regulations allow for variances and exemptions from all or part of the standard, and from the reporting and recordkeeping requirements. A variance[137] is permission to vary from one or more requirements of the standard. Upon application by a manufacturer, the Director of CDRH may grant a variance for a product if it is determined that the variance is so limited in applicability that an amendment to the standard is not justified, or is of such need that there is not sufficient time for amending the standard, and that granting the variance is in agreement with the Radiation Control Act. Specifically, a variance may be granted if:

- there are alternate but at least equal means of safety; or
- there are suitable means of safety and, further, either the product could not perform its function if it were in compliance, or one or more requirements of the standard are inappropriate for the product.

In requesting a variance, a manufacturer should carefully follow the format for submission set forth in the regulations.[138] Failure to provide all the required information may result in a delay in issuance of the variance; the variance is required before the product may be introduced into commerce. Exemptions from the standard and from the recordkeeping and reporting requirements have been granted to several Federal agencies,[139] including the Departments of Defense and Energy, some NASA facilities, and NOAA, for certain unique or classified products.

Manufacturers of certain specialized products may also be exempted from annual reporting and recordkeeping.[140] Manufacturers who wish such an exemption should apply by submitting, with the Initial Report, justification and evidence showing that:

- the product cannot under any conditions emit levels of radiation that are hazardous; or
- the product is produced in such small numbers that the need for continuous reporting and recordkeeping is negated, and the product is to be used by individuals trained and knowledgeable in the hazards of such use.

The Director of CDRH may also exempt manufacturers from any part of the reporting and recordkeeping requirements if the exemption is judged to be in keeping with the purposes of the Act.

22.8 Conclusion

The US FDA's Center for Devices and Radiological Health (CDRH) is responsible for regulating radiation-emitting electronic products. Surgical dermatologic lasers, including laser and light-based (IPL) devices used for the cosmetic dermatology indication, fall under CDRH for marketing clearance. The CDRH is responsible for regulating firms who manufacture, repackage, relabel, and/or import medical devices sold in the United States.

Medical devices are classified into one of three classes: Class I, II, and III. Regulatory control and complexity increases from Class I to Class III. The device classification regulation defines the regulatory requirements for a general device type. Most Class I devices are exempt from Premarket Notification 510(k); most Class II devices require Premarket Notification 510(k); and most Class III devices require Premarket Approval.

There of three generalized steps to obtaining Device Marketing Clearance from CDRH:

STEP ONE in the marketing process is to make absolutely sure that the product that you wish to market is a medical device, that is, does it meet the definition of a medical device in Section 201(h) of the FD&C Act. You also need to determine if the product may be an electronic radiation emitting product which has additional regulatory requirements.

STEP TWO is to determine how FDA may classify your device—which one of the three classes the device may fall into. Unless exempted, the FDA will classify your device based on its indication and market precedence. Classification identifies the level of regulatory control that is necessary to assure the safety and effectiveness of a medical device. Most important, the classification of the device will identify, unless exempt, the marketing process (either premarket notification [510(k)] or premarket approval (PMA)) the manufacturer must complete in order to obtain FDA clearance/approval for marketing.

STEP THREE is the development of data and/or information necessary to submit a marketing application, and to obtain FDA clearance to market. For some [510(k)] submissions and most PMA applications, clinical performance data is required to obtain clearance to market. In these cases, conduct of the trial must be done in accord with FDA's Investigational Device Exemption regulation.

The medical device clearance/approval pathway is well-defined by the FDA. The FDA has also made a plethora of resources available to guide the medical-device development and submission process. The reader is encouraged to consult these resources and immerse themselves in the procedural process for medical devices before initiating the pursuit to commercialize a medical device. It is strongly recommended that individuals knowledgeable in successfully navigating the intricacies and nuances impacting medical device submissions be consulted to assist the sponsor in developing the strongest submission and maximize with optimizing the likelihood of achieving a favorable FDA clearance/approval.

Notes

[1]See http://www.fda.gov/ for an overview of the FDA.

[2]21 U.S.C. 301; see http://www.fda.gov/opacom/laws/fdcact/fdctoc.htm

[3]See generally H.R. Rep. no. 853 94th Cong. 2d Session, 1976.

[4]http://thomas.loc.gov/cgi-bin/bdquery/z?d101:HR03095:@@@D&summ2=1&%7CTOM:/bss/d101query.html%7C

[5]http://thomas.loc.gov/cgi-bin/bdquery/z?d102:SN02783:@@@D&summ2=m&%7CTOM:/bss/d102query.html%7C

[6]http://www.fda.gov/cdrh/modact/modernfr.html
[7]21 U.S.C. Sec. 360e(b).
[8]21 U.S.C. Sec. 360c.
[9]General controls include: prohibition against adulteration and misbranding (FDCA §501 and §502); banned devices (§516); notification, repair, and replacement or refund (§518); records and reports (§519); and restricted devices (§520). Unless specifically exempted by regulation, general controls contain requirements for device manufacturers or other designated persons to: (i) register their establishment with FDA; (ii) list their devices with FDA; (iii) comply with labeling regulations in 21 CFR §801, 809, or 812; (iv) submit a premarket notification to FDA; and (v) design and produce devices under good manufacturing practices. See generally Lee H. Monsein, Primer on Medical Device Regulation, Part II: Classification, 205 RADIOLOGY 2 (1997); CDRH, FDA, Device Advice (see www.fda.gov/cdrh/devadvice).
[10]An example of such controls include performance standards (if adopted by the FDA) requiring the device to meet certain functional characteristics; postmarket surveillance; patient registries; development and dissemination of guidelines; recommendations and other appropriate actions. In the absence of any special controls established by regulation, only general controls apply to Class II devices. See FDCA §513(a)(1)(B) (21 U.S.C.§360c(a)(1)(B)(ii)).
[11]http://www.fda.gov/oc/pdufa/default.htm
[12]http://www.fda.gov/oc/initiatives/advance/fdaaa.html
[13]http://www.fda.gov/cdrh/devadvice/314a.html
[14]http://www.fda.gov/cdrh/devadvice/pma/userfees.html
[15]http://www.fda.gov/cdrh/centennial/milestones.html
[16]21 U.S.C. 321; see http://www.fda.gov/opacom/laws/fdcact/fdctoc.htm
[17]http://www.fda.gov/opacom/hpview.html
[18]http://www.fda.gov/cdrh/devadvice/3133.html
[19]http://www.fda.gov/cdrh/k863.html
[20]FD&C Act, Sec 513(i).
[21]21 CFR §807.3.
[22]http://www.fda.gov/cdrh/devadvice/341.html
[23]http://www.fda.gov/opacom/morechoices/fdaforms/cdrh.html
[24]http://www.fda.gov/cdrh/devadvice/342.html
[25]http://www.fda.gov/cdrh/devadvice/32.html
[26]http://www.fda.gov/cdrh/devadvice/33.html
[27]As with Class I devices, Class II device are evaluated individually under §510(k) (if applicable), and then regulated by the type of medical product.
[28]http://www.fda.gov/cdrh/devadvice/3133.html
[29]21 CFR §868.5580.
[30]21 CFR §886.3200.
[31]FDCA §513(a)(1)(C)(ii) (21 U.S.C. §360c(a)(1)(C)(ii)).
[32]21 U.S.C. §360e(b).
[33]21 CFR §870.3925.
[34]21 CFR §878.3540.
[35]21 CFR §888.3480.
[36]21 CFR §886.5925.
[37]Transitional devices are devices that were regulated as drugs prior to May 28, 1976, the date the Medical Device Amendments were signed into law. Any device that was approved by the New Drug Application process is now governed by the PMA regulations. The original NDA approval number is maintained.
[38]21 CFR §862 – §892.
[39]21 C.F.R. Sec. 860.123.
[40]http://www.accessdata.fda.gov/scripts/cdrh/cfdocs/cfpcd/classification.cfm
[41]http://www.accessdata.fda.gov/scripts/cdrh/cfdocs/cfCFR/CFRSearch.cfm?fr=878.4810
[42]http://www.accessdata.fda.gov/scripts/cdrh/cfdocs/cfcfr/CFRSearch.cfm?fr=807.25
[43]http://www.fda.gov/cdrh/devadvice/3133.html#exempt_devices

[44]http://www.accessdata.fda.gov/scripts/cdrh/cfdocs/cfpcd/315.cfm
[45]http://www.fda.gov/cdrh/devadvice/314312.html
[46]http://www.fda.gov/cdrh/devadvice/314311.html
[47]http://www.fda.gov/cdrh/ode/guidance/1567.html
[48]http://www.fda.gov/cdrh/ode/guidance/1567.html
[49]http://www.fda.gov/cdrh/ode/forms/510kchecklist.html
[50]21 CFR §807.
[51]Manufacturers of device components are not required to submit a 510(k), unless such components are promoted for sale to an end-user as replacement parts. Also, contract manufacturers, those firms assembling devices on contract according to someone else's specifications, are not required to submit a 510(k).
[52]Manufacturer must ascertain if significant changes to the labeling have occurred by modifying manuals, deleting or adding warnings, contraindications, etc., and if the packaging operation could alter the condition of the device.
[53]21 CFR §812.
[54]http://www.accessdata.fda.gov/scripts/cdrh/cfdocs/cfpcd/315.cfm
[55]The device sponsor to decide whether or not a modification could significantly affect safety or effectiveness. Whatever the conclusion, the rationale and justification should be reflected in the device master record and change control records, required under the medical device good manufacturing practices.
[56]21 CFR §807.81(a)(3).
[57]http://www.fda.gov/cdrh/devadvice/3146.html
[58]21 CFR §807.87(g).
[59]http://www.fda.gov/cdrh/devadvice/3143.html
[60]http://www.fda.gov/cdrh/devadvice/3144.html
[61]http://www.fda.gov/cdrh/devadvice/3145.html
[62]http://www.fda.gov/cdrh/ode/parad510.html
[63]John J. Smith, Physician Modification of Legally Marketed Medical Devices: Regulatory Implications Under the Federal Food, Drug, and Cosmetic Act, 55 Food & Drug L.J. 245.
[64]21 U.S.C. §360c(i)(1)(A)(ii)(I).
[65]http://www.fda.gov/cdrh/k863.html
[66]http://www.fda.gov/cdrh/devadvice/313.html
[67]http://www.fda.gov/cdrh/thirdparty/index.html
[68]http://www.accessdata.fda.gov/scripts/cdrh/cfdocs/cfPCD/classification.cfm
[69]http://www.fda.gov/cdrh/modact/classiii.html
[70]http://www.accessdata.fda.gov/scripts/cdrh/cfdocs/cfcfr/CFRSearch.cfm?FR=886.3600
[71]http://www.fda.gov/cdrh/devadvice/312.html
[72]42 U.S.C. Sec. 263b.
[73]For performance standards for lasers, see 21 CFR 1040.10 and 1040.11. In determining the applicable reporting category for a laser product, the CDRH bases its decision on the worst-case hazard present within the laser product.
[74]21 CFR §1000.15.
[75]For an overview of this area, see CDRH Device Advice on products emitting radiation at: http://www.fda.gov/cdrh/devadvice/311.html
[76]FDCA §531.
[77]The Gray Sheet, Vol.12, No. 46, pp. 9–11 (available at http://www.fdcreports.com/grayout2.shtml - Subscription required.
[78]As specified in 21 CFR §812.
[79]Such as monitoring investigations, maintaining records, making reports, and complying with prohibitions on promotion and commercialization of investigational devices.
[80]See 21 CFR §1000–1040.11 and 21 CFR §1040.10(h)(2)(ii).
[81]Reporting guides and related regulatory information are available from the CDRH website at: http://www.fda.gov/cdrh/devadvice.
[82]As specified in 21 CFR §1000–1010.

[83]See CDRH Information Sheet at: http://www.fda.gov/cdrh/lasik/what.htm
[84]21 CFR §1040.10(d)(Table VI).
[85]21 CFR §1040.10(d)(Table VI).
[86]21 CFR §812.20.
[87]http://www.fda.gov/oc/ohrt/irbs/devices.html
[88]http://www.fda.gov/CDRH/DEVADVICE/ide/application.shtml
[89]21 CFR §812.25.
[90]21 CFR §812.27.
[91]21 CFR §812.50.
[92]21 CFR §812.43(b).
[93]21 CFR §50.
[94]21 CFR §812.46.
[95]21 CFR §812.70.
[96]21 CFR §812.140 & 21 CFR §812.150.
[97]21 CFR §812.150(b)(9).
[98]21 CFR §812.2 (b).
[99]Ibid 91.
[100]Ibid 94.
[101]Ibid 96.
[102]Ibid 95.
[103]21 CFR §812.40.
[104]21 CFR §812.18.
[105]21 CFR §812.30.
[106]21 CFR §812.5.
[107]45 CFR §46.107; http://www.hhs.gov/ohrp/humansubjects/guidance/45cfr46.htm#46.107
[108]http://www.unmc.edu/ethics/evolution/evolution.html#13#13
[109]21 CFR §50.25.
[110]21 CFR §1040.10(b)(19).
[111]21 CFR §1040.10(b)(23).
[112]21 CFR §1040.10(b)(21).
[113]21 CFR §1000.3(f).
[114]21 CFR §807.87.
[115]http://www.fda.gov/opacom/morechoices/fdaforms/FDA-3514.pdf
[116]21 CFR §807.87(b).
[117]21 CFR §807.87(c).
[118]21 U.S.C. 360kk.
[119]21 CFR §1000.3(g).
[120]21 CFR §807.87(e).
[121]21 CFR §801.109(b)(1).
[122]21 U.S.C. 360c.
[123]21 CFR §807.94.
[124]"Reviewer's Guidance for Computer-Controlled Medical Devices," Rockville, MD, FDA, CDRH, August 1991.
[125]http://www.fda.gov/cdrh/ode/guidance/337.pdf.
[126]http://www.fda.gov/cdrh/ode/guidance/361.pdf.
[127]http://www.fda.gov/cdrh/devadvice/314c.html#sterility.
[128]"Kit Certification and Information for Kit 510(k)s," Rockville, MD, FDA, CDRH, ODE, June 7, 1993.
[129]Safe Medical Devices Act of 1990.
[130]FD&C Act, sect 513(i).
[131]21 CFR 807.3.
[132]"Premarket Notification Submissions; Substantial Equivalence; 510(k) Summaries and Statements, and Class III Summaries; Information Confidentiality," 59 FR:64287–64296.
[133]21 CFR 807.92(a).

[134]21 CFR 807.93.
[135]21 CFR 807.92(b).
[136]21 CFR §1040.11(a).
[137]21 CFR §1040.4.
[138]21 CFR §1040.4(b).
[139]21 CFR §1010.5.
[140]21 CFR §1002.50.

23

Dermal Safety of Laser and Light-Based Systems

J. Frank Nash[1], Melea Ward[1], and Gurpreet S. Ahluwalia[2]

[1]The Procter and Gamble Company, Cincinnati, OH, USA
[2]The Gillette Company, a wholly owned subsidiary of The P&G Company, Needham, MA, USA

Gurpreet S. Ahluwalia (ed.), *Cosmetic Applications of Laser and Light-Based Systems*, 473–498,

23.1 Background

The use of light energy as a treatment modality in dermatology is as old as civilization itself. The therapeutic properties of sunlight, alone or in combination with exogenously applied compounds have been used in the treatment of skin conditions for centuries. The earliest suspected application of light therapy was by the Egyptians, who used the interaction between sunlight and exogenously applied plant materials to treat skin disease [1]. With the development of artificial light sources, the practice of phototherapy has evolved to a point where, for example, the standard treatment of psoriasis is "PUVA" or 8-methoxypsorlen + ultraviolet (UV)-A [2]. In many cases, the therapeutic benefit of light-based treatments is achieved with a nonspecific broadband, that is, a range of wavelengths such as 320–400 nm, light source. Thus, there is a greater potential to interact with multiple chromophores and elicit both beneficial and adverse effects. Technological advances have transformed phototherapy, making diverse light sources more readily available and practical to use.

It has been over forty years since Dr. Leon Goldman working at the University of Cincinnati School of Medicine published a series of papers describing the use of lasers in dermatology [3–7]. Dr. Goldman largely recognized as the father of laser therapy in dermatology, opened the door for multiple researchers/clinicians and helped establish the principles for dermatological treatments using lasers. As with many innovations, the early years of laser treatment were filled with trial and error, refining therapy and reducing the adverse effects, particularly nontarget tissue damage, contributing to scarring and pigmentary changes. Many of the advancements have also been in the laser systems, for example, pulse-duration, waveband versatility, electronics, etc., and for some systems, skill of the operator.

Today, there are multiple lasers in use with wavelengths at the end of visible radiation and into the infrared. The breadth of electromagnetic radiation and the portion where lasers for dermatology are highlighted in Fig. 23.1. The most common lasers used in cosmetic dermatological practice is briefly reviewed.

23.1.1 Types of Lasers

Laser (Light Amplification by the Stimulated Emission of Radiation) is based on Einstein's quantum theory and the concept of stimulated emission of light. In brief, when an atom returns to ground state following excitation, it emits a photon of light at a specific wavelength. If the photon collides with a similarly excited atom, the atom will return to ground state and emit a photon that is synchronized in time and space with the incoming photon. Energy, in the form of intense flashes of light or electrical discharge, is used to excite a population of atoms. The "inversion" of the excited population of atoms creates monochromatic, coherent light which is reflected by mirrors, further stimulating the emission of radiation. One of the mirrors is half-silvered allowing some light to escape, and this is the laser light. There are several excellent reviews of lasers written for physicians/dermatologists [8–10].

There are many different lasers which can be categorized based on the type of lasing material. For medical applications, these are solid, for example, ruby, gas, (e.g., carbon dioxide) liquid (e.g., rhodamine G6), and semiconductors (e.g., diode). The lasing medium determines the wavelength of the light emitted. Further, the light can be continuous or quality switched (Q-switched). Continuous light is relatively low power with long bursts, for example, milliseconds, compared to pulsed or Q-switched, the latter of which emits a

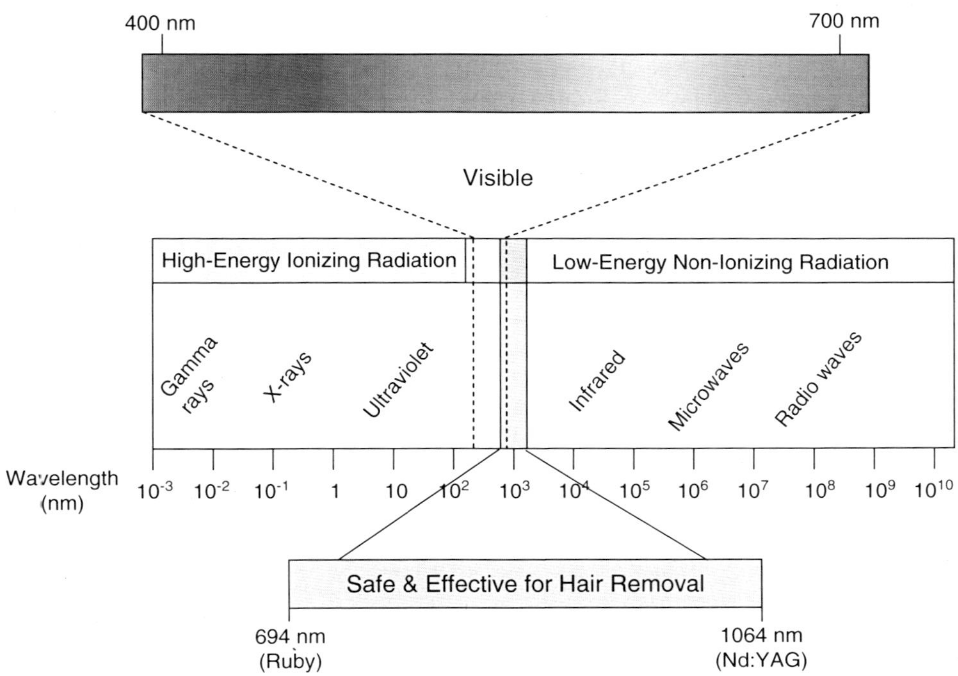

Figure 23.1 Spectrum of electromagnetic radiation.

high power burst of very short duration, for example, nanoseconds. In general, the longer the wavelength, for example, 1064 nm Nd:YAG versus 694 nm ruby, the deeper the light will penetrate into the skin [11]. Examples of commercially available lasers are presented in Table 23.1. A brief description of such lasers is summarized here.

23.1.1.1 Ruby

The ruby laser was the first of such devices to be used in dermatology [6]. The ruby laser emits red light at a wavelength of 694 nm. The depth of penetration into the skin is believed to be approximately 1 mm using clinically relevant doses [12]. This laser has been used to treat congenital melanocytic lesions [13], benign pigmented lesions [14], remove tattoos [15], and also as a hair epilating device [16]. A review of clinical history of ruby laser is presented by Anderson [17] .

23.1.1.2 Alexandrite

The alexandrite laser emits red light at a wavelength of 755 nm. Not surprisingly, the clinical uses of the alexandrite laser are similar, if not identical to the ruby laser [18–21].

23.1.1.3 Diode

The diode laser used in dermatology emits infrared light at 810 nm. The lasing medium is a semiconductor comprised of gallium–aluminum–arsenide. The depth of penetration of 810 nm diode light is greater than ruby or alexandrite lasers, whereas its energy and melanin absorption is slightly lower, resulting in fewer dermal side effects, compared to the ruby and alexandrite. The diode laser is primarily used for hair removal [22], although its use in the treatment of leg veins has also been reported [23]. Presently, in some countries there is a commercially available hand-held diode laser device intended for hair removal [23].

23.1.1.4 Neodymium:Yttrium-Aluminum-Garnet (Nd:YAG)

The Nd:YAG is a popular laser that emits light at the far end of the infrared spectrum, 1064 nm. Like the others, it has been used to remove tattoos [24], hair epilation [25], skin

Table 23.1 Examples of Commercially Available Lasers

Laser (Example)	Wavelength (nm)
Ruby lasers (Epilaser®, EpiTouch®, RubyStar®)	694
Alexandrite laser (Apogee®,GentleLASE®*, EpiTouch ALEX®)	755
Diode (LightSheer®*, Apex-800®, LaserLite, SLP1000®, MeDioStar®, EpiStar®)	800
Nd:YAG system (Softlight®)	1064
Intense pulsed-light source (EpiLight®)	590–1200

resurfacing [26] and treatment of varicose veins [27]. The penetration of light at 1064 nm is believed to be deeper than the ruby, alexandrite, or diode lasers. Further, the lower energy wavelengths have a lower potential to elicit adverse effects. However, the efficacy of Nd: YAG compared to the other lasers is thought to be "less" because of its lower absorption intensity for the melanin chromophore.

23.1.1.5 (Non-laser) Intense Pulsed Light

A brief consideration of Intense Pulsed Light (IPL) is warranted, as such devices have gained popularity because of their versatility. Intense Pulsed Light is not a laser but rather a high energy burst of noncoherent light in the range of 515–1200 nm. Like laser, IPL sources have been used in the treatment facial vascular lesions [28], hair removal [29,30], skin rejuvenation [31], and acne [32].

23.1.2 Irradiance, Fluence, and Exposure

In photobiology/phototoxicology, it is essential to understand "dose" in order to comprehend the consequences of light exposure. Perhaps the most basic property when considering electromagnetic radiation is that energy (E) is inversely proportional to wavelength or $E = hc \div \lambda$, where h equals Planck's constant, c equals the speed of light in a vacuum, and λ is the wavelength of light. Thus, when comparing lasers, for example a ruby (694 nm) to a long wavelength Nd:YAG (1064 nm), the energy of the light from the former is greater. Similarly, when considering the biological consequences of ultraviolet (UV) radiation in comparison to visible or infrared light, the energy of shorter wavelength UV is considerable greater. The net result is that the biological consequences of exposure to shorter, more energetic "doses" of light will influence the final outcome, such as a direct DNA damage produced by absorption of UV at wavelengths from 290 to 310 nm versus thermal response from vibrational energy resulting from the absorption of infrared radiation by melanin.

The unit of energy commonly used in photobiology is the Joule. The fluence, which is regularly referred to as "dose", is Joule per unit area. The irradiance is the rate of delivery per unit area of the light source. The relationship between fluence and irradiance is: Fluence (J/cm^2) = Irradiance (W/cm^2) × time (seconds). With this simple equation and knowledge of the wavelength of light, photobiological events, adverse or otherwise, can be compared, and in some cases, predicted.

The energy associated with the lasers commonly used for nonablative purposes in dermatology lie in the infrared (IR), which is between visible (400–760 nm) and microwave (>1 mm). The portion of the IR concerned with lasers is near IR, 760–1400 nm. These divisions are quite arbitrary and based on different considerations, from optical detection to different temperature ranges used in astronomy. Regardless of this fact, the energy from IR-emitting lasers is not sufficient to promote electrons in the majority of chromophores and, as such, it is incapable of producing changes in the molecular structure [33]. The energy from IR radiation is converted to vibrational energy, which is thermal. This is clearly the energy form responsible for efficacy and adverse events.

There are excellent reviews of basic photochemistry/photobiology which are geared toward clinicians/biologists, for example Kochevar [33] and Arnt et al. [34,35].

23.2 Laser–Skin Interaction

23.2.1 Chromophores in the Skin

Skin optics has been evaluated by multiple investigators with the intent of improving phototherapy [36–38]. Though the optics of human skin is extraordinarily complex, there are a few key considerations, as it relates to lasers and human safety. When considering transmission of light into tissue, there are three components: (i) reflection, (ii) scattering, and (iii) absorption. Reflectance is of minor importance compared to absorption and scattering, particularly when considering the depth of penetration of light into the skin. Scattering in the dermis is most probably the interaction of light with collagen. This event plays a large role in the depth of penetration of light with greater scattering at shorter wavelengths, in accordance with the Kubelka-Munk theory [38]. The chromophores in the skin which are the targets of lasers used in dermatology and absorb the photon energy from such devices are melanin, oxyhemaglobin/hemoglobin, and water [39]. The absorption profile of melanin and oxyhemaglobin are presented in Fig. 23.2. Depending on the wavelength of light, oxyhemaglobin or melanin can be preferentially targeted. The photon energy absorbed by melanin or oxyhemaglobin/hemoglobin is dissipated as heat.

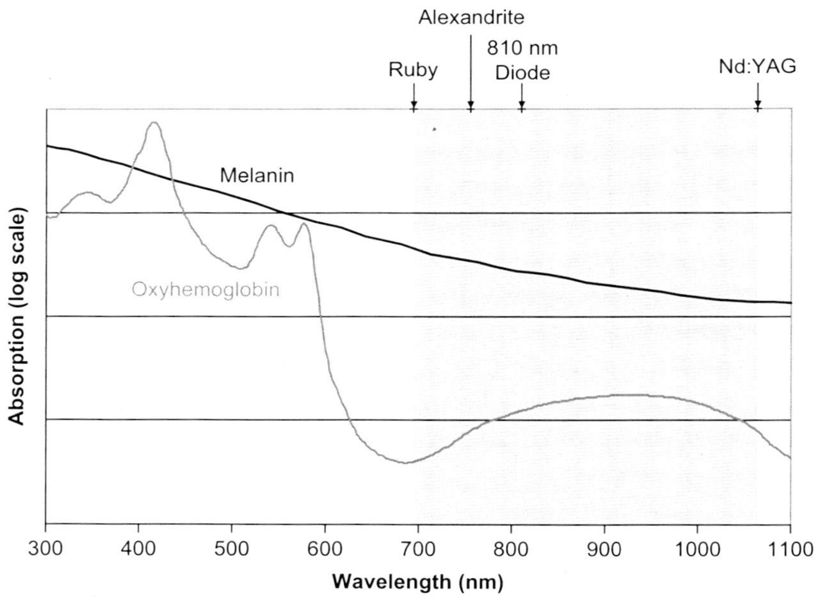

Figure 23.2 Absorption profile for melanin and oxyhemaglobin. The shaded region identifies the range of wavelengths for lasers commonly used in nonablative dermatological procedures, e.g., tattoo removal, hair epilation.

23.2.2 Mechanism of Action

The principles used in toxicological risk assessment, are: (i) hazard identification, (ii) dose-response, and (iii) exposure. In considering the hazard identification for laser devices, the mechanism or mode of action is particularly important. Specifically, by knowing the mode/mechanism of action and information around dose-responsivity, the adverse event profile can be predicted. In this regard, the work of Anderson et al. [40,41] which introduced the concept of *selective photothermolysis* as the mechanism of action of lasers, is central to our understanding of such devices and their therapeutic/adverse event profiles. Selective photothermolysis is quite straightforward, to the extent that a chromophore target is identified and its absorption profile understood. Based on this information, a device can be selected which emits a monochromatic wavelength of light to specifically target the chromophore. The aim of photothermolysis is to produce thermal damage in the target, while minimizing damage to the surrounding tissue. Nearly every laser treatment modality has a component of photothermolysis as part of its therapeutic benefit [42–54].

> A component of photothermolysis is the thermal relaxation time [55]. As stated by Choi and Welch [55], the concept of thermal relaxation time is used to determine the pulse-width for laser light in such a way that the heat generated within the target structure due to absorption of photon energy produces maximum damage without damaging the surrounding tissue. In other words, to limit damage beyond the target, the laser "dose" (i.e., energy + duration of exposure) must be sufficiently short so that absorption is limited to the chromophore at the site or target, and not the surrounding tissue. By and large, the dermal safety of laser devices commonly used in dermatology is based on photothermolysis/thermal relaxation time.

23.3 Dermatological Uses of Lasers and Light-Based Devices

A very brief overview of some nonablative clinical uses of lasers is presented. This highlights the most common therapeutic uses for which the safety/adverse events have been evaluated.

23.3.1 Acne

Lasers have been used to treat acne [32,56–58], alone or as part of a photodynamic regimen [59–61], and to reverse/correct the scarring produced by more serious cases of this condition [62–66]. It is quite likely that with advances in technology and the identification and selective targeting of key structures, that is, chromophores, a consumer device may become available for treating and/or preventing acne.

23.3.2 Cellulite

A recent application of laser devices is the cosmetic treatment of cellulite. As this is a more recent use of lasers, the clinical efficacy remains to be established. Nonetheless, it has been reported that use of laser treatment together with other agents results in an improvement in the appearance of cellulite [67–70].

23.3.3 Hair Removal

One of the oldest, established uses of lasers is the so-called permanent hair removal [71,72]. This procedure has been performed with lasers including ruby [73], alexandrite [74], diode [23], Nd:YAG [75], and IPL [76]. The procedure using these devices is generally considered safe and effective.

23.3.4 Pigmentary Lesions

It has been debated that one of the greatest benefits resulting from the use of lasers is treatment of a variety of pigmentary lesions [14,77–84]. Port wine stains and congenital nevi are a couple of dermatological conditions that have been treated with lasers. The therapeutic outcome can, in many instances, be remarkable, leading to improved quality of life for patients. Depending on the nature of the lesion, the targeted chromophore may be oxyhemaglobin or melanin.

23.3.5 Skin Rejuvenation

Treatment of photo- and chronological-aging of skin by laser resurfacing is another cosmetic use of lasers [85–88]. Generally, this is an aggressive, infrequent treatment. Treatments such as microdermabrasion, chemical peels, and laser skin rejuvenation have gained popularity as aging "baby-boomers" search for solutions to combat intrinsic and photoaging of the skin.

23.3.6 Tattoo Removal

As presented earlier in this chapter, the removal of tattoos was one, if not the first, use of lasers in dermatology [89]. The type of laser used varies and is, in part, dependent on the pigments used in the tattoo [90]. With the ever-increasing number of individuals getting tattoos [91], it will not be surprising to see more frequent use of laser treatment for tattoo removal.

23.3.7 Vascular Lesions

There are a variety of vascular lesions which are now routinely treated with lasers [92–99]. The chromophore targeted is oxyhemaglobin. Laser treatment of vascular lesions has been used in children to manage common childhood lesions, including port wine stain and haemangiomas [100]. In adults, visible veins in the lower extremities, commonly called spider veins, are treated with lasers [101,102].

23.3.8 Other

A host of cosmetic and medical applications of lasers are being explored. From the treatment of tumors, vitiligo [103,104], and wound healing [105], the application of laser light is finding novel uses.

As the professional use of lasers for treating a multitude of dermatological conditions continues to expand in terms of the number of patients and applications, it is not surprising that the devices are moving out of the dermatologists office to become consumer-type devices [23].

23.4 Dermal Safety Evaluation of Lasers

Up to this point, the types of lasers, their use in treatment of skin conditions, and the presumed mechanism/mode of action have been described. The remainder of this chapter will review the toxicological or adverse events associated with the laser treatments of nonablative dermatological conditions. Considering the mechanism/mode of action of lasers with wavelengths between visible and near IR (500–1300 nm), the areas of toxicological concern include: (i) primary effects such as photon absorption by a chromophore, and the dissipation of energy as heat and potential photochemical conversion of endogenous chromophore to a new chemical entity; (ii) secondary events such as inhibition/stimulation of biomolecules, that is, inhibition of cytochrome enzyme and formation of free radicals; and (iii) events associated with the procedure including exposure to tissue debris, fumes, etc. and the safety of the laser device, that is, electrical/device failure, that is, fire. The focus will be on the primary and secondary events at the site of treatment or exposure. Significantly, systemic toxicity is not considered to be of concern, given the localized effects of laser light in the skin or underlying fat.

Of all potential concerns at the site of exposure, the prominent one is thermal events. In general, the longer wavelength visible and infrared lasers are used at fluences incapable of promoting the majority of biomolecules to electronically excited states and, as such, photochemical conversions such as isomerization, breaking bonds, or formation of photoproducts are not observed. Infrared radiation is absorbed, and produces vibrational energy that may cause thermal damage to the target and, if there is sufficient energy, to the surrounding tissue. Thermal effects are considered in the context of acute and chronic toxicity, with special attention given to reversibility. Secondary toxicities and idiosyncratic responses are reviewed as well. Handley [106] provides an excellent review of adverse events associated with nonablative cutaneous laser treatment.

23.4.1 Primary Effects: Thermal-Related Toxicity

In general, nonablative laser treatments are targeting melanin for hair removal and pigmentary lesions, oxyhemaglobin for vascular lesions, and extrinsic chromophores, that is, ink for tattoo removal. For the group of devices discussed, namely ruby, alexandrite, diode, Nd:YAG and IPL, thermal-related skin damage is fluence- or dose-dependent. Thus, as stated by Goldberg [107] for lasers used in hair removal, "... the common theme to almost all complications is too much laser or light source thermally delivered damage to the skin". As such, the frequency and magnitude of thermal-related adverse events depends on the condition being treated which will, in turn, dictate the fluence, wavelength, and frequency of exposure.

23.4.1.1 Erythema and Edema

The most frequent adverse effect in skin from nonablative laser treatment is erythema [108–111]. Erythema may be accompanied by edema, and is a reflection of the inflammatory response. Even when the clinician or operator uses careful precautions, erythema may be observed. In fact, the presence of perifollicular erythema or edema (inflammation surrounding the hair pore) is used by some clinicians as an indication of effective treatment. Despite the high frequency, these events are reversible, and largely considered to be of no human health concern.

In the treatment of facial telangiectasisa, that is, spider veins, using a diode-pumped Nd:YAG laser, all subjects reported erythema [112]. Similarly, in another study using ruby laser to treat facial telangiectasisa, all subjects experienced erythema [113]. In a study by Wheeland [23], 33% of the subjects experienced erythema rated as minimal using a portable diode laser for hair removal. Facial resurfacing using a Nd:YAG as reported by McDaniel [114] found erythema that lasted up to five days in this aggressive treatment. Finally, using a 1450 nm diode laser as a treatment for facial acne, it was found that the side effects of erythema and edema were minimal and transient [115]. In all these examples, erythema was observed, but it was minimal to moderate and perhaps, more important, reversible.

In general, erythema resulting from laser treatment is diffuse within the treatment area. There is, however, a report of reticulated erythema in subjects following repeated laser treatment [116]. In these ten cases, subjects received high fluence (i.e., >40 J/cm^2) diode laser and after one to five treatments the reticulated erythema was noted. Whereas the authors state that the prognosis is excellent once treatment is stopped, repeated exposures may result in pigmentary changes.

23.4.1.2 Pain and Discomfort

There are few studies in which pain from laser treatments have been evaluated exclusively. Because pain is subjective, the degree of such a sensation can be quite arbitrary. Nonetheless, like erythema, pain or discomfort is a common adverse event associated with nonablative laser treatments. In one study, pain produced by alexandrite and diode lasers was compared [117]. In the same study, the size of the treatment area was evaluated. The larger the treatment area the more there was pain, when the diode laser was compared to the alexandrite in this study. However, it should be noted that the outcome was not as clear-cut as the authors conclude. Alster et al. [118] evaluated hair removal in skin types IV–VI, and reported a high incidence of discomfort/pain in these 20 female subjects using a long-pulsed Nd:YAG laser. For hair removal, the site of application plays a role in the discomfort associated with the procedure. For example, there are areas such as bikini lines and along the shin bone that are sensitive to laser hair removal. In addition, IPL can be more painful because of lower selectivity and higher energy requirement for efficacy [119]. There are interventions used to reduce the discomfort with laser procedures, including cooling, use of topical anesthetic creams, and flattening the treatment area, which are discussed in greater detail.

23.4.1.3 Pigmentary Changes

Either hypo- (lightening) or hyper- (darkening) pigmentation have been reported to occur in subjects treated with lasers. Regrettably, the rigorous determination of pigmentary changes has not been documented. The pigmentary changes seen with laser treatment are generally transient, although the appearance/disappearance period can be from days to months. The clinical perception is that pigmentary changes, hyper- or hypo-, are a consequence of postinflammatory reaction or loss of melanocytes, respectively, and are more common in darker-skinned individuals. In any case, the precise reason(s) is unclear.

Lanigan [120] reported the incidence of hyperpigmentation to be 2% in a multicenter, prospective study of 480 subjects receiving 3143 laser hair-removal procedures. In general, the incidence of hyperpigmentation associated with permanent hair removal is greater for short wavelength lasers used in dark-skinned subjects [121]. For example, treatment with a ruby laser had a 1% incidence of hyperpigmentation in Fitzpatrick skin types II/III, and nearly 10% in skin types IV–VI [117]. This may have more to do with the selection of laser, than the true hyperpigmentation observed in these patients. In a prospective clinical study evaluating the effect of a 595 nm dye pulsed laser to treat port wine stains in Japanese subjects, hyperpigmentation up to 17% was reported [122]. The response was mild and transient in this study. In a study of the effect of a short-pulse erbium:YAG laser on acquired melanocytic nevi, postinflammatory hyperpigmentation was reported in 2 out of 14 patients [123]. In a study of 87 Korean patients with Ota's nevus treated with Q-switched alexandrite laser, 14 subjects had hyperpigmentation [124]. Rogachefsky et al. [125] evaluated hair removal using a diode laser (810 nm) using long pulse durations of 200–1000 msec. delivering fluences of 23–115 J/cm^2. In five suntanned individuals, hyperpigmentation and hypopigmentation were reported at six months following treatment, primarily at the sites receiving the highest fluence (115 J/cm^2). These examples, and others (Table 23.2) provide evidence of hyperpigmentation associated with the use of different lasers for nonablative

Table 23.2 Examples of Studies Reporting Pigmentary Change Following Laser Treatment

Study	N	Fitzpatrick Skin Type	Scarring	Pigmentary Change
Lou WW, et al.	50	II & III Dark brown or black hair	No scarring seen	**At 3-month follow up:** Hyperpigmentation: 3% (Type II), 8% (Type III) Hypopigmentation: 3% (Type II), 15% (Type III)
Rogachefsy AS, et al.	5	II–IV Suntanned w/dark brown & black hair	No scarring observed	**At 6 month follow up:** Hyperpigmentation: 10% Hypopigmentation: 23%
Rogachefsy AS, et al.	10	I–VI Hair brown & black	Scarring was not observed	**At 6 month follow up:** Hyperpigmentation: 11% (9 of 80) Hypopigmentation: 9% (7 of 80)
Greppi I	8	V–VI Black pts	Not mentioned	Hyperpigmentation: 3 of 8 pts Hypopigmentation: 2 of 8 pts
Galadari I	32	IV–VI	Scarring (tiny atropheal): 6.2%	Hyperpigmentation: 31% Hypopigmentation: 5.3%

treatment of pigmentary/vascular lesions and permanent hair removal [126–128]. The conclusion is that short wavelength long-pulse in darker-skin types is associated with transient hyperpigmentation following laser treatment.

In the Lanigan study [120], the incidence of hypopigmentation was 1.2%. Kono et al. reported no hypopigmentation in a study of 18 Asians treated with a Q-switched ruby or long-pulsed 595 nm laser for facial lentigines [113]. In a prospective study of 322 subjects undergoing 3 or more permanent hair-removal treatments with a long-pulsed alexandrite laser, 2 cases of hypopigmentation were reported [21]. In the studies where hyperpigmentation [116–120] was noted, hypopigmentation was also reported with a similar incidence and severity. On the other hand, Moreno-Arias [129] indicated that long-term hypopigmentation is extremely rare. Specifically, in a 5-year retrospective analysis involving over 15,000 administered treatments for laser hair removal, only one case of hypopigmentation was noted. As with hyperpigmentation, the mechanism of laser-induced hypopigmentation is unknown. One hypothesis is that melanocytes at the epidermal/dermal junction are destroyed, leading to skin lightening.

23.4.1.4 Other Acute Primary Effects

- Folliculitis: Inflammation of the hair follicles due to irritation or infection is an adverse effect largely associated with hair removal. Depending on the laser system and skin type treated, the incidence rate for folliculitis can range from <1% to as high as 35% [106].
- Blistering and Crusting: These dermal events are generally associated with using too high a fluence level than indicated for a particular skin type [116]. With the appropriate selection of the laser parameters for the skin type of the subject being treated, blistering and crusting are considered rare events [130].
- Scarring: With nonablative laser treatments, scarring is also a rare event and associated with excessive fluences and secondary events such as infection.

23.4.2 Secondary Effects

When skin is exposed to infrared light, there is the distinct possibility that some of the energy will result in secondary molecular/biochemical events. In experimental studies, various biological systems, that is, cell-based studies have reported changes in matrix metalloproteinase (MMP) expression [131,132], perhaps as a result of free radical formation [133] and other events [134], most of which have been discussed in the context of photoaging of the skin. Where the potential for lasers to produce free radicals exists, at lower doses, such events are unlikely. This view is supported by at least one study which found no effect of ruby laser on free radical formation in human-skin biopsies [135]. Moreover, there is some evidence that IR exposure will increase antioxidant activity in skin [136]. Regardless of this fact, at the doses and frequency of treatments used in dermatological practice, the contribution of oxidative stress/damage following exposure to currently available lasers would seem to be minor.

23.4.3 Chronic Effects

23.4.3.1 Skin Cancer

Unlike the well-established mechanism by which UV radiation can induce DNA mutations [137,138] leading to skin cancers [139], there is no scientific evidence that thermal energy from nonionizing IR-laser radiation can induce, promote, or progress cancer development [140]. It has been reported that IR exposure may provide some protective benefit against cytotoxicity of shorter wavelengths of UV [141], possibly through a p53 mechanism [142].

Wulf and colleagues have investigated the effects of laser treatment alone, and before and after UV exposure on tumor formation in a well-studied animal model, lightly pigmented hairless mice. In one such study, lightly pigmented hairless mice were exposed to different fluences from a copper laser (578 nm), followed by solar-simulated radiation 4 times per week for 18 months. One laser treatment did not have any effect on skin tumor formation, alone, or in combination with UV [143]. In a similarly designed study, lightly pigmented hairless mice were treated up to 6 times with a 1060 nm CO_2 laser before or after solar-simulated UV exposure. The CO_2 laser had no effect on skin tumor formation, and did not have an effect on UV-induced skin tumor formation [144]. Finally, the effects of nonablative IPL, alone and in combination with solar-simulated UV, were studied [145]. The IPL had no effect on skin tumor formation in the lightly pigmented hairless mice, nor did it affect UV-induced tumor formation. These data are supportive of the view that laser treatment alone has no effect on the development of skin tumors nor does it have an impact on solar-simulated skin tumor formation in a well-established animal model.

The studies in animals corroborate clinical findings which, to date, have found that lasers used in cosmetic dermatology for the past two decades have no effect on the incidence of skin cancer. Nevertheless, some concerns remain, particularly as it relates to the treatment of pigmented skin lesions such as dysplastic and pre-melanocytic nevi with lasers [146]. In the course of laser treatment for hair removal or skin rejuvenation, it is likely that dysplastic nevi get inadvertently exposed on a regular basis. Again, to date, there have been no reported cases of melanoma or nonmelanoma skin cancers as a result of laser treatments.

The standard treatment modality for congenital melanocytic nevi used to be surgical excision. However, lasers treatments are now being increasingly used on these nevi to provide better cosmetic and therapeutic benefits [13,147–150]. There have been no reported cases of skin cancer from these treatments. Theoretically, reducing the number of melanocytes in these lesions reduces the number of cells that have a malignant potential. An 8-year follow up study of 85 patients with congenital nevi treated with the ruby laser showed no evidence of melanoma formation [151]. Histological analysis of the treated skin biopsy samples performed on select patients in this study also showed no evidence of melanoma, squamous cell carcinoma, or sarcoma formation. In a recent study, Goldberg et al. looked at four different markers for malignant transformation after exposing benign nevi to Q-switched Nd:YAG laser [152]. No significant increase or presence of the malignant transformation markers was seen in the exposed samples. In conclusion, the scientific evidence suggests that the risk for inducing cancer with laser treatment in either healthy skin or skin with lesions is extremely low, if not negligible.

23.4.3.2 Paradoxical Hair Growth

Though not as much a safety issue, in some patients laser hair-removal treatment has a paradoxical effect of stimulating hair growth [129,153–156]. It is considered a rare side effect of laser treatment. The most susceptible patients are the ones that have darker Fitzpatrick skin type (III–V), are of Mediterranean descent, and receiving treatments on their face. New terminal hair growth has been observed in areas untreated but close proximity to the treated ones. In most cases, this paradoxical hair growth occurs at a site that has high vellus growth and relatively free of terminal hair. In a 7-year retrospective analysis of 750 patients of Mediterranean ancestry, terminal hair induction was seen in 4% of the subjects [157]. In 28 out of 30 cases, the terminal hair growth was on the face and the subjects were skin type III or IV. Moreover, there was no relation between stimulated hair growth and the laser system, or the energy fluence used. In another study of 543 patients treated over 5 years in a dermatology clinic in Spain, the incidence rate for hair-growth stimulation on the face was somewhat higher at 10.5% [158]. Unlike these studies in Mediterranean populations, a retrospective study from Canada involving a similarly large patient group of 489, the terminal hair stimulation was seen in only 3 (0.6%) subjects who had Type IV skin [159].

Localized hypertrichosis has been observed under conditions that result in dermal trauma, stress, or inflammation. Increased localized hair growth has been reported after fracture cast application [160], bug bites [161], site surrounding a burn [162], or scar [163], and site of chronic inflammation [164]. Since the laser-induced paradoxical hair growth has mostly been associated with the Mediterranean population, it is not clear what the underlying cause(s) are for this effect.

23.4.4 Contraindications

In the dermatological setting, a careful patient history and thoughtful selection of laser including fluence, wavelength, and treatment frequency can minimize any potential adverse events. Nonetheless, there are several conditions/situations which warrant further consideration.

23.4.4.1 Compromised Skin

Common sense applies when using laser or any other device/treatment on compromised or damaged skin. In the case of laser, it should be noted that experimental studies have found that laser treatment promotes and even improves wound healing [165–169]. Nonetheless, treatment of severely compromised skin with lasers should be avoided.

23.4.4.2 Tan Skin

It is generally recommended that patients avoid getting excessive suntan before their laser treatment [125].

23.4.4.3 Pigmentary Lesions

Pigment-containing lesions on skin can be an indication of the use of a laser or a contraindication, depending on the desired outcome. If the intent is to remove hair, then the melanin-containing skin lesions such as dysplastic or congenital melanocytic nevi should be avoided from laser exposure because of the risk of burn and potential scarring.

23.4.4.4 Tattoos

The most common adverse effects resulting from the interaction between laser light and tattoo ink are burning, scarring, and transient pigmentary changes. These adverse effects are dependent on the laser being used and the kind and depth of the tattoo ink particles. As repeatedly stated, lasers are one of the treatments for tattoo removal. This does pose a problem if subjects with no intention of removing or otherwise altering existing tattoo(s) require a laser treatment of the area in question. For example, it is conceivable that permanent hair removal is desired at the location of a tattoo. The concern is twofold: (i) lasing a tattoo might result in fading and (ii) absorbance of the photons by tattoo inks could reduce the efficacy and/or increase thermal-related adverse events. In any case, it might be advisable to avoid lasing skin that has a tattoo.

23.4.4.5 Photosensitizing Drugs

It is possible that an exogenously administered drug/cosmetic might potentiate the adverse effects of laser treatment. This could happen in at least two ways: (i) drug/cosmetic absorbs light from the laser, becomes activated resulting in a phototoxicological event; and, (ii) drug/cosmetic changes the structure/function of the skin. A third possibility falls into the category of unknown, which may be populated with anecdotal reports without any mechanistic basis.

There are few, if any, reports of a phototoxicological response at the wavelengths of the lasers in question (600–1200 nm) outside of intentional photodynamic therapy. This is related to the absorption profile of chemicals where the majority of photosensitizers are activated by more energetic short wavelengths falling in the UVA (320–400 nm) region [170–172], and to a lesser extent UVB (290–320 nm). In short, a direct phototoxicological response is unlikely if not negligible, since current drugs/cosmetics do not absorb wavelengths emitted by nonablative visible/IR lasers.

The second possible contraindication related to photosensitivity is the change in structure and function of the skin produced by a drug or cosmetic which might impart enhanced sensitivity to laser light. For example, treatment of severe acne with isotretinoin has been reported to result in keloid formation and reduce wound healing following laser treatment [173]. This increased skin sensitivity is consistent with the effects of isotretinoin and other retinoic acid derivatives on skin. This potential interaction is of concern as female patients with androgen profiles including severe acne and hirsutism have a tendency to coexist. Khatri [174] followed 7 patients undergoing isotretinoin therapy treated with diode laser for hair removal. In this study, patients showed mild erythema after laser treatment, but no erythema, pigmentary change, swelling, or scarring at any of the follow-up visits. Cassano et al. [175] reported similar findings in 6 patients treated with diode laser, 4 subjects undergoing isotretinoin therapy, and the remaining 2 who had just completed isotretinoin therapy. These studies, while limited, demonstrate the safety of diode laser hair removal in patients undergoing isotretinoin therapy. Whereas these data should not be applied to all lasers, patients using isotretinoin can be safety treated.

23.5 Management of Adverse Effects

23.5.1 Selection of Laser

Laser-induced dermal adverse events depend on the properties of the laser light and laser-chromophore interactions. Depending on the target chromophore, the wavelengths of light will, in part, determine the depth of penetration and as such, the tissue interaction. In general, the penetration depth in skin increases with wavelength. For example, for hair removal, longer wavelength devices such as diode and Nd:YAG have lower absorption coefficients for melanin, thereby penetrating deeper into the epidermis/dermis where the hair bulb resides. By extending the pulse duration of such lasers, the desired efficacy can be achieved, and the thermal-mediated adverse events reduced.

23.5.2 Epidermal Cooling

As discussed, the predominant adverse effect associated with current laser devices is thermally mediated. It follows that cooling or precooling the skin has been suggested as an approach to reduce tissue damage associated with laser treatment [176], especially where melanin is the primary target chromophore. Precooling can be accomplished in several ways. The design of the laser may be such that contact with the skin either cools the skin or may act as a heat sink [177–180]. An actively cooled sapphire tip has been used on several laser hair removal systems to provide heat conduction from epidermis before, during, and after each laser pulse. For treatment, the hand piece is placed on the skin for about 0.5 seconds to cool the surface (epidermis can be cooled in 0.2 seconds) before a laser pulse is fired. Cryogenic sprays are used commonly in laser treatments to reduce thermal injury [180–187]. Cool air [188,189] has been used with success. A recent report, however, found an increase in postinflammatory hyperpigmentation in Asian skin that was attributed to the use of super-cooled air, used in combination with Nd:YAG [190]. However, it is generally accepted that cooling allows higher laser fluence to be delivered, which is important when treating the deeper dermal targets such as hair follicle, while effectively protecting the skin and reducing adverse events.

Other measures used to manage dermal events including, erythema, and skin inflammation and swelling resulting from laser treatment include the use of ice packs and topical corticosteroids. Patients are also instructed to avoid excess sun exposure and trauma, for example, scratching the treatment area, to diminish the likelihood of damaging the skin.

23.5.3 Analgesics

Attempts to reduce the pain associated with laser treatment include application of topical analgesics [191–196]. In February 2007, the Food and Drug Administration (FDA) issued a Public Health Advisory regarding the use of skin products containing numbing ingredients for cosmetic procedures causing life-threatening side effects. These included: lidocaine, tetracaine, benzocaine, and prilocaine in a cream, ointment, or gel for use as topical anesthetics used in conjunction with laser treatment. The FDA concern arose from reports of two deaths attributed to local anesthetics applied over large body surfaces to reduce pain/discomfort associated with laser hair removal. Though the topical application of analgesis/anesthetics can reduce pain and discomfort associated with laser treatment, such exposure may carry significant health risks, and should be done with caution and under medical supervision.

23.6 Conclusions

To better appreciate the toxicological evaluation of lasers, it helps to compare it to the equivalent approach used for a drug/cosmetic. For a topically applied chemical entity, the pharmacokinetics, that is, absorption, distribution, metabolism, and elimination serve as a suitable starting point. A chemical that is absorbed into the skin or general circulation will, most likely distribute to body compartments such as the blood, undergo biotransformation (metabolism), and be eliminated over time. To understand the toxicological implications of this complex ADME process, preclinical studies are needed, where the parent compound along with any metabolites is evaluated. This evaluation includes the site of application for topically administered drugs and systemic endpoints, for example, reproductive toxicity, repeat exposure toxicity, etc. In contrast, with laser light, we are concerned about the toxicological consequences of the energy (i.e., wavelength) at the site of application/treatment or local, but not systemic toxicity. Thus, for any chemical entity, the toxicological concerns are much more complex, given the ADME and local and systemic toxicity. Further, for most drugs/chemicals the mechanism or the mode of action is unknown. In contrast, for nonablative laser treatments, the mechanism of action is known, and the events are predictable, based on this mechanism. This does not mean the toxicological evaluation of laser light is trivial; rather, the specificity of such devices can be evaluated in a more precise manner.

The toxicological consequences of contemporary nonablative laser treatments are related to the production of heat following the absorption of the energy emitted from such devices. The list of common adverse events attributed to the use of lasers, such as erythema, pain, and discomfort, although frequent in occurrence, are not severe. More important, the most common side effects are reversible. There are some events including pigmentary changes which are infrequent events, largely reversible over time, although less predictive than other toxicological events. Finally, the repeated exposure to laser energy does not appear to have toxicological consequences at this point in time.

It has been 40 years since the first compact microwave ovens were introduced to the American kitchen. In the beginning, such devices were large, awkward, and expensive, and there were many health-related concerns including radiation poisoning, blindness, and sterility. However, over time these fears have been replaced by confidence and enjoyment of the benefits of microwave cooking with an estimated 90% of households in the United States having such a device. Though lasers have not yet become household items like microwaves, concerns regarding health-related fears are analogous. Currently, the majority of dermatological practices use some laser/pulsed light devices in treatment. As such, it is not unreasonable to suggest the future of laser devices most probably includes home use for diagnostic and nonmedical/cosmetic benefits. In fact, there is a hand-held hair removal device available for home use that is available in some countries [23].

To summarize, the use of lasers is in its infancy. Nonetheless, based on existing principles of photobiology/phototoxicology, widespread clinical use of lasers, fluence/spectral response and anticipated uses, there is good reason for optimism regarding human safety.

References

1. Wheeland RG. (1995) History of lasers in dermatology. *Clin. Dermatol.* **13**, 3–10.
2. Halprin KM., Stern RS, and Parrish JA. (1980) Psoriasis, skin cancer, and PUVA. *Journal of the American Academy of Dermatology* **2**, 432–434.
3. Goldman L and Rockwell, Jr RJ. (1968) Laser systems and their applications in medicine and biology. *Adv. Biomed. Eng Med. Phys.* **1**, 317–382.
4. Goldman L and Wilson RG. (1964) Treatment of basal cell epithelioma by laser radiation. *JAMA* **189**, 773–775.
5. Goldman L, Hornby P, and Long C. (1964) Effect of the laser on the skin. III Transmission of laser beams through fiber optics. *J. Invest Dermatol.* **42**, 231–234.
6. Goldman L, Blaney, DJ,, Kindel, Jr DJ, Richfield D, and Franke EK. (1963) Pathology of the effect of the laser beam on the skin. *Nature* **197**, 912–914.
7. Goldman L, Blaney DJ , Kindel, Jr DJ , and Franke EK. (1963) Effect of the laser beam on the skin. Preliminary report. *J. Invest Dermatol.* **40**, 121–122.
8. Herd RM, Dover JS, and Arndt KA. (1997) Basic laser principles. *Dermatol. Clin.* **15**, 355–372.
9. Acland KM and Barlow RJ. (2000) Lasers for the dermatologist. *Br. J. Dermatol.* **143**, 244–255.
10. Houk LD and Humphreys T. (2007) Masers to magic bullets: an updated history of lasers in dermatology. *Clinics in Dermatology* **25**, 434–442.
11. Anderson RR, Margolis RJ, Watenabe S, Flotte T, Hruza GJ, and Dover JS. (1989) Selective photothermolysis of cutaneous pigmentation by Q-switched Nd: YAG laser pulses at 1064, 532, and 355 nm. *J. Invest Dermatol.* **93**, 28–32.
12. Topping A, Gault D, Grobbelaar A, Sanders,R, Green,C, and Linge C. (2001) Does low penetration of human skin by the normal mode ruby laser account for poor permanent depilatory success rates? *Lasers Med. Sci.* **16**, 224–229.
13. Noordzij MJ, DG van den Broecke, Alting MC, and Kon M. (2004) Ruby laser treatment of congenital melanocytic nevi: a review of the literature and report of our own experience. *Plast. Reconstr. Surg.* **114**, 660–667.
14. Raulin C, Schonermark MP , Greve B, and Werner S. (1998) Q-switched ruby laser treatment of tattoos and benign pigmented skin lesions: a critical review. *Ann. Plast. Surg.* **41**, 555–565.
15. Taylor CR, Gange RW, Dover JS , Flotte TJ, Gonzalez E, Michaud N, and Anderson RR. (1990) Treatment of tattoos by Q-switched ruby laser. A dose-response study. *Arch. Dermatol.* **126**, 893–899.
16. Chana JS and Grobbelaar AO. (2002) The long-term results of ruby laser depilation in a consecutive series of 346 patients. *Plast. Reconstr. Surg.* **110**, 254–260.
17. Anderson RR. (2003) Dermatologic history of the ruby laser: the long story of short pulses. *Arch. Dermatol.* **139**, 70–74.
18. Rosenbach A, Williams CM , and Alster TS. (1997) Comparison of the Q-switched alexandrite (755 nm) and Q-switched Nd: YAG (1064 nm) lasers in the treatment of benign melanocytic nevi. *Dermatol. Surg.* **23**, 239–244.
19. Fitzpatrick RE and Goldman MP. (1994) Tattoo removal using the alexandrite laser. *Arch. Dermatol.* **130**, 1508–1514.
20. Kim S and Kang WH. (2005) Treatment of congenital nevi with the Q-switched Alexandrite laser. *Eur. J. Dermatol.* **15**, 92–96.
21. Tatlidede S, Egemen O, Saltat A, Turgut,G, Karasoy A, and Kuran I. (2005) Hair removal with the long-pulse alexandrite laser. *Aesth Surg J.* **25**, 138–143.

22. Sadick NS and Prieto VG. (2003) The use of a new diode laser for hair removal. *Dermatol. Surg.* **29**, 30–33.
23. Wheeland RG. (2007) Simulated consumer use of a battery-powered, hand-held, portable diode laser (810 nm) for hair removal: A safety, efficacy and ease-of-use study. *Lasers Surg. Med.* **39**, 476–493.
24. Kilmer SL, Lee MS, Grevelink JM, Flotte TJ, and Anderson RR. (1993) The Q-switched Nd:YAG laser effectively treats tattoos. A controlled, dose-response study. *Arch. Dermatol.* **129**, 971–978.
25. Goldberg DJ, Littler CM, and Wheeland RG. (1997) Topical suspension-assisted Q-switched Nd:YAG laser hair removal. *Dermatol. Surg.* **23**, 741–745.
26. Goldberg DJ and J Whitworth. (1997) Laser skin resurfacing with the Q-switched Nd:YAG laser. *Dermatol. Surg.* **23**, 903–906.
27. Baumler W, Ulrich H, Hartl A, Landthaler M, and Shafirstein G. (2006) Optimal parameters for the treatment of leg veins using Nd:YAG lasers at 1064 nm. *Br. J. Dermatol.* **155**, 364–371.
28. Angermeier MC. (1999) Treatment of facial vascular lesions with intense pulsed light. *J. Cutan. Laser Ther.* **1**, 95–100.
29. Gold MH, Bell MW, Foster TD, and Street S. (1999) One-year follow-up using an intense pulsed light source for long-term hair removal. *J. Cutan. Laser Ther.* **1**, 167–171.
30. Sadick NS, Weiss RA, Shea CR, Nagel H, Nicholson J , and Prieto VG. (2000) Long-term photoepilation using a broad-spectrum intense pulsed light source. *Arch. Dermatol.* **136**, 1336–1340.
31. Bitter PH. (2000) Noninvasive rejuvenation of photodamaged skin using serial, full-face intense pulsed light treatments. *Dermatol. Surg.* **26**, 835–842.
32. Chang SE, Ahn SJ, Rhee DY, Choi JH, Moon KC, Suh HS, and Soyun C. (2007) Treatment of facial acne papules and pustules in Korean patients using an intense pulsed light device equipped with a 530- to 750-nm filter. *Dermatol. Surg.* **33**, 676–679.
33. Kochevar IE. (1993) Basic Principles in Photomedicine and Photochemistry. In *Clinical Photomedicine*.(Eds HW. Lim and N. A. Soter), pp. 1–18. Marcel Dekker, New York.
34. Arndt KA, Noe JM, Northam DB, and Itzkan I. (1981) Laser therapy. Basic concepts and nomenclature. *J. Am. Acad. Dermatol.* **5**, 649–654.
35. Herd RM, MRCP U, Dover JS, and Arndt KA. (1997) Basic Laser Principles. *Dermatologic Clinics* **15**, 355–372.
36. Van Gemert MJC, Jacques SL, Sterenborg HJCM, and Star WM. (1989) Skin optics. *IEEE Transactions on Biomedical Engineering* **36**, 1146–1154.
37. Morelli JG and Parrish JA. (1985) Lasers in dermatology: a selective historical review. *Photodermatol.* **2**, 303–309.
38. Anderson RR and Parrish JA. (1981) The optics of human skin. *J Invest Dermatol* **77**, 13–19.
39. Stoker MR. (2005) Basic principles of lasers. *Anaesthesia & intensive care medicine* **6**, 402–404.
40. Anderson RR and Parrish JA. (1983) Selective photothermolysis: precise microsurgery by selective absorption of pulsed radiation. *Science* **220**, 524–527.
41. Anderson RR, Jaenicke KF., and Parrish JA. (1983) Mechanisms of selective vascular changes caused by dye lasers. *Lasers Surg. Med.* **3**, 211–215.
42. Keijzer M, J PickeringW, and van Gemert MJ. (1991) Laser beam diameter for port wine stain treatment. *Lasers Surg. Med.* **11**, 601–605.
43. Naito SK. (2007) Fractional photothermolysis treatment for resistant melasma in Chinese females. *J. Cosmet. Laser Ther.* **9**, 161–163.
44. Mezzana P and Valeriani M. (2007) Rejuvenation of the aging face using fractional photothermolysis and intense pulsed light: a new technique. *Acta Chir Plast.* **49**, 47–50.
45. Alster TS, Tanzi,EL, and Lazarus M. (2007) The use of fractional laser photothermolysis for the treatment of atrophic scars. *Dermatol. Surg.* **33**, 295–299.
46. Anderson RR, Farinelli W, Laubach H, Manstein D, Yaroslavsky AN, Gubeli J, III, Jordan K, Neil GR, Shinn M, Chandler W, Williams GP, Benson SV, Douglas DR, and Dylla HF. (2006)

Selective photothermolysis of lipid-rich tissues: a free electron laser study. *Lasers Surg. Med.* **38**, 913–919.
47. Hasegawa T, Matsukura T, Mizuno Y, Y. Suga, Ogawa H, and Ikeda S. (2006) Clinical trial of a laser device called fractional photothermolysis system for acne scars. *J. Dermatol.* **33**, 623–627.
48. Rokhsar CK and Fitzpatrick RE. (2005) The treatment of melasma with fractional photothermolysis: a pilot study. *Dermatol. Surg.* **31**, 1645–1650.
49. Fisher GH and Geronemus RG. (2005) Short-term side effects of fractional photothermolysis. *Dermatol. Surg.* **31**, 1245–1249.
50. Randeberg LL, Bonesronning JH, Dalaker M, Nelson JS, and Svaasand LO. (2004) Methemoglobin formation during laser induced photothermolysis of vascular skin lesions. *Lasers Surg. Med.* **34**, 414–419.
51. Lloyd JR and Mirkov M. (2002) Selective photothermolysis of the sebaceous glands for acne treatment. *Lasers Surg. Med.* **31**, 115–120.
52. Hern S, Allen MH, Sousa AR, Harland CC, Barker JN, Levick JR, and Mortimer PS Mortimer. (2001) Immunohistochemical evaluation of psoriatic plaques following selective photothermolysis of the superficial capillaries. *Br. J. Dermatol.* **145**, 45–53.
53. Walther T, Baumler W, Wenig M, Landthaler M, and Hohenleutner U. (1998) Selective photothermolysis of hair follicles by normal-mode ruby laser treatment. *Acta Derm. Venereol.* **78**, 443–444.
54. Eppley BL and Sadove AM. (1994) Systemic effects of photothermolysis of large port-wine stains in infants and children. *Plast. Reconstr. Surg.* **93**, 1150–1153.
55. Choi B and Welch AJ. (2001) Analysis of thermal relaxation during laser irradiation of tissue. *Lasers Surg. Med.* **29**, 351–359.
56. Perez-Maldonado A, Runger TM, and Krejci-Papa N. (2007) The 1,450-nm diode laser reduces sebum production in facial skin: a possible mode of action of its effectiveness for the treatment of acne vulgaris. *Lasers Surg. Med.* **39**, 189–192.
57. Orringer JS, Kang S, Maier L, Johnson TM, Sachs DL, Karimipour DJ, Helfrich YR, Hamilton T, and Voorhees JJ. (2007) A randomized, controlled, split-face clinical trial of 1320-nm Nd:YAG laser therapy in the treatment of acne vulgaris. *J. Am. Acad. Dermatol.* **56**, 432–438.
58. Nouri K and Ballard CJ. (2006) Laser therapy for acne. *Clin. Dermatol.* **24**, 26–32.
59. Alster TS and Surin-Lord SS. (2006) Photodynamic therapy: practical cosmetic applications. *J. Drugs Dermatol.* **5**, 764–768.
60. Alexiades-Armenakas M. (2006) Laser-mediated photodynamic therapy. *Clin. Dermatol.* **24**, 16–25.
61. Gold MH, Bradshaw VL, Boring MM, Bridges TM, Biron JA, and Carter LN. (2004) The use of a novel intense pulsed light and heat source and ALA-PDT in the treatment of moderate to severe inflammatory acne vulgaris. *J. Drugs Dermatol.* **3**, S15–S19.
62. Woo SH, Park JH, and Kye YC. (2004) Resurfacing of different types of facial acne scar with short-pulsed, variable-pulsed, and dual-mode Er:YAG laser. *Dermatol. Surg.* **30**, 488–493.
63. Batra RS, Jacob CI, Hobbs L, Arndt KA, and Dover JS. (2003) A prospective survey of patient experiences after laser skin resurfacing: results from 2 1/2 years of follow-up. *Arch. Dermatol.* **139**, 1295–1299.
64. Lent WM and David LM. (1999) Laser resurfacing: a safe and predictable method of skin resurfacing. *J. Cutan. Laser Ther.* **1**, 87–94.
65. Apfelberg DB. (1997) A critical appraisal of high-energy pulsed carbon dioxide laser facial resurfacing for acne scars. *Ann. Plast. Surg.* **38**, 95–100.
66. Alster TS and McMeekin TO. (1996) Improvement of facial acne scars by the 585 nm flashlamp-pumped pulsed dye laser. *Journal of the American Academy of Dermatology* **35**, 79–87.
67. Avram MM. (2004) Cellulite: a review of its physiology and treatment. *J. Cosmet. Laser Ther.* **6**, 181–185.
68. Sadick N and Magro C. (2007) A study evaluating the safety and efficacy of the VelaSmooth system in the treatment of cellulite. *J. Cosmet. Laser Ther.* **9**, 15–20.
69. Nootheti PK, Magpantay A, Yosowitz G, Calderon S, and Goldman MP. (2006) A single center, randomized, comparative, prospective clinical study to determine the efficacy of the VelaSmooth system versus the Triactive system for the treatment of cellulite. *Lasers Surg. Med.* **38**, 908–912.

70. Alster TS and Tanzi EL. (2005) Cellulite treatment using a novel combination radiofrequency, infrared light, and mechanical tissue manipulation device. *J. Cosmet. Laser Ther.* **7**, 81–85.
71. Polla LL, Margolis RJ, Dover JS, Whitaker D, Murphy GF, Jacques SL, and Anderson RR. (1987) Melanosomes are a primary target of Q-switched ruby laser irradiation in guinea pig skin. *J. Invest Dermatol.* **89**, 281–286.
72. Haedersdal M and Wulf HC. (2006) Evidence-based review of hair removal using lasers and light sources. *J Eur Acad Dermatol Venereol.* **20**, 9–20.
73. Allison KP, Kiernan MN, Waters RA, and Clement RM. (2003) Evaluation of the ruby 694 Chromos for hair removal in various skin sites. *Lasers Med. Sci.* **18**, 165–170.
74. Bouzari N, Nouri K, Tabatabai H, Abbasi Z, Firooz A, and Dowlati Y. (2005) The role of number of treatments in laser-assisted hair removal using a 755-nm alexandrite laser. *J. Drugs Dermatol.* **4**, 573–578.
75. Tanzi EL and Alster TS. (2004) Long-pulsed 1064-nm Nd:YAG laser-assisted hair removal in all skin types. *Dermatol. Surg.* **30**, 13–17.
76. El Bedewi AF. (2004) Hair removal with intense pulsed light. *Lasers Med. Sci.* **19**, 48–51.
77. Suh DH, Han KH, and Chung JH. (2001) The use of Q-switched Nd:YAG laser in the treatment of superficial pigmented lesions in Koreans. *J. Dermatolog. Treat.* **12**, 91–96.
78. Becker-Wegerich PM, Kuhn A, Malek L, Lehmann P, Megahed M, and Ruzicka T. (2000) Treatment of nonmelanotic hyperpigmentation with the Q-switched ruby laser. *J Am Acad of Dermatol.* **43**, 272–274.
79. Goldberg DJ. (1997) Laser Treatment of Pigmented Lesions. *Dermatol Clin.* **15**, 397–407.
80. Fiskerstrand, EJ, Dalaker M, and Norvang LT. (1995) Laser treatment of port wine stains: a study comparing therapeutic outcome with morphologic characteristics of the lesions. Preliminary results. *Acta Derm. Venereol.* **75**, 92–93.
81. Taylor CR and Anderson RR. (1993) Treatment of benign pigmented epidermal lesions by Q-switched ruby laser. *Int. J. Dermatol.* **32**, 908–912.
82. Nelson JS and Applebaum J. (1992) Treatment of superficial cutaneous pigmented lesions by melanin-specific selective photothermolysis using the Q-switched ruby laser. *Ann. Plast. Surg.* **29**, 231–237.
83. Tan OT, Morelli JG, and Kurban AK. (1992) Pulsed dye laser treatment of benign cutaneous pigmented lesions. *Lasers Surg. Med.* **12**, 538–542.
84. Wang HW, Liu YH, Zhang GK, Jin HZ, Zuo YG, Jiang GT, and Wang JB. (2007) Analysis of 602 Chinese cases of nevus of Ota and the treatment results treated by Q-switched alexandrite laser. *Dermatol. Surg.* **33**, 455–460.
85. Carruthers J and Carruthers A. (2005) Adjunctive botulinum toxin type A: fillers and light-based therapies. *Int. Ophthalmol. Clin.* **45**, 143–151.
86. Kim MG, Kim WS, Lee JH, Yang JM, Lee ES, K. H. Park, and Lee DY. (2006) Carbon dioxide laser de-epithelialization: selective removal of the epidermis with minimal change to the basement membrane zone. *Dermatol. Surg.* **32**, 329–331.
87. Sabini P. (2004) Classifying, diagnosing, and treating the complications of resurfacing the facial skin. *Facial Plast. Surg. Clin. North Am.* **12**, 357–361, vi.
88. Greeley A. (2000) Cosmetic laser surgery. A high-tech weapon in the fight against aging skin. *FDA. Consum.* **34**, 34–37.
89. Bernstein EF. (2006) Laser treatment of tattoos. *Clin. Dermatol.* **24**, 43–55.
90. Prinz BM, Vavricka SR, Graf P, Burg G, and Dummer R. (2004) Efficacy of laser treatment of tattoos using lasers emitting wavelengths of 532 nm, 755 nm and 1064 nm. *Br. J. Dermatol.* **150**, 245–251.
91. Burris K and Kim K. (2007) Tattoo removal. *Clin Dermatol* **25**, 388–392.
92. Greenwald J, Rosen S, Anderson RR, Harrist T, MacFarland F, Noe J, and Parrish JA. (1981) Comparative histological studies of the tunable dye (at 577 nm) laser and argon laser: the specific vascular effects of the dye laser. *J. Invest Dermatol.* **77**, 305–310.
93. Rosenfeld H and Sherman R. (1986) Treatment of cutaneous and deep vascular lesions with the Nd:YAG laser. *Lasers Surg. Med.* **6**, 20–21.
94. Kiyoizumi T. (1988) Low level diode laser treatment for hematomas under grafted skin and its photobiological mechanisms. *Keio J. Med.* **37**, 415–428.

95. McDaniel DH. (1990) Cutaneous vascular disorders: advances in laser treatment. *Cutis* **45**, 339–349, 354.
96. Alster T and Tan OT. (1991) Laser treatment of benign cutaneous vascular lesions. *Am. Fam. Physician* **44**, 547–554.
97. Morelli JG, Huff,JC, and Weston WL. (1993) Treatment of congenital telangiectatic vascular malformations with the pulsed-dye laser (585 nm). *Pediatrics* **92**, 603–606.
98. Ross BS, Levine VJ, and Ashinoff R. (1997) Laser treatment of acquired vascular lesions. *Dermatol. Clin.* **15**, 385–396.
99. Dohil MA, Baugh WP, and Eichenfield LF. (2000) Vascular and pigmented birthmarks. *Pediatr. Clin. North Am.* **47**, 783–787vi.
100. Mehmi M and S LaniganW. (2007) Laser treatment of skin lesions in children. *Paediatr Child Health.* **17**, 395–399.
101. Fodor L, Ramon Y, Fodor A, Carmi N, Peled IJ, and Ullmann Y. (2006) A side-by-side prospective study of intense pulsed light and Nd:YAG laser treatment for vascular lesions. *Ann. Plast. Surg.* **56**, 164–170.
102. Trelles MA, Allones I, Alvarez J, Velez M, Martin-Vazquez M, Trelles OR, Luna R, Buil C, and Mordon SR. (2006) The 800-nm diode laser in the treatment of leg veins: Assessment at 6 months. *J Am Acad Dermatol.* **54**, 282–289.
103. Forschner T, Buchholtz S, and Stockfleth E. (2007) Current state of vitiligo therapy–evidence-based analysis of the literature. *J. Dtsch. Dermatol. Ges.* **5**, 467–475.
104. Thissen M. and Westerhof W. (1997) Laser treatment for further depigmentation in vitiligo. *Int. J. Dermatol.* **36**, 386–388.
105. Al Watban FA and Zhang XY. (2004) The comparison of effects between pulsed and CW lasers on wound healing. *J. Clin. Laser Med. Surg.* **22**, 15–18.
106. Handley JM. (2006) Adverse events associated with nonablative cutaneous visible and infrared laser treatment. *J Am Acad Dermatol* **55**, 482–489.
107. Goldberg DJ. (2007) Laser- and light-based hair removal: an update. *Expert Rev. Med. Devices* **4**, 253–260.
108. Wlotzke U, Hohenleutner U, Abd-El-Raheem TA, Baumler W, and Landthaler M. (1996) Side-effects and complications of flashlamp-pumped pulsed dye laser therapy of port-wine stains. A prospective study. *Br. J. Dermatol.* **134**, 475–480.
109. Nanni CA and Alster TS. (1999) Laser-assisted hair removal: side effects of Q-switched Nd:YAG, long-pulsed ruby, and alexandrite lasers. *J. Am. Acad. Dermatol.* **41**, 165–171.
110. Bernstein EF, Ferreira M, and Anderson D. (2001) A pilot investigation to subjectively measure treatment effect and side-effect profile of non-ablative skin remodeling using a 532 nm, 2 ms pulse-duration laser. *J. Cosmet. Laser Ther.* **3**, 137–141.
111. Aldraibi MS, Touma DJ, and Khachemoune A. (2007) Hair removal with the 3-msec alexandrite laser in patients with skin types IV-VI: efficacy, safety, and the role of topical corticosteroids in preventing side effects. *J. Drugs Dermatol.* **6**, 60–66.
112. Cassuto DA, Ancona DM, and Emanuelli G. (2000) Treatment of facial telangiectasias with a diode-pumped Nd:YAG laser at 532 nm. *J Cutan Laser Ther.* **2**, 141–146.
113. Kono T, Manstein D, Chan HH, Nozaki M, and Anderson RR (2006) Q-switched ruby versus long-pulsed dye laser delivered with compression for treatment of facial lentigines in Asians. *Lasers Surg. Med.* **38**, 94–97.
114. McDaniel DH, Ash K, Lord J, Newman,J, and Zukowski M. (1997) The erbium: YAG laser: A review and preliminary report on resurfacing of the face, neck, and hands. *Aesth Sur J* **17**, 157–164.
115. Jih MH, Friedman PM, Goldberg LH, Robles M, Glaich,AS, and Kimyai-Asadi A. (2006) The 1450-nm diode laser for facial inflammatory acne vulgaris: dose-response and 12-month follow-up study. *J. Am. Acad. Dermatol.* **55**, 80–87.
116. Lapidoth M, Shafirstein G, Amitai DB, Hodak E, Waner M, and David M. (2004) Reticulate erythema following diode laser-assisted hair removal: a new side effect of a common procedure. *J. Am. Acad. Dermatol.* **51**, 774–777.
117. Eremia S and Newman N. (2000) Topical anesthesia for laser hair removal: comparison of spot sizes and 755 nm versus 800 nm wavelengths. *Dermatol. Surg.* **26**, 667–669.

118. Alster TS, Bryan H, and Williams CM. (2001) Long-pulsed Nd:YAG laser-assisted hair removal in pigmented skin: a clinical and histological evaluation. *Arch. Dermatol.* **137**, 885–889.
119. Lask G, Friedman D, Elman M, Fournier N, Shavit R, and Slatkine M. (2006) Pneumatic skin flattening (PSF): a novel technology for marked pain reduction in hair removal with high energy density lasers and IPLs. *J. Cosmet. Laser Ther.* **8**, 76–81.
120. Lanigan SW. (2003) Incidence of side effects after laser hair removal. *J. Am. Acad. Dermatol.* **49**, 882–886.
121. Lim SP and Lanigan SW (2006) A review of the adverse effects of laser hair removal. *Lasers Med. Sci.* **21**, 121–125.
122. Asahina A, Watanabe T, Kishi A, Hattori N, Shirai A, Kagami S, Watanabe R, Le Pavoux A, Maekawa T, Tamaki K, and Ohara K. (2006) Evaluation of the treatment of port-wine stains with the 595-nm long pulsed dye laser: A large prospective study in adult Japanese patients. *J Am Acad Dermatol* **54**, 487–493.
123. Baba M and Bal N. (2006) Efficacy and safety of the short-pulse erbium:YAG laser in the treatment of acquired melanocytic nevi. *Dermatol. Surg.* **32**, 256–260.
124. Suh DH, Hwang JH, Lee HS, Youn JI, and Kim PM. (2000) Clinical features of Ota's naevus in Koreans and its treatment with Q-switched alexandrite laser. *Clin. Exp. Dermatol.* **25**, 269–273.
125. Rogachefsky AS, Silapunt S, and Goldberg DJ. (2001) Evaluation of a super long pulsed 810-nm diode hair removal laser in suntanned individuals. *J. Cutan. Laser Ther.* **3**, 57–62.
126. Greppi I. (2001) Diode laser hair removal of the black patient. *Lasers Surg. Med.* **28**, 150–155.
127. Lou WW, Quintana AT, Geronemus RG, and Grossman MC. (2000) Prospective study of hair reduction by diode laser (800 nm) with long-term follow-up. *Dermatol. Surg.* **26**, 428–432.
128. Galadari I. (2003) Comparative evaluation of different hair removal lasers in skin types IV, V, and VI. *Int. J. Dermatol.* **42**, 68–70.
129. Moreno-Arias GA, Castelo-Branco C, and Ferrando J. (2002) Side-effects after IPL photodepilation. *Dermatol Surg.* **28**, 1131–1134.
130. Lepselter J and M Elman. (2004) Biological and clinical aspects in laser hair removal. *J. Dermatolog. Treat.* **15**, 72–83.
131. Schieke SM., Stege H, rten V, Grether-Beck SG, Sies H, and Krutmann J. (2002) Infrared-A radiation-induced matrix metalloproteinase 1 expression is mediated through extracellular signal-regulated kinase 1/2 activation in human dermal fibroblasts. *J Invest Dermatol.* **119**, 1323–1329.
132. Schieke SM, Schroeder P, and Krutmann J. (2003) Cutaneous effects of infrared radiation: From clinical observations to molecular response mechanisms. *Photodermatol Photoimmunol and Photomed.* **19**, 228–234.
133. Schroeder P, Pohl, Calles C, Marks C, Wild S, and Krutmann J. (2007) Cellular response to infrared radiation involves retrograde mitochondrial signaling. *Free Radic Biol Med.* **43**, 128–135.
134. Merk HF, J Abel, J BaronM, and Krutmann J. (2004) Molecular pathways in dermatotoxicology. *Toxicol Appl Pharmacol.* **195**, 267–277.
135. Haywood RM, Wardman P, Gault DT, and Linge C. (1999) Ruby laser irradiation (694 nm) of human skin biopsies: assessment by electron spin resonance spectroscopy of free radical production and oxidative stress during laser depilation. *Photochem. Photobiol.* **70**, 348–352.
136. Applegate LA, Scaletta C, Panizzon R, E. Frenk, P Hohlfeld, and Schwarzkopf S. (2000) Induction of the putative protective protein ferritin by infrared radiation: implications in skin repair. *Int J of Mol Med.* **5**, 247–251.
137. Ananthaswamy HN, Loughlin SM, Cox,P, Evans RL, Ullrich SE, and Kripke ML. (1997) Sunlight and skin cancer: inhibition of p53 mutations in UV-irradiated mouse skin by sunscreens. *Nat. Med.* **3**, 510–514.
138. Halliday GM, Agar,NS, Barnetson RS, Ananthaswamy HN, and Jones AM. (2005) UV-A fingerprint mutations in human skin cancer. *Photochem. Photobiol.* **81**, 3–8.
139. Hussein MR. (2005) Ultraviolet radiation and skin cancer: Molecular mechanisms. *J Cutan Pathol.* **32**, 191–205.

140. Alam M, Hsu TS, Dover JS, Wrone DA, and Arndt KA. (2003) Nonablative laser and light treatments: histology and tissue effects–a review. *Lasers Surg. Med.* **33**, 30–39.
141. Menezes S, Coulomb B, Lebreton C, and Dubertret L. (1998) Non-coherent near infrared radiation protects normal human dermal fibroblasts from solar ultraviolet toxicity. *J Invest Dermatol.* **111**, 629–633.
142. Frank S, Menezes S, Lebreton-De Coster C, Oster M, Dubertret L, and Coulomb B. (2006) Infrared radiation induces the p53 signaling pathway: Role in infrared prevention of ultraviolet B toxicity. *Exp Dermatol.* **15**, 130–137.
143. Haedersdal, M. and H. C. Wulf (1996) Copper vapor laser and photocarcinogenesis in hairless mice. *Lasers Surg. Med.* **18**, 391–396.
144. Hedelund L, Haedersdal M, Egekvist H, Heidenheim M, Wulf HC, and Poulsen T. (2004) CO2 laser resurfacing and photocarcinogenesis: an experimental study. *Lasers Surg. Med.* **35**, 58–61.
145. Hedelund L, Lerche C, Wulf HC, and Haedersdal M. (2006) Carcinogenesis related to intense pulsed light and UV exposure: An experimental animal study. *Lasers Med Sci.* **21**, 198–201.
146. Suzuki H and Anderson RR. (2005) Treatment of melanocytic nevi. *Dermatol. Ther.* **18**, 217–226.
147. Michel JL. (2003) Laser therapy of giant congenital melanocytic nevi. *Eur. J. Dermatol.* **13**, 57–64.
148. Nelson,JS and Kelly KM. (1999) Q-switched ruby laser treatment of a congenital melanocytic nevus. *Dermatol. Surg.* **25**, 274–276.
149. Kono T, Nozaki M, Sasaki K, Kwon SG, and Henry Chan HL. (2001) Combined use of normal mode and Q-switched ruby lasers in the treatment of congenital melanocytic naevi. *Br J Plast Surg.* **54**, 640–643.
150. Burd A. (2004) Laser treatment of congenital melanocytic nevi. *Plast. Reconstr. Surg.* **113**, 2232–2233.
151. Imayama S and Ueda S. (1999) Long- and short-term histological observations of congenital nevi treated with the normal-mode ruby laser. *Arch. Dermatol.* **135**, 1211–1218.
152. Goldberg DJ, J ZeichnerA, Hodulik SG, and Marmur E. (2006) Q-Switched laser irradiation of pigmented nevi: analysis of markers for malignant transformation. *Lasers in Surg Med.* **18 (suppl)**, 53.
153. Moreno-Arias G, Castelo-Branco C, Ferrando J, and Avram MR. (2002) Paradoxical effect after IPL photoepilation. *Dermatol Surg.* **28**, 1013–1016.
154. Marayiannis KB, Vlachos SP, Savva MP, and Kontoes PP. (2003) Efficacy of long- and short pulse alexandrite lasers compared with an intense pulsed light source for epilation: a study on 532 sites in 389 patients. *J. Cosmet. Laser Ther.* **5**, 140–145.
155. Vlachos SP and Kontoes PP. (2002) Development of terminal hair following skin lesion treatments with an intense pulsed light source. *Aesthetic Plast. Surg.* **26**, 303–307.
156. Bouzari N, Tabatabai H, Abbasi Z, Firooz A, and Dowlati Y. (2004) Laser hair removal: comparison of long-pulsed Nd:YAG, long-pulsed alexandrite, and long-pulsed diode lasers. *Dermatol. Surg.* **30**, 498–502.
157. Kontoes P, Vlachos S, Konstantinos M, Anastasia L, and Myrto S. (2006) Hair induction after laser-assisted hair removal and its treatment. *J Am Acad Dermatol.* **54**, 64–67.
158. Willey A, Torrontegui J, Azpiazu J, and Landa N. (2007) Hair stimulation following laser and intense pulsed light photo-epilation: Review of 543 cases and ways to manage it. *Lasers Surg Med.* **39**, 297–301.
159. Alajlan A, Shapiro J, Rivers,JK, Macdonald N, Wiggin J, and Lui H. (2005) Paradoxical hypertrichosis after laser epilation. *J. Am. Acad. Dermatol.* **53**, 85–88.
160. Leung AKC and Kiefer GN. (1989) Localized acquired hypertrichosis associated with fracture and cast application. *J Natl Med Assoc.* **81**, 65–67.
161. Tisocco LA, Del Campo DV, Bennin B, and Barsky S. (1981) Acquired localized hypertrichosis. *Arch Dermatol* **117**, 127.
162. Shafir R and Tsur H. (1979) Local hirsutism at the periphery of burned skin. *Br J Plast Surg.* **32**, 93.
163. Gupta SS, Gupta A, Kanwar J, and Kumar B. (2004) Hypertrichosis surrounding scar of knee replacement surgery [8]. *J Am Acad Dermatol.* **50**, 802–803.

164. Wendelin DS, Pope DN, and Mallory SB. (2003) Hypertrichosis. *J Am Acad Dermatol.* **48**, 161–179.
165. Trelles MA, Allones I, and Mayo E. (2006) Combined visible light and infrared light-emitting diode (LED) therapy enhances wound healing after laser ablative resurfacing of photodamaged facial skin. *Medical Laser Application* **21**, 165–175.
166. Hawkins D and Abrahamse H. (2007) Influence of broad-spectrum and infrared light in combination with laser irradiation on the proliferation of wounded skin fibroblasts. *Photomed. Laser Surg.* **25**, 159–169.
167. Al Watban FA, Zhang XY, and Andres BL. (2007) Low-level laser therapy enhances wound healing in diabetic rats: a comparison of different lasers. *Photomed. Laser Surg.* **25**, 72–77.
168. Rezende SB, Ribeiro MS, Nunez SC, Garcia VG, and Maldonado EP. (2007) Effects of a single near-infrared laser treatment on cutaneous wound healing: biometrical and histological study in rats. *J. Photochem. Photobiol. B* **87**, 145–153.
169. de Araujo CEN, Ribeiro MS, Favaro R, Zezell DM, and Zorn TMT. (2007) Ultrastructural and autoradiographical analysis show a faster skin repair in He-Ne laser-treated wounds. *J of Photochem Photobil. B* **86**, 87–96.
170. Spielmann H, Balls M, Brand M, Doring B, Holzhutter HG, Kalweit S, Klecak G, Eplattenier HL, Liebsch M, Lovell WW, Maurer T, Moldenhauer F, Moore L, Pape WJW, Pfanenbecker U, Potthast J, de Silva O, Steiling W, and Willshaw A. (1994) EEC/COLIPA project on in vitro phototoxicity testing: First results obtained with a Balb/c 3T3 cell phototoxicity assay. *Toxicol In Vitro* **8**, 793–796.
171. Spielmann H, Balls M, Dupuis J, Pape WJ, Pechovitch G, de Silva O, Holzhutter HG, Clothier R, Desolle P, and Gerberick F. (1998) The International EU/COLIPA In Vitro Phototoxicity Validation Study: Results of Phase II (Blind Trial). Part 1: The 3T3 NRU Phototoxicity Test. *Toxicol In vitro* **12**, 305–327.
172. Spielmann H, Muller L, Averbeck Balls DM, Brendler-Schwaab S, Castell,JV, Curren R, de Silva O, Gibbs NK, Liebsch M, Lovell WW, Merk HF, J. F. Nash, Neumann NJ, Pape WJ, Ulrich P, and Vohr HW (2000) The second ECVAM workshop on phototoxicity testing. The report and recommendations of ECVAM workshop 42. *Altern. Lab Anim* **28**, 777–814.
173. Zachariae,H. (1988) Delayed wound healing and keloid formation following Argon laser treatment or dermabrasion during isotretinoin treatment. *Br J Dermatol.* **118**, 703–706.
174. Khatri KA. (2004) Diode laser hair removal in patients undergoing isotretinoin therapy. *Dermatol. Surg.* **30**, 1205–1207.
175. Cassano N, Arpaia,N, and Vena GA. (2005) Diode laser hair removal and isotretinoin therapy [2]. Dermatol Surg. **31**, 380–381.
176. Anvari B, Milner TE, Tanenbaum BS, Kimel S, L. O. Svaasand, and J. S. Nelson (1995) Selective cooling of biological tissues: application for thermally mediated therapeutic procedures. *Phys. Med. Biol.* **40**, 241–252.
177. Altshuler GB, Zenzie HH, Erofeev AV, Smirnov MZ, Anderson RR, and Dierickx C. (1999) Contact cooling of the skin. *Phys. Med. Biol.* **44**, 1003–1023.
178. Ellis DL, Kozub J, and Reinisch L. (2006) Controlling thermal damage of incisions using diamond, copper, and sapphire heat-conducting templates with and without cooling. *Lasers Surg. Med.* **38**, 814–823.
179. Ramli R, Chung CC, Fried NM, Franco N, and Hayman MH. (2004) Subsurface tissue lesions created using an Nd:YAG laser and a sapphire contact cooling probe. *Lasers Surg. Med.* **35**, 392–396.
180. Nelson JS, Majaron B, and Kelly KM. (2000) Active skin cooling in conjunction with laser dermatologic surgery. *Semin. Cutan. Med. Surg.* **19**, 253–266.
181. Hodges GJ, Kosiba,WA, Zhao K, Alvarez,GE, and Johnson JM. (2007) The role of baseline in the cutaneous vasoconstrictor responses during combined local and whole body cooling in humans. *Am. J. Physiol Heart Circ. Physiol.*
182. Svaasand LO, Randeberg LL, Aguilar G, Majaron B, Kimel S, Lavernia EJ, and Nelson JS. (2003) Cooling efficiency of cryogen spray during laser therapy of skin. *Lasers Surg. Med.* **32**, 137–142.
183. Pikkula BM, Torres JH, Tunnell JW, and Anvari B. (2001) Cryogen spray cooling: Effects of droplet size and spray density on heat removal. *Lasers Surg. Med.* **28**, 103–112.

184. Fried NM and Walsh, Jr JT. (2000) Cryogen spray cooling during laser tissue welding. *Phys. Med. Biol.* **45**, 753–763.
185. Nelson JS and Kimel S. (2000) Safety of cryogen spray cooling during pulsed laser treatment of selected dermatoses. *Lasers Surg. Med.* **26**, 2–3.
186. Chang CJ and Nelson JS. (1999) Cryogen spray cooling and higher fluence pulsed dye laser treatment improve port-wine stain clearance while minimizing epidermal damage. *Dermatol. Surg.* **25**, 767–772.
187. Hoffman WL, Anvari B, Said S, Tanenbaum BS, Liaw LH, Milner T, and Nelson JS. (1997) Cryogen spray cooling during Nd:YAG laser treatment of hemangiomas. A preliminary animal model study. *Dermatol. Surg.* **23**, 635–641.
188. Hammes S and Raulin C. (2005) Evaluation of different temperatures in cold air cooling with pulsed-dye laser treatment of facial telangiectasia. *Lasers Surg. Med.* **36**, 136–140.
189. Raulin C, Greve B, and Hammes S. (2000) Cold air in laser therapy: first experiences with a new cooling system. *Lasers Surg. Med.* **27**, 404–410.
190. Manuskiatti W, Eimpunth S, and Wanitphakdeedecha R. (2007) Effect of cold air cooling on the incidence of postinflammatory hyperpigmentation after Q-switched Nd:YAG laser treatment of acquired bilateral nevus of Ota like macules. *Arch. Dermatol.* **143**, 1139–1143.
191. Ashinoff R and Geronemus RG. (1990) Effect of the topical anesthetic EMLA on the efficacy of pulsed dye laser treatment of port-wine stains. *J. Dermatol. Surg. Oncol.* **16**, 1008–1011.
192. Sherwood KA. (1993) The use of topical anesthesia in removal of port-wine stains in children. *J. Pediatr.* **122**, S36–S40.
193. Scheepers JH and Quaba AA. (1994) A two-year review of pain control during laser therapy using the flashlamp pulsed dye laser. *Br. J. Plast. Surg.* **47**, 112–116.
194. Alster TS and Lupton JR. (2002) Evaluation of a novel topical anesthetic agent for cutaneous laser resurfacing: a randomized comparison study. *Dermatol. Surg.* **28**, 1004–1006.
195. Kilmer SL, Chotzen V, Zelickson BD, McClaren M, Silva S, Calkin J, and No D. (2003) Full-face laser resurfacing using a supplemented topical anesthesia protocol. *Arch. Dermatol.* **139**, 1279–1283.
196. Nestor MS. (2006) Safety of occluded 4% liposomal lidocaine cream. *J. Drugs Dermatol.* **5**, 618–620.

24

Eye Safety of Laser and Light-Based Devices

David H. Sliney

Consulting Medical Physicist, 406 Streamside Drive Fallston, MD, USA

Gurpreet S. Ahluwalia (ed.), *Cosmetic Applications of Laser and Light-Based Systems*, 499–516,

24.1 Introduction

Ocular safety of both laser and intense-pulse light (IPL) devices is always important to protect both the user and patient from potential hazards. Questions also sometimes arise regarding the eye safety of the much lower power Light Emitting Diode (LED) devices. Safety has always been important for any laser application in surgical procedure, and several issues dominate. Two key issues arise: (1) wearing eye protectors by the clinical staff and (2) protecting the eyes of the patient. No one denies that lasers can pose a serious hazard to the eye, but other light-based systems may also pose similar hazards, particularly for the patient. For periorbital surgery in particular, the type of protector for the patient's eyes can vary with the wavelength and the type of laser. Since the penetration of laser light through the lid and adjacent tissues vary, the criticality of the eye protector has not been fully appreciated, and the decision to use some types of eye protectors in some procedures has frequently questioned. Fortunately, eye injuries are becoming much less likely because of the development of more contact applicators. Other ancillary safety issues include the need to properly deal with the plume of vaporized tissue during ablative procedures and controlling potential fire hazards. The choice of control measures to minimize the very serious risk from chronic breathing of vaporized tissue also requires judgment.

With the increasing variety of lasers and the number of wavelengths now available, safe laser use has become a more complex issue [1]. The extent of potentially hazardous reflections, the type of eye protection, and the ancillary hazards can vary considerably with the type of laser used and the procedure. However, in all cases, the laser operator (the clinical user) and the staff must be concerned with both the protection of the patient, and the protection of the persons in the vicinity of the procedure. Patient safety is assured by limiting needless exposure to adjacent tissues (by choice of wavelength and to a large extent by technique), using noncombustible materials adjacent to the beam, and by properly protecting the patient's eyes. Safety of the operator and assistants requires concern for both system-safety design and the means to limit potentially hazardous reflections. For ablative procedures, the environmental hazards from the smoke produced by vaporizing tissue must be minimized by local exhaust ventilation or fume extractors. The pathogenicity and chemical toxicity of vaporized tissue has been the subject of a number of investigations, as discussed later. Safety standards for medical laser applications have been issued, which consider all these potential hazards and their control measures [1–10]. The current consensus standard in the United States, that is, the American National Standard Z136.3-2005, *Safe Use of Lasers in Health Care Facilities* [4], and similar user guidelines have been crafted to provide a realistic and balanced approach to clinical laser safety, hopefully without needless control measures. Realistic safety procedures can be achieved only if the entire community of laser users participates in the development of consensus standards and codes of practice.

The one hazard that is truly unique to the laser, and that requires special attention results from the laser beam itself—the optical radiation hazard. Unlike other light sources, the laser beam may be collimated and directed over some distance; hence the area of potential hazard may not be limited to the immediate surgical site. Unwarranted fears often accompany the introduction of lasers into the clinical environment. Therefore, proper appreciation of the real laser-beam hazard is necessary for each member of the professional staff so that realistic safety precautions are followed [10–18]. In most countries, occupational safety and health regulations emphasize the critical importance of informing and educating the worker on workplace risks, and this is clearly important regarding laser use [19].

Laser hazards depend on the laser in use, the environment, and the personnel involved with the laser operation (the operator, ancillary personnel, and the patient). The laser hazard is roughly defined by the hazard classification [1–4], while the other factors must be analyzed in each situation. A basic understanding of laser biological effects and hazards is necessary to assess each laser hazard in the operating room intelligently. Once the hazards are understood, the safety measures become obvious.

24.2 Biological Hazards of Laser Beams

24.2.1 Eye Hazards

Because of its special optical properties, the human eye is considered to be most vulnerable to laser light. Apart from the oral mucosa, the only living tissue exposed to the environment is the cornea and conjunctiva. Without the comparative protective features of the stratum corneum of the skin, the eye is exposed to the harsh environment of the sun, wind, dust, ultraviolet radiation, and intense light. The eye has a natural protective mechanism in its aversion response, which limits the retinal exposure to very intense visible light by papillary constriction and eye movement or by the lid reflex and eye movement, which in turn limits exposures from intense infrared radiant energy that raise the temperature of the cornea. However, some laser-beam intensities are so great that injury can occur faster than the protective action of the aversion response, which occurs within 0.25 seconds [9].

Obviously, laser energy cannot damage the tissue, unless the light energy is able to penetrate and be absorbed in that structure. For this reason, rays in the visible and near infrared (visible and IR-A band), which can be transmitted through the clear ocular media and be absorbed in the retina, can, in sufficient intensity, damage the retina. The actual size of such a "point" at the retina is of the order of 10–20 μm (smaller than the diameter of a human hair). For this reason, lasers operating between 400 and 1400 nm are particularly dangerous to the retina. This spectral region is often referred to as the "retinal- hazard region", since the increased concentration of light after entering the eye and focused on the retina is of the order of 100,000-fold! Hence, a collimated beam of 1 W/cm^2 at the cornea will focus to a small spot with an irradiance of 100 kW/cm^2. Although damage to such a small region of the retina may seem insignificant initially, it is important to realize that certain parts of the retina, for example, the central retina, the macula, and its fovea (center of the macula), are extremely small areas responsible for critically important high-acuity vision. If these areas are damaged by laser radiation, substantial loss of vision can result. The argon, krypton, KTP, helium-neon (He-Ne), ruby, alexandrite, diode and neodymium:YAG, all emit in the retinal-hazard spectral region (~400 to ~1400 nm). The concentration of energy at the retina, make these lasers particularly hazardous to the eye, as shown in Fig. 24.1.

The retinal image area alone may not be the only site of damage, but as a result of heat flow and mechanical (acoustic) transients, the tissue surrounding the image site may also be damaged, leading to more severe consequences on visual function. For example, it has not been uncommon for an individual to lose almost the total function in an eye exposed to a very small amount of energy (several hundred microjoules) when a short-pulse (Q-switched) laser that has been accidentally imaged on the fovea. Instead of a normal visual acuity of 6/6 (20/20 in the United States), the visual acuity in such accidental situations has often been recorded as 6/60 (20/200) following an accident. Fortunately, in most

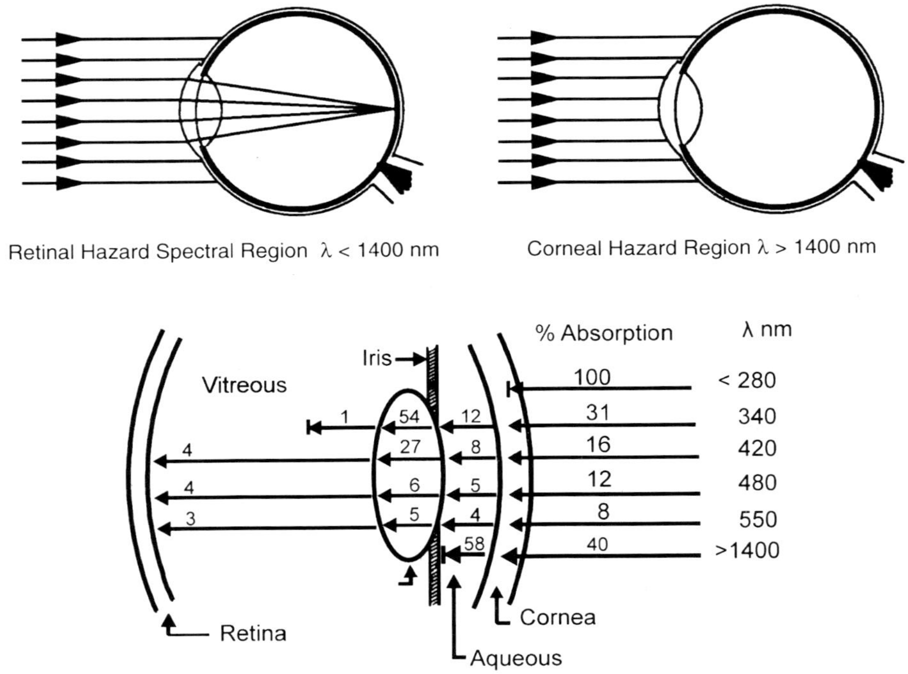

Figure 24.1 Spectral properties of the eye determine the ocular structures vulnerable to energy. The cornea, lens, iris, and retina can be injured at transitional wavelengths approaching 1400 nm. Top panel shows qualitative conditions; bottom panel gives examples of energy absorbed in different anterior structures of the eye.

accidents, only one eye is exposed to a collimated beam. However, after some recovery, any visual loss remaining after 60 days is generally permanent, since the neural tissue of the retinal has very little ability for repair [1,14,15].

At wavelengths outside of the retinal hazard region—in both the ultraviolet and far-infrared regions of the spectrum—injury to the anterior segment of the eye is possible. Certain spectral bands may injure the lens (notable at wavelengths between 295 and 320 nm in the ultraviolet region and wavelengths between 1 and 2 μm in the infrared region may actually pose a greater risk for permanent injury than the 10.6 μm CO_2 laser wavelength), which does not penetrate the cornea [15]. See Fig. 24.2 for hazards by spectral region.

Excimer lasers operating in the ultraviolet spectral region are no longer unusual in some special applications. Certain excimer lasers pose a particular hazard to the cornea, and the 308-nm Xe-Cl excimer laser can be considered additionally dangerous, as it can produce an immediate cataract of the lens. By contrast, the 193-nm ArF excimer laser wavelength used in laser refractive surgery cannot ever penetrate deep into a single cell; hence the biological consequences of scattered radiation are not at all serious, even if the conservative MPE were to be exceeded by hundredfold at this extremely short wavelength; hazard zones are therefore only a few centimeter from the ablated cornea [12]. The surface cells (wing cells) of the cornea have an average lifetime of only 48 hours, and are quickly sloughed off after being damaged, thus leading to no sequellae [13]. The holmium: YAG, hydrogen-fluoride, carbon dioxide, and carbon monoxide lasers are all potentially hazardous to the

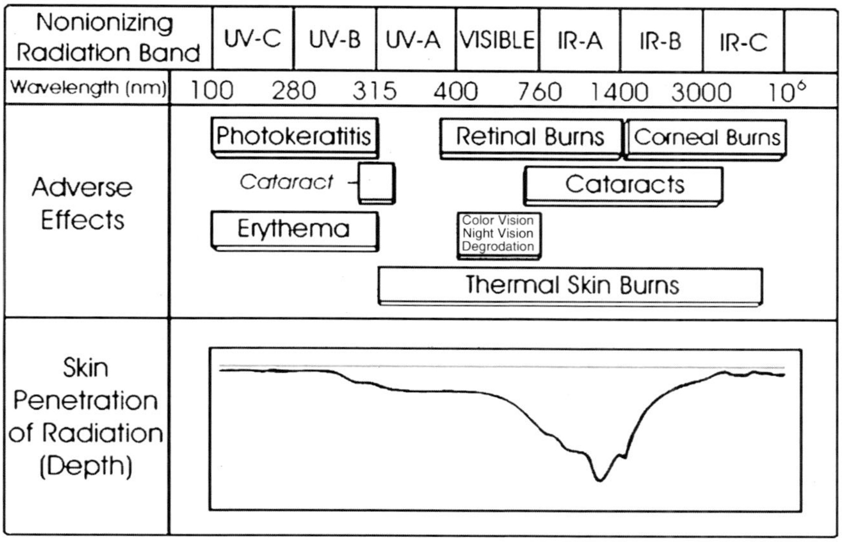

Figure 24.2 Hazards summarized by CIE spectral region. The CIE is the International Commission on Illumination.

cornea, but because wavelengths that cause corneal damage are not reconcentrated by the eyes as are wavelengths in the retinal hazard region, the thresholds for injury of the cornea are generally much higher than those that may injure the retina. Table 24.1 lists permissible occupational exposure limits for many of the commonly used dermatological lasers [2–8].

24.2.2 Periorbital Surgery and Patient Eye Protection

No one questions the need to protect the eye during periorbital surgery. However, the most appropriate means of protection and standardized requirements have been the subject of some debate. The treatment of the facial tissue near the eye can potentially expose ocular tissues to scattered laser radiation that has been internally diffusely scattered around optical tissues. This is of particular concern for the more penetrating laser wavelengths within the 750–1400 nm spectral region. The use of eye patches and occluders protect the globe from direct exposure; however, scattered energy from around the protector can reach the eye. Sensitive imaging methods and a subjective method have been employed to estimate the fluence at different internal ocular structures. Sliney measured the radiance (brightness) of scattered optical radiation in the orbital area when a low-power laser was placed at several periorbital locations to quantify the level of periorbital exposure. Subjective measures of brightness have also permitted the determination of retinal exposure levels. The measured levels scaled to 1- and 10 watt laser power levels showed that significant levels of laser radiation can be transmitted to ocular tissues despite the use of lid patches, and great caution must be exercised in the choice of appropriate eye protection for the more deeply penetrating wavelengths in the 600–1100 nm spectral region. Indeed, a number of accidental eye injuries have been reported from inadequate precautions during laser skin resurfacing and hair removal in the periorbital area [16–22]. Although accidental injuries have resulted from the CO_2 laser [16], injuries from the more deeply penetrating wavelengths in

Table 24.1 Selected Occupational Exposure Limits (MPEs) for Some Lasers[a]*#

Type of Laser	Principal Wavelengths	MPE (Eye)
Argon-fluoride laser	193 nm*	3.0 mJ/cm^2 over 8 h
Xenon-chloride laser	308 nm	40 mJ/cm^2 over 8 h
Argon ion laser	488, 514.5 nm	3.2 mW/cm^2 for 0.1 s
KTP (Nd:YAG-freq.-doubled)	532 nm	2.5 mW/cm^2 for 0.25 s
Helium–neon laser	632.8 nm	1.8 mW/cm^2 for 1.0 s
Dye-lasers	580–590 nm	1.0 mW/cm^2 for 10 s
Krypton ion laser	568, 647 nm	
		5.2 mW/cm^2 for 0.1 s
		4.1 mW/cm^2 for 0.25 s
Diode lasers	~808 nm	3.0 mW/cm^2 for 1.0 s
		1.6 mW/cm^2 for 10 s
Neodymium:YAG laser (primary λ)	1,064 nm	5.0 μJ/cm^2 for 1 ns to 50 μs No MPE for $t < 1$ ns 5 mW/cm^2 for 10 s
Neodymium:YAG laser (secondary λ)	1,334 nm	40 μJ/cm^2 for 1 ns to 50 μs 40 mW/cm^2 for 10 s
Diode 1.44, Pulsed Nd:YAG-1.44	1.41–1.44 μm	0.1 J/cm^2 for 1 ns to 1 ms
Pulsed Holmium laser	2.1 μm	100 mW/cm^2 for 10 s to 8 h, limited area
CW holmium laser	2.1 μm	
Erbium :YAG CW carbon monoxide laser	2.94 μm ~ 5 μm	10 mW/cm^2 for >10 s for most of body skin area)
Carbon dioxide laser	10.6 μm	

[a] All standards/guidelines have MPEs at other wavelengths and exposure durations.

*Sources: ANSI Standard 136-1-2007; ACGIH TLVs (2008) and ICNIRP (2000).

Note: to convert MPEs in mW/cm^2 to mJ/cm^2, multiply by exposure time t in seconds, e.g., the He-Ne or Argon MPE at 0.1 s is 0.32 mJ/cm^2.

#The retinal-hazard limits apply to a single "point" source; higher levels apply to fractionated and diffused-source laser products.

the 700–850 nm region, such as diode and alexandrite lasers have been reported more frequently [17–22]. As one example, a patient treated for a port-wine stain in the periorbital area with a 755–nm alexandrite laser with a radiant exposure of 50 J/cm^2 and 20 ms pulses having a beam diameter of 12.5 mm produced permanent injury of the iris and some reduction in vision [20]. An iris atrophy was produced during diode laser epilation [17]. The CO_2 laser injury of the eyes apparently took place, despite the use of metal eye-shields; however, in that instance, the treatment of both eyelids with two passes as well as the entire face (a 210-minute procedure) may have overheated the metal eye shields, causing thermal damage from heat conduction to the underlying ocular tissue from the 0.25–0.3 J, 6–7 mm spots (although dislocation of the eye shields cannot be ruled out). Special eye shields are available for patient protection and normally should be opaque and in contact with the cornea and conjunctiva (Fig. 24.3).

Figure 24.3 Eye Shields: Eye shields that reflect incident energy accidentally directed toward the eye are essential for laser surgery near the eye. It is important that eye shields are placed under the lid when performing periorbital resurfacing.

The thickness of soft tissues in and around the orbit varies greatly with position and facial type, therefore it is very difficult to provide absolute assurance that a specific eyelid cover or even a corneal-contact metal eye shield will provide certain protection in every case. With the deeply penetrating wavelengths, some energy will definitely reach into the globe and potentially heat-pigmented tissues—most notably the iris, and to a much lesser extent, even the retina. The findings of Pham and colleagues are notable here, as they demonstrated retinal stimulation by scattered red light, but of course this level of multiply scattered light around the globe—while sufficient to see—was far below levels required to injure the retina [23]. It is worth remembering that eye shields developed and tested in the 1990s for covering the eye and protecting the cornea against excessive CO_2 laser radiation may not necessarily be ideal for deep red and IR-A (780–1400 nm) wavelengths [24–27]. Surfacing of rhytides near the eye and blepharoplasty require great care to protect the eye [28]. Ocular injuries from IPL devices can also occur to the iris, but less likely to occur to the retina.

24.3 Safety of the Staff—Reflections and Probability of Exposure

An examination of laser-accident records indicates that the source of accidental ocular exposure of the laser operator and staff is most frequently a reflected beam. Figure 24.4 illustrates the types of mirror-like (specular) laser-beam reflections that can occur from flat or curved specular surfaces, which are characteristic of metallic instruments used in other surgical procedures, but are not frequently encountered in skin resurfacing. Skin injury of the hand holding an instrument is also possible. Normally, a collimated beam is considered the most hazardous type of reflection, but at very close range, a diverging beam may pose a greater likelihood of striking the eye [1,14,15,25,29].

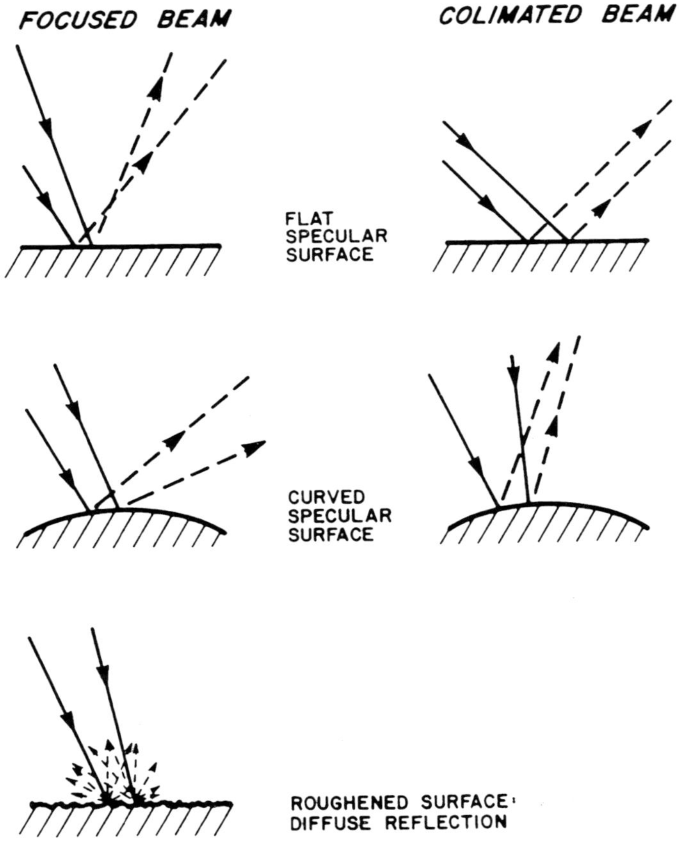

Figure 24.4 Potentially hazardous reflections from specular (mirror-like) surfaces can extend some distance if the beam and reflecting surface are flat. If the beam is convergent, divergent, or the surface curved, the reflected beam will normally diverge rapidly, limiting the hazard zone.

A number of steps can be taken to minimize the potential hazards to the staff and bystanders. Preventive measures will depend upon the type of laser. One of the most commonly employed lasers in surgical applications today may still be the CO_2 laser. Since the CO_2 laser wavelength of 10.6 μm is in the far-infrared spectral region—and invisible—the presence of hazardous secondary beams could go unnoticed. This added hazard resulting from an infrared laser beam's lack of visibility is common to other infrared lasers, such as the 2.1 μm holmium or the 1064 nm Nd:YAG laser. The 755–795 nm alexandrite lasers are visible, but these wavelengths are weakly visible, with the result that many users have mistakenly thought a high-power beam is "safe." Because there have been a number of serious retinal injuries caused by improper attention to safety with Nd: YAG lasers [15], the use of the Nd:YAG laser must be approached with even greater caution than the CO_2 laser. By contrast, the dye, argon laser, and the second-harmonic Nd:YAG (sometimes referred to as the "KTP") laser emit highly visible, blue–green (488, 514.5, 532 and 588 nm) beams, and in some ways pose a lesser potential hazard.

Many dermatological lasers, such as the CO_2, Nd:YAG, holmium, or argon, are continuous-wave (CW), or nearly so. Even the so-called super-pulse laser is quasi-CW compared to single-pulse or multiply pulsed fractional laser systems. The biological effects and potential hazards from high-peak power-pulsed lasers are quite different from those of CW lasers. This is particularly true of lasers operating in the *retinal hazard region* of the visible (400–780 nm) and near-infrared spectrum (IR-A, 780 to 1400 nm), as shown in Figs. 24.1 and 24.2. The severity of retinal lesions from a visible or near-infrared (IR-A) CW laser is normally considered to be far less than from a short-pulse laser. Another major factor that influences the potential hazard is the degree of beam collimation. Many dermatological lasers are focused, thereby limiting the hazardous area (referred to as the "nominal hazard zone" in IEC 60825-1 and ANSI Z136.1-2007 [3,9]. An exception is the highly collimated beam from the initial laser prior to the delivery system optics, which may remain hazardous at some distance from the instrument; this can be a problem during laser servicing [1].

Reflections are most serious from flat mirror-like (specular) surfaces—characteristic of many metallic surgical instruments and glass plates. Many surgical instruments now have black anodized or sandblasted and roughened surfaces to reduce (but not eliminate) potentially hazardous reflections. The strong curvature and surface roughening spread the reflected energy and greatly reduce the reflection hazard. The surface roughening is generally more effective than the black (ebonized) surface, since the beam is diffused. However, in some cases, combining a special black surface with roughening provides increased protection, and adding a black polymer finish to surgical implements placed in or near the beam has been shown by experiment measurements to offer the greatest protection at the CO_2 wavelength—despite initial skepticism by investigators [29]. However, other groups argue against blackening the surface, since the instrument will become hotter than without for visible wavelengths. Therefore, the use of the special blackened surfaces must be approached with caution for each application; fortunately, such instruments are not common in most procedures.

It should be noted that both the surface finish and reflectance seen in the visible spectrum do *not* indicate these qualities in the invisible, far-infrared spectrum. In fact, a roughened surface that appears to be quite dull and diffuse at a shorter, visible, or IR-A wavelengths, will always be more specular at far-infrared wavelengths (e.g., at 10.6 m). This results from the fact that the relative size of the microscopic structure of the surface relative to the incident wavelength determines whether the beam is reflected as a specular or diffuse reflection [1,14,29].

A specularly reflected beam with only 1% of the initial beam's power can still be quite hazardous. Hence, the rougher the surface of an instrument likely to intercept the beam, the safer the reflection. For example, even a 1% reflection of a 40 watt (40 W) laser beam is 400 mW! It is somewhat surprising that there have been few cases reported of eye injuries to residents and other persons observing Nd:YAG laser surgery without eye protectors. Hazardous specular reflections from a laser beam emerging from an optical fiber are limited in extent because the beam rapidly diverges—just as would a focused beam, as shown in Fig. 24.5.

Most invisible beam surgical lasers have a visible alignment beam. Infrared lasers most often make use of a low-power coaxial He-Ne (632.8 nm) or diode (e.g., 635 nm) red laser. It is desirable where feasible for this alignment beam to be 1 mW (milliwatt) or less, since the maximum CW, visible laser beam power that can safely enter the eye within the aversion response (i.e., within the blink reflex of 0.25 seconds) is 1 mW. This type of laser is then classified as Class 2, and poses a very low risk to the user.

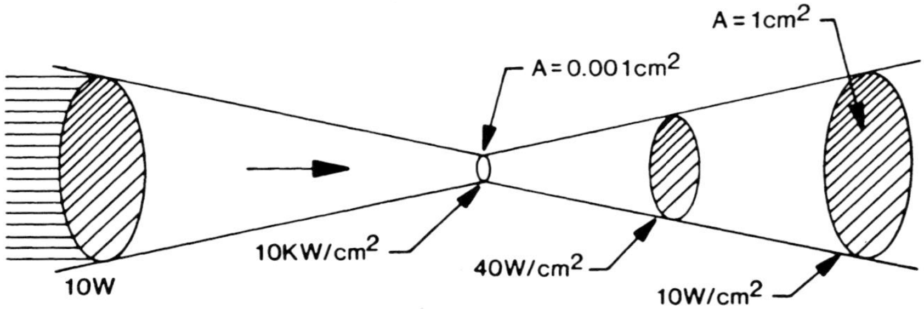

Figure 24.5 When a laser beam is focused, it is potentially hazardous only within the region around the focal zone. If the cone angle is wide, the irradiance will drop rapidly with distance.

24.3.1 Remitted Light

Remitted light is the laser or IPL energy that leaves the skin after internal scattering, and gives the special opalescent tones characteristic of skin. This light is frequently observed around contact applicators and has raised concerns that retinal hazards may exist when individuals view this light surrounding the treatment site when delivering high-intensity light or laser radiation to the skin. Certainly, there are many complaints about discomfort and even headaches arising from the remitted light, which most feel is at least noisome. Certainly, during skin resurfacing, hair removal, or other dermatological treatments, very high irradiances are delivered by contact applications, but some scattered energy is remitted from the skin surrounding the applicator after undergoing multiple internal scattering. Sliney et al. developed a method to determine the retinal hazard of viewing the remitted light using a calibrated CCD camera to measure the radiance (brightness) of the remitted optical radiation from the skin. For laser or incoherent sources, the radiance camera can be calibrated with a radiometer using the same source employed in the dermatological application to properly account for the instrument's spectral and temporal response. The camera once calibrated was used to record the brightness profile of remitted light. Using this method showed that conventional laser and IPL sources do not pose a retinal hazard, although some physical light sources may appear to be annoying to view. Since almost all applicators represent large, extended sources (greater than α_{max}), the potential retinal hazards of viewing either direct or remitted light from phototreatment devices using various sources of light, including those with nonuniform spatial distribution of optical energy are properly addressed by the CCD radiance camera.

24.3.2 Safety of the Operator

The IEC safety document [11] defines the laser operator as "the person who handles the laser equipment. In general, the laser operator controls the delivery of the laser radiation to the working area. The laser operator may appoint other person(s), who assist with the selection and/or setting of the parameters". Because some had interpreted the "laser operator" to be the assistant setting up the equipment and adjusting the settings rather than the dermatologist or other specialist directing the beam on the target tissue, the 2005 Edition of the

ANSI Z136.3 standard termed this person the "laser user" [4]. In any case, whether the laser operator is termed the "operator" or the "user," this individual often has the highest likelihood of incurring an eye injury from a laser reflection. Fortunately, many of the newer dermatological IPL and laser systems deliver energy to the treatment area by contact, and the handpiece delivery blocks any direct reflection. If the operator can view the target tissue when the optical energy is delivered in a noncontact mode, the reflections must be safely attenuated by the use of laser eye protectors. Under such indirect viewing conditions, the laser operator is normally not highly susceptible to injury due to proper design of the laser instrument. However, if the laser or IPL is accidentally actuated when the applicator is misdirected away from the target tissue, the operator will be at risk as any other person in the room. In addition, with hand-held laser delivery systems, one should remember that the operator's hand is the closest to the laser target, and therefore it is closest to potentially hazardous contact with the beam or reflections from adjacent metallic or glass surfaces.

24.3.3 Safety of the Surgical Staff

Nurses, technicians and other assisting family or staff are potentially exposed to misdirected laser beams. Lasers and IPLs have been accidentally initiated when the beam-delivery system was directed other than at the patient, a foot switch was accidentally pressed, or similar errors have occurred, and the beam has been directed at a person. Accidental firing of a laser has also occurred because of confusion created by multiple foot switches employed with other equipment positioned below the system. If a foot switch is employed, it should be covered and clearly identified. The IEC Standard 60601-22 requires that any "foot-operated laser emission control switch shall be shrouded" in order to prevent unintentional laser operation. Assistants are potentially exposed to secondary reflections from the treatment site, but these seldom extend to a distance greater than 1–2 m.

24.3.4 Safety of Other Bystanders

Bystanders in the spa facility or medical treatment facility who are present to observe or to calm the patient (e.g., a patient's relative) may be susceptible to exposure from reflected laser beams in the same manner as the professional staff. In addition, because of lack of training or knowledge about the laser surgical procedure, bystanders may be at a greater risk by inadvertently placing themselves in a dangerous position. Those individuals should always be provided with eye-protective goggles. It may also be wise to provide visitors with protective eyewear, even if not theoretically needed, since there have been incidences where individuals who discovered totally unrelated eye pathologies shortly after their visit blamed their experience in the laser-treatment facility.

24.3.5 Service Personnel

Service personnel are particularly susceptible to ocular injury since they often gain access to collimated laser beams from the laser cavity itself, or by opening up the beam-delivery optics and gaining access to collimated laser beams prior to the beam focusing optics or fiber-optic beam-delivery system. Once the laser beam leaves the delivery system and comes rapidly into a focus, it then diverges again, or if emerging from a fiber, it also rapidly diverges. Because of the nature of flashlamps, the light emitted from IPL devices rapidly diverge. The zone where the beam is concentrated to a level sufficient to pose severe hazard to the eyes or skin (the Nominal Hazard Zone or NHZ), is normally a limited zone of less than 1–2 m near the beam focal point. However, a collimated laser beam, as the raw beam for most laser cavities, or a specular reflection from a turning mirror or Brewster window in the laser console may be emitted from the laser cabinet (protective housing) when the service person gains access. Several serious eye injuries have occurred to service personnel exposed to secondary, collimated, and invisible 1064 nm Nd:YAG laser beams when the service personnel gained access to the laser cavity. According to ANSI Z136.3, a "temporary controlled area" must be established when such potential risks occur during servicing.

24.3.6 Eye Protective Goggles

Laser eye protectors provide the principal means to assure against ocular injury from the direct or reflected laser beams in the operating room [1,30,31]. Although viewing optics in some surgical specialties may inherently protect the eyes of the laser operator, there are seldom viewing optics in cosmetic applications. If viewing optics are employed during the procedure, their safety with the particular laser should always be ascertained from the manufacturer of the viewing optics. In this regard, ordinary optical glass in compound-lens systems protects substantially against all wavelengths shorter than about 300 nm and greater than approximately 2700 nm [15], although certainly at wavelengths greater than 4000 nm. Laser protective filters may be obtained for endoscopes and other viewing optics for the spectral region between these two spectral bands. Eye protectors are available as spectacles, wrap-around lenses, goggles, and related forms of eyewear. It is important that the eyewear be marked with the wavelengths and optical densities provided at those wavelengths. The markings must be clearly understood by all the operating-room staff. The proper use of eyewear and the meaning of the eye-protector markings are key subjects for laser- safety training of the staff [30–32].

Clear plastic goggles or spectacles with side shields, which are known to be made of polycarbonate, are normally suitable for use with the CO_2 laser, but should be marked by the laser-safety officer with an indication of the optical density, for example, "OD-4 at CO_2 wavelength of 10.6 μm". Some Laser Safety Officers (LSOs) may be uneasy about marking eye protectors not sold as "laser" eye protection, because of perceived (or very real) legal concerns. Studies of plastic eye protectors show clearly that polycarbonate is far superior in burn-through resistance than other plastics, and such a marking has been argued to be quite justifiable for use with CO_2 lasers having a power output up to about 100 W [29]. In some countries, the marking of eye protection by anyone other than the manufacturer is not legally recognized, and the LSO has no alternative but to obtain similar polycarbonate eye protectors that have been certified and labeled by the manufacturer. Some manufacturers of laser eye protection and some laser-safety specialists have made a major issue of the importance of damage resistance of eye protectors, and this concern is evident in eye-protection standards in Europe. However, burn-through times of plastic eye protectors appear to be of little concern in some quarters, as in the United States [1]. Those who are skeptical about these concerns of burn-through argue that with the powers generally used in laser surgery of 100 W or less, burn-through is unrealistic, since the wearer would certainly move their head within a second after detecting a flame shooting from the goggle. Indeed, the skin will incur a serious burn as would the unprotected, exposed cornea, and clothing would ignite at levels below plastic burn-through irradiances (Fig. 24.6). Goggles manufactured of special glass can frequently be designed to withstand irradiances higher than the ~100 W/cm^2 order-of-magnitude typically required to burn through polycarbonate lenses in 10 s [29]. In any case, the eye protector requirements vary from country to country, and the user is under an obligation to be informed of the legal requirements in his or her locality.

Eye protectors have been developed for use with IPL units, more to provide comfort than to merely afford protection. Because of the repeated bright light remitted from the skin or from the side of the applicator, many complain of discomfort and headaches created by looking at the repeated flashing visible light. To avoid this discomfort from repeated transient adaptation of the eyes, some manufacturers offer autodarkening lenses (such as those used in some welding helmets) that automatically become darker during the flash.

A more serious problem associated with laser eye protection occurs when more than one laser eye protector must be worn for some procedures, as in tattoo removal. Several different visible wavelengths may be required to remove the different tattoo inks, and choosing the wrong eye protector during a procedure nearing the end of a tiring day has reportedly injured some dermatologists. Unclear labeling has been one contributing factor. Again, the application of a customized label related to the specific laser has been recommended in

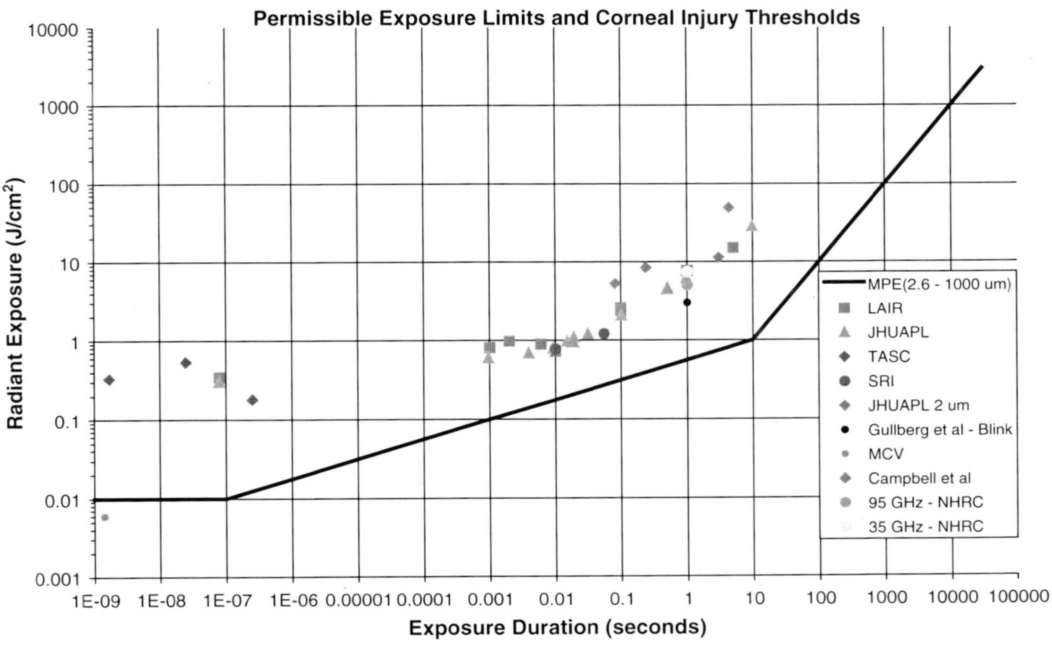

Figure 24.6 The thresholds for corneal and skin damage are of the order of 10 W/cm^2 for a 1-second laser exposure. Thresholds for 2–10.6 μm are shown (plot courtesy of B. Stuck, USAMRD, San Antonio, TX). The thresholds for laser eye-protector surface damage are higher.

some countries (as in the United States). However, in other countries, placing labels not provided by the manufacturer on eye protector may not be approved. To compound the problem, the issue of laser eye protectors having only a coded indication of the protection (e.g., "L4-1064") is strictly forbidden in the United States, because of the great importance placed upon informing the user in an understandable statement! Hopefully, future safety standards would be harmonized on this subject.

24.4 Product Safety

With a constant evolution in skin phototreatment technologies, more attention has been paid to engineering controls incorporated into the laser or lamp product itself [10]. Contact sensors that prevent firing of the device unless within contact with the skin have evolved with ever-increasing reliability. Reliability becomes essential for any products intended for home use, as in laser hair-removal products. The difficulty of determining eye safety and the NHZ is confounded by increased sophistication and complexity of modern phototreatment devices, such as devices with intentionally nonuniform output distribution of optical energy, such as fractionated beams, diode arrays, and some IPL systems. In addition, the safety levels often have to be assessed under two distinct sets of conditions: diffusely reflected light present during normal operation of the device, and direct viewing of the output surface that constitutes a misuse or abnormal condition.

24.5 Standards for Quantifying Risks

24.5.1 Occupational Exposure Limits

Relevant MPEs or ELs for lasers of interest are given in Table 24.1, and are specified at the corneal plane. If the laser beam is less than 7 mm in diameter, it is assumed that the entire beam could enter the dark-adapted pupil, and one can express the maximal safe power or energy in the beam (in the 0.4–1.4 (m retinal hazard region); it is the EL multiplied by the area of a 7 mm pupil, that is, 0.4 cm^2. For example, for the visible CW lasers, an exposure limited by the natural aversion response of 0.25 seconds is 2.5 W/cm^2, and this EL multiplied by 0.4 cm^2 results in the limiting power of 1.0 mW. This 1 mW value has a special significance in laser-safety standards, since it is the Accessible Emission Limit (AEL) of Class 2, that is, the dividing line between two laser-safety-hazard classifications: Class 2 and Class 3 [2–8]. A 3.5 mm aperture is applied with CW infrared lasers operating at wavelengths greater than 1400 nm, and a 1 mm aperture is applied in the UV spectral region for brief ocular exposures.

24.5.2 Laser Hazard Classification

As noted, any CW visible laser (400–700 nm) that has an output power <1.0 mW is termed a Class 2 (low-risk) laser, and could be considered more or less equivalent in risk with staring at the sun, at a tungsten-halogen spotlight, or at other bright lights that can cause a photic maculopathy (central retinal injury). Only if one purposely overcomes their natural aversion response to bright light, can a Class 2 laser pose a real ocular hazard. An aiming beam or alignment laser operating at a total power above 1.0 mW would fall into hazard Class 3, and could be hazardous *even if viewed momentarily within the aversion response time*. A subcategory of Class 3, termed Class 3R (formerly 3A in the United States), consists of lasers from 1–5 mW in power, and these lasers pose a moderate ocular hazard under viewing conditions, where most of the beam enters the eye. Class 3b is when the subcategory that comprises, among certain pulsed lasers, CW visible lasers that emit 5–500 mW output power. Even momentary viewing of Class 3b lasers is potentially hazardous to the eye.

Only lasers that are totally enclosed or that emit extremely low output powers fall under Class 1 and are safe to view. Any CW laser with an output power above 0.5 W (500 mW) falls under Class 4. Class 4 lasers are considered to pose skin or fire hazards as well as severe eye hazards, if not properly used. The purpose of assigning hazard classes to laser products is to simplify the determination of adequate safety measures, that is, Class 3a measures are more stringent than Class 2 measures, and Class 4 measures are more stringent than Class 3b measures. Virtually, all surgical lasers fall into Class 4, although the ophthalmic Nd:YAG photodisruptor is one example of a Class 3B surgical laser. Low-Level Laser Therapy (LLLT) lasers and LED arrays that employ nonthermal effects to stimulate or "biomodulate" cells in the dermis and epidermis are frequently Class 3B—or even Class 3R laser products, or Risk Group 2 lamp products.

24.6 Non-Beam Hazards

As with other electrical or electronic medical equipment, lasers and IPL systems in the clinical or spa environment may pose electrical safety problems as well. Potential hazards of electrical shock exist, requiring appropriate grounding, and other electrical safety laser use, and biomedical engineers and medical electronics technicians familiar with safe installation of electrical and electronic equipment in the hospital, clinical, or spa environment should have no difficulty in providing guidance for the safe electrical installation and use of laser equipment [1,10]. Only trained service personnel should access the high-voltage power supplies employed with IPL and flashlamp-pulsed lasers, since capacitor discharges can be lethal. Stored charge, even after the system is disconnected at the wall-plug, have produced lethal discharges.

Tissue-ablative laser procedures—just as with electrosurgical techniques—can produce potentially hazardous airborne contaminants from the photovaporization of tissues. Unfortunately, the vaporized tissue ("smoke") from laser surgery has often been referred to as "laser smoke" or the "laser plume," suggesting that it is unique to laser surgery. This emphasis on the laser origin has frequently led to the result that vaporized tissue fragments from pyrolysis products of tissue from electrosurgery have been overlooked as having the same degree of hazard. Vaporized tissue in sufficient quantities must receive special attention, and local exhaust ventilation almost always will be required [1,33]. The pyrolysis products are similar to those resulting from the barbecuing of meats. They contain toxic by-products and known carcinogens such as nitrosamines. A number of studies have measured the concentration of potentially hazardous airborne contaminants in conventional laser-operating rooms, with the result that concentration are shown to be kept below permissible concentrations with appropriate exhaust ventilation [1]. The studies of both the chemical toxicity of pyrolysis products and of the potential viability of infectious particulates (e.g., viral fragments) have shown real cause for concern, unless very good exhaust ventilation and respiratory protection are employed [33–35].

24.7 Safety Administration and Training

The practical implementation of a laser-safety program, which includes a laser-safety training program, cannot be treated in detail here. Other reviews of the subject treat these aspects in detail [1,4,10–18]. Clearly, the design of a safety program depends largely on the size of the institution and the variety and the number of lasers in use. An office practice might only have a safety SOP and a designated LSO; a large institution frequently benefits from a laser-safety committee. In the end, the importance of a well-trained staff cannot be over emphasized. Accidents can only be prevented by a well-trained staff and an administrative policy that encourages a sustained effort toward safe laser use.

24.8 Conclusions

The potential exposure levels to the eye and skin from scattered IPL and laser radiation from most dermatologic laser applications are substantially below a threshold for injury, and only the direct beam or specular reflections are of concern. Only with UV lasers should one be seriously concerned with chronic exposure and delayed effects. The cosmetic laser and IPL user can be assured that today a consensus exists almost worldwide regarding the appropriate laser-safety measure to preclude injury from acute or chronic effects. Procedural controls requiring the use of appropriate eye protection when needed, and control of vaporized tissue byproducts requires both a well-trained operator–user and any assisting staff. As with many other applications of lasers in industry and research, laser-safety training is of crucial importance.

References

1. Sliney DH and Trokel SL. *Medical Lasers and Their Safe Use.* New York: Springer-Verlag, 1992.
2. American Conference of Governmental Industrial Hygienists (ACGIH), ACGIH TLV's, Threshold Limit Values and Biological Exposure Indices for 2008, ACGIH, Cincinnati, OH, 2008.
3. ANSI, Safe Use of Lasers, Standard Z136.1-2007, American National Standards Institutes, Laser Institute of America, Orlando, FL, 2007.
4. ANSI, Safe Use of Lasers in Health Care Facilities, Standard Z136.3-2005, American National Standards Institute, Laser Institute of America, Orlando, FL, 2005.
5. Standards Association of Australia, *Guide to the safe use of lasers in health care*. Standard AS/NZS 417. Standards Australia, Sydney, 2004.
6. World Health Organization (WHO), Environmental Health Criteria No. 23, Lasers and Optical Radiation, joint publication of the United Nations Environmental Program, the International Radiation Protection Association and the World Health Organization, Geneva, 1982.
7. International Commission on Non-Ionizing Radiation Protection, (ICNIRP), Revision of guidelines on limits of exposure to laser radiation of wavelengths between 400 nm and 1.4 μm, *Health Phys*, 79(4):431–440, 2000.

8. Center for Devices and Radiological Health (CDRH), *Laser Product Performance Standard*, Title 21, Code of Federal Regulations, Part 1040, Washington, DC, US Food and Drug Administration (FDA), US Government Printing Office.
9. International Electrotechnical Commission (IEC), *Radiation Safety of Laser Protects, Equipment Classification, and Users Guide, Standard 60825-1*, 2nd Edn., IEC, Geneva, 2007.
10. International Electrotechnical Commission (IEC), *Medical electrical equipment - Part 2-22: Particular requirements for basic safety and essential performance of surgical, therapeutic and diagnostic laser equipment, IEC 60601-2-22* 3rd Edn., 2007.
11. International Electrotechnical Commission (IEC), Technical Report IEC TR 60825-8, *Safety of laser products –Part 8: Guidelines for the safe use of laser beams on humans,* 2nd Edn., 2006.
12. Bower KS, Burka JM, Hope RJ, Franks JK, Lyon TL, Nelson BA, and Sliney DH. Scattered laser radiation and broadband actinic ultraviolet plasma emissions during LADARVision excimer refractive surgery. *J Cataract Refract Surg*., 31(8):1506–1511, 2005.
13. Sliney DH, Krueger RR, Trokel SL, and Rappaport KD. Photokeratitis from 193 nm argon-fluoride laser radiation, *Photochemistry and Photobiology*, 53(6):739–744, 1991.
14. Sliney DH and Wolbarsht ML. *Safety with Lasers and Other Optical Sources*, New York: Plenum, 1980.
15. Sliney DH. Ocular injuries from laser accidents, *Proceedings of Laser-Inflicted Eye Injuries: Epidemiology, Prevention, and Treatment*, *Proc SPIE*, pp. 25–33 (1996).
16. Widder RA, Severin M, Kirchof B, and Krieglstein GK. Corneal injury after carbon dioxide laser skin resurfacing, *Am J Ophthalmol.*, 125:392–394, 1998.
17. Brilakis HS and Holland EHJ. Diode-laser-induced cataract and iris atrophy as a complication of eyelid hair removal, *Am J Ophthalmol.*, 137:762–763, 2004.
18. Wessely D and Lieb W. Ocular complications of diode laser epilation in the face: loss of pupillary symmetry and pigment layer defect as well as coagulation of the ciliary body with intraocular inflammation caused by laser treatment [German], *Ophthalmologe*, 99:60–61, 2002.
19. Herbold TM, Busse H, Uhlig CE. Bilateral cataract and corectopia after laser eyebrow epilation. *Ophthalmology*, 113(6):984, 2006 [with comment in *Ophthalmology*, 114(3):624; author reply 624–625, 2007.
20. Hammes S, Augustin A, Raulin C, Okenfels HM, and Fischer E. Pupil damage after periorbital laser treatment of a port-wine stain, *Arch Dermatol.*, 143:392–394, 2007.
21. Le Jeune M, Autié M, Monnet D, and Brézin AP. Ocular complications after laser epilation of eyebrows. *Eur J Dermatol.*, 17(6):553–554, 2007.
22. Spelsberg H, Hering P, Reinhard T, and Sundmacher R. Bilateral scleral thermal injury: complication after skin laser resurfacing, *Arch Ophthalmol.*, 118:846–850, 2000.
23. Pham RT, Tzekov RT, Biesman BS, and Marmor MF, Retinal evaluation after 810 nm diode laser removal of eyelashes, *Dermatol. Surg.*, Sep;28(9):836–840, 2002.
24. Nelson CC, Pasyk KA, and Dootz GI. Eye shield for patients undergoing laser treatment, *Am J Ophthalmol.*, 110(1):39–44, 1990.
25. Sliney D H. Laser safety for plastic surgery and dermatology, *Lasers in Plastic Surgery and Dermatology*, pp. 176–184, New York, Thieme Medical Publishers, 1992.
26. Ries WR, Clymer MA, and Reinisch L. Laser safety features of eye shields, *Lasers Surg Med.*, 18(3):309–315, 1996.
27. Rohrich RJ, Gyimesi IM, Clark P, and Burns AJ. CO_2 Laser safety considerations in facial skin resurfacing. *Plastic Reconstr Surg.*, 100(5):1285–1290, 1997.
28. Seckel, BR, Kovanda, CJ, Cetrulo CL. Jr, Passmore AK, Meneses PG, and White T. Laser blepharoplasty with transconjunctival orbicularis muscle/septum tightening and periocular skin resurfacing: A safe and advantageous technique. *Plastic & Reconstructive Surgery.*, 106(5):1127–1141, 2000.
29. Wood RL, Sliney DH, and Basye RA. Laser reflections from surgical instruments. *Lasers Surg Med.*, 12:675–678, 1992.
30. Sliney DH, Sparks SD, and Wood RL. The protective characteristics of polycarbonate lenses against CO_2 laser radiation. *J Laser Appl.*, 5(1):49–52, 1992.
31. Laser Institute of America (LIA). *LIA Guide for Selection of Laser Eye Protection*, Orlando, LIA, 2007.

32. Sliney D.H., Radiometric quantities and units used in photobiology and photochemistry: recommendations of the Commission Internationale de L'Eclairage (International Commission on Illumination). *Photochem Photobiol.* 83(2):425–432, 2007.
33. Ball KA. *Lasers: The Perioperative Challenge*. 3rd Edn., Denver, AORN, 2004.
34. US Department of Labor, title 29, Codes of Federal Regulations, Occupational Health and Safety.
35. Kokasa JM and Eugene J. Chemical composition of laser tissue interaction smoke plume. *J Laser Appl.*, 1:59–63, 1989.

25

Light-Based Devices for At-Home Use

Michael Moretti

Medical Insight, Inc., Aliso Viejo, CA, USA

25.1 Introduction

Consumers have long sought products that would help them look their best, conveniently and affordably. Not only must the prices of professional procedures performed in a salon, spa, or medical office incorporate costs and profit margins associated with these businesses, but also, traveling to these establishments for treatment typically entails an additional commitment of time. Therefore, products developed for use at home often meet this need best. In fact, consumers are quite comfortable selecting and using products at home, as most are accustomed to purchasing skin-care and hair-removal products, cosmetics, hair-care products, and other personal-care items. Until recently, however, the ability of these products to offer more efficacious results was limited, and consumers seeking more profound results had no choice but to utilize professional services. As aesthetic products and equipment manufacturers increasingly expand beyond professional markets into consumer markets, however, this scenario is changing.

Gurpreet S. Ahluwalia (ed.), *Cosmetic Applications of Laser and Light-Based Systems*, 517–526,

In the infancy stage, this expansion from professional to consumer markets is the result of three important trends that are poised to continue through the foreseeable future:

1. advancing technology that allows the development of highly effective, yet extremely compact equipment;
2. rising competition and market saturation in the professional device sector, leading manufacturers to seek growth in the larger consumer sector;
3. rising interest in aesthetic treatments among the general public.

While the emerging new aesthetic devices for home use will offer significantly more benefit than home-use products available earlier because they will be based on technology developed for professional aesthetic practitioners, they will nonetheless not compete directly with professional treatments. This is because the new home-use devices will deliver lower levels of energy to provide the highest possible level of safety. Further, as professional equipment continues to advance, the capabilities available to physicians and other practitioners will continue to expand.

> "Home devices will have less energy," says Syneron's Chief Marketing Officer, Mark Tager, MD. "They're working at a different scale than professional treatments, at which the energy is at a very therapeutic level for immediate efficacy. We'll see these products being very supportive of professional treatments, such as in maintenance programs. People get into regimens, many of which are approved by their physicians, for skin rejuvenation and skin tightening to keep them looking young. We see home devices as being supportive of good skin maintenance and enhancing awareness of the professionals' role in the skin care process."

Therefore, home devices are expected to address a different need than professional services.

This type of differentiation has also been observed in the relatively recent introduction of home microdermbrasion kits, which appeal in large part to consumers who purchase drug-store cosmetics, and are highly price sensitive. This group also includes teens and senior citizens, many of whom like to experiment with new skin-care products, but are not comfortable visiting a professional for treatment. However, James Bartholomeusz, Syneron's Director of Product Development, notes that home microdermabrasion kits have a problem differentiating products since they cannot show enhanced benefits. Because of this, he says, they are less likely to be recommended by physicians.

25.2 First-Generation Home-Use Devices

The home aesthetic device market is either extremely large or quite small, depending on how it is defined. A large number of manufacturers have long offered a variety of relatively low-technology products intended to improve skin condition, remove unwanted hair, and address cellulite. Most of these products, however, were not based on medical technology, so the benefit they could provide was limited, and their acceptance has been slow. Table 25.1 shows some of these devices; many have been available for several years, while others are relatively new. Quite a few of the home hair-removal devices have been sold for more than

Table 25.1 Low Tech Home Aesthetic Devices

Type of Device	Application	Selected Manufacturers	Details
Electric clippers	Hair removal	Braun, Emjoi, Panasonic, Philips/Norelco, Wahl, others	Available for many years, widely used
Electric razor	Hair removal	Braun, Panasonic, Philips/Norelco, Remington, others	Available for many years, widely used
Home electrolysis (RF)	Hair removal	Emjoi, Tactica International, others	Newer, limited acceptance
Rotating coils	Hair removal	Braun, Emjoi, Epilady, Tactica International, others	Newer, limited acceptance
Ionic blemish remover	Acne	Tactica International, others	Newer, limited acceptance
Massager	Cellulite	Guitay*, Tactica International, others	Newer, gaining some acceptance
Ionic facial toner	Skin rejuvenation	Tactica International, Ya-Man, others	Newer, limited acceptance
Microcurrent facial toner	Skin rejuvenation	Skin Star, SLC, others	Newer, limited acceptance but growing
Microdermabrasion	Skin rejuvenation	Guthy Renker, Johnson & Johnson, Mary Kay, L'Oreal, Procter & Gamble, Zia, others	Relatively new but gaining some acceptance; many new products entering the market
Ultrasonic facial toner	Skin rejuvenation	Tactica International, Ya-Man, others	Newer, limited acceptance

Source: Medical Insight, Inc.
*Device is FDA-cleared for muscular pain relief.

a decade with a few changes. Most are not subject to Food & Drug Administration (FDA), approval and therefore are not required to submit clinical studies demonstrating safety and efficacy. It can be reasonably assumed that such devices provide little significant benefit and in fact, most of the home hair-removal systems offer only temporary hair removal.

25.3 Next Generation Home-Use Devices

The new generation of home-use devices incorporates advanced technologies such as phototherapy, radio frequency (RF), ultrasound, and so on, and therefore requires FDA clearance to market, supported by rigorous clinical studies demonstrating the safety and efficacy in the intended application. Leading devices that meet these criteria, both available and under development, are shown in Table 25.2.

Table 25.2 Next Generation Home Aesthetic Devices

Company	Product Name	Application	Development Status	Expected Launch
Palomar	n/a	Hair removal	FDA cleared	2008
Radiancy	No! No!	Hair removal	Launched	n/a
Spectragenics	i-epi	Hair removal	Launched (Japan)	2008 (US, EU)
Ya-Man	n/a	Hair removal	Launched (Asia)	2008 (US, EU)
Radiancy	CTL	Acne	Launched	n/a
Tyrell	Zeno	Acne	Launched	n/a
Palomar	n/a	Skin rejuvenation	Early development	2009
Aesthera	n/a	Skin rejuvenation	Early development	2009
Candela	n/a	Skin rejuvenation	Early development	2009
CyDen	n/a	Skin rejuvenation	Early development	2008
Light BioScience	n/a	Skin rejuvenation	FDA clearance pending	2008
Photo Therapeutics	New-U	Skin rejuvenation, acne	FDA clearance pending	2008
Radiancy	FSD	Skin rejuvenation	Launched	n/a
Syneron	n/a	Skin rejuvenation	Early development	2009
Xthetix	n/a	Skin rejuvenation	Early development	2009

Source: Medical Insight, Inc.

As of mid-2007, few of these were available. One exception to this was a set of home-use devices offered by Israel-based Radiancy. A leader in the professional aesthetic device market, Radiancy manufactures equipment that delivers Light–Heat Energy (LHE), a proprietary technology utilizing thermal energy (heat), as well as photothermolysis to achieve intended effects.

In the home-device market, Radiancy was an early pioneer with the introduction of its No!No! hair-removal device. First launched in 2004, the system uses a patented proprietary technology that Radiancy calls "thermicon." In this process, thermal energy is delivered to the hair follicle with a high-temperature thermodynamic wire that glides just above the skin and singes hair at the skin surface, while conducting thermal energy through the hair shaft down the follicle. Heat energy from the device, transiently stored in the hair shaft, completes the thermolysis process. An electromechanical system monitors the device's movement over the skin and controls energy delivery; when the sensors detect that the device is being moved too slowly, heating of the wire stops, and a mechanical mechanism rapidly raises the heating wire away from the skin. Radiancy says that such repeated treatments weaken the hair follicle and lead to reduced hair growth. In clinical studies, women who used the device two to three times per week on various body areas experienced a sustained 48% average reduction in hair count after six weeks of use. Studies show these results to be as efficacious and safe as professional light-based systems.

Consumer response to the No!No! has been strong. In May 2006, Radiancy's distributor in Europe ran a trial late night television commercial to test consumer interest. The launch of the product via televison sales was later expanded to retail, where, according to the company, during the first week of the test, 10,000 units were sold. By September, more than 100,000 orders had been taken in Spain alone with a run rate of 20,000 devices per week. The unit is priced at $250 to $300, with replacement heads (which are required roughly every 2 months) priced at $10. No!No! is currently available in Chile, Mexico, Argentina, Peru, Uruguay, Spain, Portugal, Germany, Greece, Belgium, Australia, Taiwan, Korea, and the United States, with another 12 planned launches by the end of 2007. More than 700,000 devices have been sold as of May 2007, according to Radiancy's President and CEO, Dr. Dolev Rafaeli.

Two other home-use devices incorporate a modified version of the LHE (a proprietary and patented combination of Light-and-Heat Energy) technology that Radiancy calls LHE Micro Phototherapy. This adaptation utilizes lower-energy fluence, providing greater safety for consumer use. The Facial Skincare Device (FSD) was cleared in 2006 by FDA for skin rejuvenation, and Clear Touch Lite (CTL) was cleared in 2006 for acne treatment. Both are being sold worldwide. The FSD is a continuous pulsed-light system that delivers light in the 400–1200 nm wavelength range to a large 14 mm × 27 mm spot-size treatment area. Fluence ranges from 6 to 12 J/cm^2 with 1.6 ms pulse duration. The 2.8 lb device is easily held in the hand, and plugs into a standard electrical outlet. Radiancy says the FSD (which is also called Facial Toning Device or FTD outside the United States) can stimulate collagen production to soften fine lines, homogenize skin tone, decrease pore size, and improve skin texture. Similarly, the CTL is a continuous pulsed-light system that delivers light in the 430 nm–1100 nm wavelength range with a 14 mm × 27 mm spot size. Fluence is 6 J/cm^2 with a 1 ms pulse duration. It weighs 2.2 lb and plugs into a standard electrical outlet. Like Radiancy's professional acne-treatment devices, CTL helps reduce the appearance of inflammatory acne lesions.

Radiancy also has other home-device product initiatives.

In Asia, two companies currently sell laser systems for home hair removal: Spectragenics and Ya-Man.

Although Spectragenics is relatively new, its founders have a long history in the aesthetics industry, having developed the highly successful and technologically innovative LightSheer diode laser for professional hair removal in the early 1990s. They subsequently sold the technology to Palomar Medical, and Palomar divested to Coherent Medical, which in turn merged with ESC to form Lumenis. LightSheer remains one of the industry's top-selling professional laser hair-removal systems, although Spectragenics focuses entirely on the emerging home market, selling a diode laser home hair-removal device in Japan. Available for roughly three years, the product generated approximately $3.0 million in sales in 2006. The company is working to expand distribution for the i-epi, but as of July 2007, the product had not obtained clearance to market either in the United States or in Europe.

Tokyo-based Ya-Man offers a variety of home-use devices for skin care, hair care, and analysis of body fat. Founded in 1969, the privately held company was the first Japanese manufacturer to introduce professional electrolysis for hair removal in Japan. With more than 400 patents pending, Ya-Man's technology-based products are sold worldwide in retail stores, salons, mail order, and television shopping channels. Ya-Man has introduced a light-based home hair-removal device in Asia, and is expected to eventually obtain FDA approval to sell the product in the United States.

Palomar Medical Technologies, one of the industry's oldest manufacturers of professional aesthetic laser equipment and an early technology pioneer, is also working on a home-use device for hair removal. In February 2003, Palomar signed a development agreement with razor maker Gillette (which has since been acquired by Procter & Gamble), wherein Gillette would provide $7 million in funding and work with Palomar over a 30-month period to develop a patented self-use device for women. In June 2004, the companies extended their relationship, with Gillette agreeing to fund a further $2.1 million through a planned project completion in August 2006. The collaboration extends Palomar's own efforts to develop a home-use hair-removal device, which date back to the late 1990s. In December 2006, Palomar announced that FDA had cleared its home device for marketing. While the product had not been introduced as of mid-2007, a high-profile launch is expected shortly that will draw national attention to these new types of devices.

In February 2007, Palomar and Gillette extended their agreement to develop an additional home use, light-based hair-removal device for women. Under the 11-month agreement, Gillette will provide Palomar with development payments of $1.2 million and an additional $300,000 upon the completion of certain deliverables.

Palomar is also working with Johnson & Johnson to develop home-use devices for skin rejuvenation, acne treatment, and cellulite reduction. In September 2004, the two companies signed an agreement wherein Johnson & Johnson would fund development and then (presumably) market a product upon FDA approval. While the program is ongoing, Palomar has not publicly released any details regarding its progress.

Texas-based Tyrell was founded in 2002 to market a novel device for home treatment of acne. The company's product, Zeno, utilizes proprietary ClearPoint technology to kill acne-causing bacteria. It does this through the delivery of precisely controlled heat in a replaceable treatment tip that heats to 118.5 degrees. In one clinical trial, 90% of pimples treated with the device demonstrated improvement or resolution within 24 hours. When Zeno was first launched in 2005, it was sold exclusively through dermatologists' offices and medispas. The following year, however, Tyrell introduced two new versions, Zeno PRO and Zeno MD, which contained higher treatment counts. The full product line now includes original Zeno, a 60-treatment count device priced at $149; Zeno PRO, a 90-treatment count device with carrying case priced at $185; and Zeno MD, a 150-treatment count device with carrying case costing $200. The new models also feature new product colors.

Photo Therapeutics Limited, a UK-based company, is also getting ready to launch an advanced home-use device. Like its professional equipment, the company's home device would be based upon light emitting diode (LED) technology to stimulate collagen regeneration. Photo Therapeutics' President, Sue D'Arcy, says that its first device will address skin rejuvenation and acne, with a scheduled launch in 2007. It will be priced at few hundred dollars and will be distributed by a group of marketing partners. The company is also planning to investigate the ability of its home-use device to address cellulite. Photo Therapeutics has obtained FDA approval for a similar home-use product for wound healing.

Although it is still a relatively young company, Syneron has quickly established a leading position in the market for professional light-based aesthetic devices with a series of multifunctional platforms that combine the company's unique broad spectrum light/bi-polar RF (ELOS) technology. The devices address all the leading aesthetic applications, including hair removal, skin rejuvenation, skin tightening, acne reduction, treatment of vascular lesions, removal of tattoos, and pigmented lesions, as well as body shaping and skin tightening.

In early 2007, the company announced a partnership with Procter & Gamble (P&G) to develop skin-rejuvenation products for the home market, and is currently conducting early stage clinical studies with a prototype device. Syneron will be responsible for product development while P&G will handle marketing and distribution. Although P&G will ultimately set the prices, Syneron has stated that the device will probably be priced less than $500 and there could be a disposable component.

Aesthera, the technology pioneer which developed photopneumatic therapy for professional aesthetic treatment, is also working on a home-use device. Company president Alon Maor says a home system would use photopneumatic technology, but unlike other home units, would not merely represent a scaled-down version of professional devices. As of mid-2007, the company has demonstrated proof of concept and has submitted data in pursuit of intellectual property protection. Its first application will be treatment of the skin in a comprehensive procedure that Maor says goes beyond skin rejuvenation to address both medical and cosmetic deficiencies. The company expects to begin commercialization toward late 2008 or early 2009; devices would be priced at "a few hundred dollars" and as with Aesthera's professional systems, would include a disposable component.

Photopneumatic technology combines broadband light with pneumatic (vacuum) energy; the vacuum stretches the skin, increasing its surface area as well as the amount of energy transmitted to the target. This enables treatment for hair removal, skin rejuvenation, reduction of pigmented lesions, and vascular lesions that the company says is uniquely safe, painless, and up to seven times faster than conventional light-based hair-removal therapies.

Candela is one of the oldest and most established aesthetic laser companies, with a broad range of professional products that address hair removal, skin rejuvenation, skin tightening, acne reduction, treatment of vascular lesions, as well as removal of tattoos, and pigmented lesions. The company's interest in the home-use market is relatively recent but its significant resources could lead to the quick development of a product.

In March 2007, Candela acquired Inolase, a developer of Serenity Pneumatic Skin Flattening (PSF) technology. This technology allows painless, more effective, and safer laser and intense pulsed light (IPL) treatments even at high energy levels. The PSF provides multiple advantages in a single step: painless treatments by activation of a natural pain blocking mechanism; blood expulsion from treatment zone for better light penetration, and improved efficacy; no erythema. The PSF's vacuum system works with or without gel, and can be mounted on any laser/IPL handpiece for fast and convenient operation. Candela has indicated that this approach may have applications in professional as well as home-use devices. The company appears to be committed to this application, reportedly allocating $10.5 million of the $16.8 million it paid, to acquire Inolase to "goodwill" associated with the technology's applicability to home use.

Spun out of medical ultrasound developer Guided Therapy Systems, Ulthera is focused on ultrasonic body-shaping devices for the professional market. The company has developed the Ultrasite GT Device, a platform of ultrasound instrumentation that utilizes disposable probes to apply energy below the skin's surface, while sparing the intervening tissue such as the epidermis. The depth and precision of energy deposition is reportedly unprecedented, and the device's imaging capability provides a significant advantage over alternative technologies. Ulthera is expected to introduce its first professional product in 2008, and its researchers are also working on a smaller home-use device which would be launched after the professional system. To launch this product, another company has been spun out from the Guided Therapy Systems group.

Xthetix will focus on commercial applications of Ulthera's technology for the home market. The company's first product will be for acne treatment and prevention, says Ulthera's President, Brian O'Connor. It has the ability to prevent the occurrence of acne by inhibiting sebaceous gland function, using heat, he says, without undesirable side effects. Additional devices in the pipeline target skin rejuvenation, hair removal, and inflammation-related skin diseases such as rosacea. Like Ulthera's professional device, home-use units produced by Xthetix will include both platforms and disposable components.

25.4 Market Growth

These ongoing product development activities are expected to result in the launch of a plethora of devices, with a strong related promotion. As this occurs, the sale of home-use devices will rise by 38.3% per year on average; from an estimated $33.4 million in 2006 to $771.7 million in 2011, according to industry research firm Medical Insight. The most sustained growth will occur when Palomar and Syneron enter the market with strong marketing support from Gillette and Procter & Gamble, respectively. It should be noted, however, that these estimates reflect the total retail dollars spent on home use aesthetic devices and *not* necessarily the revenues earned by manufacturers. Because retail markups can be as much as 50% or more, device sales revenues represent total dollars spent on each device which will be divided between device developers and their distribution partners.

Radiancy's products accounted for 67.4% of all retail dollars spent on advanced home-use devices in 2006, according to Medical Insight. By 2011 Palomar and Syneron are expected to stake out more significant positions, as are other manufacturers, as shown in Fig. 25.1.

Over and above the sales of devices, consumers will also purchase disposable tips and cartridges, as they currently do for razors, since most, although not all the home-use devices will include a disposable component. This is the case, for example, with Radiancy's No! No! As this occurs, sales of its disposables will rise by more than 31% per year from about $10 million in 2006 to $48 million in 2011, according to Medical Insight.

25.5 Consumer Response

As of today, consumers have demonstrated a strong desire for efficacious home-use products and equally strong comfort with the technologies introduced. Far from being intimidated by words like "laser," "radio frequency," "microcurrent," "chemical peel," "microdermabrasion," on the boxes of these new products, consumers have embraced them as a means to access treatments that were once only available through physicians. This offers the potential of greater convenience at a lower cost. As these products have generally been developed so that consumers can easily and safely use them at home, they have not yet been proven too complex or sophisticated for the average value-conscious consumer

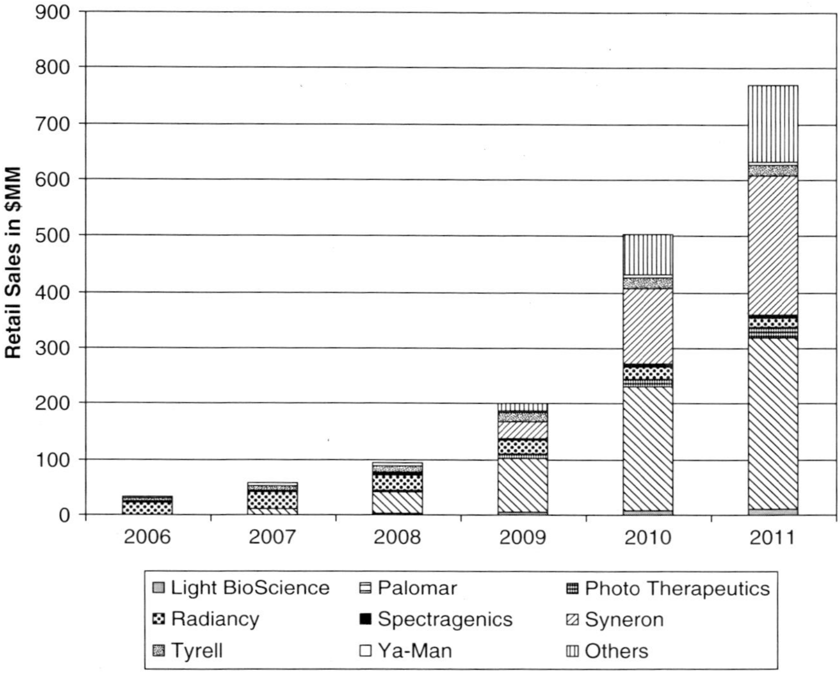

Figure 25.1 Aesthetic home device retail sales by company, 2006–2011.

to try. For aesthetic procedure patients, this confidence is also fueled in part by the rising usage by physicians of nurses, aestheticians and others to perform treatments, leading consumers to believe that with proper instruction, they could also operate advanced aesthetic devices.

That said, not all the emerging home-use products have provided a high cost-benefit. Some have proven relatively ineffective, or have simply not lived up to the consumers' expectations. This has largely been the result of unbalanced marketing campaigns, wherein the benefits of new products are exaggerated to justify their higher prices, with the inevitable consequence of consumer dissatisfaction. Among the more spectacular examples of this phenomenon has been rotating coil hair-removal systems, which were initially promoted as a quick and painless method of long-term hair removal. After rushing out to try the EpiLady, which launched the category, women of all ages found that this was not the case at all, and the resulting negative publicity drove the product's distributors into bankruptcy. Marketers of newer home-use devices appear to be more cautious, so that acceptance of the products is proceeding in a more controlled and sustainable manner.

25.6 Impact on Physician Practices

The impact of aesthetic home-use devices on physician practices has been modest as of mid-2007; however, this is expected to change with the next generation of considerably more efficacious devices. First-generation home-use systems have largely provided only incremental improvements over conventional home-use aesthetic products. For example, home electrolysis units offer only marginally longer-lasting hair removal than waxing and tweezing; and other types of hair-removal devices such as electric shavers and rotating coil units are not substantially different from shaving. Therefore, these devices have posed effectively no threat to physicians offering light-based hair removal. In fact, they may have emphasized the efficacy of office-based treatments, for consumers disappointed in the results from these supposed "advancements."

However, the next generation of home hair-removal devices, including lasers from Palomar/Gillette, Spectrageneics, and others, is expected to dramatically shift this dynamic by making substantially more efficacious treatments available to the consumer at home. Although home-use devices will never offer the results possible from professional equipment, they will nonetheless provide significantly greater benefit than topical and other conventional products; for some consumers, particularly those who are cost-conscious, this will be sufficient to wean them from professional treatments to home-use products.

For consumers who are more loyal to professional treatments, new home-use devices will offer a means to maintain professional results for longer periods of time and/or try a new type of procedure at home before investing in a professional procedure. A small but growing proportion of savvy physicians will market to this clientele by:

- explaining to them the benefits and limitations of new home-use devices vis-à-vis professional treatments;
- selling them complimentary topical, and other products that will boost the results delivered by home-use devices.

However, many physicians will lose business to home-use devices, particularly in the short run, as consumers try the new products. Practices that stand the most to lose include those that market mainly on the basis of price, those with suboptimal marketing programs, and those that focus on hair removal.

Index